# Medicinal Herbs and Fungi

Dinesh Chandra Agrawal •
Muralikrishnan Dhanasekaran
Editors

# Medicinal Herbs and Fungi

## Neurotoxicity vs. Neuroprotection

Springer

*Editors*
Dinesh Chandra Agrawal
Department of Applied Chemistry
Chaoyang University of Technology
Taichung, Taiwan

Muralikrishnan Dhanasekaran
Department of Drug Discovery and Development
Harrison School of Pharmacy, Auburn University
Auburn, USA

ISBN 978-981-33-4140-1 ISBN 978-981-33-4141-8 (eBook)
https://doi.org/10.1007/978-981-33-4141-8



This Springer imprint is published by the registered company Springer Nature Singapore Pte Ltd.
The registered company address is: 152 Beach Road, #21-01/04 Gateway East, Singapore 189721, Singapore

*Editor Professor Agrawal dedicates this book to his beloved spouse*
Manju
*and*
*Editor Professor Dhanasekaran dedicates this book to his beloved spouse*
Madhumalini M. Nadar

# Preface

This book is the continuation of Professor Agrawal's previous Springer books: *Medicinal Plants—Recent Advances in Research and Development* (Springer link: http://www.springer. com/in/book/9789811010842); *Medicinal Plants and Fungi—Recent Advances in Research and Development* (Springer link: https://www.springer.com/gp/book/9789811059773); and *Medicinal Mushrooms—Recent Progress in Research and Development* (Springer link: https://www.springer.com/la/book/9789811363818).

The ever-rising increase in the consumption of medicinal herbs and its products and its exposure in the human population have generated concerns about the potential neurotoxicity of several new and existing botanicals. This book on "Medicinal Herbs and Fungi—Neurotoxicity and Neuroprotection" offers an accurate, relevant, and comprehensive coverage of a wide variety of medicinal herbs and fungi, which are associated with neurological diseases (central and peripheral nervous system disorders). It includes chapters (review articles) that thoroughly describe the benefits and adverse effects associated with the use of some of the most commonly used medicinal herbs and fungi, and the pathophysiological mechanisms underlying them. The rich compilation aims to deliver thorough and extensive research updates on the advances in medicinal herbs and fungi related to neurotoxicity and neuroprotection, ranging from discussions on cellular and molecular processes and pathology to clinical aspects. The chapters in the book have been contributed by the experienced and eminent academicians, researchers, and scientists working in the field across the globe.

Chapter "Mitosis Inhibitors and Medicinal Plants: Neurotoxicity and Neuroprotection" constitutes an exhaustive review on factors involved in the pathogenesis of chemotherapy-induced peripheral neuropathy (CIPN), neurotoxic mitosis inhibitors, and natural products as neuroprotective and/or neuropreventive agents. Besides, future perspectives have been manifested for the prevention or treatment of CIPN occurrence. Chapter "The Neurotrophic and Neuroprotective Potential of Macrofungi" reviews recent advances in research on the neuroprotective potential of macrofungi and perspectives for their application as neuroprotectants in biomedicine to prevent, support, or cure neurodegenerative disorders. Chapter

"Andrographolide, a Diterpene from *Andrographis paniculata*, and Its Influence on the Progression of Neurodegenerative Disorders" describes several biologically important derivatives of andrographolide and covers a comprehensive discussion on the upregulation or downregulation of specific signaling pathways targeted by andrographolide and their derivatives in the pathogenesis of important neurodegenerative diseases, including conditions such as pain and depression. Chapter "Ginseng: A Boon or a Curse to Neurodegenerative Diseases" provides a theoretical basis for the treatment of neurodegenerative diseases by ginseng and its extracts. Chapter "Insights into Mechanisms and Models for Studying Neurological Adverse Events Mediated by Pharmacokinetic Interactions Between Clinical Drugs and Illicit Substances of Herbal and Fungal Origin" discusses novel insights into the potential mechanisms of pharmacokinetics-based interactions between clinical drugs and illicit substances of herbal/fungal origin that may be responsible for neurological and related adverse events. Also, the chapter provides insights into potential experimental models that can be used in studying these pharmacokinetic interactions that lead to neurological adverse events. Chapter "Cannabis-Induced Neuroactivity: Research Trends and Commercial Prospects" deals with the concise yet broad review of chemical, medicinal (neuroprotection), and adverse psychotic aspects of cannabis. The ancient and traditional use of cannabis leaves (bhang) in India for medical as well as cultural purposes has been discussed in the scientific perspective. Further, the trends in scientific research, intellectual property (patents), and commercial prospects related to cannabis are discussed. Chapter "Neurotoxicity of Polyherbal Formulations: Challenges and Potential Solutions" focuses on potentially toxic substances present in the polyherbal products and discusses various general toxicity and neurotoxicity tests of the herbal products. Chapter "Balancing the Neuroprotective Versus Neurotoxic Effects of Cannabis" reviews the current neuropharmacological and neurotoxicological properties of cannabinoids. Chapter "Alpha-Synuclein: Biomarker for Parkinson's Disease, It's Estimation Methods, and Targeted Medicinal Therapies" provides an overview of the role of α-synuclein in Parkinson's disease, its estimation methods, and the use of phytochemicals in targetting α-synuclein for preventing the neurotoxicity. Chapter "Screening of Herbal Medicines for Neurotoxicity: Principles and Methods" reviews screening methods of herbal medicines for neurotoxicity. Chapter "Plants with Phytomolecules Recognized by Receptors in the Central Nervous System" describes the traditional, medicinal, and recreational uses of the plants, phytomolecules in these plants recognized by receptors in the Central Nervous System. Chapter "Reserpine-Induced Depression and Other Neurotoxicity: A Monoaminergic Hypothesis" summarizes depression and the monoamine depletion hypothesis, focusing on the drug reserpine and its role in establishing the hypothesis. Chapter "Traditional Medicinal Plants of Sri Lanka and Their Derivatives of Benefit to the Nervous System" describes the traditional medicinal plants of Sri Lanka and their derivatives of benefit to the nervous system. Chapter "Ameliorative Effects of Shodhana (Purification) Procedures on Neurotoxicity Caused by Ayurvedic Drugs of Mineral and Herbal Origin" discusses the ameliorative effects of shodhana (purification) procedures on neurotoxicity caused by ayurvedic drugs of mineral and herbal origin. Chapter "St. John's Wort: A Therapeutic Herb to Be Cautioned for Its Potential Neurotoxic Effects and Major Drug Interactions" provides a brief history of St. John's Wort (SJW), a summary of its active constituents, current and potential therapeutic uses, adverse drug effects, and drug interactions that should be considered when prophylactically and/or therapeutically using SJW with prescribed medications. Chapter "Neurotoxic Potential of Alkaloids from Thorn Apple (*Datura stramonium* L.): A Commonly Used Indian Folk Medicinal Herb" describes the neurotoxic potential of alkaloids from thorn apple (*Datura stramonium* l.)—a commonly used source of folklore medicinal herb known for its mental stimulation and curative properties. It discusses the noteworthy pharmacological potential of this plant utilized by Ayurvedic practitioners in the traditional system of Indian medicine. Chapter "Medicinal Plants in Uganda as Potential Therapeutics Against Neurological Disorders" presents the complementary and alternative therapies that could potentially narrow the treatment gap in the management of neurological disorders in Uganda. Specifically, plant species from the Ugandan context are presented from ethnobotanical studies. Chapter "Ayurvedic Ideology on *Rasapanchak*-Based Cognitive Drug Intervention" explores the rationale of Ayurveda in the management of cognitive disorders. Chapter "Neurotoxic Medicinal Plants of Indian Himalayan Regions: An Overview" provides an overview of neurotoxic medicinal plants of the Indian Himalayan region and their neurotoxins, mechanism of neurotoxicity, and a few case studies explaining the adverse effects of neurotoxins. Chapter "Neuroprotective Effects of *Portulaca oleracea* and *Portulaca quadrifida* Linn" includes the important neuroprotective activity and other therapeutic benefits of two *Portulaca* species.

The editors hope that this compendium of review articles will be useful as a reference book for advanced students, researchers, academics, business houses, and all individuals concerned with medicinal herbs and fungi.

Taichung, Taiwan — Dinesh Chandra Agrawal
Auburn, AL, USA — Muralikrishnan Dhanasekaran
23 September 2020

# Acknowledgments

The editors thank all the invited authors to this book for preparing their valuable manuscripts. Without their contributions, this book would not have been possible.

The co-editor Professor Dhanasekaran wishes to place on record special appreciation and thanks to Professor Agrawal for handling the entire correspondence with the Springer and authors and dealing with the editing, reviewing, and revision process of manuscripts and managing them from start to finish. Without his untiring efforts, this book would not have become a reality.

Editor Professor Agrawal thanks Professor Tao-Ming Cheng, President of the Chaoyang University of Technology (CYUT); Professor Wen-Goang Yang, Vice-President, CYUT; Professor Sung-Chi Hsu, Dean, R&D office and Assistant Vice-President, CYUT; Professor Hsi-Hsien Yang, Dean, College of Science and Engineering, CYUT; Professor Wei-Jyun Chien, Chairperson, Department of Applied Chemistry, CYUT, Taichung, Taiwan for their constant support and encouragement during the progress of the book.

Editor Professor Dhanasekaran thanks the administrators, faculty, and staff in Harrison School of Pharmacy, Auburn University, and all his beloved students for their relentless dedication and inspiration.

Editors thank Dr. Satyanarayana R. Pondugula, Auburn University, USA; Dr. Jeyabalan Govindasamy, Rajasthan University of Health Sciences, Jaipur, India; and Dr. Rajiv Kumar Chaturvedi, QLeap Academy, Pune, India, for their help in peer-reviewing some of the manuscripts.

The editors sincerely thank the entire Springer Nature Singapore Pte Ltd. team concerned with the publication of this book.

Editors thank and appreciate their respective families (Manju, Somya, Neha, and Mihir—Family Agrawal; Madhu and Rishi—Family Dhanasekaran) for the encouragement and wholehearted support during the progress of the book.

Editors express profound gratitude towards "God, the Infinite Being" for providing strength to accomplish the arduous task of handling this book.

# Contents

# Editors and Contributors

## About the Editors

**Dinesh Chandra Agrawal** graduated in 1976 from Aligarh Muslim University (National university) and obtained his Ph.D. degree in 1982 from HNB Garhwal University (National university). Professor Agrawal has more than 38 years of research experience in plant biotechnology of diverse species, including medicinal plants and medicinal mushrooms. After serving for more than 31 years, in 2013, he superannuated as a chief scientist and professor of biological sciences at the CSIR-National Chemical Laboratory, Pune, the top-ranking institute in chemical sciences under the umbrella of the Council of Scientific and Industrial Research (CSIR), Ministry of Science and Technology, Govt. of India. Currently, he is working as a professor in the Department of Applied Chemistry, Chaoyang University of Technology (CYUT), Taiwan. While in CSIR-NCL, Prof. Agrawal worked as a coordinator and project leader of several research projects funded by the Govt. of India. He has more than 180 publications, including five books (3 by Springer Nature) to his credit on different aspects of plant biotechnology, including medicinal plants and medicinal mushrooms. More than 35 M.Tech./M.Sc. and 7 Ph.D. students have completed their thesis work under his guidance.

Professor Agrawal has been bestowed several prestigious awards and fellowships such as the Alexander von Humboldt Fellowship (Germany), DBT Overseas Associateship (USA), British Council Scholar (UK),

European Research Fellow (UK), and INSA Visiting Scientist (India). During these fellowships, he had opportunities to work in the USA, Germany, and the UK. He had a research collaboration with UMR Vigne et Vins, INRA, Centre de Recherché Colmar, France. For more than 10 years, he was a member of the executive committee of the Humboldt Academy, Pune Chapter, and held the position of treasurer. Professor Agrawal has reviewed a large number of research papers for several SCI journals on plant biotechnology and served as a member of the editorial board of *Medicinal and Aromatic Plant Abstracts*, NISCAIR, Govt. of India. Presently, he is on the editorial board of the *International Journal of Applied Science and Engineering* (Scopus), serving as associate editor in chief of the journal.

**Muralikrishnan Dhanasekaran** completed his Bachelor of Pharmacy from Annamalai University and Master of Pharmacy from Jadavpur University, West Bengal, India. He obtained his Ph.D. degree from the Indian Institute of Chemical Biology, Kolkata, India, under the guidance of Dr. K.P. Mohanakumar. Following which he attained his post-doctoral training from renowned scientists Dr. Manuchair Ebadi (Prof. University of North Dakota, Grand Forks, ND, and Dr. Bala Manyam (Scott & White Clinic/Texas A & M, Temple, TX). Dr. Dhanasekaran joined Auburn University in the year 2005 and currently working as a full Professor at Harrison School of Pharmacy, Auburn University, USA. Dr. Dhanasekaran's area of research and interest focuses on neuropharmacology, toxicology, dietary and natural products. Dr. Dhanasekaran completed the New Investigator Research Grant from Alzheimer's Association, several Auburn University grants, and several other research projects from a Pharmaceutical Company. He has graduated 14 students (as a mentor) and currently has 2 graduate and trained more than 50 undergraduate students in his lab. Dr. Dhanasekaran has received several teaching awards from Auburn University for teaching Pharm D. and graduate students. He has published more than 200 scientific abstracts, 85 peer-reviewed publications, a book, and several book chapters.

## Contributors

**Kodye L. Abbott** Department of Anatomy, Physiology and Pharmacology, Auburn University, Auburn, AL, USA

**Rabinarayan Acharya** Department of Dravyaguna Vigyan, Institute for Post Graduate Teaching & Research in Ayurveda, Gujarat Ayurved University, Jamnagar, Gujarat, India

**Dinesh Chandra Agrawal** Department of Applied Chemistry, Chaoyang University of Technology, Taichung, Taiwan

**Nabil A. Alhakamy** Department of Pharmaceutics, Faculty of Pharmacy, King Abdulaziz University, Jeddah, Saudi Arabia

**Waleed S. Alharbi** Department of Pharmaceutics, Faculty of Pharmacy, King Abdulaziz University, Jeddah, Saudi Arabia

**Mohammed Almaghrabi** Department of Drug Development and Discovery, Harrison School of Pharmacy, Auburn University, Auburn, AL, USA

**Alshaimaa M. Almehmady** Department of Pharmaceutics, Faculty of Pharmacy, King Abdulaziz University, Jeddah, Saudi Arabia

**Godwin U. Anywar** Department of Plant Sciences, Microbiology & Biotechnology, Makerere University, Kampala, Uganda

**Savina Asiimwe** Department of Plant Sciences, Microbiology & Biotechnology, Makerere University, Kampala, Uganda

**Susanna M. Badalyan** Laboratory of Fungal Biology and Biotechnology, Department of Biomedicine, Institute of Pharmacy, Yerevan State University, Yerevan, Armenia

**Vinicia C. Biancardi** Department of Anatomy, Physiology and Pharmacology, Auburn University, Auburn, AL, USA

**Dawn M. Boothe** Department of Anatomy, Physiology and Pharmacology, Auburn University, Auburn, AL, USA

**Prashanth Boralingaiah** ESIC Medical College, Bengaluru, Karnataka, India

**Manal Buabeid** College of Pharmacy and Health Sciences, Ajman University, Ajman, UAE

**Nupe Hewage Akila Chamikara** Management and Science University, Shah Alam, Malaysia

**Aneesh Chandran** Molecular Biophysics Unit, Indian Institute of Science, Bangalore, India

**Rajiv Kumar Chaturvedi** QLeap Academy, Pune, India

**Malvi Choudhary** School of Biotechnology, University of Jammu, Jammu, Jammu & Kashmir, India

**Randall C. Clark** Department of Drug Development and Discovery, Harrison School of Pharmacy, Auburn University, Auburn, AL, USA

**Jack Deruiter** Department of Drug Discovery and Development, Harrison School of Pharmacy, Auburn University, Auburn, AL, USA

**Muralikrishnan Dhanasekaran** Department of Drug Discovery and Development, Harrison School of Pharmacy, Auburn University, Auburn, AL, USA

**Manoj K. Dhar** School of Biotechnology, University of Jammu, Jammu, Jammu & Kashmir, India

**Shivani V. Dhokne** Dr. D.Y. Patil Institute of Pharmaceutical Sciences and Research, Pimpri, Pune, Maharashtra, India

**Thomas Efferth** Department of Pharmaceutical Biology, Institute of Pharmacy and Biomedical Sciences, Johannes Gutenberg University, Mainz, Germany

**Mary Fabbrini** Department of Drug Discovery and Development, Harrison School of Pharmacy, Auburn University, Auburn, AL, USA

**Patrick C. Flannery** Department of Anatomy, Physiology and Pharmacology, Auburn University, Auburn, AL, USA
Rocky Vista University College of Osteopathic Medicine, Parker, CO, USA

**Ayaka Fujihashi** Department of Drug Development and Discovery, Harrison School of Pharmacy, Auburn University, Auburn, AL, USA

**Sabrina Ait Gacem** College of Pharmacy and Health Sciences, Ajman University, Ajman, UAE

**Sudesh Gaidhani** Central Council for Research in Ayurvedic Sciences (Ministry of AYUSH, Govt. of India), New Delhi, India

**Kristina S. Gill** Department of Anatomy, Physiology and Pharmacology, Auburn University, Auburn, AL, USA
University of Massachusetts Memorial Medical Center, Worcester, MA, USA

**Manoj Govindarajulu** Department of Drug Discovery and Development, Harrison School of Pharmacy, Auburn University, Auburn, AL, USA

**Chen-Che J. Huang** Department of Anatomy, Physiology and Pharmacology, Auburn University, Auburn, AL, USA

**Esezah Kyomugisha Kakudidi** Department of Plant Sciences, Microbiology & Biotechnology, Makerere University, Kampala, Uganda

**Ruckmani Kandasamy** Department of Pharmaceutical Technology, University College of Engineering, Bharathidasan Institute of Technology Campus, Anna University, Tiruchirappalli, Tamil Nadu, India

**Sanjana Kaul** School of Biotechnology, University of Jammu, Jammu, Jammu & Kashmir, India

**Hinanit Koltai** Institute of Plant Science, Agriculture Research Organization, Rishon LeZion, Israel

**Meng-Jen Lee** Applied Biomedical Sciences Lab, Department of Applied Chemistry, Chaoyang University of Technology, Taichung, Taiwan, Republic of China

**Sridhar Mani** Department of Molecular Pharmacology, Albert Einstein College of Medicine, Bronx, NY, USA

**Amulya Manohar** Department of Pediatrics, Ramaiah Medical College and Hospital, Bengaluru, Karnataka, India

**Adrian Cuda Banda Meedeniya** Griffith Health, Griffith University, Brisbane, Australia

**Timothy Moore** Department of Drug Discovery and Development, Harrison School of Pharmacy, Auburn University, Auburn, AL, USA

**Shravanthi Mouli** Department of Drug Development and Discovery, Harrison School of Pharmacy, Auburn University, Auburn, AL, USA

**Saraswathy Nachinuthu** Department of Biotechnology, Kumaraguru College of Technology, Coimbatore, India

**Rishi M. Nadar** Department of Drug Discovery and Development, Harrison School of Pharmacy, Auburn University, Auburn, USA

**Suneel K. Onteru** Animal Biochemistry Division, National Dairy Research Institute, ICAR-NDRI, Karnal, Haryana, India

**Nadire Özenver** Department of Pharmacognosy, Faculty of Pharmacy, Hacettepe University, Ankara, Turkey
Department of Pharmaceutical Biology, Institute of Pharmacy and Biomedical Sciences, Johannes Gutenberg University, Mainz, Germany

**Shriya Patel** Department of Drug Discovery and Development, Harrison School of Pharmacy, Auburn University, Auburn, AL, USA

**Sharad D. Pawar** Regional Ayurveda Institute for Fundamental Research, Pune, Maharashtra, India

**Pathirage Kamal Perera** Institute of Indigenous Medicine, University of Colombo, Colombo, Sri Lanka

**Satyanarayana Pondugula** Department of Anatomy, Physiology and Pharmacology, Auburn University, Auburn, AL, USA

**Ramalingam Ponnusamy** Department of Drug Discovery and Development, Harrison School of Pharmacy, Auburn University, Auburn, AL, USA

**Ashish Raina** School of Biotechnology, University of Jammu, Jammu, Jammu & Kashmir, India

**Sindhu Ramesh** Department of Drug Discovery and Development, Harrison School of Pharmacy, Auburn University, Auburn, AL, USA

**Anagha Ranade** Regional Ayurveda Institute for Fundamental Research, Pune, Maharashtra, India

**Sylvie Rapior** Laboratoire de Botanique, Phytochimie et Mycologie, Faculté des Sciences Pharmaceutiques et Biologiques, CEFE UMR 5175, CNRS—Université de Montpellier—Université Paul-Valéry Montpellier—EPHE—IRD, Montpellier Cedex 5, France

**Julia M. Salamat** Department of Anatomy, Physiology and Pharmacology, Auburn University, Auburn, AL, USA

**Tharanath Shankar** Department of Internal Medicine, Ramaiah Medical College, and Hospital, Bengaluru, Karnataka, India

**Itika Sharma** School of Biotechnology, University of Jammu, Jammu, Jammu & Kashmir, India

**Sreelakshmi Sharma** ESIC Medical College, Bengaluru, Karnataka, India

**Supriya Sharma** School of Biotechnology, University of Jammu, Jammu, Jammu & Kashmir, India

**Badrinathan Sridharan** Applied Biomedical Sciences Lab, Department of Applied Chemistry, Chaoyang University of Technology, Taichung, Taiwan, Republic of China

**Mayur Surana** Vardhayu Ayurved, Nashik, Maharashtra, India

**Sivasudha Thilagar** Department of Environmental Biotechnology, School of Environmental Sciences, Bharathidasan University, Tiruchirappalli, India

**Ayeh Trish** College of Science and Mathematics, Auburn University, Auburn, USA

**Patience Tugume** Department of Plant Sciences, Microbiology & Biotechnology, Makerere University, Kampala, Uganda

**Vaishali R. Undale** Pharmacology, Dr. D.Y. Patil Institute of Pharmaceutical Sciences and Research, Pune, Maharashtra, India

**Saraswathi Vishveshwara** Molecular Biophysics Unit, Indian Institute of Science, Bangalore, India

**Zohara Yaniv** Institute of Plant Science, Agriculture Research Organization, Rishon LeZion, Israel

# Mitosis Inhibitors and Medicinal Plants: Neurotoxicity and Neuroprotection

**Nadire Özenver and Thomas Efferth**

## Contents



**Abstract** Cancer is one of the devastating diseases worldwide, causing desperate outcomes and high mortality rates. Despite undeniable improvements in cancer treatment, many patients with malignancies still suffer from adverse drug reactions, among which peripheral neurotoxicity holds great importance. Peripheral neuropathy as a representation of peripheral neurotoxicity is a usual complication of chemotherapy, reducing the life quality of individuals since it can adversely induce sensory and motor dysfunctions influencing patients' life. Mitosis inhibitors are substantially administered drugs during chemotherapy. However, these drugs may induce the occurrence of chemotherapy-induced peripheral neuropathy (CIPN).

N. Özenver
Department of Pharmacognosy, Faculty of Pharmacy, Hacettepe University, Ankara, Turkey

Department of Pharmaceutical Biology, Institute of Pharmacy and Biomedical Sciences, Johannes Gutenberg University, Mainz, Germany

T. Efferth (✉)
Department of Pharmaceutical Biology, Institute of Pharmacy and Biomedical Sciences, Johannes Gutenberg University, Mainz, Germany
e-mail: efferth@uni-mainz.de

 
D. C. Agrawal, M. Dhanasekaran (eds.), *Medicinal Herbs and Fungi*,
https://doi.org/10.1007/978-981-33-4141-8_1

Efficient therapy with fewer side effects specifically focusing on the eradication of malignant cells without affecting healthy cells adversely constitutes today's approach for chemotherapy. Therefore, to target the discovery of functional chemotherapeutics with minimized adverse effects is a rational approach. In this context, medicinal plants and phytochemicals may come into the focus to avoid or treat the complications of peripheral neurotoxicity, if combined with standard chemotherapy regimens, or as complementary and alternative therapy interventions. In the present chapter, we review the factors involved in the pathogenesis of CIPN, neurotoxic mitosis inhibitors, and natural products as neuroprotective and/or neuropreventive agents. Future perspectives will be further manifested for the prevention or treatment of CIPN occurrence.

**Keywords** Cancer · Medicinal plant · Mitosis inhibitor · Neuroprotective · Neurotoxic · Phytochemical

## Abbreviations

| | |
|---|---|
| 5-FU | 5-Fluorouracil |
| Adrs | Adverse drug reactions |
| AUC | Area under the curve |
| AUC12 | Arca under the curve of 12 |
| AUC6 | Area under the curve of 6 |
| CAM | Complementary and alternative medicine |
| CAMKK1 | Calcium/calmodulin dependent protein kinase 1CIPN |
| *CBS* | Cystathionine-β-synthase |
| Chıs | Chinese herb injections |
| CHM | Chinese herbal medicines |
| CIPN | Chemotherapy-induced peripheral neuropathy |
| CNS | Central nervous system |
| CRC | Colorectal cancer |
| Crel | Cremophor-EL |
| CYP | Cytochrome P450 |
| CYP2C8 | Cytochrome P450 Form 1 |
| CYP2C9 | Cytochrome P450 PB-1 |
| DHA | Docosahexaenoic acid |
| DRG | Dorsal root ganglia |
| FDA | Food and Drug Administration |
| GFAP | Glial fibrillary acidic protein |
| HGWD | *Huangqi Guizhi Wuwu* decoction |
| ID3 | Inhibitor of differentiation 3 |
| LJZT | *Liu Jun Zi Tang* |
| LV | Leucovorin |
| MTD | Maximum tolerated dose |
| MTE | *Marsdenia tenacissima* extract |

| | |
|---|---|
| NCI | National Cancer Institute |
| NFATC2 | Nuclear factor of activated T-cells 2 |
| NF-κβ | Nuclear factor kappa B |
| OATP1B2 | Organic anion–transporting polypeptide B2 |
| OCT2 | Organic cation transporter 2 |
| OCTN2 | Organic cation transporter novel 2 |
| PN | Peripheral neuropathy |
| Pt | Platinum |
| RCT | Randomized controlled trial |
| ROS | Reactive oxygen species |
| SNP | Single nucleotide polymorphism |
| TCM | Traditional Chinese Medicine |
| VIPN | Vincristine-induced peripheral neuropathy |

## 1 Introduction

Cancer is a group of diseases in which cells lose their ability to control cell proliferation and gain properties to invade and metastasize distant organs of the body. Cancer is the second-leading cause of death and is predicted to induce 9.6 million deaths globally in 2018, an incidence that is steadily increasing throughout the world (WHO 2020).

Currently, the 5-year survival rate of patients receiving cancer therapy is 67%, and the number of patients is considered to reach up to 23.6 million by 2030 (Abe et al. 2016; NCI 2020). Many anticancer agents in the clinic display adverse drug reactions (ADRs) such as hematological toxicities, hepatotoxicity, nephrotoxicity, neurotoxicity, and so on (Adachi et al. 1983; King and Perry 2001; Perazella and Moeckel 2010; Yang and Moon 2013). Today's concept on chemotherapy is to perform effective treatment with fewer side effects, which means selective therapy with a specific focus on the eradication of malignant cells without affecting healthy cells adversely since ADRs are the fourth cause of mortality with 6.7% incidence in hospitalized patients (Lazarou et al. 1998). Despite their severe side effects, chemotherapeutics are unavoidable elements of cancer management, and minimizing their unwanted effects is of great importance to improve survival rates and life quality of individuals.

Peripheral neuropathy (PN), an expression of peripheral neurotoxicity, is a common complication resulting from chemotherapeutic- and anti-HIV medication-associated toxicity (Fig. 1). Although olfactory neurons and taste receptors can regenerate, many cells in the nervous system divide slowly or not at all. Therefore, neurotoxicity is quite an exciting feature of chemotherapeutics, since they generally target rapidly dividing cells. Besides, the existence of blood–brain, blood–cerebrospinal fluid, and blood–nerve barriers should theoretically prohibit the admission of antineoplastic agents to the nervous system. Recent investigations have demonstrated that chemotherapeutics may harm the nervous system through other

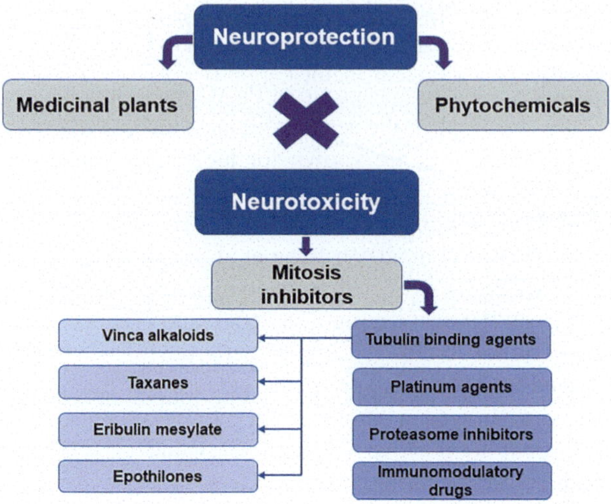


**Fig. 1** Schematic representation of neurotoxic and neuroprotective agents induced by mitosis inhibitors and natural products

mechanisms (Dietrich et al. 2006). Peripheral neuropathy, usually inclined by direct involvement of peripheral nerves, is the most common complication (Magge and DeAngelis 2015). Medications may adversely affect different elements of the peripheral nervous system inclining neuropathy, which is mostly linked to axonal degeneration by a "dying back" type. Peripheral neuropathy develops weeks to months following medication exposure and may prolong even after the cessation of the drug. Commonly used sensitive electrophysiology studies have identified the existence of neuropathy, enabling early diagnosis and treatment of peripheral neurotoxicity (Peltier and Russell 2002).

Neurotoxic side effects are frequent outcomes of well-known chemotherapeutics along with bone marrow suppression and renal toxicity requiring termination of the antitumor therapy or alteration of the dose regimen. Many chemotherapeutics may even incline polyneuropathy, while only a few produce peripheral neuropathy (Quasthoff and Hartung 2002), which results in disrupted sensory and motor symptoms (Fig. 2; Table 1). Numbness, tingling, increased sensitivity to heat and cold, and pain, particularly in the hands and feet, are sensory symptoms. Motor dysfunctions include symptoms of muscle weakness and deteriorative balance (Windebank and Grisold 2008).

Chemotherapy-induced peripheral neuropathy (CIPN) is a usual long-term side effect of antineoplastic agents during or after treatment, decreasing the life quality of patients with cancer. CIPN occurrence affects oncologic treatment resultants adversely by reducing the patient's adhesion to therapy, inclining dose modifications, and therapy disruptions. Evaluations regarding demographic and clinical properties revealed that patients with neurotoxicity were usually older, not employed, and had less annual family income (Miaskowski et al. 2018). CIPN arises approximately in two-thirds (68.1%) of individuals in the first month after chemotherapy. At the same time, the prevalence of which is 60% at 3 months and 30% at 6 months or more among patients (Seretny et al. 2014). The peripheral nervous

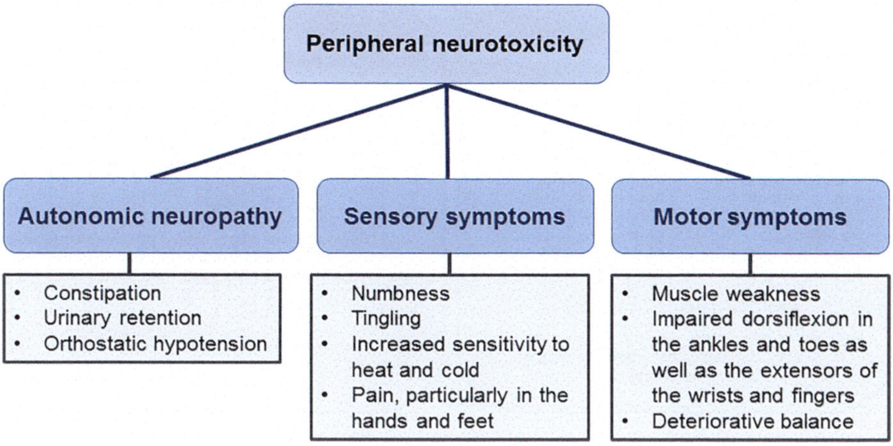


**Fig. 2** Classification of the general symptoms of peripheral neurotoxicity

**Table 1** Type of peripheral neurotoxicity and clinical pattern of chemotherapy-induced peripheral neuropathy (CIPN) by neurotoxic drug classification

| Mitosis inhibitors | Type of peripheral neurotoxicity | Clinical pattern | References |
|---|---|---|---|
| Vincristine | Sensory, motor, and autonomic | Repression of the Achilles–tendon reflex<br>Paraesthesia in the feet and/or hands | Bradley et al. (1970), Gomber et al. (2010), Weiss et al. (1974) |
| Vinorelbine | Sensory and motor | Distal paraesthesia<br>Decrease or abolition of tendon jerks | Pace et al. (1996) |
| Paclitaxel | Sensory | Decline in vibration sense in the hands and feet | Hershman et al. (2011) |
| Docetaxel | Sensory | Tingling in hands or feet<br>Numbness in fingers or toes | Katsumata (2003) |
| Eribulin mesylate | Sensory and motor | Decrease in motor activities<br>Reduction in the sense in distal toes | Wozniak et al. (2011), Vahdat et al. (2013) |
| Ixabepilone | Sensory and motor | Numbness and paraesthesias | Vahdat et al. 2012 |
| Platinum agents | Sensory | Cold-induced syndrome<br>Reduced vibratory sensitivity | Calls et al. (2020), Velasco and Bruna (2014) |
| Bortezomib | Sensory and motor | Distal paraesthesias, numbness, a burning sensation, and neuropathic pain | García-Sanz et al. (2017), Thawani et al. (2015), Zajaczkowska et al. (2019) |
| Thalidomide | Sensory and motor | Reduction in physical activity<br>Distal weakness in the lower limbs | García-Sanz et al. (2017), Mohty et al. (2010) |

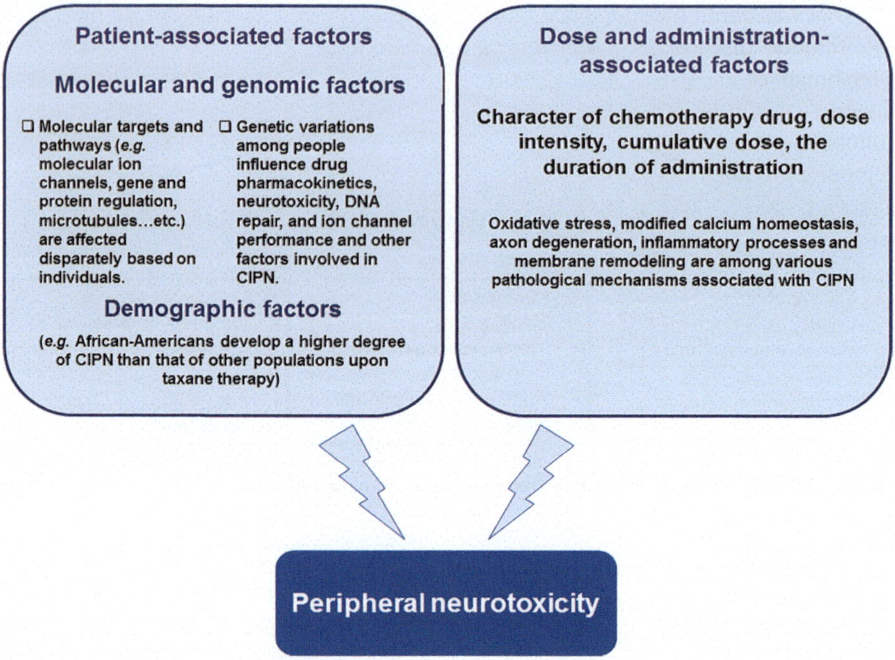


**Fig. 3** Factors involved in the development of peripheral neurotoxicity

system generally can regenerate itself in response to injury. Reformation requires adequate time without damage caused by the chemotherapeutic agent. Based on the drug type and dosage, CIPN may be either reversible or irreversible, determining life quality accordingly (Quasthoff and Hartung 2002). Chemotherapeutic agents involved in the occurrence of CIPN comprise anti-tubulin drugs, platinum agents, taxanes, *Vinca* alkaloids, bortezomib, and thalidomide analogs (Cavaletti and Marmiroli 2010; Chan et al. 2019), most of which function against the dorsal root ganglia (DRG) neurons or the peripheral nerve axons due to lower efficacy of the blood–nerve barrier at these sites (Cavaletti and Marmiroli 2010).

Factors involved in the occurrence of peripheral neurotoxicity mainly consist of two groups: (1) Patient-associated factors and (2) Dose and administration-associated factors (Fig. 3).

Patient-associated factors comprise molecular, genomic, and demographic predictors of CIPN. As the response to the specific drug differs from patient to patient, CIPN development and intensity may vary based on the particular genetic variation of an individual by affecting drug pharmacokinetics, neurotoxicity, DNA repair, and ion channel performance (Chan et al. 2019). Elderly patients were at higher risk for neurotoxicity in some studies (Bulls et al. 2019; Raphael et al. 2017), while no association between age and higher CIPN incidence has been reported so far (Argyriou et al. 2006). The variance among individuals may arise from their accompanying comorbidities and demographic factors. For instance, a higher rank of CIPN was common in diabetic patients, in contrast to the reports of patients with autoimmune diseases, which were related to the decreased odds of neuropathy (Hershman et al. 2016). Furthermore, in the case of obesity co-occurrence, the severity of CIPN symptoms may enhance reducing life quality compared to nonobese patients (Cox-Martin et al. 2017). Another example is the impact of demographic features that African Americans represented a higher degree of CIPN than that of other populations developed CIPN upon taxane therapy (Schneider et al. 2015).

Regarding dose and administration-associated factors, pathophysiology, clinical and biological predictors influenced the occurrence of CIPN. The type and intensity of peripheral neurotoxicity mainly rely on the character of chemotherapy drug, dose intensity, cumulative dose, and the duration of administration forming clinical predictors. CIPN may vary based on the type of cytotoxic drugs. Since these drugs can affect either directly sensory neurons or other cell types inducing off-target effects. Various pathological mechanisms may lead to CIPN development, including oxidative stress, modified calcium homeostasis, axon degeneration, inflammatory processes, and membrane remodeling (Argyriou et al. 2008; Mironov et al. 2005; Starobova and Vetter 2017).

The existence of genetic polymorphisms in patients may also lead up to the development of CIPN in the case of chemotherapy exposure by affecting molecular targets and pathways (Chan et al. 2019). Apart from drug type and dose, the presence of existing diseases (e.g., diabetes, alcohol neuropathy, etc.) may also increase the grade of neuropathy (Quasthoff and Hartung 2002).

CIPN is a significant adverse effect of cancer therapy, reducing life quality. In the present overview, we will review the factors involved in the pathogenesis of CIPN as a representation of neurotoxicity. Conventional mitosis inhibitors may usually act as neurotoxic chemotherapeutic agents. In contrast, medicinal plants and their phytochemical constituents mostly exhibit neuroprotective and/or neuropreventive properties (Fig. 1), which will be exemplified and discussed. Future perspectives for the prevention or treatment of CIPN occurrence will be further put forward.

## 2 Peripheral Neurotoxicity of Mitosis İnhibitors

### *2.1 Tubulin-Binding Agents*

Owing to the highly dynamic character of the microtubule system of eukaryotic cells linked to cell division and cell function, it is an interesting target for drug discovery (Dumontet and Jordan 2010; Karsenti and Vernos 2001). To date, agents interfering with tubulin represent a wide range of classes of agents with significant antitumor activity. Tubulin-binding agents include both naturally occurring and semisynthetic agents, which inhibit cell division by blocking microtubule dynamics (Dumontet and Jordan 2010). The *Vinca* alkaloids, discovered more than 60 years ago (Noble et al. 1958), and the taxanes firstly identified more than 40 years ago (Wani et al. 1971) are among the commonly administered agents in various malignancies (Minev 2011). These drugs have been usually incorporated into multi-agent chemotherapy regimens (Islam et al. 2019; Reck et al. 2020) to achieve more effective treatment outcomes.

Tubulin-binding agents are usually derived from natural sources and can bind to tubulin and/or microtubule (Dumontet and Jordan 2010). Microtubules consist of a backbone of tubulin dimers and microtubule-related proteins (Dustin 1980). Each tubulin molecule is formed by two globular subunits α- and β-tubulin and is represented by a sequence of nearly 450 amino acids and a molecular weight of 50 kD (Downing and Nogales 1998). α- and β-tubulin monomers come together as heterodimers to form a head to tail arrangement for the construction of protofilaments (Amos and Baker 1979). Microtubules are constituted by 12 or 13 protofilaments adjusted in parallel with similar polarity and have a role in intracellular transport, signaling, and mitosis (Perez 2009). Microtubule-targeting agents present highly structural diversity and structural complexity and are frequently obtained from medicinal plants or marine organisms in quite low amounts (Amador et al. 2003; Moudi et al. 2013). Taxanes, *Vinca* alkaloids, epothilones, halichondrins, maytansinoids are among the tubulin-binding agents, which influence microtubule dimerization and dynamics in different ways (Dumontet and Jordan 2010). All of these compounds are antimitotic agents, which block cell proliferation by interfering with microtubules and inhibiting microtubule dynamics during specifically the mitotic stage of the cell cycle. Mainly, the microtubule-targeted antimitotic drugs are categorized into two principal groups the microtubule-destabilizing and microtubule-stabilizing agents. Microtubule-destabilizing agents repress microtubule polymerization at high concentrations and interact with one of two domains of tubulin, which are the "*Vinca*" and "colchicine" domains. *Vinca* alkaloids, the dolastatins, eribulin, and maytansinoids are among the *Vinca*-site binders. Colchicine-site binders comprise various molecules of different origins such as podophyllotoxin, combretastatin colchicine, and its analogs (Dumontet and Jordan 2010). The microtubule-stabilizing agents increase microtubule polymerization at high drug concentrations, among which paclitaxel, docetaxel, the epothilones, and ixabepilone are well-known drugs. The stabilizing agents usually interact with the exact position of the taxoid binding site on beta-tubulin (Buey et al. 2005).

### 2.1.1 Vinca Alkaloids

*Vinca* alkaloids comprise a subset of drugs obtained from the Madagascar periwinkle plant. The substantial *Vinca* alkaloids in clinical use are either naturally occurring (vincristine and vinblastine) or semisynthetic (vindesine and vinorelbine) agents obtained from the pink periwinkle plant *Catharanthus roseus* G. Don (formerly: *Vinca rosea*) (Islam et al. 2019; Moudi et al. 2013). They bind intracellular tubulin and obstruct microtubule polymerization leading to the interruption of mitotic spindle formation and prohibiting cell division. *Vinca* alkaloids are commonly used against hematological malignancies, such as pediatric acute lymphoblastic leukemia. Besides, they can be included in multidrug chemotherapy regimens in a wide range of cancers in adults (Islam et al. 2019; Weiss et al. 1974). *Vinca* alkaloids have usually been applied as direct intravenous injection or continuous infusion and highly metabolized and eliminated by the hepatobiliary system through cytochrome P450 3A (CYP3A) enzyme system (Saba et al. 2015).

Vincristine is the most neurotoxic one among the *Vinca* alkaloids with extensive distribution in body tissues apart from the central nervous system (CNS) due to hindering property of blood–brain barrier. However, it may lead to the development of vincristine-induced peripheral neuropathy (VIPN) even at lower cumulative dosages (Lavoie Smith et al. 2015).

Indicators of VIPN are substantially divided into three categories as sensory, motor, and autonomic neuropathy (Mora et al. 2016; Lavoie Smith et al. 2015). Numbness and tingling, neuropathic pain are common features of sensory neuropathy in the upper and lower extremities (Mora et al. 2016). Motor involvement is also a well-known disabling manifestation of vincristine neurotoxicity, impairing the dorsiflexion in the ankles and toes as well as the extensors of the wrists and fingers (Bradley et al. 1970; Casey et al. 1973). The earliest and universal indication of vincristine neurotoxicity is the repression of the Achilles–tendon reflex (Weiss et al. 1974), which can be either asymptomatic or a stage subsequently giving rise to depression of other deep tendon reflexes (Casey et al. 1973). Paraesthesia in the feet and/or hands is another subjective manifestations of VIPN, usually arises in the early weeks of therapy (Bradley et al. 1970). Hyporeflexia, a decline in deep tendon reflexes are the signs and symptoms of both sensory and motor VIPN. Indications of autonomic neuropathy mainly comprise constipation, urinary retention, and orthostatic hypotension (Gomber et al. 2010).

A number of mechanisms induce VIPN, among which axonal degeneration in peripheral nerves (Gottschalk et al. 1968) and axonal transport dysfunction (Topp et al. 2000) are of significance that axonal degeneration is a SARM1 dependent cellular process (Gerdts et al. 2016) and mice were reported to alleviate vincristine-induced neuropathy in case of its genetic deficiency (Geisler et al. 2016). Larger doses or smaller time intervals may markedly increase VIPN (Diouf et al. 2015). Vincristine is administered intravenously through bolus injections or prolonged infusions. The way of administration also affects the VIPN development and intensity that vincristine (if administered intravenously as bolus injection in comparison to prolonged infusion) caused increased VIPN occurrence in children due to reaching out high concentration in plasma (Kellie et al. 2004). Pharmacokinetic profiles and genetic factors of patients are further factors determining the risk and intensity of VIPN. The general concept about vincristine plasma clearance is that the children have higher vincristine clearance than the adults (Crom et al. 1994), representing a reduced risk of VIPN occurrence in children. Another study conducted by Egbelakin et al. (2011) unraveled that children with precursor B cell acute lymphoblastic leukemia faced less VIPN in case they had a higher CYP3A5 expression genotype than CYP3A5 non-expressers (Egbelakin et al. 2011). Many researchers have pointed out the relationship between the VIPN phenomenon and DNA single nucleotide polymorphisms (SNPs). For instance, an SNP in the centrosome protein encoded by *CEP72* gene was involved in the VIPN development through improving the susceptibility of neuronal cells to vincristine damage (Diouf et al. 2015). Moreover, other SNPs such as CAMKK1 (calcium/calmodulin-dependent protein kinase 1), CYP2C8 (cytochrome P450 Form 1) and CYP2C9 (cytochrome P450 PB-1), NFATC2 (nuclear factor of activated T-cells 2), ID3 (inhibitor of differentiation 3), and SLC10A2 (apical sodium-dependent bile acid transporter) have been proposed to participate in vincristine-induced neuropathy (Johnson et al. 2011). Newly developed lipid-coated or liposomal vincristine has improved the pharmacokinetics and pharmacodynamics of vincristine easing its neurotoxicity (Shah et al. 2016; Silverman and Deitcher 2013).

A semisynthetic *Vinca* alkaloid vinorelbine (5′-nor-anhydro-vinblastine) is used as a single agent or in combination with other drugs against breast cancer, non-small cell lung carcinoma and other malignancies (Furuse et al. 1996; Toso and Lindley 1995). Similar to other *Vinca* alkaloids, vinorelbine blocks axonal transport but at higher concentrations compared to vincristine and vinblastine. Besides, it has a higher selective affinity for tubulin and lower activity on axonal microtubules, leading to less neurotoxicity than other *Vinca* alkaloids (Pace et al. 1996; Toso and Lindley 1995). A randomized study of vinorelbine versus vindesine showed that the peripheral neurotoxic effects of vinorelbine were milder than that of vindesine (Furuse et al. 1996). Despite vinorelbine itself did not exert clinically relevant neurotoxicity in many studies, the neurotoxic injury may intensify in case of a combination of which with other chemotherapeutics such as platinum compounds and paclitaxel (Pace et al. 1996).

### 2.1.2 Taxanes

Taxanes are spindle poisons inducing PN, which mostly include sensory or motor neuropathy based on the type of nerve fibers affected (Swain and Arezzo 2008). The common aspect of taxane-induced PN is that taxanes bind to the $\beta$-tubulin subunit of microtubules, causing the stabilization of microtubules and the disruption of microtubule function, which influences the structures and functions of neurons resulting in neuropathy (Rivera and Cianfrocca 2015). Taxanes usually gather in the soma of sensory neurons of DRG. The process contributing to the neurotoxicity is vice versa to the general proceeding that it often initiates at distal nerve endings followed with the Schwann cell, neuronal body, or axonal transport alterations (Argyriou et al. 2008; Chan et al. 2019).

The level of neuronal damage relies on various factors such as cumulative dose, duration of the agent, and the chemotherapeutic used (Guo et al. 2019; Wolf et al. 2008). The impact of agents in toxicity profiles is diverse based on their formulation. For instance, polyoxyethylated castor oil, or Cremophor® EL (recently renamed Kolliphor® EL), is used in the formulation of paclitaxel. At the same time, docetaxel is formulated with polysorbate 80 (or TWEEN® 80), and solvent-free *nab*-paclitaxel is formed with paclitaxel and human serum albumin at a concentration equivalent to the concentration of albumin in the blood (Summit 2014).

Paclitaxel, a member of taxanes, is a microtubule-stabilizing agent and efficient in the treatment of assorted tumor types including breast, lung, and ovarian cancer (Armstrong et al. 2006; Camidge et al. 2014; Tolaney et al. 2015). However, apart from its non-hematological toxicity, peripheral neurotoxicity is the substantial drawback of paclitaxel (Gornstein and Schwarz 2014; Mielke et al. 2006). The reason for axonal degeneration, secondary demyelination, nerve fiber loss, oxidative stress, and abnormalities in sphingolipids was reported to be linked to peripheral neurotoxicity (Duggett et al. 2016; Gornstein and Schwarz 2014; Kramer et al. 2015). Many investigations uncovered that paclitaxel-induced neurotoxicity is associated with SNPs in cytochrome P450 (CYP) CYP2C8 (Boora et al. 2016), ABCB1 (Abraham et al. 2014; Boora et al. 2016), and TUBB2A (Abraham et al. 2014). However, due to varying replication outcomes, their application in the clinic has not been confirmed. Many studies pointed out the correlation of paclitaxel pharmacokinetics with PN occurrence (Hertz et al. 2018; Scripture et al. 2006). A meta-analysis conducted by Guo et al. (2019) unraveled the dosage and the type of administration that may affect the severity and incidence of PN. Since solvent (Cremophor EL)-based paclitaxel-induced PN inclined lower rate of peripheral neurotoxicity in patients receiving monochemotherapy if compared with *nab*-paclitaxel (an albumin-bound formulation of paclitaxel free from Cremophor EL) (Guo et al. 2019). On the other hand, some clinical studies showed that intense neuropathy remained a long time in patients with metastatic breast cancer following the termination of paclitaxel and docetaxel therapy in comparison to *nab*-paclitaxel treatment (Cortes and Saura 2010; Gradishar et al. 2012).

Docetaxel is a member of the taxoid family obtained via a semisynthetic procedure from the needles of *Taxus baccata* (Chabner and Longo 2011). Docetaxel has been exhibited convincing in vitro and in vivo cytotoxic activity toward various tumor types such as breast, lung, and ovarian cancers (Katsumata 2003). Like paclitaxel, docetaxel functions as a spindle poison inducing the blockage of microtubule dynamics and cell cycle arrest (Ringel and Horwitz 1991). Despite having shared tubulin binding sites and identical characters, the mechanistic and pharmacological variations in between are available. To exemplify, docetaxel is a more potent promoter of tubulin polymerization in vitro with a longer intracellular half-life and exhibits better activity in several tumors (Bissery et al. 1996; Katsumata 2003; Ringel and Horwitz 1991). More effective cytotoxic profile of docetaxel ranges between 1.2- and 2.6-fold than paclitaxel and more than 1000-fold than cisplatin or etoposide in ovarian carcinoma cells were reported previously (Engblom et al. 1997; Kelland and Abel 1992). Likewise, docetaxel and paclitaxel have mainly different toxicity profiles. Noteworthy, docetaxel is associated with minimalized neurotoxicity and suggested as an alternative therapy to paclitaxel for the combination therapy with platinum-based regimens against advanced ovarian cancer (Katsumata 2003; Vasey 2002).

The incidence and intensity of taxane-associated neurotoxicity depend on dose levels, the cumulative dose, and probably the use of paclitaxel with other chemotherapeutics such as cisplatin. Furthermore, predisposing factors such as preexisting PNs contribute to the emergence of neurotoxicity (Mielke et al. 2006). Mielke et al.

(2003) performed research to contrast weekly 1 h infusion of paclitaxel to 3 h infusion in patients with progressive cancer of different origin (substantially breast and lung) placing neurotoxicity to the forefront. A high rate of peripheral neurotoxicity with a significant difference was noted in both infusions, suggesting the occurrence of a strong cumulative property of peripheral neurotoxicity (Mielke et al. 2003). A dose-escalation study to determine pharmacodynamics of non-break weekly paclitaxel and pharmacokinetics of the vehicle Cremophor-EL (CrEL) demonstrated that clinically significant PN usually occurs at around 1500 mg/m$^2$ cumulative dosage at the weekly interval and CrEL levels do not significantly accumulate at doses up to 90 mg/m$^2$ (Briasoulis et al. 2002).

A phase III trial study to determine the optimal duration of chemotherapy showed that later courses of therapy were associated with persistent neuropathy that patients with advanced non-small cell lung cancer receiving first-line paclitaxel plus carboplatin exhibited higher grades of peripheral neuropathy from cycle 4 (20%) to cycle 8 (43%) (Socinski et al. 2002). In another randomized trial, comparative pharmacokinetics of unbound paclitaxel during 1- and 3-h infusions were evaluated. The attenuation in the infusion duration time from 3 to 1 h decreased in the area under the curve (AUC) for unbound paclitaxel in contrast to the AUC of CrEL increased. The study pointed out shorter infusion regimes may be attributable to the occurrence of less severe paclitaxel-associated peripheral neurotoxicity but potential CrEL-related adverse effects, suggesting a challenge to foresee which infusion model (1 or 3 h) is beneficial due to the characters of both paclitaxel and CrEL inducing PN (Gelderblom et al. 2002). Still, the outcomes of many studies mostly reached an agreement about the fact that exposure to paclitaxel might be intimately related to the peripheral neurotoxicity development rather than CrEL. Since, peripheral neurotoxicity developed in patients with advanced malignancies in case of the administration of two CrEL-free formulations including ABI-007 (a novel CrEL-free, protein-stabilized, a nanoparticle formulation of paclitaxel) and Genexol-PM (a polymeric micelle formulated paclitaxel free of CrEL) (Ibrahim et al. 2002; Kim et al. 2004). Surprisingly, in another clinical study, docosahexaenoic acid (DHA)-paclitaxel, a conjugate produced by covalently binding of the natural fatty acid DHA to paclitaxel containing CrEL introduced no cases of severe PNP, which was probably due to extended exposure to very low concentrations of paclitaxel (Wolff et al. 2003). Long-term complications of taxane-associated peripheral neuropathy also differ among patients based on the time after drug exposure discontinuation. For instance, paclitaxel and docetaxel-related PN appeared in 64% of patients who were in 1–13 years post-taxane therapy after the end of the last cure and ceased 14% of them despite the symptoms, which were well-tolerated (Osmani et al. 2012). In another study, among patients receiving 6 months to 2 years of post-taxane therapy, 81% demonstrated symptoms of PN. Among these patients, 27% exhibited severe symptoms in their hands in addition to 25% in their feet (Hershman et al. 2011).

The impact of taxane-induced neuropathy may vary among other chemotherapeutics targeting tubulin. Eribulin mesylate, a microtubule-targeting antineoplastic agent presented relatively lower neuropathy in mice than paclitaxel or ixabepilone at the equivalent maximum tolerated dose (MTD)-based doses. Notable loss of caudal nerve conduction velocity, caudal amplitude, and digital nerve amplitudes, as well as severe degenerative pathologic changes in DRG and sciatic nerve, were observed in patients receiving paclitaxel or ixabepilone. Conversely, eribulin mesylate did not cause critical adverse effects on any nerve conduction parameter, but only contributed to the emergence of milder, rare effects on morphology (Wozniak et al. 2011). Subsequently, another preclinical study directed by the same group evaluated further peripheral neuropathy occurrence in mice with preexisting paclitaxel-induced PN in the case of paclitaxel or eribulin treatment. Eribulin mesylate-administered mice with preexisting paclitaxel-induced PN displayed reduced deleterious outcomes in contrast to additional paclitaxel treatment (Wozniak et al. 2013). These findings led the authors to the conclusion that the severity of taxane-related neuropathies are variable and linked to various factors (Guo et al. 2019; Mielke et al. 2006; Rivera and Cianfrocca 2015).

### 2.1.3 Eribulin Mesylate

Eribulin mesylate is a synthetic derivative of halichondrin B, a macrolide obtained from Japanese marine sponge *Halichondria okadai*, and used for the treatment of metastatic breast cancers (Cavaletti and Marmiroli 2015; Huyck et al. 2011). Eribulin acts as an anti-tubulin agent, the binding sites of which does not collide with the taxane and *Vinca* binding sites. A study conducted by Wozniak et al. (2011) compared the neuropathy-inducing effects of eribulin mesylate, paclitaxel, and ixabepilone based on their MTDs in mice and revealed eribulin mesylate to cause less neuropathy at corresponding MTD-based doses (Wozniak et al. 2011). On the other hand, a randomized phase II clinical trial study pointed out no significant differences between eribulin and ixabepilone in terms of the overall incidence of neuropathy (Vahdat et al. 2013).

### 2.1.4 Epothilones

Epothilones are a relatively new class of natural microtubule-stabilizing agents with probable activity against numerous tumor types, which do not respond to the therapy or weaken after taxane medication (Argyriou et al. 2011). Like taxanes, epothilones bind to $\beta$-tubulin at a molecular epitope apart from that of conventional microtubule-stabilizing agents (Cortes and Baselga 2007). Despite having different chemical structures compared to that of taxanes, PN is a significant adverse effect of epothilones (Cheng et al. 2008).

Epothilones are substantially represented by ixabepilone, which is approved by the US Food and Drug Administration (FDA) in the United States but not in Europe, especially for the treatment of breast cancer, on which other chemotherapeutics are not efficient (Lopus et al. 2015). Sagopilone is another epothilone applied for the treatment of numerous cancers such as non-small cell lung cancer, ovarian cancer, and prostate cancer. Nevertheless, the FDA has not approved it as yet (Heigener et al.

2013). Ixabepilone-induced PN was of primary sensory and cumulative severity, which was correlated with the dose applied during the treatment cycle, the duration of infusion and cumulative dose, and preexisting neuropathy. The incidence rates varied from 1% in early untreated breast cancer up to 24% in heavily pretreated metastatic breast cancer in addition to the grade 4 PN less than 1%. The complications of PN were reversible in many patients resolving within 5–6 weeks (Vahdat et al. 2012).

The clinical disclosure of neuropathy induced by epothilones is mostly characterized by mild and moderate sensory neuropathy. Motor involvement is barely available. The autonomic manifestation was extremely rare as less than 1% of patients (Vahdat et al. 2012; Zajaczkowska et al. 2019). Due to representing a new group of anticancer drugs, the studies assessing epothilones-associated peripheral neurotoxicity are limited. Therefore, the mechanism behind their neurotoxicity has not been entirely enlightened. Microtubule disruption and mitochondrial dysfunction take part in the neurotoxicity occurrence, but the intensity of epothilones-related peripheral neurotoxicity was usually lower than that of taxanes (Vahdat et al. 2012).

## 2.2 *Platinum Agents*

Peripheral neurotoxicity is a common dose-limiting, unwanted effect of platinum compounds approved for clinical use. The incidence and severity of peripheral neurotoxicity vary among platinum (Pt) compounds, the predominant factors deciding which are the cumulative doses, dose intensity, and the type of chemotherapeutic coadministered. Cisplatin ($>350$ mg/m$^2$) and oxaliplatin ($>510$–$765$ mg/m$^2$) may cause high levels of neurotoxicity in contrast to the less neurotoxic carboplatin at standard doses. Still, higher doses of carboplatin with $\text{AUC} > 6$ might generate comparable neurotoxicity (Argyriou et al. 2012).

Sensory neuropathy and cold-induced syndrome are the consequences of platinum exposure over time, among which sensory neuropathy develops in the case of periodical platinum medication over time and is characterized by a reduced vibratory sensitivity. Cold-induced syndrome induced by oxaliplatin caused paresthesias in the perioral part and distal extremities (Calls et al. 2020; Velasco and Bruna 2014). Peripheral neurotoxicities may occur during or at the end of the medication; however, signs and complications can deteriorate or even emerge during the 2–6 months after the termination of treatment. This phenomenon was called "the coasting effect", in which recovery is usually not accomplished (Argyriou 2015).

Mechanism of action of platinum compounds mainly consists of four steps, including cellular uptake, aquation and activation, DNA platination, and cellular processing of Pt-DNA lesions, inducing cell survival or apoptosis. Besides, platinum compounds generate oxidative stress if the aquated platinum interacts with cytoplasmic nucleophiles resulting in peroxidation of proteins and lipids, which is another mechanism associated with the cytotoxicity of platinum compounds (Calls et al. 2020). Moreover, the inadequacy of the cytoplasmic antioxidant mechanisms further enhances reactive oxygen species (ROS) production, boosting neuronal apoptosis. Inflammatory mechanisms mediated by the P2X7 receptor activation and paracrine cytokines release are other mechanisms involved particularly in oxaliplatin-induced neurotoxicity (Calls et al. 2020; Massicot et al. 2013).

Cisplatin, a synthetic inorganic and water-soluble platinum complex and the first platinum-derived compound, was approved by the FDA in 1978 for the treatment of numerous solid tumor malignancies. Since then, platinum compounds have been increasingly applied in the clinic due to their confirmed efficacy (Calls et al. 2020). However, it has limited use due to inclining nephrotoxicity, ototoxicity, and sensory neuropathy. Cisplatin-induced neurotoxicity can persist for decades. A previous investigation reported that approximately 15 years after cisplatin treatment, patients with non-symptomatic neuropathy, symptomatic neuropathy, and disabling neuropathy was observed in 38%, 28%, and 6% of patients, respectively (Strumberg et al. 2002). A cumulative cisplatin dose of 225–500 $mg/m^2$ inclined any grade of peripheral nerve hurt in nearly 60% of patients (Argyriou et al. 2012). Still, solely 10% of them suffer from grade 3 to 4 neurotoxicity (Gatzemeier et al. 2000; Sutton et al. 2000). Furthermore, peripheral neurotoxicity improved if cisplatin was coadministered with paclitaxel in comparison to cisplatin medication itself (Berger et al. 1997). The adverse effects profile and development of inherent and/or acquired tumor resistance at standard rationale doses encouraged scientists to the discovery of other platinum-based drug alternatives. Therefore, carboplatin and oxaliplatin have been the only drugs approved in 1989 and 1996, respectively, globally to date (Calls et al. 2020).

Carboplatin, the second-generation platinum drug, is used either to treat bladder, breast, endometrial, head and neck, lung and ovarian cancers or as adjuvant therapy in the treatments of germ cells, bladder, endometrial, and head and neck tumors (Chu and DeVita 2018). Carboplatin is a less neurotoxic agent than cisplatin and oxaliplatin (Argyriou et al. 2012). Carboplatin monotherapy was nearly linked to no peripheral neuropathy at an area under the curve of 6 (AUC6), whereas grade 3–4 neurotoxicity occurred at an area under the curve of 12 (AUC12) (Gore et al. 1998).

Oxaliplatin, a third generation of platinum, inhibits tumor growth as an alkylating agent and is administered in the treatment of numerous solid tumors, particularly against metastatic colon cancer (Chu and DeVita 2018; de Gramont et al. 2012). The major side effect of oxaliplatin is the underlying neurotoxicity mechanism which is still not determined. Oxaliplatin-treated patients encountered different levels of neurotoxicity (Kemeny et al. 2004). Patients who received oxaliplatin-based regimens experienced chronic-oxaliplatin induced neurotoxicity in the range of 60–75% (Argyriou et al. 2007; Argyriou et al. 2013). Oxaliplatin was reported to gather in the DRG inducing axonal hyperexcitability and repetitive discharges (Pasetto et al. 2006). The cumulative dose, time of infusion, the presence of PN prior to the onset of chemotherapy hold the most prominent parts in the occurrence of oxaliplatin-associated PN (Grothey 2005). Despite the lack of efficacious therapy to tackle oxaliplatin-associated neurotoxicity, a number of Chinese herbal medicines (CHM) or Chinese herbal extracts were attenuated oxaliplatin-associated neurotoxicity (Chen et al. 2018).

Based on the data available until now, the prevention and treatment of platinum compounds-induced peripheral neurotoxicity are still not fully achieved. Clinical settings can only present indirect information. Since it is usually not possible to obtain the tissue of interest from patients, preclinical studies are of vital importance and provide elucidation of underlying mechanisms associated with platinum compounds-related peripheral neurotoxicity. Therefore, future research should primarily focus on method and/or model development to be performed in preclinical studies.

## *2.3 Proteasome Inhibitors*

Bortezomib and carfilzomib are reversible proteasome inhibitors applied against multiple myeloma and certain types of lymphoma. Bortezomib, a boronic acid dipeptide, blocks the chymotryptic site of the 26S proteasome (Broyl et al. 2010). The development of a painful, sensory PN is one of the most common adverse effects of bortezomib medication, which may be accompanied by predominant axonal neuropathy as well as demyelinating polyneuropathy (Thawani et al. 2015; Zajaczkowska et al. 2019). Several factors decide the incidence and severity of PN induced by proteasome inhibitors. For instance, bortezomib, if administered subcutaneously, was morc tolerable than its intravenous form (Hu et al. 2017). Oral proteasome inhibitor ixazomib exhibited less PN (Zajaczkowska et al. 2019). Numerous factors are involved in the development of bortezomib-induced PN including damage to mitochondria and endoplasmic reticulum in DRG (Cavaletti et al. 2007), dysregulation of mitochondrial calcium homeostasis (Landowski et al. 2005), autoimmune factors, and inflammation (Ravaglia et al. 2008), activation of neuronal nuclear factor kappa B (NF-κB) (Alé et al. 2016), genetic factors (e.g., inherent SNPs, or a genetic locus mapping to *PKNOX1* and in cystathionine-β-synthase (*CBS*) at 21q22.3) (Broyl et al. 2010; Magrangeas et al. 2016).

## *2.4 Immunomodulatory Drugs (Thalidomide Analogues)*

The immunomodulatory agent thalidomide prohibits cell division by inhibiting angiogenesis and modifying the response of the immune system by various mechanisms (Chan et al. 2019; Seretny et al. 2014). FDA has approved thalidomide to cure several malignant diseases, including multiple myeloma and prostate cancer (Melchert and List 2007). Thalidomide-associated PN is substantially sensory axonal neuropathy (García-Sanz et al. 2017; Mohty et al. 2010). The cumulative dose and duration of therapy determine the risk of neurotoxicity (Cavaletti et al. 2004), which may be enhanced by other factors including older age, other chemotherapeutics, and vitamin $B_{12}$/folate insufficiency (Chaudhry et al. 2008; Mohty et al. 2010).

Lenalidomide and pomalidomide are other thalidomide analogs, whose neurotoxic mechanisms are not well known at present (Zajaczkowska et al. 2019). Principally, the hypotheses focus on three assumptions: (a) Due to its antiangiogenic property, thalidomide may result in ischemia-linked neuropathy (Kirchmair et al. 2007), (b) Thalidomide metabolites may incline the generation of ROS, causing neuronal damage (Wani et al. 2017), and (c) Thalidomide inhibits NF-κB activity and TNF-α production, which may increase neuronal fragility (Islam et al. 2019).

## 3 Neuroprotection of Medicinal Plants/Phytochemicals and Treatment Alternatives

Cancer ranks as one of the substantial causes of death in humans and is regarded as a significant threat to the enhancement of life quality and survival in every country worldwide. Theoretical research and clinical practice have enabled human beings to illuminate facts associated with cancer. Considering cancer as a genetic illness in the early 1990s due to the relation between p53 mutation and cancer has moved to the identification of many genes involved in cancer during the last 30 years (Turnbull et al. 2018). Subsequently, investigations put forward epigenetics along with genetic changes in the initiation, progress, and other steps of cancer. By the isolation of cancer stem cells in 1994 (Lapidot et al. 1994), the characteristics of cancer stem cells contributing to the tumor initiation, development, migration, invasion, and metastasis were further proven (Jiang et al. 2012). Another additional concept, that is, tumor microenvironment, is also a prominent factor accounting for intratumoral heterogeneity and chemotherapeutical resistance affecting the biology of tumors (Son et al. 2017).

During recent years, a great interest in the use of complementary and alternative medicine (CAM) approaches in patients with cancer have come to the attention in recent years, reaching up to 80% in Western countries (Richardson et al. 2000). Dobos et al. reported that approximately 1.5 billion people use CHM globally (Dobos et al. 2005). Traditional Chinese Medicine (TCM), including herbal and Chinese patented drugs, has been recognized as a fashionable trend of CAM above all in China (Chen et al. 2015). The reason behind the use of traditional herbal medicines as an alternative therapy by a vast number of patients has arisen from its effectiveness and absence of adverse effects. In cancer therapy, traditional herbal medicines were disclosed to reduce chemotherapy-related adverse effects, enhance life quality of patients (Molassiotis et al. 2009), improve cellular immunity of cancer patients undergoing chemo−/radiotherapy (Wang et al. 2017), alleviate cancer pain (Smith and Bauer-Wu 2012), mitigate cancer-related fatigue (Zee-Cheng 1992), and curing anorexia and cachexia (Ming-Hua et al. 2016). For instance, Mcculloch et al. (2006) reported that *Astragalus*-based CHM might enhance the efficacy of platinum-based chemotherapy in the case of its combination with chemotherapy (McCulloch et al. 2006). A systematic review by Li et al. (2013) further suggested that the inclusion of CHM as an adjuvant therapy may attenuate chemotherapy toxicity, extend survival rate and improve tumor response in advanced non-small cell lung cancer (Li et al. 2013). Moreover, numerous CHM-derived phytochemicals such as quercetin, resveratrol, berberine, curcumin, and so on, have been studied to demonstrate and display antitumor activity in many cancers (Diogo et al. 2011; Rauf et al. 2018a; Rauf et al. 2018b; Wilken et al. 2011). In TCM philosophy, cancers occur due to a disturbance of the smooth flow of vital energy and/or yin-yang balance (the balance sustains health) and are attributed to the occurrence of incoordination among the body–mind–environment network. Therefore, TCM has a holistic approach with a special focus on the correction of imbalances. However, the lack of formal documentation and little high-quality scientific data limit the administration of TCM as conventional therapy. In this context, paying attention to the TCM as a complementary therapy in addition to the traditional therapy regimens may provide more effective treatment outcomes with fewer side effects in cancer (Xu et al. 2006).

Several medicinal plants, as well as phytochemical constituents, have been reported to reduce peripheral neurotoxicity as a well-known side effect of chemotherapy until now. Examples are summarized as follows and in Table 2.

*Marsdenia tenacissima* is represented as an anticancer herb in TCM and documented in Medicinal Plants in Southern Yunnan in Ming Dynasty 600 years ago (Zhou et al. 2019). A systematic review of the clinical practice of *Marsdenia tenacissima* extract (MTE) from the stems was generated to form a base of the randomized controlled trial (RCT) used against gastric cancer. According to the literature search by Zhou et al. (2019) suggested MTE as a potential adjuvant therapy enhancing the response toward anticancer therapy and decreasing numerous chemotherapy-induced adverse effects including peripheral neurotoxicity (Risk ratio 0.77, 95% Confidence interval 0.59–1.01). Oral administration of MTE was reported to be the favored choice (Zhou et al. 2019).

Chen et al. (2018) compiled and reviewed the data about the probable functions of CHMs in the recovery of chemotherapy-induced adverse effects in patients with colorectal cancer (CRC). Assorted Astragali radix extracts were reported to prevent nervous tissue from oxaliplatin-induced nerve injury without altering chemotherapeutic activity (Chen et al. 2018; Deng et al. 2016a). In particular, 50% hydroalcoholic extract of Astragali radix alleviated neuro-damage-induced pain in oxaliplatin-receiving cells. Those effects were attributed to the inactivation of caspase-3 in the DRG (Di Cesare Mannelli et al. 2015).

Another example *goshajinkigan* (TJ-107) is a traditional Japanese medicine (Kampo), including several medicinal plant parts in its formulation (Tsumura Co 2014). A phase II clinical trial study unraveled TJ-107 has a convincing activity in retarding the initiation of grade 2 or advanced oxaliplatin-induced neurotoxicity with pleasing safety and tolerability in colorectal cancer patients receiving oxaliplatin (Kono et al. 2013; Matsui et al. 2011).

*Huangqi Guizhi Wuwu* decoction (HGWD), comprising Astragali radix, Cinnamomi ramulus, Paeonia radix Alba, Jujubae fructus, and Zingiberis rhizoma, was documented in "Synopsis of the Golden Chamber" to treat some neurological symptoms (Wiseman et al. 2009). Cheng et al. (2017) tested the impact of AC591, a

**Table 2** Neuroprotective medicinal plants and phytochemicals from natural origin

| Natural product | Animal model or types of cancer in patients exposed to natural agents | Mechanism | Method or Preclinical/ Clinical study | References |
|---|---|---|---|---|
| *Goshajinkigan* | Patients with colorectal cancer | Increasing nitric oxide production and induction of dynorphin release in the spinal cord | Clinical trial | Hosokawa et al. (2012), Ohnishi and Takeda (2015), Yoshida et al. (2013) |
| *Wen-luo-tong* decoction | Rat model | Averting hypertrophy and activation of glial fibrillary acidic protein (GFAP)-positive astrocytes in the spinal dorsal horn Downregulation of GFAP and TNF-$\alpha$ | Preclinical study (in vivo) | Deng et al. (2016b) |
| *Guilongtongluofang* | Patients with colorectal cancer | ns | Clinical trial | Liu et al. (2013) |
| *Keishikajutsubutou* (TJ-18) | Patients with colon cancer | ns | Clinical trial | Fu et al. (2018), Yamada et al. (2012) |
| *Liu Jun Zi Tang* (LJZT) | Human neuroblastoma cells and mice model | Exerting antioxidative effect and controlling mitochondrial function | Preclinical study (in vitro and in vivo) | Chiou et al. (2018) |
| *Aidi* and compound matrine injection | Patients with colorectal cancer | ns | Network meta-analysis to evaluate the clinical efficacy and safety of *Aidi* and compound matrine injection combined with FOLFOX | Fujii et al. (2004), Ge et al. (2016) |
| *Shakuyaku-kanzo-to* | Patients with epithelial ovarian carcinoma | ns | Clinical trial | Fujii et al. (2004), Imai et al. (2012), Yamamoto et al. (2001) |

(continued)

**Table 2** (continued)

| Natural product | Animal model or types of cancer in patients exposed to natural agents | Mechanism | Method or Preclinical/ Clinical study | References |
|---|---|---|---|---|
| *Curcumin* | Murine model | Activation of NGF/Akt and Nrf2/HO-1 pathways and blockage of oxidative stress | Preclinical study (in vivo) | Dai et al. (2020) |
| Cyanidin | PC12 cells from neural crest origin | Blocking ROS-mediated DNA damage and apoptosis | Preclinical study (in vitro) | Li et al. (2015) |
| Berberine | Rat model | Upregulation of the *nrf2* gene | Preclinical study (in vivo) | Singh et al. (2019) |

*ns* not stated

standardized extract of HGWD, in rodents and clinical trials in which oxaliplatin-associated PN occurred. The study uncovered not only the decline in cold hyperalgesia, mechanical allodynia as well as morphological damage of DRG, but also an increment in antitumor response in a rat model oxaliplatin-induced peripheral neuropathy developed. Subsequently, the AC591-treated group ($n = 36$) exhibited grades 1–2 neurotoxicity as 25%, while in the control group the incidence was 55.55% ($P < 0.01$) ($n = 36$). No significant differences in terms of antitumor activity between the two groups were observed, suggesting AC591 as a potential adjuvant in oxaliplatin-dependent therapy through reducing PN symptoms in a clinical application (Cheng et al. 2017).

Another study conducted by Hosokawa et al. (2012) demonstrated the preventive effects of *goshajinkigan*, a traditional Japanese medicine including ten herbal constituents, and *shakuyakukanzoto*, an extract of a mixture of *Glycyrrhiza* and *Paeony* root, on oxaliplatin-induced neurotoxicity. Both the traditional Japanese medicines eased oxaliplatin-induced neurotoxicity without causing any adverse effect on antitumor activity (Hosokawa et al. 2012). In a clinical trial, *goshajinkigan* prohibited aggravation of oxaliplatin-induced peripheral neuropathy in colorectal cancer patients (Ohnishi and Takeda 2015; Yoshida et al. 2013). On the other hand, some clinical investigations and a systematic review concluded the fact that the administration of *goshajinkigan* as standard care is not currently advocated due to the low quality and inadequate amount of confirmed data (Hoshino et al. 2018; Kuriyama and Endo 2018; Oki et al. 2015; Zhang et al. 2018).

*Wen-luo-tong* decoction is a herbal medicine belonging to the TCM and contains *Epimedium brevicornum, Geranium qilfordii, Cinnamomum cassia*, and *Carthamus tinctorius. The decoction was reported to be used externally for the treatment of* *chemotherapy-induced neuropathic pain in China (*Chen et al. 2018*)*. Deng et al. (2016a, b) *studied the impact of Wen-luo-tong* decoction in a rat model of oxaliplatin-related chronic neuropathic pain. *Wen-luo-tong* decoction lessened oxaliplatin-induced mechanical allodynia and mechanical hyperalgesia, prohibited the alterations in the somatic, nuclear, and nucleolar areas of neurons in DRG as well as hypertrophy and activation of glial fibrillary acidic protein (GFAP)-positive astrocytes in the spinal dorsal horn. Furthermore, it downregulated GFAP and TNF-*α, all proposing Wen-luo-tong* decoction inverted either glial activation in the spinal dorsal horn or nociceptive sensitization of oxaliplatin-resultant chronic neuropathic pain in rats (Deng et al. 2016b).

It has been shown in a randomized double-blind and placebo-controlled clinical trial that, compared with controls, *Guilongtongluofang* remarkably alleviated the development of grades 1–2 neurotoxicities after 6 cycles of adjuvant oxaliplatin-based chemotherapy in colorectal cancer patients (70.0 vs. 51.7%, $P < 0.05$). Besides, *Guilongtongluofang* also delayed the onset time of grades 1–4 neurotoxicities relative to the control group (9.4 vs. 6.5 weeks in the trial vs. control groups, $P < 0.05$). Most importantly, no statistically significant difference was found in tumor response rate between the trial and the control groups, indicating that *Guilongtongluofang* is a promising prescription for preventing oxaliplatin-induced neurotoxicity in CRC patients without reducing the efficacy of oxaliplatin-based chemotherapy. However, the underlying mechanisms need to be further investigated (Liu et al. 2013).

*Keishikajutsubutou* (TJ-18) is a herbal formula containing cassia twig, monkshood, and rhizoma Atractylodis and usually administered to treat arthralgia and neuralgia. A clinical study demonstrated that the combined use of TJ-18 with powdered refined aconite root (TJ-3023) might decrease oxaliplatin-induced PN in patients with colon cancer (Fu et al. 2018; Yamada et al. 2012).

*Liu Jun Zi Tang* (LJZT), a popular Chinese herbal formula and Kampo medicine, has been used in folkloric medicine for several intentions since ancient times (Yang et al. 2013). Chiou et al. (2018) investigated the effect of LJZT in a mice model of cisplatin-induced neurotoxicity, which indicated LJZT alleviated cisplatin-associated thermal hyperalgesia, cytosolic and mitochondrial-free radical formation in mice as well as apoptosis in human neuroblastoma SH-SY5Y cells. Moreover, it also reverted the cisplatin-induced decrease in mitochondrial membrane potential and enhanced the release of mitochondrial pro-apoptotic factors. To sum up, LJZT prevented cisplatin-induced neurotoxicity through antioxidative effects and controlling mitochondrial function (Chiou et al. 2018).

Ge et al. (2016) performed a meta-analysis of Chinese herb injections (CHIs) combined with FOLFOX chemotherapy (a regimen containing ranging doses and schedules of 5-fluorouracil (5-FU), leucovorin (LV) combined with oxaliplatin) to assess clinical efficacy and safety of CHIs in patients with advanced CRC. Combinations of FOLFOX with *Aidi* and compound matrine injection remarkably attenuated the incidence of peripheral neurotoxicity when compared with FOLFOX alone. Still, due to insufficient evidence, the necessity of further studies was emphasized and warranted to prove the efficacy of CHIs plus FOLFOX therapy in patients with

CRC (Ge et al. 2016). Likewise, another review compiled by Chen et al. (2016) evaluated the efficacy and adverse effects of CHM in case of inclusion of which to the chemo- or radiotherapy in cases with oesophageal cancer. No precise proof but only probable predictions were obtained, proposing the requirement for additional future trials in high quality (Chen et al. 2016).

A meta-analysis performed by Chen et al. (2014) appraised coadministration of herbal medicines and a kind of FOLFOX therapy FOLFOX 4. Based on the outcomes of meta-analysis, if herbal medicines were added to FOLFOX 4 regimen in advanced colorectal cancer patients, tumor response rate and 1-year survival increased in contrast to the decline in peripheral neurotoxicity induced by chemotherapeutics. Still, pursuance of further RCTs with a strong methodological property was recommended to obtain precise consequences (Chen et al. 2014).

*Shakuyaku-kanzo-to* is herbal medicine and one of the interventions applied to relieve the symptoms of taxane-based neurotoxicity *in Japan and China* (Yoshida et al. 2009). *Shakuyaku-kanzo-to* was demonstrated to be effective in the prevention or development of CIPN-associated muscle pain. Besides, no *Shakuyaku-kanzo-to-related* adverse reactions and no troubles about its safety were monitored (Yamamoto et al. 2001). Another investigation by Fujii et al. (2004) studied 21 cases with arthralgia and muscular pain arisen from carboplatin and paclitaxel chemotherapy. All patients with ovarian epithelial carcinoma received *Shakuyaku-kanzo-to* per os and among which 43% was observed to reduce pain (Fujii et al. 2004). Imai et al. (2012) further reviewed the extensively used and investigated medications in case studies as well as clinical trials to lessen taxane-induced neurotoxicity, which reported *Shakuyaku-kanzo-to* as being of importance. All of the performed investigations suggested coadministration of conventional chemotherapeutics with *Shakuyaku-kanzo-to* may be crucial due to providing optional strategy to the conventional medical agents reducing peripheral neurotoxicity (Imai et al. 2012).

Curcumin, a constituent of *Curcuma longa*, has been traditionally used for several purposes in Asia, among which counteracting carcinogenesis, sensitizing cancer cells to chemotherapeutics, and protecting the healthy cells from chemotherapy-associated toxicities were reported (Liu et al. 2018; Mendonça et al. 2013; Zhou et al. 2011). A recent study conducted by Dai et al. (2020) demonstrated that joint use of oral curcumin and colistin notably enhanced colistin-induced impaired sensory and motor dysfunctions in a dose-dependent manner. Furthermore, the preventive effect of oral curcumin on colistin induced peripheral neurotoxicity was attributed to the activation of NGF/Akt and Nrf2/HO-1 pathways and blockage of oxidative stress, proposing the potential clinical importance of curcumin as an oral neuroprotective agent coadministered along with colistin therapy (Dai et al. 2020).

Having a convincing antioxidant property, the possible protective influence of the natural flavonoid cyanidin was studied on PC12 cells with cisplatin-induced neurotoxicity. Cyanidin inhibited cisplatin-induced neurotoxicity through blocking ROS-mediated DNA damage and apoptosis, suggesting its potential use to prohibit the emergence of neurotoxicity by cisplatin (Li et al. 2015).

Berberine, a naturally occurring alkaloid, possesses various effects (Imanshahidi and Hosseinzadeh 2008), among which anti-inflammatory property has led scientists to examine its effect on paclitaxel-induced neuropathy. Mice treated with berberine significantly attenuated paclitaxel-associated thermal hyperalgesia, which is a measure of neuropathic pain in comparison to the non-treated group (Rezaee et al. 2019). Moreover, a recent study established the neuropathy preventive effect of berberine in paclitaxel-induced peripheral neuropathy model in rats. Berberine upregulated the *nrf2* gene, which is assumed to mediate the neuropathy preventive role of berberine (Singh et al. 2019).

Many examples of natural products have come into prominence to avoid or treat the complications of peripheral neurotoxicity so far, and among which some are exemplified in the present chapter. Although a number of those exhibit convincing activity against PN, still more robust, precise, and dependable preclinical and clinical studies are warranted to convince scientists about their efficacy and safety. Yet, based on the comprehensive literature search, it can be concluded that overall and severe CIPN incidence may be eased by the combination of chemotherapeutics with traditional medicine applications in cancer patients.

## 4 Discussion and Future Perspectives

CIPN is a devastating outcome of cancer therapy affecting the life quality of individuals adversely. Although the exact pathophysiology behind CIPN is not known, improvements in molecular genetics and pathobiological mechanisms have been achieved. In this context, few efficient pharmacological and nonpharmacological interventions are in existence to treat CIPN, although no treatments are still available to prevent CIPN. The American Society of Clinical Oncology Guidelines for CIPN put adequate proof to advocate duloxetine for the treatment of existing CIPN (Dorsey et al. 2019; Hershman et al. 2014).

Prevention alternatives are restricted, and various preclinical phases have remained under investigation. To exemplify, genetic lack of SARM1 in mice inhibited VIPN development; thus intense striving to obtain SARM-1 inhibitors for clinical use has been continuing (Geisler et al. 2016). Besides, other genetic deficiencies in DRG comprising the murine solute carrier organic anion–transporting polypeptide B2 (OATP1B2), organic cation transporter novel 2 (OCTN2) and organic cation transporter 2 (OCT2) avoid vincristine-, taxane- and platin-associated PN (Fujita et al. 2019; Leblanc et al. 2018).

An RCT drew attention to behavioral interposition that exercises for cancer patients along with chemotherapy attenuated CIPN complications of hot/coldness in hands/feet and numbness/tingling in comparison to the control group, emphasizing the prominence of physical exercise (Keckner et al. 2018).

A number of evidence exists about the influence of natural products reducing chemotherapy and radiotherapy-induced side effects. Thus, these agents may be used as adjuvant therapy regimens for prophylaxis and treatment of chemotherapy and radiotherapy-associated neurotoxic effects in cancer patients. However, this conception still requires more investigation using preclinical and clinical evaluations in high quality together with the determination underlying mechanism of CIPNs and the effect of natural products on CIPN. New therapies in clinical validation phases specifically focus on the research about gut microbiota due to their prebiotic potential and, thus, fighting against many diseases with particular interest to the targets of CIPN. These studies will enable us to make the explicit preventive process of chemotherapy-associated side effects by natural products and develop improved treatment approaches (Zhang et al. 2018).

Majithia et al. (2016) assessed the outcomes of CIPN trials funded by National Cancer Institute and summarized those with importance for CIPN trial design in the following: (a) Preclinical studies should be strengthened, (b) Patient-associated factors of CIPN are of more significance than clinical ones, (c) Standard traditional medications (e.g., calcium and magnesium coadministered with chemotherapy) are not used anymore for the prevention of CIPN, (d) Neuropathy and its related response differ based on various factors (e.g., the preexistence of neuropathy, paclitaxel-induced, taxane-induced PN, etc.), and (e) Genetics of individuals rank prominence (Majithia et al. 2016).

To conclude, treatment and/or prevention choices for CIPN are still missing despite numerous investigations. PN is induced by multiple causative agents such as demographic and clinical properties of the patient as well as genetics. The data should be collected from large scale patients in clinical trials and analyzed adequately in terms of physiological, environmental, clinical, and demographic properties of patients to clarify the CIPN phenotype. A comprehensive examination of genetics and different phenotypes should be the main focus for future studies, and the development of standardized, more robust, and accurate tools are needed for this purpose.

Accumulating evidence suggests that Kampo medicine, together with TCM and other traditional therapy regimens, may appear to hold useful importance for the treatment or prophylaxis of CIPN (Ohnishi and Takeda 2015; Zhang et al. 2018). Natural products involved in traditional therapy regimens have been proven to be effective in many studies. They include crude extracts, bioactive components-enriched fractions, and pure compounds and usually have several substances rather than only one compound in their composition. Therefore, further studies should focus on clarifying, which constituent has a substantial beneficial effect related to the treatment or prevention of PN and what is the background mechanism behind. Placebo-controlled, double-blind, RCTs will allow scientists to form a basis similar to those for Western medicine, providing decisive evidence. After that, these folkloric approaches may present alternative interventions against CIPN.

**Acknowledgments** We are grateful for a postdoc stipend given to N.Ö. by the Germany Academic Exchange Service (DAAD) (Funding programme/-ID: Research Grants- Short-Term Grants, 2019 (57440917)) for a research stay at the Johannes Gutenberg University, Mainz, Germany.

*Conflict of interest*: The authors declare that there is no conflict of interest.

## References


Abe JI, Martin JF, Yeh ETH (2016) The future of onco-cardiology: we are not just "side effect hunters.". Circ Res 119:896–899. https://doi.org/10.1161/CIRCRESAHA.116.309573

Abraham JE, Guo Q, Dorling L et al (2014) Replication of genetic polymorphisms reported to be associated with taxane-related sensory neuropathy in patients with early breast cancer treated with paclitaxel. Clin Cancer Res 20:2466–2475. https://doi.org/10.1158/1078-0432.CCR-13-3232

Adachi T, Sezaki T, Ishii H et al (1983) Clinical responses and side effects of aclarubicin (ACR) in the treatment of multiple myeloma. Gan To Kagaku Ryoho 10:1653–1658. (Article in Japanese)

Alé A, Bruna J, Calls A et al (2016) Inhibition of the neuronal NF-κB pathway attenuates bortezomib-induced neuropathy in a mouse model. Neurotoxicology 55:58–64. https://doi.org/10.1016/j.neuro.2016.05.004

Amador ML, Jimeno J, Paz-Ares L et al (2003) Progress in the development and acquisition of anticancer agents from marine sources. Ann Oncol 14:1607–1615. https://doi.org/10.1093/annonc/mdg443

Amos LA, Baker TS (1979) The three-dimensional structure of tubulin protofilaments. Nature 279 (5714):607–612. https://doi.org/10.1038/279607a0

Argyriou AA (2015) Updates on oxaliplatin-induced peripheral neurotoxicity (OXAIPN). Toxics 3 (2):187–197. https://doi.org/10.3390/toxics3020187

Argyriou AA, Bruna J, Marmiroli P et al (2012) Chemotherapy-induced peripheral neurotoxicity (CIPN): an update. Crit Rev Oncol Hematol 82(1):51–77. https://doi.org/10.1016/j.critrevonc.2011.04.012

Argyriou AA, Cavaletti G, Antonacopoulou A et al (2013) Voltage-gated sodium channel polymorphisms play a pivotal role in the development of oxaliplatin-induced peripheral neurotoxicity: results from a prospective multicenter study. Cancer 119(19):3570–3577. https://doi.org/10.1002/cncr.28234

Argyriou AA, Koltzenburg M, Polychronopoulos P et al (2008) Peripheral nerve damage associated with administration of taxanes in patients with cancer. Crit Rev Oncol Hematol 66(3):218–228. https://doi.org/10.1016/j.critrevonc.2008.01.008

Argyriou AA, Marmiroli P, Cavaletti G et al (2011) Epothilone-induced peripheral neuropathy: a review of current knowledge. J Pain Symptom Manag 42(6):931–940. https://doi.org/10.1016/j.jpainsymman.2011.02.022

Argyriou AA, Polychronopoulos P, Iconomou G et al (2007) Incidence and characteristics of peripheral neuropathy during oxaliplatin-based chemotherapy for metastatic colon cancer. Acta Oncol 46(8):1131–1137. https://doi.org/10.1080/02841860701355055

Argyriou AA, Polychronopoulos P, Koutras A et al (2006) Is advanced age associated with increased incidence and severity of chemotherapy-induced peripheral neuropathy? Support Care Cancer 14(3):223–229. https://doi.org/10.1007/s00520-005-0868-6

Armstrong DK, Bundy B, Wenzel L et al (2006) Intraperitoneal cisplatin and paclitaxel in ovarian cancer. N Engl J Med 354(1):34–43. https://doi.org/10.1056/NEJMoa052985

Berger T, Malayeri R, Doppelbauer A et al (1997) Neurological monitoring of neurotoxicity induced by paclitaxel/cisplatin chemotherapy. Eur J Cancer 33(9):1393–1399. https://doi.org/10.1016/s0959-8049(97)00103-2

Bissery M, Vrignaud P, Lavelle F (1996) Preclinical profile of docetaxel (Taxotere): efficacy as a single agent and in combination. Semin Oncol 22:3–16

Boora GK, Kanwar R, Kulkarni AA et al (2016) Testing of candidate single nucleotide variants associated with paclitaxel neuropathy in the trial NCCTG N08C1 (Alliance). Cancer Med 5 (4):631–639. https://doi.org/10.1002/cam4.625

Bradley WG, Lassman LP, Pearce GW et al (1970) The neuromyopathy of vincristine in man. Clinical, electrophysiological and pathological studies. J Neurol Sci 10(2):107–131. https://doi.org/10.1016/0022-510X(70)90013-4

Briasoulis E, Karavasilis V, Tzamakou E et al (2002) Pharmacodynamics of non-break weekly paclitaxel (Taxol) and pharmacokinetics of Cremophor-EL vehicle: results of a dose-escalation study. Anticancer Drug Des 13(5):481–489. https://doi.org/10.1097/00001813-200206000-00006

Broyl A, Corthals SL, Jongen JLM et al (2010) Mechanisms of peripheral neuropathy associated with bortezomib and vincristine in patients with newly diagnosed multiple myeloma: a prospective analysis of data from the HOVON-65/GMMG-HD4 trial. Lancet Oncol 11 (11):1057–1065. https://doi.org/10.1016/S1470-2045(10)70206-0

Buey RM, Barasoain I, Jackson E et al (2005) Microtubule interactions with chemically diverse stabilizing agents: thermodynamics of binding to the paclitaxel site predicts cytotoxicity. Chem Biol 12(12):1269–1279. https://doi.org/10.1016/j.chembiol.2005.09.010

Bulls HW, Hoogland AI, Kennedy B et al (2019) A longitudinal examination of associations between age and chemotherapy-induced peripheral neuropathy in patients with gynecologic cancer. Gynecol Oncol 152(2):310–315. https://doi.org/10.1016/j.ygyno.2018.12.002

Calls A, Carozzi V, Navarro X et al (2020) Pathogenesis of platinum-induced peripheral neurotoxicity: insights from preclinical studies. Exp Neurol 325:113141. https://doi.org/10.1016/j.expneurol.2019.113141

Camidge DR, Berge EM, Doebele RC et al (2014) A phase II, open-label study of ramucirumab in combination with paclitaxel and carboplatin as first-line therapy in patients with stage IIIB/IV non-small-cell lung cancer. J Thorac Oncol 9(10):1532–1539. https://doi.org/10.1097/JTO.0000000000000273

Casey EB, Jellife AM, Le Quesne PM et al (1973) Vincristine neuropathy. Clinical and electrophysiological observations. Brain 96(1):69–86. https://doi.org/10.1093/brain/96.1.69

Cavaletti G, Beronio A, Reni L et al (2004) Thalidomide sensory neurotoxicity: a clinical and neurophysiologic study. Neurology 62(12):2291–2293. https://doi.org/10.1212/wnl.62.12.2291

Cavaletti G, Gilardini A, Canta A et al (2007) Bortezomib-induced peripheral neurotoxicity: a neurophysiological and pathological study in the rat. Exp Neurol 204(1):317–325. https://doi.org/10.1016/j.expneurol.2006.11.010

Cavaletti G, Marmiroli P (2010) Chemotherapy-induced peripheral neurotoxicity. Nat Rev Neurol 6 (12):657–666. https://doi.org/10.1038/nrneurol.2010.160

Cavaletti G, Marmiroli P (2015) Chemotherapy-induced peripheral neurotoxicity. Curr Opin Neurol 28(5):500–507. https://doi.org/10.1097/WCO.0000000000000234

Chabner BA, Longo DL (2011) Cancer chemotherapy and biotherapy: principles and practice. Sultan Qaboos Univ Med J 11(3):424–425

Chan A, Hertz DL, Morales M et al (2019) Biological predictors of chemotherapy-induced peripheral neuropathy (CIPN): MASCC neurological complications working group overview. Support Care Cancer 27(10):3729–3737. https://doi.org/10.1007/s00520-019-04987-8

Chaudhry V, Cornblath DR, Polydefkis M et al (2008) Characteristics of bortezomib- and thalidomide-induced peripheral neuropathy. J Peripher Nerv Syst 13(4):275–282. https://doi.org/10.1111/j.1529-8027.2008.00193.x

Chen D, Zhao J, Cong W (2018) Chinese herbal medicines facilitate the control of chemotherapy-induced side effects in colorectal cancer: progress and perspective. Front Pharmacol 9:1442–1442. https://doi.org/10.3389/fphar.2018.01442

Chen G, Qiao TT, Ding H et al (2015) Use of Chinese herbal medicine therapies in comprehensive hospitals in Central China: a parallel survey in cancer patients and clinicians. J Huazhong Univ Technol Med Sci 35(6):808–814. https://doi.org/10.1007/s11596-015-1511-5

Chen M, May BH, Zhou IW et al (2014) FOLFOX 4 combined with herbal medicine for advanced colorectal cancer: a systematic review. Phytother Res 28(7):976–991. https://doi.org/10.1002/ptr.5092

Chen X, Deng L, Jiang X et al (2016) Chinese herbal medicine for oesophageal cancer. Cochrane Database Syst Rev (1.):Cd004520. doi:https://doi.org/10.1002/14651858.CD004520.pub7

Cheng KL, Bradley T, Budman DR (2008) Novel microtubule-targeting agents–the epothilones. Biologics 2(4):789. https://doi.org/10.2147/btt.s3487

Cheng X, Huo J, Wang D et al (2017) Herbal medicine ac591 prevents oxaliplatin-induced peripheral neuropathy in animal model and cancer patients. Front Pharmacol 8:344–344. https://doi.org/10.3389/fphar.2017.00344

Chiou C-T, Wang K-C, Yang Y-C et al (2018) Liu Jun Zi tang-a potential, multi-herbal complementary therapy for chemotherapy-induced neurotoxicity. Int J Mol Sci 19(4):1258. https://doi.org/10.3390/ijms19041258

Chu E, DeVita Jr VT (2018) Physicians' cancer chemotherapy drug manual 2019. Burlington, the United States of America

Cortes J, Baselga J (2007) Targeting the microtubules in breast cancer beyond taxanes: the epothilones. Oncologist 12(3):271. https://doi.org/10.1634/theoncologist.12-3-271

Cortes J, Saura C (2010) Nanoparticle albumin-bound (nab™)-paclitaxel: improving efficacy and tolerability by targeted drug delivery in metastatic breast cancer. Eur J Cancer 8(1):1–10. https://doi.org/10.1016/S1359-6349(10)70002-1

Cox-Martin E, Trahan LH, Cox MG et al (2017) Disease burden and pain in obese cancer patients with chemotherapy-induced peripheral neuropathy. Support Care Cancer 25(6):1873–1879. https://doi.org/10.1007/s00520-017-3571-5

Crom WR, de Graaf SS, Synold T et al (1994) Pharmacokinetics of vincristine in children and adolescents with acute lymphocytic leukemia. J Pediatr 125(4):642–649. https://doi.org/10.1016/s0022-3476(94)70027-3

Dai C, Xiao X, Zhang Y et al (2020) Curcumin attenuates colistin-induced peripheral neurotoxicity in mice. ACS Infect Dis 6(4):715–724. https://doi.org/10.1021/acsinfecdis.9b00341

de Gramont A, Van Cutsem E, Schmoll H-J et al (2012) Bevacizumab plus oxaliplatin-based chemotherapy as adjuvant treatment for colon cancer (AVANT): a phase 3 randomised controlled trial. Lancet Oncol 13(12):1225–1233. https://doi.org/10.1016/S1470-2045(12)70509-0

Deng B, Jia L, Cheng Z (2016a) Radix astragali-based Chinese herbal medicine for oxaliplatin-induced peripheral neuropathy: a systematic review and meta-analysis. Evid Based Complement Alternat Med 2016:2421876. https://doi.org/10.1155/2016/2421876

Deng B, Jia L, Pan L et al (2016b) Wen-Luo-Tong prevents glial activation and nociceptive sensitization in a rat model of oxaliplatin-induced neuropathic pain. Evid Based Complement Alternat Med 2016:3629489. https://doi.org/10.1155/2016/3629489

Di Cesare Mannelli L, Zanardelli M, Bartolucci G et al (2015) In vitro evidence for the use of *Astragali* radix extracts as adjuvant against oxaliplatin-induced neurotoxicity. Planta Med 81 (12-13):1045–1055. https://doi.org/10.1055/s-0035-1546117

Dietrich J, Han R, Yang Y et al (2006) CNS progenitor cells and oligodendrocytes are targets of chemotherapeutic agents in vitro and in vivo. J Biol 5(7):22. https://doi.org/10.1186/jbiol50

Diogo C, Machado N, Barbosa I et al (2011) Berberine as a promising safe anti-cancer agent- is there a role for mitochondria? Curr Drug Targets 12:850–859. https://doi.org/10.2174/138945011795528930

Diouf B, Crews KR, Lew G et al (2015) Association of an inherited genetic variant with vincristine-related peripheral neuropathy in children with acute lymphoblastic leukemia. JAMA 313 (8):815–823. https://doi.org/10.1001/jama.2015.0894

Dobos GJ, Tan L, Cohen MH et al (2005) Are national quality standards for traditional Chinese herbal medicine sufficient? Current governmental regulations for traditional Chinese herbal medicine in certain Western countries and China as the eastern origin country. Complement Ther Med 13(3):183–190. https://doi.org/10.1016/j.ctim.2005.06.004

Dorsey SG, Kleckner IR, Barton D et al (2019) The National Cancer Institute clinical trials planning meeting for prevention and treatment of chemotherapy-induced peripheral neuropathy. J Natl Cancer Inst 111(6):531–537. https://doi.org/10.1093/jnci/djz011

Downing KH, Nogales E (1998) New insights into microtubule structure and function from the atomic model of tubulin. Eur Biophys J 27(5):431–436. https://doi.org/10.1007/s002490050153

Duggett NA, Griffiths LA, McKenna OE et al (2016) Oxidative stress in the development, maintenance and resolution of paclitaxel-induced painful neuropathy. Neuroscience 333:13–26. https://doi.org/10.1016/j.neuroscience.2016.06.050

Dumontet C, Jordan MA (2010) Microtubule-binding agents: a dynamic field of cancer therapeutics. Nat Rev Drug Discov 9(10):790–803. https://doi.org/10.1038/nrd3253

Dustin P (1980) Microtubules. Sci Am J 243(2):66–76. https://doi.org/10.1038/scientificamerican0880-66

Egbelakin A, Ferguson MJ, MacGill EA et al (2011) Increased risk of vincristine neurotoxicity associated with low CYP3A5 expression genotype in children with acute lymphoblastic leukemia. Pediatr Blood Cancer 56(3):361–367. https://doi.org/10.1002/pbc.22845

Engblom P, Rantanen V, Kulmala J et al (1997) Taxane sensitivity of ovarian carcinoma in vitro. Anticancer Res 17(4a):2475–2479

Fu B, Wang N, Tan H-Y et al (2018) Multi-component herbal products in the prevention and treatment of chemotherapy-associated toxicity and side effects: a review on experimental and clinical evidences. Front Pharmacol 9:1394–1394. https://doi.org/10.3389/fphar.2018.01394

Fujii K, Okamoto S, Saitoh K et al (2004) The efficacy of Shakuyaku-Kanzo-to for peripheral nerve dysfunction in paclitaxel combination chemotherapy for epithelial ovarian carcinoma. Gan To Kagaku Ryoho 31:1537–1540

Fujita S, Hirota T, Sakiyama R et al (2019) Identification of drug transporters contributing to oxaliplatin-induced peripheral neuropathy. J Neurochem 148(3):373–385. https://doi.org/10.1111/jnc.14607

Furuse K, Fukuoka M, Kuba M et al (1996) Randomized study of vinorelbine (VRB) versus vindesine (VDS) in previously untreated stage IIIB or IV non-small-cell lung cancer (NSCLC). Ann Oncol 7(8):815–820. https://doi.org/10.1093/oxfordjournals.annonc.a010760

García-Sanz R, Corchete LA, Alcoceba M et al (2017) Prediction of peripheral neuropathy in multiple myeloma patients receiving bortezomib and thalidomide: a genetic study based on a single nucleotide polymorphism array. Hematol Oncol 35(4):746–751. https://doi.org/10.1002/hon.2337

Gatzemeier U, von Pawel J, Gottfried M et al (2000) Phase III comparative study of high-dose cisplatin versus a combination of paclitaxel and cisplatin in patients with advanced non-small-cell lung cancer. J Clin Oncol 18(19):3390–3399. https://doi.org/10.1200/JCO.2000.18.19.3390

Ge L, Wang Y-f, Tian J-h et al (2016) Network meta-analysis of Chinese herb injections combined with FOLFOX chemotherapy in the treatment of advanced colorectal cancer. J Clin Pharm Ther 41(4):383–391. https://doi.org/10.1111/jcpt.12410

Geisler S, Doan RA, Strickland A et al (2016) Prevention of vincristine-induced peripheral neuropathy by genetic deletion of SARM1 in mice. Brain 139(12):3092–3108. https://doi.org/10.1093/brain/aww251

Gelderblom H, Mross K, ten Tije AJ et al (2002) Comparative pharmacokinetics of unbound paclitaxel during 1- and 3-hour infusions. J Clin Oncol 20(2):574–581. https://doi.org/10.1200/JCO.2002.20.2.574

Gerdts J, Summers DW, Milbrandt J et al (2016) Axon self-destruction: new links among SARM1, MAPKs, and NAD+ metabolism. Neuron 89(3):449–460. https://doi.org/10.1016/j.neuron.2015.12.023

Gomber S, Dewan P, Chhonker D (2010) Vincristine induced neurotoxicity in cancer patients. Indian J Pediatr 77(1):97–100. https://doi.org/10.1007/s12098-009-0254-3

Gore M, Mainwaring P, A'Hern R et al (1998) Randomized trial of dose-intensity with single-agent carboplatin in patients with epithelial ovarian cancer. London gynaecological oncology group. J Clin Oncol 16(7):2426–2434. https://doi.org/10.1200/JCO.1998.16.7.2426

Gornstein E, Schwarz TL (2014) The paradox of paclitaxel neurotoxicity: mechanisms and unanswered questions. Neuropharmacology 76:175–183. https://doi.org/10.1016/j.neuropharm.2013.08.016

Gottschalk PG, Dyck PJ, Kiely JM (1968) *Vinca* alkaloid neuropathy: nerve biopsy studies in rats and in man. Neurology 18(9):875–882. https://doi.org/10.1212/wnl.18.9.875

Gradishar WJ, Krasnojon D, Cheporov S et al (2012) Phase II trial of nab-paclitaxel compared with docetaxel as first-line chemotherapy in patients with metastatic breast cancer: final analysis of overall survival. Clin Breast Cancer 12(5):313–321. https://doi.org/10.1016/j.clbc.2012.05.001

Grothey A (2005) Clinical management of oxaliplatin-associated neurotoxicity. Clin Colorectal Cancer 5(1):38–46. https://doi.org/10.3816/CCC.2005.s.006

Guo X, Sun H, Dong J et al (2019) Does nab-paclitaxel have a higher incidence of peripheral neuropathy than solvent-based paclitaxel? Evidence from a systematic review and meta-analysis. Crit Rev Oncol Hematol 139:16–23. https://doi.org/10.1016/j.critrevonc.2019.04.021

Heigener DF, von Pawel J, Eschbach C et al (2013) Prospective, multicenter, randomized, independent-group, open-label phase II study to investigate the efficacy and safety of three regimens with two doses of sagopilone as second-line therapy in patients with stage IIIB or IV non-small-cell lung cancer. Lung Cancer 80(3):319–325. https://doi.org/10.1016/j.lungcan.2013.02.007

Hershman D, Weimer L, Wang A et al (2011) Association between patient reported outcomes and quantitative sensory tests for measuring long-term neurotoxicity in breast cancer survivors treated with adjuvant paclitaxel chemotherapy. Breast Cancer Res Treat 125:767–774. https://doi.org/10.1007/s10549-010-1278-0

Hershman DL, Lacchetti C, Dworkin RH et al (2014) Prevention and management of chemotherapy-induced peripheral neuropathy in survivors of adult cancers: American Society of Clinical Oncology clinical practice guideline. J Clin Oncol 32(18):1941–1967. https://doi.org/10.1200/JCO.2013.54.0914

Hershman DL, Till C, Wright JD et al (2016) Comorbidities and risk of chemotherapy-induced peripheral neuropathy among participants 65 years or older in southwest oncology group clinical trials. J Clin Oncol 34(25):3014–3022. https://doi.org/10.1200/JCO.2015.66.2346

Hertz D, Kidwell K, Vangipuram K, Sun D, Henry N (2018) Association of systemic paclitaxel concentrations with severity and progression of paclitaxel-induced peripheral neuropathy. In: Proceedings of the 2017 San Antonio Breast Cancer Symposium, San Antonio, TX, December 2017. Cancer Research, vol 78(4). AACR, Philadelphia. doi:https://doi.org/10.1158/1538-7445.SABCS17-P6-11-03

Hoshino N, Ganeko R, Hida K et al (2018) Goshajinkigan for reducing chemotherapy-induced peripheral neuropathy: a systematic review and meta-analysis. Int J Clin Oncol 23(3):434–442. https://doi.org/10.1007/s10147-017-1229-4

Hosokawa A, Ogawa K, Ando T et al (2012) Preventive effect of traditional Japanese medicine on neurotoxicity of FOLFOX for metastatic colorectal cancer: a multicenter retrospective study. Anticancer Res 32(7):2545–2550. https://doi.org/10.1016/S1359-6349(09)70696-2

Hu B, Zhou Q, Wu T et al (2017) Efficacy and safety of subcutaneous versus intravenous bortezomib in multiple myeloma: a meta-analysis. Int J Clin Pharmacol Ther 55(4):329. https://doi.org/10.5414/CP202714

Huyck TK, Gradishar W, Manuguid F et al (2011) Eribulin mesylate. Nat Rev Drug Discov 10 (3):173–174. https://doi.org/10.1038/nrd3389

Ibrahim NK, Desai N, Legha S et al (2002) Phase I and pharmacokinetic study of ABI-007, a Cremophor-free, protein-stabilized, nanoparticle formulation of paclitaxel. Clin Cancer Res 8 (5):1038–1044

Imai A, Matsunami K, Takagi H et al (2012) Proposed medications for taxane-induced myalgia and arthralgia. Oncol Lett 3(6):1181–1185. https://doi.org/10.3892/ol.2012.651

Imanshahidi M, Hosseinzadeh H (2008) Pharmacological and therapeutic effects of *Berberis vulgaris* and its active constituent, berberine. Phytother Res 22(8):999–1012. https://doi.org/10.1002/ptr.2399

Islam B, Lustberg M, Staff NP et al (2019) *Vinca* alkaloids, thalidomide and eribulin-induced peripheral neurotoxicity: from pathogenesis to treatment. J Peripher Nerv Syst 24(S2):63–73. https://doi.org/10.1111/jns.12334

Jiang W, Peng J, Zhang Y et al (2012) The implications of cancer stem cells for cancer therapy. Int J Mol Sci 13(12):16636–16657. https://doi.org/10.3390/ijms131216636

Johnson DC, Corthals SL, Walker BA et al (2011) Genetic factors underlying the risk of thalidomide-related neuropathy in patients with multiple myeloma. J Clin Oncol 29 (7):797–804. https://doi.org/10.1200/JCO.2010.28.0792

Karsenti E, Vernos I (2001) The mitotic spindle: a self-made machine. Science 294(5542):543–547. https://doi.org/10.1126/science.1063488
Katsumata N (2003) Docetaxel: an alternative taxane in ovarian cancer. Br J Cancer 89(3):9–15. https://doi.org/10.1038/sj.bjc.6601495
Keckner IR, Kamen C, Gewandter JS et al (2018) Effects of exercise during chemotherapy on chemotherapy-induced peripheral neuropathy: a multicenter, randomized controlled trial. Support Care Cancer 26(4):1019–1028. https://doi.org/10.1007/s00520-017-4013-0
Kelland LR, Abel G (1992) Comparative in vitro cytotoxicity of taxol and taxotere against cisplatin-sensitive and-resistant human ovarian carcinoma cell lines. Cancer Chemother Pharmacol 30 (6):444–450. https://doi.org/10.1007/BF00685595
Kellie SJ, Koopmans P, Earl J et al (2004) Increasing the dosage of vincristine: a clinical and pharmacokinetic study of continuous-infusion vincristine in children with central nervous system tumors. Cancer 100(12):2637–2643. https://doi.org/10.1002/cncr.20220
Kemeny N, Garay CA, Gurtler J et al (2004) Randomized multicenter phase II trial of bolus plus infusional fluorouracil/leucovorin compared with fluorouracil/leucovorin plus oxaliplatin as third-line treatment of patients with advanced colorectal cancer. J Clin Oncol 22 (23):4753–4761. https://doi.org/10.1200/JCO.2004.03.119
Kim TY, Kim DW, Chung JY, Shin SG, Kim SC, Heo DS, Kim NK, Bang YJ (2004) Phase I and pharmacokinetic study of Genexol-PM, a cremophor-free, polymeric micelle-formulated paclitaxel, in patients with advanced malignancies. Clin Cancer Res 10(11):3708–3716. https://doi.org/10.1158/1078-0432.CCR-03-0655
King PD, Perry MC (2001) Hepatotoxicity of chemotherapy. Oncologist 6(2):162–176. https://doi.org/10.1634/theoncologist.6-2-162
Kirchmair R, Tietz AB, Panagiotou E et al (2007) Therapeutic angiogenesis inhibits or rescues chemotherapy-induced peripheral neuropathy: taxol- and thalidomide-induced injury of vasa nervorum is ameliorated by VEGF. Mol Ther 15(1):69–75. https://doi.org/10.1038/sj.mt.6300019
Kono T, Hata T, Morita S et al (2013) Goshajinkigan oxaliplatin neurotoxicity evaluation (GONE): a phase 2, multicenter, randomized, double-blind, placebo-controlled trial of goshajinkigan to prevent oxaliplatin-induced neuropathy. Cancer Chemother Pharmacol 72(6):1283–1290. https://doi.org/10.1007/s00280-013-2306-7
Kramer R, Bielawski J, Kistner-Griffin E et al (2015) Neurotoxic 1-deoxysphingolipids and paclitaxel-induced peripheral neuropathy. FASEB J 29(11):4461–4472. https://doi.org/10.1096/fj.15-272567
Kuriyama A, Endo K (2018) Goshajinkigan for prevention of chemotherapy-induced peripheral neuropathy: a systematic review and meta-analysis. Support Care Cancer 26(4):1051–1059. https://doi.org/10.1007/s00520-017-4028-6
Landowski TH, Megli C, Nullmeyer KD et al (2005) Mitochondrial-mediated disregulation of $Ca^2+$ is a critical determinant of Velcade (PS-341/bortezomib) cytotoxicity in myeloma cell lines. Cancer Res 65(9):3828–3836. https://doi.org/10.1158/0008-5472.CAN-04-3684
Lapidot T, Sirard C, Vormoor J et al (1994) A cell initiating human acute myeloid leukaemia after transplantation into SCID mice. Nature 367(6464):645–648. https://doi.org/10.1038/367645a0
Lavoie Smith EM, Li L, Chiang C et al (2015) Patterns and severity of vincristine-induced peripheral neuropathy in children with acute lymphoblastic leukemia. J Peripher Nerv Syst 20 (1):37–46. https://doi.org/10.1111/jns.12114
Lazarou J, Pomeranz BH, Corey PN (1998) Incidence of adverse drug reactions in hospitalized patients: a meta-analysis of prospective studies. JAMA 279(15):1200–1205. https://doi.org/10.1001/jama.279.15.1200
Leblanc AF, Sprowl JA, Alberti P et al (2018) OATP1B2 deficiency protects against paclitaxel-induced neurotoxicity. J Clin Investig 128(2):816–825. https://doi.org/10.1172/JCI96160
Li DW, Sun JY, Wang K et al (2015) Attenuation of cisplatin-induced neurotoxicity by cyanidin, a natural inhibitor of ros-mediated apoptosis in PC12 cells. Cell Mol Neurobiol 35(7):995–1001. https://doi.org/10.1007/s10571-015-0194-6

Li SG, Chen HY, Ou-Yang CS et al (2013) The efficacy of Chinese herbal medicine as an adjunctive therapy for advanced non-small cell lung cancer: a systematic review and meta-analysis. PLoS One 8(2):e57604. https://doi.org/10.1371/journal.pone.0057604

Liu Y, Zhu G, Han L et al (2013) Clinical study on the prevention of oxaliplatin-induced neurotoxicity with guilongtongluofang: results of a randomized, double-blind, placebo-controlled trial. Evid Based Complement Alternat Med 2013:541217. https://doi.org/10.1155/2013/541217

Liu Z, Huang P, Law S et al (2018) Preventive effect of curcumin against chemotherapy-induced side-effects. Front Pharmacol 9:1374–1374. https://doi.org/10.3389/fphar.2018.01374

Lopus M, Smiyun G, Miller H et al (2015) Mechanism of action of ixabepilone and its interactions with the βIII-tubulin isotype. Cancer Chemother Pharmacol 76(5):1013–1024. https://doi.org/10.1007/s00280-015-2863-z

Magge RS, DeAngelis LM (2015) The double-edged sword: neurotoxicity of chemotherapy. Blood Rev 29(2):93–100. https://doi.org/10.1016/j.blre.2014.09.012

Magrangeas F, Kuiper R, Avet-Loiseau H et al (2016) A genome-wide association study identifies a novel locus for bortezomib-induced peripheral neuropathy in European patients with multiple myeloma. Clin Cancer Res 22(17):4350. https://doi.org/10.1158/1078-0432.CCR-15-3163

Majithia N, Temkin SM, Ruddy KJ et al (2016) National Cancer Institute-supported chemotherapy-induced peripheral neuropathy trials: outcomes and lessons. Support Care Cancer 24 (3):1439–1447. https://doi.org/10.1007/s00520-015-3063-4

Massicot F, Hache G, David L et al (2013) P2X7 cell death receptor activation and mitochondrial impairment in oxaliplatin-induced apoptosis and neuronal injury: cellular mechanisms and in vivo approach. PLoS One 8(6):e66830. https://doi.org/10.1371/journal.pone.0066830

Matsui T, Kono T, Hata T et al (2011) Neuroprotective effects of goshajinkigan (TJ-107) in colorectal cancer patients receiving oxaliplatin-based chemotherapy: a double-blind, placebo-controlled, randomized phase II trial from the GONE study group. J Clin Oncol 29(15):e14007–e14007. https://doi.org/10.1200/jco.2011.29.15_suppl.e14007

McCulloch M, See C, X-j S et al (2006) *Astragalus*-based Chinese herbs and platinum-based chemotherapy for advanced non-small-cell lung cancer: meta-analysis of randomized trials. J Clin Oncol 24:419–430. https://doi.org/10.1200/JCO.2005.03.6392

Melchert M, List A (2007) The thalidomide saga. Int J Biochem Cell Biol 39(7):1489–1499. https://doi.org/10.1016/j.biocel.2007.01.022

Mendonça LM, da Silva Machado C, Teixeira CC et al (2013) Curcumin reduces cisplatin-induced neurotoxicity in NGF-differentiated PC12 cells. Neurotoxicology 34:205–211. https://doi.org/10.1016/j.neuro.2012.09.011

Miaskowski C, Mastick J, Paul SM et al (2018) Impact of chemotherapy-induced neurotoxicities on adult cancer survivors' symptom burden and quality of life. J Cancer Surviv 12(2):234–245. https://doi.org/10.1007/s11764-017-0662-8

Mielke S, Mross K, Gerds TA et al (2003) Comparative neurotoxicity of weekly non-break paclitaxel infusions over 1 versus 3 h. Anticancer Drug Des 14(10):785–792. https://doi.org/10.1097/00001813-200311000-00003

Mielke S, Sparreboom A, Mross K (2006) Peripheral neuropathy: a persisting challenge in paclitaxel-based regimes. Eur J Cancer 42(1):24–30. https://doi.org/10.1016/j.ejca.2005.06.030

Minev B (2011) Cancer management in man: chemotherapy, biological therapy, hyperthermia and supporting measures. Springer, Netherlands

Ming-Hua C, Bao-Hua Z, Lei Y (2016) Mechanisms of anorexia cancer cachexia syndrome and potential benefits of traditional medicine and natural herbs. Curr Pharm Biotechnol 17 (13):1147–1152. https://doi.org/10.2174/1389201017666161018123311

Mironov SL, Ivannikov MV, Johansson M (2005) [$Ca^2$+] signaling between mitochondria and endoplasmic reticulum in neurons is regulated by microtubules. From mitochondrial permeability transition pore to Ca2+−induced Ca2+ release. J Biol Chem 280(1):715–721. https://doi.org/10.1074/jbc.M409819200

Mohty B, El-Cheikh J, Yakoub-Agha I et al (2010) Peripheral neuropathy and new treatments for multiple myeloma: background and practical recommendations. Haematologica 95(2):311–319. https://doi.org/10.3324/haematol.2009.012674

Molassiotis A, Potrata B, Cheng KK (2009) A systematic review of the effectiveness of Chinese herbal medication in symptom management and improvement of quality of life in adult cancer patients. Complement Ther Med 17(2):92–120. https://doi.org/10.1016/j.ctim.2008.11.002

Mora E, Smith EML, Donohoe C et al (2016) Vincristine-induced peripheral neuropathy in pediatric cancer patients. Am J Cancer Res 6(11):2416–2430

Moudi M, Go R, Yien CYS et al (2013) *Vinca* alkaloids. Int J Prev Med 4(11):1231–1235. https://doi.org/10.2165/00128415-200711380-00080

National Cancer Institute (NCI) (2020) Cancer Statistics. https://www.cancer.gov/about-cancer/understanding/statistics. Accesssed 7 July 2020

Noble RL, Beer CT, Cutts JH (1958) Role of chance observations in chemotherapy: *Vinca rosea*. Ann N Y Acad Sci 76(3):882–894. https://doi.org/10.1111/j.1749-6632.1958.tb54906.x

Ohnishi S, Takeda H (2015) Herbal medicines for the treatment of cancer chemotherapy-induced side effects. Front Pharmacol 6:14–14. https://doi.org/10.3389/fphar.2015.00014

Oki E, Emi Y, Kojima H et al (2015) Preventive effect of Goshajinkigan on peripheral neurotoxicity of FOLFOX therapy (GENIUS trial): a placebo-controlled, double-blind, randomized phase III study. Int J Clin Oncol 20(4):767–775. https://doi.org/10.1007/s10147-015-0784-9

Osmani K, Vignes S, Aissi M et al (2012) Taxane-induced peripheral neuropathy has good long-term prognosis: a 1- to 13-year evaluation. J Neurol 259:1936–1943. https://doi.org/10.1007/s00415-012-6442-5

Pace A, Bove L, Nisticò C et al (1996) Vinorelbine neurotoxicity: clinical and neurophysiological findings in 23 patients. J Neurol Neurosurg Psychiatr 61(4):409–411. https://doi.org/10.1136/jnnp.61.4.409

Pasetto LM, D'Andrea MR, Rossi E et al (2006) Oxaliplatin-related neurotoxicity: how and why? Crit Rev Oncol Hematol 59(2):159–168. https://doi.org/10.1016/j.critrevonc.2006.01.001

Peltier AC, Russell JW (2002) Recent advances in drug-induced neuropathies. Curr Opin Neurol 15 (5):633–638. https://doi.org/10.1097/00019052-200210000-00015

Perazella MA, Moeckel GW (2010) Nephrotoxicity from chemotherapeutic agents: clinical manifestations, pathobiology, and prevention/therapy. Semin Nephrol 30(6):570–581. https://doi.org/10.1016/j.semnephrol.2010.09.005

Perez EA (2009) Microtubule inhibitors: differentiating tubulin-inhibiting agents based on mechanisms of action, clinical activity, and resistance. Mol Cancer Ther 8(8):2086–2095. https://doi.org/10.1158/1535-7163.MCT-09-0366

Quasthoff S, Hartung HP (2002) Chemotherapy-induced peripheral neuropathy. J Neurol 249 (1):9–17. https://doi.org/10.1007/PL00007853

Raphael MJ, Fischer HD, Fung K et al (2017) Neurotoxicity outcomes in a population-based cohort of elderly patients treated with adjuvant oxaliplatin for colorectal cancer. Clin Colorectal Cancer 16(4):397–404. https://doi.org/10.1016/j.clcc.2017.03.013

Rauf A, Imran M, Butt MS et al (2018a) Resveratrol as an anti-cancer agent: a review. Crit Rev Food Sci Nutr 58(9):1428–1447. https://doi.org/10.1080/10408398.2016.1263597

Rauf A, Imran M, Khan IA et al (2018b) Anticancer potential of quercetin: a comprehensive review. Phytother Res 32(11):2109–2130. https://doi.org/10.1002/ptr.6155

Ravaglia S, Corso A, Piccolo G et al (2008) Immune-mediated neuropathies in myeloma patients treated with bortezomib. Clin Neurophysiol 119(11):2507–2512. https://doi.org/10.1016/j.clinph.2008.08.007

Reck M, Shankar G, Lee A et al (2020) Atezolizumab in combination with bevacizumab, paclitaxel and carboplatin for the first-line treatment of patients with metastatic non-squamous non-small cell lung cancer, including patients with EGFR mutations. Expert Rev Respir Med 14 (2):125–136. https://doi.org/10.1080/17476348.2020.1701439

Rezaee R, Monemi A, SadeghiBonjar M et al (2019) Berberine alleviates paclitaxel-induced neuropathy. J Pharmacopuncture 22:90–94. https://doi.org/10.3831/KPI.2019.22.011

Richardson MA, Sanders T, Palmer JL et al (2000) Complementary/alternative medicine use in a comprehensive cancer center and the implications for oncology. J Clin Oncol 18 (13):2505–2514. https://doi.org/10.1200/JCO.2000.18.13.2505

Ringel I, Horwitz SB (1991) Studies with RP 56976 (Taxotere): a semisynthetic analogue of taxol. J Natl Cancer Inst 83(4):288–291. https://doi.org/10.1093/jnci/83.4.288

Rivera E, Cianfrocca M (2015) Overview of neuropathy associated with taxanes for the treatment of metastatic breast cancer. Cancer Chemother Pharmacol 75(4):659–670. https://doi.org/10.1007/s00280-014-2607-5

Saba N, Bhuyan R, Nandy SK et al (2015) Differential interactions of cytochrome p450 3a5 and 3a4 with chemotherapeutic agent-vincristine: a comparative molecular dynamics study. Anti Cancer Agents Med Chem 15(4):475–483. https://doi.org/10.2174/1871520615666150129213510

Schneider BP, Li L, Radovich M et al (2015) Genome-wide association studies for taxane-induced peripheral neuropathy in ECOG-5103 and ECOG-1199. Clin Cancer Res 21(22):5082–5091. https://doi.org/10.1158/1078-0432.CCR-15-0586

Scripture CD, Figg WD, Sparreboom A (2006) Peripheral neuropathy induced by paclitaxel: recent insights and future perspectives. Curr Neuropharmacol 4(2):165–172. https://doi.org/10.2174/157015906776359568

Seretny M, Currie GL, Sena ES et al (2014) Incidence, prevalence, and predictors of chemotherapy-induced peripheral neuropathy: a systematic review and meta-analysis. Pain 155 (12):2461–2470. https://doi.org/10.1016/j.pain.2014.09.020

Shah NN, Merchant MS, Cole DE et al (2016) Vincristine sulfate liposomes injection (VSLI, Marqibo(R)): results from a phase I study in children, adolescents, and young adults with refractory solid tumors or leukemias. Pediatr Blood Cancer 63(6):997–1005. https://doi.org/10.1002/pbc.25937

Silverman JA, Deitcher SR (2013) Marqibo® (vincristine sulfate liposome injection) improves the pharmacokinetics and pharmacodynamics of vincristine. Cancer Chemother Pharmacol 71 (3):555–564. https://doi.org/10.1007/s00280-012-2042-4

Singh J, Saha L, Singh N et al (2019) Study of nuclear factor-2 erythroid related factor-2 activator, berberine, in paclitaxel induced peripheral neuropathy pain model in rats. J Pharm Pharmacol 71 (5):797–805. https://doi.org/10.1111/jphp.13047

Smith ME, Bauer-Wu S (2012) Traditional Chinese medicine for cancer-related symptoms. Semin Oncol Nurs 28(1):64–74. https://doi.org/10.1016/j.soncn.2011.11.007

Socinski MA, Schell MJ, Peterman A et al (2002) Phase III trial comparing a defined duration of therapy versus continuous therapy followed by second-line therapy in advanced-stage IIIB/IV non-small-cell lung cancer. J Clin Oncol 20(5):1335–1343. https://doi.org/10.1200/JCO.2002.20.5.1335

Son B, Lee S, Youn H et al (2017) The role of tumor microenvironment in therapeutic resistance. Oncotarget 8(3):3933–3945. https://doi.org/10.18632/oncotarget.13907

Starobova H, Vetter I (2017) Pathophysiology of chemotherapy-induced peripheral neuropathy. Front Mol Neurosci 10:174–174. https://doi.org/10.3389/fnmol.2017.00174

Strumberg D, Brügge S, Korn M et al (2002) Evaluation of long-term toxicity in patients after cisplatin-based chemotherapy for non-seminomatous testicular cancer. Ann Oncol 13 (2):229–236. https://doi.org/10.1093/annonc/mdf058

Summit N (2014) ABRAXANE® for Injectable Suspension (paclitaxel protein-bound particles for injectable suspension) (albumin-bound). https://www.accessdata.fda.gov/drugsatfda_docs/label/2014/021660s040lbl.pdf. Accessed 15 April 2020

Sutton G, Brunetto VL, Kilgore LS et al (2000) A phase III trial of ifosfamide with or without cisplatin in carcinosarcoma of the uterus: A Gynecologic Oncology Group Study. Gynecol Oncol 79(2):147–153. https://doi.org/10.1006/gyno.2000.6001

Swain SM, Arezzo JC (2008) Neuropathy associated with microtubule inhibitors: diagnosis, incidence, and management. Clin Adv Hematol Oncol 6(6):455–467

Thawani SP, Tanji K, De Sousa EA et al (2015) Bortezomib-associated demyelinating neuropathy—clinical and pathologic features. J Clin Neuromuscul Dis 16(4):202–209. https://doi.org/10.1097/CND.0000000000000077

Tolaney SM, Barry WT, Dang CT et al (2015) Adjuvant paclitaxel and trastuzumab for node-negative, HER2-positive breast cancer. New Engl J Med 372(2):134–141. https://doi.org/10.1056/NEJMoa1406281

Topp KS, Tanner KD, Levine JD (2000) Damage to the cytoskeleton of large diameter sensory neurons and myelinated axons in vincristine-induced painful peripheral neuropathy in the rat. J Comp Neurol 424(4):563–576. https://doi.org/10.1002/1096-9861(20000904)424:4<563::AID-CNE1>3.0.CO;2-U

Toso C, Lindley C (1995) Vinorelbine: a novel vinca alkaloid. Am J Health Syst Pharm 52(12):1287–1304. https://doi.org/10.1093/ajhp/52.12.1287

Tsumura Co (2014) TSUMURA Goshajinkigan Extract Granules for Ethical Use. https://www.tsumura.co.jp/english/products/pi/JPR_T107.pdf. Accessed 14 April 2020

Turnbull C, Sud A, Houlston RS (2018) Cancer genetics, precision prevention and a call to action. Nat Genet 50(9):1212–1218. https://doi.org/10.1038/s41588-018-0202-0

Vahdat LT, Garcia AA, Vogel C et al (2013) Eribulin mesylate versus ixabepilone in patients with metastatic breast cancer: a randomized phase II study comparing the incidence of peripheral neuropathy. Breast Cancer Res Treat 140(2):341–351. https://doi.org/10.1007/s10549-013-2574-2

Vahdat LT, Thomas ES, Roché HH et al (2012) Ixabepilone-associated peripheral neuropathy: data from across the phase II and III clinical trials. Support Care Cancer 20(11):2661–2668. https://doi.org/10.1007/s00520-012-1384-0

Vasey PA (2002) Survival and longer-term toxicity results of the SCOTROC study: Docetaxel-carboplatin vs. paclitaxel-carboplatin in epithelial ovarian cancer. In Abstracts of the OncoLink Scientific Meetings Coverage ASCO, University of Pennsylvania, Pennsylvania, 21 May 2002

Velasco R, Bruna J (2014) Oxaliplatin neurotoxicity. Curr Colorectal Cancer Rep 10(3):303–312. https://doi.org/10.1007/s11888-014-0230-9

Wang Q, Wang Q, Wang SF et al (2017) Oral Chinese herbal medicine as maintenance treatment after chemotherapy for advanced non-small-cell lung cancer: a systematic review and meta-analysis. Curr Oncol 24(4):e269–e276. https://doi.org/10.3747/co.24.3561

Wani MC, Taylor HL, Wall ME et al (1971) Plant antitumor agents. VI. The isolation and structure of taxol, a novel antileukemic and antitumor agent from *Taxus brevifolia*. J Am Chem Soc 93(9):2325–2327. https://doi.org/10.1021/ja00738a045

Wani TH, Chakrabarty A, Shibata N et al (2017) The dihydroxy metabolite of the teratogen thalidomide causes oxidative DNA damage. Chem Res Toxicol 30(8):1622–1628. https://doi.org/10.1021/acs.chemrestox.7b00127

Weiss HD, Walker MD, Wiernik PH (1974) Neurotoxicity of commonly used antineoplastic agents (second of two parts). N Engl J Med 291(3):127–133. https://doi.org/10.1056/NEJM197407182910305

WHO (2020) Cancer. https://www.who.int/health-topics/cancer#tab=tab_1. Accessed 1 April 2020

Wilken R, Veena MS, Wang MB et al (2011) Curcumin: a review of anti-cancer properties and therapeutic activity in head and neck squamous cell carcinoma. Mol Cancer 10:12–12. https://doi.org/10.1186/1476-4598-10-12

Windebank AJ, Grisold W (2008) Chemotherapy-induced neuropathy. J Peripher Nerv Syst 13(1):27–46. https://doi.org/10.1111/j.1529-8027.2008.00156.x

Wiseman N, Willms S, Ye F (2009) Jin Gui Yao Lue-essential prescriptions of the golden coffer. Paradigm Publications, Summer

Wolf S, Barton D, Kottschade L et al (2008) Chemotherapy-induced peripheral neuropathy: prevention and treatment strategies. Eur J Cancer 44(11):1507–1515. https://doi.org/10.1016/j.ejca.2008.04.018

Wolff AC, Donehower RC, Carducci MK et al (2003) Phase I study of docosahexaenoic acid-paclitaxel: a taxane-fatty acid conjugate with a unique pharmacology and toxicity profile. Clin Cancer Res 9:3589–3597

Wozniak KM, Nomoto K, Lapidus RG et al (2011) Comparison of neuropathy-inducing effects of eribulin mesylate, paclitaxel, and ixabepilone in mice. Cancer Res 71(11):3952–3962. https://doi.org/10.1158/0008-5472.CAN-10-4184

Wozniak KM, Wu Y, Farah MH et al (2013) Neuropathy-inducing effects of eribulin mesylate versus paclitaxel in mice with preexisting neuropathy. Neurotox Res 24(3):338–344. https://doi.org/10.1007/s12640-013-9394-3

Xu W, Towers AD, Li P et al (2006) Traditional Chinese medicine in cancer care: perspectives and experiences of patients and professionals in China. Eur J Cancer Care 15(4):397–403. https://doi.org/10.1111/j.1365-2354.2006.00685.x

Yamada T, Kan H, Matsumoto S (2012) Reduction in oxaliplatin-related neurotoxicity by the administration of Keishikajutsubuto (TJ-18) and powdered processed aconite root. Gan To Kagaku Ryoho 39(11):1687–1691

Yamamoto K, Hoshiai H, Noda K (2001) Effects of shakuyaku-kanzo-to on muscle pain from combination chemotherapy with paclitaxel and carboplatin. Gynecol Oncol 81(2):333–334. https://doi.org/10.1006/gyno.2001.6168

Yang J, Huang H, Zhu L (2013) World century compendium to TCM: volume 3: introduction to Chinese Materia Medica. World Scientific, United States

Yang M, Moon C (2013) Neurotoxicity of cancer chemotherapy. Neural Regen Res 8 (17):1606–1614. https://doi.org/10.3969/j.issn.1673-5374.2013.17.009

Yoshida N, Hosokawa T, Ishikawa T et al (2013) Efficacy of goshajinkigan for oxaliplatin-induced peripheral neuropathy in colorectal cancer patients. J Oncol 2013:139740–139740. https://doi.org/10.1155/2013/139740

Yoshida T, Sawa T, Ishiguro T et al (2009) The efficacy of prophylactic shakuyaku-kanzo-to for myalgia and arthralgia following carboplatin and paclitaxel combination chemotherapy for non-small cell lung cancer. Support Care Cancer 17(3):315–320. https://doi.org/10.1007/s00520-008-0508-z

Zajaczkowska R, Kocot-Kepska M, Leppert W et al (2019) Mechanisms of chemotherapy-induced peripheral neuropathy. Int J Mol Sci 20(6):1451. https://doi.org/10.3390/ijms20061451

Zee-Cheng RK (1992) Shi-quan-da-bu-tang (ten significant tonic decoction), SQT. A potent Chinese biological response modifier in cancer immunotherapy, potentiation and detoxification of anticancer drugs. Methods Find Exp Clin Pharmacol 14(9):725–736

Zhang Q-Y, Wang F-X, Jia K-K, Kong L-D (2018) Natural product interventions for chemotherapy and radiotherapy-induced side effects. Front Pharmacol 9:1253. https://doi.org/10.3389/fphar.2018.01253

Zhou QM, Wang XF, Liu XJ et al (2011) Curcumin improves MMC-based chemotherapy by simultaneously sensitising cancer cells to MMC and reducing MMC-associated side-effects. Eur J Cancer 47(14):2240–2247. https://doi.org/10.1016/j.ejca.2011.04.032

Zhou X, Liu M, Ren Q et al (2019) Oral and injectable *Marsdenia tenacissima* extract (MTE) as adjuvant therapy to chemotherapy for gastric cancer: a systematic review. BMC Complement Altern Med 19(1):366. https://doi.org/10.1186/s12906-019-2779-y

# The Neurotrophic and Neuroprotective Potential of Macrofungi

**Susanna M. Badalyan and Sylvie Rapior**

## Contents



**Abstract** Diversity of wild and cultivated macrofungi as edible and medicinal mushrooms has long been known by humans as a source of valuable food and medicines used by tradipraticians. In the fungal kingdom, macrofungi taxonomically belong to two phyla, the Basidiomycota (class Agaricomycetes) and Ascomycota

---

S. M. Badalyan (✉)
Laboratory of Fungal Biology and Biotechnology, Department of Biomedicine, Institute of Pharmacy, Yerevan State University, Yerevan, Armenia
e-mail: s.badalyan@ysu.am

S. Rapior
Laboratoire de Botanique, Phytochimie et Mycologie, Faculté des Sciences Pharmaceutiques et Biologiques, CEFE, Univ Montpellier, CNRS, EPHE, IRD, Univ Paul Valéry Montpellier 3, Montpellier, France
e-mail: sylvie.rapior@umontpellier.fr

 
D. C. Agrawal, M. Dhanasekaran (eds.), *Medicinal Herbs and Fungi*,
https://doi.org/10.1007/978-981-33-4141-8_2

(class Pezizomycetes). Macrofungi have been used in traditional Asian and European Medicines, and based on 90,000 known worldwide distributed mushroom species, are considered an important resource for modern clinical and pharmacological research. They are regarded as a source of high- and low-molecular-weight bioactive compounds (alkaloids, lipids, phenolics, polysaccharides, proteins, steroids, terpenoids, etc.) with more than 130 therapeutic effects (anti-inflammatory, antimicrobial, antioxidant, antitumor, antiviral, cytotoxic, hepatoprotective, hypocholesterolemic, hypoglycemic, hypotensive, immunomodulatory, etc.). There is also scientific evidence of using macrofungi as neuroprotectants, that is, *Agaricus blazei* (= *Agaricus subrufescens*), *Ganoderma lucidum*, *Grifola frondosa*, *Hericium erinaceus*, *Pleurotus ostreatus*, and *Trametes versicolor*. However, their neuroprotective effects have not been fully explored. This review discusses recent advances in research on the neuroprotective potential of macrofungi and perspectives for their application as neuroprotectants in biomedicine to prevent, support, or cure neurodegenerative disorders.

**Keywords** Anti-inflammatory · Antioxidant · Macrofungi · Neurodegenerative · Neuroprotective · Polysaccharides

## Abbreviations

| | |
|---|---|
| ACh | Acetylcholine |
| AChE | Acetylcholinesterase |
| AD | Alzheimer's disease |
| AE | Aqueous extract |
| AIF | Apoptosis-inducing factor |
| AInA | Anti-inflammatory activity |
| ANDA | Anti-neurodegenerative activity |
| AOA | Antioxidant activity |
| APP | Amyloid precursor protein |
| ASD | Autism spectrum disorder |
| Aβ | Amyloid-β |
| BACE1 | β-Site APP-cleaving enzyme 1 |
| BDNF | Brain-derived neurotrophic factor |
| ChAT | Choline acetyltransferase |
| COX-2 | Cyclooxygenase-2 |
| CREB | C-AMP response element-binding protein |
| DA | Dopamine |
| DPPH | 2,2-Diphenyl-1-picrylhydrazyl |
| DRG | Dorsal root ganglia |
| EE | Ethanolic extract |
| ERK1/2 | Extracellular signal-regulated kinase ½ |
| FRAP | Ferric reducing power |
| GABA | γ-Aminobutyric acid |

| | |
|---|---|
| GFAP | Glial fibrillary acidic protein |
| GMI | *Ganoderma microsporum* immunomodulatory |
| hMAOB | Human monoamine oxidase B |
| Hsp | Heat shock proteins |
| HWE | Hot water extract |
| IL | Interleukin |
| iNOS | Inducible nitric oxide synthase |
| LB | Lewy bodies |
| LPS | Lipopolysaccharide |
| LXA4 | Lipoxin A4 |
| MD | Meniere's disease |
| MDA | Malondialdehyde |
| ME | Methanolic extract |
| MM | Medicinal macrofungi |
| MMP | Mitochondrial membrane potential |
| MPTP | 1-Methyl-4-phenyl-1,2,3,6-tetrahydropyridine |
| N2a | Neuroblastoma-2a |
| NAC | Non-amyloid-β component |
| NDD | Neurodegenerative disease |
| NF-κB | Nuclear factor kappa B |
| NGF | Nerve growth factor |
| NMDA | N-methyl-D-aspartate |
| NO | Nitric oxide |
| NPE | Neuroprotective effect |
| NRA | Neuroregenerative activity |
| NRF2 | Nuclear factor erythroid-2-related factor 2 |
| NTA | Neurotrophic activity |
| NTF | Neurotrophic factor |
| PARP1 | Poly (ADP-ribose) polymerase 1 |
| PBS | Phosphate-buffered saline |
| PC12 | Pheochromocytoma 12 |
| PD | Parkinson's disease |
| PE | Pyrocatechol equivalent |
| $PGE_2$ | Prostaglandin $E_2$ |
| PhAC | Pharmacologically active compounds |
| PI3K-AKT | Phosphoinositide-3-kinase-AKT |
| PNI | Peripheral nerve injury |
| PNR | Peripheral nerve regeneration |
| PSAM | Protoilludane sesquiterpenoid aromatic |
| PSPC | Protein-bound polysaccharide complex |
| PUFA | Polyunsaturated fatty acids |
| ROS | Reactive oxygen species |
| SE | Status epilepticus |
| SOD | Superoxide dismutase |

| | |
|---|---|
| TCM | Traditional Chinese medicine |
| TEAC | Trolox equivalent antioxidant |
| TNF | Tumor necrosis factor |
| VDM | Vitamin $D_2$-enriched mushrooms |

## 1 Introduction

In recent years, neurodegenerative diseases (NDD), including age-related Alzheimer (AD), Parkinson (PD), and Meniere's diseases (MD), are affecting more than 36 million people worldwide. Understanding mechanisms of aging and determinants of life span will help to reduce age-associated diseases and morbidity and facilitate healthy cognitive aging (Gorman 2008; Deary et al. 2009). Mitochondria, as critical regulators of cell death, play important role in the development of aging-related NDD. Mutations in mitochondrial DNA and oxidative stress are contributed to the aging process—a risk factor for NDD (Lin and Beal 2006).

Recent literature data support the opinion that the oxidative stress-derived neuroinflammation is an early pathological feature in the development of NDD. An anti-stress lifestyle, healthy diet, and caloric restriction appear to extend healthy life by reducing reactive oxygen species (ROS)-mediated oxidative damage. The functional role of mitochondria and ROS formation are positively implicated in cellular stress response mechanisms, and in highly regulated processes controlled by several intracellular signaling pathways, including vitagenes, as an intracellular redox system involved in neuroprotection (Cornelius et al. 2013; Chen et al. 2016b; Uddin and Ashraf 2018). The vitagenes encode for cytoprotective heat shock proteins (Hsp) Hsp32 and Hsp70, heme oxygenase-1, sirtuin protein systems, thioredoxin, and lipoxin A4 (LXA4) and are involved in preserving cellular homeostasis during stress. Current research focuses on biomolecules that activate the vitagene system as novel targets to minimize processes associated with free radical-induced cell damage, such as neurodegeneration (Cornelius et al. 2013).

Within the kingdom of Fungi, macrofungi are a group of 90,000 known mushroom species that form visible cap-like structures (namely known as fruiting bodies or sporocarps). Based on a large number of chemical and myco-pharmacological studies, the macrofungi (fruiting bodies and mycelium) are producers of different pharmacologically active compounds (PhAC) with neuroprotective effect (NPE) to prevent the development of different neurodegenerative processes in the human brain (Kim et al. 2014; Mahmoud et al. 2014; Phan et al. 2015, 2017; Zengin et al. 2015; Cheng et al. 2016; Trovato et al. 2016a, b; Zhang et al. 2016a, b; Ahuja et al. 2017; He et al. 2017; Sabaratnam and Phan 2018; Knežević et al. 2018; Trovato Salinaro et al. 2018; Bai et al. 2019; Ćilerdžić et al. 2019; Lai et al. 2019; Varghese et al. 2019; Wang et al. 2019a, b; Liang et al. 2020; Lucius 2020; Yadav et al. 2020). Mushroom-derived LXA4 is an emerging endogenous eicosanoid (based on the enzymatic or nonenzymatic of polyunsaturated fatty acids, PUFA) able to prevent an inflammatory process (Cornelius et al. 2013; Trovato et al. 2016a). Although the mechanism of neuroprotective action of macrofungi-derived PhAC has not been thoroughly investigated, recent studies have revealed their potential to develop novel myco-pharmaceuticals to prevent and mitigate the development of various NDD (Bennett et al. 2013b; Zajac et al. 2016; Cardwell et al. 2018; Sabaratnam and Phan 2018; Badalyan et al. 2019; Yadav et al. 2020). Mushrooms-based dietary biotech products with different formulation have also been demonstrated to be neuroprotective (Barros et al. 2008; Palacios et al. 2011; Phan et al. 2012, 2014a, b, 2015, 2017; Cornelius et al. 2013; Wang et al. 2014; Bandara et al. 2015; Friedman 2015; Zhang et al. 2016a, b, 2017; Brandalise et al. 2017; Solayman et al. 2017; Lemieszek et al. 2018; Rossi et al. 2018; Yin et al. 2018; Bell et al. 2019; Dhakal et al. 2019; Jang et al. 2019; Ho et al. 2020; Lucius 2020).

The current review discusses recent advances in research on the neuroprotective potential of macrofungi and the perspectives of their application in biomedicine to prevent or cure neurodegenerative disorders.

## 2 Etiopathogenesis of Neurodegenerative Diseases

### *2.1 Age-Related Alzheimer's, Parkinson's, and Meniere's Diseases*

Aging is an inevitable biological process and the greatest risk factor for different neurodegenerative disorders, such as AD and Lewy body (LB) dementia (mental dysfunction), PD, MD, and Huntington's disease, and multiple sclerosis (Gorman 2008; Deary et al. 2009; Olanow and Brundin 2013; Theillet et al. 2016; Uddin and Ashraf 2018; Luryi et al. 2019). The process of aging is caused by changes in cells to loss of nutrient sensing, cellular homeostasis, and genomic instability, disrupted cellular functions, increased oxidative stress, accumulation of misfolded protein, impaired cellular defenses, and telomere shortening. The perturbation of cellular processes in neuronal cells can lead to life-threatening neurological disorders, which are the most frequent cause of death in elderly people (Uddin and Ashraf 2018).

Oxidative stress and antioxidant systems, as well as mitochondrial dysfunction and neuro-inflammation, are considered to play a very important role in the etiology and pathogenesis of major NDD (Lin and Beal 2006; Kim et al. 2014; Phan et al. 2015; Chen et al. 2016b; Sabaratnam and Phan 2018; Trovato Salinaro et al. 2018; Uddin and Ashraf 2018; Jiang et al. 2020; Yadav et al. 2020). Under oxidative and inflammatory pathological conditions, the development of different NDD, including ocular neural degeneration or neurosensory degeneration occurring in glaucoma and MD, is taking place, respectively (Chen et al. 2016b; Luryi et al. 2019). Reducing the level of stress that would produce protective responses against pathogenic processes is an innovative area of neurobiology to understand the basics of neurodegeneration and develop new approaches to treat different NDDs.

Alzheimer's disease (AD) is the most common neurodegenerative disorder, which annually affects the daily life of more than 5% of the population worldwide. This disease has a poor prognosis, and new therapeutic approaches are required to improve the quality of life of AD patients (Gorman 2008; Uddin and Ashraf 2018). As a potentially innovative approach in AD therapy associated with mitochondrial dysfunction and neuro-inflammation, the endogenous cellular defense mechanism modulation and neurohormesis were documented (Lin and Beal 2006; Chen et al. 2016b; Trovato Salinaro et al. 2018).

Parkinson's disease (PD) is another neurodegenerative disorder with unknown origin. Progressive functional impairment of the nigral dopaminergic neurons, intraneuronal aggregation of amyloid protein α-synuclein—a principal component of Lewy pathology is linked to PD. Except for protein aggregation, mitochondrial dysfunction, iron accumulation, both neuro-inflammation, and oxidative stress also play an important role in the etiopathogenesis of the disease. Moreover, mutations in the α-synuclein gene cause rare familial forms of PD (Olanow and Brundin 2013; Theillet et al. 2016). The duplication/triplication of the wild type α-synuclein gene is also considered the etiology of PD, indicating that increased levels of normal α-synuclein protein are sufficient for the development of the disease. α-Synuclein protein can transfer from affected to unaffected nerve cells to promote misfolding of the host α-synuclein, which leads to the formation of larger aggregates, neuronal dysfunction, and neurodegeneration. This mechanism plays an important role in the pathogenesis of PD and allows to develop novel neuroprotective therapies (Olanow and Brundin 2013; Theillet et al. 2016).

The anti-Parkinson drugs, such as levodopa, carbidopa, dopamine (DA) agonists, monoamine oxidase type B inhibitors, and anticholinergics to replace DA, are associated with numerous side effects. Therefore, the search for new therapeutic approaches that regulate pathways leading to neuronal dysfunction and death is warranted.

Meniere's disease (MD) represents a clinical syndrome mainly characterized by episodes of spontaneous vertigo, associated with fluctuating sensorineural hearing loss and tinnitus, affecting one or both ears (Sajjadi and Paparella 2008; Luryi et al. 2019). The cause of MD is still unknown. Increasing evidence shows that oxidative stress and neuro-inflammation may be important causes of developing endolymphatic hydrops and consequent otolithic degeneration (Luryi et al. 2019).

Cellular pathways, such as vitagenes conferring protection against oxidative stress, are not sufficient to prevent full neuroprotection, which can be reinforced by exogenous nutritional approaches. The search for innovative approaches can promote the development of therapies able to enhance the intrinsic reserve of vulnerable neurons, such as ganglion cells to maximize anti-degenerative stress responses and neuroprotection (Trovato Salinaro et al. 2018; Scuto et al. 2020).

Various therapeutic molecules have been designed to overcome the social, economic, and healthcare problems caused by NDD; however, almost all compounds in clinical practice are being limited to palliative care. The antioxidant polyphenolics may potentially be the most effective preventative strategy against NDD (Dhakal et al. 2019).

The age-related diseases also involve epigenetic changes in the epigenome influenced by lifestyle and diet; therefore, dietary components could accelerate or prevent age-related NDD. Thus, the combined dietary and therapeutic approaches are required to treat these diseases. The research interest in new anti-neurodegenerative drugs is high. For many years, natural products derived from medicinal plants, fruits, and vegetables have been regarded as primary resources for the discovery of potential therapeutic agents. They have been effective as anti-PD agents due to their neuroprotective properties, antioxidative and anti-inflammatory activities, as well as inhibitory effects regarding iron accumulation, protein misfolding, maintenance of proteasomal degradation, and mitochondrial homeostasis (Solayman et al. 2017; Rossi et al. 2018; Yin et al. 2018; Bell et al. 2019; Dhakal et al. 2019; Jang et al. 2019; Ho et al. 2020).

## *2.2 Autism, Epilepsy, and Depression*

Autism Spectrum Disorder (ASD) is a largely unknown neurological disease, including a condition related to brain development that impacts how a person leading to problems in social interaction and communication. ASD is an incurable systemic neurological disability in the pathogenesis of which inflammation plays an important role. Previous studies showed that gut microbiota may assist in the development of ASD symptoms. Autistic patients may benefit from a balanced diet rich with antioxidants, improvement of gut microbiota, and immunity. Mushroom-derived dietary supplementation may decrease gastrointestinal inflammation and improve health conditions in patients (Bell et al. 2019).

Status Epilepticus (SE) or epilepsy is accompanied by continuous or rapidly repeating seizures persisting for 20–30 min and cause injury to the neurons. SE can also be caused by penicillin and related compounds that antagonize the effects of γ-aminobutyric acid (GABA)—the main inhibitory neurotransmitter of the central nervous system. Although the danger of seizure activity has been recognized since ancient times, the pathophysiology of SE is not completely understood (Lowenstein and Alldredge 1998).

Depression is a common and severe neuropsychiatric disorder in humans. The symptoms of depression include feelings of intense sadness and hopelessness, which may occur after a specific event or in response to a gradual decline in health associated with aging. The treatments of depression include antidepressants and behavioral therapy. However, antidepressant drugs are associated with mild to severe side effects.

Recent studies focus on the pharmacology and feasibility of bioactive herbal and fungal compounds as a potential strategy to target a variety of human metabolic and brain disorders. Natural bioactive ingredients are reported to produce both neuroprotective and psychotropic activities and may help to combat mental disorders, including depression, anxiety, sleep disturbances, and cognitive alterations (Nagano et al. 2010; Zhang et al. 2019; Huang et al. 2020; Lew et al. 2020).

## 3 Neuroprotective and Psychotropic Compounds of Macrofungi

Macrofungi are ascomycetous and basidiomycetous mushrooms that fruit above or below the ground. Mushroom "any fungus with a distinctive fruiting body that is large enough to be ... picked by hand" (Chang and Wasser 2017). Macrofungi are taxonomically placed in two phyla, the Basidiomycota (class Agaricomycetes) and Ascomycota (class Pezizomycetes) of the subkingdom Dikarya (Hibbet and Taylor 2013). From an estimated 0.5–(1.5) – (5.1) million fungal species, about 140,000–160,000 are macrofungi from which around 10% (14,000–16,000) have been taxonomically identified (Hawksworth 2012). Macrofungi are widely distributed worldwide and have been appreciated for their nutritional and medicinal properties in traditional medicine for thousands of years (Chang and Miles 2004; Barros et al. 2008; Badalyan 2012; Chang and Wasser 2012; Gupta et al. 2018; Badalyan et al. 2019). Many species of macrofungi are industrially cultivated at a large scale, however, their medicinal and biotechnological potential and perspectives of usage in biomedicine and bio-industry have not been fully exploited, yet (Kües and Badalyan 2017; Pleszczyńska et al. 2017; Badalyan and Zambonelli 2019; Badalyan et al. 2019; Hyde et al. 2019; Diallo et al. 2020).

Macrofungi are increasingly recognized as rich sources of PhAC (alkaloids, fatty acids, lectins, lipids, phenolics, polyketides, polysaccharides, proteins, peptides, steroids, terpenoids, etc.) possessing more than 130 therapeutic effects, that is, antimicrobial, anti-inflammatory, antioxidant, antiviral, cytotoxic, hepatoprotective, hypocholesterolemic, hypoglycemic, hypotensive, immunomodulatory, mitogenic/regenerative, and so on (Mizuno 1999; Poucheret et al. 2006; Ma et al. 2010; Palacios et al. 2011; Badalyan 2012, 2016; De Silva et al. 2013; Friedman 2015; Zengin et al. 2015; Wang et al. 2017; Gupta et al. 2018; Morel et al. 2018; Badalyan et al. 2019; Akiba et al. 2020; Kosanić et al. 2020a, b). New screening strategies based on innovative biological, biochemical, and genetic approaches have identified novel macrofungi—metabolites-derived products widely applicable in biomedicine (Schueffler and Anke 2014; Kües and Badalyan 2017).

The evaluation of mushroom resources and establishment of specialized culture collections will have an incredible impact on myco-pharmacological and biotechnological research that will assist in developing novel mushroom-based health-enhancing biotech products (Badalyan 2012, 2016, 2020; Bandara et al. 2015; Badalyan and Gharibyan 2016, 2017, 2020; Badalyan and Borhani 2019a, b; Hyde et al. 2019; Badalyan and Zambonelli 2019; Badalyan and Rapior 2020; Diallo et al. 2020; Badalyan et al. 2021).

The mushrooms are widely used in Asian countries as antitumor, anti-inflammatory, antioxidative, and antimicrobial agents and are becoming popular in the Western hemisphere (Hobbs 2004; Grienke et al. 2014; Chen et al. 2016a). They have also been reported as antidepressant and neuroprotective agents (Park et al. 2007; Aguirre-Moreno et al. 2013; Sabaratnam and Phan 2018; Ćilerdžić et al. 2018; Chen et al. 2018a, b, 2019; Lemieszek et al. 2018; Chong et al. 2019; Yadav et al.

2020). However, further interdisciplinary collaboration and myco-pharmacological study with the involvement of new medicinal species, as well as valid clinical and preclinical trials are still required for the comprehensive evaluation of their therapeutic potential not only in the form of dietary supplements but also in the form of approved clinical drugs (Lew et al. 2020; Lucius 2020).

The recent advances, perspectives, and major challenges of medicinal macrofungi concerning their nutraceutical and pharmaceutical properties, as well as dietary value, the biotechnological production of fruiting bodies and mycelial biomass, isolation, purification, and characterization of various bioactive compounds have been recently reviewed (Gargano et al. 2017; Kües and Badalyan 2017; Badalyan and Zambonelli 2019; Badalyan et al. 2019; Hyde et al. 2019; Gründemann et al. 2020).

### *3.1 Polysaccharides*

Indeed, polysaccharides and proteins from macrofungi are potential therapeutics for aging and age-related neurodegeneration. Previous myco-pharmacological studies have shown that polysaccharides are considered one of the major bioactive compounds with immunomodulatory and antitumor effects without any toxicity (Meng et al. 2016; Wang et al. 2017; Badalyan et al. 2019). Mushroom-derived β-glucans also possess significant antioxidant, anti-inflammatory, and neuroprotective activities (Wasser and Didukh 2005; Khan et al. 2014; Kozarski et al. 2014; He et al. 2017; Bandara et al. 2019).

*Hericium erinaceus* is among the most appraised edible medicinal agaricomycetous mushroom regarded as a producer of different bioactive metabolites with the potential to treat different pathological conditions, including NDD. The polysaccharides are supposed to be one of the major bioactive compounds of *H. erinaceus*. It possesses immunomodulating, antitumor, antioxidant, gastroprotective, neuroprotective, hepatoprotective, hypoglycemic, and hypolipidemic activities. The current advancements in extraction, purification, structural characteristics, and bioactivities of polysaccharides obtained from the fruiting bodies, mycelium, and culture broth of *H. erinaceus* have been reported, and new prospects for their biomedical usage has been proposed (He et al. 2017; Wang et al. 2019a).

The neuroprotective activity of polysaccharide-enriched aqueous extract (AE) from the mycelium of *H. erinaceus* was also reported (Zhang et al. 2016a). In the AD mouse model, AE administration enhanced the horizontal and vertical movements, improved the endurance time, and decreased the escape latency time, as well as enhanced the central cholinergic system function, demonstrated dose-dependent enhancement of acetylcholine (ACh), and choline acetyltransferase (ChAT) concentrations in the serum and hypothalamus in mice. Thus, the NPE of *H. erinaceus* is useful for the prevention and treatment of NDD (Zhang et al. 2016a).

The observed in vitro dose-dependent inhibitory effect of Chaga mushroom (*Inonotus obliquus*) polysaccharide extracts (25, 50, 100, 200, and 500 μg/mL) on U251 human neurogliocytoma cells was related to the downregulation of anti-apoptotic Bcl-2 and upregulation of caspase-3 proteins (Ning et al. 2014). The water-soluble polysaccharide extracted from *Agaricus bisporus*, composed of glucose and galactose, showed potent antioxidant and acetylcholinesterase (AChE) inhibitory activity (Mahmoud et al. 2014; Kozarski et al. 2020). The polysaccharide extracts of two medicinal coprini mushrooms *Coprinus comatus* and *Coprinellus truncorum* contain β-glucans, proteins, and polyphenolics and showed AChE inhibitory effect that may allow using these species in the palliative treatment of AD (Pejin et al. 2019; Badalyan 2020).

Study of NPE of $1 \rightarrow 2, 3$ and $1 \rightarrow 3$-glucans isolated from edible medicinal mushroom *Cantharellus cibarius* in different in vitro models of neurodegeneration revealed the beneficial effect of *C. cibarius* polysaccharide fractions CC2a and CC3 on neuron viability and neurite outgrowth under normal and stress conditions. Both fractions showed antioxidant activity (AOA) and effectively neutralized the negative changes induced by glutamatergic system activators. They can be suggested as an effective and safe therapeutic strategy to prevent or mitigate neurodegenerative pathologies (Lemieszek et al. 2018).

## 3.2 Terpenoids and Steroids

Mushrooms are considered sources of different bioactive terpenoids, steroids, and sterols with NPE (Rupcic et al. 2018; Tang et al. 2019; Wang et al. 2019a, b; Yin et al. 2019; Akiba et al. 2020; Lee et al. 2020; Yadav et al. 2020).

The NPE of *H. erinaceus* has been attributed to terpenoids that can stimulate the production of NGF or brain-derived neurotrophic factor (BDNF). Along with six previously identified cyathane diterpenes, the novel erinacines possess neurotrophin-inducing effects and act on NGF expression (Ma et al. 2010; Rupcic et al. 2018).

Erinacine A derived from ethanol extract (EE) of *H. erinaceus* mycelium shows effects on 1-methyl-4-phenyl-1,2,3,6-tetrahydropyridine (MPTP)-induced neurotoxicity with posttreatment regimens (Lee et al. 2020). A novel mechanism for posttreatment with erinacine A to protect from neurotoxicity through regulating neuronal survival and cell death pathways was suggested by Lee and coauthors (Lee et al. 2020).

Evaluation of neurotrophic activity (NTA) of new cyathane diterpenoids cyafricanins and cyahookerins isolated from mycelia of two agaricoid mushrooms, *Cyathus africanus* and *Cyathus hookeri* in PC-12 cells and anti-neuroinflammatory activity in BV2 microglia cells have been reported (Tang et al. 2019; Yin et al. 2019; Yadav et al. 2020). These compounds showed NGF-induced neurite outgrowth-promoting activity and strong inhibitory effects on inducible iNOS and COX-2 expression (Yin et al. 2019). Cyahookerins and its known analogs showed differential NGF-induced neurite outgrowth-promoting activity in PC-12 cells, while cyahookerin, cyathin, cyathin, and cyathin inhibitory effect on nitric oxide (NO) production in lipopolysaccharide (LPS)-activated BV-2 microglial cells (Tang et al. 2019).

Isolation and structure elucidation of ten meroterpenoids from chloroform extracts of russoloid fungus *Albatrellus yasudae* (Agaricomycetes) and their Aβ aggregation inhibitory activity has been reported (Akiba et al. 2020). Three compounds were novel whereas seven were previously identified as grifolin, grifolic acid, neogrifolin, confluentin, 2-hydroxyneogrifolin, daurichromenic acid, and a cerebroside derivative.

Seven secondary metabolites, that is, a new lanostane triterpene, four known triterpenes, and two known aromatic meroterpenoids were isolated from fruiting bodies of medicinal polypore mushroom *Ganoderma lucidum*; they showed in vitro AOA and NPE against $H_2O_2$ and aged Aβ-induced cell death in SH-SY5Y cells (Wang et al. 2019b).

Sixteen secondary metabolites, including 3 new lanostane triterpenes, 3 ganoleucoins, and 13 known compounds with NPE tested on PC12 cells, were isolated from *Ganoderma leucocontextum,* cultivated in China. Two ganoleucoins showed NPE against $H_2O_2$-induced damage of PC12 cells at 200 μM with a survival rate of 83.19% and 73.37%, respectively, and induced neurite outgrowth at 50–200 μM (Chen et al. 2018a). Thus, *G. lucidum* and *G. leucocontextum* metabolites, particularly meroterpenoids, may be suggested as potential antioxidants and neuroprotective functional food ingredients to prevent the development of NDD (Chen et al. 2018a; Wang et al. 2019b).

### 3.3 Phenolics and Other Compounds

Phenolics are a diverse group of PhAC, including a large number of subclasses, such as flavonoids, phenolic acids, quinones, tocopherols, tannins, and so on. Mushroom-derived phenolics are known for their diverse pharmacological effects, including anti-inflammatory, antioxidant, and neuroprotective (Palacios et al. 2011; Khatua et al. 2013; Kozarski et al. 2015; Zengin et al. 2015; Islam et al. 2016; Pop et al. 2018; Dhakal et al. 2019; Jiang et al. 2020; Yadav et al. 2020).

Polyphenols extracted from hymenochaetoid fungus *Phellinus baumii* showed strong DPPH-scavenging (78.76%) and Trolox equivalent antioxidant (TEAC) (32.28 μmol Trolox/g sample) activities. The phenolic compound hispidin was isolated and identified from the ethanolic extract (EE) of mycelia of *Ph. baumii.* Hispidin showed a strong ability to scavenge DPPH free radicals and TEAC, equivalent to positive (vitamin C) value of 89.41% and 75.98%, respectively. Furthermore, hispidin protected $H_2O_2$-induced PC12 cells injured by the decreased oxidative stress level. These results indicated that *Ph. baumii* is a potential source to develop new natural antioxidants for food or medicines (Jiang et al. 2020).

The total phenolic content of methanolic extract (ME) and AE from *Ganoderma applanatum,* as well as their antioxidant, antimicrobial, and inhibitory effects against cholinesterase, tyrosinase, α-amylase, and α-glucosidase enzymes support to be considered this fungus as a source of new food supplements and represent a model for the development of new drug formulations (Zengin et al. 2015).

The total phenolic contents and AOA of wild polypore species *Trametes versicolor* and *T. gibbosa* were evaluated using ME and AE. The highest AOA was observed in ME, whereas the highest polyphenol and flavonoid contents were detected in AE of screened species (Pop et al. 2018).

High concentrations of tryptamine alkaloids psilocybin and psilocin (dephosphorylated psilocybin) as well as a small amount of baeocystin were detected in cultivated fruiting bodies of *Psilocybe samuiensis*. The psilocybin amount varied from 0.23% to 0.90% dry weight and was the highest in the caps. It was also found in the mycelium of *P. samuiensis* (0.24–0.32%). The relative content of psilocybin, psilocin, and baeocystin found in *P. samuiensis* was similar to that measured in many other psychoactive fungi but completely different from that found in *Psilocybe semilanceata* (Gartz et al. 1994).

A novel ergosterol conjunction-type alkaloid, hericirine, and ergosterol were isolated from the dried fruiting bodies of *H. erinaceum*. The hericirine significantly inhibited the protein expression of iNOS and COX-2 and reduced NO, $PGE_2$, TNF-α, IL-6, and IL-1β production in RAW264.7 cells exposed to LPS (Li et al. 2014).

Vitamin D deficiency, particularly in elderly people, plays an important role in the development of neurological and psychiatric disorders. However, interventional clinical evidence is lacking. Mushrooms are a nonanimal source of vitamin D and a rich source of ergosterol—the precursor of vitamin $D_2$. Vitamin $D_2$-enriched mushrooms (VDM) could prevent the cognitive and pathological abnormalities associated with dementia. However, randomized clinical trials to determine whether VDM consumption improves cognitive performance in the broader population are required (Bennett et al. 2013b; Zajac et al. 2016; Cardwell et al. 2018).

## 4 Macrofungi as Neuroprotectants

Mushrooms have been part of human culture for thousands of years as food, medicine, and religious attribute (Hobbs 2004; Badalyan 2012; Thongbai et al. 2015; Badalyan and Zambonelli 2019). Edible and medicinal mushrooms are an excellent source of polysaccharides (β-glucans), proteins (lectins), phenolics, unsaturated fatty acids, ergosterol (precursor of vitamin $D_2$), minerals, fiber, as well as bioactive metabolites (alkaloids, lectins, phenolics, polysaccharides, terpenoids, etc.). Possessing different medicinal properties, they may play an important role in the prevention, mitigation, and treatment of many diseases, including age-associated neurological dysfunctions, including AD and PD (Chang and Miles 2004; Cheung 2010; Chang and Wasser 2012; Thangthaeng et al. 2015; Valverde et al. 2015; Zengin et al. 2015; Gargano et al. 2017; Phan et al. 2015, 2017; Rathore et al. 2017; Badalyan and Zambonelli 2019; Badalyan et al. 2019; Chen et al. 2019; Hyde et al. 2019; Ho et al. 2020; Lucius 2020; Yadav et al. 2020) (Table 1).

**Table 1** Macrofungi with potential neuroprotective activity

| Mushroom species | Bioactive compounds | Therapeutic effects | References |
|---|---|---|---|
| *Agaricus bisporus* (J.E. Lange) Imbach | Polysaccharides | Against AD, improves memory, AChE inhibitory | Bennett et al. (2013b), Mahmoud et al. (2014), Kozarski et al. (2020) |
| *Agaricus blazei* Murrill (= *Agaricus brasiliensis* Wasser, M. Didukh, Amazonas & Stamets (= *Agaricus subrufescens* Peck.) | Polysaccharide (WSP-AbM) | Antidepressant, anti-inflammatory, neuroprotective, against diabetic neuropathy | Ji et al. (2014), Qin and Han (2014) |
| *Albatrellus yasudae* (Lloyd) Pouzar | Meroterpenoids | Aβ-aggregation inhibitory activity | Akiba et al. (2020) |
| *Amanita caesarea* (Scop.) Pers. | Fatty acids, phenolics | Antioxidant, neuroprotective | Doğan and Akbaş (2013), Li et al. (2017) |
| *Amanita muscaria* (L.) Lam. | Muscimol | Neuroprotective | Kondeva-Burdina et al. (2019) |
| *Antrodia cinnamomea* T.T. Chang & W.N. Chou [= *Taiwanofungus camphoratus* (M. Zang & C.H. Su) Sheng H. Wu, Z.H. Yu, Y.C. Dai & C.H. Su] | Polysaccharides, terpenoids | Anti-inflammatory, antioxidant, neuroprotective | Lu et al. (2013), Hsiao and Weng (2019) |
| *Armillaria mellea* (Vahl) P. Kumm. | Protoilludane sesquiterpenoids, aromatic esters | Antidepressant-like activity | Zhang et al. (2019) |
| *Calocybe indica* Purkayastha & A. Chandra | Polyphenols | Antioxidant, neuroprotective | Cardoso et al. (2015), Rathore et al. (2018) |
| *Cantharellus cibarius* Fr. | Polysaccharides | Neuroprotective | Lemieszek et al. (2018) |
| *Clitocybe geotropa* (Bull. ex DC.) Quél. [= *Infundibulicybe geotropa* (Bull. Ex DC.) Harmaja] | Phenolics | Anti-AChE activity, antioxidant, strong neuroprotective | Kosanić et al. (2020a) |
| *Clitocybe nebularis* (Batsch) P. Kumm. | Phenolics | Antioxidant, anti-AChE activity, strong neuroprotective | Kosanić et al. (2020a) |
| *Coprinellus truncorum* (Scop.) Redhead, Vilgalys & Moncalvo | Polysaccharides | Antioxidant, anti-AChE activity | Pejin et al. (2019) |
| *Coprinus comatus* (O.F. Müll.) Pers. | Polysaccharides | Antioxidant, anti-AChE activity | Pejin et al. (2019) |

(continued)

**Table 1** (continued)

| Mushroom species | Bioactive compounds | Therapeutic effects | References |
|---|---|---|---|
| *Cyathus africanus* H.J. Brodie | Cyathane diterpenoids cyafricanins | Anti-neuroinflammatory activity, NGF-induced neurite outgrowth-promoting activity | Yin et al. (2019), Yadav et al. (2020) |
| *Cyathus hookeri* Berkeley | Cyathane diterpenoids cyahookerins | Anti-neuroinflammatory NGF-induced neurite outgrowth-promoting and neutrophic activity | Tang et al. (2019), Yadav et al. (2020) |
| *Dictyophora indusiata* (Vent.) Desv. (= *Phallus indusiatus* Vent.) | Polysaccharides | Anti-inflammatory, antioxidant, neuroprotective | Zhang et al. (2016b) |
| *Flammulina velutipes* (Curtis) Singer | Polysaccharides | Neuroprotective, mitigate neurodegeneration | Phan et al. (2017), Sabaratnam and Phan (2018) |
| *Fomitopsis officinalis* (Vill.) Bondartsev & Singer | Polysaccharides, lanostane triterpenoids officimalonic and eburicoic acids, flavonoids, organic acids, coumarins, phenolic compounds | Antioxidant, anti-inflammatory, neuroprotective, neuroregenerative, antidepressant, anti-epileptic | Muszyńska et al. (2020) |
| *Ganoderma applanatum* (Pers.) Pat. [=*Ganoderma lipsiense* (Batsch) G.F.] | Polysaccharides, triterpenes, aromatic meroterpenoids, proteins, peptides, sterols | Neuroprotective, against AD and PD, modulation of neurogenesis, therapeutic effect on epilepsy, protective effect on neural cells in stroke injury | Zengin et al. (2015), Zhao et al. (2019) |
| *Ganoderma leucocontextum* T.H. Li, W.Q. Deng, Sheng H. Wu, D.M. Wang & H.P. Hu | Lanostane triterpenes, meroterpenoids, ganoleucoins | Antioxidant, neuroprotective | Chen et al. (2018a) |
| *Ganoderma lucidum* (Curtis) P. Karst. | Polysaccharides, triterpenes, aromatic meroterpenoids, proteins, peptides, sterols | Antioxidant, anti-AChE activity, neuroprotective against AD and PD, analgesic, antidepressant, antiepileptic, antinociceptive, hypnotic, neuroprotective, sedative | Qin and Han (2014), Zengin et al. (2015), Diling et al. (2017), Ćilerdžić et al. (2018), Cui and Zhang (2019), Lai et al. (2019), Wang et al. (2019b), Zhao et al. (2019), Yadav et al. (2020) |
| *Ganoderma microsporum* R.S. Hseu | Proteins | Inhibition of neuronal cell death | Chen et al. (2018b) |

| | | | |
|---|---|---|---|
| *Ganoderma neo-japonicum* Imazeki | Polysaccharides | Antidepressant, neuroprotective | Tan et al. (2015) |
| *Grifola frondosa* (Dicks.) Gray | Protein-bound polysaccharide (PGM) | Antidepressant, neuroprotective, ameliorates AD-like pathology and cognitive impairments by enhancing microglial amyloid-β clearance | Bai et al. (2019), Fan et al. (2019) |
| *Hericium erinaceus* (Bull.: Fr.) Pers. | Polysaccharides, cyathane diterpenoids, hericenones and erinacines | Antioxidant, antidepressant, memory enhancer, neuroprotective and neurostimulating | Ma et al. (2010), Nagano et al. (2010), Phan et al. (2014a), Zengin et al. (2015), Cheng et al. (2016), He et al. (2017), Rossi et al. (2018), Rupcic et al. (2018), Chong et al. (2019), Kushairi et al. (2019), Saitsu et al. (2019), Üstün and Ayhan (2019), Wang et al. (2019a), Limanaqi et al. (2020), Yadav et al. (2020) |
| *Hygrophorus eburneus* (Bull.) Fr. | Phenolics | Antioxidant, anti-AChE activity, strong NPE | Kosanić et al. (2020b) |
| *Inonotus obliquus* (Ach. ex Pers.) Pilát | Polysaccharides | Inhibitory effects on the proliferation of human neurogliocytoma cells | Ning et al. (2014) |
| *Laetiporus sulphureus* (Bull.) Murrill | Polysaccharides, phenolics | Antioxidant, neuroprotective, against AD and PD, AChE, and tyrosinase inhibitory activities | Ćilerdžić et al. (2019) |
| *Lentinus edodes* (Berk.) Pegler | Polysaccharides | Antioxidant, anti-inflammatory, antidepressant, neuroprotective | Badalyan et al. (2019), Diallo et al. (2020) |
| *Lignosus rhinocerotis* (Cooke) Ryvarden | Glucans, proteins, proteases, phenolics | Antidepressant, anti-inflammatory, antioxidant, enhances motor and sensory functional recovery after nerve injury and has no adverse effects on nervous tissues, stimulation of neurite outgrowth, neuroprotective, neuroregenerative | Phan et al. (2013), Nallathamby et al. (2018), Farha et al. (2019) |

(continued)

**Table 1** (continued)

| Mushroom species | Bioactive compounds | Therapeutic effects | References |
|---|---|---|---|
| *Phellinus linteus* (Berk. & M.A. Curtis) Teng, Zhong Guo De Zhen Jun [= *Tropicoporus linteus* (Berk. & M.A. Curtis) L.W. Zhou & Y.C. Dai] | Aromatic acids, polysaccharides, flavones, phenolics, triterpenes | Antidepressant, anti-inflammatory, neuroprotective | Hsieh et al. (2013), Chen et al. (2016a), Chen et al. (2019) |
| *Phellinus pini* (Brot.) A. Ames [= *Porodaedalea pini* (Brot.) Murrill] | Polysaccharides, phenolic compound hispidin, phenolic acids | Anti-AChE activity, anti-inflammatory | Im et al. (2016a) |
| *Pleurotus eryngii* (DC.) Quél. | Polysaccharides, phenolic compounds | Ameliorates memory and learning deficit, against AD | Liang et al. (2020), Zhang et al. (2020) |
| *Pleurotus giganteus* (Berk.) S.C. Karunarathna & K.D. Hyde | Polysaccharides, phenolics, triterpenoids, proteins | Antioxidant, antidepressant, neuroprotective, neurite stimulation | Phan et al. (2012), Bennett et al. (2013a), Yadav et al. (2020) |
| *Pleurotus ostreatus* (Jacq.) P. Kumm. | Polysaccharides pleuran | Antioxidant, against AD and PD | Bobek and Galbavy (2001), Ćilerdžić et al. (2019) |
| *Polyporus umbellatus* (Pers.) Fr. | Exopolysaccharides | Antioxidant | Bandara et al. (2015) |
| *Poria cocos* F.A. Wolf [= *Wolfiporia cocos* (F.A. Wolf) Ryvarden & Gilb.] | *P. cocos* water extract (PCW) | Antidepressant-like, antioxidant, anti-inflammatory, sedative and tonic agent, regulation of monoaminergic neurotransmission | Huang et al. (2020) |
| *Psilocybe samuiensis* Guzmán, Bandala & J.W. Allen | Alkaloids psilocybin, psilocin and baeocystin | Psychoactive | Gartz et al. (1994) |
| *Trametes gibbosa* (Pers.) Fr. | Polysaccharides, phenolics | Antioxidative, anti-neurodegenerative | Knežević et al. (2018) |
| *Trametes pubescens* (Schumach.) Pilát | Polysaccharides, phenolics | Antioxidant, anti-neurodegenerative | Im et al. (2016b), Knežević et al. (2018) |
| *Trametes versicolor* (L.) Lloyd | Polysaccharides, phenolics | Antioxidant, anti-dementia, anti-inflammatory | Knežević et al. (2018), Pop et al. (2018) |
| *Tremella fuciformis* Berk. | Fatty acids, proteins, polysaccharides, enzymes, phenols, flavonoids, volatile oil | Antioxidant, antiaging, neuroprotective, neurotrophic, induces neurite outgrowth via activation of CREB transcription and cholinergic systems | Park et al. (2007), Park et al. (2012), Liu et al. (2019), Wu et al. (2019) |

*AChE* acetylcholinesterase, *AD* Alzheimer's disease, *CREB* C-AMP response element-binding protein, *NGF* nerve growth factor, *NPE* neuroprotective effect, *PD* Parkinson's disease, *WSP-AbM* water-soluble polysaccharide of *Agaricus blazei* Murill

The NDD will continue to rise steadily and is expected to reach 42 million cases in 2020 worldwide. The oxidative stress and inflammation in the neuron–glia system are key factors in the pathogenesis of NDD, the main challenges to modern medicine (Uddin and Ashraf 2018). The general strategy to prevent the development of NDD is a stress-free lifestyle, physical activity, and a healthy diet, enriched with different natural supplements. Despite the advancement of pharmacological treatment, the management of these diseases remains largely ineffective. Moreover, available drugs have always been associated with several side effects; natural products have gained recognition for the prevention or management of diseases. Therefore, it is urgent to explore novel neuroprotective agents and myco-pharmaceuticals to mitigate the development of age-related neurodegenerative conditions.

In vivo and in vitro studies revealed a high nutritional value and nutraceutical properties of macrofungi and have provided evidence for their protective effects against oxidation and inflammation, as well as various lifestyle- and age-related chronic diseases, including AD, PD, arterial hypertension, cancer, and high risk of stroke (Chang and Buswell 1996; Poucheret et al. 2006; Barros et al. 2008; Palacios et al. 2011; Badalyan 2012; Phan et al. 2015, 2017).

The association between dietary patterns with cognitive function has not been thoroughly investigated (Nagano et al. 2010; Lucius 2020). A recent study in a population of elderly Japanese patients suggests that frequent mushroom consumption is associated with a lower risk of incident dementia (Zhang et al. 2017). It has been suggested that both "mushroom, vegetable, and fruits" and "meat and soybean products" patterns were associated with better cognitive function among adults aged more than 60 years old (Yin et al. 2018). Mushroom-derived bioactive compounds and mushroom-based daily diets can improve the cognitive abilities in aging people, inhibit AChE and tyrosinase activity, and prevent the development of NDD (Cardwell et al. 2018; Rossi et al. 2018; Yin et al. 2018; Dhakal et al. 2019). However, scientific validation is required to consider macrofungi as neuroprotective agents, to understand the molecular and biochemical mechanisms involved in the stimulation of neurite outgrowth in in vitro and in vivo studies (Sabaratnam et al. 2013; El Sayed and Ghoneum 2020).

Macrofungi, such as *Agaricus brasiliensis* (Ji et al. 2014; Qin and Han 2014), *Cantharellus cibarius* (Lemieszek et al. 2018), *Laetiporus sulphureus* and *Pleurotus ostreatus* (Bobek and Galbavy 2001; Ćilerdžić et al. 2019), *Fomitopsis betulina* (Pleszczyńska et al. 2017), *Fomitopsis officinalis* (Muszyńska et al. 2020), *Polyporus umbellatus* (Bandara et al. 2015), *Amanita caesarea* (Li et al. 2017), *Hericium erinaceus* (Ma et al. 2010; Üstün and Ayhan 2019), *Phellinus linteus* (Chen et al. 2016a), *Ganoderma lucidum* (Diling et al. 2017), *Ganoderma neo-japonicum* (Tan et al. 2015), *Trametes* (= *Coriolus*) species (Im et al. 2016b; Knežević et al. 2018), *Lignosus rhinocerotis* (Phan et al. 2013; Nallathamby et al. 2018; Farha et al. 2019) and others species, that is, *Agaricus bisporus, Auricularia polytricha, Flammulina velutipes, Grifola frondosa, Lentinus edodes* and *Pleurotus giganteus* (Phan et al. 2012, 2015; Bennett et al. 2013a; Fan et al. 2019; Kozarski et al. 2020) have been used in traditional medicine as neuroprotective and antidepressant agents against age-related NDD (Table 1, Figs. 1 and 2). Among these,

**Fig. 1** Fruiting bodies of edible potential neuroprotective mushrooms: (**a**) *Hericium erinaceus* (Photo Courtesy of Angelini C), (**b**) *Grifola frondosa* (Photo Courtesy of Angelini C), (**c**) *Lentinus edodes* (Photo Courtesy of Angelini C), and (**d**) *Tremella fuciformis* (Photo Courtesy of Callac P)

**Fig. 2** Wild-growing fruiting bodies of agaricoid and polyporoid neuroprotective mushrooms: (**a**) *Ganoderma lucidum* (Photo Courtesy of Angelini C), (**b**) *Pleurotus ostreatus* (Photo Courtesy of Angelini C) (**c**) *Trametes versicolor* (Photo Courtesy of Angelini C), (**d**) *Flammulina velutipes* (Photo Courtesy of Moingeon JM, (**e**) *Amanita muscaria* (Photo Courtesy of Callac P and Guinberteau J), (**f**) wild specimen CA918 *Agaricus subrufescens* (Photo Courtesy of Callac P)

*H. erinaceus, G. lucidum* and *L. edodes* are widely used as bio-ingredients in the formulation of cholesterol-free functional food products which are of great interest to the modern dietary nutrition industry (Chang and Buswell 1996; Ma et al. 2010; Valverde et al. 2015; Ho et al. 2020; Lucius 2020; Yadav et al. 2020).

Recent literature data have provided information about neuroprotective, psychotropic, and antidepressant effects of macrofungi attributed to their antioxidant, anti-neuroinflammatory, and cholinesterase inhibitory properties (Grienke et al. 2014; Phan et al. 2015, 2017; Badalyan and Zambonelli 2019; Badalyan et al. 2019; Hyde et al. 2019; Zhang et al. 2020). More than 20 different brain-improving culinary–medicinal mushrooms and their around 80 bioactive compounds from basidiocarps and mycelia have been reported to reduce β-amyloid-induced neurotoxicity, to show anti-AChE and neurite outgrowth stimulatory effects, as well as to assist NGF synthesis. Moreover, the in vitro and in vivo studies on the molecular mechanisms of meuroprotection and possible clinical trials are also discussed (Phan et al. 2015). Nevertheless, there is a gap between traditional knowledge, experimental evidence, and clinical studies, which are restricted by the quality of trials and other important criteria that may affect their success to become clinically verified drugs. Moreover, before the clinical application of mushroom-derived neuroprotectants, the therapeutic potential and synergistic effects of bioactive ingredients and stabilization for the administration of the drug needs to be evaluated appropriately (Lindequist 2013; Money 2016; Badalyan et al. 2019; Lew et al. 2020; Lucius 2020).

## *4.1* Hericium erinaceus

The cultivated edible mushroom, *Hericium erinaceus* (Lion's mane or Monkey's head) known as "Houtou" or "Shishigashira" in Chinese and "Yamabushitake" in Japanese), has been frequently prescribed in TCM, as an important medicinal fungus with immunomodulatory, antioxidant, anti-inflammatory, antitumor, and neuroprotective properties (Thongbai et al. 2015; Badalyan et al. 2019; El Sayed and Ghoneum 2020).

The modern myco-pharmacological study has attracted considerable attention on *H. erinaceus* as a neuroprotector to prevent and mitigate NDD, including AD, PD, and other forms of dementia, anxiety, or depression (Mizuno 1999; Wong et al. 2007, 2011, 2012, 2014; Ma et al. 2010; Nagano et al. 2010; Mori et al. 2011; Kim et al. 2014; Phan et al. 2014a, b, 2019; Thongbai et al. 2015; Cheng et al. 2016; Kuo et al. 2016; Zhang et al. 2016a; Spelman et al. 2017; Chong et al. 2019; Jang et al. 2019; Kushairi et al. 2019; Saitsu et al. 2019; Üstün and Ayhan 2019; Limanaqi et al. 2020; Yadav et al. 2020).

*H. erinaceus* contains high amounts of antioxidants and polysaccharides (β-glucans); in addition, a potent catalyst for brain tissue regeneration helps to improve memory and cognitive functions. The fruiting bodies and mycelium of *H. erinaceus* possess immunomodulating, antitumor, hypoglycemic, and antiaging properties. This fungus can be considered as useful therapeutic agents in the management and/or treatment of NDD (Ma et al. 2010; Wong et al. 2012; Kim et al. 2014; Cheng et al. 2016; Zhang et al. 2016a; Diling et al. 2017; Chong et al. 2019).

The fruiting bodies and fermented mycelia of *H. erinaceus* have been reported to produce different groups of bioactive compounds (polysaccharides, proteins, lectins, phenolic derivatives, and terpenoids) among which two classes of terpenoids—hericenones and erinacines stimulate the synthesis of nerve growth factor (NGF)—a neurotrophic factor and neuropeptide primarily involved in the regulation of growth, maintenance, proliferation, and survival of certain target neurons (Kim et al. 2014; Thongbai et al. 2015). Isolation, structural elucidation, and bioactivity of hericenones and erinacines from the fruiting body and mycelium of *H. erinaceus* have been reviewed (Kawagishi et al. 1991; Ma et al. 2010; Phan et al. 2014a; Li et al. 2018). Hericenones and erinacines have been found to promote the expression of the neurotrophic factor (NTF) associated with cell proliferation. However, only erinacine A has revealed therapeutic properties in the central nervous system of experimental rats (Li et al. 2018). The screening of 58 biomolecules isolated from *H. erinaceus* and their neurotropic activity was recently reported (Aparicio-Razo 2020). Four benzyl alcohol derivatives of hericenones B–E exerts NTA in PC12 cells by extracellular signal-regulated 21 kinase 1/2 (ERK1/2) and phosphoinositide-3-kinase/AKT or protein kinase B (PI3K/AKT) signaling pathways (Phan et al. 2014a). The hericenones, an inhibitor of the β-secretase enzyme improved mitochondrial dysfunction, intracellular $Ca^{2+}$ levels, inhibition of the production of ROS, increase in the mitochondrial membrane potential (MMP) and ATP levels, as well as regulation of the expression of genes encoding for p21, COX I, COX II, PARP1, and NF-κB proteins can be suggested as neuroprotectants for the treatment of various NDD (Diling et al. 2017).

The hericipins A, B, C, E, F and hericenones C, D, E, and Y-A-8-c, promoted the induction of NGF synthesis. The NGF could prolong neuronal axons and regulate the formation of neurons, as well as promote their regeneration in animals. The activity of hericipins was suspected to be more potent compared to the activity of adrenaline. This finding should have opened a new perspective in the treatment of Alzheimer-type dementia and neurasthenia (Kawagishi et al. 1990, 1991; Mizuno 1999; Ma et al. 2010). Bioactive compounds erinaceolactones with plant-growth regulatory activity were isolated from the culture broth of *H. erinaceus* (Wu et al. 2014).

Although antidepressant effects of *H. erinaceus* have not been scientifically validated and compared to conventional antidepressants, based on the neurotrophic and neurogenic pathophysiology of depression, the medicinal properties of this fungus may allow developing a potential alternative drug for prevention and treatment of depression (Nagano et al. 2010; Chong et al. 2019).

Although *H. erinaceus* has shown therapeutic potential in many neurological diseases, its role in SE-mediated neuronal death remains unclear. The NPE of crude extracts obtained from *H. erinaceus* before and after SE was observed. At 7 d after SE, animals treated with 60 mg/kg and 120 mg/kg of *H. erinaceus* revealed improved hippocampal neuronal survival, whereas those treated with 300 mg/kg showed similar neuronal death to that of vehicle-treated controls. Thus,

*H. erinaceus,* as a neuroprotectant can be used for preventing neuronal death after epileptic seizures (Jang et al. 2019).

The neurobiological activities of *H. erinaceus,* such as its effect on neurite outgrowth and differentiation in PC12 cells, have been revealed. The NPE of two high molecular weight of polysaccharides (1.7 × 10(5)Da, 1.1 × 10(5)Da) obtained from EE of *H. erinaceus* showed that 250 μg/mL of polysaccharide prevented Aβ-induced shrinkage and nuclear degradation of PC12 cells. The polysaccharides, except NPE, also showed AOA (Cheng et al. 2016).

*H. erinaceus* has been extensively tested in vivo and in vitro as a stimulator of neurite outgrowth in cultured cells of the neural hybrid clone NG108–15 and rat PC12 cells (Phan et al. 2014a; Wong et al. 2007, 2011, 2012, 2014, 2016). The enhancement of peripheral neuro-regenerative activity (NRA) by AE from cultivated *H. erinaceus* after crush injury in vivo, as well as the physiological mechanisms of its effect on growth and development of neuronal cell cultures, improvement of functional recovery after nerve injury, together with its neurotropic and neurotrophic effects, have been previously reported (Wong et al. 2011, 2012, 2014, 2016). *H. erinaceus* did not cause neuron damage, toxic effect, or suppression of cellular respiration, improving the myelination process in the mature myelinating fibers. However, further studies are warranted to elucidate molecular mechanisms to promote the growth and regeneration of axons by this fungus. The development of *H. erinaceus* in alternative therapies is in progress (Wang et al. 2014; Wong et al. 2014, 2016).

HWE and EE from fruiting bodies of *H. erinaceus* were investigated for their NPE. The EE showed potent NPE leading to a significant increase in the viability of $H_2O_2$-treated neurons accompanied by a reduction in ROS and improvement of the catalase (CAT) and glutathione (GSH) content. It also increased the production of MMP and ATP, while reducing mitochondrial toxicity, Bcl-2-associated X (Bax) gene expression, and nuclear apoptosis, as well as reduced NO level in LPS-treated BV2, indicating anti-inflammatory activity (AInA) in microglia. Thus, the EE of *H. erinaceus* may be considered as a potential neuroprotective and anti-inflammatory agent (Phan et al. 2019; Kushairi et al. 2019).

The oral administration of mycelial biomass of *H. erinaceus* during 3 months showed NPE in an experimental animal model. The upregulation of LXA4 was associated with increased content of redox-sensitive proteins. The maximal induction of LXA4 was observed in the cortex, and hippocampus followed by substantia nigra, striatum, and cerebellum (Trovato et al. 2016b).

Peripheral nerve injury (PNI) is an important health problem. Insights into this process are important for the development of novel effective therapies. NGF plays a significant role in the survival, growth, and maintenance of various neurons in the nervous system. The study of NPE of *H. erinaceus* and NGF on a mouse PNI model showed that *H. erinaceus* exhibits a higher NPE compared to the NGF. The combination of both increases the axonal regeneration ability of axotomized neurons in mice. Moreover, *H. erinaceus* prevents the death of neurons and regenerates their axons, therefore, may serve as a neuroprotective and neuro-regenerative agent for treating PNI. Further studies of *H. erinaceus* as a potential source of biocompounds to cure PNI are warranted (Üstün and Ayhan 2019).

The neuroprotective properties of *H. erinaceus* mycelia enriched with erinacines may contribute to further research on the therapeutic role of this fungus. Preclinical studies have shown that daily consumption of *H. erinaceus* enriched with erinacines can prevent the occurrence of ischemic stroke, AD, PD, and depression (Li et al. 2018). Oral supplementation with *H. erinaceus* results in specific and selective improvements in novelty-seeking behavior and object recognition in mice (Rossi et al. 2018). Studies related to the impact of *H. erinaceus* dietary oral supplementation on brain function are few; however, the effect on cognitive skills and hippocampal neurotransmission in wild-type mice has been reported (Nagano et al. 2010; Brandalise et al. 2017; Lucius 2020).

Several bioactive compounds of *H. erinaceus* have been developed into food supplements and alternative medicines (Kawagishi et al. 1990, 1991; Hobbs 2004; Jiang et al. 2014; Wang et al. 2014, 2019a, b; Friedman 2015; Thongbai et al. 2015; He et al. 2017; Li et al. 2018; Rossi et al. 2018; Trovato Salinaro et al. 2018). A new mushroom product available in the market for the treatment of AD and dementia has been developed on standardized extracts, containing hericinones and amiloban from *H. erinaceus*. However, the correspondence of active mushroom components that cause the observed effects is often not clear (Chang and Wasser 2012; Lindequist 2013; Wang et al. 2014; Money 2016).

Since NDD is associated with oxidative stress, antioxidant therapy has been suggested for its prevention and treatment. The antioxidant product *Antia* developed from *H. erinaceus* and several plants (*Centella asiatica, Dioscorea villosa, Salacia reticulata* and *Phyllanthus emblica*) showed a NPE in AD-induced mice. Moreover, the treatment with *Antia* showed a protective effect on malondialdehyde (MDA), NF-κB, IL-6, TNF-α, and amyloid-β and other compounds (El Sayed and Ghoneum 2020).

A recent study has also explored *H. erinaceus* as potential fortified foods enriched with lithium. Co-cultivation of *H. erinaceus* with lithium chloride results in a concentration-dependent uptake of lithium and its accumulation in *H. erinaceus* fruiting bodies, to be useful as supplementation for daily dietary intake of lithium underlying the beneficial effects in the brain (Limanaqi et al. 2020). However, further investigation should be carried out in psychiatric disorders.

## *4.2* Ganoderma *Species*

The age-related oxidative damage and DNA methylation generated in the human body can cause different neuropathological disorders, including dementia. Therefore, the modulation of these conditions may be an effective strategy to delay the progression of NDD (Cornelius et al. 2013; Uddin and Ashraf 2018; Lai et al. 2019).

Numerous studies have reported AOA and NPE of *Ganoderma* mushrooms during stroke injury, modulation of neurogenesis, as well as treatment of AD, dementia and epilepsy (Chen et al. 2018a; Ćilerdžić et al. 2018; Lai et al. 2019; Wang et al. 2019b; Zhao et al. 2019; Rahman et al. 2020; Yadav et al. 2020).

The ganoderic acid and lucidone A isolated from alcoholic extracts of *Ganoderma lucidum* delay AD progression, improve learning and memory function, ameliorate neuronal apoptosis, and brain atrophy, and downregulate the expression of AD intracellular marker, Ab1–42, in animals (Lai et al. 2019). It has been shown that the biotechnological cultivation of *G. lucidum* is promising, and the cultivation substrate greatly affects not only the chemical profile but also the neuroprotective capacity of basidiocarps (Ćilerdžić et al. 2018). The meroterpenoids isolated from *G. lucidum* have been suggested as potential antioxidant and neuroprotective functional food ingredients (Wang et al. 2019b). Hypercholesterolemia is a risk factor in the development of AD. HWE from *G. lucidum* showed spatial learning and memory-related behavioral amelioration and has an important role in the pathogenesis of AD in hypercholesterolemic rats (Rahman et al. 2020).

The antioxidant, antimicrobial, and cholinesterase, tyrosinase, α-amylase, and α-glucosidase inhibitory activities of ME and AE from other medicinal *Ganoderma* mushrooms, such as *G. applanatum,* and *G. resinaceum* have been reported. The highest AOA and enzyme inhibitory effects were detected in ME of tested species with the highest amount of phenolics. These macrofungi can be potentially used to develop new food supplements and drug formulations (Zengin et al. 2015). Furthermore, anti-inflammatory and neuroprotective effects of *Ganoderma microsporum* immunomodulatory protein (GMI) involving microglial inhibition are reported (Chen et al. 2018b).

According to Cui and Zhang (2019), *Ganoderma* species, and mainly *G. lucidum* have a broad spectrum of neuropharmacological effects as an analgesic, antidepressant, antiepileptic, antinociceptive, hypnotic, neuroprotective, and sedative. The authors summarize among other things rare preclinical and clinical trials of *Ganoderma* and its ingredients in the patients with these disorders.

## *4.3* **Pleurotus** *Species*

One of the major etiological factors of AD is oxidative stress, which accelerates Aβ peptide plaque and neurofibrillary tangle accumulation in the brain. Oyster mushrooms (*Pleurotus* spp.) possess a high quantity of antioxidants, including ergothioneine, adenosine, and polyphenol, which reduce oxidative stress-related aging (Badalyan 2012; Phan et al. 2014b; Liang et al. 2020).

A recent study has shown that consumption of edible medicinal mushroom *Pleurotus eryngii* delayed the development of brain atrophy, ameliorated memory deficit in mice, and significantly decreased the levels of brain phosphorylated τ-protein, Aβ plaque deposition, MDA, and protein carbonyl therefore may improve memory and learning capacity in an Aβ-induced AD mouse model (Zhang et al. 2020).

In vitro and in vivo studies of neuroprotective and antidepressant properties of polysaccharides isolated from *P. eryngii* on Aβ-induced neurotoxicity in PC12 cells showed, that 28-weeks treatment by polysaccharides resulted in the significant elevation of cell viability, decreased in the levels of intracellular calcium, and attenuation of the Aβ-mediated cell apoptosis. In aging rats, *P. eryngii* polysaccharides may decrease the production of amyloid precursor protein (APP) in the brain by a mechanism associated with the lowering of iNOS and COX-2 levels showing neuroprotective and antidepressant effects (Zhang et al. 2020).

*Pleurotus* (= *Panus*) *giganteus* contain a high number of polysaccharides, phenolics, triterpenoids, proteins, dietary fiber, potassium as well as vitamins $B_1$, $B_2$, and $B_3$. Both AE and EE of *P. giganteus* induced neurite outgrowth in rat PC12 and mouse N2a cells, as well as stimulates neuronal differentiation and elongation. Linoleic acid present in the EE promoted NGF biosynthesis, whereas the phenolics positively correlated to the AOA. Thus, *P. giganteus* contains bioactive compounds that mimic NGF and are responsible for neurite stimulation. They can be used in a well-balanced diet as a source of antioxidants to promote neuronal health (Phan et al. 2014a, b; Yadav et al. 2020). The AChE inhibitory activity was detected in HWE from fruiting bodies of *Pleurotus ostreatus*, however, the effect was weaker than the effect of the commercial anti-AD preparation, galantamine (Ćilerdžić et al. 2019).

## *4.4* Trametes (= Coriolus) *Species*

Several white-rot polypore *Trametes* (= *Coriolus*) species, such as *Trametes versicolor, T. hirsute, T. gibbosa,* and *T. pubescens* have been used for centuries in the traditional medicine of East Asian cultures, however, only *T. versicolor* has been comprehensively studied (Hobbs 2004; Im et al. 2016b; Knežević et al. 2018; Pop et al. 2018; Kıvrak et al. 2020). The high-molecular-weight fractions, especially polysaccharide Krestin (PSK) derived from mycelium, have been studied in human clinical trials in Japan with several chemotherapy protocols. Numerous in vitro and in vivo studies and clinical trials of bioproducts obtained from *T. versicolor* have shown enhancement of immune response and improvement of the quality of life (Wasser 2017; Badalyan et al. 2019).

*T. versicolor* mycelium extract was the most effective inhibitor of AChE activity, but twice weaker than galantamine. The mycelium extract of *T. gibbosa* more significantly inhibited tyrosinase activity than kojic acid. The chemical screening revealed strong synergistic action of the content of bioactive compounds, such as triterpenes, sugars, and polyphenols produced by studied *Trametes* species (Knežević et al. 2018).

Basidiocarp extracts from *Trametes* species (*T. gibbosa*, *T. hirsuta*, *T. versicolor*) were more effective ABTS+ scavengers and $Fe^{2+}$ reducers in comparison with mycelium extract, however, they were less effective than L-ascorbic acid (Knežević et al. 2018).

The LXA4 upregulation was associated with increased content of redox-sensitive proteins involved in cellular stress response (Trovato et al. 2016a). The maximum induction of LXA4 by *Coriolus* (= *Trametes*) *versicolor* was observed in the cortex and hippocampus of the brain in experimental rats. The supplementation with *C. versicolor* maintained a response to counteract intracellular pro-oxidant status in MD patients (Scuto et al. 2020).

The emerging role of the inflammasome and the importance of *Coriolus* and *Hericium* medicinal mushroom-based nutra- and nutriceuticals in neuroprotection have recently been considered (Trovato Salinaro et al. 2018).

In vitro evaluation of ME and HWE from fruiting bodies of *Trametes pubescens* revealed antioxidant related antidiabetes, anti-dementia, and anti-inflammatory activities (Im et al. 2016b). Eleven phenolic compounds were detected in tested extracts. The ME showed NPE against glutamate-induced PC-12 cell cytotoxicity at 2–40 μg/mL. The AChE and BChE inhibitory effects by both tested extracts at 2.0 mg/mL were moderate and comparable with galanthamine (CAS 357–70-0), the standard drug to treat AD. The extracts also possess AInA and significantly suppressed the carrageenan-induced rat paw edema.

## *4.5* Amanita *Species*

The bioactivity, chemical composition, as well as AOA and NPE of edible mushroom *Amanita caesarea* (Caesar's mushroom) have been reported (Doğan and Akbaş 2013; Li et al. 2017). The AE of *A. caesarea* improve cell viability, restored mitochondrial function, inhibited the overproduction of intracellular ROS and $Ca^{2+}$, and suppressed the high expression levels of cleaved-caspase-3 and calpain 1 enzymes, apoptosis-inducing factor (AIF) in the AD–mouse model, as well as alleviated the deposition of Aβ in the brain. The extract improved the central cholinergic system function, as indicated by an increase in ACh and ChAT concentrations and a reduction in AChE levels, as well as reduced ROS and increased SOD levels in the brain of experimental mice. The results provide evidence that *A. caesarea* may be used as a potential neuroprotectant to prevent or mitigate different neurodegenerative disorders (Li et al. 2017).

Muscimol (known also as agarin or pantherine) is one of the principal psychoactive constituents of *Amanita* mushrooms. Muscimol is a potent, selective agonist for the $GABA_A$ receptors and possesses neurotropic, sedative, and hallucinogenic activity. Muscimol is the main hallucinogenic compound found in toxic *Amanita muscaria* mushroom (Kupka and Wieczorek 2016; Kupka et al. 2020). Several studies have described the suppressive effect of muscimol on tremor, without impairing speech and coordination in Parkinson-affected patients. The extract of *A. muscaria* containing high amounts of muscimol showed statistically significant in vitro neuroprotective and antioxidant effects on different neurotoxicity models tested at subcellular and cellular levels in rat brain microsomes, mitochondria, synaptosomes, as well as on neuroblastoma cell line SH-SY5Y (Kondeva-Burdina et al. 2019).

## 4.6 Agaricus blazei (= Agaricus subrufescens)

Well-known medicinal mushroom *Agaricus blazei* (= *Agaricus subrufescens*) contains several bioactive compounds with various pharmacological effects (Val et al. 2015; Badalyan et al. 2019; Rahmani-Nezhad et al. 2019). For taxonomy and synonymy of that species we recommend the following report as mentioned in Badalyan et al. (2019): the main synonyms of *A. subrufescens* are *Agaricus blazei* Murrill sensu Heinemann, *A. rufotegulis* Nauta, *A. brasiliensis* Wasser, M. Didukh, Amazonas and Stamets, *A. albopersistens* Zuccher, and *A. bambusae* Beeli var. *bambusae*.

The severe clinical presentation of cerebral malaria has been associated with poor treatment access, therapeutic complexity, and drug resistance; therefore, alternative therapies are required. Mice treated with AE or fraction C from *A. blazei* (= *A. subrufescens*) showed AOA, lower parasitemia, increased survival, reduction in weight loss, decrease in brain lesions, and protection against cerebral malaria caused by *Plasmodium berghei*. Thus, *A. blazei* was effective in improving the consequences of cerebral malaria in mice and may provide bases for the development of novel therapeutic strategies (Val et al. 2015).

Different extracts of *A. subrufescens* obtained from Iranian and French strains showed selective AChE inhibitory activity with $IC_{50}$ values of 154.63 and 145.43 μg/mL, respectively. However, the extracts were not demonstrated BChE inhibitory activity whereas its anti-Aβ aggregation activity was comparable to donepezil, as a reference drug. Thus, the extracts induced moderate AOA by DPPH-radical scavenging activity and weak neuroprotective activity against A-β-induced damage (Rahmani-Nezhad et al. 2019).

## 4.7 Grifola frondosa

Maitake mushroom, *Grifola frondosa,* possesses nutritional and medicinal value and contains a high amount of health-enhancing bioactive compounds (Badalyan and Zambonelli 2019; Badalyan et al. 2019; Bai et al. 2019; Fan et al. 2019). It has been reported that proteoglucan isolated from *G. frondosa* (PGM) possesses strong immunomodulatory effects and can improve learning and memory, decrease the loss of neurons and histopathological abnormalities in mice (Bai et al. 2019; Fan et al. 2019) Moreover, PGM treatment could activate microglia, astrocytes, promote microglial recruitment to the Aβ plaques, and enhance Aβ phagocytosis, thereby alleviating Aβ burden and pathological changes in the cortex and hippocampus. The administration of PGM as a dietary supplement may provide potential benefits on brain aging-related memory dysfunction (Bai et al. 2019; Fan et al. 2019).

## 4.8 Other Mushroom Species as Potential Neuroprotectants

In addition to species as mentioned earlier, other Agaricomycetes mushrooms, such as *Antrodia cinnamomea, Armillaria mellea, Calocybe indica, Clitocybe geotropa, Cyathus hookeri, Dictyophora indusiata, Fomitopsis betulina, Hygrophorus eburneus, Laetiporus sulphurous, Lignosus rhinocerotis, Paxillus panuoides, Phellinus linteus, Poria cocos,* and *Tremella fuciformis* have also been reported to possess NPE (Park et al. 2007; Hsieh et al. 2013; Lu et al. 2013; Phan et al. 2013; Bandara et al. 2015; Chen et al. 2016a; Zhang et al. 2016b; Nallathamby et al. 2018; Rathore et al. 2018; Ćilerdžić et al. 2019; Farha et al. 2019; Hsiao and Weng 2019; Lee et al. 2019; Wu et al. 2019; Kosanić et al. 2020a, b; Yadav et al. 2020).

The NPE of mushrooms acts by several biological pathways, including inhibiting β-site APP-cleaving enzyme (BACE1), responsible for releasing toxic Aβ peptide from the brain. Several medicinal mushrooms, such as *Agaricus bisporus, Auricularia polytricha, Flammulina velutipes,* and *Lentinus edodes* have been tested for the regulation of BACE1. Both BACE1 inhibitory and stimulatory effects were observed. The inhibitory effect was detected in *A. polytricha*, whereas *A. bisporus, F. velutipes*, and *L. edodes* were activators of BACE1. The inhibitory effect was attributed to lipophilic hispidin-derived polyphenols with possible brain bioavailability, whereas the stimulatory effect—to polysaccharides (Bennett et al. 2013a).

Protoilludane sesquiterpenoid aromatic (PSAM) esters isolated from edible medicinal mushroom *A. mellea* are the main active components with antibacterial and anticancer activities. However, 1 mg/kg intraperitoneal injection of PSAM esters showed also significant antidepressant-like activity, which could be reversed by pretreatment with haloperidol (a nonselective $D_2$ dopamine receptor antagonist), bicuculline—a competitive GABA antagonist and N-methyl-D-aspartate (NMDA)—an agonist at the glutamate site. PSAM esters also effectively increased the hippocampus DA and GABA and decreased the hippocampus glutamate (Glu) levels of mice, indicating that the antidepressant-like effect of PSAM ester might be mediated by the DAergic, GABAergic, and Gluergic systems (Zhang et al. 2019).

The medicinal mushroom *Dictyophora indusiata (= Phallus indusiata)* has been traditionally used in China to cure different inflammatory and neurological diseases. *D. indusiata* polysaccharides are shown to possess in vitro and in vivo antioxidant-related NPE in *Caenorhabditis elegans* nematode. The fungus was shown not only to increase survival rate and reduce stress level but also to decrease ROS and MDA levels and to increase SOD activity. Moreover, *D. indusiata* has restored the functional parameters of mitochondria in the nematode. These findings demonstrate AOA and NPE of *D. indusiata* polysaccharide and suggest further pharmacological usage of this mushroom in the treatment of NDD (Zhang et al. 2016b).

The hymenochaetoid medicinal mushroom *Phellinus linteus* contains polysaccharides, flavones, triterpenes, aromatic acids, phenylpropanoids, furans, amino acids and has been widely used in Asian countries to treat hemorrhage and blood coagulation disorders. Antitumor, hypoglycemic, anti-inflammatory, anti-obesity, and neuroprotective activities of *Ph. linteus* have also been reported (Hsieh et al. 2013; Chen et al. 2016a, 2019).

The highest potency to induce neurite outgrowth in PC12 cells compared to other previously reported natural substances of HWE derived from *Tremella fuciformis* was reported (Park et al. 2007). The *T. fuciformis*-treated PC12 cells before β-amyloid peptide treatment have significantly diminished toxicity. Therefore, *T. fuciformis* may potentially be used as a neuroprotectant in the therapy of NDD. Park et al. (2012) also reported that on the one hand, *T. fuciformis* enhances the neurite outgrowth of PC12 cells and, on the other hand, restores trimethyltin-induced impairment of memory in rats via activation of CREB transcription and cholinergic systems. Strong AOA ($IC_{50} = 0.176$ mg/mL) detected in *T. fuciformis* was positively correlated with the concentration of volatile oil extracted from this fungus (Liu et al. 2019).

The neuroprotective, hepatoprotective, anti-inflammatory, antioxidant, and antineoplastic activities were reported in a medicinal polypore mushroom, *A. cinnamomea* (Lu et al. 2013; Hsiao and Weng 2019). It was shown that the oral administration of a low dosage of AE from another polypore species *L. rhinocerotis* has improved motor and functional sensory recovery after nerve injury and had no adverse effect on nervous tissues, unlike mecobalamin, used for the treatment of peripheral neuropathies (Phan et al. 2013; Nallathamby et al. 2018; Farha et al. 2019).

A medicinal polypore mushroom *Poria cocos* (= *Wolfiporia cocos*) possesses antioxidant and anti-inflammatory activities and was used as a sedative, diuretic, and tonic agent in traditional medicine. A potent antidepressant-like effect of AE of *P. cocos* via regulation of monoaminergic neurotransmission and inactivation of inflammation in a rodent animal model have been reported (Huang et al. 2020).

The study of neuroprotective, antioxidant, antimicrobial, and cytotoxic activities of acetone extracts from agaricoid mushrooms *Hygrophorus eburneus, Clitocybe geotropa,* and *Clitocybe nebularis* showed dose-dependent AChE inhibitory activity and a strong NPE. Estimated as pyrocatechol equivalent (PE), total phenolic content of *H. eburneus* was 9.27 μg PE/mg, of *C. geotropa*—95.71 μg PE/mg, and of *C. nebularis*—93.94 μg PE/mg. These mushrooms can be regarded as a source of nutraceuticals and neuroprotective functional food (Kosanić et al. 2020a, b).

In vitro study of antioxidant and neuroprotective properties of extracts from *P. ostreatus* and *Laetiporus sulphureus* showed the highest reducing power in *L. sulphureus*. In comparison to α-kojic acid, the tested extracts showed a weaker tyrosinase inhibitory activity. Fungal extracts were rich in phenolics, which were in positive correlation with AOA, AChE, and tyrosinase inhibition. Thus, mushroom-derived phenolic compounds are the potential carriers of NPE possessing significant antioxidant and anti-neurodegenerative capacity and can be suggested as novel nutraceuticals and myco-pharmaceuticals (Ćilerdžić et al. 2019).

The study of a neuroprotective mechanism by *p*-terphenyl leucomentins derived from *Paxillus panuoides* showed potent inhibition of lipid peroxidation and $H_2O_2$-induced neurotoxicity, but free from any role as ROS scavengers. The leucomentins can chelate iron when DNA is present with iron and $H_2O_2$, thereby inhibiting DNA single-strand breakage. These results suggest that the NPE of leucomentins is dependent on their ability to chelate iron (Lee et al. 2003).

The bracket fungus *Fomitopsis betulina* has been traditionally used in folk medicine as antimicrobial, anticancer, and anti-inflammatory agents. Due to its therapeutic properties, the pieces of its fruiting body were carried by Ötzi the Iceman (Pöder 2005). Modern myco-pharmacological studies confirm the health-enhancing effects of *F. betulina* and provided evidence supporting the antibacterial, antiparasitic, antiviral, anti-inflammatory, anticancer, and immunomodulatory activities, as well as NPE. *F. betulina* is a source of bioactive triterpenoids, valuable enzymes, (1 $\rightarrow$ 3)-α-D-glucan, and can be considered as a promising source for the development of new healthcare bioproducts (Pleszczyńska et al. 2017).

The qualities of medicinal preparations obtained from the *Fomitopsis officinalis* fruiting bodies are determined by the unique composition of its bioactive compounds, such as triterpenoids, polysaccharides, organic acids, coumarins, and phenolic compounds. It has been proved that both crude extracts and the compounds isolated from *F. officinalis* have a wide spectrum of therapeutic effects, including anti-inflammatory, cytotoxic, and antimicrobial effects. The potential mechanism of action of bioactive compounds, such as flavonoids of *F. officinalis* on the central nervous system has been discussed. Dietary products originated from *F. officinalis*, such as *Agarikon* capsules, and powdered *F. officinalis* mycelium are already available in the market (Muszyńska et al. 2020).

The total phenolic content, free radical DPPH-scavenging activity, and ferric reducing power (FRAP) of ME from Se-enriched fruiting bodies of *Calocybe indica* have almost doubled. The correlations amongst the biomass yield, polyphenols, and AOA at 5 mg/mL concentrations of the Se was demonstrated (Rathore et al. 2018). Indeed, further investigation should be carried out on Se-enriched *C. indica* biomass, as a potential novel food supplement with improved Se bioavailability in neurological disorders (Cardoso et al. 2015).

The hypogeal sclerotia of edible medicinal fungus *Polyporus umbellatus* contain antitumor, anticancer, immune-modulating, antimicrobial, and antioxidant compounds and currently used as an ingredient in many healthcare products and food supplements (Bandara et al. 2015). Further investigation could be carried out on antioxidant capacity, and free radical scavenging activity of *P. umbellatus* to develop natural health food from mycelium and sclerotia and exopolysaccharides-based food supplement to prevent, support, or cure neurodegenerative diseases.

## 5 Conclusion and Future Prospects

Presently, human neurodegenerative and neurological diseases, such as Alzheimer's, Parkinson's, Meniere diseases, as well as epilepsy and depression, are affecting the adult population worldwide. Therefore, scientists have been attempting for more than 20 years to discover new resources of medicines, including mushrooms-derived neuroactive compounds, to delay the progression of these diseases.

This review discusses the current state of knowledge and the findings of recent studies on the neuroprotective potential of macrofungi. It offers a preliminary roadmap for both physicians and researchers interested in expanding their knowledge about neurodegenerative disorders. However, the investigated list of macrofungi is far from being complete.

Future research should be carried out using an interdisciplinary approach involving physicians, biologists, chemists, pharmacologists, and mycologists partnering with social scientists to create a scientific framework that incorporates traditional knowledge, biochemical, and biomedical data.

**Acknowledgments** This review arises from a long-standing collaboration between two authors (S.M.B. and S.R.) on research directed to the identification of bioactive compounds from mushrooms and develops their medicinal properties. The realization of research work was performed in collaboration between the Institute of Pharmacy, Yerevan State University, Armenia; and Faculty of Pharmacy of the University of Montpellier/UMR 5175 CNRS, France. The research project on medicinal mushrooms was supported by the MESCS Republic of Armenia (grant number #18 T-1F115). The authors are thankful to Dr. A. Barkhudaryan for a critical review of the manuscript.

We would like to thank our colleagues Claudio Angelini (Pordenone, Italy), Philippe Callac (Villenave d'Ornon, France), Jacques Guinberteau (France) and Jean-Marc Moingeon (Goux-les-Usiers, France) for kindly providing photos of medicinal mushrooms (Figs. 1 and 2).

## References

Aguirre-Moreno A, Campos-Pena V, del Rio-Portilla F et al (2013) Anticonvulsant and neuroprotective effects of oligosaccharides from lingzhi or reishi medicinal mushroom, *Ganoderma lucidum* (higher Basidiomycetes). Int J Med Mushrooms 15(6):555–568. https://doi.org/10.1615/intjmedmushr.v15.i6.40

Ahuja M, Patel M, Majrashi M et al (2017) *Centella asiatica*, an ayurvedic medicinal plant, prevents the major neurodegenerative and neurotoxic mechanisms associated with cognitive impairment. In: Agrawal DC, Tsay HS, Shyur LF, Wu YC, Wang SY (eds) Medicinal plants and fungi: recent advances in research and development, medicinal and aromatic plants of the world, vol 4. Springer Nature, Singapore, pp 3–48. https://doi.org/10.1007/978-981-10-5978-0_1

Akiba M, Kinoshita K, Kino Y et al (2020) Isolation of three new meroterpenoids and seven known compounds from *Albatrellus yasudae* and their Aβ-aggregation inhibitory activity. Bioorg Med Chem Lett 30(2):126808. https://doi.org/10.1016/j.bmcl.2019.126808

Aparicio-Razo M (2020) Analysis of biomolecules of the fungus *Hericium erinaceus* through the theory of electron transfer of quantum chemistry and its relationship with the primary amino acids. World J Pharm Sci 9(6):139–147. https://doi.org/10.20959/wjpps20206-16343

Badalyan SM (2012) Medicinal aspects of edible ectomycorrhizal mushrooms. In: Zambonelli A, Bonito G (eds) Edible ectomycorrhizal mushrooms, current knowledge and future prospects, vol 34. Springer, Berlin/Heidelberg, pp 317–334. https://doi.org/10.1007/978-3-642-33823-6

Badalyan SM (2016) Fatty acid composition of different collections of coprinoid mushrooms (Agaricomycetes) and their nutritional and medicinal values. Int J Med Mushrooms 18 (10):883–893. https://doi.org/10.1615/IntJMedMushrooms.v18.i10.40

Badalyan SM (2020) Medicinal coprinoid mushrooms (Agaricomycetes) distributed in Armenia (review). Int J Med Mushrooms 22(3):257–267. https://doi.org/10.1615/intjmedmushrooms.2020033981

Badalyan SM, Barkhudaryan A, Rapior S (2019) Recent progress in research on the pharmacological potential of mushrooms and prospects for their clinical application. In: Agrawal DC, Dhanasekaran M (eds) Medicinal mushrooms: recent progress in research and development. Springer Nature, Singapore, pp 1–70. https://doi.org/10.1007/978-981-13-6382-5_1

Badalyan SM, Barkhudaryan A, Rapior S (2021) The cardioprotective properties of agaricomycetes mushrooms growing in the Territory of Armenia (review). Int J Med Mushrooms 23(4) in press

Badalyan SM, Borhani A (2019a) The diversity and distribution of edible and medicinal mushrooms from Mazandaran province of northern Iran. Proc YSU Chem Biol 53(1):33–41

Badalyan SM, Borhani A (2019b) Medicinal, nutritional and cosmetic values of macrofungi distributed in Mazandaran province of northern Iran (review). Int J Med Mushroom 21 (11):1099–1106. https://doi.org/10.1615/IntJMedMushrooms.2019032743

Badalyan SM, Gharibyan NG (2016) Diversity of polypore bracket mushrooms, Polyporales (Agaricomycetes) recorded in Armenia and their medicinal properties. Int J Med Mushrooms 18(4):347–354. https://doi.org/10.1615/intjmedmushrooms.v18.i4.80

Badalyan SM, Gharibyan NG (2017) Characteristics of mycelial structures of different fungal collections. YSU Press, Yerevan

Badalyan SM, Gharibyan NG (2020) Pharmacological properties and resource value of hymenochaetoid fungi (Agaricomycetes) distributed in Armenia: review. Int J Med Mushrooms 22(12):1133–1145. https://doi.org/10.1615/IntJMedMushrooms.2020037092

Badalyan SM, Rapior S (2020) Perspectives of biomedical application of macrofungi. Curr Trends Biomed Eng Biosci 19(5):556024. https://doi.org/10.19080/CTBEB.2020.19.556024

Badalyan SM, Zambonelli A (2019) Biotechnological exploitation of macrofungi for the production of food, pharmaceuticals and cosmeceuticals. In: Sridhar KR, Deshmukh SK (eds) Advances in macrofungi: diversity, ecology and biotechnology. CRC Press, pp 199–230. https://doi.org/10.1201/9780429504075

Bai Y, Chen L, Chen Y (2019) A Maitake (*Grifola frondosa*) polysaccharide ameliorates Alzheimer's disease-like pathology and cognitive impairments by enhancing microglial amyloid-β clearance. RSC Adv 9(64):37127–37135. https://doi.org/10.1039/C9RA08245J

Bandara AR, Rapior S, Bhat DJ et al (2015) *Polyporus umbellatus*, an edible-medicinal cultivated mushroom with multiple developed health-care products as food, medicine and cosmetics: a review. Cryptogam Mycol 36(1):3–42. https://doi.org/10.7872/crym.v36.iss1.2015.3

Bandara AR, Rapior S, Mortimer PE et al (2019) A review of the polysaccharide, protein and selected nutrient content of *Auricularia*, and their potential pharmacological value. Mycosphere 10(1):579–607. https://doi.org/10.5943/mycosphere/10/1/10

Barros L, Cruz T, Baptista P (2008) Wild and commercial mushrooms as source of nutrients and nutraceuticals. Food Chem Toxicol 46:2742–2747. https://doi.org/10.1016/j.fct.2008.04.030

Bell V, Ferrão J, Chaquisse E et al (2019) Role of mushrooms in autism. Austin J Nutri Food Sci 7 (6):id1128

Bennett L, Kersaitis C, Macaulay SL et al (2013b) Vitamin $D_2$-enriched button mushroom (*Agaricus bisporus*) improves memory in both wild type and APPswe/PS1dE9 transgenic mice. PLoS One 8(10):e76362. https://doi.org/10.1371/journal.pone.0076362

Bennett L, Sheean P, Zabaras D et al (2013a) Heat-stable components of wood ear mushroom, *Auricularia polytricha* (higher basidiomycetes), inhibit in vitro activity of beta secretase (BACE1). Int J Med Mushrooms 15(3):233–249. https://doi.org/10.1615/intjmedmushr.v15.i3.20

Bobek P, Galbavy S (2001) Effect of pleuran (β-glucan from *Pleurotus ostreatus*) on the antioxidant status of the organism and on dimethylhydrazine-induced precancerous lesions in rat colon. Br J Biomed Sci 58(3):164–168

Brandalise F, Cesaroni V, Gregori A et al (2017) Dietary supplementation of *Hericium erinaceus* increases mossy fiber-CA3 hippocampal neurotransmission and recognition memory in wild-type mice. Evid Based Complement Alternat Med 2017:3864340. https://doi.org/10.1155/2017/3864340

Cardoso BR, Roberts BR, Bush AI et al (2015) Selenium, selenoproteins and neurodegenerative diseases. Metallomics 7:1213–1228. https://doi.org/10.1039/c5mt00075k

Cardwell G, Bornman JF, James AP et al (2018) A review of mushrooms as a potential source of dietary vitamin D. Nutrients 10(10):1498. https://doi.org/10.3390/nu10101498

Chang ST, Buswell JA (1996) Mushroom nutriceuticals. World J Microbiol Biotechnol 12 (5):473–476. https://doi.org/10.1007/BF00419460

Chang ST, Miles PG (2004) Mushrooms: cultivation, nutritional value, medicinal effect, and environmental impact, 2nd edn. CRC Press, Boca Raton. https://doi.org/10.1021/np058221b

Chang ST, Wasser SP (2012) The role of culinary-medicinal mushrooms on human welfare with a pyramid model for human health. Int J Med Mushrooms 14(2):95–134. https://doi.org/10.1615/IntJMedMushr.v14.i2.10

Chang ST, Wasser SP (2017) The cultivation and environmental impact of mushrooms. In: Oxford research encyclopedia. Environmental science—agriculture and the environment, Oxford University press USA, 2016, pp 1–39. doi:https://doi.org/10.1093/acrefore/9780199389414.013.231

Chen H, Tian T, Miao H et al (2016a) Traditional uses, fermentation, phytochemistry and pharmacology of *Phellinus linteus*: a review. Fitoterapia 113:6–26. https://doi.org/10.1016/j.fitote.2016.06.009

Chen H, Zhang J, Ren J et al (2018a) Triterpenes and meroterpenes with neuroprotective effects from *Ganoderma leucocontextum*. Chem Biodivers 15(5):e1700567. https://doi.org/10.1002/cbdv.201700567

Chen W, Tan H, Liu Q et al (2019) A review: the bioactivities and pharmacological applications of *Phellinus linteus*. Molecules 24(10):E1888. https://doi.org/10.3390/molecules24101888

Chen WW, Zhang X, Huang WJ (2016b) Role of neuroinflammation in neurodegenerative diseases (review). Mol Med Rep 13(4):3391–3396. https://doi.org/10.3892/mmr.2016.4948

Chen WY, Chang CY, Li JR et al (2018b) Anti-inflammatory and neuroprotective effects of fungal immunomodulatory protein involving microglial inhibition. Int J Mol Sci 19(11):3678. https://doi.org/10.3390/ijms19113678

Cheng JH, Tsai CL, Lien YY et al (2016) High molecular weight of polysaccharides from *Hericium erinaceus* against amyloid beta-induced neurotoxicity. BMC Complement Altern Med 16:170. https://doi.org/10.1186/s12906-016-1154-5

Cheung PCK (2010) The nutritional and health benefits of mushrooms. Nutr Bull 35(4):292–299. https://doi.org/10.1111/j.1467-3010.2010.01859.x

Chong PS, Fung ML, Wong KH et al (2019) Therapeutic potential of *Hericium erinaceus* for depressive disorder. Int J Mol Sci 21(1):E163. https://doi.org/10.3390/ijms21010163

Ćilerdžić J, Galić M, Vukojević J et al (2019) *Pleurotus ostreatus* and *Laetiporus sulphureus* (Agaricomycetes): possible agents against Alzheimer and Parkinson diseases. Int J Med Mushrooms 21(3):275–289. https://doi.org/10.1615/IntJMedMushrooms.2019030136

Ćilerdžić JL, Sofrenić IV, Tešević VV et al (2018) Neuroprotective potential and chemical profile of alternatively cultivated *Ganoderma lucidum* basidiocarps. Chem Biodivers 15(5):e1800036. https://doi.org/10.1002/cbdv.201800036

Cornelius C, Perrotta R, Graziano A et al (2013) Stress responses, vitagenes and hormesis as critical determinants in aging and longevity: mitochondria as a "chi". Immun Ageing 10(1):15. https://doi.org/10.1186/1742-4933-10-15

Cui X, Zhang Y-H (2019) Neuropharmacological effect and clinical applications of *Ganoderma* (Lingzhi). In: Lin Z, Yang B (eds) *Ganoderma* and health. Advances in experimental medicine and biology book series 1182. Springer Nature Singapore Pte Ltd, pp 143–157. https://doi.org/10.1007/978-981-32-9421-9_5

De Silva DD, Rapior S, Sudarman E et al (2013) Bioactive metabolites from macrofungi: ethnopharmacology, biological activities, chemistry. Fungal Divers 62:1–40. https://doi.org/10.1007/s13225-013-0265-2

Deary IJ, Corley J, Gow AJ et al (2009) Age-associated cognitive decline. Br Med Bull 92 (1):135–152. https://doi.org/10.1093/bmb/ldp033

Dhakal S, Kushairi N, Phan CW et al (2019) Dietary polyphenols: a multifactorial strategy to target Alzheimer's disease. Int J Mol Sci 20(20):E5090. https://doi.org/10.3390/ijms20205090

Diallo I, Boudard F, Morel S et al (2020) Antioxidant and anti-inflammatory potential of shiitake culinary-medicinal mushroom, *Lentinus edodes* (Agaricomycetes), sporophores from various culture conditions. Int J Med Mushrooms 22(6):535–546. https://doi.org/10.1615/IntJMedMushrooms.2020034864

Diling C, Tianqiao Y, Jian Y et al (2017) Docking studies and biological evaluation of a potential β-secretase inhibitor of 3-hydroxyhericenone F from *Hericium erinaceus*. Front Pharmacol 8:219. https://doi.org/10.3389/fphar.2017.00219

Doğan HH, Akbaş G (2013) Biological activity and fatty acid composition of Caesar's mushroom. Pharm Biol 51(7):863–871. https://doi.org/10.3109/13880209.2013.768272

El Sayed NS, Ghoneum MH (2020) Antia, a natural antioxidant product, attenuates cognitive dysfunction in streptozotocin-induced mouse model of sporadic Alzheimer's disease by targeting the amyloidogenic, inflammatory, autophagy, and oxidative stress pathways. Oxidative Med Cell Longev 2020:4386562. https://doi.org/10.1155/2020/4386562

Fan L, Chen L, Liang Z et al (2019) A polysaccharide extract from Maitake culinary-medicinal mushroom, *Grifola frondosa* (Agaricomycetes) ameliorates learning and memory function in aluminum chloride-induced amnesia in mice. Int J Med Mushrooms 21(11):1065–1074. https://doi.org/10.1615/IntJMedMushrooms.2019032914

Farha M, Parkianathan L, Abdul Amir NAI et al (2019) Functional recovery enhancement by tiger milk mushroom, *Lignosus rhinocerotis,* in a sciatic nerve crush injury model and morphological study of its neurotoxicity. J Animal Plant Sci 29(4):930–942

Friedman M (2015) Chemistry, nutrition, and health-promoting properties of *Hericium erinaceus* (lion's mane) mushroom fruiting bodies and mycelia and their bioactive compounds. J Agric Food Chem 63(32):7108–7123. https://doi.org/10.1021/acs.jafc.5b02914

Gargano ML, van Griensven LJLD, Isikhuemhen OS et al (2017) Medicinal mushrooms: valuable biological resources of high exploitation potential. Plant Biosyst 151(3):548–565. https://doi.org/10.1080/11263504.2017.1301590

Gartz J, Allen JW, Merlin MD et al (1994) Ethnomycology, biochemistry, and cultivation of *Psilocybe samuiensis* Guzmán, Bandala and Allen, a new psychoactive fungus from Koh Samui, Thailand. J Ethnopharmacol 43(2):73–80. https://doi.org/10.1016/0378-8741(94)90006-x

Gorman AM (2008) Neuronal cell death in neurodegenerative diseases: recurring themes around protein handling. J Cell Mol Med 12(6A):2263–2280. https://doi.org/10.1111/j.1582-4934.2008.00402.x

Grienke U, Zöll M, Peintner U et al (2014) European medicinal polypores—a modern view on traditional uses. J Ethnopharmacol 154:564–583. https://doi.org/10.1016/j.jep.2014.04.030

Gründemann C, Reinhardt JK, Lindequist U (2020) European medicinal mushrooms: do they have potential for modern medicine?—An update. Phytomedicine 66:153131. https://doi.org/10.1016/j.phymed.2019.153131

Gupta S, Summuna B, Gupta M et al (2018) Edible mushrooms: cultivation, bioactive molecules, and health benefits. In: Mérillon JM, Ramawat KG (eds) Bioactive molecules in food, reference series in Phytochemistry. Springer, pp 1–33. https://doi.org/10.1007/978-3-319-54528-8_86-1

Hawksworth DL (2012) Global species numbers of fungi: are tropical studies and molecular approaches contributing to a more robust estimate? Biodivers Conserv 21:2425–2433. https://doi.org/10.1007/s1531-012-0335-x

He X, Wang X, Fang J et al (2017) Structures, biological activities, and industrial applications of the polysaccharides from *Hericium erinaceus* (Lion's mane) mushroom: a review. Int J Biol Macromol 97:228–237. https://doi.org/10.1016/j.ijbiomac.2017.01.040

Hibbet DS, Taylor JW (2013) Fungal systematics: is a new age of enlightenment at hand? Nat Rev Microbiol 11:129–133. https://doi.org/10.1038/nrmicro2963

Ho LH, Zulkifli NA, Tan TC (2020) Edible mushroom: Nutritional properties, potential nutraceutical values, and its utilisation in food product development. In: An introduction to mushroom. IntechOpen. doi:https://doi.org/10.5772/intechopen.91827

Hobbs C (2004) Medicinal value of Turkey tail fungus *Trametes versicolor* (L.:Fr.) Pilát (Aphyllophoromycetideae). Int J Med Mushroom 6:195–218. https://doi.org/10.1615/IntJMedMushrooms.v7.i3.100

Hsiao HY, Weng YM (2019) Repeated dose 90-day oral toxicity study of an *Antrodia cinnamomea* product via a novel process in Sprague Dawley rats. International Conference on Biotechnology and Bioengineering (8th ICBB), 24–26 October 2018. AIP Conf Proceedings 2079:020027, Budapest. doi:https://doi.org/10.1063/1.5092405

Hsieh PW, Wu JB, Wu YC (2013) Chemistry and biology of *Phellinus linteus*. BioMed 3 (3):106–113. https://doi.org/10.1016/j.biomed.2013.01.002

Huang YJ, Hsu NY, Lu KH et al (2020) *Poria cocos* water extract ameliorates the behavioural deficits induced by unpredictable chronic mild stress in rats by down-regulating inflammation. J Ethnopharmacol 258:112566. https://doi.org/10.1016/j.jep.2020.112566

Hyde KD, Xu J, Rapior S et al (2019) The amazing potential of fungi: 50 ways we can exploit fungi industrially. Fungal Divers 97:1–136. https://doi.org/10.1007/s13225-019-00430-9

Im KH, Nguyen TK, Kim JK et al (2016a) Evaluation of anticholinesterase and inflammation inhibitory activity of medicinal mushroom Phellinus pini (Basidiomycetes) fruiting bodies. Int J Med Mushrooms 18(11):1011–1022. https://doi.org/10.1615/IntJMedMushrooms.v18.i11.60

Im KH, Nguyen TK, Choi J et al (2016b) *In vitro* antioxidant, anti-diabetes, anti-dementia, and inflammation inhibitory effect of *Trametes pubescens* fruiting body extracts. Molecules 21(5): E639. https://doi.org/10.3390/molecules21050639

Islam T, Yu X, Xu B (2016) Phenolic profiles, antioxidant capacities and metal chelating ability of edible mushrooms commonly used in China. LWT Food Sci Technol 72:423–431. https://doi.org/10.1016/j.lwt.2016.05.005

Jang HJ, Kim JE, Jeong KH et al (2019) The neuroprotective effect of *Hericium erinaceus* extracts in mouse hippocampus after pilocarpine-induced status epilepticus. Int J Mol Sci 20(4):E859. https://doi.org/10.3390/ijms20040859

Ji W, Huang H, Chao J et al (2014) Medicinal and edible fungi as an alternative medicine for treating age-related disease. Evid Based Complement Alternat Med 2014:ID679259. https://doi.org/10.1155/2014/679259

Jiang F, Zhang HN, Zhang L et al (2020) Antioxidant and neuroprotector influence of endo-polyphenol extract from magnesium acetate multi-stage addition in the oak bracket medicinal mushroom, *Phellinus baumii* (Agaricomycetes). Int J Med Mushrooms 22(2):183–195. https://doi.org/10.1615/IntJMedMushrooms.2020033699

Jiang S, Wang S, Sun Y et al (2014) Medicinal properties of *Hericium erinaceus* and its potential to formulate novel mushroom-based pharmaceuticals. Appl Microbiol Biotechnol 98 (18):7661–7670. https://doi.org/10.1007/s00253-014-5955-5

Kawagishi H, Ando M, Mizuno T (1990) Hericenone A and B as cytotoxic principles from the mushroom *Hericium erinaceum*. Tetrahedron Lett 31:373–376. https://doi.org/10.1016/S0040-4039(00)94558-1

Kawagishi H, Ando M, Sakamoto H et al (1991) Hericenones C, D and E, stimulators of nerve growth factor (NGF)-synthesis, from the mushroom *Hericium erinaceum*. Tetrahedron Lett 32:4561–4564. https://doi.org/10.1016/0040-4039(91)80039-9

Khan MS, Zhang X, You L (2014) Structure and bioactivities of fungal polysaccharides. In: Ramawat K, Mérillon JM (eds) Polysaccharides—bioactivity and biotechnology. Springer, Cham, pp 1851–1866. https://doi.org/10.1007/978-3-319-03751-6_28-1

Khatua S, Paul S, Acharya K (2013) Mushroom as the potential source of new generation of antioxidant– a review. Res J Pharm Technol 6(5):496–505

Kim YO, Lee SW, Kim JS (2014) A comprehensive review of the therapeutic effects of *Hericium erinaceus* in neurodegenerative disease. J Mushroom 12(2):77–81. https://doi.org/10.14480/JM.2014.12.2.77

Kıvrak I, Kıvrak S, Karababa E (2020) Assessment of bioactive compounds and antioxidant activity of Turkey tail medicinal mushroom *Trametes versicolor* (Agaricomycetes). Int J Med Mushrooms 22(6):559–571. https://doi.org/10.1615/IntJMedMushrooms.2020035027

Knežević A, Stajić M, Sofrenić I et al (2018) Antioxidative, antifungal, cytotoxic and anti-neurodegenerative activity of selected *Trametes* species from Serbia. PLoS One 13(8): e0203064. https://doi.org/10.1371/journal.pone.0203064

Kondeva-Burdina M, Voynova M, Shkondrov A et al (2019) Effects of *Amanita muscaria* extract on different in vitro neurotoxicity models at sub-cellular and cellular levels. Food Chem Toxicol 132:110687. https://doi.org/10.1016/j.fct.2019.110687

Kosanić M, Petrović N, Stanojković T (2020a) Bioactive properties of *Clitocybe geotropa* and *Clitocybe nebularis*. J Food Meas Charact 14(2):1046–1053. https://doi.org/10.1007/s11694-019-00354-7

Kosanić MM, Šeklic DS, Jovanovic MM et al (2020b) *Hygrophorus eburneus*, edible mushroom, a promising natural bioactive agent. EXCLI J 19:442–457. https://doi.org/10.17179/excli2019-2056

Kozarski M, Klaus A, Jakovljevic D et al (2015) Antioxidants of edible mushrooms. Molecules 20:19489–19525. https://doi.org/10.3390/molecules201019489

Kozarski M, Klaus A, Vunduk JD et al (2020) Health impact of the commercially cultivated mushroom *Agaricus bisporus* and the wild-growing mushroom *Ganoderma resinaceum*—a comparative overview. J Serb Chem Soc 85(6):721–735. https://doi.org/635.8+631:576+615.9:577.15:616–01

Kozarski MS, Klaus AS, Nikšić MP et al (2014) Polysaccharides of higher fungi: biological role, structure and antioxidative activity. Hem Ind 68(3):305–320. https://doi.org/10.2298/HEMIND121114056K

Kües U, Badalyan SM (2017) Making use of genomic information to explore the biotechnological potential of medicinal mushrooms. In: Agrawal DC, Tsay HS, Shyur LF, Wu YC, Wang SY (eds) Medicinal plants and fungi: recent advances in research and development, medicinal and aromatic plants of the world, vol 4. Springer, New York, pp 397–458. https://doi.org/10.1007/978-981-10-5978-0_13

Kuo HC, Lu CC, Shen CH et al (2016) *Hericium erinaceus* mycelium and its isolated erinacine a protection from MPTP-induced neurotoxicity through the ER stress, triggering an apoptosis cascade. J Transl Med 14:78. https://doi.org/10.1186/s12967-016-0831-y

Kupka T, Broda MA, Wieczorek PP (2020) What is the form of muscimol from fly agaric mushroom (*Amanita muscaria*) in water? An insight from NMR experiment supported by molecular modeling. Magn Reson Chem 58(6):584–593. https://doi.org/10.1002/mrc.4990

Kupka T, Wieczorek PP (2016) Theoretical and experimental NMR studies on muscimol from fly agaric mushroom (*Amanita muscaria*). Spectrochim Acta A Mol Biomol Spectrosc 153:216–225. https://doi.org/10.1016/j.saa.2015.08.026

Kushairi N, Phan CW, Sabaratnam V et al (2019) Lion's mane mushroom, *Hericium erinaceus* (bull.:Fr.) Pers. suppresses $H_2O_2$-induced oxidative damage and LPS-induced inflammation in HT22 hippocampal neurons and BV2 microglia. Antioxidants 8(8):E261. https://doi.org/10.3390/antiox8080261

Lai G, Guo Y, Chen D et al (2019) Alcohol extracts from *Ganoderma lucidum* delay the progress of Alzheimer's disease by regulating DNA methylation in rodents. Front Pharmacol 10:272. https://doi.org/10.3389/fphar.2019.00272

Lee IK, Yun BS, Kim JP et al (2003) Neuroprotective activity of *p*-terphenyl leucomentins from the mushroom *Paxillus panuoides*. Biosci Biotechnol Biochem 67(8):1813–1816. doi.https://doi.org/10.1271/bbb.67.1813

Lee KF, Tung SY, Teng CC et al (2020) Post-treatment with erinacine A, a derived diterpenoid of *H. erinaceus*, attenuates neurotoxicity in MPTP model of Parkinson's disease. Antioxidants 9 (2):E137. https://doi.org/10.3390/antiox9020137

Lee W, Fujihashi A, Govindarajulu M et al (2019) Role of mushrooms in neurodegenerative diseases. In: Agrawal DC, Dhanasekaran M (eds) Medicinal mushrooms: recent progress in research and development. Springer Nature, Singapore, pp 223–249. https://doi.org/10.1007/978-981-13-6382-5_8

Lemieszek MK, Nunes FM, Cardoso C et al (2018) Neuroprotective properties of *Cantharellus cibarius* polysaccharide fractions in different in vitro models of neurodegeneration. Carbohydr Polym 197:598–607. https://doi.org/10.1016/j.carbpol.2018.06.038

Lew SY, Teoh SL, Lim SH et al (2020) Discovering the potentials of medicinal mushrooms in combating depression. A review. Mini Rev Med Chem 20(15):1518–1531. https://doi.org/10.2174/1389557520666200526125534

Li IC, Lee LY, Tzeng TT et al (2018) Neurohealth properties of *Hericium erinaceus* mycelia enriched with erinacines. Behav Neurol 2018:5802634. https://doi.org/10.1155/2018/5802634

Li W, Zhou W, Lee DS et al (2014) Hericirine, a novel anti-inflammatory alkaloid from *Hericium erinaceum*. Tetrahedron Lett 55(30): 4086–4090. doi:https://doi.org/10.1016/j.tetlet.2014.05.117

Li Z, Chen X, Lu W et al (2017) Anti-oxidative stress activity is essential for *Amanita caesarea* mediated neuroprotection on glutamate-induced apoptotic HT22 cells and an Alzheimer's disease mouse model. Int J Mol Sci 18(8):E1623. https://doi.org/10.3390/ijms18081623

Liang C-H, Huang P-C, Mau J-L et al (2020) Effect of the king oyster culinary-medicinal mushroom *Pleurotus eryngii* (Agaricomycetes) basidiocarps powder to ameliorate memory and learning deficit in ability in Aβ-induced Alzheimer's disease C57BL/6J mice model. Int J Med Mushrooms 22(2):145–159. https://doi.org/10.1615/IntJMedMushrooms.2020033766

Limanaqi F, Biagioni F, Busceti CL et al (2020) Potential antidepressant effects of *Scutellaria baicalensis*, *Hericium erinaceus* and *Rhodiola rosea*. Antioxidants 9(3):234. https://doi.org/10.3390/antiox9030234

Lin MT, Beal MF (2006) Mitochondrial dysfunction and oxidative stress in neurodegenerative diseases. Nature 443:787–795. https://doi.org/10.1038/nature05292

Lindequist U (2013) The merit of medicinal mushrooms from a pharmaceutical point of view. Int J Med Mushrooms 15:517–523. https://doi.org/10.1615/intjmedmushr.v15.i6.10

Liu W, Tang Q, Wei YT et al (2019) Chemical compounds and antioxidant activity of volatile oil from the white jelly mushroom, *Tremella fuciformis* (Tremellomycetes). Int J Med Mushrooms 21(3):207–214. https://doi.org/10.1615/IntJMedMushrooms.2019030099

Lowenstein DH, Alldredge BK (1998) Status epilepticus. N Engl J Med 338(14):970–976. https://doi.org/10.1056/NEJM199804023381407

Lu MC, El-Shazly M, Wu TY et al (2013) Recent research and development of *Antrodia cinnamomea*. Pharmacol Ther 139(2):124–156. https://doi.org/10.1016/j.pharmthera.2013.04.001

Lucius K (2020) Medicinal mushrooms: current use in clinical practice. Altern Complement Ther 26(3):119–126. https://doi.org/10.1089/act.2020.29275.kha

Luryi AL, Morse E, Michaelides E (2019) Pathophysiology and diagnosis of Meniere's disease. In: Babu S, Schutt C, Bojrab D (eds) Diagnosis and treatment of vestibular disorders. Springer, Cham, pp 165–188. https://doi.org/10.1007/978-3-319-97858-1_13

Ma BJ, Shen JW, Yu HY et al (2010) Hericenones and erinacines: stimulators of nerve growth factor (NGF) biosynthesis in *Hericium erinaceus*. Mycology 1(2):92–98. https://doi.org/10.1080/21501201003735556

Mahmoud MG, Ibrahim AY, Asker MM et al (2014) Therapeutic potential and structural elucidation of a water-soluble polysaccharide of a wild edible mushroom *Agaricus bisporus* against neurodegenerative disease, Alzheimer. World J Pharm Sci 2(10):1136–1145

Meng X, Liang H, Luo L (2016) Antitumor polysaccharides from mushrooms: a review on the structural characteristics, antitumor mechanisms and immunomodulating activities. Carbohydr Res 424:30–41. https://doi.org/10.1016/j.carres.2016.02.008

Mizuno T (1999) Bioactive substances in *Hericium erinaceus* (bull. : Fr.). Pers. (Yamabushitake), and its medicinal utilization. Int J Med Mushrooms 1(2):105–119. https://doi.org/10.1615/IntJMedMushrooms.v1.i2.10

Money NP (2016) Are mushrooms medicinal? Fungal Biol 120:449–453. https://doi.org/10.1016/j.funbio.2016.01.006

Morel S, Arnould S, Vitou M et al (2018) Antiproliferative and antioxidant activities of wild Boletales mushrooms from France. Int J Med Mushrooms 20(1):13–29. https://doi.org/10.1615/IntJMedMushrooms.2018025329

Mori K, Obara Y, Moriya T et al (2011) Effects of *Hericium erinaceus* on amyloid β (25-35) peptide-induced learning and memory deficits in mice. Biomed Res 32(1):67–72. https://doi.org/10.2220/biomedres.32.67

Muszyńska B, Fijałkowska A, Sułkowska-Ziaja K et al (2020) *Fomitopsis officinalis*: a species of arboreal mushroom with promising biological and medicinal properties. Chem Biodivers 17: e2000213. https://doi.org/10.1002/cbdv.202000213

Nagano M, Shimizu K, Kondo R et al (2010) Reduction of depression and anxiety by 4 weeks *Hericium erinaceus* intake. Biomed Res 31:231–237. https://doi.org/10.2220/biomedres.31.231

Nallathamby N, Phan CW, Seow SL et al (2018) A status review of the bioactive activities of Tiger milk mushroom *Lignosus rhinocerotis* (Cooke) Ryvarden. Front Pharmacol. https://doi.org/10.3389/fphar.2017.00998

Ning X, Luo Q, Li C et al (2014) Inhibitory effects of a polysaccharide extract from the Chaga medicinal mushroom, *Inonotus obliquus* (higher basidiomycetes), on the proliferation of human neurogliocytoma cells. Int J Med Mushrooms 16(1):29–36. https://doi.org/10.1615/intjmedmushr.v16.i1.30

Olanow CW, Brundin P (2013) Parkinson's disease and alpha synuclein: is Parkinson's disease a prion-like disorder? Mov Disord 28(1):31–40. https://doi.org/10.1002/mds.25373

Palacios I, Lozano M, Moro C et al (2011) Antioxidant properties of phenolic compounds occurring in edible mushrooms. Food Chem 128:674–678. https://doi.org/10.1016/j.foodchem.2011.03.085

Park HJ, Shim HS, Ahn YH et al (2012) *Tremella fuciformis* enhances the neurite outgrowth of PC12 cells and restores trimethyltin-induced impairment of memory in rats via activation of CREB transcription and cholinergic systems. Behav Brain Res 229:82–90. https://doi.org/10.1016/J.BBR.2011.11.017

Park KJ, Lee SY, Kim HS et al (2007) The neuroprotective and neurotrophic effects of *Tremella fuciformis* in PC12h cells. Mycobiology 35(1):11–15. https://doi.org/10.4489/MYCO.2007.35.1.011

Pejin B, Tešanović K, Jakovljević D et al (2019) The polysaccharide extracts from the fungi *Coprinus comatus* and *Coprinellus truncorum* do exhibit AChE inhibitory activity. Nat Prod Res 33(5):750–754. https://doi.org/10.1080/14786419

Phan CW, David P, Naidu M et al (2013). Neurite outgrowth stimulatory effects of culinary-medicinal mushrooms and their toxicity assessment using differentiating neuro-2a and embryonic fibroblast BALB/3T3. BMC Complement Altern Med 13:261. doi: https://doi.org/10.1186/1472-6882-13-261

Phan CW, David P, Naidu M et al (2015) Therapeutic potential of culinary-medicinal mushrooms for the management of neurodegenerative diseases: diversity, metabolite, and mechanism. Crit Rev Biotechnol 35(3):355–368. https://doi.org/10.3109/07388551.2014.887649

Phan CW, David P, Sabaratnam V (2017) Edible and medicinal mushrooms: emerging brain food for the mitigation of neurodegenerative diseases. J Med Food 20(1):1–10. https://doi.org/10.1089/jmf.2016.3740

Phan CW, David P, Tan YS et al (2014b) Intrastrain comparison of the chemical composition and antioxidant activity of an edible mushroom, *Pleurotus giganteus*, and its potent neuritogenic properties. Sci World J 2014:378651. https://doi.org/10.1155/2014/378651

Phan CW, Lee GS, Hong SL et al (2014a) *Hericium erinaceus* (Bull.: Fr.) Pers. cultivated under tropical conditions: isolation of hericenones and demonstration of NGF-mediated neurite outgrowth in PC12 cells via MEK/ERK and PI3K-Akt signaling pathways. Food Funct 5 (12):160–169. https://doi.org/10.1039/c4fo00452c

Phan CW, Wong WL, David P et al (2012) *Pleurotus giganteus* (Berk.) Karunarathna & KD Hyde: nutritional value and *in vitro* neurite outgrowth activity in rat pheochromocytoma cells. BMC Complement Altern Med 12:102. https://doi.org/10.1186/1472-6882-12-102

Phan KN, Wei C, Vikineswary S et al (2019) Lion's mane mushroom, *Hericium erinaceus* (bull.: Fr.) Pers. suppresses $H_2O_2$-induced oxidative damage and LPS-induced inflammation in HT22 hippocampal neurons and BV2 microglia. Antioxidants 8(8):261. https://doi.org/10.3390/antiox8080261

Pleszczyńska M, Lemieszek MK, Siwulski M et al (2017) *Fomitopsis betulina* (formerly *Piptoporus betulinus*): the iceman's polypore fungus with modern biotechnological potential. World J Microbiol Biotechnol 33(5):83. https://doi.org/10.1007/s11274-017-2247-0

Pöder R (2005) The ice man's fungi: facts and mysteries. Int J Med Mushrooms 7(3):357–359. https://doi.org/10.1615/IntJMedMushrooms.v7.i3.190

Pop RM, Puia IC, Puia A et al (2018) Characterization of *Trametes versicolor*: medicinal mushroom with important health benefits. Not Bot Horti Agrobo 46(2):343–349. https://doi.org/10.15835/nbha46211132

Poucheret P, Fons F, Rapior S (2006) Biological and pharmacological activity of higher fungi: 20-year retrospective analysis. Cryptogam Mycol 27(4):311–333

Qin DW, Han C (2014) Medicinal and edible fungi as an alternative medicine for treating age-related disease. Evid Based Complement Alternat Med 2014:ID638561. https://doi.org/10.1155/2014/638561

Rahman MA, Hossain S, Abdullah N et al (2020) Reishi medicinal mushroom, *Ganoderma lucidum* (Agaricomycetes) ameliorates spatial learning and memory deficits in rats with hypercholesterolemia and Alzheimer's disease. Int J Med Mushrooms 22(1):93–103. https://doi.org/10.1615/IntJMedMushrooms.2020033383

Rahmani-Nezhad S, Dianat S, Mahdizadeh V et al (2019) Investigation of polysaccharide extracts from Iranian and French strains of *Agaricus subrufescens* against enzymes involved in Alzheimer's disease. Bol latinoam Caribe plantas med aromát 18(6):544–554. https://doi.org/10.35588/blacpma.19.18.6.37

Rathore H, Prasad S, Sharma S (2017) Mushroom nutraceuticals for improved nutrition and better human health: a review. Pharma Nutrition 5(2):35–46. https://doi.org/10.1016/j.phanu.2017.02.001

Rathore H, Sharma A, Prasad S et al (2018) Selenium bioaccumulation and associated nutraceutical properties in *Calocybe indica* mushroom cultivated on se-enriched wheat straw. J Biosci Bioeng 126(4):482–487. https://doi.org/10.1016/j.jbiosc.2018.04.010

Rossi P, Cesaroni V, Brandalise F et al (2018) Dietary supplementation of lion's mane medicinal mushroom, *Hericium erinaceus* (Agaricomycetes), and spatial memory in wild-type mice. Int J Med Mushroom 20(5):485–494. https://doi.org/10.1615/IntJMedMushrooms.2018026241

Rupcic Z, Rascher M, Kanaki S et al (2018) Two new cyathane diterpenoids from mycelial cultures of the medicinal mushroom *Hericium erinaceus* and the rare species, *Hericium flagellum*. Int J Mol Sci 19(3):E740. https://doi.org/10.3390/ijms19030740

Sabaratnam V, Kah-Hui W, Naidu M et al (2013) Neuronal health—can culinary and medicinal mushrooms help? J Tradit Complement Med 3(1):62–68. https://doi.org/10.4103/2225-4110.106549

Sabaratnam V, Phan CW (2018). Neuroactive components of culinary and medicinal mushrooms with potential to mitigate age-related neurodegenerative diseases. In: Discovery and development of neuroprotective agents from natural products. Brahmachari G (ed), 1st edn. Elsevier, p 401–413. doi:https://doi.org/10.1016/B978-0-12-809593-5.00010-0

Saitsu Y, Nishide A, Kikushima K et al (2019) Improvement of cognitive functions by oral intake of *Hericium erinaceus*. Biomed Res 40(4):125–131. https://doi.org/10.2220/biomedres.40.125

Sajjadi H, Paparella MM (2008) Meniere's disease. Lancet 372(9636):406–414. https://doi.org/10.1016/S0140-6736(08)61161-7

Schueffler A, Anke T (2014) Fungal natural products in research and development. Nat Prod Rep 31(10):1425–1448. https://doi.org/10.1039/c4np00060a

Scuto M, Di Mauro P, Ontario ML et al (2020) Nutritional mushroom treatment in Meniere's disease with *Coriolus versicolor*: a rationale for therapeutic intervention in neuroinflammation and antineurodegeneration. Int J Mol Sci 21(1):284. https://doi.org/10.3390/ijms21010284

Solayman M, Islam MA, Alam F et al (2017) Natural products combating neurodegeneration: Parkinson's disease. Curr Drug Metab 18(1):50–61. https://doi.org/10.2174/1389200217666160709204826

Spelman K, Sutherland E, Bagade A (2017) Neurological activity of lion's mane (*Hericium erinaceus*). J Restorative Med 6(1):19–26. https://doi.org/10.14200/jrm.2017.6.0108

Tan W-C, Kuppusamy UR, Phan C-W et al (2015) *Ganoderma neo-japonicum* Imazeki revisited: domestication study and antioxidant properties of its basidiocarps and mycelia. Sci Rep 5:12515. https://doi.org/10.1038/srep12515

Tang D, Xu Y-Z, Wang W-W et al (2019) Cyathane diterpenes from cultures of the bird's nest fungus *Cyathus hookeri* and their neurotrophic and anti-neuroinflammatory activities. J Nat Prod 82(6):1599–1608. https://doi.org/10.1021/acs.jnatprod.9b00091

Thangthaeng N, Miller MG, Gomes SM et al (2015) Daily supplementation with mushroom (*Agaricus bisporus*) improves balance and working memory in aged rats. Nutr Res 35 (12):1079–1084. https://doi.org/10.1016/j.nutres.2015.09.012

Theillet FX, Binolfi A, Bekei B et al (2016) Structural disorder of monomeric α-synuclein persists in mammalian cells. Nature 530(7588):45–50. https://doi.org/10.1038/nature16531

Thongbai B, Rapior S, Hyde KD et al (2015) *Hericium erinaceus*, an amazing medicinal mushroom. Mycol Progress 14(10):91. https://doi.org/10.1007/s11557-015-1105-4

Trovato A, Siracusa R, Di Paola R et al (2016a) Redox modulation of cellular stress response and lipoxin A4 expression by *Coriolus versicolor* in rat brain: relevance to Alzheimer's disease pathogenesis. Neurotoxicology 53:350–358. https://doi.org/10.1016/j.neuro.2015.09.012

Trovato A, Siracusa R, Di Paola R et al (2016b) Redox modulation of cellular stress response and lipoxin A4 expression by *Hericium erinaceus* in rat brain: relevance to Alzheimer's disease pathogenesis. Immun Ageing 13:23. https://doi.org/10.1186/s12979-016-0078-8

Trovato Salinaro A, Pennisi M, Di Paola R et al (2018) Neuroinflammation and neurohormesis in the pathogenesis of Alzheimer's disease and Alzheimer-linked pathologies: modulation by nutritional mushrooms. Immun Ageing 15:8. https://doi.org/10.1186/s12979-017-0108-1

Uddin MS, Ashraf GM (2018) Introductory chapter: Alzheimer's disease—the most common cause of dementia. In: Ashraf GM, Uddin MS (eds) Advances in dementia research. IntechOpen. https://doi.org/10.5772/intechopen.82196

Üstün R, Ayhan P (2019) Regenerative activity of *Hericium erinaceus* on axonal injury model using in vitro laser microdissection technique. Neurol Res 41(3):265–274. https://doi.org/10.1080/01616412.2018.1556494

Val CH, Brant F, Miranda AS et al (2015) Effect of mushroom *Agaricus blazei* on immune response and development of experimental cerebral malaria. Malar J 14(1):311. https://doi.org/10.1186/s12936-015-0832-y

Valverde ME, Hernández-Pérez T, Paredes-López O (2015) Edible mushrooms: improving human health and promoting quality of life. Int J Microbiol 2015:376387. https://doi.org/10.1155/2015/376387

Varghese R, Dalvi YB, Lamrood PD et al (2019) Historical and current perspectives on therapeutic potential of higher basidiomycetes: an overview. 3 Biotech 9:362–377. https://doi.org/10.1007/s13205-019-1886-2

Wang C, Liu X, Lian C et al (2019b) Triterpenes and aromatic meroterpenoids with antioxidant activity and neuroprotective effects from *Ganoderma lucidum*. Molecules 24(23):4353. https://doi.org/10.3390/molecules24234353

Wang M, Gao Y, Xu D et al (2014) *Hericium erinaceus* (yamabushitake): a unique resource for developing functional foods and medicines. Food Funct 5(12):3055–3064. https://doi.org/10.1039/c4fo00511b

Wang Q, Wang F, Xu Z, Ding Z (2017) Bioactive mushroom polysaccharides: a review on monosaccharide composition, biosynthesis and regulation. Molecules 22:955. https://doi.org/10.3390/molecules22060955

Wang XY, Zhang DD, Yin JY et al (2019a) Recent developments in *Hericium erinaceus* polysaccharides: extraction, purification, structural characteristics and biological activities. Crit Rev Food Sci Nutr 59(1):S96–S115. https://doi.org/10.1080/10408398.2018.152137

Wasser SP (2017) Medicinal mushrooms in human clinical studies. Part I. anticancer, oncoimmunological, and immunomodulatory activities: a review. Int J Med Mushrooms 19 (4):279–317. https://doi.org/10.1615/IntJMedMushrooms.v19.i4.10

Wasser SP, Didukh MY (2005) Mushroom polysaccharides in human health care. In: Deshmukh SK, Rai MK (eds) Biodiversity of fungi: their role in human life. Oxford & IBH Publishing, pp 289–328

Wong KH, Grygansky AP, Cheng PG et al (2016) Lion's mane mushroom—the natural healer for nerve damage. In: Gabriel J (ed) Macromycetes: medicinal properties and biological peculiarities, vol 2. Nash Format, Kiev, pp 69–104

Wong KH, Kanagasabapathy G, Naidu M et al (2014) *Hericium erinaceus* (bull.: Fr.) Pers., a medicinal mushroom, activates peripheral nerve regeneration. Chin J Integr Med 22 (10):759–767. https://doi.org/10.1007/s11655-014-1624-2

Wong KH, Naidu M, David P et al (2011) Peripheral nerve regeneration following crush injury to rat peroneal nerve by aqueous extract of medicinal mushroom *Hericium erinaceus* (bull.: Fr) Pers. (Aphyllophoromycetideae). Evid Based Complement Alternat Med 2011:580752. https://doi.org/10.1093/ecam/neq062

Wong KH, Naidu M, David P et al (2012) Neuroregenerative potential of Lion's mane mushroom, *Hericium erinaceus* (bull.: Fr.) Pers. (higher basidiomycetes), in the treatment of peripheral nerve injury (review). Int J Med Mushrooms 14(5):427–446. https://doi.org/10.1615/IntJMedMushr.v14.i5.10

Wong KH, Sabaratnam V, Abdullah N et al (2007) Activity of aqueous extracts of lion's mane mushroom *Hericium erinaceus* (Bull.: Fr.) Pers. (Aphyllophoromycetideae) on the neural cell line NG108-15. Int J Med Mushrooms 9(1):57–65. https://doi.org/10.1615/IntJMedMushr.v9.i1.70

Wu JT, Tokunaga T, Kondo M et al (2014) Erinaceolactones a to C, from the culture broth of *Hericium erinaceus*. J Nat Prod 78(1):155–158. https://doi.org/10.1021/np500623s

Wu YJ, Wei ZX, Zhang FM et al (2019) Structure, bioactivities and applications of the polysaccharides from *Tremella fuciformis* mushroom: a review. Int J Biol Macromol 121:1005–1010. https://doi.org/10.1016/j.ijbiomac.2018.10.117

Yadav SK, Ir R, Jeewon R et al (2020) A mechanistic review on medicinal mushrooms-derived bioactive compounds: potential mycotherapy candidates for alleviating neurological disorders. Planta Med 86:1–15. https://doi.org/10.1055/a-1177-4834

Yin X, Wei J, Wang WW et al (2019) New cyathane diterpenoids with neurotrophic and anti-neuroinflammatory activity from the bird's nest fungus *Cyathus africanus*. Fitoterapia 134:201–219. https://doi.org/10.1016/j.fitote.2019.02.013

Yin Z, Chen J, Zhang J et al (2018) Dietary patterns associated with cognitive function among the older people in underdeveloped regions: finding from the NCDFaC study. Nutrients 10(4):E464. https://doi.org/10.3390/nu10040464

Zajac I, Cavuoto P, Danthiir V et al (2016) Study protocol: a randomised, double blinded, placebo-controlled clinical trial testing the effects of a vitamin D-enriched mushroom supplement on cognitive performance and mood in healthy elderly adults. Healthy Aging Res 5:1–4. https://doi.org/10.1097/01.HXR.0000511866.70301.d6

Zengin G, Sarikurkcu C, Gunes E et al (2015) Two *Ganoderma* species: profiling of phenolic compounds by HPLC–DAD, antioxidant, antimicrobial and inhibitory activities on key enzymes linked to diabetes mellitus, Alzheimer's disease and skin disorders. Food Funct 6 (8):2794–2802. https://doi.org/10.1039/c5fo00665a

Zhang CJ, An S, Hu W et al (2016a) The neuroprotective properties of *Hericium erinaceus* in glutamate-damaged differentiated PC12 cells and an Alzheimer's disease mouse model. Int J Mol Sci 17(11):E1810. https://doi.org/10.3390/ijms17111810

Zhang CJ, Guo JY, Cheng H et al (2020) Protective effects of the king oyster culinary-medicinal mushroom, *Pleurotus eryngii* (Agaricomycetes), polysaccharides on β-amyloid-induced neurotoxicity in PC12 cells and aging rats, in vitro and in vivo studies. Int J Med Mushrooms 22 (4):325–333. https://doi.org/10.1615/IntJMedMushrooms.2020033990

Zhang CJ, Shi R, Li H et al (2016b) Antioxidant and neuroprotective effects of *Dictyophora indusiata* polysaccharide in *Caenorhabditis elegans*. J Ethnopharmacol 192:413–422. https://doi.org/10.1016/j.jep.2016.09.031

Zhang S, Tomata Y, Sugiyama K et al (2017) Mushroom consumption and incident dementia in elderly japanese: the Ohsaki cohort 2006 study. J Am Geriatr Soc 65(7):1462–1469. https://doi.org/10.1111/jgs.14812

Zhang T, Du Y, Liu X et al (2019) Study on antidepressant-like effect of protoilludane sesquiterpenoid aromatic esters from *Armillaria mellea*. Nat Prod Res 22:1–4. https://doi.org/10.1080/14786419.2019.1614577

Zhao C, Zhang C, Xing Z et al (2019) Pharmacological effects of natural *Ganoderma* and its extracts on neurological diseases: a comprehensive review. Int J Biol Macromol 121:1160–1178. https://doi.org/10.1016/j.ijbiomac.2018.10.076

# Andrographolide, a Diterpene from *Andrographis paniculata*, and its Influence on the Progression of Neurodegenerative Disorders

**Badrinathan Sridharan and Meng-Jen Lee**

## Contents



**Abstract** Traditional medicines play a significant role in our lives against mild infections to acute, chronic, and traumatic disorders. A large number of secondary metabolites from medicinal plants have been identified and explored for their therapeutic properties, including anticancer, anti-HIV, hepatic, neuroprotective, antidiabetic, antimicrobial, and so on. Neurodegeneration is a phenomenon that occurs due to various disease conditions leading to the loss of neurons and its

B. Sridharan · M.-J. Lee (✉)
Department of Applied Chemistry, Chaoyang University of Technology, Taichung, Taiwan, Republic of China
e-mail: mjlee@cyut.edu.tw

 
D. C. Agrawal, M. Dhanasekaran (eds.), *Medicinal Herbs and Fungi*,
https://doi.org/10.1007/978-981-33-4141-8_3

connections. Some common neurodegenerative diseases, as well as neuropathological conditions like depression and pain, exert high pressure on the public health resources and pose a socioeconomic burden on patients and their families. Traditional medicines became popular due to the lack of the appropriate therapeutic strategies to target multiple factors involved in disease initiation and progression. Among various plant-derived compounds successfully used for the treatment of some of these disorders, leaves, and stem of *Andrographis paniculata* and its key bioactive compound, andrographolide has been extensively studied for its neuroprotective property. In this chapter, we have described several biologically important derivatives of andrographolide, which are either natural or synthetic analogs. Furthermore, we have comprehensively discussed the upregulation or downregulation of specific signaling pathways targeted by andrographolide and their derivatives in the pathogenesis of important neurodegenerative diseases, as well as conditions such as pain and depression that concern the nervous system.

**Keywords** Alzheimer's disease · Andrographolide · Anti-inflammation · Ischemic stroke · Neurodegeneration · Parkinson's disease

## Abbreviations

| | |
|---|---|
| $^{-}$OH | Hydroxyl radical |
| 14-DA | 14-deoxyandrographolide |
| 14-DDA | 14-deoxy-11,12-didehydroandrographolide |
| *A. paniculata* | *Andrographis paniculata* |
| AD | Alzheimer's disease |
| Akt | Protein kinase B |
| Andro | Andrographolide |
| APP | Amyloid precursor protein |
| ARE | Antioxidant responsive element |
| ATP | Adenosine triphosphate |
| Aβ | Amyloid beta (β) peptide |
| Aδ | A-Delta (δ) fibers |
| Bax | Bcl-2-associated X protein |
| BBB | Blood–brain barrier |
| BDNF | Brain-derived neurotrophic factor |
| C1q | Complement 1q |
| C3 | Complement 3 |
| $Ca^{2+}$ | Calcium ions |
| $CD3^{+}$ | Cluster differentiation $3^{+}$ |
| $CD4^{+}$ | Cluster differentiation $4^{+}$ |
| $CD68^{+}$ | Cluster differentiation $68^{+}$ |
| $CD8^{+}$ | Cluster differentiation $8^{+}$ |
| Cdk5 | Cyclin-dependent kinase-5 |
| CEC | Cerebral endothelial cells |

| | |
|---|---|
| CGRP | Calcitonin gene-related peptide |
| COX-2 | Cyclooxygenase-2 |
| CREB | cAMP response element-binding protein |
| CUMS | Chronic unpredictable mild stress |
| CXCL12 | C–X–C motif chemokine 12 |
| CXCR4 | C–X–C chemokine receptor type 4 |
| DAI | Diffused axonal injury |
| DAT | Dopamine transporter |
| DC | Dendritic cells |
| DNA | Deoxyribo nucleic acid |
| Drp-1 | Dynamin-related protein-1 |
| EA.hy926 | Type of human endothelial cells |
| EAE | Experimental autoimmune encephalomyelitis |
| ED-1 | Microglail marker |
| eIF4E | Eukaryotic translation initiation factor 4E |
| ER | Endoplasmic reticulum |
| ERK | Extracellular signal-regulated kinases |
| FGF | Fibroblast growth factor |
| Fz | Frizzled receptor |
| GP-91PHOX | Subunit of NADPH oxidase |
| GSK-3β | Glycogen synthase kinase-3β |
| $H_2O_2$ | Hydrogen peroxide |
| HO-1 | Heme oxygenase-1 |
| HUVEC | Human umbilical vein endothelial cells |
| ICAM-1 | Intercellular adhesion molecule 1 |
| IFN-β | Interferon-β |
| IL-1 | Interleukin-1 |
| IL-10 | Interleukin-10 |
| IL-1β | Interleukin-1β |
| IL-6 | Interleukin −6 |
| iNOS | Inducible nitric oxide synthase |
| JNK | c-Jun N-terminal kinases |
| Keap-1 | Kelch-like ECH-associated protein 1 |
| LEF | Lymphoid enhancer-binding factor |
| LPS | Lipopolysaccharide |
| MAPK | Mitogen-activated protein kinase |
| MCA | Middle cerebral artery |
| MDA | Malondialdehyde |
| MOG | Myelin oligodendrocyte glycoprotein |
| $MPP^+$ | 1-methyl-4-phenylpyridinium |
| MPTP | 1-methyl-4-phenyl-1,2,3,6-tetrahydropyridine |
| MS | Multiple sclerosis |
| mTOR | Mammalian target of rapamycin |
| mTORC1 | Mammalian target of rapamycin complex 1 |
| mTORC2 | Mammalian target of rapamycin complex 2 |

| | |
|---|---|
| NADPH | Nicotinamide adenine–dinucleotide phosphate-reduced |
| NeuroD1 | Neurogenic differentiation 1 |
| NF-κB | Nuclear factor-κB |
| NGF | Nerve growth factor |
| NO | Nitric oxide |
| NOX2 | NADPH oxidase-2 |
| NPC | Neural progenitor cells |
| Nrf-2 | Nuclear erythroid factor-2 |
| NSC | Neural stem cells |
| $O_2^-$ | Oxygen radical |
| $ONOO^-$ | Peroxy nitrate radical |
| OX-42 | Microglial marker |
| p70S6K | Ribosomal protein S6 kinase beta-1 |
| PD | Parkinson's disease |
| $PGE_2$ | Prostaglandin-E2 |
| PGF | Placental growth factor |
| PI3K | Phosphoinositol-3 kinase |
| PKC | Protein kinase C |
| PS-1 | Presinilin-1 |
| PTEN | Phosphatase and tensin homolog |
| RNS | Reactive nitrogen species |
| ROS | Reactive oxygen species |
| SDF-1α | Stromal cell-derived factor-1alpha |
| Ser 9 | Serine-9 phosphorylation |
| SOD | Superoxide dismutase |
| STAT3 | Signal transducer and activator of transcription 3 |
| TBI | Traumatic brain injury |
| TCF | Transcription factor |
| Tg | Transgenic |
| TGF-β | Transforming growth factor-β |
| TH | Tyrosine hydroxylase |
| TNF-α | Tumor necrosis factor-α |
| TrkB | Tropomyosin receptor kinase B |
| VEGF | Vascular endothelial growth factor |
| Wnt | Wingless-related integration site |
| τ | Tau protein |

## 1 Introduction

Medicinal plants and their metabolites are being used in Asian, African, and other traditional medicine for centuries, and their usage is increasing worldwide. Compounds from the medicinally valuable plants started gaining the attention of the researchers who explored their ability to influence various disease conditions and pertaining signaling pathways (Akinyemi et al. 2018). Among various diseases that were successfully managed by herbal extracts and phytocompounds, neurodegenerative disorders are considered as one of the severe disorders with high mortality and morbidity, which are being successfully treated (Junsathian et al. 2018; van der Schyf 2011). Neurodegenerative diseases such as Alzheimer's and Parkinson's diseases, trauma, and so on, can be acute or chronic depends on the pathophysiology, and molecular signaling pathways differ according to the disease etiology, which makes therapeutic management of these disorders a challenging task (van der Schyf 2011). It was considered that some of the important signaling pathways that are being altered during neurodegenerative disorders include, nuclear factor-κB (NF-κB)-mediated inflammation (Kumar and Singh 2015; Luo et al. 2019), nuclear erythroid factor 2 (Nrf-2)-mediated anti-oxidation (Magesh et al. 2012; Wong et al. 2018), mitochondrial dysregulation and apoptosis (Martinez et al. 2018), mammalian target of rapamycin (mTOR)-mediated autophagy (Oddo 2012), and regulation of neurotrophic factors (Yu and Chen 2011; Neto et al. 2011).

Some of the common nutraceutical plant species that include medicinal herbs and functional foods are *Azadirachta indica* (Neem) (Alzohairy 2016), *Cinnamomum cassia* (Zhang et al. 2019a), *Curcuma longa* (Turmeric) (Krup et al. 2013), *Gymnema sylvestre* (Fabio et al. 2014), *Moringa oleifera* (Igado and Olopade 2016), *Phyllanthus emblica* (Indian gooseberry) (Husain et al. 2019), *Syzygium cumini* (Vardhan et al. 2018), *Terminalia arjuna* (Suganthy et al. 2018), and *Withania somnifera* (Bhatnagar et al. 2009). These plants' parts and their active secondary metabolites were extensively explored for its neuroprotective and neuroregenerative potential by influencing various signaling pathways, including anti-inflammatory activity.

## 2 *Andrographis paniculata* and Andrographolide

Among many potential neuroprotective plant species, *Andrographis paniculata* was extensively studied and well-established. It is commonly known as the King of bitters and is distributed almost widely in Southern Asia (Dai et al. 2019). Andrographolide (Andro) is the terpenoid compound isolated from leaves and stem of *A. paniculata*. It is considered as the major constituent that makes the plant a potent neuroprotective agent (Lu et al. 2019). Andro has been studied extensively for its ability to regulate certain important signaling pathways that accelerate or reduce disease progression (Tan et al. 2017). Derivatives of andro naturally synthesized by the plant that are illustrated in Fig. 1 (a–h) and their biological activities are primarily targeted toward certain mechanisms like activation of NOS signaling (Zhang and Tan 1998), inhibition of macrophage-derived $PGE_2$ and COX-2 signaling (Batkhuu et al. 2002; Liu et al. 2007), NF-κB-dependent signaling during tumor formation and other inflammation-mediated disorders (Chao et al. 2010), p38/MAPK signaling (Liu et al. 2008), SDF-1α upregulation

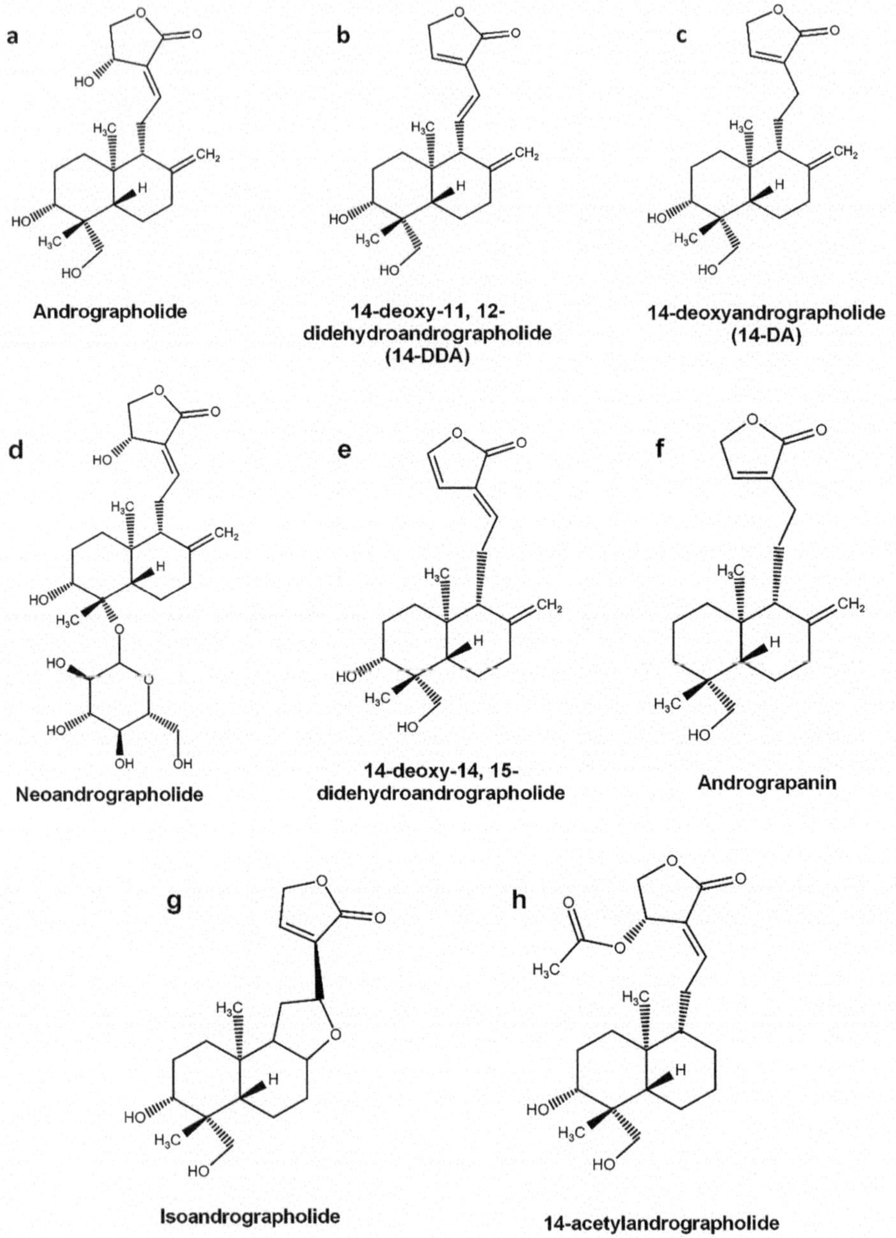


**Fig. 1** (**a–h**) Natural analogs of andrographolide

**Table 1** Effect of andro and its natural and synthetic analogs on various signaling pathways

| Compound name | Figure number | Signaling pathway targeted | Reference |
|---|---|---|---|
| *Natural analogs* | | | |
| Andrographolide | Fig. 1a | Antioxidant (Nrf-2 signaling), anti-inflammation (NF-κB signaling) | Yang et al. (2017); Mussard et al. (2019) |
| 14-deoxy-11,12-didehydroandrographolide (14-DDA) | Fig. 1b | Activation of NOS signaling and vasorelaxation | Zhang and Tan (1998) |
| 14-deoxyandrographolide (14-DA) | Fig. 1c | $PGE_2$, iNOS, and COX-2 secretion from activated macrophages | Batkhuu et al. (2002); Liu et al. (2007) |
| Neoandrographolide | Fig. 1d | | |
| 14-deoxy-14,15-didehydro andrographolide | Fig. 1e | NF-κB activation and nuclear translocation | Chao et al. (2010) |
| Andrograpanin | Fig. 1f | p38/MAPKs pathway, SDF-1α signaling | Liu et al. (2008) |
| Isoandrographolide | Fig. 1g | Anti-proliferation and anti- inflammatory | Ji et al. (2005) |
| 14-acetyandrographolide | Fig. 1a | NF-κB activation and nuclear translocation | Yoopan et al. (2007) |
| *Synthetic analogs* | | | |
| 3, 19-diacetylandrographolide | Fig. 2a | Downregulation of iNOS, TNF-α, and IL-6; reduced NO production | Xu et al. (2019) |
| 3, 12, 19-triacetylandrographolide | Fig. 2b | | |
| 12-hydroxy-14-dehydroandrographolide (Acylated at 3 and 9 positions) | Fig. 3a | Downregulation of TNF-α and IL-6 in LPS induced inflammation | Li et al. (2007) |
| 12-hydroxy-14-dehydroandrographolide (Phenylated at 3 and 9 positions) | Fig. 3b | | |
| Isoandrographolide (Acylated at position 12) | Fig. 3c | | |
| Isoandrographolide (Phenylated at position 12) | Fig. 3d | | |
| Isoandrographolide (nitro acylated at position 12) | Fig. 3e | | |
| Isoandrographolide (aromatic nitro group at position 12) | Fig. 3f | | |
| 3, 19-diacetyl-14-deoxy-14,15-didehydroandrographolide | Fig. 4a | NF-κB inhibition | Chao et al. (2010) |
| 19-acetyl-14-deoxy-14,15-didehydroandrographolide | Fig. 4b | | |

(Ji et al. 2005), upregulation of neural circuitry during muscle relaxation (Yoopan et al. 2007), and so on (Tables 1 and 2).

**Table 2** Signaling pathways influenced by andrographolide

| Disease | Signaling pathway | Molecular markers | References |
|---|---|---|---|
| Alzheimer's disease | Inflammatory pathway | ↓ TLR-4, ↓ NF-κB, ↓ TNF-α, ↓ IL-6, ↓ IL-1β, ↓ iNOS, ↓ COX-2, ↓ MIP-1α, ↓ HMGB1 ↑ TGF- β, ↑IL-10 | Das et al. (2017); Xia et al. (2004) |
| | Anti-oxidation pathway | ↓ ICAM-1, ↑ Nrf2, ↑ HO-1, ↑ AREs ↑ Keap-1, ↑ p-ERK, ↑ p38 | Wong et al. (2016); Yu et al. (2010); Gu et al. (2018) |
| | m-TOR | ↓ Bax, ↓ activated caspase 3, ↓ PI3K, ↓ Akt, ↓ PTEN ↓ mTOR | Kumar et al. (2015); Zhang et al. (2019b) |
| | Wnt/β-catenin | ↓ GSK-3β, ↑ β-catenin, ↑ Wnt, ↑ NeuroD1 | Tapia-Rojas et al. (2015); Varela-Nallar et al. (2015) |
| Parkinson's disease | Inflammatory pathway | ↓ NF-κB, ↓ IκBα, ↓ TNF-α, ↓ IL-6, ↓ IL-1β, ↓ COX-2, ↓ nitrite, ↓ $PGE_2$, ↑ tyrosine hydroxylase (TH) | Wang et al. (2004); Xia et al. (2004) |
| | Oxidative stress, anti-oxidant pathway | ↓ ROS ($^{-}OH$, $ONOO^{-}$, $^{-}O_2$), ↓ cytochrome C, ↓ MDA, ↑ SOD, ↑ catalase | Geng et al. (2019b) |
| | DRP-1 | ↓ DRP-1-mediated GTPase activity, ↓ mitochondrial fission | Geng et al. (2019b) |
| | Pathological markers | ↑ TH, ↑DAT, ↓ striatal apoptosis | |
| Cerebral ischemia | Inflammatory pathway | ↓ NF-κB, ↓ p65 nuclear translocation, ↓ TNF-α, ↓ IL-1β, ↓ $PGE_2$, ↓pAkt, ↓ HIF-1α | Chan et al. (2010); Yang et al. (2017); Xia et al. (2004) |
| | Anti-oxidation pathway | ↑ Nrf2, ↑ HO-1, ↑ AREs, ↓ Erk1/2, ↓iNOS, ↓ GP-91PHOX-NOX2, ↓ p38/MAPK, ↓ Jnk1/2, ↓$^{-}OH$ free radical | Yang et al. (2019); Yen et al. (2016); Yang et al. (2017) |
| | Glutamate/$Ca^{2+}$-mediated excitotoxicity | ↓ $Ca^{2+}$ influx, inhibition of voltage-gated ion channels (Na and Ca channels) | Yang and Song (2014) |
| | Apoptosis (cerebral endothelial cells, vascular smooth muscle cells) | ↓ activated caspase 3, ↑ Bax, ↑ p38/MAPK, ↓pAkt, ↓PTEN | Yen et al. (2013); Chen et al. (2014) |
| Traumatic brain injury | Inflammatory pathway | ↓ p65 nuclear translocation, ↓ TNF-α, ↓ IL-1β, ↓ IL-6, ↓ p-Erk, ↓ p-p38, ↓ MAPK | Tao et al. (2018); Xia et al. (2004) |
| Multiple sclerosis | Immunosuppression | ↓ T cells activation, ↓ Dendritic cell maturation, | Iruretagoyena et al. (2005) |
| Pain | Inflammatory pathway | ↓ NF-κB, ↓ COX-2, ↓ $PGE_2$, ↓ substance P | Wang et al. (2018); Sulaiman et al. (2010) |

(continued)

**Table 2** (continued)

| Disease | Signaling pathway | Molecular markers | References |
|---|---|---|---|
| Depression | BDNF | ↓ BDNF, ↓ p-TrkB, ↓ p-ERK 1 and 2, ↓ p-AKT, ↓ p-CREB | Zhang et al. (2019c) |
| | Inflammatory pathway | ↓ TLR-4, ↓ NF-κB, ↓ p65 nuclear translocation, ↓ TNF-α, ↓ IL-6, ↓ IL-1β, ↓ iNOS, ↓ COX-2 | Geng et al. (2019a); Xia et al. (2004) |

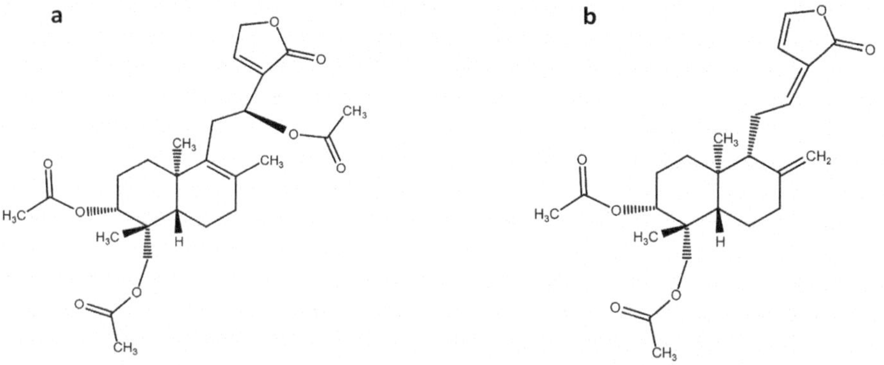


**3, 19-diacetylandrographolide**    **3, 12, 19-triacetylandrographolide**

**Fig. 2** (**a, b**) Synthetic analogs of andrographolide having anti-inflammatory activities

To improve the efficacy of andro, certain derivatives were synthesized, and its anticancer, antiviral, antipyretic, neuroprotective, hepatoprotective, antidiabetic, anti-inflammatory properties were evaluated and reported. Antitumor, anti-inflammatory, and neuroprotective properties of andro and its derivatives were widely reported (Kandanur et al. 2019; Lu et al. 2019).

Inflammation critically influences most of the disease conditions and predominantly influences the progression of tumors and neurodegeneration. Xu et al. (2019) had synthesized up to 12 derivatives of andro and checked the efficacy of all the compounds against the major signaling pathways of inflammation and neurodegeneration. Out of that synthesized derivatives, two compounds (Fig. 2a, b) showed a significant reduction of microglial-mediated inflammation, especially by downregulation of TNF-α, IL-6, iNOS, and NO production and reduced activation of microglia. (Xu et al. 2019).

Natural analogs of andro mainly isoandrographolide and 12-hydroxy-14-dehydroandrographolide are the potent anti-inflammatory agent, and its efficacy was improved after derivatization with acyl and aromatic groups. It was found that the synthesized derivatives (Fig. 3a–f) rendered much better action against LPS induced upregulation of IL-6 and TNF-α than andro (Li et al. 2007). Similarly, acetylation at 3 and 19 or only at 19 positions (Fig. 4a, b) showed significantly

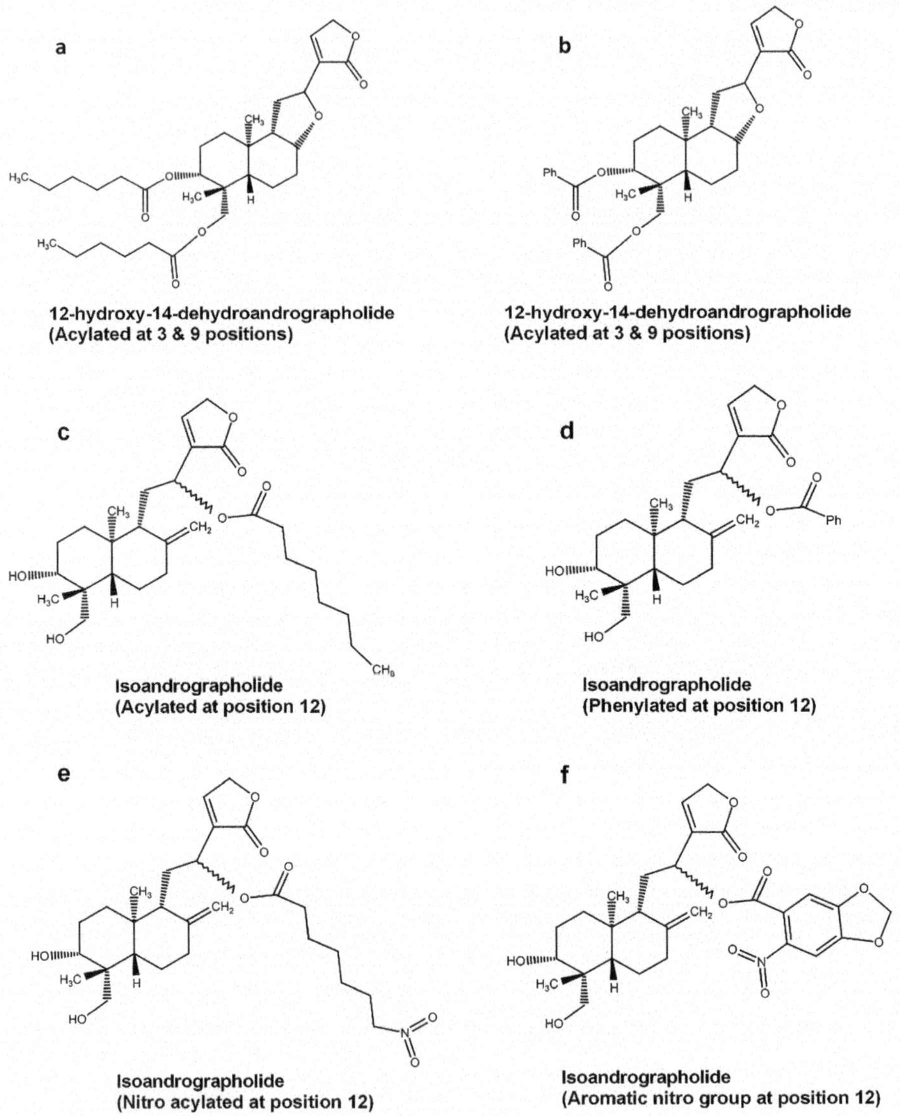


**Fig. 3** **(a–f)** Synthetic analogs of andrographolide having inhibitory activity on TNF-α and IL-6 expression induced by LPS

higher activity in inhibition of NF-κB transactivation than its parent compound 14-deoxy-14,15-didehydroandrographolide (Chao et al. 2010).

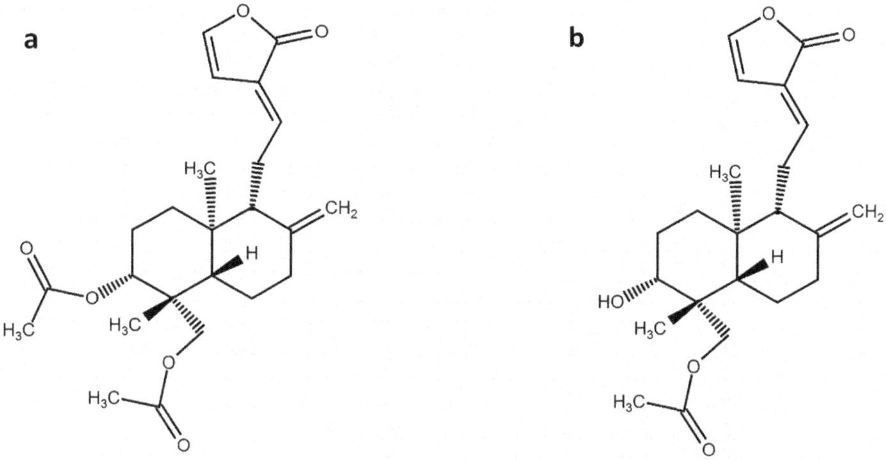


**3, 19-diacetyl-14-deoxy-14,15-didehydroandrographolide**

**19-diacetyl-14-deoxy-14,15-didehydroandrographolide**

**Fig. 4** (**a, b**) Synthetic analogs of andrographolide that inhibit NF-κB transactivation

# 3 Influence of *Andrographis paniculata* and Andrographolide on Neurodegenerative Diseases

## *3.1 Alzheimer's Disease*

Neurodegeneration characterized by a decline in cognitive ability with dementia and loss of neuron is termed as Alzheimer's disease (AD), and it accounts for 70% of the reported cases of dementia. Many hypotheses have been discussed for years regarding the initiation and progression of AD, such as Aβ peptide/Amyloid cascade hypothesis, Tau (τ) protein hypothesis, and cholinergic hypothesis, and so on (Choi 1995). So far Aβ hypothesis was prevailing as the best-explained mechanism of AD progression where Aβ aggregation, τ protein hyperphosphorylation are considered as the hallmarks that results in neuroinflammation mediated by activated microglia and astrocytes and the coexisting mechanisms (oxidative stress, apoptosis, etc.) that accelerate the disease progression (Hardy and Higgins 1992; Hardy and Allsop 1991; Choi 1995). Here we have illustrated (Fig. 5) and discussed various significant pathways that are responsible for AD pathogenesis and the ability of andro to influence these mechanisms in reducing AD progression.

### 3.1.1 Inhibition of Glial-Mediated Inflammation

Activation of microglial and macrophage population resulting in the neuroinflammatory process through hypersecretion of pro-inflammatory cytokines

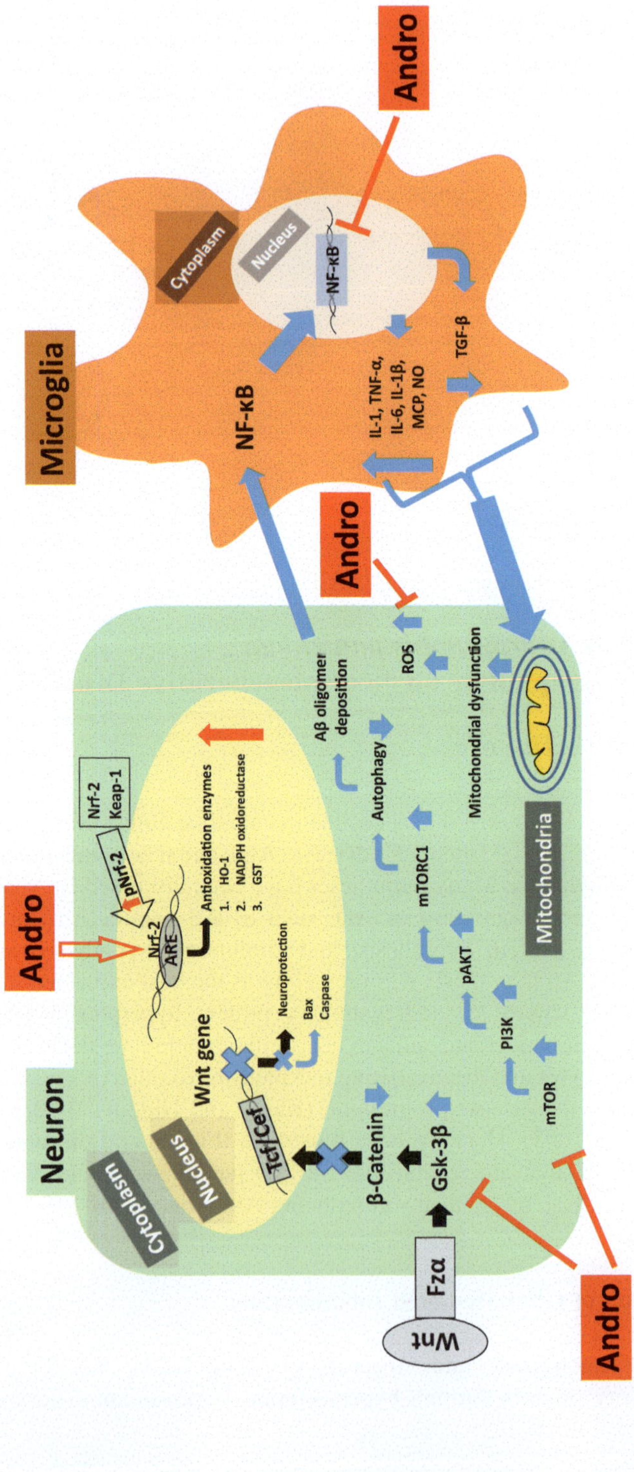


**Fig. 5** Drug targets influenced by andrographolide during Alzheimer's disease

starting with activation and nuclear translocation of NF-κB, and it is one of the key events in inflammatory signaling as various cytokines like TNF-α, IL-6, IL-1β are upregulated followed by this step (Streit et al. 2004; 1998). Therefore, a raise in Aβ levels in the neurons leads to the hyperexpression of various cytokine and chemokines but not limited to IL-6, TNF-α, and so on, and NF-κB (Fig. 5), mediated by active microglia and macrophages (Welcome 2020). Neuroinflammation further initiates a cascade of mitochondrial dysregulation, ROS/RNS-mediated oxidative stress, and corresponding apoptotic steps, end up in neuronal damage (Chen et al. 2020). Andrographolide is well known for its anti-inflammatory activity especially in downregulation of NF-κB (Yang et al. 2017) through TLR-4 reduction (Das et al. 2017), and this significantly resulted in hampered expression of other downstream pro-inflammatory cytokines like TNF-α, IL-1β and, and so on (Fig. 5) (Jing et al. 2019; Varela-Nallar et al. 2015). Nitric oxide secretion by glial significantly influences the upregulation of inflammation, oxidative stress, and apoptosis, while andro had shown it inhibitory activity against NO secretion by downregulation of iNOS. It was followed by COX-2 inhibition and decreased production of $PGE_2$, leading to inhibition of neuroinflammation (Jing et al. 2019). Apart from the downregulation of pro-inflammatory cytokines, it was observed that andro had shown upregulation of specific anti-inflammatory cytokines such as TGF-β and IL-10 (Wang et al. 2004). Andro was also effective in reducing AD pathogenesis in APPswe/PS-1 Tg mice through inhibition of mitochondria swelling and neuronal death (Geng et al. 2018). A study with a South American rodent, *Octodon degus* that spontaneously develop AD at old age, showed that andro provided neuroprotection by inhibited pro-inflammatory and oxidative stress-mediated apoptotic processes (Rivera et al. 2016; Lindsay et al. 2020).

### 3.1.2 Effect of Andro on GSK-3β activity and Wnt/β-Catenin Pathway

Another independent signaling involved in AD pathogenesis is Aβ-oligomers and Tau (τ) protein hyperphosphorylation-mediated downregulation of Wnt/β-catenin signaling by increasing of Glycogen Synthase Kinase-3β (GSK-3β) activity (Inestrosa and Toledo 2008; Arrázola et al. 2017). Under physiological condition, Wnt binding to the Frizzled (Fz) receptor complex inhibits the GSK-3β activity by phosphorylating it at Ser 9 residue. It results in increased β-catenin levels in the cytoplasm, which activates its entry into the nucleus and binding to factors (TCF/LEF) and upregulate expression of *Wnt* genes that regulate cellular homeostasis and neuroprotection (Gordon and Nusse 2006). However, during AD pathogenesis, Wnt signaling is inhibited by Aβ-oligomers deposition leading to phosphorylation of β-catenin and its degradation processes. Reduced levels of β-catenin results in downregulation of Wnt target proteins and various pathological conditions such as oxidative stress and inflammation (Fig. 5) (Arrázola et al. 2017; 2009). Hence, reduced activity of GSK-3β enzyme and upregulation of downstream signals pertaining to Wnt/ β-catenin could be an important strategy for the management of AD (Inestrosa and Toledo 2008; Cerpa et al. 2009). Andro significantly inhibits GSK-3β activity and activates Wnt/β-catenin pathway and neuroprotection in vitro and in vivo (Fig. 5). Andro specifically acts on GSK-3β and not on other physiologically active kinases such as JNK, ERK, PKC, Cdk5, and so on. Therefore, inhibition of active GSK-3β formation by andro could be a possible therapeutic target for the improvement of synaptic plasticity and memory behavior (Tapia-Rojas et al. 2015). Inhibition of Aβ aggregation process with the administration of andro in transgenic mice (APPswe/PS-1 Tg mice) was successfully carried out, and it was observed that Tg mice showed a significant reduction of GSK-3β activity (Serrano et al. 2014).

### 3.1.3 Inhibition of mTOR Pathway

The mammalian target of rapamycin (mTOR)-mediated pathway triggers a cascade of cellular signaling to activate an important cellular process called autophagy (Wullschleger et al. 2006; Loewith et al. 2002). In a normal cell, mTOR regulates protein synthesis and degradation through its complexes (mTORC1 and mTORC2) (Quinsay et al. 2010). The brain of the deceased AD patient showed increased expression of mTOR and its downstream signaling molecules (p70S6K and eIF4E) (Oddo 2012; Pei and Hugon 2008). mTOR-mediated autophagy consists of Akt activation through phosphorylation by PI3K. Akt activates mTORC1, which regulates autophagy (Rajapakse et al. 2011; Alessi et al. 1996). Reduced autophagy results in increased steady-state Aβ oligomers, hence create a vicious cycle leading to more Aβ oligomers deposition. Neurons undergo extensive biochemical alterations leading to death. Dysregulated mTOR/autophagy is one of the mechanisms that accelerate the neurodegeneration and AD progression (Zhang et al. 2019b; Wang et al. 2014; Shen et al. 2017). Andro had shown to halt the upregulation of mTOR and their signaling molecules both in vitro and in vivo in various disease conditions, including neurodegeneration (Fig. 5) (Sharifi-Rad et al. 2020; Zhang et al. 2019b; Kumar et al. 2015).

### 3.1.4 Upregulation of Nrf-2-Related Molecules

Nuclear erythroid-related factor-2 (Nrf-2) is responsible for counteraction against mitochondrial dysfunction and oxidative stress (Cole and Frautschy 2007; Johnson et al. 2008; Wu et al. 1999). As shown in Fig. 5 during the oxidative injury, Nrf-2 gets activated and translocated to the nuclear region, where it binds to antioxidant responsive element (ARE) for upregulation of certain important antioxidant mediators such as NADPH-quinone oxidoreductase 1, glutathione-S-transferase, and heme oxygenase-1 (HO-1), and so on (Magesh et al. 2012; Cole and Frautschy 2007). These signaling molecules target oxidative stress and free radicals ($O_2^-$, $^-OH$, $ONOO^-$, $H_2O_2$, and NO), generated due to mitochondrial dysregulation and inflammatory response (Chen et al. 2020; Herrera-Arozamena et al. 2016). Various researchers have reported the downregulation of the Nrf-2/ARE pathway during all the neurodegenerative conditions of AD. Hence, this pathway should be the prime target to reduce the progression of AD. Andro has demonstrated significantly increased nuclear translocation of Nrf-2 in EA.hy926 cells treated with TNF-α and further the upregulation of antioxidant signaling molecules was also reported, which clearly showed the ability of andro to trigger the antioxidant process to counteract the oxidative stress (Wong et al. 2016; Yu et al. 2010; Herrera-Arozamena et al. 2016).

## 3.2 Parkinson's Disease

Parkinson's disease (PD) is an age-related neurodegenerative disorder that occurs due to the continuous loss of dopaminergic neurons at the pars compacta region of the brain (Savica et al. 2010). PD can be characterized by reduced motor ability, including bradykinesia, posture instability, and resting tremors apart from nonmotor symptoms such as hyposmia, cognitive dysfunction, sleep, and mental disorders (Poewe et al. 2017). Multiple factors predisposing the disease initiation that starts with the formation of Lewy bodies, leading to a degeneration of dopaminergic neurons and consequent depletion of dopamine. Lewy bodies deposition is the hallmark of PD that occurs due to the development of protein aggregates containing misfolded proteins (α-Synuclein) and ubiquitin due to improper proteasome formation. They play a significant role in neuronal loss through various neurodegenerative signaling pathways triggered through activation of astrocytes and microglia (Beitz 2014; Poewe et al. 2017; Dexter and Jenner 2013).

As depicted in Fig. 6, a typical PD mechanism is initiated by multiple factors that lead to protein aggregates resulted in the rapid elevation of oxidative stress and a decline in mitochondrial function (Martinez et al. 2018; Lu et al. 2019). These alterations result in neuronal damage carried out by inflammation and apoptosis through astroglial activation (Tansey and Goldberg 2010). Several research articles had provided evidence for the inflammatory process and oxidative stress in the dopaminergic neuron activated by Lewy bodies. These pieces of evidence make those two processes as the main therapeutic targets for the management of PD. Lewy bodies which trigger the pro-inflammatory process is achieved by NF-κB activation and followed by upregulation of the primary mediators like TNF-α, IL-1 beta, and IL-6 (Tansey and Goldberg 2010; Welcome 2020). ROS/RNS-mediated neurodegeneration was reported to be caused by mitochondrial dysfunction leading to ionic imbalance. Free radical-mediated macromolecular damage processes like lipid peroxidation, protein, and DNA oxidation can lead to further pathological processes in PD (Dexter and Jenner 2013). Apart from free radical-mediated neuronal death, another vital hypothesis supported excessive mitochondrial fission mediated by Dynamin-related protein-1 (Drp-1), leading to a change in mitochondrial dynamics and severe depolarization and increased ROS production (Martinez et al. 2018; Ni et al. 2015). Andro had successfully reduced the PD progression by reducing the neuronal damage firstly by quenching the oxidative stress through its

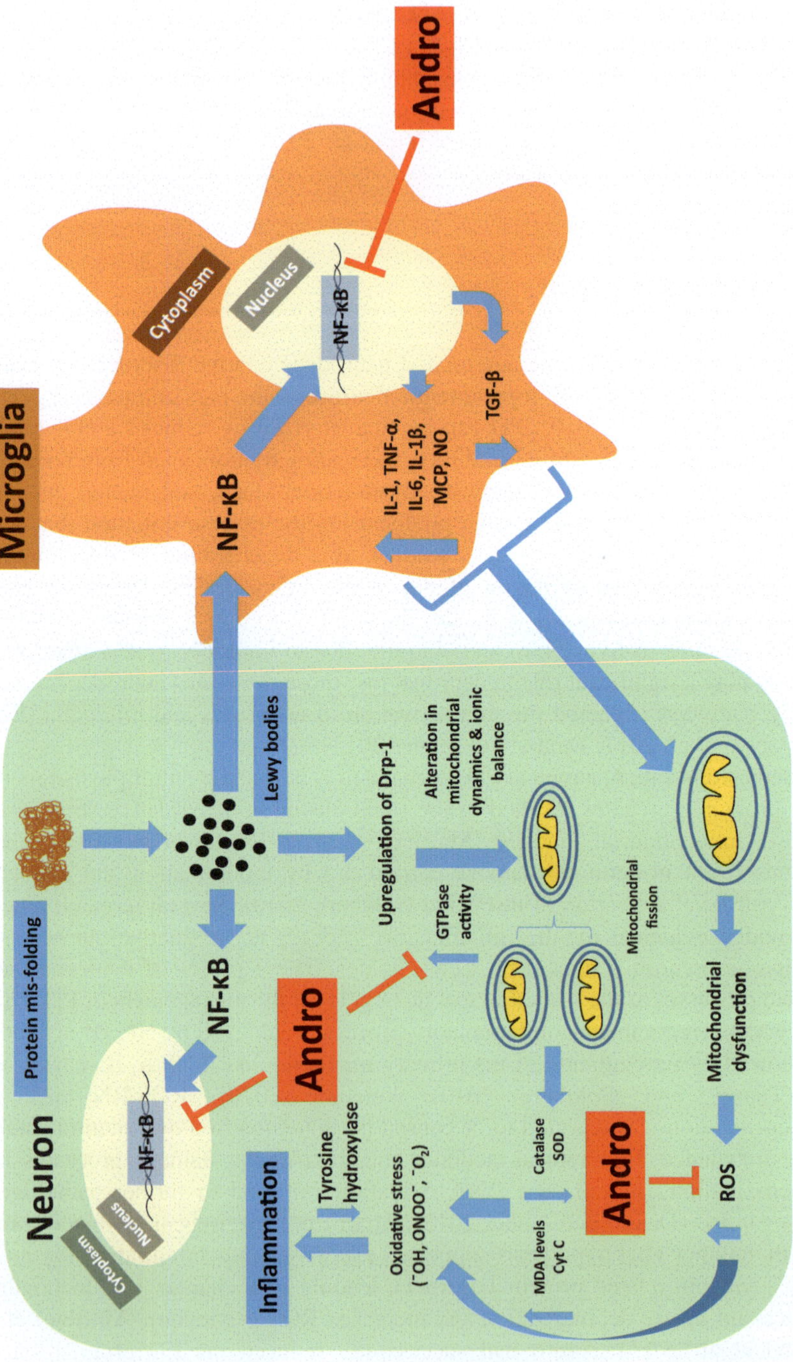


**Fig. 6** Drug targets influenced by andrographolide during Parkinson's disease

anti-oxidative property, and further downregulation of Drp-1 by andro was also successfully reported that resulted in the improvement of motor and nonmotor functions in MPTP induced PD in mice (Geng et al. 2019b).

Downregulation of NF-κB activation and its nuclear translocation helps in the reduction of inflammatory cascades in the neurons (Lu et al. 2019; Tansey and Goldberg 2010). As reported by many researchers, andrographolide is a potent anti-inflammatory agent and can unarguably be considered for therapeutic management of PD (Lindsay et al. 2020). However, there is supporting evidence in mice model by $MPP^+$/MPTP-induced neurotoxicity and behavioral impairment mimicking the PD pathogenesis in humans (Zhang et al. 2014). Suppression of microglial-mediated neuroinflammation and neuroprotection is another precise target strategy observed by Wang et al. (2004), where andro significantly avoided the activation of microglia through the reduction of ROS, and that initiated secondary messengers like $PGE_2$ and TNF-α (Fig. 6). Furthermore, the study also provided evidence about andro successfully activating protein degradation through the downregulation of iNOS and COX-2 mechanisms (Chiou et al. 2000). It is considered a novel strategy exerted by andro compared to the traditional mechanism of iNOS-mediated anti-PD activity (Wang et al. 2004).

## 3.3 *Ischemic Brain Injury*

Cerebral ischemia is characterized by neuronal insult caused due to interrupted blood supply to the brain in the form of thrombus, occlusion, and atherosclerotic plaques (Lee et al. 2018; Mehta et al. 2007). It also leads to reduced motor and cognitive functions of the brain as the energy supply to the motor and sensory neurons was interrupted, usually at the middle cerebral artery (MCA) (Traystman 2003). The patients with occluded MCA displayed symptoms like hemiparesis, deficiency in hemi-sensory and visuals, dysarthria, and ataxia (Lee et al. 2018).

A complex cascade of pathophysiological mechanisms (Fig. 7) is triggered due to depleted blood supply and energy to the neurons, which are completely dependent on the severity and duration of ischemia. As a result of an ischemic injury-depleted energy source, ATP dependent ion pumps were not executed in neurons followed by an imbalance in ionic gradient (Lee et al. 2018; Luo et al. 2019). It leads to neuron swelling and necrosis due to excess intracellular calcium extracellular potassium (Neumann et al. 2013; Harukuni and Bhardwaj 2006). Excitatory neurotransmitters toxicity, free radical-mediated neuronal damage, inflammatory processes, and disruption of blood–brain barrier are further downstream neurodegenerative pathways (Sugawara and Chan 2003; Leker and Shohami 2002; Mehta et al. 2007). Ischemic injury, if not checked within a short period, forms an ischemic infarct surrounded by ischemic penumbra (Mehta et al. 2007; Lee et al. 2018).

Reperfusion period after initial ischemia distributes sufficient oxygen to the cells for normal function. However, the biochemical reactions involving oxygen release free radicals, and that can further aid in neuron damage. Apart from this, cells tend to

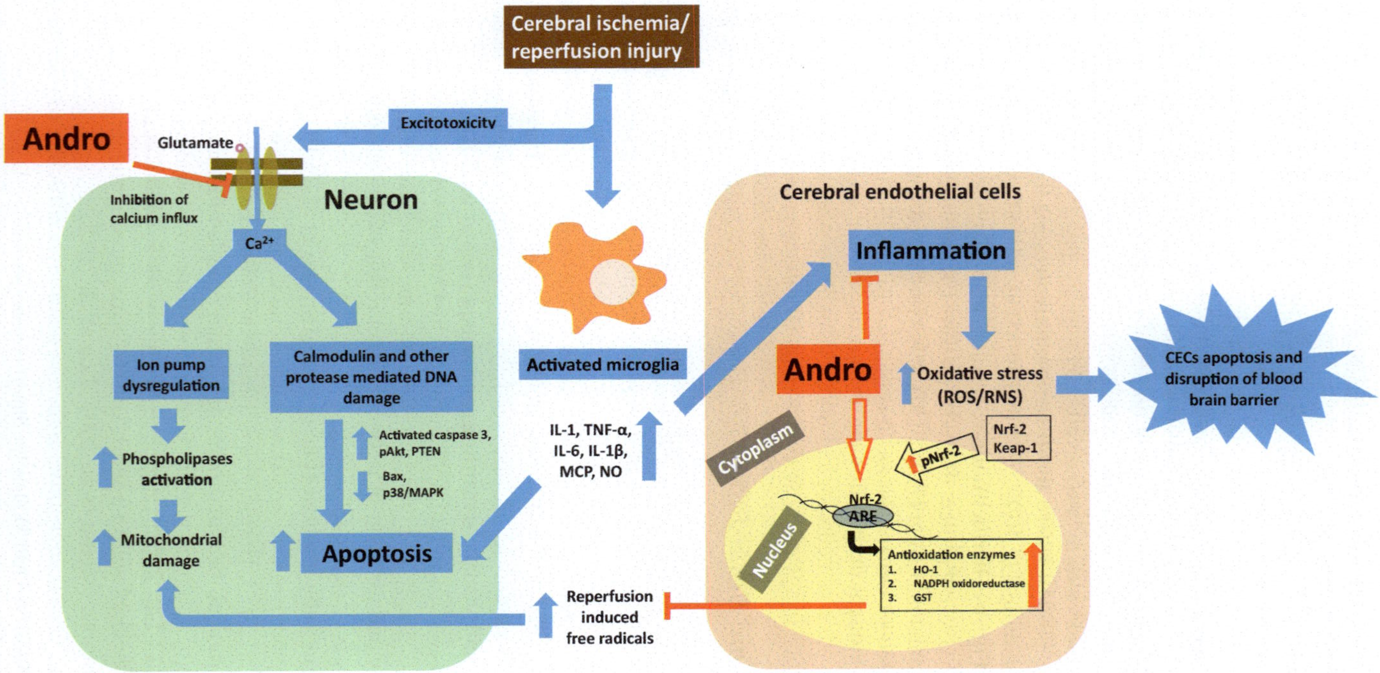


**Fig. 7** Drug targets influenced by andrographolide during ischemic brain injury

cope up with the ischemic injury by eliminating excess calcium. It leads to the activation of phospholipase A, which can trigger damage in mitochondrial and cellular membranes (Traystman 2003; Mehta et al. 2007; Wu et al. 2020).

Drug treatment is highly essential for restoring the normal function of the penumbra partially or wholly so that the diffusion of neuronal damage can be avoided to evade severe brain damage (Wu et al. 2020; Lee et al. 2018). In vitro and in vivo experiments have shown the ability of andrographolide to inhibit neuronal damage through several mechanisms (Chan et al. 2010; Adedayo et al. 2020; Lu et al. 2019; Wang et al. 2020; Hou et al. 2010; Yang et al. 2019). Various cellular processes are targeted by andro during the management of ischemic damage such as ionic imbalance created by calcium influx, cerebral endothelial cells (CECs) apoptosis, oxidative stress due to increased mitochondrial and ER stress, activation of microglia, and mediated inflammation (Yen et al. 2013; Mouw et al. 2003; Khan et al. 2004). The phenotypic changes observed as a result of regulating these cellular signals were reduced disruption of blood–brain barrier and hippocampus, reduced brain infarct volume, improved motor, and other behavioral functions, and reduced infiltration of leucocytes (Traystman 2003; Lee et al. 2018).

### 3.3.1 Blockage of Calcium Channel

The blockade of calcium channel is one of the successful targets for drugs to reduce ischemia-mediated neuron injury. Intracellular calcium-activated signaling leads to depolarization of the presynaptic membrane and the increase in excitatory amino acids like glutamate and aspartate, which in turn upregulated ligand-gated ion channel and excessive calcium, sodium, and chloride ions along with water. The loss of ionic homeostasis and neurodegeneration is primarily triggered by calcium. Calcium channel blockers are selective calcium antagonists that obstruct the voltage-dependent ion channels, which significantly inhibit the calcium influx to the cells and dilate the blood vessels to increase the blood flow (Leker and Shohami 2002; Luo et al. 2019). Andrographolide had successfully helped in the calcium channel blocking process (Fig. 7) thereby significantly abolished the subsequent depolarization (Zeng et al. 2017; Yang and Song 2014).

### 3.3.2 Anti-Oxidative Molecules in CEC and Anti-Oxidation of Neuron Cells

Blood–brain barrier (BBB) primarily constituted of CECs interacting with capillaries through tight junctions. They protect the brain from various toxins and foreign bodies in the circulation. Hence, micro/macromolecules protecting the CECs can be considered to possess a better therapeutic effect on ischemia-induced neuronal damage (Mark and Davis 2002). CECs protect the blood–brain barrier with the well-equipped signaling machinery in the form of Nrf-2 and HO-1 (Ding et al. 2014). In vivo studies have been reported for the neuroprotective effect of andro on CECs, and

MCAO-induced injury was successfully avoided by upregulation of Nrf2 nuclear translocation and HO-1 activation (Fig. 7). These factors are the primary source of the antioxidant mechanism by CECs, where Nrf-2 triggers the upregulation of detoxification and antioxidant genes, while HO-1, predominantly activated by Nrf-2 render anti-oxidative property through heme catabolic pathway (Yen et al. 2013; 2016).

Role Nrf-2 and HO-1 have clearly indicated that oxidative stress can be a vital target mechanism for neuroprotection from ischemia as brain tissues are highly susceptible to oxidative injury organized by $O_2^-$, $^-OH$, $ONOO^-$, $H_2O_2$, and NO which are the outcome of reperfusion and mitochondrial alterations (Sugawara and Chan 2003; Wu et al. 2020). These highly reactive radicals, apart from damaging the cellular macromolecules, trigger microglia activation-mediated inflammation and neurodegeneration, leading to damaged CECs and disintegrated BBB (Lee et al. 2018; Mehta et al. 2007). Several studies have shown the anti-oxidative property of andro and support a strong argument for andro to be considered as a potential neuroprotective agent (Mussard et al. 2019; Thakur et al. 2016).

### 3.3.3 Anti-Inflammation in CEC and Glial Cells

Inflammation is another vital mechanism that progresses the focal/transient ischemic/reperfusion injury into significant neurodegeneration. The upregulation of most of the signaling cascade that carries out neuroinflammation is focused on the activation of NF-κB and successful translocation to the nucleus (Wu et al. 2020; Harukuni and Bhardwaj 2006; Sugawara and Chan 2003). Andrographolide was reported for its inhibition of TNF-α and NF-κB signaling cascade and reduced expression of ICAM-1 in human endothelial cells (EA.hy926 cells) (Chao et al. 2011). NF-κB-p65 subunit activation through phosphorylation of Ser536, which leads to successful translocation of active p65 subunit for mediating the inflammatory process (Yang et al. 2019). These steps were potentially inhibited by andro and successfully downregulated the neuroinflammation (Fig. 7). Microglial activation and subsequent inflammation were another typical signaling in most of the acute/chronic neurodegenerative disorders (Yang et al. 2017; 2019). This process was also potentially inhibited by andro and further inflammatory cascade, which was exhibited through reduced expression of OX-2 and EG-1, the markers for activated microglia (Chan et al. 2010).

## 3.4 *Multiple Sclerosis*

Autoimmunity is a potential disorder that can affect any part of the system. Neurological degeneration due to autoimmunity is termed as Multiple sclerosis (MS) and is characterized by inflammation in the central nervous system through the infiltration of immune cells (Loma and Heyman 2011; Goodin 2015). Auto antigen-presenting cells (mostly dendritic cells) upon abnormal activation along with $CD4^+/CD8^+$ T cells lead to gradual demyelination and subsequent axonal damage (van der Star et al. 2012; Steinman and Zamvil 2003). DCs and T-lymphocytes are potential targets for drugs to attenuate the progression of MS (Farjam et al. 2015).

MS activates the majority of the immune cells involved in inflammation and secretion of pro-inflammatory cytokines such as cytotoxic ($CD8^+$) and helper ($CD4^+$), $CD3^+$ T cells, $CD68^+$ microglia/macrophages. The disease progresses at the perivascular cuff, parenchyma of the brain, and spinal cord (Steinman and Zamvil 2003; Farjam et al. 2015). Complement proteins such as C1q and C3 were also reported to be released during MS (Lu et al. 2019).

Inflammation-mediated demyelination leading to the loss of neuron/glial cells was observed in Experimental Autoimmune encephalomyelitis (EAE). It is identical to the pathogenesis of MS. Upcoming and approved drugs for the treatment of MS have been validated based on EAE animal model studies (Baker and Amor 2014; Constantinescu et al. 2011). The major evidence for andro to be a significant candidate for MS treatment is its ability to inhibit key therapeutic targets such as T-cells activation, the release of cytokines, and the maturation of DCs, which organizes the antigen presentation (Farjam et al. 2015). In a synthesized MOG peptide stimulated EAE model, the activation of $CD4^+$ T-cells was inhibited, which reduce the EAE symptoms that are manifested in human during multiple sclerosis. The ability of andro-treated DCs to activate T-cells was wholly reduced, indicating andro's interference in the maturation of DCs (Iruretagoyena et al. 2005).

Although andro was not significantly explored, however, clinically, *A. paniculata* has been shown in subsiding the symptoms of MS, and in the normalization of related phenotypic alterations (Bertoglio et al. 2016). The study reported the reduction of fatigue in MS patients in synergy with interferon-beta (IFN-β), significantly better than IFN-β supplementation without *A. paniculata*. The study also made a point that the combination of supplements does not change the clinical parameters, and only MS-related fatigue was improved up to 44% in a period of 12 months (Bertoglio et al. 2016). This composition developed by Burgos and coworkers was patented in the year 2015 (Orozco and Burgos 2015).

## 3.5 Traumatic Brain Injury

Brain injury due to mechanical force can lead to trauma, and a set of neurological damage caused by cellular and molecular signaling processes occurring as a result of primary injury is collectively known as Traumatic Brain Injury (TBI). More than 70% of the primary damage results in diffused axonal injury (DAI) that results in the disruption of microtubular and membrane structure (Lee and Ng 2019). It leads to physiological impairments such as varicosity formation and axonal disconnection that impairs the synaptic function (Blennow et al. 2016). On the other hand, secondary injury is described as a delayed cellular process that occurs as a result of a primary injury such as excitotoxicity, mitochondrial dysfunction, neuroinflammation, and apoptosis leading to axonal and endothelial degeneration leading to cerebral edema, and disruption of BBB (Prins et al. 2013; Lee and Ng 2019).

The management strategies of TBI should be timely and effective since the extent of the injury will increase if prolonged for a longer time. Therapeutic intervention can significantly inhibit the molecular and cellular signaling cascades so that secondary injury after the initial damage can be avoided, and cognitive and other behavioral functions can be recovered (Galgano et al. 2017; Leker and Shohami 2002). The inhibitory role of andro in microglial activation inflammation, apoptosis, ionic imbalance, and so on, makes it an apparent candidate for therapeutic management of secondary traumatic injury (Lee and Ng 2019). Experimental animal model study on TBI had addressed the ability of andro in inhibiting pathways like microglial activation-mediated inflammation, nuclear translocation of activated p56 subunit of NF-κB, downregulation of cytokines, phospho-ERK, p38-MAPK signaling-mediated inflammation (Tao et al. 2018).

## *3.6 Antidepressant-like Property of Andrographolide*

Depression is a neuropsychological disorder that is characterized by disinterest in almost all activities (anhedonia), feelings of guilt, anxiety, and unhappy mood, which ends up with patients getting recurring thoughts about death and suicide. Monoamine hypothesis is one of the well-established mechanisms of the pathobiology of depression, which says deficiency of serotonin, norepinephrine, and dopamine leads to depressive behavior. Also, the role of neurotrophic factors and its involvement in the pathophysiology of depression was discussed (Yu and Chen 2011). Various studies have reported evidence on BDNF and its influence on depressive behavior. The significance of BDNF expression and depression was clarified by observing a minor variant in the human BDNF gene leading to altered BDNF levels, which resulted in neuropsychiatric disorders like anxiety-mediated neural dysfunction, bipolar disorders, and so on, apart from depression (Neto et al. 2011). It was reported that BDNF expression has decreased in the hippocampus and cortex, while the striatum and amygdala showed upregulated BDNF expression. This evidence suggests the heterogeneity of BDNF expression during the depression, and the therapeutic strategy should be sight specific according to its inhibitory or activating influence of the gene (Yu and Chen 2011). Different researchers reported where the hippocampal expression of BDNF was reduced in an experimental model of chronic unpredictable mild stress (CUMS) by supplementation of andrographolide. Another study discussed the regulation of BDNF signaling protein by andro and resulted in a significant improvement in the cognitive and motor abilities in CUMS induced mice (Zhang et al. 2019c; Geng et al. 2019a).

## *3.7 Analgesic Property of Andrographolide*

Noxious stimulus leading to activation of Aδ and C fibers in the peripheral nervous system, which responds to thermal, chemical, and mechanical stimuli, and the entire flow of neurological signal is called Nociception. Action potential and pain sensor molecules were processed in the spinal cord. Nociceptors conduct the pain in various tissues through Substance P and calcitonin gene-related peptide (CGRP) (Lynn 1996; Foreman 1987). The pain blocker mainly blocks the voltage-gated sodium and calcium channels that conduct the action potential caused by the noxious stimuli. Most of the first and second lines of analgesic drugs include some antidepressants, voltage-gated channel blockers for NMDA receptors, opioids, and muscle relaxants (Fornasari 2017).

Several papers have reported the inhibition of acid writhing and hot plate test for andrographolide, although somewhat contradictory to each other (Madav et al. 1995; Suebsasana et al. 2009; Sulaiman et al. 2010). Acid writhing and hot plate tests examine nociceptive pain, which results from an encounter with a noxious stimulus such as heat, chemical stimulus, and pressure. Inhibition to neuropathic pain is reported (Wang et al. 2018), which associated reduced IL-1 and astrocytic activity, which is part of an inflammatory reaction. IL-1 is part of the signaling transduction downstream to NF-κB. Activations of NF-κB/MAPKs in astrocytes and microglia can enforce a positive feedback loop for further upregulation of glial receptors and pro-nociceptive mediators (Ji et al. 2009). Recently there have been several reports on andro-reducing pain in different pain models. These include generalized hyperalgesia in induced adenomyosis (Mao et al. 2011), HIV induced pain (Yi et al. 2018), and arthritis-related pain (Gupta et al. 2020). These mechanisms are associated with the anti-inflammatory reaction, although other mechanisms like glutamate-related activity and channel blocking remain to be investigated.

## *3.8 Influence of Andrographolide on Angiogenesis and Stem Cell Infiltration during Neurodegeneration*

Angiogenesis in the brain is regulated by a balance between pro/antiangiogenic factors (Dameron et al. 1994). Deterioration of the blood vessels was observed in neurodegenerative conditions, while embryonic development, malignant tumor, brain repair after traumatic injury, or ischemia require extensive neoangiogenesis (Vallon et al. 2014; Kim and Lee 2009; Plate 1999). Placental growth factor (PGF), fibroblast growth factor (FGF), and vascular endothelial growth factor (VEGF) are pro-angiogenic, while thrombospondins, angiostatin, and endostatin show the antiangiogenic property (Harrigan 2003; Fukumura et al. 1998; Folkman 2006; O'Reilly et al. 1994).

During AD, Aβ deposition occurs even in the capillaries and trigger anti-angiogenesis. Also, it was presumed that VEGF aggregates along with Aβ deposits leading to the downregulation of angiogenesis (Paris et al. 2004b; Vallon et al. 2014; Yang et al. 2004). It leads to increased expression of VEGF and other pro-angiogenic factors and activation of endothelial cells. Increased VEGF might further help in co-deposition with Aβ, while endothelial activation leads to the upregulation of various proteases and inflammatory factors that aid in neuroinflammation-mediated neurodegeneration (Kalaria et al. 1998; Paris et al. 2004a). Andro has been positively reported for its antiangiogenic activity on various disease models (Li et al. 2020; Dai et al. 2017; Li et al. 2016; Huang et al. 2017; Sheng et al. 2016). On the contrary, Duan et al. (2019) have reported that andro increased expression of VEGF and stimulation of HUVEC cell migration, which are critical steps toward angiogenesis. Andro improved cell migration and angiogenesis by upregulating PI3K/Akt/eNOS signaling pathway. They proposed that the pro-angiogenic activity of andro depends primarily on its concentration (Duan et al. 2019). Cytotoxicity of andro on CECs was reported by the same research group with increasing concentration, while 50 μM andro rendered cytoprotective activity to CECs and significantly increased the expression of VEGF, cell migration, and tube formation (Dai et al. 2017). This study provides a new perspective on the biological activity of andro, as angiogenesis is an important step during the recovery of the brain from trauma/ischemia. Active angiogenesis occurs after acute phase injury not only for increasing the oxygen and nutrient supply to the brain but also for neuro and synaptogenesis.

Activation of CECs occurs during brain injury, and this leads to the secretion of stromal cell-derived factors (SDF-1α) or CXCL12, which attracts the CXCR4 positive neural stem cells (NSCs) (Vallon et al. 2014; Itoh et al. 2009). SDF-1α plays a significant role in the maintenance of embryonic and adult NSCs, acts as a regulatory signal for peripheral hematopoietic stem cells, and hence is hypothesized to recruit the NSCs to the site of brain damage to carry out stem cell-based tissue repair (Li et al. 2012). Inhibition of CXCL4, a receptor of SDF-1α, results in impaired differentiation and remyelination in multiple sclerosis patients (Patel et al. 2010; Carbajal et al. 2010). CECs and perivascular astrocytes are thought to play a pivotal role in SDF-1α expression and recruitment of NPCs and NSCs to the damage site for tissue repair during an ischemic stroke (Imitola et al. 2004; Wang et al. 2008). SDF-1α levels were decreased in both experimental models and AD patients, which implies that its upregulation by therapeutic strategy might provide a significant improvement in AD management (Parachikova and Cotman 2007; Shin et al. 2011). Intracerebral administration of SDF-1α had helped in the migration of Mesenchymal stem cells (MSCs) and lead to tissue repair that improved memory and cognitive abilities. At the same time, the level of Aβ deposits was unaltered (Shin et al. 2011). Ji et al. (2005) reported that chemotaxis of Jurkat cells enhanced by SDF-1α when supplemented with andrographanin (a derivative of andrographolide). SDF-1α and CXCR4-mediated chemotaxis was activated clearly due to reduced SDF-1α induced CXCR4 receptor internalization by andrographanin (Ji et al. 2005).

Based on the set of studies discussed and the proposed hypothesis, it is clear that andrographolide and its derivatives provide antiangiogenic effects at higher concentrations, but a pro-angiogenic property at a lower concentration. Also, the ability of andro based compounds to induce the upregulation of SDF-1α and CXCL4 shows that it can be a potential candidate for the migration of chemically induced stem cells to the damaged site, and the stem cell-based therapy.

## 4 Conclusions

In summary, the biological properties of andro and its derivatives have been studied widely on various diseases. Concerning neurodegenerative diseases, andro has received special attention in recent years, mainly because it targets different molecular signaling pathways and affects neuroprotection in multiple routes. Further, andro shows the ability to treat these diseases not only by downregulation of pathological signaling (NF-κB-mediated inflammation and caspase 3-mediated apoptosis) but through upregulation of physiological signaling (Nrf-2-mediated anti-oxidation and Wnt/β-catenin-mediated anti-apoptosis). Therefore, we can conclude that andro is a potential multi-targeted drug for the treatment of various chronic and life-threatening ailments, including neurodegenerative diseases.

## References

Adedayo BC, Ogunsuyi OB, Akinniyi ST et al (2020) Effect of *Andrographis paniculata* and *Phyllanthus amarus* leaf extracts on selected biochemical indices in Drosophila melanogaster model of neurotoxicity. Drug Chem Toxicol:1. https://doi.org/10.1080/01480545.2019.1708377

Akinyemi O, Oyewole S, Jimoh K (2018) Medicinal plants and sustainable human health: a review. Hortic Int J 2(4):194–195. https://doi.org/10.15406/hij.2018.02.00051

Alessi DR, Andjelkovic M, Caudwell B et al (1996) Mechanism of activation of protein kinase B by insulin and IGF-1. EMBO J 15(23):6541–6551. https://doi.org/10.1002/j.1460-2075.1996.tb01045.x

Alzohairy MA (2016) Therapeutics role of *Azadirachta indi*ca (Neem) and their active constituents in diseases prevention and treatment. Evid Based Complement Alternat Med. https://doi.org/10.1155/2016/7382506

Arrázola MS, Ramos-Fernandez E, Cisternas P et al (2017) Wnt signaling prevents the Aβ oligomer-induced mitochondrial permeability transition pore opening preserving mitochondrial structure in hippocampal neurons. PLoS One 12(1):e0168840. https://doi.org/10.1371/journal.pone.0168840

Arrázola MS, Varela-Nallar L, Colombres M et al (2009) Calcium/calmodulin-dependent protein kinase type IV is a target gene of the Wnt/β-catenin signaling pathway. J Cell Physiol 221(3):658–667. https://doi.org/10.1002/jcp.21902

Baker D, Amor S (2014) Experimental autoimmune encephalomyelitis is a good model of multiple sclerosis if used wisely. Mult Scler Relat Disord 3(5):555–564. https://doi.org/10.1016/j.msard.2014.05.002

Batkhuu J, Hattori K, Takano F et al (2002) Suppression of NO production in activated macrophages *in vitro* and *ex vivo* by neoandrographolide isolated from *Andrographis paniculata*. Biol Pharm Bull 25(9):1169–1174. https://doi.org/10.1248/bpb.25.1169

Beitz JM (2014) Parkinson's disease: a review. Front Biosci (Schol Ed) 6:65–74. https://doi.org/10.2741/s415

Bertoglio J, Baumgartner M, Palma R et al (2016) *Andrographis panicul*ata decreases fatigue in patients with relapsing-remitting multiple sclerosis: a 12-month double-blind placebo-controlled pilot study. BMC Neurol 16(1):77. https://doi.org/10.1186/s12883-016-0595-2

Bhatnagar M, Sharma D, Salvi M (2009) Neuroprotective effects of *Withania somnifera* Dunal.: a possible mechanism. Neurochem Res 34(11):1975–1983. https://doi.org/10.1007/s11064-009-9987-7

Blennow K, Brody DL, Kochanek PM et al (2016) Traumatic brain injuries. Nat Rev Dis Primers 2 (1):1–19. https://doi.org/10.1038/nrdp.2016.84

Carbajal KS, Schaumburg C, Strieter R et al (2010) Migration of engrafted neural stem cells is mediated by CXCL12 signaling through CXCR4 in a viral model of multiple sclerosis. Proc Natl Acad Sci 107(24):11068–11073. https://doi.org/10.1073/pnas.1006375107

Cerpa W, Toledo EM, Varela-Nallar L et al (2009) The role of Wnt signaling in neuroprotection. Drug News Perspect 22(10):579–591. https://doi.org/10.1358/dnp.2009.10.1436817

Chan SJ, Wong WF, Wong PT et al (2010) Neuroprotective effects of andrographolide in a rat model of permanent cerebral ischemia. Br J Pharmacol 161(3):668–679. https://doi.org/10.1111/j.1476-5381.2010.00906.x

Chao C-Y, Lii C-K, Tsai I-T et al (2011) Andrographolide inhibits ICAM-1 expression and NF-κB activation in TNF-α-treated EA.hy926 cells. J Agric Food Chem 59(10):5263–5271. https://doi.org/10.1021/jf104003y

Chao W-W, Kuo Y-H, Lin B-F (2010) Anti-inflammatory activity of new compounds from *Andrographis paniculata* by NF-κB transactivation inhibition. J Agric Food Chem 58 (4):2505–2512. https://doi.org/10.1021/jf903629j

Chen S-Y, Gao Y, Sun J-Y et al (2020) Traditional Chinese medicine: role in reducing β-amyloid, apoptosis, autophagy, neuroinflammation, oxidative stress, and mitochondrial dysfunction of Alzheimer's disease. Front Pharmacol 11:497. https://doi.org/10.3389/fphar.2020.00497

Chen Y-Y, Hsieh C-Y, Jayakumar T et al (2014) Andrographolide induces vascular smooth muscle cell apoptosis through a SHP-1-PP2A-p38MAPK-p53 cascade. Sci Rep 4:5651. https://doi.org/10.1038/srep05651

Chiou WF, Chen CF, Lin JJ (2000) Mechanisms of suppression of inducible nitric oxide synthase (iNOS) expression in RAW 264.7 cells by andrographolide. Br J Pharmacol 129(8):1553–1560. https://doi.org/10.1038/sj.bjp.0703191

Choi BH (1995) Oxidative stress and Alzheimer's disease. Neurobiol Aging 16(4):675–678. https://doi.org/10.1016/0197-4580(95)00065-M

Cole GM, Frautschy SA (2007) The role of insulin and neurotrophic factor signaling in brain aging and Alzheimer's disease. Exp Gerontol 42(1–2):10–21. https://doi.org/10.1016/j.exger.2006.08.009

Constantinescu CS, Farooqi N, O'Brien K et al (2011) Experimental autoimmune encephalomyelitis (EAE) as a model for multiple sclerosis (MS). Br J Pharmacol 164(4):1079–1106. https://doi.org/10.1111/j.1476-5381.2011.01302.x

Dai J, Lin Y, Duan Y et al (2017) Andrographolide inhibits angiogenesis by inhibiting the Mir-21-5p/TIMP3 signaling pathway. Int J Biol Sci 13(5):660–668. https://doi.org/10.7150/ijbs.19194

Dai Y, Chen S-R, Chai L et al (2019) Overview of pharmacological activities of *Andrographis paniculata* and its major compound andrographolide. Crit Rev Food Sci Nutr 59(sup1):S17–S29. https://doi.org/10.1080/10408398.2018.1501657

Dameron KM, Volpert OV, Tainsky MA et al (1994) Control of angiogenesis in fibroblasts by p53 regulation of thrombospondin-1. Science 265(5178):1582–1584. https://doi.org/10.1126/science.7521539

Das S, Mishra K, Ganju L et al (2017) Andrographolide-a promising therapeutic agent, negatively regulates glial cell derived neurodegeneration of prefrontal cortex, hippocampus and working memory impairment. J Neuroimmunol 313:161–175. https://doi.org/10.1016/j.jneuroim.2017.11.003

Dexter DT, Jenner P (2013) Parkinson's disease: from pathology to molecular disease mechanisms. Free Radic Biol Med 62:132–144. https://doi.org/10.1016/j.freeradbiomed.2013.01.018

Ding Y, Zhang B, Zhou K et al (2014) Dietary ellagic acid improves oxidant-induced endothelial dysfunction and atherosclerosis: role of Nrf2 activation. Int J Cardiol 175(3):508–514. https://doi.org/10.1016/j.ijcard.2014.06.045

Duan M-X, Zhou H, Wu Q-Q et al (2019) Andrographolide protects against HG-induced inflammation, apoptosis, migration, and impairment of angiogenesis via PI3K/AKT-eNOS signalling in HUVECs. Mediat Inflamm. https://doi.org/10.1155/2019/6168340

Fabio GD, Romanucci V, De Marco A et al (2014) Triterpenoids from *Gymnema sylvestre* and their pharmacological activities. Molecules 19(8):10956–10981. https://doi.org/10.3390/molecules190810956

Farjam M, Zhang G-X, Ciric B et al (2015) Emerging immunopharmacological targets in multiple sclerosis. J Neurol Sci 358(1–2):22–30. https://doi.org/10.1016/j.jns.2015.09.346

Folkman J (2006) Antiangiogenesis in cancer therapy – endostatin and its mechanisms of action. Exp Cell Res 312(5):594–607. https://doi.org/10.1016/j.yexcr.2005.11.015

Foreman JC (1987) Peptides and neurogenic inflammation. Br Med Bull 43(2):386–400. https://doi.org/10.1093/oxfordjournals.bmb.a072189

Fornasari D (2017) Pharmacotherapy for neuropathic pain: a review. Pain Ther 6(Suppl 1):25–33. https://doi.org/10.1007/s40122-017-0091-4

Fukumura D, Xavier R, Sugiura T et al (1998) Tumor induction of VEGF promoter activity in stromal cells. Cell 94(6):715–725. https://doi.org/10.1016/s0092-8674(00)81731-6

Galgano M, Toshkezi G, Qiu X et al (2017) Traumatic brain injury: current treatment strategies and future endeavors. Cell Transplant 26(7):1118–1130. https://doi.org/10.1177/0963689717714102

Geng J, Liu J, Yuan X et al (2019a) Andrographolide triggers autophagy-mediated inflammation inhibition and attenuates chronic unpredictable mild stress (CUMS)-induced depressive-like behavior in mice. Toxicol Appl Pharmacol 379:114688. https://doi.org/10.1016/j.taap.2019.114688

Geng J, Liu W, Gao J et al (2019b) Andrographolide alleviates Parkinsonism in MPTP-PD mice via targeting mitochondrial fission mediated by dynamin-related protein 1. Br J Pharmacol 176 (23):4574–4591. https://doi.org/10.1111/bph.14823

Geng J, Liu W, Xiong Y et al (2018) Andrographolide sulfonate improves Alzheimer-associated phenotypes and mitochondrial dysfunction in APP/PS1 transgenic mice. Biomed Pharmacother 97:1032–1039. https://doi.org/10.1016/j.biopha.2017.11.039

Goodin DS (2015) The pathogenesis of multiple sclerosis. Clin Exp Neuroimmunol 6:2–22. https://doi.org/10.1111/cen3.12261

Gordon MD, Nusse R (2006) Wnt signaling: multiple pathways, multiple receptors, and multiple transcription factors. J Biol Chem 281(32):22429–22433. https://doi.org/10.1074/jbc.R600015200

Gu L, Yu Q, Li Q et al (2018) Andrographolide protects PC12 cells against β-amyloid-induced autophagy-associated cell death through activation of the Nrf2-mediated p62 signaling pathway. Int J Mol Sci 19(9):2844. https://doi.org/10.3390/ijms19092844

Gupta S, Mishra KP, Kumar B et al (2020) Andrographolide attenuates complete freund's adjuvant induced arthritis via suppression of inflammatory mediators and pro-inflammatory cytokines. J Ethnopharmacol 261:113022. https://doi.org/10.1016/j.jep.2020.113022

Hardy J, Allsop D (1991) Amyloid deposition as the central event in the aetiology of Alzheimer's disease. Trends Pharmacol Sci 12(10):383–388. https://doi.org/10.1016/0165-6147(91)90609-v

Hardy JA, Higgins GA (1992) Alzheimer's disease: the amyloid cascade hypothesis. Science 256 (5054):184–186. https://doi.org/10.1126/science.1566067

Harrigan MR (2003) Angiogenic factors in the central nervous system. Neurosurgery 53 (3):639–661. https://doi.org/10.1227/01.neu.0000079575.09923.59

Harukuni I, Bhardwaj A (2006) Mechanisms of brain injury after global cerebral ischemia. Neurol Clin 24(1):1–21. https://doi.org/10.1016/j.ncl.2005.10.004

Herrera-Arozamena C, Martí-Marí O, Estrada M et al (2016) Recent advances in neurogenic small molecules as innovative treatments for neurodegenerative diseases. Molecules 21(9):1165. https://doi.org/10.3390/molecules21091165

Hou Y-C, Wang Y-H, Liou K-T et al (2010) Neuroprotective effect of andrographolide against ischemic stroke in rats through reducing iNOS and gp91phox/NOX2 expression. J Chin Med 21 (3,4):85–98

Huang B, Peng Y, Li J et al (2017) An andrographolide derivative AGP-26b exhibiting anti-angiogenic activity in HUVECs and zebrafish via blocking the VEGFA/VEGFR2 signaling pathway. Mol BioSyst 13(3):525–536. https://doi.org/10.1039/c6mb00641h

Husain I, Zameer S, Madaan T et al (2019) Exploring the multifaceted neuroprotective actions of *Emblica officinalis* (Amla): a review. Metab Brain Dis 34:957–965. https://doi.org/10.1007/s11011-019-00400-9

Igado O, Olopade J (2016) A review on the possible neuroprotective effects of *Moringa oleifera* leaf extract. Niger J Physiol Sci 31(2):183–187

Imitola J, Raddassi K, Park KI et al (2004) Directed migration of neural stem cells to sites of CNS injury by the stromal cell-derived factor 1α/CXC chemokine receptor 4 pathway. Proc Natl Acad Sci 101(52):18117–18122. https://doi.org/10.1073/pnas.0408258102

Inestrosa NC, Toledo EM (2008) The role of Wnt signaling in neuronal dysfunction in Alzheimer's disease. Mol Neurodegener 3(1):9. https://doi.org/10.1186/1750-1326-3-9

Iruretagoyena MI, Tobar JA, González PA et al (2005) Andrographolide interferes with T cell activation and reduces experimental autoimmune encephalomyelitis in the mouse. J Pharmacol Exp Ther 312(1):366–372. https://doi.org/10.1124/jpet.104.072512

Itoh T, Satou T, Ishida H et al (2009) The relationship between SDF-1 α/CXCR4 and neural stem cells appearing in damaged area after traumatic brain injury in rats. Neurol Res 31(1):90–102. https://doi.org/10.1179/174313208X332995

Ji LL, Wang Z, Dong F et al (2005) Andrograpanin, a compound isolated from anti-inflammatory traditional Chinese medicine *Andrographis paniculata*, enhances chemokine SDF-1α-induced leukocytes chemotaxis. J Cell Biochem 95(5):970–978. https://doi.org/10.1002/jcb.20464

Ji R-R, Gereau IV RW, Malcangio M, Strichartz GR (2009) MAP kinase and pain. Brain Res Rev 60(1):135–148. https://doi.org/10.1016/j.brainresrev.2008.12.011

Jing M, Wang Y, Xu L (2019) Andrographolide derivative AL-1 ameliorates dextran sodium sulfate-induced murine colitis by inhibiting NF-κB and MAPK signaling pathways. Oxid Med Cell Longev:6138723. https://doi.org/10.1155/2019/6138723

Johnson JA, Johnson DA, Kraft AD et al (2008) The Nrf2-ARE pathway: an indicator and modulator of oxidative stress in neurodegeneration. Ann N Y Acad Sci 1147:61–69. https://doi.org/10.1196/annals.1427.036

Junsathian P, Yordtong K, Corpuz HM et al (2018) Biological and neuroprotective activity of Thai edible plant extracts. Ind Crop Prod 124:548–554. https://doi.org/10.1016/j.indcrop.2018.08.008

Kalaria R, Cohen D, Premkumar D et al (1998) Vascular endothelial growth factor in Alzheimer's disease and experimental cerebral ischemia. Brain Res Mol Brain Res 62(1):101–105. https://doi.org/10.1016/s0169-328x(98)00190-9

Kandanur SGS, Tamang N, Golakoti NR et al (2019) Andrographolide: a natural product template for the generation of structurally and biologically diverse diterpenes. Eur J Med Chem 176:513–533. https://doi.org/10.1016/j.ejmech.2019.05.022

Khan M, Sekhon B, Jatana M et al (2004) Administration of N-acetylcysteine after focal cerebral ischemia protects brain and reduces inflammation in a rat model of experimental stroke. J Neurosci Res 76(4):519–527. https://doi.org/10.1002/jnr.20087

Kim WY, Lee HY (2009) Brain angiogenesis: mechanism and therapeutic intervention in brain tumors. FEBS J 276(17):4653–4664. https://doi.org/10.1111/j.1742-4658.2009.07177.x

Krup V, Prakash L, Harini A (2013) Pharmacological activities of turmeric (*Curcuma longa* Linn): a review. J Homeop Ayurv Med 2(133). https://doi.org/10.4172/2167-1206.1000133

Kumar A, Singh A (2015) A review on Alzheimer's disease pathophysiology and its management: an update. Pharmacol Rep 67(2):195–203. https://doi.org/10.1016/j.pharep.2014.09.004

Kumar D, Das B, Sen R et al (2015) Andrographolide analogue induces apoptosis and autophagy mediated cell death in U937 cells by inhibition of PI3K/Akt/mTOR pathway. PLoS One 10(10): e0139657. https://doi.org/10.1371/journal.pone.0139657

Lee AYW, Ng SY (2019) Traumatic brain injuries: pathophysiology and potential therapeutic targets. Front Cell Neurosci 13:528. https://doi.org/10.3389/fncel.2019.00528

Lee RH, Lee MH, Wu CY et al (2018) Cerebral ischemia and neuroregeneration. Neural Regen Res 13(3):373–385. https://doi.org/10.4103/1673-5374.228711

Leker RR, Shohami E (2002) Cerebral ischemia and trauma—different etiologies yet similar mechanisms: neuroprotective opportunities. Brain Res Rev 39(1):55–73. https://doi.org/10.1016/s0165-0173(02)00157-1

Li J, Huang W, Zhang H et al (2007) Synthesis of andrographolide derivatives and their TNF-α and IL-6 expression inhibitory activities. Bioorg Med Chem Lett 17(24):6891–6894. https://doi.org/10.1016/j.bmcl.2007.10.009

Li J, Li F, Tang F et al (2020) AGS-30, an andrographolide derivative, suppresses tumor angiogenesis and growth *in vitro* and *in vivo*. Biochem Pharmacol 171:113694. https://doi.org/10.1016/j.bcp.2019.113694

Li J, Peng Y, Li S et al (2016) Anti-angiogenic activity of a new andrographolide derivative in zebrafish and HUVECs. Eur J Pharmacol 789:344–353. https://doi.org/10.1016/j.ejphar.2016.07.043

Li M, Hale JS, Rich JN et al (2012) Chemokine CXCL12 in neurodegenerative diseases: an SOS signal for stem cell-based repair. Trends Neurosci 35(10):619–628. https://doi.org/10.1016/j.tins.2012.06.003

Lindsay CB, Zolezzi JM, Rivera DS et al (2020) Andrographolide reduces neuroinflammation and oxidative stress in aged *Octodon degus*. Mol Neurobiol 57(2):1131–1145. https://doi.org/10.1007/s12035-019-01784-6

Liu J, Wang Z-T, Ge B-X (2008) Andrograpanin, isolated from Andrographis paniculata, exhibits anti-inflammatory property in lipopolysaccharide-induced macrophage cells through down-regulating the p38 MAPKs signaling pathways. Int Immunopharmacol 8(7):951–958. https://doi.org/10.1016/j.intimp.2007.12.014

Liu J, Wang Z-T, Ji L-L et al (2007) Inhibitory effects of neoandrographolide on nitric oxide and prostaglandin E 2 production in LPS-stimulated murine macrophage. Mol Cell Biochem 298 (1–2):49–57. https://doi.org/10.1007/s11010-006-9349-6

Loewith R, Jacinto E, Wullschleger S et al (2002) Two TOR complexes, only one of which is rapamycin sensitive, have distinct roles in cell growth control. Mol Cell 10(3):457–468. https://doi.org/10.1016/s1097-2765(02)00636-6

Loma I, Heyman R (2011) Multiple sclerosis: pathogenesis and treatment. Curr Neuropharmacol 9 (3):409–416. https://doi.org/10.2174/157015911796557911

Lu J, Ma Y, Wu J et al (2019) A review for the neuroprotective effects of andrographolide in the central nervous system. Biomed Pharmacother 117:109078. https://doi.org/10.1016/j.biopha.2019.109078

Luo Y, Tang H, Li H et al (2019) Recent advances in the development of neuroprotective agents and therapeutic targets in the treatment of cerebral ischemia. Eur J Med Chem 162:132–146. https://doi.org/10.1016/j.ejmech.2018.11.014

Lynn B (1996) Neurogenic inflammation caused by cutaneous polymodal receptors. In: Progress in brain research, vol 113. Elsevier, pp 361-368. doi: https://doi.org/10.1016/S0079-6123(08)61098-5

Madav S, Tripathi H, Mishra S (1995) Analgesic, anti-pyretic and antiulcerogenic effects of andrographolide. Ind J Pharma Sci 57(3):121–125

Magesh S, Chen Y, Hu L (2012) Small molecule modulators of Keap1-Nrf2-ARE pathway as potential preventive and therapeutic agents. Med Res Rev 32(4):687–726. https://doi.org/10.1002/med.21257

Mao X, Wang Y, Carter AV et al (2011) The retardation of myometrial infiltration, reduction of uterine contractility, and alleviation of generalized hyperalgesia in mice with induced adenomyosis by levo-tetrahydropalmatine (l-THP) and andrographolide. Reprod Sci 18 (10):1025–1037. https://doi.org/10.1177/1933719111404610
Mark KS, Davis TP (2002) Cerebral microvascular changes in permeability and tight junctions induced by hypoxia-reoxygenation. Am J Physiol Heart Circ Physiol 282(4):H1485–H1494. https://doi.org/10.1152/ajpheart.00645.2001
Martinez JH, Alaimo A, Gorojod RM et al (2018) Drp-1 dependent mitochondrial fragmentation and protective autophagy in dopaminergic SH-SY5Y cells overexpressing alpha-synuclein. Mol Cell Neurosci 88:107–117. https://doi.org/10.1016/j.mcn.2018.01.004
Mehta SL, Manhas N, Raghubir R (2007) Molecular targets in cerebral ischemia for developing novel therapeutics. Brain Res Rev 54(1):34–66. https://doi.org/10.1016/j.brainresrev.2006.11.003
Mouw G, Zechel JL, Gamboa J et al (2003) Activation of caspase-12, an endoplasmic reticulum resident caspase, after permanent focal ischemia in rat. Neuroreport 14(2):183–186. https://doi.org/10.1097/00001756-200302100-00004
Mussard E, Cesaro A, Lespessailles E et al (2019) Andrographolide, a natural antioxidant: an update. Antioxidants 8(12):571. https://doi.org/10.3390/antiox8120571
Neto FL, Borges G, Torres-Sanchez S et al (2011) Neurotrophins role in depression neurobiology: a review of basic and clinical evidence. Curr Neuropharmacol 9(4):530–552. https://doi.org/10.2174/157015911798376262
Neumann JT, Cohan CH, Dave KR et al (2013) Global cerebral ischemia: synaptic and cognitive dysfunction. Curr Drug Targets 14(1):20–35. https://doi.org/10.2174/138945013804806514
Ni H-M, Williams JA, Ding W-X (2015) Mitochondrial dynamics and mitochondrial quality control. Redox Biol 4:6–13. https://doi.org/10.1016/j.redox.2014.11.006
Oddo S (2012) The role of mTOR signaling in Alzheimer disease. Front Biosci (Schol Ed) 4:941–952. https://doi.org/10.2741/s310
O'Reilly MS, Holmgren L, Shing Y et al (1994) Angiostatin: a novel angiogenesis inhibitor that mediates the suppression of metastases by a Lewis lung carcinoma. Cell 79(2):315–328. https://doi.org/10.1016/0092-8674(94)90200-3
Orozco JLH, Burgos R (2015) Combination therapy with interferon and andrographolides for multiple sclerosis. US patent, 9,060,994 B2, 23 Jun 2015
Parachikova A, Cotman C (2007) Reduced CXCL12/CXCR4 results in impaired learning and is downregulated in a mouse model of Alzheimer disease. Neurobiol Dis 28(2):143–153. https://doi.org/10.1016/j.nbd.2007.07.001
Paris D, Patel N, DelleDonne A et al (2004a) Impaired angiogenesis in a transgenic mouse model of cerebral amyloidosis. Neurosci Lett 366(1):80–85. https://doi.org/10.1016/j.neulet.2004.05.017
Paris D, Townsend K, Quadros A et al (2004b) Inhibition of angiogenesis by Aβ peptides. Angiogenesis 7(1):75–85. https://doi.org/10.1023/B:AGEN.0000037335.17717.bf
Patel JR, McCandless EE, Dorsey D et al (2010) CXCR4 promotes differentiation of oligodendrocyte progenitors and remyelination. Proc Natl Acad Sci 107(24):11062–11067. https://doi.org/10.1073/pnas.1006301107
Pei JJ, Hugon J (2008) mTOR-dependent signalling in Alzheimer's disease. J Cell Mol Med 12 (6b):2525–2532. https://doi.org/10.1111/j.1582-4934.2008.00509.x
Plate KH (1999) Mechanisms of angiogenesis in the brain. J Neuropathol Exp Neurol 58 (4):313–320. https://doi.org/10.1097/00005072-199904000-00001
Poewe W, Seppi K, Tanner CM et al (2017) Parkinson disease. Nat Rev Dis Primers 3(1):1–21. https://doi.org/10.1038/nrdp.2017.13
Prins M, Greco T, Alexander D et al (2013) The pathophysiology of traumatic brain injury at a glance. Dis Model Mech 6(6):1307–1315. https://doi.org/10.1242/dmm.011585
Quinsay MN, Thomas RL, Lee Y et al (2010) Bnip3-mediated mitochondrial autophagy is independent of the mitochondrial permeability transition pore. Autophagy 6(7):855–862. https://doi.org/10.4161/auto.6.7.13005

Rajapakse AG, Yepuri G, Carvas JM et al (2011) Hyperactive S6K1 mediates oxidative stress and endothelial dysfunction in aging: inhibition by resveratrol. PLoS One 6(4):e19237. https://doi.org/10.1371/journal.pone.0019237

Rivera DS, Lindsay C, Codocedo JF et al (2016) Andrographolide recovers cognitive impairment in a natural model of Alzheimer's disease (*Octodon degus*). Neurobiol Aging 46:204–220. https://doi.org/10.1016/j.neurobiolaging.2016.06.021

Savica R, Rocca WA, Ahlskog JE (2010) When does Parkinson disease start? Arch Neurol 67 (7):798–801. https://doi.org/10.1001/archneurol.2010.135

Serrano FG, Tapia-Rojas C, Carvajal FJ et al (2014) Andrographolide reduces cognitive impairment in young and mature AβPPswe/PS-1 mice. Mol Neurodegener 9(1):61. https://doi.org/10.1186/1750-1326-9-61

Sharifi-Rad M, Lankatillake C, Dias DA et al (2020) Impact of natural compounds on neurodegenerative disorders: from preclinical to pharmacotherapeutics. J Clin Med 9(4):1061. https://doi.org/10.3390/jcm9041061

Shen CY, Jiang JG, Yang L et al (2017) Anti-ageing active ingredients from herbs and nutraceuticals used in traditional Chinese medicine: pharmacological mechanisms and implications for drug discovery. Br J Pharmacol 174(11):1395–1425. https://doi.org/10.1111/bph.13631

Sheng D, Li J, Wang K et al (2016) Differential *in vitro* and *in vivo* anti-angiogenic activities of acetal and ketal andrographolide derivatives in HUVEC and zebrafish models. RSC Adv 6 (105):102831–102842. https://doi.org/10.1039/C6RA16758F

Shin JW, Lee JK, Lee JE et al (2011) Combined effects of hematopoietic progenitor cell mobilization from bone marrow by granulocyte colony stimulating factor and AMD3100 and chemotaxis into the brain using stromal cell-derived factor-1α in an Alzheimer's disease mouse model. Stem Cells 29(7):1075–1089. https://doi.org/10.1002/stem.659

Steinman L, Zamvil S (2003) Transcriptional analysis of targets in multiple sclerosis. Nat Rev Immunol 3(6):483–492. https://doi.org/10.1038/nri1108

Streit WJ, Mrak RE, Griffin WST (2004) Microglia and neuroinflammation: a pathological perspective. J Neuroinflammation 1(1):1–4. https://doi.org/10.1186/1742-2094-1-14

Streit WJ, Semple-Rowland SL, Hurley SD et al (1998) Cytokine mRNA profiles in contused spinal cord and axotomized facial nucleus suggest a beneficial role for inflammation and gliosis. Exp Neurol 152(1):74–87. https://doi.org/10.1006/exnr.1998.6835

Suebsasana S, Pongnaratorn P, Sattayasai J et al (2009) Analgesic, anti-pyretic, anti-inflammatory and toxic effects of andrographolide derivatives in experimental animals. Arch Pharm Res 32 (9):1191–1200. https://doi.org/10.1007/s12272-009-1902-x

Suganthy N, Ramkumar VS, Pugazhendhi A et al (2018) Biogenic synthesis of gold nanoparticles from *Terminalia arjuna* bark extract: assessment of safety aspects and neuroprotective potential via antioxidant, anticholinesterase, and antiamyloidogenic effects. Environ Sci Pollut Res Int 25 (11):10418–10433. https://doi.org/10.1007/s11356-017-9789-4

Sugawara T, Chan PH (2003) Reactive oxygen radicals and pathogenesis of neuronal death after cerebral ischemia. Antioxid Redox Signal 5(5):597–607. https://doi.org/10.1089/152308603770310266

Sulaiman M, Zakaria Z, Abdul Rahman A et al (2010) Antinociceptive and antiedematogenic activities of andrographolide isolated from *Andrographis paniculata* in animal models. Biol Res Nurs 11(3):293–301. https://doi.org/10.1177/1099800409343311

Tan WD, Liao W, Zhou S et al (2017) Is there a future for andrographolide to be an anti-inflammatory drug? Deciphering its major mechanisms of action. Biochem Pharmacol 139:71–81. https://doi.org/10.1016/j.bcp.2017.03.024

Tansey MG, Goldberg MS (2010) Neuroinflammation in Parkinson's disease: its role in neuronal death and implications for therapeutic intervention. Neurobiol Dis 37(3):510–518. https://doi.org/10.1016/j.nbd.2009.11.004

Tao L, Zhang L, Gao R et al (2018) Andrographolide alleviates acute brain injury in a rat model of traumatic brain injury: possible involvement of inflammatory signaling. Front Neurosci 12:657. https://doi.org/10.3389/fnins.2018.00657

Tapia-Rojas C, Schüller A, Lindsay CB et al (2015) Andrographolide activates the canonical Wnt signalling pathway by a mechanism that implicates the non-ATP competitive inhibition of GSK-3β: autoregulation of GSK-3β *in vivo*. Biochem J 466(2):415–430. https://doi.org/10.1042/BJ20140207

Thakur AK, Rai G, Chatterjee SS et al (2016) Beneficial effects of an *Andrographis paniculata* extract and andrographolide on cognitive functions in streptozotocin-induced diabetic rats. Pharm Biol 54(9):1528–1538. https://doi.org/10.3109/13880209.2015.1107107

Traystman RJ (2003) Animal models of focal and global cerebral ischemia. ILAR J 44(2):85–95. https://doi.org/10.1093/ilar.44.2.85

Vallon M, Chang J, Zhang H et al (2014) Developmental and pathological angiogenesis in the central nervous system. Cell Mol Life Sci 71(18):3489–3506. https://doi.org/10.1007/s00018-014-1625-0

van der Schyf CJ (2011) The use of multi-target drugs in the treatment of neurodegenerative diseases. Expert Rev Clin Pharmacol 4(3):293–298. https://doi.org/10.1586/ecp.11.13

van der Star BJ, Vogel DYS, Kipp M et al (2012) *In vitro* and *in vivo* models of multiple sclerosis. CNS Neurol Disord Drug Targets 11(5):570–588. https://doi.org/10.2174/187152712801661284

Vardhan AN, Senthil Kumar S, Vijay Raghavan R (2018) Neuro vigilance of *Syzygium cumini* plant phytochemicals. Int J Pharm Sci Res 9(3):806–815. https://doi.org/10.26452/ijrps.v9i3.1572

Varela-Nallar L, Arredondo SB, Tapia-Rojas C et al (2015) Andrographolide stimulates neurogenesis in the adult hippocampus. Neural Plast 2015:935403. https://doi.org/10.1155/2015/935403

Wang C, Yu J-T, Miao D et al (2014) Targeting the mTOR signaling network for Alzheimer's disease therapy. Mol Neurobiol 49(1):120–135. https://doi.org/10.1007/s12035-013-8505-8

Wang D-P, Chen S-H, Wang D et al (2020) Neuroprotective effects of andrographolide on chronic cerebral hypoperfusion-induced hippocampal neuronal damage in rats possibly via PTEN/AKT signaling pathway. Acta Histochem 122(3):151514. https://doi.org/10.1016/j.acthis.2020.151514

Wang H-C, Tsay H-S, Shih H-N et al (2018) Andrographolide relieved pathological pain generated by spared nerve injury model in mice. Pharm Biol 56(1):124–131. https://doi.org/10.1080/13880209.2018.1426614

Wang T, Liu B, Zhang W et al (2004) Andrographolide reduces inflammation-mediated dopaminergic neurodegeneration in mesencephalic neuron-glia cultures by inhibiting microglial activation. J Pharmacol Exp Ther 308(3):975–983. https://doi.org/10.1124/jpet.103.059683

Wang Y, Deng Y, Zhou G-Q (2008) SDF-1α/CXCR4-mediated migration of systemically transplanted bone marrow stromal cells towards ischemic brain lesion in a rat model. Brain Res 1195:104–112. https://doi.org/10.1016/j.brainres.2007.11.068

Welcome MO (2020) Neuroinflammation in CNS diseases: molecular mechanisms and the therapeutic potential of plant derived bioactive molecules. Pharma Nutr:100176. https://doi.org/10.1016/j.phanu.2020.100176

Wong DPW, Ng MY, Leung JY et al (2018) Regulation of the Nrf2 transcription factor by andrographolide and organic extracts from plant endophytes. PLoS One 13(10):e0204853. https://doi.org/10.1371/journal.pone.0204853

Wong SY, Tan MG, Wong PT et al (2016) Andrographolide induces Nrf2 and heme oxygenase 1 in astrocytes by activating p38 MAPK and ERK. J Neuroinflammation 13(1):251. https://doi.org/10.1186/s12974-016-0723-3

Wu L, Xiong X, Wu X et al (2020) Targeting oxidative stress and inflammation to prevent ischemia-reperfusion injury. Front Mol Neurosci 13:28. https://doi.org/10.3389/fnmol.2020.00028

Wu Z, Puigserver P, Andersson U et al (1999) Mechanisms controlling mitochondrial biogenesis and respiration through the thermogenic coactivator PGC-1. Cell 98(1):115–124. https://doi.org/10.1016/S0092-8674(00)80611-X

Wullschleger S, Loewith R, Hall MN (2006) TOR signaling in growth and metabolism. Cell 124 (3):471–484. https://doi.org/10.1016/j.cell.2006.01.016

Xia Y-F, Ye B-Q, Li Y-D et al (2004) Andrographolide attenuates inflammation by inhibition of NF-κB activation through covalent modification of reduced cysteine 62 of p50. J Immunol 173 (6):4207–4217. https://doi.org/10.4049/jimmunol.173.6.4207

Xu Y, Wei H, Wang J et al (2019) Synthesis of andrographolide analogues and their neuroprotection and neurite outgrowth-promoting activities. Bioorg Med Chem 27 (11):2209–2219. https://doi.org/10.1016/j.bmc.2019.04.025

Yang C-H, Yen T-L, Hsu C-Y et al (2017) Multi-targeting andrographolide, a novel NF-κB inhibitor, as a potential therapeutic agent for stroke. Int J Mol Sci 18(8):1638. https://doi.org/10.3390/ijms18081638

Yang EJ, Song KS (2014) Andrographolide, a major component of *Andrographis paniculata* leaves, has the neuroprotective effects on glutamate-induced HT22 cell death. J Funct Foods 9:162–172. https://doi.org/10.1016/j.jff.2014.04.023

Yang MY, Yu QL, Huang YS et al (2019) Neuroprotective effects of andrographolide derivative CX-10 in transient focal ischemia in rat: involvement of Nrf2/AE and TLR/NF-κB signaling. Pharm Res 144:227–234. https://doi.org/10.1016/j.phrs.2019.04.023

Yang SP, Bae DG, Kang HJ et al (2004) Co-accumulation of vascular endothelial growth factor with β-amyloid in the brain of patients with Alzheimer's disease. Neurobiol Aging 25 (3):283–290. https://doi.org/10.1016/S0197-4580(03)00111-8

Yen TL, Chen RJ, Jayakumar T et al (2016) Andrographolide stimulates p38 mitogen-activated protein kinase–nuclear factor erythroid-2-related factor 2–heme oxygenase 1 signaling in primary cerebral endothelial cells for definite protection against ischemic stroke in rats. Transl Res 170:57–72. https://doi.org/10.1016/j.trsl.2015.12.002

Yen TL, Hsu WH, Huang SKH et al (2013) A novel bioactivity of andrographolide from *Andrographis paniculata* on cerebral ischemia/reperfusion-induced brain injury through induction of cerebral endothelial cell apoptosis. Pharm Biol 51(9):1150–1157. https://doi.org/10.3109/13880209.2013.782051

Yi Z, Ouyang S, Zhou C et al (2018) Andrographolide inhibits mechanical and thermal Hyperalgesia in a rat model of HIV-induced neuropathic pain. Front Pharmacol 9:593. https://doi.org/10.3389/fphar.2018.00593

Yoopan N, Thisoda P, Rangkadilok N et al (2007) Cardiovascular effects of 14-deoxy-11, 12-didehydroandrographolide and *Andrographis paniculata* extracts. Planta Med 73 (06):503–511. https://doi.org/10.1055/s-2007-967181

Yu AL, Lu CY, Wang TS et al (2010) Induction of heme oxygenase 1 and inhibition of tumor necrosis factor α-induced intercellular adhesion molecule expression by andrographolide in EA. hy926 cells. J Agric Food Chem 58(13):7641–7648. https://doi.org/10.1021/jf101353c

Yu H, Chen ZY (2011) The role of BDNF in depression on the basis of its location in the neural circuitry. Acta Pharmacol Sin 32(1):3–11. https://doi.org/10.1038/aps.2010.184

Zeng M, Jiang W, Tian Y et al (2017) Andrographolide inhibits arrhythmias and is cardioprotective in rabbits. Oncotarget 8(37):61226. https://doi.org/10.18632/oncotarget.18051

Zhang C, Fan L, Fan S et al (2019a) *Cinnamomum cassia* Presl: a review of its traditional uses, phytochemistry, pharmacology and toxicology. Molecules 24(19):3473. https://doi.org/10.3390/molecules24193473

Zhang C, Tan B (1998) Vasorelaxation of rat thoracic aorta caused by 14-deoxyandrographolide. Clin Exp Pharmacol Physiol 25(6):424–429. https://doi.org/10.1111/j.1440-1681.1998.tb02226.x

Zhang H, Wang W, Du Q (2019b) Andrographolide attenuates bupivacaine-induced cytotoxicity in SH-SY5Y cells through preserving Akt/mTOR activity. Drug Des Devel Ther 13:1659–1666. https://doi.org/10.2147/DDDT.S201122

Zhang JJ, Gao TT, Wang Y et al (2019c) Andrographolide exerts significant antidepressant-like effects involving the hippocampal BDNF system in mice. Int J Neuropsychopharmacol 22 (9):585–600. https://doi.org/10.1093/ijnp/pyz032

Zhang Z, Lai D, Wang L et al (2014) Neuroprotective effects of the andrographolide analogue AL-1 in the MPP+/MPTP-induced Parkinson's disease model *in vitro* and in mice. Pharmacol Biochem Behav 122:191–202. https://doi.org/10.1016/j.pbb.2014.03.028

# Ginseng: A Boon or a Curse to Neurodegenerative Diseases

**Sindhu Ramesh, Manoj Govindarajulu, Shriya Patel, Rishi M. Nadar, Mary Fabbrini, Randall C. Clark, Jack Deruiter, Timothy Moore, Dinesh Chandra Agrawal, and Muralikrishnan Dhanasekaran**

## Contents



**Abstract** The herbal medication ginseng has been utilized worldwide for the treatment of a variety of conditions and has currently become readily available for domestic purposes. Acute use of ginseng in recommended dosages is reported by many to be reliable and effective. However, its usefulness and long-term safety in various neurodegenerative diseases remain to be documented. This chapter provides an overview of ginseng and its effect on neurodegenerative diseases. Several constituents of ginseng have been established to show neurotrophic effects, improving

---

Sindhu Ramesh and Manoj Govindarajulu contributed equally to this work.

---

S. Ramesh · M. Govindarajulu · S. Patel · R. M. Nadar · M. Fabbrini · R. C. Clark · J. Deruiter · T. Moore · M. Dhanasekaran (✉)
Department of Drug Discovery and Development, Harrison School of Pharmacy, Auburn University, Auburn, AL, USA
e-mail: dhanamu@auburn.edu

D. C. Agrawal (✉)
Department of Applied Chemistry, Chaoyang University of Technology, Taichung, Taiwan
e-mail: dcagrawal@gm.cyut.edu.tw

 
D. C. Agrawal, M. Dhanasekaran (eds.), *Medicinal Herbs and Fungi*,
https://doi.org/10.1007/978-981-33-4141-8_4

cognition, memory, and learning and neuroprotective actions for the prevention of neurodegenerative diseases. Besides, further investigation of ginseng and its metabolites is necessary to establish a therapeutic outcome. This chapter would provide a theoretical basis for the treatment of neurodegenerative diseases by ginseng and its extracts.

**Keywords** Alzheimer's disease · Ginseng · Ginsenosides · Neurodegenerative diseases · Parkinson's disease · Signaling pathways

## Abbreviations

| | |
|---|---|
| AD | Alzheimer's disease |
| ADAS-Cog | Alzheimer's Disease Assessment Scale Cognitive Subscale |
| AMPA | α-Amino-3-hydroxy-5-methyl-4-isoxazole propionic acid |
| APP | Amyloid precursor protein |
| Aβ | Amyloid-beta |
| BSSG | β-sitosterol β-D-glucoside |
| CAG | Cytosine–adenine–guanine |
| CaMKII | Calcium–calmodulin-*dependent* protein kinase II |
| CCl4 | Carbon tetrachloride |
| ChAT | Choline acetyltransferase |
| CHOP | C/EBP homologous protein |
| CNS | Central nervous system |
| CYP | Cytochromes P450 |
| CYP1A2 | Cytochrome P450 family 1 subfamily A member 2 |
| CYP2B1 | Cytochrome P450, family 2, subfamily B, polypeptide 1 |
| CYP3A23 | Cytochrome P450, family 3, subfamily A, polypeptide 23 |
| CYP3A4 | Cytochrome P450 family 3 subfamily A member 4 |
| ER | Endoplasmic reticulum |
| ERK | Extracellular signal-regulated kinase 1/2 |
| GSK-3 | Glycogen synthase kinase 3 |
| HTT | Huntingtin |
| MAPKs | Mitogen-activated protein kinases |
| MMSE | Mini-mental state examination |
| MPP+ | 1-Methyl-4-phenylpyridinium. |
| MPTP | 1-Methyl-4-phenyl-1,2,3,6-tetrahydropyridine |
| mTOR | Mammalian target of rapamycin |
| NMDAR | N-Methyl-D-aspartate receptor |
| Nrf2 | Nuclear factor erythroid 2-related factor |
| OATP | Organic-anion-transporting polypeptides |
| PERK | Protein kinase R-like endoplasmic reticulum kinase |
| PD | Parkinson's disease |
| PI3K/Akt | Phosphatidylinositol 3-kinase/protein kinase B |
| PKA | Protein kinase A |

| | |
|---|---|
| PNS | *Panax notoginseng* saponins |
| PPARγ | Peroxisome proliferator-activated receptor-γ |
| PPD | Protopanaxadiol |
| PPT | Protopanaxatriol |
| PSD-95 | Postsynaptic density protein 95 |
| RNS | Reactive nitrogen species |
| ROS | Reactive oxygen species |
| SIRT1 | Silent information regulator 1 |
| TrkA | Tropomyosin receptor kinase A |
| UDP-UGT | Uridine 5′-diphospho-glucuronosyltransferase |

## 1 Introduction

The increase in life expectancy has intensified the attention in exploring approaches to prevent age-related diseases. One of the common manifestations in the aging population is memory decline (Small 2002), which can be attributed to altered cellular activity, synaptic dysfunction, and plasticity (Kim et al. 2014a, b). Despite extensive research, there have been limited therapeutic interventions to treat cognitive disorders. Interestingly, herbal medicines offer several benefits in treating aging-related diseases, which has attracted wide consideration. However, the determination of the protective signaling mechanisms involved and potential target(s) of herbal drugs are complex due to the multiple components in these preparations. Plication of modern biological and pharmacological approaches may be of value for determining the underlying biochemical mechanism of herbal medicines.

*Panax ginseng* is a traditional herbal medicine widely used in China, Korea, and Japan (Fig. 1). The word Panax means "all-healing" and indicating that ginseng can heal all illnesses (Lee and Kim 2014). Studies suggest that long-term use of ginseng

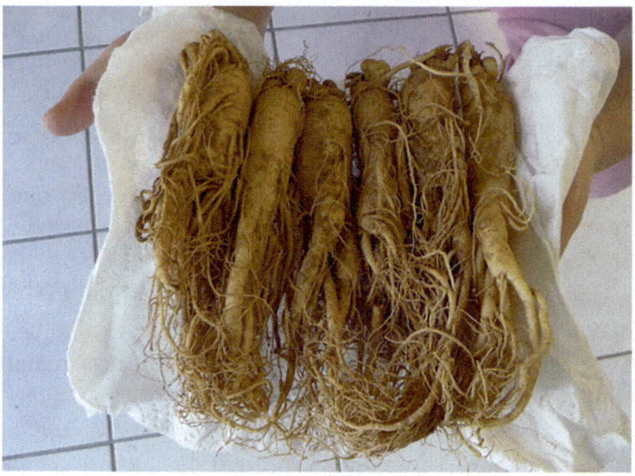

**Fig. 1** "Fresh Korean Ginseng": Attrition: Eugene Kim licensed under CC BY 2.0 (https://www.flickr.com/photos/eekim/4145898809/)

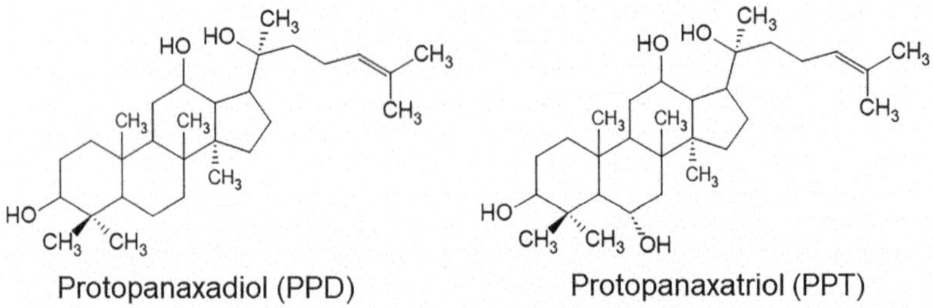


**Fig. 2** The chemical structure of protopanaxadiol (PPD) and protopanaxatriol (PPT)

promotes the health and well-being of aged populations (Yang et al. 2014). Worldwide there are two common forms of ginseng recognized, and they are the Asian ginseng, *Panax ginseng*, and the American ginseng, *Panax quinquefolius* (Qi et al. 2011). *P. ginseng* is a traditional form of herbal medicine that has been shown to have extensive pharmacological activity in the central nervous system and the potential to enhance immune function, concentration, and memory (Qi et al. 2011). It is one of the most popular herbal medicine used in many Asian countries such as China and Japan, dating as far back as 5000 years. It has now become a household item all over the world because of reported benefits, including reduced stress and depression, as well as anti-inflammatory activity. The American ginseng products have not been as extensively investigated as the Asian ginsengs for their therapeutic potential.

The two different varieties—the Asian ginseng and the American ginseng differ in terms of total ginsenoside content as well as composition. The American ginseng contains 40–60 g/kg ginsenoside, while Asian ginseng contains 20–40 g/kg. Rb1, Re, Rd. are the major ginsenosides present in American ginseng, whereas Rb1, Rg1, and Rb2 are the major ginsenosides in Asian ginseng. American ginseng is grown in the temperate regions of North America, from southern Quebec, Minnesota, and Wisconsin in the North, to Oklahoma, the Ozark Plateau, and Georgia in the south (Assinewe et al. 2003).

The main component in *Panax* ginseng is ginsenosides, which are a group of saponins and are considered to be the main active ingredient (Jin et al. 2019). They are classified into two major groups, protopanaxadiols (PPD) and protopanaxatriols (PPT), based on the nature of their core aglycone structures, such as (Fig. 2). Both PPD and PPT have dammarane triterpenoid skeletons with sugar moieties but differ in the degree of substitution at the 6-position (Shibata et al. 1985). The structural diversities of ginsenosides based on a varying degree of glycosylation are responsible for the therapeutic and pharmacological effects upon intake of the root in its various forms (Table 1). Interestingly, PPD- and PPT-type ginsenosides in ginseng may be associated with a biphasic effect that can both stimulate and sedate the CNS. The health benefits of ginsenoside are from its metabolites, which are known to enhance important body mechanisms such as phagocytosis. Phagocytosis is the natural defense mechanisms in the body where cells can ingest microbes or molecules and help to maintain normal healthy tissues. This process results in an

**Table 1** Structures of the ginsenosides

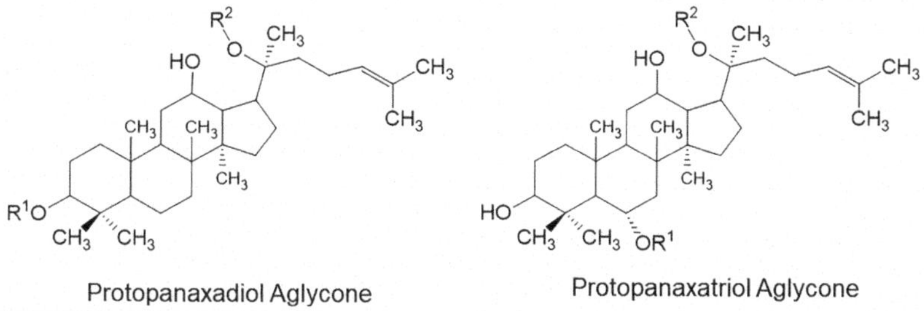


| Aglycone | Ginsenoside | *R*[1] | *R*[2] |
|---|---|---|---|
| PPD | Rb | Glu (1–2)Glu- | Glu (1–6)Glu- |
| PPD | Rd | Glu (1–2)Glu- | Glu- |
| PPD | F2 | Glu- | Glu- |
| PPD | Rg3 | Glu (1–2)Glu- | H- |
| PPD | Rh2 | Glu- | H- |
| PPT | Re | Rha(1–2)Glu- | Glu- |
| PPT | Rg1 | Glu- | Glu- |
| PPT | Rh1 | Glu- | H- |
| PPT | F1 | H- | Glu- |

improved immune system by an efficient defensive mechanism, allowing the body to build resistance to stress (Günther and Seyfert 2018). Ginseng also has been marked to have neuroprotective effects in many neurodegenerative diseases (Alzheimer's disease, Parkinson's, or Huntington's disease) that work to prevent neuronal dysfunction by antioxidant and antiapoptotic effects, which is an important observation when considering treatments for neurodegenerative diseases (Cho 2012).

## 2 Pharmacokinetics

Ginseng's pharmacokinetic profile has not been well characterized due to several chemical components of the ginsenoside and their varying structures. Following oral administration, the absorption rate of ginseng is low, as several studies have shown that high doses are required to detect ginseng levels in the plasma. Several theories have been proposed to account for the low oral bioavailability of ginseng saponins, including low solubility of deglycosylated products (Gu et al. 2009), low enterocyte membrane permeability (Liu et al. 2009), and extensive metabolism in the gastrointestinal tract (Tawab et al. 2003; Cai et al. 2003). Furthermore, the bioavailability of the protopanaxadiol (PPD) group (Li et al. 2007; Wang et al. 2007; Xie et al. 2005; Qian et al. 2005) and protopanaxatriol (PPT) group (Sun et al. 2005; Joo et al. 2010; Lai et al. 2009) is <5%. PPT saponins degrade slower than PPD saponins and hence show better bioavailability than PPD saponins (Xu et al. 2003). Saturated kinetics are noted on high oral doses, and changing the pharmaceutical formulation may also improve bioavailability (Paek et al. 2006).

Several studies involving cell culture, animal, and human studies have studied the impact of ginseng constituents and their metabolites on multiple CYP enzymes (Yu et al. 2005; Henderson et al. 1999; He et al. 2006). However, the in vitro studies about the influence of ginsenoside on CYP3A4 activity have yielded conflicting results with some studies reporting CYP inhibition, other studies noting CYP induction, and still, others claiming no CYP effect. These results were similar to those observed with respect to different CYP isoforms (Henderson et al. 1999; He et al. 2006). The impact of *P. ginseng* on CYP activity has been assessed in several animal models with varying results. CYP activity has been shown to increase after a single dose of *P. ginseng* (10–30 mg/kg) in adult male rats (Lee et al. 1987).

Similarly, a 20 mg/kg oral dose or a 100 mg/kg oral or the intraperitoneal dose of *P. ginseng* for 4 days did not alter hepatic CYP2B1, CYP3A23, and CYP1A2 in adult male Sprague–Dawley rats (Yu et al. 2005). Furthermore, the effect of *Panax notoginseng* saponins (PNS) on CYP activity using specific probe drugs such as caffeine was investigated. Administration of PNS increased clearance of caffeine, which could be attributed to PNS induced CYP1A2 activity and protein expression. Intriguingly, the protein content or the activity of other CYP enzymes were not altered by PNS administration (Liu et al. 2012). Currently, it is not possible to apply the results from these studies to the clinical setting due to conflicting data from in vitro and in vivo studies assessing the effect of *P ginseng* on CYP activity.

UDP-glucuronosyltransferase (UGT) enzymes UGTs are involved in drug metabolism as well as conjugation of endogenous substances, namely bilirubin, bile acids, and fat-soluble vitamins (Ritter 2000). Studies demonstrate that medications that are metabolized by UGT isoforms (UGT1A1 and UGT2B7) show elevated drug concentrations when concomitantly treated with ginseng (He et al. 2013). In contrast, human liver microsomes treated with *P. ginseng* did not show significant inhibition of UGT isoforms (Mohamed and Frye 2011). Furthermore, the administration of ginsenoside Rg3 increased the bioavailability of co-administrated substrates, which are metabolized by UGT isoforms (Fang et al. 2013). Ginseng and its metabolites may alter Phase 1 and/or Phase 2 metabolic activity through either its additive, synergistic, or antagonistic effects. However, evaluating the ginsenosides and their metabolites' role in regulating drug metabolism is not likely to give an accurate depiction of the true drug interaction potential of *P. ginseng* administration in a clinical setting.

Drug transport proteins such as P-glycoprotein (P-gp), Organo Anion Transporter (OAT) Family, and others are important in the uptake or efflux of drugs in several tissues. *P. ginseng* and/or its metabolites have been shown to influence drug transport proteins such as P-gp and OAT family (Ramanathan and Penzak 2017). P-glycoprotein (P-gp), the major transport protein, and the effect of ginseng and/or its metabolites have been extensively investigated (Choi et al. 2003; Li et al. 2014; You et al. 2015). Specifically, ginseng extract and ginsenosides have been shown to inhibit P-gp activity and expression (Choi et al. 2003; Li et al. 2014). With respect to the OAT family, ginsenosides (Rb1, Rc, and Rd) inhibited OATP1B3 and OATP1BI significantly, resulting in elevated concentrations of substrates resulting in toxicity (Jeon et al. 2020). Carefully monitored clinical studies are required to evaluate this alleged drug interaction further.

The biliary system utilizes active transport to excrete most ginsenosides and their deglycosylated products. (Liu et al. 2009). After an oral administration, variable bioavailability has been demonstrated, indicating the involvement of enterohepatic recirculation (Paek et al. 2006). Finally, around 0.2–1.2% of ginsenosides are excreted in human urine (Cui et al. 1997).

## 3 Effects of Ginseng on the CNS

Evidence for the therapeutic benefits of *P. ginseng* has significantly expanded over the past decade. Specifically, the neuroprotective effects of ginsenosides have gained widespread interest with respect to its utility in mediating several intracellular neuronal signaling pathways important in regulating cell survival and death. The following sections describe the effects of ginseng on various neurodegenerative diseases such as Alzheimer's disease, Parkinson's disease, multiple sclerosis, and Huntington's disease.

### *3.1 Neuroprotective Effects of Ginseng on Alzheimer's Disease*

Alzheimer's disease (AD), the most cause of dementia worldwide, is a neurodegenerative disorder characterized by the accumulation of extracellular amyloid plaques and neurofibrillary tangles in the form of hyperphosphorylated tau. Furthermore, several other pathologies such as mitochondrial dysfunction (Ramesh et al. 2018a), oxidative stress (Tönnies and Trushina 2017), neuroinflammation (Hampel et al. 2020), excitotoxicity and calcium dyshomeostasis (Hynd et al. 2004), brain energy dysfunction (Yin et al. 2016), and so on, are associated with AD.

### *3.2 Effect of Ginseng on Cognition*

Several studies have shown that ginseng and its constituents improve cognition in AD patients as assessed by standard cognitive tests such as mini-mental state examination (MMSE) and Alzheimer's Disease Assessment Scale Cognitive Subscale (ADAS-Cog) tests (Heo et al. 2008; Heo et al. 2011). A recent study by Lho et al. researched the effects of lifetime cumulative ginseng intake on late-life cognitive function. This study concluded that ginseng use for 5 years could be favorable in cognitive function in late life (Lho et al. 2018), similar to the results published from several other studies.

Gintonin treatment in cognitively impaired patients improved scores in MMSE and ADAS-Cog, thereby indicating gintonin potential as an anti-AD drug. Interestingly, the adverse effects typically are seen with whole ginseng extracts, such as flushing, dizziness, and gastrointestinal symptoms were not seen with gintonin (Moon et al. 2018).

Similarly, animal models of AD have shown the cognitive-enhancing effects of ginseng. For instance, ginseng extract improved memory and attenuated Aβ levels in an AD mouse model (Kim et al. 2013).

Mice injected with oligomeric Aβ injection into the hippocampus manifest with memory dysfunction, and this effect was mitigated by ginseng extract (Lee et al. 2011). Another study showed ginseng extract improved synaptophysin and choline acetyltransferase activity in Aβ oligomer injected mice (Choi et al. 2017). These studies indicate that ginseng and its constituents improve cognitive deficits in AD by acting through various signaling pathways such as prevent amyloid formation, interfering with tau hyperphosphorylation, attenuating oxidative stress and inflammation, or by modulating neurotransmission. The further sections describe the effects of ginseng on several pathologies of AD.

## 3.3 *Effect of Ginseng on Amyloid and* tau *Pathology*

The amyloid precursor protein (APP) belongs to a family of conserved type I membrane proteins and has a large extracellular and a short cytoplasmic domain. The processing of APP involves two pathways. The first pathway, termed as non-amyloidogenic pathway mediated by α- and γ-secretases, involves the generation of sAPPα, which is considered to be beneficial to the neurons (Ramesh et al. 2018b). On the contrary, the amyloidogenic pathway involving β- and γ-secretase results in the generation of toxic Aβ, which aggregate and form amyloid plaques (Hooper 2005). Hence, any efforts to inhibit β- and γ-secretase or activate α-secretase could act as a therapeutic strategy to attenuate Aβ formation.

Ginseng and its constituents have been shown to inhibit amyloidogenesis by modulating the amyloidogenic pathway. In in vitro and molecular docking studies have shown ginsenosides (Rb1 and Rb2) to exhibit inhibitory activity against β-secretase (Choi et al. 2016; Karpagam et al. 2013). Similarly, ginsenoside Re reduced Aβ levels by inhibiting β-secretase (mRNA, protein expression, and activity) in N2a/APP695 cells. Furthermore, the β-secretase inhibitory effect was mediated via activation of nuclear peroxisome proliferator-activated receptor-γ (PPARγ) (Cao et al. 2016). Interestingly, ginsenoside Rg1 treatment in in vitro (B103-APP cells) and in vivo (Tg mAPP mouse) inhibited γ-secretase activity through activation of protein kinase A/CREB (cAMP response element-binding protein) pathway (Fang et al. 2012). Gintonin, a derivative of ginseng, has been shown to act through lysophosphatidic acid receptor and facilitate sAPPα production by increasing α-secretase activity (Hwang et al. 2012b). Similarly, in the same study, gintonin reduced Aβ formation and attenuated Aβ-induced neurotoxicity in ShSy5y neuroblastoma cells indicating the potential anti-Aβ effect of gintonin.

The anti-Aβ effects of ginseng were investigated in several animal models. Ginsenoside Rh2 treatment improved memory and decreased the size of Aβ plaques in a transgenic mouse model of AD (Hou et al. 2013). Another study utilizing D-galactose/AlCl3 induced AD rats showed improved memory, reduction of Aβ, and p-tau. These effects were attributed to the activation of PI3K/Akt signaling (Li et al. 2016). Similar to cell culture studies, gintonin administration improved behavioral and cognitive deficits, decreased amyloid plaque deposits in a mouse model of AD (Hwang et al. 2012a). These in vitro and in vivo studies indicate that ginseng and its constituents could attenuate Aβ pathology by modulating various signaling mechanisms.

Only limited studies have been done to investigate the effects of ginseng on tau hyperphosphorylation. For instance, Aβ induced tau hyperphosphorylation was attenuated by treatment with ginsenoside Rb1, and this effect was mediated by inhibition of GSK-3β (Zhao et al. 2013). Similarly, okadaic acid tau hyperphosphorylation was attenuated by ginsenoside Rd. in cultured cortical neurons (Li et al. 2011).

### 3.4 *Effect of Ginseng on Neurotransmission*

The two major neurotransmitter abnormalities noted in AD include the cholinergic and glutamatergic systems. In AD patients and animal models of AD, reduced acetylcholine levels in the brain are noted, which could result from a decrease in choline acetyltransferase (ChAT), a major enzyme involved in acetylcholine synthesis, or an increase in acetylcholinesterase, an enzyme involved in degrading acetylcholine.

Similarly, alterations in the glutamatergic system have been noted in AD. Neuronal homeostasis and synaptic plasticity are maintained by excitatory glutamatergic neurotransmission mediated by N-methyl-D-aspartate receptor (NMDAR). During the initial stages of AD, excessive NMDAR activity mediated by Aβ leads to excitotoxicity and neuronal death. The NMDAR's located in the synaptic region (also termed as synaptic NMDAR) leads to synaptic plasticity changes and promotes cell survival. However, activation of NMDARs located in the extrasynaptic space (extrasynaptic NMDARs) promotes cell death. As the disease progresses, synaptic depression, a phenomenon noted due to a decrease in the glutamate receptors in the synapses and eventual synaptic loss, occurs in AD. Amyloid-beta has been shown to induce endocytosis of AMPA-type glutamate receptors. The alterations in the glutamate receptor trafficking and functioning lead to impaired ion channel functioning leading to altered action potential changes in the neurons. Furthermore, downstream signaling is impaired, leading to long-term changes in the synapses.

Ginseng and its derivatives have been shown to alter both cholinergic and glutamatergic neurotransmission by modulating several pathways. An overview of ginseng and its constituents on various signaling pathways in cholinergic and glutamatergic neurotransmission is reviewed in Table 2.

### 3.5 *Effect of Ginseng on Oxidative Stress and Neuroinflammation*

Normal neuronal homeostasis is characterized by a balance between reactive species (oxygen/nitrogen) and the antioxidant defense system. Oxidative stress occurs when there is an imbalance between reactive oxygen species/reactive nitrogen species (ROS/RNS) and the antioxidant system (Uttara et al. 2009). Markers of oxidative stress (oxidized proteins, lipid peroxidation, free protein carbonyls, oxidative modifications in RNA, and DNA) have been noted in postmortem brains from patients with preclinical or early stages of AD (Sultana et al. 2011). Conversely, the levels of antioxidants are also reduced in affected brain regions of AD patients (Benzi and Moretti 1995; Sultana et al. 2008).

Several studies indicate that immune-mediated alterations in the brain characterize AD. For instance, Aβ deposition bind to several receptors on the glial cells such as microglia and astrocytes, eliciting an immune response. This immune response over a protracted period leads to pro-inflammatory mediators, which induces neuroinflammation. The effect of ginseng constituents on oxidative stress and neuroinflammation, apoptosis, and neurogenesis with respect to AD has been summarized in Table 3.

## 4 Ginseng in Parkinson's Disease

Parkinson's disease is another major neurodegenerative disorder characterized by movement problems such as tremor, rigidity, and bradykinesia. This is attributed to the loss of dopaminergic neurons in the basal ganglia, specifically substantia nigra pars compacta (Sherer et al. 2001).

The effects of ginseng in attenuating the pathologies of PD has been investigated in several animal models of PD. In β-sitosterol β-D-glucoside (BSSG) induced PD animal model, ginseng extract G115 attenuated microglial activation, α-synuclein accumulation, and dopaminergic cell loss. Also, improvement in locomotor deficits was noted (Van Kampen et al. 2014). Similarly, in 1-methyl-4-phenyl-1,2,3,6-tetrahydropyridine (MPTP) mice, a valid animal model of PD, treatment with G115 improved locomotor function and prevented tyrosine hydroxylase-positive

**Table 2** Effect of Ginseng on neurotransmission

| Pathology | Specific constituent of Ginseng | Effects | Signaling mechanism | References |
|---|---|---|---|---|
| Glutamatergic neurotransmission | Ginsenoside Rb1 | Increase glutamate release | PKA signaling pathway | Liu et al. (2010) |
| | Ginsenoside Rg1 | Increase glutamate release | CaMKII dependent | Liu et al. (2010) |
| | Gintonin | Increase glutamate release | LPA1 receptor activation | Nah (2012) |
| | Ginsenoside Rg1 | Increase in synaptophysin, NMDAR, PSD95, and CaMKIIα | mTOR pathway | Yang et al. (2014) |
| | Gintonin | Synaptic enhancement | Activation of the LPA1 receptor | Park et al. (2015) |
| | Notoginsenoside R1 | Inhibits voltage-gated K+ currents | N/A | Meng et al. (2014) |
| | Ginsenoside Rb1 | Inhibits L-type voltage-gated calcium channels | N/A | Lin et al. (2012b) |
| Cholinergic neurotransmission | Ginsenoside Rd and Re | Increase the expression of choline acetyltransferase (ChAT).<br>Increase vesicular acetylcholine transporter | Cholinergic neurotransmission | Kim et al. (2014a) |
| | Ginsenoside Rb1 | Increased ChAT, microtubule-associated protein-2, and synaptophysin | Cholinergic neurotransmission | Shin et al. (2016) |
| | Gintonin | Increased ChAT expression in the hippocampus of an AD mouse model<br>Reduced acetylcholinesterase | Cholinergic neurotransmission | Kim et al. (2015) |
| | Ginsenoside Rd. and Re | Increase microtubule-associated protein-2 expression, nerve growth factor receptor (p75), p21, and TrkA (tropomyosin receptor kinase A) | TrkA signaling pathway | Kim et al. (2014a) |
| | Gintonin | Differentiate hippocampal neural progenitor cells into cholinergic cells | N/A | Kim et al. (2015) |

**Table 3** Effect of ginseng on other pathologies in AD

| Pathology | Specific constituent of ginseng | Effects | Signaling mechanism | References |
|---|---|---|---|---|
| Oxidative stress and inflammation | Protopanaxtriol | Increase in antioxidant enzymes | Increased Nrf2 | Gao et al. (2015) |
| | Notoginsenoside R1 | Increase in antioxidant enzymes | Increased Nrf2 | Meng et al. (2014) |
| | Ginsenoside Rb1 or Rg1 | A decrease in ROS, lipid peroxidation, LDH release, and MDA production | Scavengers of ROS, MAPK/ERK, and PI3K/Akt pathway | Shi et al. (2012); Wu et al. (2016) |
| | Ginsenoside Rg1 | Increase in mitochondrial membrane potential (MMP), ATP, and cytochrome C activity | Suppressed intracellular mitochondrial oxidative stress | Huang et al. (2012) |
| | Ginsenoside Rh3 | Inhibition of NF-κB | Increased Nrf2 and SIRT1 | Lee et al. (2015) |
| | Ginsenoside Rh1 | Inhibit pro-inflammatory cytokines and increase in anti-inflammatory cytokines | PKA and heme oxygenase-1 | Lee et al. (2015) |
| | Ginsenoside Rb1 | Decrease in neuroinflammatory markers | Direct and indirect effects | Wang et al. (2011) |
| | Ginsenoside Rh3 | Microglial phagocytosis of Aβ and clearance | Upregulation of type A macrophage scavenger receptor | Wang et al. (2011) |
| | Gintonin | Decrease microglia activation by inhibiting expression of allograft inflammatory factor I | Inhibits microglial activation and promotes microglial phagocytosis | Hwang et al. (2012a) |
| | Gintonin | Inhibition of LPS induced pro-inflammatory cytokines | Inhibition of MAPK and NF-kB pathway | Saba et al. (2015) |
| Apoptosis | Ginsenoside Rg | Reduction of calpain I and caspase-3 activity | N/A | He et al. (2012) |
| | Ginsenoside Rb1 | Decreased ER stress proteins-PERK and CHOP | Inhibition of ER stress-mediated apoptosis | Liu et al. (2013) |

(continued)

**Table 3** (continued)

| Pathology | Specific constituent of ginseng | Effects | Signaling mechanism | References |
|---|---|---|---|---|
| Neurodegeneration | Ginsenoside Rb1 | Increase in neurotrophin expression | N/A | Liu et al. (2013) |
| | Ginsenoside Rd | Increase in neural stem cell proliferation and neurogenesis | N/A | Wang et al. (2013); Lin et al. (2012a) |
| | Gintonin | Hippocampal neural progenitor proliferation | LPA receptor | Kim et al. (2016b) |
| | Panax notoginseng saponins | Decrease neurite inhibitory molecules Nogo-A and NgR | N/A | Liu et al. (2014) |

cell loss in the substantia nigra (Van Kampen et al. 2003). High dose administration of carbon tetrachloride (CCl4) induced loss of nigrostriatal dopaminergic nerve cells, which was improved by the administration of ginsenosides Rd. and Re (Zhang et al. 2016). Ginsenoside Rg1 reduced dopaminergic neuronal loss through the Wnt/β-catenin pathway (Zhou et al. 2016) or by modulating the glucocorticoid receptor signaling (Sun et al. 2016). Proteomic analysis of mouse model of PD treated with Korean Red Ginseng showed several proteins involved in glycolysis, gluconeogenesis, and oxidative phosphorylation to be altered (Kim et al. 2016a). Furthermore, Korean Red Ginseng repressed MPTP-induced behavioral dysfunction and nigrostriatal cell death (Ryu et al. 2018). Ginseng extract administration in a rotenone-based rat model of PD showed therapeutic benefit indicating that ginseng could be beneficial in attenuating pathologies in PD (Zaafan et al. 2019). The results from animal models of PD are supplemented with in vitro studies.

MPP+ induced apoptosis in dopaminergic neuronal (SH-SY5Y) cells was prevented by ginseng extracts treatment. This effect was mediated by a reduction in the levels of Bax, Bcl-2, cytochrome c, and cleaved caspase-3 apoptotic proteins (Hu et al. 2011). A similar study utilizing MPP+ induced apoptosis in PC12 neuroblastoma cells showed decreased Bcl-xL and cleaved caspase-3 expression after ginsenoside Rg1 treatment (Zhou et al. 2016). The mechanism by which ginseng offers neuroprotective benefit might be due to its potent antioxidant effects and improvement of mitochondrial function (Liu et al. 2015). Although there is a wide range of cell culture and animal studies done to determine the utility of ginseng in PD, clinical trials in humans need to be done to investigate the efficacy and safety of ginseng in humans.

## 5 Ginseng in Huntington's Disease

Huntington's disease (HD) is a fatal dominantly inherited neurodegenerative disease characterized by progressive neurodegeneration. It is caused by repeated cytosine–adenine–guanine (CAG) triplet expansion in the huntingtin (HTT) gene on chromosome 4. This results in abnormal protein accumulation, which is toxic to neurons. These patients manifest with abnormal dance-like body movements (chorea), mood disorders, and cognitive impairments (Rüb et al. 2015).

Gintonin administrated in a cellular and animal model of HD alleviated mitochondrial dysfunction, apoptosis, microglial activation, and inflammation. These effects were attributed to the stimulation of LPA receptors and nuclear factor erythroid 2-related factor 2 (Nrf2) signaling pathway. Interestingly, gintonin inhibited mitogen-activated protein kinases (MAPKs) signaling and nuclear factor-κB (NF-κB) signaling pathways. Furthermore, gintonin attenuated neurological deficits in mice overexpressing mutant HTT in the striatum (Jang et al. 2019). The administration of 3-nitropropionic acid in rodents is considered a model to mimic HD characteristics. Ginseng extract protopanaxatriol treatment in rats injected with 3-nitropropionic acid prevented oxidative stress through Nrf2 signaling (Gao et al. 2015). Finally, an in vitro model of HD utilizing glutamate for inducing apoptosis showed attenuation of apoptosis in neurons with ginsenoside treatment (Wu et al. 2009).

## 6 Other Neurodegenerative Diseases

The neuroprotective effects of ginsenosides in an animal model of multiple sclerosis termed experimental autoimmune encephalomyelitis (EAE) was evaluated. Ginsenoside Rd. given intraperitoneally (40 mg/kg/day) in EAE mice, improved the severity of EAE. This improvement was attributed to the effect of ginsenoside on cytokines, specifically interferon-gamma and interleukin-4 (IL-4) (Zhu et al. 2014). In another study, EAE mice treated with Panax quinquefolium (American ginseng) at a dose of 150 mg/kg resulted in the improvement of demyelination scores (Bowie et al. 2012).

Finally, ginseng treatment in an experimental amyotrophic lateral sclerosis model improved motor impairment and survival (Jiang et al. 2000). These studies indicate that ginseng could offer therapeutic benefits in other neurodegenerative diseases. However, more mechanistic studies need to be performed to validate the findings for its effectiveness.

## 7 Adverse Effects and Toxicity of Ginseng

Several clinical trials and case reports in humans have addressed the adverse events and toxicities associated with ginseng use. Some of the mild adverse effects associated with ginseng include gastrointestinal symptoms such as diarrhea and loss of appetite; CNS symptoms such as nervousness, insomnia, and depression; cardiovascular such as edema, hypertension, and hypotension; and skin eruptions (Siegel 1979). Most of these symptoms are attributed to high dosage and long-term administration. Human studies report that a comparatively small incidence of toxicity has been correlated with ginseng application (Chang et al. 2003). Furthermore, other studies report that *P. ginseng* is occasionally associated with adverse drug interactions (Coon and Ernst 2002). Generally, it is thought that ginseng is a safe and nontoxic substance even though the safety issues of which remain to be investigated.

Several instances of anaphylactic reactions have been reported with ginseng use. Asian ginseng caused respiratory distress, angioedema, skin eruptions, and hypotension in a 20-year-old patient (Wiwanitkit and Taungjaruwinai 2004). Similarly, respiratory distress (Lee et al. 2006) and occupational asthma have been reported in individuals exposed to ginseng (Kim et al. 2008).

Toxicities due to long term abuse of ginseng have been reported. Cardiac arrhythmias in the form of *torsades de pointes* have been reported in a patient without preexisting cardiovascular risk factors on long term ginseng intake (Torbey et al. 2011). The risk of cardiovascular disease is increased by chronic renal insufficiency (Go et al. 2004). Atrial fibrillation in an elderly patient with preexisting renal disease was noted after 1 week of ginseng administration (Liao et al. 2010). Therefore, it is required for such patients to reveal idiosyncrasy toward drugs and particular food to use ginseng with great vigilance. Ginseng should be avoided in patients who have hypertension, hypotension, and cardiovascular disorders.

Gynecomastia has been reported following ginseng use (Palop et al. 1999; Kakisaka et al. 2012). Uterine bleeding and menometrorrhagia have resulted following misuse of ginseng in middle-aged women (Palop-Larrea et al. 2000; Kabalak et al. 2004). These effects are attributed to the estrogen-like effects of ginseng. Since ginsenosides structurally resemble estradiol, ginsenosides bind to estrogen receptors and cause estrogen-like effects. Based on the above pharmacological action of *P. ginseng,* it can be used as an alternative therapy in treating postmenopausal signs and symptoms while its effects are still contentious. (Cheema et al. 2007; Wong et al. 2009). Furthermore, ginseng is contraindicated in breast and uterine cancer and endometriosis. *Panax ginseng* should be consumed under caution during pregnancy, especially during the first trimester and during lactation (Seely et al. 2008; Holst et al. 2008).

Concerning drug interactions and its associated toxicity, hepatotoxicity with concomitant use of ginseng along with imatinib has been reported. Since ginseng can act as an inhibitor of CYP3A4, the primary enzyme, which is involved in the metabolism of imatinib, the plasma concentrations of imatinib increased, resulting in hepatotoxicity. On the contrary, ginseng can interact with warfarin and reduce the anticoagulant effects of warfarin. This is attributed to the induction of CYP450 enzyme activity and thereby warfarin elimination. (Yuan et al. 2004). The differences in the ability of ginseng to either induce or inhibit the CYP family of metabolizing enzymes could be due to the type of ginseng constituents and the clinical settings. Additional research needs to be conducted to evaluate the practical significance of these prospective interactions. Further analysis, in more cases, is necessary to explain these above-mentioned adverse effects and toxicities.

## 8 Conclusion

Ginseng has been shown to exert distinct pharmacological actions in the CNS. Specifically, ginseng shows beneficial effects in neurodegenerative diseases through its impact on neurotransmission, oxidative stress, mitochondrial function, neuroinflammation, apoptosis, and other intracellular signaling mechanisms. Further studies to determine the structure–activity relationship and target specificity, pharmacokinetic, and toxicological studies need to be performed. The pharmacodynamic and pharmacokinetic effects of ginseng in neurodegenerative diseases have predominantly focused on AD based on animal and human studies. More studies to study the efficacy and safety of ginseng in other neurodegenerative disorders needs to be performed. Moreover, the adverse effects associated with ginseng abuse and misuse needs to be further investigated to understand the mechanisms related to the cause.

## References

Assinewe VA, Baum BR, Gagnon D, Arnason JT (2003) Phytochemistry of wild populations of Panax quinquefolius L. (north American ginseng). J Agric Food Chem 51(16):4549–4553. https://doi.org/10.1021/jf030042h

Benzi G, Moretti A (1995) Age-and peroxidative stress-related modifications of the cerebral enzymatic activities linked to mitochondria and the glutathione system. Free Radic Biol Med 19(1):77–101. https://doi.org/10.1016/0891-5849(94)00244-e

Bowie LE, Roscoe WA, Lui EM, Smith R, Karlik SJ (2012) Effects of an aqueous extract of north American ginseng on MOG (35–55)-induced EAE in mice. Can J Physiol Pharmacol 90 (7):933–939. https://doi.org/10.1139/y2012-092

Cai ZW, Qian TX, Wong RNS, Jiang ZH (2003) Liquid chromatography–electrospray ionization mass spectrometry for metabolism and pharmacokinetic studies of ginsenoside Rg3. Anal Chim Acta 492(1–2):283–293. https://doi.org/10.1016/S0003-2670(03)00719-0

Cao G, Su P, Zhang S, Guo L, Zhang H, Liang Y, Qin C, Zhang W (2016) Ginsenoside Re reduces Aβ production by activating PPARγ to inhibit BACE1 in N2a/APP695 cells. Eur J Pharmacol 793:101–108

Chang YS, Seo EK, Gyllenhaal C, Block KI (2003) Panax ginseng: a role in cancer therapy? Integr Cancer Ther 2(1):13–33. https://doi.org/10.1177/1534735403251167

Cheema D, Coomarasamy A, El-Toukhy T (2007) Non-hormonal therapy of post-menopausal vasomotor symptoms: a structured evidence-based review. Arch Gynecol Obstet 276 (5):463–469. https://doi.org/10.1007/s00404-007-0390-9

Cho IH (2012) Effects of Panax ginseng in neurodegenerative diseases. J Ginseng Res 36 (4):342–353. https://doi.org/10.5142/jgr.2012.36.4.342

Choi CH, Kang G, Min YD (2003) Reversal of P-glycoprotein-mediated multidrug resistance by protopanaxatriol ginsenosides from Korean red ginseng. Planta Med 69(03):235–240. https://doi.org/10.1055/s-2003-38483

Choi JG, Kim N, Huh E et al (2017) White ginseng protects mouse hippocampal cells against amyloid-beta oligomer toxicity. Phytother Res 31(3):497–506. https://doi.org/10.1002/ptr.5776

Choi RJ, Roy A, Jung HJ et al (2016) BACE1 molecular docking and anti-Alzheimer's disease activities of ginsenosides. J Ethnopharmacol 190:219–230. https://doi.org/10.1016/j.jep.2016.06.013

Coon JT, Ernst E (2002) Panax ginseng: a systematic review of adverse effects and drug interactions. Drug Saf 25(5):323–344. https://doi.org/10.2165/00002018-200225050-00003

Cui J-F, Björkhem I, Eneroth P (1997) Gas chromatographic-mass spectrometric determination of 20 (S)-protopanaxadiol and 20 (S)-protopanaxatriol for study on human urinary excretion of ginsenosides after ingestion of ginseng preparations. J Chromatogr B Biomed Sci Appl 689 (2):349–355. https://doi.org/10.1016/s0378-4347(96)00304-0

Fang F, Chen X, Huang T et al (2012) Multi-faced neuroprotective effects of Ginsenoside Rg1 in an Alzheimer mouse model. Biochim Biophys Acta 1822(2):286–292. https://doi.org/10.1016/j.bbadis.2011.10.004

Fang ZZ, Cao YF, Hu CM et al (2013) Structure–inhibition relationship of ginsenosides towards UDP-glucuronosyltransferases (UGTs). Toxicol Appl Pharmacol 267(2):149–154. https://doi.org/10.1016/j.taap.2012.12.019

Gao Y, Chu S, Li JP et al (2015) Protopanaxtriol protects against 3-nitropropionic acid-induced oxidative stress in a rat model of Huntington's disease. Acta Pharmacol Sin 36(3):311–322. https://doi.org/10.1038/aps.2014.107

Go AS, Chertow GM, Fan D et al (2004) Chronic kidney disease and the risks of death, cardiovascular events, and hospitalization. N Engl J Med 351(13):1296–1305. https://doi.org/10.1056/NEJMoa041031

Gu Y, Wang GJ, Sun JG et al (2009) Pharmacokinetic characterization of ginsenoside Rh2, an anticancer nutrient from ginseng, in rats and dogs. Food Chem Toxicol 47(9):2257–2268. https://doi.org/10.1016/j.fct.2009.06.013

Günther J, Seyfert H-M (2018) The first line of defence: insights into mechanisms and relevance of phagocytosis in epithelial cells. Semin Immunopathol 40(6):555–565. https://doi.org/10.1007/s00281-018-0701-1

Hampel H, Caraci F, Cuello AC et al (2020) A path toward precision medicine for neuroinflammatory mechanisms in Alzheimer's disease. Front Immunol 11:456–456. https://doi.org/10.3389/fimmu.2020.00456

He B, Chen P, Yang J et al (2012) Neuroprotective effect of 20 (R)-ginsenoside Rg3 against transient focal cerebral ischemia in rats. Neurosci Lett 526(2):106–111. https://doi.org/10.1016/j.neulet.2012.08.022

He N, Xie HG, Collins X et al (2006) Effects of individual ginsenosides, ginkgolides and flavonoids on CYP2C19 and CYP2D6 activity in human liver microsomes. Clin Exp Pharmacol Physiol 33 (9):813–815. https://doi.org/10.1111/j.1440-1681.2006.04445.x

He YJ, Fang ZZ, Ge GB, Jiang P, Jin HZ, Zhang WD, Yang L (2013) The inhibitory effect of 20 (S)-Protopanaxatriol (ppt) towards UGT1A1 and UGT2B7. Phytother Res 27(4):628–632. https://doi.org/10.1002/ptr.4755

Henderson G, Harkey M, Gershwin ME et al (1999) Effects of ginseng components on c-DNA-expressed cytochrome P450 enzyme catalytic activity. Life Sci 65(15):PL209–PL214. https://doi.org/10.1016/s0024-3205(99)00407-5

Heo JH, Lee ST, Chu K, Oh MJ, Park HJ, Shim JY, Kim M (2008) An open-label trial of Korean red ginseng as an adjuvant treatment for cognitive impairment in patients with Alzheimer's disease. Eur J Neurol 15(8):865–868. https://doi.org/10.1111/j.1468-1331.2008.02157.x

Heo JH, Lee ST, Oh MJ et al (2011) Improvement of cognitive deficit in Alzheimer's disease patients by long term treatment with Korean red ginseng. J Ginseng Res 35(4):457–461. https://doi.org/10.5142/jgr.2011.35.4.457

Holst L, Nordeng H, Haavik S (2008) Use of herbal drugs during early pregnancy in relation to maternal characteristics and pregnancy outcome. Pharmacoepidemiol Drug Saf 17(2):151–159. https://doi.org/10.1002/pds.1527

Hooper NM (2005) Roles of proteolysis and lipid rafts in the processing of the amyloid precursor protein and prion protein. Biochem Soc Trans 33(Pt 2):335–338. https://doi.org/10.1042/BST0330335

Hou J, Xue J, Lee M et al (2013) Ginsenoside Rh2 improves learning and memory in mice. J Med Food 16(8):772–776. https://doi.org/10.1089/jmf.2012.2564

Hu S, Han R, Mak S, Han Y (2011) Protection against 1-methyl-4-phenylpyridinium ion (MPP+)-induced apoptosis by water extract of ginseng (Panax ginseng CA Meyer) in SH-SY5Y cells. J Ethnopharmacol 135(1):34–42. https://doi.org/10.1016/j.jep.2011.02.017

Huang T, Fang F, Chen L et al (2012) Ginsenoside Rg1 attenuates oligomeric Aβ1-42-induced mitochondrial dysfunction. Curr Alzheimer Res 9(3):388–395. https://doi.org/10.2174/156720512800107636

Hwang SH, Shin EJ, Shin TJ et al (2012a) Gintonin, a ginseng-derived lysophosphatidic acid receptor ligand, attenuates Alzheimer's disease-related neuropathies: involvement of non-amyloidogenic processing. J Alzheimers Dis 31(1):207–223. https://doi.org/10.3233/JAD-2012-120439

Hwang SH, Shin TJ, Choi SH et al (2012b) Gintonin, newly identified compounds from ginseng, is novel lysophosphatidic acids-protein complexes and activates G protein-coupled lysophosphatidic acid receptors with high affinity. Mol Cells 33(2):151–162. https://doi.org/10.1007/s10059-012-2216-z

Hynd MR, Scott HL, Dodd PR (2004) Glutamate-mediated excitotoxicity and neurodegeneration in Alzheimer's disease. Neurochem Int 45(5):583–595. https://doi.org/10.1016/j.neuint.2004.03.007

Jang M, Choi JH, Chang Y et al (2019) Gintonin, a ginseng-derived ingredient, as a novel therapeutic strategy for Huntington's disease: activation of the Nrf2 pathway through lysophosphatidic acid receptors. Brain Behav Immun 80:146–162. https://doi.org/10.1016/j.bbi.2019.03.001

Jeon J-H, Lee S, Lee W et al (2020) Herb-drug interaction of red ginseng extract and ginsenoside rc with valsartan in rats. Molecules 25(3):622. https://doi.org/10.3390/molecules25030622

Jiang F, DeSilva S, Turnbull J (2000) Beneficial effect of ginseng root in SOD-1 (G93A) transgenic mice. J Neurol Sci 180(1–2):52–54. https://doi.org/10.1016/s0022-510x(00)00421-4

Jin Y, Cui R, Zhao L et al (2019) Mechanisms of Panax ginseng action as an antidepressant. Cell Prolif 52(6):e12696. https://doi.org/10.1111/cpr.12696

Joo KM, Lee JH, Jeon HY et al (2010) Pharmacokinetic study of ginsenoside re with pure ginsenoside re and ginseng berry extracts in mouse using ultra performance liquid chromatography/mass spectrometric method. J Pharm Biomed Anal 51(1):278–283. https://doi.org/10.1016/j.jpba.2009.08.013

Kabalak AA, Soyal OB, Urfalioglu A et al (2004) Menometrorrhagia and tachyarrhythmia after using oral and topical ginseng. J Women's Health (Larchmt) 13(7):830–833. https://doi.org/10.1089/jwh.2004.13.830

Kakisaka Y, Ohara T, Tozawa H et al (2012) Panax ginseng: a newly identified cause of gynecomastia. Tohoku J Exp Med 228(2):143–145. https://doi.org/10.1620/tjem.228.143

Karpagam V, Sathishkumar N, Sathiyamoorthy S et al (2013) Identification of BACE1 inhibitors from Panax ginseng saponins—an *in silico* approach. Comput Biol Med 43(8):1037–1044. https://doi.org/10.1016/j.compbiomed.2013.05.009

Kim D, Jeon H, Ryu S et al (2016a) Proteomic analysis of the effect of Korean red ginseng in the striatum of a Parkinson's disease mouse model. PLoS One 11(10):e0164906. https://doi.org/10.1371/journal.pone.0164906

Kim HJ, Kim DJ, Shin EJ et al (2016b) Effects of gintonin-enriched fraction on hippocampal cell proliferation in wild-type mice and an APPswe/PSEN-1 double Tg mouse model of Alzheimer's disease. Neurochem Int 101:56–65. https://doi.org/10.1016/j.neuint.2016.10.006

Kim HJ, Shin EJ, Lee BH et al (2015) Oral administration of gintonin attenuates cholinergic impairments by scopolamine, amyloid-β protein, and mouse model of Alzheimer's disease. Mol Cells 38(9):796–805. https://doi.org/10.14348/molcells.2015.0116

Kim J, Kim SH, Lee DS et al (2013) Effects of fermented ginseng on memory impairment and β-amyloid reduction in Alzheimer's disease experimental models. J Ginseng Res 37 (1):100–107. https://doi.org/10.5142/jgr.2013.37.100

Kim KM, Kwon HS, Jeon SG et al (2008) Korean ginseng-induced occupational asthma and determination of IgE binding components. J Korean Med Sci 23(2):232–235. https://doi.org/10.3346/jkms.2008.23.2.232

Kim MS, Yu JM, Kim HJ et al (2014a) Ginsenoside re and Rd enhance the expression of cholinergic markers and neuronal differentiation in neuro-2a cells. Biol Pharm Bull 37 (5):826–833. https://doi.org/10.1248/bpb.b14-00011

Kim S, Park JL, Hwang HS, Kim YP (2014b) Correlation between frailty and cognitive function in non-demented community dwelling older Koreans. Korean J Fam Med 35(6):309–320. https://doi.org/10.4082/kjfm.2014.35.6.309

Lai L, Hao H, Liu Y et al (2009) Characterization of pharmacokinetic profiles and metabolic pathways of 20 (S)-ginsenoside Rh1 in vivo and in vitro. Planta Med 75(08):797–802. https://doi.org/10.1055/s-0029-1185400

Lee CH, Kim JH (2014) A review on the medicinal potentials of ginseng and ginsenosides on cardiovascular diseases. J Ginseng Res 38(3):161–166. https://doi.org/10.1016/j.jgr.2014.03.001

Lee F, Park J, Ko J et al (1987) Effects of Panax ginseng extract on the benzo (a) pyrene metabolizing enzyme system. Drug Chem Toxicol 10(3–4):227–236. https://doi.org/10.3109/01480548709042984

Lee JY, Lee YD, Bahn JW, Park HS (2006) A case of occupational asthma and rhinitis caused by Sanyak and Korean ginseng dusts. Allergy 61(3):392–393. https://doi.org/10.1111/j.1398-9995.2006.01032.x

Lee MR, Yun BS, In OH, Sung CK (2011) Comparative study of Korean white, red, and black ginseng extract on cholinesterase inhibitory activity and cholinergic function. J Ginseng Res 35 (4):421–248. https://doi.org/10.5142/jgr.2011.35.4.421

Lee YY, Park JS, Lee EJ et al (2015) Anti-inflammatory mechanism of ginseng saponin metabolite Rh3 in lipopolysaccharide-stimulated microglia: critical role of 5′-adenosine monophosphate-activated protein kinase signaling pathway. J Agric Food Chem 63(13):3472–3480. https://doi.org/10.1021/jf506110y

Lho SK, Kim TH, Kwak KP et al (2018) Effects of lifetime cumulative ginseng intake on cognitive function in late life. Alzheimers Res Ther 10(1):50. https://doi.org/10.1186/s13195-018-0380-0

Li H, Kang T, Qi B, Kong L et al (2016) Neuroprotective effects of ginseng protein on PI3K/Akt signaling pathway in the hippocampus of D-galactose/AlCl3 inducing rats model of Alzheimer's disease. J Ethnopharmacol 179:162–169. https://doi.org/10.1016/j.jep.2015.12.020

Li L, Liu J, Yan X et al (2011) Protective effects of ginsenoside Rd against okadaic acid-induced neurotoxicity in vivo and in vitro. J Ethnopharmacol 138(1):135–141. https://doi.org/10.1016/j.jep.2011.08.068

Li N, Wang D, Ge G et al (2014) Ginsenoside metabolites inhibit P-glycoprotein *in vitro* and *in situ* using three absorption models. Planta Med 80(4):290–296. https://doi.org/10.1055/s-0033-1360334

Li X, Wang G, Sun J et al (2007) Pharmacokinetic and absolute bioavailability study of total panax notoginsenoside, a typical multiple constituent traditional Chinese medicine (TCM) in rats. Biol Pharm Bull 30(5):847–851. https://doi.org/10.1248/bpb.30.847

Liao WI, Lin YY, Chu SJ et al (2010) Bradyarrhythmia caused by ginseng in a patient with chronic kidney disease. Am J Emerg Med 28(4):538. https://doi.org/10.1016/j.ajem.2009.08.006

Lin T, Liu Y, Shi M et al (2012a) Promotive effect of ginsenoside Rd on proliferation of neural stem cells *in vivo* and *in vitro*. J Ethnopharmacol 142(3):754–761. https://doi.org/10.1016/j.jep.2012.05.057

Lin ZY, Chen LM, Zhang J et al (2012b) Ginsenoside Rb1 selectively inhibits the activity of L-type voltage-gated calcium channels in cultured rat hippocampal neurons. Acta Pharmacol Sin 33 (4):438–444. https://doi.org/10.1038/aps.2011.181

Liu D, Zhang H, Gu W, Liu Y, Zhang M (2013) Neuroprotective effects of ginsenoside Rb1 on high glucose-induced neurotoxicity in primary cultured rat hippocampal neurons. PLoS One 8(11): e79399. https://doi.org/10.1371/journal.pone.0079399

Liu H, Yang J, Du F et al (2009) Absorption and disposition of ginsenosides after oral administration of Panax notoginseng extract to rats. Drug Metab Dispos 37(12):2290–2298. https://doi.org/10.1124/dmd.109.029819

Liu L, Zhu L, Zou Y et al (2014) Panax notoginseng saponins promotes stroke recovery by influencing expression of Nogo-a, NgR and p75NGF, *in vitro* and *in vivo*. Biol Pharm Bull 37(4):560–568. https://doi.org/10.1248/bpb.b13-00770

Liu R, Qin M, Hang P et al (2012) Effects of Panax notoginseng saponins on the activities of CYP1A2, CYP2C9, CYP2D6 and CYP3A4 in rats in vivo. Phytother Res 26(8):1113–1118. https://doi.org/10.1002/ptr.3688

Liu Y, Zhang RY, Zhao J et al (2015) Ginsenoside Rd protects SH-SY5Y cells against 1-methyl-4-phenylpyridinium induced injury. Int J Mol Sci 16(7):14395–14408. https://doi.org/10.3390/ijms160714395

Liu ZJ, Zhao M, Zhang Y et al (2010) Ginsenoside Rg1 promotes glutamate release via a calcium/calmodulin-dependent protein kinase II-dependent signaling pathway. Brain Res 1333:1–8. https://doi.org/10.1016/j.brainres.2010.03.096

Meng X, Sun G, Ye J et al (2014) Notoginsenoside R1-mediated neuroprotection involves estrogen receptor-dependent crosstalk between Akt and ERK1/2 pathways: a novel mechanism of Nrf2/ARE signaling activation. Free Radic Res 48(4):445–460. https://doi.org/10.3109/10715762.2014.885117

Mohamed ME, Frye RF (2011) Inhibitory effects of commonly used herbal extracts on UDP-glucuronosyltransferase 1A4, 1A6, and 1A9 enzyme activities. Drug Metab Dispos 39 (9):1522–1528. https://doi.org/10.1124/dmd.111.039602

Moon J, Choi SH, Shim JY et al (2018) Gintonin administration is safe and potentially beneficial in cognitively impaired elderly. Alzheimer Dis Assoc Disord 32(1):85–87. https://doi.org/10.1097/wad.0000000000000213

Nah SY (2012) Gintonin: a novel ginseng-derived ligand that targets G protein-coupled lysophosphatidic acid receptors. Curr Drug Targets 13(13):1659–1664. https://doi.org/10.2174/138945012803529947

Paek IB, Moon Y, Kim J et al (2006) Pharmacokinetics of a ginseng saponin metabolite compound K in rats. Biopharm Drug Dispos 27(1):39–45. https://doi.org/10.1002/bdd.481

Palop V, Catalán C, Rubio E, Martínez-Mir I (1999) Ginecomastia en un varón y ginseng [gynecomastia in a male and ginseng]. Med Clin (Barc) 112:758

Palop-Larrea V, Gonzálvez-Perales JL, Catalán-Oliver C et al (2000) Metrorrhagia and ginseng. Ann Pharmacother 34(11):1347–1348. https://doi.org/10.1345/aph.10013

Park H, Kim S, Rhee J et al (2015) Synaptic enhancement induced by gintonin via lysophosphatidic acid receptor activation in central synapses. J Neurophysiol 113(5):1493–1500. https://doi.org/10.1152/jn.00667.2014

Qi LW, Wang CZ, Yuan CS (2011) Ginsenosides from American ginseng: chemical and pharmacological diversity. Phytochemistry 72(8):689–699. https://doi.org/10.1016/j.phytochem.2011.02.012

Qian T, Cai Z, Wong RN, Jiang ZH (2005) Liquid chromatography/mass spectrometric analysis of rat samples for in vivo metabolism and pharmacokinetic studies of ginsenoside Rh2. Rapid Commun Mass Spectrom 19(23):3549–3554. https://doi.org/10.1002/rcm.2232

Ramanathan MR, Penzak SR (2017) Pharmacokinetic drug interactions with Panax ginseng. Eur J Drug Metab Pharmacokin 42(4):545–557. https://doi.org/10.1007/s13318-016-0387-5

Ramesh S, Govindarajulu M, Jones E, Suppiramaniam V, Moore T, Dhanasekaran M (2018a) Mitochondrial dysfunction and the role of mitophagy in Alzheimer's disease. Alzheimer's Disease & Treatment, MedDocs Publishers LLC

Ramesh S, Govindarajulu M, Lynd T et al (2018b) SIRT3 activator Honokiol attenuates β-amyloid by modulating amyloidogenic pathway. PLoS One 13(1):e0190350. https://doi.org/10.1371/journal.pone.0190350

Ritter JK (2000) Roles of glucuronidation and UDP-glucuronosyltransferases in xenobiotic bioactivation reactions. Chem Biol Interact 129(1–2):171–193. https://doi.org/10.1016/s0009-2797(00)00198-8

Rüb U, Vonsattel JPG, Heinsen H, Korf HW (2015) The neuropathology of Huntington's disease: classical findings, recent developments and correlation to functional neuroanatomy. Adv Anat Embryol Cell Biol 217:1–146

Ryu S, Jeon H, Koo S, Kim S (2018) Korean red ginseng enhances neurogenesis in the subventricular zone of 1-methyl-4-phenyl-1,2,3,6-tetrahydropyridine-treated mice. Front Aging Neurosci 10:355–355. https://doi.org/10.3389/fnagi.2018.00355

Saba E, Jeon BR, Jeong DH et al (2015) A novel Korean red ginseng compound gintonin inhibited inflammation by MAPK and NF-κB pathways and recovered the levels of mir-34a and mir-93 in RAW 264.7 cells. Evid Based Complement Alternat Med 2015:624132. https://doi.org/10.1155/2015/624132

Seely D, Dugoua JJ, Perri D et al (2008) Safety and efficacy of Panax ginseng during pregnancy and lactation. Can J Clin Pharmacol 15(1):e87–e94

Sherer T, Betarbet R, Greenamyre J (2001) Pathogenesis of Parkinson's disease. Curr Opin Investig Drugs 2(5):657–662

Shi C, Zheng DD, Fang L et al (2012) Ginsenoside Rg1 promotes nonamyloidgenic cleavage of APP via estrogen receptor signaling to MAPK/ERK and PI3K/Akt. Bioch Biophys Acta 1820 (4):453–460

Shibata S, Tanaka O, Shoji J, Saito H (1985) Chemistry and pharmacology of Panax. In: Wagner et al (eds) Economic and medicinal plant research. Academic Press

Shin K, Guo H, Cha Y et al (2016) Cereboost™, an American ginseng extract, improves cognitive function via up-regulation of choline acetyltransferase expression and neuroprotection. Regul Toxicol Pharmacol 78:53–58. https://doi.org/10.1016/j.yrtph.2016.04.006

Siegel RK (1979) Ginseng abuse syndrome: problems with the panacea. JAMA 241 (15):1614–1615

Small GW (2002) What we need to know about age related memory loss. BMJ 324 (7352):1502–1505. https://doi.org/10.1136/bmj.324.7352.1502

Sultana R, Mecocci P, Mangialasche F et al (2011) Increased protein and lipid oxidative damage in mitochondria isolated from lymphocytes from patients with Alzheimer's disease: insights into the role of oxidative stress in Alzheimer's disease and initial investigations into a potential biomarker for this dementing disorder. J Alzheimers Dis 24(1):77–84. https://doi.org/10.3233/JAD-2011-101425

Sultana R, Piroddi M, Galli F, Butterfield DA (2008) Protein levels and activity of some antioxidant enzymes in hippocampus of subjects with amnestic mild cognitive impairment. Neurochem Res 33(12):2540–2546. https://doi.org/10.1007/s11064-008-9593-0

Sun J, Wang G, Haitang X et al (2005) Simultaneous rapid quantification of ginsenoside Rg1 and its secondary glycoside Rh1 and aglycone protopanaxatriol in rat plasma by liquid chromatography–mass spectrometry after solid-phase extraction. J Pharm Biomed Anal 38 (1):126–132. https://doi.org/10.1016/j.jpba.2004.12.007

Sun XC, Ren XF, Chen L et al (2016) Glucocorticoid receptor is involved in the neuroprotective effect of ginsenoside Rg1 against inflammation-induced dopaminergic neuronal degeneration in substantia nigra. J Steroid Biochem Mol Biol 155:94–103. https://doi.org/10.1016/j.jsbmb.2015.09.040

Tawab MA, Bahr U, Karas M et al (2003) Degradation of ginsenosides in humans after oral administration. Drug Metab Dispos 31(8):1065–1071. https://doi.org/10.1124/dmd.31.8.1065

Tönnies E, Trushina E (2017) Oxidative stress, synaptic dysfunction, and Alzheimer's disease. J Alzheimers Dis 57(4):1105–1121. https://doi.org/10.3233/JAD-161088

Torbey E, Abi Rafeh N, Khoueiry G et al (2011) Ginseng: a potential cause of long QT. J Electrocardiol 44(3):357–358. https://doi.org/10.1016/j.jelectrocard.2010.08.007

Uttara B, Singh AV, Zamboni P, Mahajan R (2009) Oxidative stress and neurodegenerative diseases: a review of upstream and downstream antioxidant therapeutic options. Curr Neuropharmacol 7(1):65–74. https://doi.org/10.2174/157015909787602823

Van Kampen J, Robertson H, Hagg T, Drobitch R (2003) Neuroprotective actions of the ginseng extract G115 in two rodent models of Parkinson's disease. Exp Neurol 184(1):521–529. https://doi.org/10.1016/j.expneurol.2003.08.002

Van Kampen JM, Baranowski DB, Shaw CA, Kay DG (2014) Panax ginseng is neuroprotective in a novel progressive model of Parkinson's disease. Exp Gerontol 50:95–105. https://doi.org/10.1016/j.exger.2013.11.012

Wang B, Feng G, Tang C et al (2013) Ginsenoside Rd maintains adult neural stem cell proliferation during lead-impaired neurogenesis. Neurol Sci 34(7):1181–1188. https://doi.org/10.1007/s10072-012-1215-6

Wang W, Wang GJ, Xie HT et al (2007) Determination of ginsenoside Rd in dog plasma by liquid chromatography–mass spectrometry after solid-phase extraction and its application in dog pharmacokinetics studies. J Chromatogr B Analyt Technol Biomed Life Sci 852(1–2):8–14. https://doi.org/10.1016/j.jchromb.2006.12.046

Wang Y, Liu J, Zhang Z et al (2011) Anti-neuroinflammation effect of ginsenoside Rbl in a rat model of Alzheimer's disease. Neurosci Lett 487(1):70–72. https://doi.org/10.1016/j.neulet.2010.09.076

Wiwanitkit V, Taungjaruwinai W (2004) A case report of suspected ginseng allergy. MedGenMed 6(3):9

Wong VC, Lim CE, Luo X et al (2009) Current alternative and complementary therapies used in menopause. Gynecol Endocrinol 25(3):166–174. https://doi.org/10.1080/09513590802549866

Wu J, Jeong HK, Bulin SE et al (2009) Ginsenosides protect striatal neurons in a cellular model of Huntington's disease. J Neurosci Res 87(8):1904–1912. https://doi.org/10.1002/jnr.22017

Wu J, Yang H, Zhao Q et al (2016) Ginsenoside Rg1 exerts a protective effect against Aβ25-35-induced toxicity in primary cultured rat cortical neurons through the NF-κB/NO pathway. Int J Mol Med 37(3):781–788. https://doi.org/10.3892/ijmm.2016.2485

Xie HT, Wang GJ, Sun JG et al (2005) High performance liquid chromatographic–mass spectrometric determination of ginsenoside Rg3 and its metabolites in rat plasma using solid-phase extraction for pharmacokinetic studies. J Chromatogr B Analyt Technol Biomed Life Sci 818(2):167–173. https://doi.org/10.1016/j.jchromb.2004.12.028

Xu QF, Fang XL, Chen DF (2003) Pharmacokinetics and bioavailability of ginsenoside Rb1 and Rg1 from Panax notoginseng in rats. J Ethnopharmacol 84(2–3):187–192. https://doi.org/10.1016/s0378-8741(02)00317-3

Yang L, Zhang J, Zheng K et al (2014) Long-term ginsenoside Rg1 supplementation improves age-related cognitive decline by promoting synaptic plasticity associated protein expression in C57BL/6J mice. J Gerontol A Biol Sci Med Sci 69(3):282–294. https://doi.org/10.1093/gerona/glt091

Yin F, Sancheti H, Patil I, Cadenas E (2016) Energy metabolism and inflammation in brain aging and Alzheimer's disease. Free Radic Biol Med 100:108–122. https://doi.org/10.1016/j.freeradbiomed.2016.04.200

You Q, Ma ZC, Wang YG et al (2015) Panax ginseng inhibits intestinal absorption of toxic Aconitum carmichaeli alkaloids in vitro. Chin J Integr Med:1–9. https://doi.org/10.1007/s11655-015-2282-3

Yu CT, Chen J, Teng XW et al (2005) Lack of evidence for induction of CYP2B1, CYP3A23, and CYP1A2 gene expression by *Panax ginseng* and *Panax quinquefolius* extracts in adult rats and primary cultures of rat hepatocytes. Drug Metab Dispos 33(1):19–22. https://doi.org/10.1124/dmd.104.001917

Yuan CS, Wei G, Dey L et al (2004) Brief communication: American ginseng reduces warfarin's effect in healthy patients: a randomized, controlled trial. Ann Intern Med 141(1):23–27. https://doi.org/10.7326/0003-4819-141-1-200407060-00011

Zaafan MA, Abdelhamid AM, Ibrahim SM (2019) The protective effect of Korean red ginseng against rotenone-induced Parkinson's disease in rat model: modulation of nuclear factor-κβ and Caspase-3. Curr Pharm Biotechnol 20(7):588–594. https://doi.org/10.2174/1389201020666190611122747

Zhang X, Wang Y, Ma C et al (2016) Ginsenoside Rd and ginsenoside re offer neuroprotection in a novel model of Parkinson's disease. Am J Neurodegener Dis 5(1):52–61

Zhao HH, Di J, Liu WS et al (2013) Involvement of GSK3 and PP2A in ginsenoside Rb1's attenuation of aluminum-induced tau hyperphosphorylation. Behav Brain Res 241:228–234. https://doi.org/10.1016/j.bbr.2012.11.037

Zhou T, Zu G, Zhang X et al (2016) Neuroprotective effects of ginsenoside Rg1 through the Wnt/β-catenin signaling pathway in both *in vivo* and *in vitro* models of Parkinson's disease. Neuropharmacology 101:480–489. https://doi.org/10.1016/j.neuropharm.2015.10.024

Zhu D, Liu M, Yang Y et al (2014) Ginsenoside Rd ameliorates experimental autoimmune encephalomyelitis in C57BL/6 mice. J Neurosci Res 92(9):1217–1226. https://doi.org/10.1002/jnr.23397

# Insights into Mechanisms and Models for Studying Neurological Adverse Events Mediated by Pharmacokinetic Interactions between Clinical Drugs and Illicit Substances of Herbal and Fungal Origin

**Julia M. Salamat, Kodye L. Abbott, Kristina S. Gill, Patrick C. Flannery, Vinicia C. Biancardi, Dawn M. Boothe, Chen-Che J. Huang, Sridhar Mani, Aneesh Chandran, Saraswathi Vishveshwara, Suneel K. Onteru, Muralikrishnan Dhanasekaran, and Satyanarayana R. Pondugula**

## Contents



---

Julia M. Salamat, Kodye L. Abbott, Kristina S. Gill, Patrick C. Flannery contributed equally to this work.

---

J. M. Salamat · K. L. Abbott · V. C. Biancardi · D. M. Boothe · C.-C. J. Huang · S. R. Pondugula (✉)
Department of Anatomy, Physiology and Pharmacology, Auburn University, Auburn, AL, USA
e-mail: srp0010@auburn.edu

K. S. Gill
Department of Anatomy, Physiology and Pharmacology, Auburn University, Auburn, AL, USA

University of Massachusetts Memorial Medical Center, Worcester, MA, USA

P. C. Flannery
Department of Anatomy, Physiology and Pharmacology, Auburn University, Auburn, AL, USA

Rocky Vista University College of Osteopathic Medicine, Parker, CO, USA

S. Mani
Department of Molecular Pharmacology, Albert Einstein College of Medicine, Bronx, NY, USA

A. Chandran · S. Vishveshwara
Molecular Biophysics Unit, Indian Institute of Science, Bangalore, India

S. K. Onteru
Animal Biochemistry Division, National Dairy Research Institute, ICAR-NDRI, Karnal, Haryana, India

M. Dhanasekaran
Department of Drug Discovery and Development, Auburn University, Auburn, AL, USA

 
D. C. Agrawal, M. Dhanasekaran (eds.), *Medicinal Herbs and Fungi*,
https://doi.org/10.1007/978-981-33-4141-8_5

**Abstract** Coadministration of herbs and fungi with clinical drugs can result in herb–drug and fungi–drug interactions, leading to adverse effects, including neurological and related adverse events. Many of the illicit substances are derived from herbs or fungi. Hence, adverse neurological events that arise from pharmacokinetics-based interactions of illicit substances of herbal/fungal origin with clinical drugs are of significant health burden. Despite the knowledge that concomitant usage of clinical drugs with illicit substances can often result in neurological and related adverse events, the underlying pharmacokinetics-based mechanisms of these interactions are poorly understood. This chapter briefly highlights the neurological and related adverse events associated with pharmacokinetics-based interactions between clinical drugs and commonly used illicit substances of herbal or fungal origin, such as ayahuasca, cannabis (marijuana), cocaine, psilocybin, khat, salvia, kratom, and mescaline. This chapter discusses a great deal of novel insights into the potential mechanisms of pharmacokinetics-based interactions between clinical drugs and illicit substances of herbal/fungal origin that may be responsible for neurological and related adverse events. Finally, this chapter provides insights into potential experimental models that can be used in studying these pharmacokinetic interactions that lead to neurological adverse events.

**Keywords** Adverse drug interactions · Adverse neurological events · Adverse pharmacokinetic interactions · Drug-metabolizing enzymes · Drug transporters · Illicit substances

## Abbreviations

| | |
|---|---|
| AHR | Aryl hydrocarbon receptor |
| AIDS | Acquired immune deficiency syndrome |
| ALDH | Aldehyde dehydrogenase |
| BBB | Blood–brain barrier |
| BPA | Bisphenol A |
| CAR | Constitutive androstane receptor |
| ChAT | Choline acetyl transferase |
| CNS | Central nervous system |
| CYPs | Cytochromes P450 |
| DEHP | Di-2-ethylhexyl phthalate |
| DMT | N-dimethyltryptamine |
| GAD67 | Glutamate decarboxylase |
| LSD | Lysergic acid diethylamide |
| MAO | Monoamine oxidase |
| MDMA | 3, 4-Methylenedioxymethamphetamine |
| NIDA | National Institute on Drug Abuse |
| P-gp | P-glycoprotein |
| PXR | Pregnane X receptor |
| SSRI | Selective serotonin reuptake inhibitor |
| SULT | Sulfotransferase |
| UGT | UDP glucuronosyltransferases |

# 1 Introduction

## *1.1 Illicit Substances of Herbal/Fungal Origin*

Illicit substances can show stimulatory or inhibitory effects within the central nervous system (CNS). These substances can also exhibit hallucinogenic effects, thus considered dangerous to consume and illegal to possess. This chapter focuses on illicit substances of herbal or fungal origin, as reported by the National Institute on Drug Abuse (NIDA) (NIDA 2018) (Fig. 1).

Illicit substances of herbal or fungal origin alone may cause stimulation, depression, or hallucinogenic effects (Graziano et al. 2017). However, these adverse neurological events caused by illicit substances are further exasperated with concomitant use of clinical drugs, as many of these psychoactive substances are used in combination with clinical drugs. This simultaneous usage can lead to adverse neurological interactions. This is because of the altered response of an illicit substance or the clinical drug following the combinatorial usage. Indeed, many clinically relevant adverse neurological interactions have been observed accompanying the simultaneous usage of these compounds (Baumeister et al. 2015).

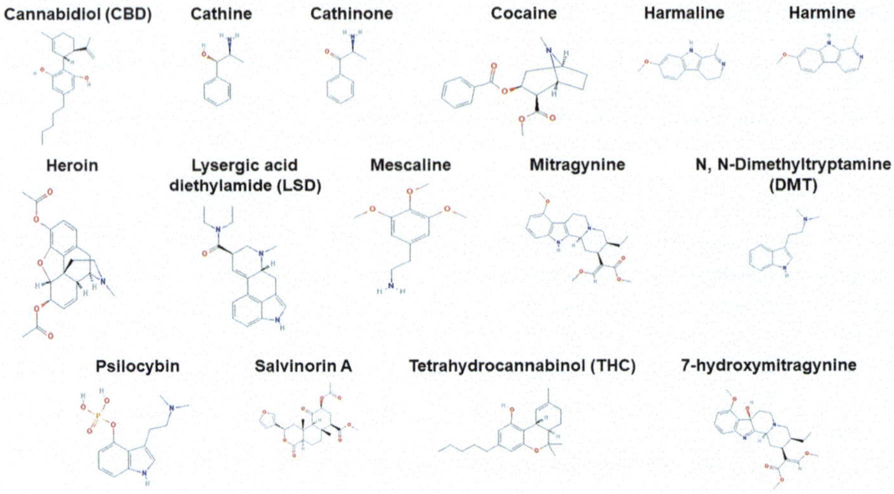


**Fig. 1** Chemical structures of the illicit substances of herbal/fungal origin

## *1.2 Mechanisms of Adverse Pharmacokinetic Interactions*

In pharmacokinetics, the absorption, distribution, metabolism, and elimination of xenobiotics, such as clinical drugs or illicit substances, is primarily determined by drug-metabolizing enzymes and drug transporters. Cytochromes P450 (CYPs), sulfotransferases (SULTs), and UDP-glucuronosyltransferases (UGTs) are predominant drug-metabolizing enzymes. ATP-binding cassette transporters and organic anion and cation transporters are the major drug transporters. A combined regimen of clinical drugs and illicit substances of herbal or fungal origin can result in alterations in the expression/activity of drug-metabolizing enzymes and/or drug transporters. These changes can result in interactions at the pharmacokinetic level, leading to adverse neurological events.

Similar to clinical drugs, illicit substances are also metabolized and transported by drug-metabolizing enzymes and drug transporters (refer to Abbott et al. (2020a) for details). At the same time, clinical drugs, as well as illicit substances, may alter the expression/activity of drug-metabolizing enzymes and/or drug transporters (refer to Abbott et al. (2020a) for details). As a result, during coadministration, illicit substances can negatively affect the outcome of clinical drugs. Similarly, clinical drugs may influence the outcome of illicit substances.

In this book chapter, we briefly discuss the neurological and related adverse events associated with pharmacokinetics-based interactions between clinical drugs and illicit substances of herbal/fungal origin (Fig. 1) in human patients. This chapter then highlights insights into the potential pharmacokinetic mechanisms of adverse neurological and related adverse events. Furthermore, this chapter discusses insights into experimental models that can be used in studying these pharmacokinetic interactions that lead to adverse neurological drug interactions.

# 2 Neurological and Related Adverse Events Associated with Pharmacokinetics-Based Interactions between Illicit Substances of Herbal and Fungal Origin and Clinical Drugs

## *2.1 Ayahuasca*

Ayahuasca contains *Banisteriopsis caapi* stalks, which consist of β-carbolines. The major β-carbolines in ayahuasca include harmine and harmaline. Ayahuasca also consists of *Psychotria Viridis* leaves, which contain DMT (NIDA 2018; Palhano-Fontes et al. 2018; Lobao-Soares et al. 2018). A previous study reported that ayahuasca adversely interacts with SSRIs (Callaway and Grob 1998). Notably, a patient undergoing depression and prescribed with the Selective serotonin reuptake inhibitor (SSRI) fluoxetine while using ayahuasca presented with adverse interactions, including tremors, shivering, and confusion (Callaway and Grob 1998). A possible mechanism underlying these adverse effects is the ability of fluoxetine to inhibit CYP2C19, CYP3A4, and CYP2D6 (Sager et al. 2014).

## *2.2 Cannabis (Marijuana)*

Cannabis is derived from the cannabis plant that is used both medicinally and recreationally (NIDA 2018). Clinical reports have shown that the concomitant intake of cannabis with clinical drugs can lead to adverse neurotoxic interactions. For instance, coadministration of cannabis with clobazam, an antiepileptic drug, leads to a decreased metabolism of clobazam. This leads to adverse neurological events such as restless sleep, ataxia, drowsiness, and tremors (Geffrey et al. 2015). These observed adverse events were likely because of decreased clobazam metabolism owing to the inhibition of CYP3A4 and CYP2C19 by CBD (Geffrey et al. 2015).

## *2.3 Cocaine (Benzoylmethylecgonine)*

Cocaine is found in the leaves of *Erythroxylon coca* (NIDA 2018). Mortalities were reported in instances of simultaneous usage of cocaine with several classes of clinical drugs (Molina and Hargrove 2011). For example, cocaine toxicity was shown to be intensified by neuroleptics and antihistamines. This is because of the upregulation of dopamine during the concomitant usage (Molina and Hargrove 2011).

## 2.4 Khat

The psychoactive compounds of khat are found in the shrub, *Catha edulis* (NIDA 2018). Although it is possible that adverse neurological events can occur when khat is simultaneously taken with clinical drugs, there are no reported adverse neurological events mediated by pharmacokinetic interactions.

## 2.5 Kratom

Kratom is scientifically known as *Mitragyna speciosa* (NIDA 2018). Adverse neurological interactions at the pharmacokinetic level could occur when kratom is taken along with clinical drugs. For instance, concomitant usage of kratom and modafinil, a CNS stimulant, and CYP3A4 substrate lead to seizures (Boyer et al. 2008).

## 2.6 Lysergic Acid Diethylamide (LSD)

LSD is found in the fungus ergot (NIDA 2018). LSD is a hallucinogenic substance, and its hallucinogenic actions can be altered when taken along with clinical drugs such as antidepressants. For example, the tricyclic antidepressant medications, imipramine, desipramine, and clomipramine, were revealed to intensify the hallucinatory responses of LSD. In contrast, patients chronically using antidepressants with inhibitory effects on serotonin reuptake (fluoxetine, paroxetine, sertraline, trazodone), or monoamine oxidase (MAO) inhibitors, reported a decreased impact of LSD response (Bonson and Murphy 1996). Possible mechanisms of these adverse interactions include the alterations in serotonin receptors, the concentration of circulating serotonin, and brain catecholamine systems (Bonson and Murphy 1996).

## 2.7 Mescaline

Mescaline, a powerful hallucinogenic substance, is found in several cacti species (NIDA 2018). To date, there is no evidence of adverse neurological interactions between mescaline and clinical drugs.

### 2.8 *N, N-Dimethyltryptamine (DMT)*

DMT is a naturally occurring indole alkaloid found in many plants (Carbonaro and Gatch 2016). However, there is no evidence of DMT interaction with clinical drugs resulting in adverse neurological events.

### 2.9 *Psilocybin*

Psilocybin is found in certain mushrooms (Johnson and Griffiths 2017). Inhibitors of aldehyde dehydrogenase (ALDH) or MAO can escalate the hallucinatory actions of psilocybin. For example, haloperidol enhances the hallucinatory actions of psilocybin (Vollenweider et al. 1998). This could be because of the inhibition of MAO by haloperidol (Fang et al. 1995). Therefore, it may be possible for other clinical drugs that inhibit the enzymes responsible for psilocybin metabolism from contributing to adverse neurological events.

### 2.10 *Salvia*

*Salvia divinorum,* a species of sage, has potent hallucinogenic properties (Serra et al. 2015; Tsujikawa et al. 2009). Salvinorin A is the major known active component of salvia. Currently, there is no data on salvia's ability to cause adverse drug interaction when taken in combination with clinical medications.

## 3 Insights into Novel Mechanisms of Pharmacokinetics-Based Interactions

As described earlier, the coadministration of illicit substances of herbal/fungal origin with clinical drugs can often result in adverse neurological and related interactions. However, the responsible pharmacokinetics-based mechanisms for these interactions are not fully understood.

### 3.1 *Xenobiotic Receptors*

Xenobiotic receptors are known to contribute to pharmacokinetic interactions facilitated by drug-metabolizing enzymes and/or drug transporters (Pondugula and Mani 2013; Pondugula et al. 2016). Xenobiotic receptors, including the pregnane X receptor (PXR), constitutive androstane receptor (CAR), and aryl hydrocarbon receptor (AHR), regulate gene expression of the major drug-metabolizing enzymes and drug transporters (Chen et al. 2012; Malaplate-Armand et al. 2005). It is noteworthy that these xenobiotic receptors are activated or inhibited by diverse groups of xenochemicals, including clinical drugs, environmental chemicals, supplements, and illicit substances. Therefore, it is highly plausible that these xenobiotic receptors play an important role in pharmacokinetic interactions of illicit substances with clinical drugs. As an example, kratom was shown to activate PXR and increase the expression and activity of PXR target genes such as P-glycoprotein (P-gp) and CYP3A4 (Manda et al. 2017). However, to which extent illicit substances of herbal/fungal origin modulate xenobiotic receptors-regulated drug-metabolizing enzymes and/or drug transporters to induce adverse neurological events is not fully understood.

## *3.2 Sites of Pharmacokinetics-Based Interactions*

The liver and intestine are the major detoxifying organs with an abundance of xenobiotics receptors, enzymes, and transporters. As a result, pharmacokinetic aspects of xenobiotics are often studied using liver and intestinal models. Another primary organ that comes in contact with xenobiotics is the lung, as it is the primary organ of the human respiratory tract. This organ serves as a major site of entry for various xenobiotics, including clinical drugs and illicit substances that are taken via the respiratory tract. Other detoxifying organs include kidneys and skin (Fig. 2).

As mentioned earlier, the xenobiotic receptors PXR, CAR, and AHR are widely expressed in the liver and gastrointestinal tract. Still, their expression in other detoxifying organs is also expected as xenobiotic chemicals also come in contact with these organs, such as the kidneys, skin, and lungs (Bikle 2015) (Fig. 2). In the kidneys, xenobiotic receptors PXR and AHR have been reported to be present, although to a lesser extent compared to the liver and intestine (Miki et al. 2005). In the kidney, drug-metabolizing enzymes and drug transporters that are regulated by these xenobiotic receptors function as eliminators of xenobiotics (Radford et al.

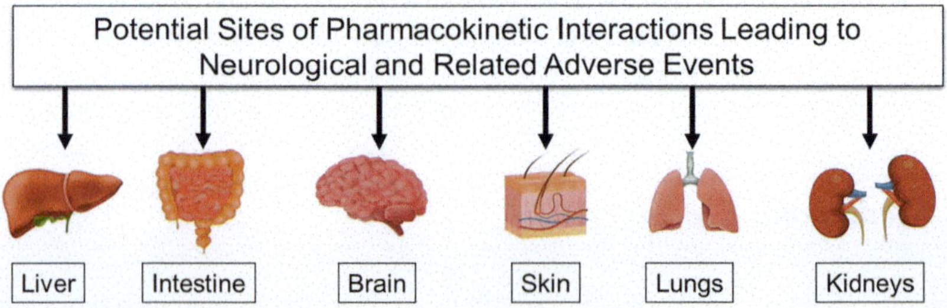


**Fig. 2** Major sites of pharmacokinetics-based interactions that can lead to adverse neurological events

2013; Dou et al. 2018). The skin was reported to harbor PXR, CAR, and AHR (Elentner et al. 2015; Di Meglio et al. 2014; Tigges et al. 2013; Avouac et al. 2014). Notably, these receptors were shown to respond to xenochemicals by regulating the enzymes and transporters in the skin (Elentner et al. 2015; Fritsche et al. 2007; Villard et al. 2002). Lungs also express xenobiotic receptors, enzymes, and transporters (Ding and Kaminsky 2003) (Anttila et al. 2011; Esser and Rannug 2015; Beamer and Shepherd 2013). While it is very likely that pharmacokinetics-based mechanisms in the kidneys, skin, and lungs also contribute to adverse neurological events associated with concomitant usage of clinical drugs with illicit substances, the mechanisms have yet to be studied in these organs.

### 3.3 Xenobiotic Receptors, Drug-Metabolizing Enzymes, and Drug Transporters in the Brain and Blood–Brain Barrier

Notably, these pathways are also expressed in sanctuary organs like the brain despite the blood–brain barrier (BBB) as these organs are also expected to come in contact with xenobiotic chemicals. Recently, expression of xenobiotic receptors, particularly PXR, CAR, and AHR, has been reported within the brain and in the BBB (Wang et al. 2012; Bauer et al. 2006; Juricek and Coumoul 2018; You et al. 2019; Kimura and Tohyama 2017). Furthermore, it was also shown that these receptors, upon activation by ligands, regulate the expression of the enzymes and transporters in the brain and BBB (Wang et al. 2012; Bauer et al. 2006; Zastre et al. 2009; You et al. 2019; Xu et al. 2016).

Moreover, plasma extracellular vesicles containing CYP enzymes produced by the liver and nearby organs can circulate through the plasma and migrate to distant organs, such as the brain, influencing drug metabolism and excretion, which can also play a role in the likelihood of inducing adverse drug events (Gerth et al. 2019). Still, limited information is available on their critical role in mediating pharmacokinetic interactions that lead to adverse neurological events (Alavijeh et al. 2005) (Fig. 3).

In the brain and within the BBB, xenobiotic receptors can significantly influence the adverse neurological events induced by xenochemicals such as clinical drugs, environmental chemicals, and illicit substances (Wang et al. 2010). For example, Upadhya et al. has provided evidence for a probable correlation between CAR activation by phenobarbital and neurotoxicity. Rats exposed to phenobarbital exhibited enhanced CYP2B1 expression and increased neurotoxic effects of chemotherapy drugs (Upadhya et al. 2002). Neurotoxicity was also observed in neuroblastoma cell lines treated with the chemotherapy drug paclitaxel, after drug-induced P-gp activity inhibition, as this amplified the accumulation of paclitaxel in sensory neurons (Stage et al. 2020; Gornstein and Schwarz 2014)

Xenobiotic receptors can also impact the efficacy of therapeutic drugs in the brain. For instance, brain endothelial cells of epileptic patients exhibiting resistance

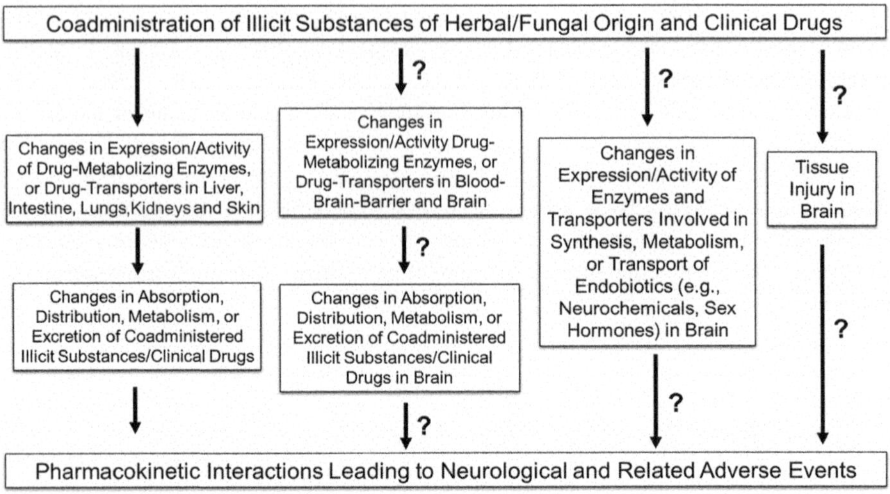


**Fig. 3** Suggested mechanisms of adverse neurological interactions at the pharmacokinetic level when illicit substances are coadministered with clinical drugs

to antiepileptic drugs increased PXR levels resulted in the upregulation of the enzymes CYP3A4 and CYP2C9 (Ghosh et al. 2017). Another study by Yu et al. showed the role of PXR in drug response in the brain. This study found that drug-resistant epilepsy is attributed to PXR-mediated P-gp overexpression in the chronic phase of epilepsy (Yu et al. 2017). Other drugs, such as phenobarbital and carbamazepine, which are also ligands of PXR and CAR, upregulated the expression of P-gp and MDRs in microvascular endothelial cells within the brain, causing resistance to antiepileptic drugs (Lombardo et al. 2008).

## *3.4 Disruption of Endobiotic Homeostasis*

Pharmacokinetic interactions could also take place because of the disruption of the mechanisms involved in the homeostasis of endobiotics. This can lead to an excessive buildup or depletion of endobiotics, resulting in toxicity or tissue injury (Fig. 3). For example, the co-treatment of the antibiotics, isoniazid and rifampicin, has been shown to amplify the liver injury when compared with either drug (Li et al. 2013). This hepatotoxicity is due to the disruption of the homeostatic mechanisms of porphyrin, leading to an excessive buildup of protoporphyrin. The key mechanisms responsible for the disruption of homeostasis of protoporphyrin in the liver include the upregulation of aminolevulinic acid synthase-1 and downregulation of ferrochelatase in a PXR-dependent manner (Li et al. 2013; Fraser et al. 2003).

Several other studies have shown that xenobiotic chemicals, including clinical drugs via xenobiotic receptors, enzymes, and transporters, contribute to the disruption of endobiotic homeostasis. PXR activation using PXR agonist rifampicin, an antituberculosis medication, can result in the disturbance of adrenal steroid homeostasis in rodents. The dyshomeostasis, characterized by an increase in corticosterone and aldosterone in PXR activated animals, maybe because of the upregulation of adrenal steroidogenic enzymes genes such as Cyp11a1, Cyp11b1, and Cyp11b2 (Zhai et al. 2007). Bisphenol A (BPA) was shown to inhibit follicular growth and diminish estradiol levels via xenobiotic receptors (Ziv-Gal et al. 2013). In one study, coadministration of the PXR-agonist BPA, and ketoconazole, antifungal medication, and known PXR antagonist (Ji et al. 2019) lead to less efficient metabolism of BPA in zebrafish, which may increase the adverse effects of BPA (Fuchs et al. 2013; Ji et al. 2019). Indeed, there was a significant change in sex hormone levels, resulting in a decrease in egg number and gonad weight in the zebrafish models (Fuchs et al. 2013; Ji et al. 2019). Another example is exposure to estrogen found in contraceptive pills and trans-nonachlor (TNC), a pesticide. It was discovered that individually, both substances have low binding affinity and efficacy to PXR. However, when combined, they efficaciously bind to PXR, resulting in synergistic activation of the receptor (Delfosse et al. 2015). Such occurrence can result in the production of inactive metabolites or potentiate the toxic effect of TNC in the human body.

Xenobiotic receptors may affect endobiotic homeostasis in the brain, leading to changes in behavior. It has been shown that the activation of PXR alters the metabolism of sex hormones, resulting in altered levels of the metabolites of these hormones involved in the regulation of behavior (Torres-Vergara et al. 2020). For example, knocking down PXR expression in the ventral tegmental area of the midbrain resulted in attenuation of allopregnanolone, a neuroactive metabolite of progesterone, and an allosteric modulator of GABA, influencing sexual receptivity and neural plasticity (Frye et al. 2014; Schule et al. 2014). Additionally, the deletion of PXR and CAR in mice led to impaired recognition and memory associated with anxiety-like behavior. Treatment with meclizine, a PXR agonist, in dementia-induced mice resulted in improved cognition. However, coadministration of the PXR-antagonist ketoconazole inhibited this beneficial effect of meclizine (Singh et al. 2020). Another example by Kaur et al. showed that forskolin, another PXR agonist, improved memory function in dementia-induced Wistar rats. Conversely, concomitant administration of ketoconazole hindered the observed positive changes, suggesting the potential role of PXR in the preservation of neuronal structure and function of meclizine and forskolin in dementia patients (Kaur et al. 2018).

Toxic substances, such as cypermethrin, affect the activity of CYP1A, CYP2B, and CYP2E1 in a PXR-dependent manner in different regions of the brain. These changes were also associated with altered expression of neurotransmitter-synthesizing enzymes such as glutamate decarboxylase (GAD67) and choline acetyltransferase (ChAT), as well as brain-derived neurotrophic factors. Altered expression of these enzymes involved in endobiotic metabolism resulted in undesirable changes in learning and memory (Singh et al. 2016). It has also been reported that chemicals such as Di-2-ethylhexyl phthalate (DEHP), a widely used plasticizer in many products like medical devices and furniture, induces neurotoxicity, particularly in the cerebellum, and this is closely connected to the disruption of cytochrome P450 homeostasis regulated by the xenobiotic receptors, CAR, and PXR (Du et al. 2017).

The above-cited information demonstrates that xenobiotics, including clinical drugs, can dysregulate endobiotic homeostasis in the brain via the xenobiotic receptors, enzymes, and transporters. However, it is unknown whether illicit substances of herbal/fungal origin can induce dysregulation of endobiotic homeostasis in a similar manner. Therefore, whether pharmacokinetics-based interactions between clinical drugs and illicit substances of herbal/fungal origin result in disruption of homeostasis of endobiotics in the brain, resulting in adverse neurological events that remain to be studied (Fig. 3).

It is likely that the combined administration of illicit substances with clinical drugs can lead to inflammation and oxidative stress in the brain. This can eventually result in the production of endogenous toxins such as reactive oxygen species, causing tissue injury, or toxicity in the brain (Fig. 3). Modulation of the enzymes and transporters in the brain via CAR and PXR can result in oxidative stress leading to neurotoxicity (Du et al. 2017). However, it remains to be understood to what degree the concomitant usage of illicit substances with clinical drugs can lead to tissue injury/toxicity in the brain because of disruption of the homeostasis of endobiotics (Fig. 3).

## 4 Insights into In Vivo, In Vitro, and In Silico Models for Studying Pharmacokinetic Interactions

### *4.1 In Vivo Models*

Different animal models have been used for preclinical in vivo studies involving xenobiotics, including illicit substances. However, xenobiotics can modulate the enzymes, transporters, and receptors in a species-dependent manner. Following this theme, the enzymes and transporters also exhibit differential species specificity for substrates. These factors confound our ability to extrapolate preclinical assessment of drug safety studies from animal models to humans.

Many of the preclinical studies are based on the anticipated impact on drug metabolism, particularly via CYPs. Interspecies differences in CYPs are not yet fully characterized but can be profound (Table 1). A number of studies involving different model species such as rat, mouse, dog, monkey, and mini pig have attempted to identify the most relevant and thus best predictor of human CYP activity. The CYP enzymes with the most similar activities are CYP1A, CYP1B, and CYP2E1 (Table 1). However, the substrates for these enzymes, their rate, extent, a pathway of metabolism, and the impact of substrates on the induction or inhibition of the specific enzymes may also vary (Campion and Dowell 2019; Court 2013; Martignoni et al. 2006). According to Martignoni, CYP2E1 does not dramatically

**Table 1** Orthologues of cytochrome P450 enzymes (CYPs) among species used to assess human drug metabolism

| Human CYP | Monkey CYP | Dog CYP | Mouse CYP | Rat CYP |
|---|---|---|---|---|
| 1A1 | 1A1 | 1A1 | 1A1 | 1A1 |
| 1A2 | 1A2 | 1A2 | 1A2 | 1A2 |
| 1B1 | 1B1 | 1B1 | 1B1 | 1B1 |
| 2A6, 2A7, 2A13 | 2A23,2A24 | 2A13, 2A25 | 2A4,2A5,2A12,2A22 | 2A1,2A2,2A3 |
| 2B6, 2B7 | 2B17 | 2B11 | 2B9,10,13,19,23 | 2B1,2B2,2B3 |
| 2C8, 2C9, 2C18,2C19 | 2C20,2C43 | 2C8, 2C9 (2C21/41), 2C38 | 2C29, 2C37 | 2C6,2C7,2C11–13,2C22,2C23 |
| 2D6, 2D7, 2D8 | 2D17,2D19,2D29,2D30 | 2D15 | 2D9–13,22, 26,34,40 | 2D1-D5, 2D14 |
| 2E1 | 2E1 | 2E1 | 2E1 | 2E1 |
| 2F1 | | | 2F2,2G1 | |
| 3A4 (3A4/5/7, 43) | 3A8 | 3A4 (3A12/26) | 3A11,13,16,25,41,44 | 3A1/ 3A23,3A2,3A9,3A18,3A62 |
| | Martignoni et al. (2006) | Antonovic and Martinez (2011), Court (2013) | Martignoni et al. (2006), Nelson et al. (2004) | Martignoni et al. (2006) |

differ among the species in terms of substrate metabolism, suggesting extrapolation among the species is reasonable. Yet, this enzyme is relatively not involved, inhibited, or induced by illicit substances. Polymorphisms in CYP2D6 and CYP2D19 also play a critical role in determining the response during treatment with medications in psychiatric patients (Kropp et al. 2006).

Court et al. and Nishimuta et al. (Court 2013; Nishimuta et al. 2013) have specifically addressed CYP differences among the dog, human, and rat, including similarities in substrate metabolism (Table 1). Nishimuta's study, based on 43 human CYP450 substrates for CYP1A2, CYP2C, CYP2D6, CYP2J2, and CYP3A, found that intrinsic hepatic clearance (based on microsomal activity) was generally within ten-fold of beagle dogs. In contrast, CYP2D6 substrates were more rapidly cleared in rats, resulting in a low correlation with human activity. The authors concluded that dog metabolic activity was most comparable to humans. Sakai et al. (Sakai et al. 2014) also explored the issue of the most appropriate animal model to predict CYP-mediated metabolism. They coupled in vitro microsomal studies with in vivo pharmacokinetic studies using typical CYP450 substrates after both oral and intravenous administration in commonly studied species (mice, rats, dogs, monkeys, and mini pigs). Similar to previous investigations, they found most substrates to be more rapidly metabolized in rodents. The activity of CYP1A2 (based on caffeine) was found to be similar among the species. CYP2C9 activity (based on losartan) in the rat, dog, and mini pig was most similar to humans. For CYP2D, the intrinsic clearance of dextromethorphan was closest between dogs and humans, whereas human CYP3A4 activity (based on midazolam) was best correlated with the mini pig. Similar to Phase I enzymes, interspecies differences in Phase II enzymes also complicate the extrapolation of preclinical data obtained in animal studies to humans (Taylor et al. 2017; Fujiwara et al. 2018).

Transport proteins impact every step during the drug movement (absorption, distribution, metabolism, and excretion). As such, understanding their impact on drug movement should facilitate the prediction of illicit substance–clinical drug interactions. Although significant progress had been made in terms of transporters (Giacomini et al. 2010; Giacomini et al. 2018), limited data are available regarding species differences (Fallon et al. 2016; Kim et al. 2015) (Table 1). Yet, a similar degree of variability in presence, activity, and substrates should be anticipated.

As discussed earlier, because of the notable interspecies differences, humanized animal models could be useful to address the species-specific drug interactions involving the enzymes, transporters, and receptors (Fig. 4). Humanized mice models are available for the xenobiotic receptors such as CAR, PXR, and AHR (Xie et al. 2000; Cheng et al. 2009; Ross et al. 2010; Hasegawa et al. 2012; Flaveny and Perdew 2009; Flaveny et al. 2010; Moriguchi et al. 2003). Humanized mice models are also available for the major drug-metabolizing enzymes and drug transporters (Cheng et al. 2009; Yamasaki et al. 2018; Holmstock et al. 2013; Hasegawa et al. 2012; Fujiwara et al. 2018).

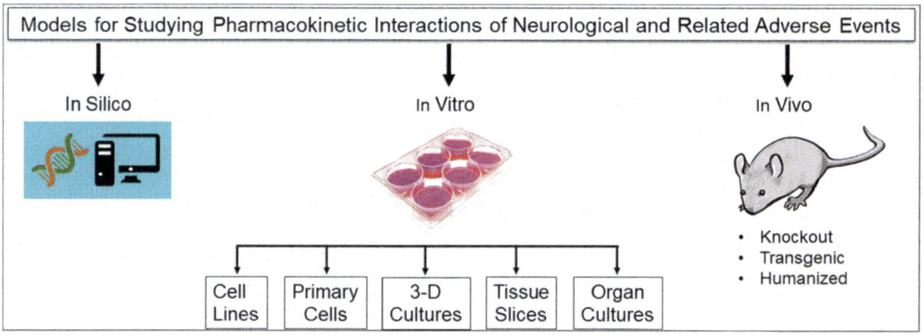


**Fig. 4** Proposed pharmacokinetic models for studying adverse neurological and related drug interactions

## *4.2 In Vitro Models*

Advanced in vitro models (Issa et al. 2017) such as engineered cell lines (Abbott et al. 2019; Pondugula et al. 2015) and primary cells (Abbott et al. 2019; Pondugula et al. 2015) of the liver, intestine, BBB, and brain are also useful preclinical tools (Fig. 4). For example, a study by Moyano et al. showed that primary hippocampal cell lines could be utilized in showing CYP3A4 induction resulting in neurotoxicity by amitraz (Moyano et al. 2019). Sandwich and 3D cell culture models can be used to mitigate the disadvantages of 2D cell models (Hoffmaster et al. 2004; De Bruyn et al. 2013; Papageorgiou et al. 2013; Issa et al. 2017). For instance, 3D brain organoids have been suggested to be better models for understanding the mechanisms behind neurological disorders (Heide et al. 2018; Chukwurah et al. 2019). BBB models of rodent, porcine, and human origin are available to study xenobiotic receptors, enzymes, and transporters (Torres-Vergara et al. 2020). These models include cell lines, primary cultures, isolated capillaries, and so on. These models have already been validated for studying the primary xenobiotic receptors, enzymes, and transporters (Torres-Vergara et al. 2020).

## *4.3 In Silico Models*

In addition to in vivo and in vitro models, in silico models can be useful to study molecular dynamic interactions with xenobiotic receptors (Issa et al. 2017; Abbott et al. 2019; Chandran and Vishveshwara 2016) (Fig. 4). Computational methods and tools are available to predict not only metabolism but also the transport of xenobiotics by drug-metabolizing enzymes and drug transporters (Tyzack and Kirchmair 2019). In silico models have also been proven to be valuable tools to predict xenobiotics with the potential to serve as substrates and inhibitors of drug-metabolizing enzymes and drug transporters (Martiny et al. 2015; Demel et al.

2008). Moreover, these tools are useful to predict pharmacokinetic interactions with enzymes/transporters (Estrada-Tejedor and Ecker 2018; Ecker et al. 2008). Taken together, computer models can be used to understand the underlying mechanisms of pharmacokinetics-based interactions between clinical drugs and illicit substances (Fig. 4). While the field of illicit substances is a growing research area, it faces many regulatory challenges. These regulations preclude the efforts of scientists to procure illicit substances for research purposes (Abbott et al. 2020b). Therefore, molecular modeling studies would certainly help fill the gap in this area of understanding of mechanisms of clinical drug interactions.

**Acknowledgments** This project was supported by Animal Health and Disease Research Grant and Auburn University Research Initiative in Cancer Grant.

## References

Abbott KL, Chaudhury CS, Chandran A et al (2019) Belinostat, at its clinically relevant concentrations, inhibits rifampicin-induced CYP3A4 and MDR1 gene expression. Mol Pharmacol 95 (3):324–334. https://doi.org/10.1124/mol.118.114587

Abbott KL, Flannery PC, Gill KS et al (2020a) Adverse pharmacokinetic interactions between illicit substances and clinical drugs. Drug Metab Rev 52(1):44–65. https://doi.org/10.1080/03602532.2019.1697283

Abbott KL, Gill KS, Flannery PC et al (2020b) Nothing ventured, nothing gained: regulations cripple potentially life-saving research of illicit substances. ACS Chem Neurosci 11 (10):1382–1384. https://doi.org/10.1021/acschemneuro.0c00241

Alavijeh MS, Chishty M, Qaiser MZ et al (2005) Drug metabolism and pharmacokinetics, the blood-brain barrier, and central nervous system drug discovery. NeuroRx 2(4):554–571. https://doi.org/10.1602/neurorx.2.4.554

Antonovic L, Martinez M (2011) Role of the cytochrome P450 enzyme system in veterinary pharmacokinetics: where are we now? Where are we going? Future Med Chem 3(7):855–879. https://doi.org/10.4155/fmc.11.37

Anttila S, Raunio H, Hakkola J (2011) Cytochrome P450-mediated pulmonary metabolism of carcinogens: regulation and cross-talk in lung carcinogenesis. Am J Respir Cell Mol Biol 44 (5):583–590. https://doi.org/10.1165/rcmb.2010-0189RT

Avouac J, Palumbo-Zerr K, Ruzehaji N et al (2014) The nuclear receptor constitutive androstane receptor/NR1I3 enhances the profibrotic effects of transforming growth factor beta and contributes to the development of experimental dermal fibrosis. Arthritis Rheumatol 66 (11):3140–3150. https://doi.org/10.1002/art.38819

Bauer B, Yang X, Hartz AM et al (2006) In vivo activation of human pregnane X receptor tightens the blood-brain barrier to methadone through P-glycoprotein up-regulation. Mol Pharmacol 70 (4):1212–1219. https://doi.org/10.1124/mol.106.023796

Baumeister D, Tojo LM, Tracy DK (2015) Legal highs: staying on top of the flood of novel psychoactive substances. Ther Adv Psychopharmacol 5(2):97–132. https://doi.org/10.1177/2045125314559539

Beamer CA, Shepherd DM (2013) Role of the aryl hydrocarbon receptor (AhR) in lung inflammation. Semin Immunopathol 35(6):693–704. https://doi.org/10.1007/s00281-013-0391-7

Bikle DD (2015) Pregnane X receptor expression in skin: the good and the bad. Exp Dermatol 24 (11):829–830. https://doi.org/10.1111/exd.12829

Bonson KR, Murphy DL (1996) Alterations in responses to LSD in humans associated with chronic administration of tricyclic antidepressants, monoamine oxidase inhibitors or lithium. Behav Brain Res 73(1–2):229–233

Boyer EW, Babu KM, Adkins JE et al (2008) Self-treatment of opioid withdrawal using kratom (Mitragynia speciosa Korth). Addiction 103(6):1048–1050. https://doi.org/10.1111/j.1360-0443.2008.02209.x

Callaway JC, Grob CS (1998) Ayahuasca preparations and serotonin reuptake inhibitors: a potential combination for severe adverse interactions. J Psychoactive Drugs 30(4):367–369. https://doi.org/10.1080/02791072.1998.10399712

Campion DP, Dowell FJ (2019) Translating pharmacogenetics and pharmacogenomics to the clinic: progress in human and veterinary medicine. Front Vet Sci 6:22. https://doi.org/10.3389/fvets.2019.00022

Carbonaro TM, Gatch MB (2016) Neuropharmacology of N,N-dimethyltryptamine. Brain Res Bull 126(Pt 1):74–88. https://doi.org/10.1016/j.brainresbull.2016.04.016

Chandran A, Vishveshwara S (2016) Exploration of the conformational landscape in pregnane X receptor reveals a new binding pocket. Protein Sci 25(11):1989–2005. https://doi.org/10.1002/pro.3012

Chen Y, Tang Y, Guo C et al (2012) Nuclear receptors in the multidrug resistance through the regulation of drug-metabolizing enzymes and drug transporters. Biochem Pharmacol 83 (8):1112–1126. https://doi.org/10.1016/j.bcp.2012.01.030

Cheng J, Ma X, Krausz KW et al (2009) Rifampicin-activated human pregnane X receptor and CYP3A4 induction enhance acetaminophen-induced toxicity. Drug Metab Dispos 37 (8):1611–1621. https://doi.org/10.1124/dmd.109.027565

Chukwurah E, Osmundsen A, Davis SW et al (2019) All together now: modeling the interaction of neural with non-neural systems using organoid models. Front Neurosci 13:582. https://doi.org/10.3389/fnins.2019.00582

Court MH (2013) Canine cytochrome P-450 pharmacogenetics. Vet Clin North Am Small Anim Pract 43(5):1027–1038. https://doi.org/10.1016/j.cvsm.2013.05.001

De Bruyn T, Chatterjee S, Fattah S et al (2013) Sandwich-cultured hepatocytes: utility for in vitro exploration of hepatobiliary drug disposition and drug-induced hepatotoxicity. Expert Opin Drug Metab Toxicol 9(5):589–616. https://doi.org/10.1517/17425255.2013.773973

Delfosse V, Dendele B, Huet T et al (2015) Synergistic activation of human pregnane X receptor by binary cocktails of pharmaceutical and environmental compounds. Nat Commun 6:8089. https://doi.org/10.1038/ncomms9089

Demel MA, Schwaha R, Kramer O et al (2008) In silico prediction of substrate properties for ABC-multidrug transporters. Expert Opin Drug Metab Toxicol 4(9):1167–1180. https://doi.org/10.1517/17425255.4.9.1167

Di Meglio P, Duarte JH, Ahlfors H et al (2014) Activation of the aryl hydrocarbon receptor dampens the severity of inflammatory skin conditions. Immunity 40(6):989–1001. https://doi.org/10.1016/j.immuni.2014.04.019

Ding X, Kaminsky LS (2003) Human extrahepatic cytochromes P450: function in xenobiotic metabolism and tissue-selective chemical toxicity in the respiratory and gastrointestinal tracts. Annu Rev Pharmacol Toxicol 43:149–173. https://doi.org/10.1146/annurev.pharmtox.43.100901.140251

Dou L, Poitevin S, Sallee M et al (2018) Aryl hydrocarbon receptor is activated in patients and mice with chronic kidney disease. Kidney Int 93(4):986–999. https://doi.org/10.1016/j.kint.2017.11.010

Du ZH, Xia J, Sun XC et al (2017) A novel nuclear xenobiotic receptors (AhR/PXR/CAR)-mediated mechanism of DEHP-induced cerebellar toxicity in quails (Coturnix japonica) via disrupting CYP enzyme system homeostasis. Environ Pollut 226:435–443. https://doi.org/10.1016/j.envpol.2017.04.015

Ecker GF, Stockner T, Chiba P (2008) Computational models for prediction of interactions with ABC-transporters. Drug Discov Today 13(7–8):311–317. https://doi.org/10.1016/j.drudis.2007.12.012

Elentner A, Ortner D, Clausen B et al (2015) Skin response to a carcinogen involves the xenobiotic receptor pregnane X receptor. Exp Dermatol 24(11):835–840. https://doi.org/10.1111/exd.12766

Esser C, Rannug A (2015) The aryl hydrocarbon receptor in barrier organ physiology, immunology, and toxicology. Pharmacol Rev 67(2):259–279. https://doi.org/10.1124/pr.114.009001

Estrada-Tejedor R, Ecker GF (2018) Predicting drug resistance related to ABC transporters using unsupervised consensus self-organizing maps. Sci Rep 8(1):6803. https://doi.org/10.1038/s41598-018-25235-9

Fallon JK, Smith PC, Xia CQ et al (2016) Quantification of four efflux drug transporters in liver and kidney across species using targeted quantitative proteomics by isotope dilution NanoLC-MS/MS. Pharm Res 33(9):2280–2288. https://doi.org/10.1007/s11095-016-1966-5

Fang J, Yu PH, Gorrod JW et al (1995) Inhibition of monoamine oxidases by haloperidol and its metabolites: pharmacological implications for the chemotherapy of schizophrenia. Psychopharmacology 118(2):206–212

Flaveny CA, Murray IA, Perdew GH (2010) Differential gene regulation by the human and mouse aryl hydrocarbon receptor. Toxicol Sci 114(2):217–225. https://doi.org/10.1093/toxsci/kfp308

Flaveny CA, Perdew GH (2009) Transgenic humanized AHR mouse reveals differences between human and mouse AHR ligand selectivity. Mol Cell Pharmacol 1(3):119–123. https://doi.org/10.4255/mcpharmacol.09.15

Fraser DJ, Zumsteg A, Meyer UA (2003) Nuclear receptors constitutive androstane receptor and pregnane X receptor activate a drug-responsive enhancer of the murine 5-aminolevulinic acid synthase gene. J Biol Chem 278(41):39392–39401. https://doi.org/10.1074/jbc.M306148200

Fritsche E, Schafer C, Calles C et al (2007) Lightening up the UV response by identification of the arylhydrocarbon receptor as a cytoplasmatic target for ultraviolet B radiation. Proc Natl Acad Sci U S A 104(21):8851–8856. https://doi.org/10.1073/pnas.0701764104

Frye CA, Koonce CJ, Walf AA (2014) Novel receptor targets for production and action of allopregnanolone in the central nervous system: a focus on pregnane xenobiotic receptor. Front Cell Neurosci 8:106. https://doi.org/10.3389/fncel.2014.00106

Fuchs I, Hafner-Blumenstiel V, Markert C et al (2013) Effect of the CYP3A inhibitor ketoconazole on the PXR-mediated induction of CYP3A activity. Eur J Clin Pharmacol 69(3):507–513. https://doi.org/10.1007/s00228-012-1388-1

Fujiwara R, Yoda E, Tukey RH (2018) Species differences in drug glucuronidation: humanized UDP-glucuronosyltransferase 1 mice and their application for predicting drug glucuronidation and drug-induced toxicity in humans. Drug Metab Pharmacokinet 33(1):9–16. https://doi.org/10.1016/j.dmpk.2017.10.002

Geffrey AL, Pollack SF, Bruno PL et al (2015) Drug-drug interaction between clobazam and cannabidiol in children with refractory epilepsy. Epilepsia 56(8):1246–1251. https://doi.org/10.1111/epi.13060

Gerth K, Kodidela S, Mahon M et al (2019) Circulating extracellular vesicles containing xenobiotic metabolizing CYP enzymes and their potential roles in extrahepatic cells via cell-cell interactions. Int J Mol Sci 20(24). https://doi.org/10.3390/ijms20246178

Ghosh C, Hossain M, Solanki J et al (2017) Overexpression of pregnane X and glucocorticoid receptors and the regulation of cytochrome P450 in human epileptic brain endothelial cells. Epilepsia 58(4):576–585. https://doi.org/10.1111/epi.13703

Giacomini KM, Galetin A, Huang SM (2018) The international transporter consortium: summarizing advances in the role of transporters in drug development. Clin Pharmacol Ther 104(5):766–771. https://doi.org/10.1002/cpt.1224

Giacomini KM, Huang SM, Tweedie D et al (2010) Membrane transporters in drug development. Nat Rev Drug Discov 9:215–236. https://doi.org/10.1038/nrd3028

Gornstein E, Schwarz TL (2014) The paradox of paclitaxel neurotoxicity: mechanisms and unanswered questions. Neuropharmacology 76(PtA):175–183. https://doi.org/10.1016/j.neuropharm.2013.08.016

Graziano S, Orsolini L, Rotolo MC et al (2017) Herbal highs: review on psychoactive effects and neuropharmacology. Curr Neuropharmacol 15(5):750–761. https://doi.org/10.2174/1570159X14666161031144427

Hasegawa M, Tahara H, Inoue R et al (2012) Investigation of drug-drug interactions caused by human pregnane X receptor-mediated induction of CYP3A4 and CYP2C subfamilies in chimeric mice with a humanized liver. Drug Metab Dispos 40(3):474–480. https://doi.org/10.1124/dmd.111.042754

Heide M, Huttner WB, Mora-Bermudez F (2018) Brain organoids as models to study human neocortex development and evolution. Curr Opin Cell Biol 55:8–16. https://doi.org/10.1016/j.ceb.2018.06.006

Hoffmaster KA, Turncliff RZ, LeCluyse EL et al (2004) P-glycoprotein expression, localization, and function in sandwich-cultured primary rat and human hepatocytes: relevance to the hepatobiliary disposition of a model opioid peptide. Pharm Res 21(7):1294–1302

Holmstock N, Gonzalez FJ, Baes M et al (2013) PXR/CYP3A4-humanized mice for studying drug-drug interactions involving intestinal P-glycoprotein. Mol Pharm 10(3):1056–1062. https://doi.org/10.1021/mp300512r

Issa NT, Wathieu H, Ojo A et al (2017) Drug metabolism in preclinical drug development: a survey of the discovery process, toxicology, and computational tools. Curr Drug Metab 18(6):556–565. https://doi.org/10.2174/1389200218666170316093301

Ji K, Seo J, Kho Y et al (2019) Co-exposure to ketoconazole alters effects of bisphenol A in Danio rerio and H295R cells. Chemosphere 237:124414. https://doi.org/10.1016/j.chemosphere.2019.124414

Johnson MW, Griffiths RR (2017) Potential therapeutic effects of psilocybin. Neurotherapeutics 14(3):734–740. https://doi.org/10.1007/s13311-017-0542-y

Juricek L, Coumoul X (2018) The aryl hydrocarbon receptor and the nervous system. Int J Mol Sci 19(9). https://doi.org/10.3390/ijms19092504

Kaur J, Sodhi RK, Madan J et al (2018) Forskolin convalesces memory in high fat diet-induced dementia in wistar rats-plausible role of pregnane x receptors. Pharmacol Rep 70(1):161–171. https://doi.org/10.1016/j.pharep.2017.07.009

Kim TW, Samet JH, Cheng DM et al (2015) The spectrum of unhealthy drug use and quality of care for hypertension and diabetes: a longitudinal cohort study. BMJ Open 5(12):e008508. https://doi.org/10.1136/bmjopen-2015-008508

Kimura E, Tohyama C (2017) Embryonic and postnatal expression of aryl hydrocarbon receptor mRNA in mouse brain. Front Neuroanat 11:4. https://doi.org/10.3389/fnana.2017.00004

Kropp S, Lichtinghagen R, Winterstein K et al (2006) Cytochrome P-450 2D6 and 2C19 polymorphisms and length of hospitalization in psychiatry. Clin Lab 52(5–6):237–240

Li F, Lu J, Cheng J, Wang L et al (2013) Human PXR modulates hepatotoxicity associated with rifampicin and isoniazid co-therapy. Nat Med 19(4):418–420. https://doi.org/10.1038/nm.3104

Lobao-Soares B, Eduardo-da-Silva P, Amarilha H et al (2018) It's tea time: interference of ayahuasca brew on discriminative learning in zebrafish. Front Behav Neurosci 12:190. https://doi.org/10.3389/fnbeh.2018.00190

Lombardo L, Pellitteri R, Balazy M et al (2008) Induction of nuclear receptors and drug resistance in the brain microvascular endothelial cells treated with antiepileptic drugs. Curr Neurovasc Res 5(2):82–92. https://doi.org/10.2174/156720208784310196

Malaplate-Armand C, Ferrari L, Masson C et al (2005) Down-regulation of astroglial CYP2C, glucocorticoid receptor and constitutive androstane receptor genes in response to cocaine in human U373 MG astrocytoma cells. Toxicol Lett 159(3):203–211. https://doi.org/10.1016/j.toxlet.2005.04.005

Manda VK, Avula B, Dale OR et al (2017) PXR mediated induction of CYP3A4, CYP1A2, and P-gp by Mitragyna speciosa and its alkaloids. Phytother Res 31(12):1935–1945. https://doi.org/10.1002/ptr.5942

Martignoni M, Groothuis GM, de Kanter R (2006) Species differences between mouse, rat, dog, monkey and human CYP-mediated drug metabolism, inhibition and induction. Expert Opin Drug Metab Toxicol 2(6):875–894. https://doi.org/10.1517/17425255.2.6.875

Martiny VY, Carbonell P, Chevillard F et al (2015) Integrated structure- and ligand-based in silico approach to predict inhibition of cytochrome P450 2D6. Bioinformatics 31 (24):3930–3937. doi:https://doi.org/10.1093/bioinformatics/btv486

Miki Y, Suzuki T, Tazawa C et al (2005) Steroid and xenobiotic receptor (SXR), cytochrome P450 3A4 and multidrug resistance gene 1 in human adult and fetal tissues. Mol Cell Endocrinol 231 (1–2):75–85. https://doi.org/10.1016/j.mce.2004.12.005

Molina DK, Hargrove VM (2011) Fatal cocaine interactions: a review of cocaine-related deaths in Bexar County, Texas. Am J Forensic Med Pathol 32(1):71–77

Moriguchi T, Motohashi H, Hosoya T et al (2003) Distinct response to dioxin in an arylhydrocarbon receptor (AHR)-humanized mouse. Proc Natl Acad Sci U S A 100(10):5652–5657. https://doi.org/10.1073/pnas.1037886100

Moyano P, Ruiz M, Garcia JM et al (2019) Oxidative stress and cell death induction by amitraz and its metabolite BTS-27271 mediated through cytochrome P450 and NRF2 pathway alteration in primary hippocampal cell. Food Chem Toxicol 129:87–96. https://doi.org/10.1016/j.fct.2019.04.042

Nelson DR, Zeldin DC, Hoffman SM et al (2004) Comparison of cytochrome P450 (CYP) genes from the mouse and human genomes, including nomenclature recommendations for genes, pseudogenes and alternative-splice variants. Pharmacogenetics 14(1):1–18

NIDA (2018) Commonly abused drugs charts. https://www.drugabuse.gov/drugs-abuse/commonly-abused-drugs-charts

Nishimuta H, Nakagawa T, Nomura N et al (2013) Species differences in hepatic and intestinal metabolic activities for 43 human cytochrome P450 substrates between humans and rats or dogs. Xenobiotica 43(11):948–955. https://doi.org/10.3109/00498254.2013.787155

Palhano-Fontes F, Barreto D, Onias H et al (2018) Rapid antidepressant effects of the psychedelic ayahuasca in treatment-resistant depression: a randomized placebo-controlled trial. Psychol Med 49(4):655–663. https://doi.org/10.1017/s0033291718001356

Papageorgiou I, Grepper S, Unadkat JD (2013) Induction of hepatic CYP3A enzymes by pregnancy-related hormones: studies in human hepatocytes and hepatic cell lines. Drug Metab Dispos 41(2):281–290. https://doi.org/10.1124/dmd.112.049015

Pondugula SR, Ferniany G, Ashraf F et al (2015) Stearidonic acid, a plant-based dietary fatty acid, enhances the chemosensitivity of canine lymphoid tumor cells. Biochem Biophys Res Commun 460(4):1002–1007. https://doi.org/10.1016/j.bbrc.2015.03.141

Pondugula SR, Mani S (2013) Pregnane xenobiotic receptor in cancer pathogenesis and therapeutic response. Cancer Lett 328(1):1–9. https://doi.org/10.1016/j.canlet.2012.08.030

Pondugula SR, Pavek P, Mani S (2016) Pregnane X receptor and cancer: context-specificity is key Nucl Receptor Res. doi:https://doi.org/10.11131/2016/101198

Radford R, Frain H, Ryan MP et al (2013) Mechanisms of chemical carcinogenesis in the kidneys. Int J Mol Sci 14(10):19416–19433. https://doi.org/10.3390/ijms141019416

Ross J, Plummer SM, Rode A et al (2010) Human constitutive androstane receptor (CAR) and pregnane X receptor (PXR) support the hypertrophic but not the hyperplastic response to the murine nongenotoxic hepatocarcinogens phenobarbital and chlordane in vivo. Toxicol Sci 116 (2):452–466. https://doi.org/10.1093/toxsci/kfq118

Sager JE, Lutz JD, Foti RS et al (2014) Fluoxetine- and norfluoxetine-mediated complex drug-drug interactions: in vitro to in vivo correlation of effects on CYP2D6, CYP2C19, and CYP3A4. Clin Pharmacol Ther 95(6):653–662. https://doi.org/10.1038/clpt.2014.50

Sakai C, Iwano S, Yamazaki Y et al (2014) Species differences in the pharmacokinetic parameters of cytochrome p450 probe substrates between experimental animals, such as mice, rats, dogs, monkeys, and microminipigs, and humans. J Drug Metab Toxicol 5(6):1–12. https://doi.org/10.4172/2157-7609.1000173

Schule C, Nothdurfter C, Rupprecht R (2014) The role of allopregnanolone in depression and anxiety. Prog Neurobiol 113:79–87. https://doi.org/10.1016/j.pneurobio.2013.09.003

Serra V, Fattore L, Scherma M et al (2015) Behavioural and neurochemical assessment of salvinorin a abuse potential in the rat. Psychopharmacology 232(1):91–100. https://doi.org/10.1007/s00213-014-3641-z

Singh A, Mudawal A, Maurya P et al (2016) Prenatal exposure of cypermethrin induces similar alterations in xenobiotic-metabolizing cytochrome p450s and rate-limiting enzymes of neurotransmitter synthesis in brain regions of rat offsprings during postnatal development. Mol Neurobiol 53(6):3670–3689. https://doi.org/10.1007/s12035-015-9307-y

Singh H, Sodhi RK, Chahal SK et al (2020) Meclizine ameliorates memory deficits in streptozotocin-induced experimental dementia in mice: role of nuclear pregnane X receptors. Can J Physiol Pharmacol 98(6):383–390. https://doi.org/10.1139/cjpp-2019-0421

Stage TB, Mortensen C, Khalaf S et al (2020) P-glycoprotein inhibition exacerbates paclitaxel neurotoxicity in neurons and patients with Cancer. Clin Pharmacol Ther 108:671. https://doi.org/10.1002/cpt.1847

Taylor AB, Roberts KM, Cao X et al (2017) Structural and enzymatic insights into species-specific resistance to schistosome parasite drug therapy. J Biol Chem 292(27):11154–11164. https://doi.org/10.1074/jbc.M116.766527

Tigges J, Weighardt H, Wolff S et al (2013) Aryl hydrocarbon receptor repressor (AhRR) function revisited: repression of CYP1 activity in human skin fibroblasts is not related to AhRR expression. J Invest Dermatol 133(1):87–96. https://doi.org/10.1038/jid.2012.259

Torres-Vergara P, Ho YS, Espinoza F et al (2020) The constitutive androstane receptor and pregnane X receptor in the brain. Br J Pharmacol 177(12):2666–2682. https://doi.org/10.1111/bph.15055

Tsujikawa K, Kuwayama K, Miyaguchi H et al (2009) In vitro stability and metabolism of salvinorin a in rat plasma. Xenobiotica 39(5):391–398. https://doi.org/10.1080/00498250902769967

Tyzack JD, Kirchmair J (2019) Computational methods and tools to predict cytochrome P450 metabolism for drug discovery. Chem Biol Drug Des 93(4):377–386. https://doi.org/10.1111/cbdd.13445

Upadhya SC, Chinta SJ, Pai HV, Boyd MR, Ravindranath V (2002) Toxicological consequences of differential regulation of cytochrome p450 isoforms in rat brain regions by phenobarbital. Arch Biochem Biophys 399(1):56–65. https://doi.org/10.1006/abbi.2001.2727

Villard PH, Sampol E, Elkaim JL et al (2002) Increase of CYP1B1 transcription in human keratinocytes and HaCaT cells after UV-B exposure. Toxicol Appl Pharm 178(3):137–143. https://doi.org/10.1006/taap.2001.9335

Vollenweider FX, Vollenweider-Scherpenhuyzen MF, Babler A et al (1998) Psilocybin induces schizophrenia-like psychosis in humans via a serotonin-2 agonist action. Neuroreport 9 (17):3897–3902

Wang X, Sykes DB, Miller DS (2010) Constitutive androstane receptor-mediated up-regulation of ATP-driven xenobiotic efflux transporters at the blood-brain barrier. Mol Pharmacol 78 (3):376–383. https://doi.org/10.1124/mol.110.063685

Wang YM, Ong SS, Chai SC et al (2012) Role of CAR and PXR in xenobiotic sensing and metabolism. Expert Opin Drug Metab Toxicol 8(7):803–817. https://doi.org/10.1517/17425255.2012.685237

Xie W, Barwick JL, Downes M et al (2000) Humanized xenobiotic response in mice expressing nuclear receptor SXR. Nature 406(6794):435–439. https://doi.org/10.1038/35019116

Xu K, Yang Z, Shi R et al (2016) Expression of aryl hydrocarbon receptor in rat brain lesions following traumatic brain injury. Diagn Pathol 11(1):72. https://doi.org/10.1186/s13000-016-0522-2

Yamasaki Y, Kobayashi K, Okuya F et al (2018) Characterization of p-glycoprotein humanized mice generated by chromosome engineering technology: its utility for prediction of drug distribution to the brain in humans. Drug Metab Dispos 46(11):1756–1766. https://doi.org/10.1124/dmd.118.081216

You D, Wen X, Gorczyca L et al (2019) Increased MDR1 transporter expression in human brain endothelial cells through enhanced histone acetylation and activation of aryl hydrocarbon receptor signaling. Mol Neurobiol 56(10):6986–7002. https://doi.org/10.1007/s12035-019-1565-7

Yu N, Zhang YF, Zhang K et al (2017) Pregnane X receptor not nuclear factor-kappa b up-regulates p-glycoprotein expression in the brain of chronic epileptic rats induced by kainic acid. Neurochem Res 42(8):2167–2177. https://doi.org/10.1007/s11064-017-2224-x

Zastre JA, Chan GN, Ronaldson PT et al (2009) Up-regulation of P-glycoprotein by HIV protease inhibitors in a human brain microvessel endothelial cell line. J Neurosci Res 87(4):1023–1036. https://doi.org/10.1002/jnr.21898

Zhai Y, Pai HV, Zhou J, Amico JA et al (2007) Activation of pregnane X receptor disrupts glucocorticoid and mineralocorticoid homeostasis. Mol Endocrinol 21(1):138–147. https://doi.org/10.1210/me.2006-0291

Ziv-Gal A, Craig ZR, Wang W et al (2013) Bisphenol a inhibits cultured mouse ovarian follicle growth partially via the aryl hydrocarbon receptor signaling pathway. Reprod Toxicol 42:58–67. https://doi.org/10.1016/j.reprotox.2013.07.022

# Cannabis-Induced Neuroactivity: Research Trends and Commercial Prospects

**Rajiv Kumar Chaturvedi and Dinesh Chandra Agrawal**

## Contents



**Abstract** The worldwide interest and push for the legalization of cannabis/marijuana, especially in the United States, are increasing with each passing day. The present article deals with the concise yet broad review of chemical, medicinal (neuroprotection), and adverse psychotic aspects of cannabis (marijuana or

R. K. Chaturvedi (✉)
QLeap Academy, Pune, India
e-mail: rajiv.k@qleap.in

D. C. Agrawal (✉)
Department of Applied Chemistry, Chaoyang University of Technology, Taichung, Taiwan
e-mail: dcagrawal@gm.cyut.edu.tw

 
D. C. Agrawal, M. Dhanasekaran (eds.), *Medicinal Herbs and Fungi*,
https://doi.org/10.1007/978-981-33-4141-8_6

marihuana). The emphasis is made to understand the influence of tetrahydrocannabinol (THC) on a broad spectrum of properties ranging from psychosis, neuroprotection, neurotoxicity to medicinal. The reason why THC shows psychoactivity, but cannabidiol (CBD) does not, has been elucidated based on the minor difference in their chemical structures inhibiting CBD to bind with cannabinoid receptors due to steric hindrance. The distribution of cannabinoid receptors (namely, CB1 and CB2) in the human body and the role of endocannabinoids (namely, anandamide and 2-arachidonoyl glycerol) throughout the human system are described. The effect of the method of consumption (inhalation vs. ingestion) on the psychotropicity of cannabis/THC has also been discussed. Additionally, the effect of the use of synthetic endocannabinoid receptor blocker (antagonist) as a drug molecule for a specific purpose, such as for reducing the appetite, to treat obesity, or for the treatment of tobacco, alcohol, and other hard drugs induced addiction, and their potential adverse effects are also the focus of the article. Both the benefits and the risks of consuming cannabinoids are mainly dose-dependent, just like any other legal or prescription pharma products or regulated/unregulated psychotropic substances. Moderation is the right old prescription for a healthy and long productive life, and it applies to the use of medicinal, cultural, and/or recreational products like cannabis/cannabinoids. The traditional use of cannabis leaves (bhang) in India for medical as well as cultural purposes has been discussed from the modern scientific perspective. Lastly, the rapidly growing trend of the number of the publication of both the scientific research papers and the patent applications on cannabis, along with the market trend of cannabis-derived products, has been provided, showing quite high and promising growth.

**Keywords** Cannabis · Marihuana · Marijuana · Medicinal plant · Neuroprotection · Psychosis

## Abbreviations

| | |
|---|---|
| 11-COOH-THC | 11-nor-9-carboxy-delta 9-tetrahydrocannabinol |
| 11-OH-THC | 11-hydroxy-delta 9-tetrahydrocannabinol |
| 2-AG | 2-Arachidonoyl glycerol |
| AIDS | Acquired immune deficiency syndrome |
| BCE | Before Common Era |
| Bn | Billion |
| CAGR | Compound Annual Growth Rate |
| CB | Cannabinoid |
| CBD | Cannabidiol |
| CBN | Cannabinol |
| CE | Common Era |
| EC | Endocannabinoid |
| FDA | Food and Drug Administration |
| NDPS | Narcotic Drugs and Psychotropic Substances |

| | |
|---|---|
| NIAAA | National Institute of Alcohol Abuse and Alcoholism |
| NIH | National Institute of Health |
| SCND | Single Convention on Narcotic Drugs |
| THC | Tetrahydrocannabinol |
| THCA | Tetrahydrocannabinolic Acid |
| USA | United States of America |
| USD | United States Dollar |

# 1 Introduction

Cannabis is among the very first few plants domesticated by humans as early as ca. 5000 BCE, mainly in Central Asia, China, and India (Tikka and D'Souza 2019; Russo 2005). Similarly, cannabis or marijuana/marihuana has been in common use in the Americas, both north and south, and Europe for fiber making and recreational purposes (Russo 2005). However, in the last century, there were some concerns raised about the potential abuse of marijuana, and the United States led the worldwide crusade against its quite prevalent use/abuse (Courtwright 1992). Because of this, cannabis/marijuana was banned and criminalized globally. It was only after the extraction and identification of the main psychoactive component of cannabis/marijuana, tetrahydrocannabinol (THC) (Gaoni and Mechoulam 1964), the interest in researchers grew significantly, leading to the discoveries of endocannabinoid (EC) receptors and EC agonists like anandamide and EC antagonist virodhamine. Furthermore, it was found that there are at least two main cannabinoids (CB) receptors (CB1 and CB2) in the human body. While CB1 receptors are predominantly located in different parts of the brain (Matsuda et al. 1990), the CB2 receptors are mainly found in the immunological and peripheral systems (Munro et al. 1993). In the last three decades, the explosion in the evidenced-based research on the medicinal as well as psychotoxic/adverse aspects of cannabinoids, be it phytocannabinoids (derived from cannabis like THC) or synthetic cannabinoids (like HU210 and many more) (Mechoulam et al. 1990).

There are numerous reviews and books available in the scientific literature on cannabis, and the trend is increasing (Devane et al. 1992; Nahas et al. 1999; Labigalini Jr et al. 1999; Wright et al. 2003; Hanuš and Mechoulam 2005; Pertwee 2006; Montoya and Weiss 2019; Finn 2020), which indicates its importance. In the last few decades, there have been quite evidenced-based discoveries covering various aspects of cannabinoid interaction with the endocannabinoid (EC) system and understanding the role they play in medicinal effects on one hand and the neurotoxic effects on the other. Sometimes, the evidenced-based reports seem even contradictory. Additionally, users' experience is also quite varying.

The present chapter deals with the concise yet broad review of chemical, medicinal (neuroprotection), and adverse psychotic aspects of cannabis. The emphasis will be made to understand the effect of THC on a broad spectrum of properties ranging from medicinal to psychosis and neuroprotection to neurotoxicity. The reason why THC shows psychoactivity, and CBD does not, will be emphasized. The effect of the method of consumption (inhalation vs. ingestion) on the psychotropicity of cannabis/THC will also be discussed. Additionally, the effect of the use of synthetic EC receptor blocker (antagonist) as a drug molecule for a specific purpose, such as for reducing the appetite, to treat obesity, or for the treatment of tobacco, alcohol, and other hard drugs induced addiction, and their potential adverse effects are also the focus of the present article. The ancient and traditional use of cannabis leaves (bhang) in India for medical as well as cultural purposes will be discussed from the modern scientific perspective. Furthermore, the trends in scientific research, intellectual property (patents) as well as commercial/business prospects related to cannabis are also discussed.

## 2 Botanical Aspects

Genus *Cannabis*, a member of the Cannabaceae family, has three main species, namely *C. sativa, C. indica,* and *C. ruderalis*. While *C. ruderalis* possesses hardly any psychotropicity, both *C. indica* and *C. sativa* exhibit varying degrees of psychosomatic effects when consumed. Significant details about the botanical aspects of cannabis are given elsewhere and can be referred to for more information (Farag and Kayser 2017). The differences between these species in morphology and their effects (Anonymous 2017) are as follows:

***Cannabis sativa***: Plants can grow up to 5–6 m in height is loosely branched. Leaves are narrow and long (Fig. 1a). Its consumption causes the feeling of high, energizing, more of psychic/cerebral buzz, motivation, focusing and increasing creativity, reduction in nausea, relieving depression, a sense of well-being, and appetite stimulation. It is more suited for use in the day time.

***Cannabis indica***: Plants grow to the medium height of about 2 m, thickly branched, and the leaves are quite broad (Fig. 1b). Its effects are mainly a feeling of body and cerebral buzz, muscle relaxation, sedative, appetite enhancement,

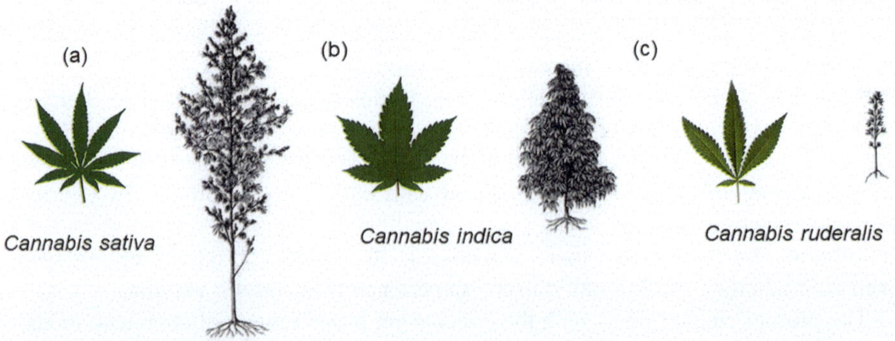


**Fig. 1** Morphological features of (**a**) *Cannabis sativa*, (**b**) *Cannabis indica*, (**c**) *Cannabis ruderalis* (Anonymous 2017) (https://www.leafscience.com/2017/10/16/indica-vs-sativa-whats-difference/)

pain-relieving, stress and anxiety-relieving, and sleep-inducing. It is more suited for evening consumption.

***Cannabis ruderalis***: The plant is smaller and grows about 1 m high. Leaves are moderately broad (Fig. 1c). It is almost non-psychotropic. It is auto-flowering (does not depend on photoperiod for flowering).

The hybridization or cross-pollination of these species produces several strains that may have variable psychoactivity as desired by the user. Furthermore, it may also produce certain strains having psychoactivity and auto-flowering. Hybrid strains can help cannabis growers requiring little attention, and the crop can be recultivated immediately after harvesting.

## 3 Chemical Aspects: Origin of Psychoactivity of Cannabis

Cannabis is one of the most studied and debated psychotropic plants, mainly due to its vast spectrum of psychosomatic effects. It is believed that the psychotropic effects of this plant were discovered in India (Tikka and D'Souza 2019), specifically in the Himalayan regions. Like medicinal plants, the psychoactive plants derive the activity from the chemicals produced by such plants. It is their chemistry that imparts the unique properties, be it medicinal or psychoactive. There is instead a thin line between psychoactive and therapeutic properties on the one hand and between psychoactive and neurotoxic properties on the other. However, in the last few decades, the psychotropic activity of cannabis attracted significant attention of researchers.

### *3.1 Cannabinoids*

The origin of the term "cannabinoids" has its roots in "cannabis," which in turn seems to have been derived from "cane-like," and is used for a group of chemicals naturally present in cannabis. Although the cannabis plant has around 100 cannabinoids, the most common are delta$^9$-tetrahydrocannabinol (from now on referred to as tetrahydrocannabinol or THC), cannabidiol (CBD), and tetrahydrocannabinolic acid (THCA). However, THCA, a nonpsychoactive substance, is decarboxylated to THC, the main psychoactive compound, when cannabis leaves/flowers are dried/heated. Further, THC and CBD are interconvertible under certain conditions.

THC was first isolated by Gaoni and Mechoulam (1964). The identification of the psychoactive constituent THC gave significant impetus to research in this area. Furthermore, it was found that when acquired immune deficiency syndrome (AIDS), patients smoked marijuana, their appetite increased, resulting in less weight loss. Experiments on animals showed that THC promotes appetite, and clinical trials confirmed its activity. This natural product was introduced clinically under the generic name Dronabinol (Mechoulam et al. 2006).

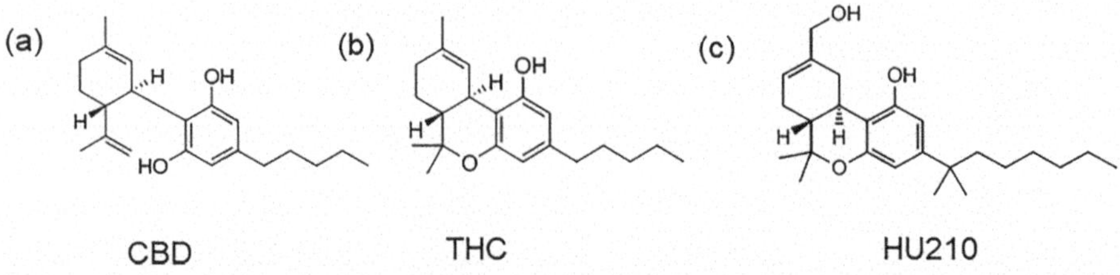


**Fig. 2** Chemical structure of (**a**) CBD, (**b**) THC, and (**c**) HU210

THC, being psychoactive, is associated with the "high" experienced when cannabis is consumed. On the other hand, CBD does not show any appreciable psychoactivity and is known to have a moderation effect of THC. The THC/CBD ratio can define the psychoactive potency in cannabis, where the higher the value of this ratio, the higher the psychoactive potency. The chemical structure of these two main components CBD and THC of cannabis, along with a synthetic cannabinoid HU210, is shown in Fig. 2 (a–c). The HU210 was first synthesized from (1R,5S)-myrtenol (Mechoulam et al. 1990). This synthetic cannabinoid was prepared in the laboratory, and its chemical structure is quite similar to that of THC. However, compared to THC, the psychoactive potency of HU210 is significantly higher (100–800 times) than that of THC. HU210 is the (−)-1,1-dimethyl heptyl analog of 11-hydroxy- $\Delta^8$- tetrahydrocannabinol. It is also called as 1,1-dimethyl heptyl-11-hydroxytetrahydrocannabinol.

## 3.2 *Why THC Is Psychoactive and CBD Is Nonpsychoactive?*

There is only a minor difference in the chemical structures of THC and CBD (Fig. 2a, b). Both are functional isomers having the same molecular formula (($C_{21}H_{30}O_2$). In CBD, there is an alcoholic group in the vicinity of the double bond (>C=$CH_2$) in an open moiety. At the same site, in THC, there is an etheric group making an additional six-membered heterocyclic ring. Now the question of why the only THC binds to the receptors, called cannabinoid (CB) receptors, present in the body, imparting high or psychoactive influence and not CBD, has puzzled researchers. Recently, Bragança et al. (2020), based on conformational analysis using quantum chemical analysis, found that the CBD structure is quite flexible and has different conformations (geometries) while THC is a rigid and planer structure, like another psychoactive compound (LSD, Heroin), and can bind the receptors present in the central nervous system. While a non-planer structure of CBD experiences a steric repulsion between hydroxyl and alkene moieties, and the structure is not planer, unlike that of THC. This aspect of CBD structure may be causing a steric hindrance not allowing to bind the EC receptors in the body and thereby shows no psychoactive effect.

### *3.3 Cannabinoids and Endocannabinoid System*

While trying to understand the psychoactive and other medicinal effects observed after consuming cannabis, it is evident that there must be a specific receptor system in the body. After the isolation and identification of the main psychoactive component in cannabis, THC, the endocannabinoid receptor system was discovered by Gaoni and Mechoulam (1986). If there are EC receptors in the body, there must be naturally occurring cannabinoid produced in the body. In 1992, scientists at Hebrew University (Devane et al. 1992), isolated and identified an endogenously produced cannabinoid (endocannabinoid), named anandamide (arachidonoylethanolamide, an arachidonic acid derivative) (Fig. 3a). The term anandamide is derived from the Sanskrit word "Ananda" meaning bliss, aptly describing the effect of this endocannabinoid in the brain. The isomer of anandamide, an amine, is termed as "virodhamine" (Fig. 3b) derived from the Sanskrit word "virodh" meaning opposite or against. Later on, another endocannabinoid was found in the body called 2-arachidonoylglycerol (2-AG) (Fig. 3c) (Mechoulam et al. 1995), which can also "lock" into EC receptors (Stella et al. 1997). Structurally, anandamide (3a) and 2-AG (3c) are quite similar, and both have long aliphatic lipid chains that facilitate slipping these molecules in the cell. It was speculated that 2-AG might be playing a complementary function in the activity of anandamide. 2-AG is found in certain parts of the brain in as large as 170 times the concentration of anandamide. Obviously, when the function of these endocannabinoids is understood, it may be quite helpful in understanding their medicinal effects. It will also help in exploiting the therapeutic applications of THC and CBD while avoiding its potentially adverse consequences.

Based on the origin of cannabinoids, these can be classified into three categories as endocannabinoids, phytocannabinoids, and synthetic cannabinoids, as depicted below (Fig. 4).

## 4 Neurotransmission and Endocannabinoids

Generally, communication between neurons happens through chemical signaling to coordinate and regulate our feelings, thinking, and actions. These chemical messengers are commonly called neurotransmitters. The transmitting neurons (presynaptic

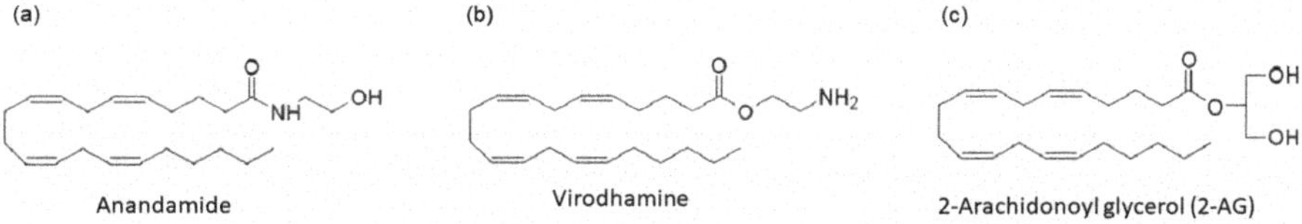


**Fig. 3** The structures of (**a**) Anandamide, (**b**) Virodhamine (**c**) 2-Arachidonoyl glycerol

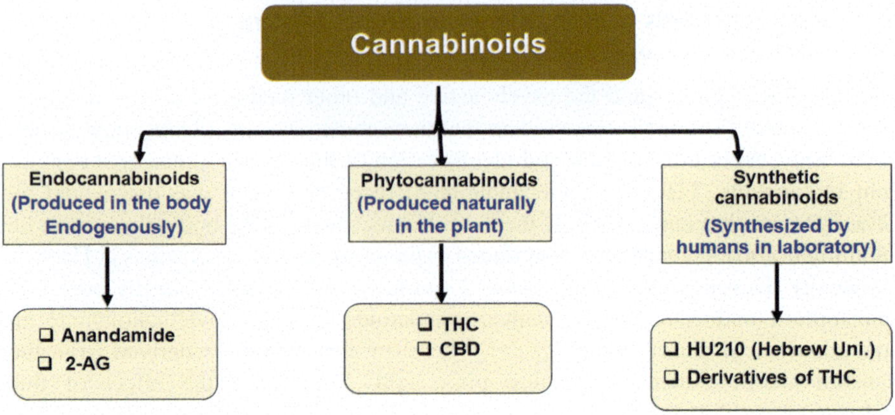


**Fig. 4** Types of cannabinoids

cells) send neurotransmitters to specific receiving neurons (receptors, also called postsynaptic cells) through a tiny gap (synapse) to trigger a series of actions to achieve the desired activity. There is a sort of "lock and key" arrangement between neurotransmitter molecules and specific receptor cells (receiving neurons), where a chemical molecule (key) attaches itself to the receptor (lock) in an unique way.

However, the endocannabinoids follow different signaling systems. In a way, the "backward route," where postsynaptic neuron, once activated, triggers an endocannabinoid messenger molecule, for example, anandamide, which is produced "on-demand" via fatty acid precursor, already present in the form of fat cells at the cell wall of neurons. Then this endocannabinoid is released and travels to presynaptic cells and attaches to the CB receptor. This unique communication way of the EC system is very important as endocannabinoids act on transmitting neurons (presynaptic cells). It can regulate/moderate the next course of action, like the controlled release of neurotransmitters and thereby controlling the manner of neuro-communication and processing the message.

For example, when neurons experience electrolyte (chloride ions) imbalance between internal (inside the cell) and external (outside the cell) fluid, there is an excessive neural firing. It triggers the production of endocannabinoid, anandamide, which then attaches to CB receptors on these cells and opens a small window in the cell wall. This leads to quick attainment of the electrolyte balance between inside and outside the cell wall, and then the uncontrolled firing is stopped. This could be the reason why the feeling of bliss and calm is experienced. However, once the electrolytic balance is achieved, this endocannabinoid ligand/agonist, like anandamide, has to be removed from the cell wall to restore the normal cell neurotransmission. It is achieved by secreting an enzyme that quickly breaks the cannabinoid by cleaving very fragile and highly active amide linkage via hydrolysis.

## *4.1 Cannabinoid Receptors (CB1 and CB2) in the Human Body*

The endocannabinoid receptors, or simply cannabis receptors, are target cells in the body that attach with cannabinoid agonists to trigger quite important psychophysiological actions. There are two main cannabinoid receptors, namely CB1 and CB2. The CB1 receptors, first cloned by Matsuda et al. (1990), are mainly located in the brain and are supposed to be involved in the psychoactive properties of cannabis. However, to understand the nonpsychoactive effects of cannabinoids, the cloning of a receptor for cannabinoids is not expressed mainly in the brain but macrophages in the marginal zone of the spleen and immunological system (Munro et al. 1993).

The distribution of CB1 and CB2 receptors throughout the human body, along with their respective target functions/activities, has been depicted in Fig. 5. Although both CB1 and CB2 EC receptors are distributed throughout the human body, CB1 types of receptors are mainly concentrated in the brain, and CB2 receptors are mostly present in the immunological system. Since these cannabinoid receptors are widely distributed in the human body, it is anticipated that they will affect a variety of psychosomatic functions.

It is no wonder that cannabis has been traditionally used for stimulating appetite. Since it is now known that CB1 receptors are present in the hypothalamus part of the brain, which regulates appetite and sexual behavior, hence, by administering a CB1 antagonist, which blocks the CB1 receptor, one can control food intake and thereby it can be used as an anti-obesity drug. A CB1 antagonist, Rimonabant (Rinaldi-

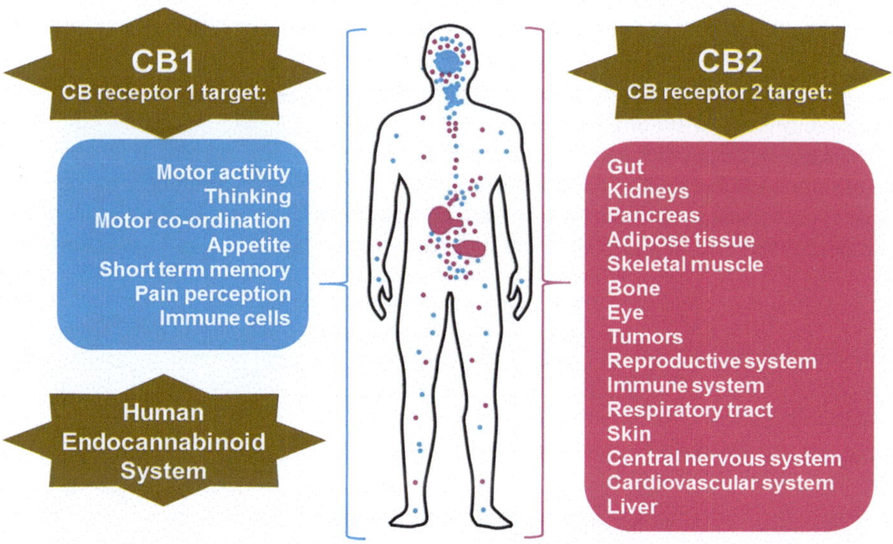


**Fig. 5** CB1 and CB2 receptors and their major role in different human activities/functions in the human body (Adapted from Komnenov (2019); https://www.rxleaf.com)

Carmona et al. 1994), was developed as an anti-obesity molecule and was marketed, mainly in Europe. Similarly, another CB1 antagonist, Surinabant (Rinaldi-Carmona et al. 2004), was developed for treating smoking addiction. The structure of these both CB1 antagonists, namely Rimonabant ((SR141716), 5-(4-chlorophenyl)-1-(2,4-dichlorophenyl)-3-methyl- *N*-(piperidin-1-yl)-1*H*-pyrazole-4-carboxamide) and Surinabant ((SR147778), 5-(4-bromophenyl)-1-(2,4-dichlorophenyl)-4-ethyl-*N*-(1-piperidinyl)-1*H*-pyrazole-3-carboxamide) is shown in Fig. 6. While the Rimonabant was approved as an anti-obesity drug, the Surinabant could not clear the clinical test and was not approved. However, due to some very severe life-threatening side effects, including the development of suicidal tendencies, Rimonabant was withdrawn (Sam et al. 2011).

This example highlights the complexity of the endocannabinoid system, which regulates a vast number of interdependent body functions. Hence, any such CB receptor antagonists may impact a variety of body functions, which may be interdependent. Therefore, utmost care must be taken while trying to develop desired therapeutic applications of CB receptors antagonist-based drugs.

## 4.2 Endocannabinoids (Anandamide and 2-AG) and their Functions

Anandamide plays a vital role in humans and other mammals, as can be seen from the distribution of EC receptors throughout the human body. Since the receptors (both CB1 and CB 2) are present in different parts of the brain and immunological system, these regulate a plethora of body functions, as mentioned in Fig. 5. In short, endocannabinoids may determine how we think, feel, and behave as they regulate our memory and cognition, emotions as well as actions.

Outside the neuro system, anandamide plays a very significant and critical role in reproduction by facilitating early pregnancy in mammals, where it acts as a chemical messenger between the embryo and placental wall of the uterus. It is the first communication between "child" and mother. The largest concentration of anandamide receptors is in the uterus during the early pregnancy at the implantation stage, and not in the brain. It was found that the mouse uterus has the highest concentration

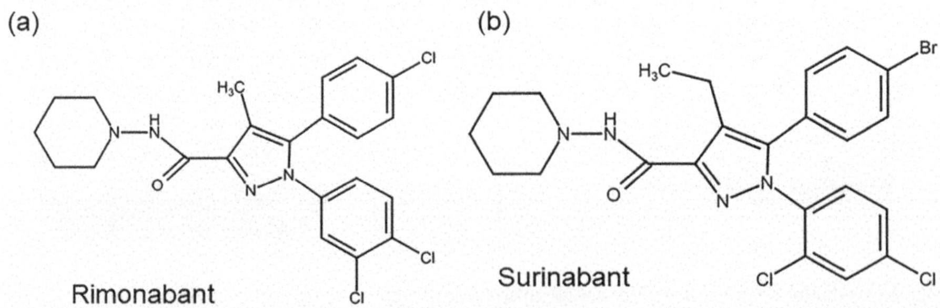


**Fig. 6** Structures of (**a**) Remonabant and (**b**) Surinabant

of cannabinoid receptors than any other part of the body (Yang et al. 1996) and endocannabinoid signaling in female reproduction (Sun and Dey 2002).

Another significant effect of endocannabinoids is to impart suckling ability to a newborn mammalian child. If the deficiency of anandamide is present in the newly born child, it has a problem in sucking the milk from the mother leading to the low survival rate in such babies (Mechoulam et al. 2006) where it was shown that when CB1 receptor antagonist, SR141716A (20 mg/kg), administered to newly borne mouse pups (day 1), it completely inhibited the physical growth of mouse pups and caused death within 1 week due to the deprivation of the essential nourishment. It was shown, in the words of Mechoulam et al. (2006), that "the first 24 h of life seem to be most critical for the putative endocannabinoid-induced growth-promoting effect, which is compatible with the peak levels of 2-AG. This is also consistent with our observation that 2-AG is present in mammalian milk from the 1st day after birth ('early' milk or colostrum)."

## *4.3 THC: A Natural Mimic of Endocannabinoid Agonist (Anandamide)*

THC, being a naturally occurring mimic of endocannabinoid agonist anandamide, when inhaled (smoking) or ingested (oral) gets absorbed in the blood and reaches the brain as well as in the rest of the body. Since CB1 receptors are mainly present in different parts of the brain, it interacts with CB receptors impacting the functions of the brain parts, which are responsible for regulating various body functions, including thinking, feeling, and behavior.

Whether the endocannabinoid receptors present in different parts of the brain or the rest of the body responsible for regulating various functions have the same or different receptivity concerning the THC concentration is not well understood. This understanding will be of great help in optimizing THC dose for beneficial effects while avoiding the adverse effects. Furthermore, it is a quite common observation that different individuals experience quite different effects (both beneficial and harmful) with comparable doses. Based on detailed observations and study on the optimal dose of cannabis for its medicinal benefits, Sulak (2018) opined that: "Dosage is the key factor in achieving the most benefits and least adverse effects of cannabis. After following thousands of patients using medical cannabis for eight years, I've observed that dosing cannabis is unlike any therapeutic agent to which I was exposed in my medical training. A basic understanding of the key characteristics of cannabis dosing can empower you to make the most of this incredibly versatile, safe, and effective herb."

## 4.4 THC Oxidation and Metabolism in Human System

THC, when subjected to humid and hot conditions or under the sun (UV light), it undergoes oxidative dehydrogenation to get cannabinol (CBN), which induces very mild psychoactivity. Although it is found in traces in fresh cannabis, its concentration increases with the aging of cannabis. However, the metabolism of THC proceeds via its oxidation once inhaled or ingested. Initially, THC gets oxidized to 11-hydroxy tetrahydrocannabinol (11-OH-THC) and then finally to 11-carboxy tetrahydrocannabinol (11-carboxy THC) (Huestis et al. 1992) as depicted in Fig. 7. While 11-OH THC is highly psychoactive (ca. 3–7 times more) as it binds with CB1 receptors and crosses the blood–brain barrier more effectively, compared to THC (Grotenhermen 2003), 11 COOH THC is nonpsychoactive.

### 4.4.1 Inhalation Versus Ingestion

The pharmacokinetics of THC depends on the mode of administration, like inhalation and ingestion. Inhalation causes a maximum absorption in the blood within minutes with almost immediate psychotropic effects reaching a maximum within 15–30 min before tapering off in a couple of hours. However, in the case of oral ingestion, the psychotropic effects have a delayed start (30–90 min) and reach their

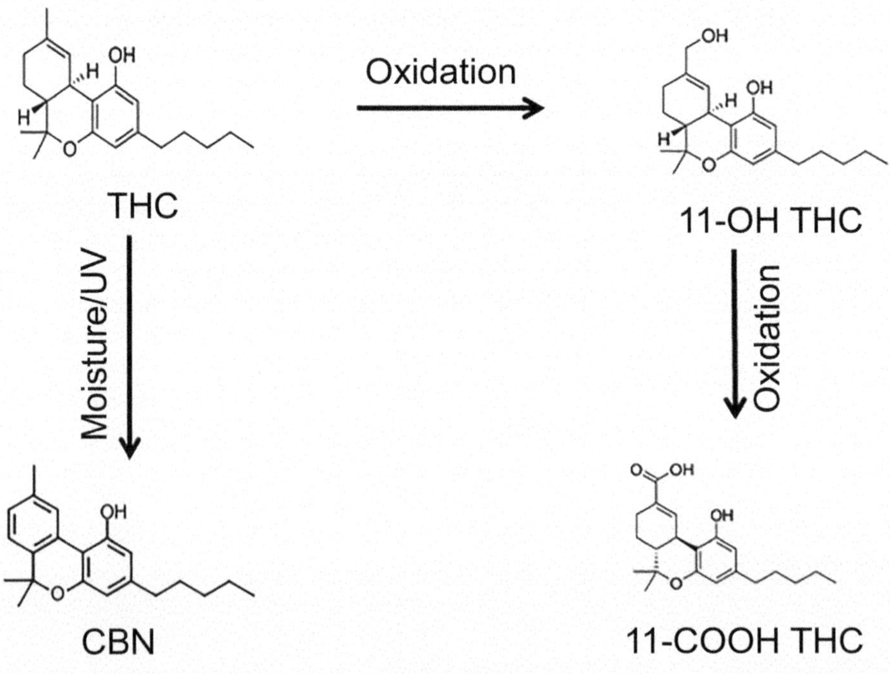


Fig. 7 Oxidative dehydrogenation and metabolism of THC

maximum after 2–3 h and last for about 4–12 h, depending on dose and specific effect (Grotenhermen 2003).

The bioavailability of THC in the blood/plasma (Sharma et al. 2012) for inhaled and ingested THC is between 10–35% and 4–12%, respectively. However, the inhalation of cannabis shows a 10:1 THC:11-OH-THC ratio; the same proportion is 1:1 for ingested THC (Wall and Perez-Reyes 1981). It is quite expected as more time for the oxidation of THC to 11-OH-THC is available when ingested. Hence, the lower bioavailability of THC when ingested compared to that with inhalation gets compensated and balanced out by the more extensive occurrence of 11-OH-THC and its higher potency in the brain. Therefore, no significant difference is experienced by the user at a peak, irrespective of the method of cannabis/THC consumption.

Furthermore, it is a common observation that cannabis-infused edibles are growing in demand. In the inhalation of cannabis/marijuana, the leaves/flowers have to be burnt, which may lead to the conversion of several cannabinoids/terpenoids to some other unknown chemicals, at least partially, and hence could be riskier compared to oral ingestion (edibles) which get absorbed in the blood through the digestive system, without causing any smoke-related stress to the lungs. Furthermore, while smoking floods the system with a substantial quantity of THC in a short time, it may cause an overdosing for a short period. While in ingestion, the release and absorption of THC and CBD are relatively slow and gradual, it will have a smoother effect on the EC system. Compared to ubiquitous legal/regulated psychotropic substances like alcohol, cannabis is much less risky and safer. However, much more research is needed to develop a detailed understanding of a rather extensive and complex endocannabinoid receptor system and the CB agonists. The relative specificity and CB agonists' relative binding affinity to different CB or simply cannabis receptors present in various parts/organs pose challenges. This understanding may lead to the better use of the cannabinoid agonists (natural/synthetic) as well as cannabinoid antagonists for specific indications/symptoms.

## 5 Modern Research on the Cannabinoids-Induced Neuroprotection and Neurotoxicity

### *5.1 Neuroprotection*

Zuardi et al. (2006) have reviewed the antianxiety and antipsychotic activity of CBD, supporting the traditional and ancient claims of cannabis for treating anxiety. Since THC is known to be useful in treating Alzheimer's disease (Volicer et al. 1997; Milano and Capasso 2018), cannabis can be used in treating dementia. Cannabis has also been extensively claimed in ancient and traditional uses to alleviate pain and induce sleep. It has been corroborated by modern clinical trial-based studies on the use of cannabis extract in treating insomnia, mainly in patients who have multiple sclerosis and neuropathic pain (Russo 2003). Also, other researchers have studied the use of cannabis extract in improving intractable neurogenic symptoms and pain (Wade et al. 2003; Notcutt et al. 2004), the anti-hyperalgesic effects of cannabinoids (Richardson et al. 1998). Furthermore, other group studied the effects of THC and CBD on nocturnal sleep and early-morning behavior in young adults (Nicholson et al. 2004).

The benefits of cannabis as antidepressant (Deyo and Musty 2003) and mood regulator were studied, where it was found that the EC system controls the extinction of aversive memories (Marsicano et al. 2002). Additionally, THC and CBD exhibit neuroprotective antioxidant properties (Hampson et al. 1998). Cannabis has also been found useful for treating migraine (Russo 1998, 2001). Carlini and Cunha (1981) reported the hypnotic and antiepileptic effects of CBD and THC exhibiting anticonvulsant activity and are useful for treating epileptic seizures. It has been observed that cannabis can benefit in reducing addiction is also supported by modern studies of alcoholics and cocaine users for decreased use rates in clinical experiments (Labigalini Jr et al. 1999; Mello et al. 1978). Table 1 broadly summarizes the cannabis indications along with a respective physiological basis (Russo 2005).

Apart from neuroprotective benefits of cannabis, cannabis extract, THC, and/or CBD, their several other uses are: (a) dermatological (anti-psoriatic (Malfait et al. 2000), anti-puritive (Richardson et al. 1998); (b) rheumatoid arthritis (Malfait et al. 2000); (c) appetite stimulation (Abrams et al. 2003); (d) oncological (anti-nausea) (Abrahamov et al. 1995), tumour reduction (Musty and Rossi 2001), apoptosis promotion (Guzman 2003; Maccarone 2005), antiangiogenesis (Guzman 2003), anti-prolactin effect (De Petrocellis et al. 1998), anti-pulmonary carcinogenesis (Roth et al. 2001); (e) pulmonary/asthma (inducing bronchodilation (Pertwee and Rossi 2002); (f) gastroenterological/intestinal spasm (causing smooth muscle relaxation) (Russo 1998), secretory diarrhoea (Pertwee 2001), gastritis (Tambe et al. 1996); (g) gynecological (Russo 2002; Brady et al. 2004); (h) premature ejaculation (Ferrari et al. 2000); (i) fertility (Battistust et al. 2008); and (j) infections (insecticidal) (Downs et al. 2000), anti-malarial (Campbell et al. 1997).

It is quite interesting to note that the above-mentioned modern scientific and clinical investigations corroborate, to a very large extent, the ancient and traditional Indian claims made for the use of this venerable herb. However, considering the rather extensive distribution of cannabinoid receptors throughout human body parts, albeit, in different proportions (Fig. 5) and their psychosomatic regulatory functions, there is no wonder that the use of cannabinoids, be it phyto-, endo-, or synthetic-cannabinoid agonists, exhibits such a wide range of applications, mentioned above, both in traditional and modern scientific observations.

## 5.2 Neurotoxicity

Recently, Tikka and D'Souza (2019) have reviewed the association between cannabinoids and psychosis and found that there is evidence that excessive use of cannabinoids (mainly THC) may lead to schizophrenic symptoms. The risk is higher for those users having such psychotic history and such premorbid condition. They

**Table 1** Cannabis indications and their physiological basis (adapted from Russo (2005))

| Cannabis indication | Physiological basis | References |
|---|---|---|
| **Psychological** | | |
| Anxiety | CBD reduces anxiety in humans | Zuardi et al. (2006) |
| Extinction of aversive memories | EC controls hippocampus | Marsicano et al. (2002) |
| Insomnia | Increased sleep in pain/multiple sclerosis patients | Wade et al. (2003); Notcutt et al. (2004) |
| Addiction treatment | Decreased usage of cocaine/alcohol | Labigalini Jr et al. (1999); Mello and Mendelson (1978) |
| **Neurological** | | |
| Neuropathic pain | Endocannabinoid modulation of central nervous system pathways Clinical pain reduction | Richardson et al. (1998); Zajicek et al. (2003); Wade et al. (2003); Notcutt et al. (2004) |
| Neuroprotection | THC/CBD antioxidant/NMDA antagonism | Hampson et al. (1998) |
| Migraine | Effects on periaqueductal grey, 5-HT, inflammation | Russo (1998, 2001, 2004) |
| Seizures/ epilepsy | CBD anticonvulsant, THC anticonvulsant EC modulation of seizure thresh hold | Carlini and Cunha (1981); Wallace et al. (2002, 2003); Pertwee (2005) |

concluded that "While these cross-sectional studies suggest that cognitive test performance is better in cannabis-using patients with schizophrenia, whether this reflects better premorbid function or is secondary to cannabis use is not clear. Future studies taking into account premorbid differences in cognitive test performance may need to be conducted."

Another common effect of cannabinoid (ab)use includes memory reduction (Solowij and Battisti 2008). Based on the study of the cannabis effect on the adolescent brain, Jacobus and Tapert (2014) found that chronic cannabis exposure is associated with diminished neuronal integrity in the dorsolateral prefrontal cortex. Anxiety and hallucinations, impaired motion, and cognition are some other effects of cannabinoid/cannabis exposure. However, the beneficial or adverse effects of cannabis/cannabinoid exposure are dose and probably exposure mode (ingestion vs. inhalation) dependent. Like any other psychoactive substances, alcohol included, personalized dose calibration is required for advantageous effects. Unlike alcohol, which is the most common and almost universally legalized and socially accepted psychotropic substance, intelligent use of cannabis/cannabinoids offers a basket of medicinal advantages apart from being used for recreational purposes.

Death is ultimate in toxicity. According to the National Institute on Alcohol Abuse and Alcoholism (NIAAA), a body of the National Institute of Health (NIH), USA, the number of death certificates mentioning alcohol as a cause of death was 72,558 in 2017, accounting for 2.6% of all deaths in the United States

(NIAAA 2020). The deaths attributed to prescription drugs are also very high. However, there are hardly any reports on deaths attributed directly to cannabis/ marijuana use. One has to take at least 1000 times more than a normal dose of 25 mg compared to ca. 25–50 g of THC as a probable lethal dose. This amount is practically impossible to be taken by anyone. When compared to alcohol, it is just three times more than a normal dosage. Considering three drinks as normal than 9–10 drinks could be poisonous and can kill the person (Wing 2018). The report states, "Study results vary, but some suggest a lethal dose of marijuana may be even higher. Extrapolating from cannabis toxicity research on monkeys, a human would have to orally ingest more than 2.9 g of THC per kilogram of body mass to be at risk of poisoning. At 88 kg, the average American man would have to down more than 9 ounces of 100 percent THC extract. That's almost certainly not possible, but if it were, it might involve taking this 'Dragon Ball' to the face" (Wing 2018).

However, this does not mean that cannabinoid/cannabis/marijuana overdose is not dangerous to humans. It may cause damage to the brain and lead to hallucinations, panic, paranoia, and so on. Our main purpose is just to mention that cannabinoids/cannabis/marijuana is far safer than other legal and illegal psychoactive substances, including alcohol, which is perhaps the most dangerous, mainly because the difference between the safer and lethal dose is very narrow compared to cannabinoids. The deaths due to accidents caused by the psychoactive substance, cannabinoids, alcohol included, are certainly a cause of concern, and hence driving should be avoided after their consumption.

## 6 Cannabis in Traditional Indian Medical System in Modern Scientific Perspective

There has been a large volume of ancient literature on the plethora of uses of cannabis, which was and is being traditionally used for medicinal, cultural, and recreational purposes in India (Russo 2005). Based on the findings of the Indian Hemp Commission, constituted by the British government, in 1893 the use of Indian hemp (cannabis), it was concluded that consumption of Indian hemp was quite widespread among almost all the sections (social, mystic, and religious) of Indian society as an integral part of the culture (Report of Indian Hemp and Drug Commission 1893–94) (Hall 2019).

It may be interesting to note that the 1961 international treaty "Single Convention on Narcotic Drugs" (SCND) included cannabis as a hard drug. However, according to the final treaty, "Cannabis means the flowering or fruiting tops of the cannabis plant (excluding the leaves and seeds) whatever name they may be designated." Since cannabis leaves and seeds are traditionally used in India as bhang (cannabis leaves), it is excluded from India's Narcotic Drugs, and Psychotropic Substances (NDPS) Act (Anonymous 2018a), where the cannabis is defined as: "(iii) cannabis (hemp) means: (a) charas, that is, the separated resin, in whatever form, whether crude or purified, obtained from the cannabis plant and also includes concentrated preparation and resin known as hashish oil or liquid hashish; (b) ganja, that is, the flowering or fruiting tops of the cannabis plant (excluding the seeds and leaves when not accompanied by the tops), by whatever name they may be known or designated; and (c) any mixture, with or without any neutral material, of any of the above forms of cannabis or any drink, prepared therefrom." Although the NDPS Act of India excluded the consumption of "bhang" and its preparations, it was left to the states to allow or ban the use of bhang. Many northern Indian states have legalized the use of bhang, but some have banned it. However, considering the medicinal and industrial applications of cannabis, some other Indian states are now considering legalizing cannabis cultivation for its industrial and medicinal uses.

Russo (2005), in a very comprehensive review on cannabis in the Indian medical system and lore, has provided details of the extensive use of cannabis in India from thousands of years in a modern research-based scientific medical perspective. In the Atharvaveda, ca. 2000–1500 BCE, the health benefits of cannabis leaves (bhang or bhanga in Hindi) are mentioned, where bhanga, along with few more herbs, are used for alleviating anxiety and providing overall good health. Sushruta Samhita (treatise of Sushruta, a renowned physician, ca. 500–800 BCE) describes the use of cannabis for phlegm and diarrhea (Grierson 1894). In Ayurveda, an anti-phlegmatic is interpreted as possessing a variety of effects. Cannabis was used in Indian folk medicine in aphrodisiacs and treatments for pain in the same era (Dwarakanath 1965). In fact, in an anonymous work Anandakand (the root of bliss) provided ca. 50 odd Sanskrit/Hindi cannabis attributes/synonyms. In Sanskrit, the same material is defined by several names, which, in a way, describe the *Guna* (properties) of the materials. Hence various names used for cannabis describe the different effects of cannabis consumption. These can be categorized into four parts for the sake of clarity as psycho- or neuro-related attributes, spiritual/high thinking attributes, somatic/immunological-related attributes, and general/common names. However, the terms related only to psycho-neuro attributes adapted from Dash (1999), and Russo (2005), are given in Table 2. Many of these ascribe to the remarkable rejuvenating and bliss giving effects of cannabis. Since bhang (cannabis leaves) has been traditionally used in Himalayan regions, which is and has been the main active ground of mystics of India and Tibet for ages, spiritual aspects of cannabis were supposed to have been discovered by such users as they might have found that the use of cannabis leaves/seeds is helpful in meditation. Besides, cannabis also provides heat in the colder regions and overall health benefits, as is mentioned in Table 2.

Based on a clinical survey of cannabis users in the Holy city Varanasi of India, where bhang (cannabis leaves) has been used for ages, Chaturvedi et al. (1991) have reported that bhang use is quite prevalent. Most commonly, bhang is consumed orally after grinding the leaves along with black pepper and almonds to make a very fine paste either as neat or preferably mixed with milk. Most users in the age group of 30–40 years employed it for anxiety or mood disorders for the resulting pleasure. The elderly population uses it for the benefits of improved appetite and digestion, as well as bowel habits and/or for alleviating insomnia. Ninety percent of subjects (out

**Table 2** The names and attributes of cannabis in Sanskrit and Hindi related to psycho-neuro attributes (Adapted from Dash 1999; Russo 2005)

| Psycho- or Neuro-related attributes | |
|---|---|
| Indian name | Meaning |
| Ananda | The joyful, joyous, laughter moving, bliss |
| Bahuvadini | Causing excessive talkativeness, garrulousness |
| Bhangini | Breaks three kinds of misery |
| Capala | Lighthearted |
| Capata | Agile, capricious, mischievous, scatterbrained |
| Chapala | The lighthearted, causer of reeling gait, causer of vacillation |
| Chidalhada | Gives happiness to the mind |
| Divyaka | Gives pleasure, luster, intoxication, beauty |
| Harshani | Joy-giver |
| Harshini | The exciter of sexual desire, the delight-giver, causer of elation |
| Madini | The intoxicator, sex intoxicator |
| Manonmana | Accomplishes the objects of the mind |
| Ranjika | Causer of excitement, grief remover |
| Samvida manjarika | Flower causes garrulousness |
| Suknidhan | Fountain of pleasures |
| Tandrakrit | Causer of drowsiness |
| Ununda | The laughter mover |
| Vrijapata | Strong nerved |

of 100) reported improved sleep without any daytime fatigue. Improvement in "marital adjustment" was also mentioned. Chaturvedi et al. (1991) reported that "all employed bhang orally, generally 1.5 g/day, for gastrointestinal indications, but 56% employed 4–10 g/day, without evidence of associated toxic adverse events." Gehlot et al. (1996) reported that cannabis (bhang) extract (17% THC) was used to treat cancer pain in 42 human beings, where it was found that in 11.9%, there was no analgesia with doses of 25 mg, but 64.3% had up to 50% pain reduction, and 9.5% had greater than 75% pain relief with no use of adjunctive medicine. In light of the details provided in the previous section (Sect. 5), it can be stated that most of the traditional medical applications mentioned above are corroborated by modern evidence-based research.

# 7 Trends of Scientific Publications, Patenting, and Commercial Aspects of Cannabis

The past two decades have witnessed increasing social acceptance and approval of medical and recreational cannabis/marijuana, which is also reflected in its legalization at the state level in the United States. This has opened the flood gates for research and development (publications, patents), propelling commercial/business activities to cater to rapidly growing demand for medicinal and also recreational cannabis uses. The same is becoming true for several other countries in the world. Although the cannabis legalization started slowly during the 1990s, the legalization of recreational cannabis/marijuana in Colorado and Washington in 2012 was a major development, followed by several other states, leading to the opening up of new opportunities in recreational cannabis markets also.

## 7.1 *Scientific Publications*

The trend of scientific publications, obtained using the PubMed database, is shown in Fig. 8, where the total number of publications on "all cannabis/marijuana" and the publications only on "medical cannabis/marijuana" are compared over 20 years (2000–2019) in the blocks of 4 years. It is clear that there is a significant spike in publications, both in "all cannabis/marijuana" and "medical cannabis/marijuana" during this period, specifically after 2011 (2012 onwards), synchronized with increasing legalization in and after 2012.

Recently, Treister-Goltzman et al. (2019) studied the trends in publications on cannabis/marijuana and medicinal cannabis/marijuana from the year 2000 till 2017 using PubMed and Web of Science databases. They observed that while the overall number of scientific publications increased ca. 2.5-fold, the same on cannabis increased 4.5-fold. However, interestingly, the number of publications on "medical cannabis" increased almost 9–10 times where publications in the field of psychiatry were the highest with a spike in cannabis publication post-2013, where ca. 50% of "all cannabis" publications were from the USA followed by Canada (Treister-Goltzman et al. 2019).

## 7.2 *Patents*

With an increasing expansion in cannabis and related research and industrial development, the stakeholders naturally seek the protection of their inventions through patents to get a competitive advantage. Figure 9 exhibits the trend of worldwide patent publications, both "Filed" and "Granted" during the year 2000–2019 (in the block of 4 years). The trend is quite similar to that observed for scientific

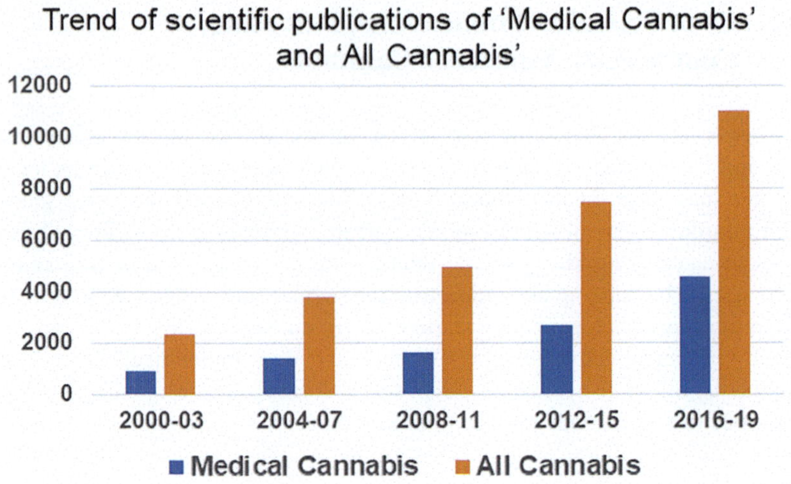


**Fig. 8** The trend of the number of scientific publications on "Medical Cannabis" and "All Cannabis" using the PubMed database. The search was made using keywords such as medical cannabis, medical marijuana, medical tetrahydrocannabinol, medical cannabidiol for "Medical Cannabis"; and cannabis, marijuana, tetrahydrocannabinol, cannabidiol for "All Cannabis" publications

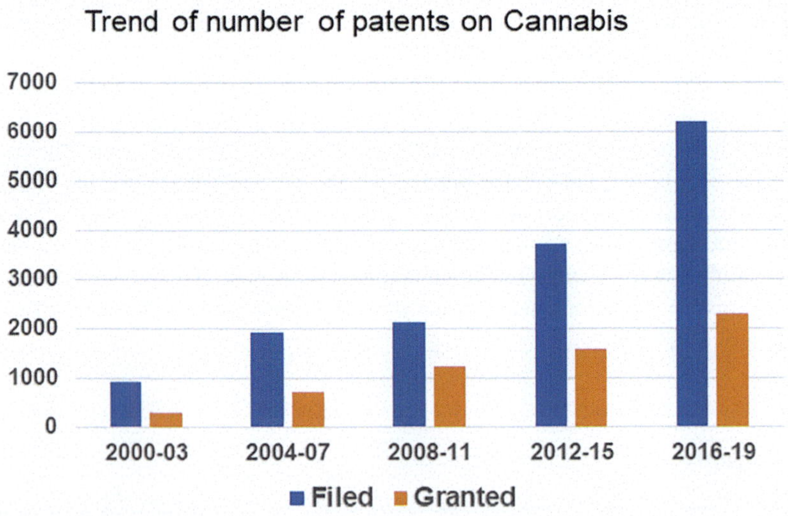


**Fig. 9** Trends of the number of patents on cannabis during 2000 and 2019. The patent search was conducted using a subscription-based advanced patent search database "Patseer." The keywords used for the search were cannabi* or marijuana or tetrahydrocannabinol or cannabidiol

publications (Fig. 8). Here also there was a significant increase in the patent applications filed post-2012, and the same is true for granted patents post 2015, as the time taken for grant of a patent takes around two and a half years on an average.

In a recent report on the trends of US patents on cannabis from the year 2000 to the end of 2019 (Johnson 2020), it was observed that the cannabis-related patent activity has increased significantly over the last decade, where the filings of cannabis-related patent applications increased 96% between 2012 and 2019. However, the patent filings related to recreational and medical cannabis increased 257% over the same time. Clearly, the legalization is synchronized with cannabis patent activity, as expected. It is quite obvious, that the investments, both for research and for innovation, are expected to increase further in this rather quiet, rapidly growing area. This will lead to significant upward trends in the publication of the research papers and the patents with the time as the legalization of medical and industrial cannabis increases in different parts of the world.

### 7.3 *Commercial Prospects*

As has been mentioned above, the rapid development in the area of the medicinal use of cannabis/marijuana for treating a plethora of ailments/indications (such as anxiety, depression, insomnia, cancer, neuropathic pain, etc.) along with increasing acceptability and legalization is driving the commercialization of medical cannabis. Some recent market reports (Conway 2020; Anonymous 2020; Grand View Research 2020) about the present and future business size of the products based on legal cannabis or medical cannabis show that the legal cannabis market is growing very fast and expected to grow further quite significantly. Most recently (2 Dec. 2020), the UN Commission on Narcotic Drugs (CND) has reclassified Cannabis out of the most dangerous category of drugs (UN News 2020). This may hasten up the legalization and commercialization of cannabis globally.

The global spending on legal cannabis increased from USD 3.3 Bn in 2014 to USD 14.9 Bn in 2019, with a CAGR of whopping 35.19%. The global legal cannabis spending is expected to be of the tune of ca. 43 Bn USD by 2024 (Conway 2020). However, according to the IMARC report, the medical cannabis market size was USD 16.5 Bn (Anonymous 2020). Furthermore, as per the recent Grand View Research market report on legal marijuana/cannabis in the United States and world for a period from 2016 to 2027 (Grand View Research 2020), the legal cannabis business in 2019 was USD 17.7 Bn and expected to reach ca. USD 73 Bn in 2027 at a healthy CAGR of 18.1%. This report shows the market size and expected growth of cannabis-based products like buds, oil (mainly THC/CBD) and tinctures in the USA. Presently, the United States accounts for ca. 85% of the global market.

Although these above-mentioned market research reports vary in absolute figures, say for 2019, the market size was reported to be in the range of between USD 14.9 and 17.7 Bn, the commonality is in the very high growth profile of this very exciting area. The difference in figures may be due to the range of types and products of

**Table 3** FDA approved commercial cannabinoid-based drugs

| Drug name | Active component | FDA approval for | Other conditions | Company/supplier |
|---|---|---|---|---|
| Sativex/ Nabiximols | THC + CBD (extracted) (THC: CBD in 2.75:2.5 mg per spray) | Multiple sclerosis, induced spasticity (muscle spasms), migraine, analgesic | Pain-induced insomnia | GW Pharmaceuticals plc, UK Novartis |
| Dronabinol/ Marinol/ Syndros | Synthetic Delta-9 THC | HIV-induced anorexia, chemotherapy-induced nausea/ vomiting | Appetite stimulant, antiemetic, and sleep apnea reliever | Unimed Pharmaceuticals, a subsidiary of Solvay Pharmaceuticals, Brussels, Belgium |
| Epidiolex | CBD | Epilepsy-induced seizures, Dravet syndrome, Lennox–Gastaut syndrome | Anti-inflammatory, anti-acne | GW Pharmaceuticals plc, UK |

Adapted from https://medicalmarijuana.procon.org/pharmaceutical-drugs-based-on-cannabis/

cannabis/marijuana, like legal, medical, cultural, and recreational, which are covered in the market research. However, our main purpose in mentioning the commercial and economic aspects of cannabis-related products is to emphasize the fact that this fascinating area of cannabis offers very promising commercial opportunities.

Major commercial cannabinoid-based drugs approved by the Food and Drug Administration (FDA), USA, have been listed in Table 3.

There are many other cannabis-based drugs that are under trials at various stages in different countries and may be applied for approval after successful completion of the required clinical trials before their launch.

Recently, in India, the industrial applications of cannabis (hemp) have been legalized, and Indian Institute for Integrative Medicine (IIIM) in Jammu and Kashmir, India, has been given a license to cultivate cannabis for research and developing phytocannabinoid based medicines. (Anonymous 2018b).

## 8 Conclusions

Owing to the modern evidence-based finding of the beneficial effects of cannabis/ cannabinoids consumption for a variety of common ailments like insomnia, pain, multiple sclerosis, anxiety, depression, epilepsy, Alzheimer's disease, decreased appetite (post-chemotherapy), and so on, on one hand, and cancer (antitumor/anti-angiogenesis), aging (antioxidant), sexual and gynecological problems, asthma, rheumatoid arthritis, infections, and many more on the other, the demand for its legalization and decriminalization of cannabis/cannabinoids is growing rapidly. The lawmakers are also responding, albeit cautiously and perhaps rightly so, to the social needs. Otherwise, it will be a classic case of "throwing the baby with the bathwater."

Concerning the question of whether cannabis-induced neuroactivity is a great boon or vice, both benefits and risks of consuming cannabinoids are mostly dose-dependent, just like any other legal or prescription pharma products or regulated/unregulated psychotropic substances like alcohol-based drinks/edibles. Moderation is the right old prescription for a healthy and long productive life, and it applies to the use of both medicinal, cultural, and/or recreational products like cannabis/cannabinoids. The importance of cannabis-related products for a variety of applications is aptly reflected by legalization and scientific research driven growth in research publications, intellectual property generation, and protection and economic activity. This is a highly promising area not only for the researchers and the companies but also for the entrepreneurs.

**Acknowledgments** The authors thank Abhishek Choudhury, Cocreate Consulting Private Limited, Pune, for providing the patent data used to draw Fig. 9.

## References

Abrahamov A, Abrahamov A, Mechoulam R (1995) An efficient new cannabinoid antiemetic in pediatric oncology. Life Sci 56:2097–2102. https://doi.org/10.1016/0024-3205(95)00194-b

Abrams DI, Hilton JF, Leiser RJ et al (2003) Short-term effects of cannabinoids in patients with HIV-1 infection. A randomized, placebo-controlled clinical trial. Ann Intern Med 139:258–266. https://doi.org/10.7326/0003-4819-139-4-200308190-00008

Anonymous (2017) Indica Vs Sativa: What's the Difference? https://www.leafscience.com/2017/10/16/indica-vs-sativa-whats-difference. Accessed 14 August 2020

Anonymous (2018a) The Narcotic Drugs and Psychotropic Substances Act, 1985. https://www.indiankanoon.org/doc/1727139/, section 2 iii, Accessed on 14 August 2020

Anonymous (2018b) CSIR-IIIM, BOHECO to develop cannabis-based drugs for cancer, epilepsy, sickle cell anemia. https://www.business-standard.com/article/pti-stories/csir-iiim-boheco-to-develop-cannabis-based-drugs-for-cancer-epilepsy-sickle-cell-anemia-118112300749_1.html, Accessed 14 August 2020

Anonymous (2020) Medical cannabis market: global industry trends, share, size, growth, opportunity and forecast 2020–2025. https://www.imarcgroup.com/medical-cannabis-market, Accessed on 31 August 2020

Battistust N, Rapino C, Di Tommaso M, Bari M, Pasquariello N, Maccarrone M (2008) Regulation of male fertility by the endocannabinoid system. Mol Cell Endocrinol 286(S1):S17–S23. https://doi.org/10.1016/j.mce.2008.01.010

Brady CM, Das Gupta R, Dalton C et al (2004) An open-label pilot study of cannabis based extracts for bladder dysfunction in advanced multiple sclerosis. Mult Scler 10:425–433. https://doi.org/10.1191/1352458504ms1063oa

Bragança VAN, França TG, de Jesus ACSPS et al (2020) Impact of conformational and solubility properties on psycho-activity of cannabidiol (CBD) and tetrahydrocannabinol (THC). Chem Data Coll 26:100345–100350. https://doi.org/10.1016/j.cdc.2020.100345

Campbell WE, Gammon DW, Smith P et al (1997) Composition and anti-malarial activity in vitro of the essential oil of *Tetradenia riparia*. Planta Med 63:270–272. https://doi.org/10.1055/s-2006-957672

Carlini EA, Cunha JM (1981) Hypnotic and antiepileptic effects of cannabidiol. J Clin Pharmacol 21:417S–427S. https://doi.org/10.1002/j.1552-4604.1981.tb02622.x

Chaturvedi GN, Rai NP, Pandey US, Sing KP, Tiwari SK (1991) Clinical survey of cannabis users in Varanasi. Anc Sci Life 10:194–198
Conway J (2020) Global legal cannabis market size from 2014 to 2024. https://www.statista.com/statistics/1005176/global-legal-cannabis-market-size/ accessed on 31 august 2020
Courtwright DT (1992) A century of American narcotic policy. In: Gerstein DR, Harwood HJ (eds) Treating drug problems, vol 2. National Academies Press, Washington (DC)
Dash VB (1999) Fundamentals of ayurvedic medicine. Sri Satguru Publications, Delhi, India
De Petrocellis L, Melck D, Palmisano A et al (1998) The endogenous cannabinoid anandamide inhibits human breast cancer cell proliferation. Proc Natl Acad Sci U S A 95:8375–8380. https://doi.org/10.1073/pnas.95.14.8375
Devane WA, Hanus L, Breuer A et al (1992) Isolation and structure of a brain constituent that binds to the cannabinoid receptor. Science 258:1946–1949. https://doi.org/10.1126/science.1470919
Deyo RA, Musty RE (2003) A cannabichromene (CBC) extract alters behavioural despair on the mouse tail suspension test of depression. In: Symposium on the Cannabinoids. International Cannabinoid Research Society, June 24–29, 2003, Cornwall, ON, Canada p 146
Downs AM, Stafford KA, Coles GC (2000) Monoterpenoids and tetralin as pediculicides. Acta Derm Venereol 80:69–70. https://doi.org/10.1080/000155500750012676
Dwarakanath C (1965) Use of opium and cannabis in the traditional systems of medicine in India. Bull Narc 17:15–19
Farag S, Kayser O (2017) The Cannabis plant: botanical aspects. In: Preddy PR (ed) Handbook of Cannabis and related pathologies, biology, pharmacology, diagnosis and treatment, 1st edn. Academic Press, London, pp 3–12
Ferrari F, Ottani A, Giuliani D (2000) Inhibitory effects of the cannabinoid agonist HU 210 on rat sexual behaviour. Physiol Behav 69:547–554. https://doi.org/10.1016/s0031-9384(00)00214-6
Finn K (ed) (2020) Cannabis in medicine: an evidence-based approach. Springer International Publishing, Springer Nature, Switzerland AG. https://doi.org/10.1007/978-3-030-45968-0
Gaoni Y, Mechoulam R (1964) Isolation, structure, and partial synthesis of an active constituent of hashish. J Am Chem Soc 86:1646–1647. https://doi.org/10.1021/ja01062a046
Gaoni Y, Mechoulam R (1986) The pharmacohistory of Cannabis sativa. In: Mechoulam R (ed) The cannabinoids as therapeutic agents. CRC Press Inc., Boca Raton, FL, pp 1–19
Gehlot S, Rastogi V, Dubby GP (1996) Role of cannabis extract (tetrahydrocannabinol) for relief of cancer pain. Sachitra Ayurveda 49:138–140
Grand View Research (2020). https://www.grandviewresearch.com/industry-analysis/legal-marijuana-market. Accessed 11 June 11 2020
Grierson GA (1894) The hemp plant in Sanskrit and Hindi literature. Indian Antiquary:260–262
Grotenhermen F (2003) Clinical pharmacokinetics of cannabinoids. J Cannabis Ther 3(1):3–51. https://doi.org/10.1300/J175v03n01_02
Guzman M (2003) Cannabinoids: potential anticancer agents. Nat Rev Cancer 3:745–755. https://doi.org/10.1038/nrc1188
Hall W (2019) The Indian hemp drugs commission 1893–1894. Addiction Classics, Wiley Online Library 114:1679. https://doi.org/10.1111/add.14640
Hampson AJ, Grimaldi M, Axelrod J, Wink D (1998) Cannabidiol and (−)Delta-9-tetrahydrocannabinol are neuroprotective anti-oxidants. Proc Natl Acad Sci U S A 95:8268–8273. https://doi.org/10.1073/pnas.95.14.8268
Hanuš LO, Mechoulam R (2005) Cannabinoid chemistry: an overview. In: Mechoulam R (ed) Cannabinoids as therapeutics, series: milestones in drug therapy. Birkhäuser Verlag, Basel – Boston – Berlin, pp 23–46
Huestis MA, Henningfield JE, Cone EJ (1992) Blood cannabinoids I. absorption of THC and formation of 11-OH-THC and THCCOOH during and after smoking marijuana. J Anal Toxicol 16(5):276–282. https://doi.org/10.1093/jat/16.5.276
Jacobus J, Tapert SF (2014) Effects of cannabis on the adolescent brain. Curr Pharm Des 20:2186–2193. https://doi.org/10.2174/13816128113199990426

Johnson I (2020). https://plantlaw.com/2020/01/20/cannabis-trends-medical-recreational/ accessed on august 10, 2020
Komnenov D (2019). https://www.rxleaf.com/four-types-of-cannabinoid-receptors-for-killing-pain-and-stopping-inflammation/
Labigalini E Jr, Rodrigues LR, Da Silveira DX (1999) Therapeutic use of cannabis by crack addicts in Brazil. J Psychoactive Drugs 31:451–455. https://doi.org/10.1080/02791072.1999.10471776
Maccarone M (2005) Endocannabinoids and regulation of fertility. In: Mechoulam R (ed) Cannabinoids as therapeutics. Birkhäuser Verlag, Switzerland, pp 67–78
Malfait AM, Gallily R, Sumariwalla PF et al (2000) The non-psychoactive cannabis constituent cannabidiol is an oral anti-arthritic therapeutic in murine collagen-induced arthritis. Proc Natl Acad Sci USA 97:9561–9566. https://doi.org/10.1073/pnas.160105897
Marsicano G, Wotjak CT, Azad SC et al (2002) The endogenous cannabinoid system controls extinction of aversive memories. Nature 418:530–534. https://doi.org/10.1038/nature00839
Matsuda LA, Lolait SJ, Brownstein MJ et al (1990) Structure of a cannabinoid receptor and functional expression of the cloned cDNA. Nature 346:561–564. https://doi.org/10.1038/346561a0
Mechoulam R, Ben-Shabat S, Hanus L et al (1995) Identification of an endogenous 2-monoglyceride, present in canine gut, that binds to cannabinoid receptors. Biochem Pharmacol 50:83–90. https://doi.org/10.1016/0006-2952(95)00109-D
Mechoulam R, Berry EM, Avraham Y et al (2006) Endocannabinoids, feeding and suckling—from our perspective. Int J Obes 30:S24–S28. https://doi.org/10.1038/sj.ijo.0803274
Mechoulam R, Lander N, University A, Zahal J (1990) Synthesis of the individual, pharmacologically distinct, enantiomers of a tetrahydrocannabinol derivative. Tetrahedron: asymmetry 1:315–318. https://doi.org/10.1016/S0957-4166(00)86322-3
Mello NK, Mendelson JH (1978) Marihuana, alcohol, and polydrug use: human self-administration studies. NIDA Res Monogr:93–127
Mello NK, Mendelson JH, Kuehnle JC, Sellers ML (1978) Human polydrug use: marihuana and alcohol. J Pharmacol Exp Ther 207:922–935
Milano W, Capasso A (2018) Neuroprotection by cannabinoids in neurodegenerative diseases. Alzheimers dement. Cogn Neurol 2(1):1–7. https://doi.org/10.15761/ADCN.1000120
Montoya ID, Weiss SRB (eds) (2019) Cannabis use disorders. Springer International Publishing, Springer Nature, Switzerland AG. https://doi.org/10.1007/978-3-319-90365-1
Munro S, Thomas K, Abu-Shaar M (1993) Molecular characterization of a peripheral receptor for cannabinoids. Nature 365:61–65. https://doi.org/10.1038/365061a0
Musty RE, Rossi R (2001) Effects of smoked cannabis and oral delta-9-tetrahydrocannabinol on nausea and emesis after cancer chemotherapy: A review of state clinical trials. J Cannabis Ther 1:29–42. https://doi.org/10.1300/J175v01n01_03
Nahas GG, Sutin KM, Harvey DJ, Agurell S (eds) (1999) Marihuana and medicine, 1st edn. Springer Science+Business Media, New York, pp 1–792
NIAAA (2020). https://www.niaaa.nih.gov/news-events/news-releases/alcohol-related-deaths-increasing-united-states, Accessed 20 July 2020
Nicholson AN, Turner C, Stone BM, Robson PJ (2004) Effect of delta-9-tetrahydrocannabinol and cannabidiol on nocturnal sleep and early-morning behaviour in young adults. J Clin Psychopharmacol 24:305–313. https://doi.org/10.1097/01.jcp.0000125688.05091.8f
Notcutt W, Price M, Miller R et al (2004) Initial experiences with medicinal extracts of cannabis for chronic pain: results from 34 "N of 1" studies. Anesthesia 59:440–452. https://doi.org/10.1111/j.1365-2044.2004.03674.x
Pertwee RG (2001) Cannabinoids and the gastrointestinal tract. Gut 48:859–867. https://doi.org/10.1136/gut.48.6.859
Pertwee RG (2005) Cannabidiol as potential medicine. In: Mechoulam R (ed) Cannabinoids as therapeutics, 1st edn. Birkhäuser Verlag, Switzerland, pp 47–65
Pertwee RG (2006) Cannabinoid pharmacology: the first 66 years. Br J Pharmacol 147(S1):S163–S171. https://doi.org/10.1038/sj.bjp.0706406

Pertwee RG, Rossi RA (2002) Cannabinoid receptors and their ligands. Prostaglandins Leukot Essent Fatty Acids 66:101–121. https://doi.org/10.1054/plef.2001.0341

Richardson JD, Aanonsen L, Hargreaves KM (1998) Antihyperalgesic effects of spinal cannabinoids. Eur J Pharmacol 345:145–153. https://doi.org/10.1016/s0014-2999(97)01621-x

Rinaldi-Carmona M, Barth F, Congy C et al (2004) SR147778 [5-(4-bromophenyl)-1-(2,4-dichlorophenyl)-4-ethyl-N-(1-piperidinyl)-1H-pyrazole-3-carboxamide], a new potent and selective antagonist of the CB1 cannabinoid receptor: biochemical and pharmacological characterization. J Pharmacol Exp Ther 310:905–914. https://doi.org/10.1124/jpet.104.067884

Rinaldi-Carmona M, Barth F, Héaulme M et al (1994) SR141716A, a potent and selective antagonist of the brain cannabinoid receptor. FEBS Lett 350:240–244. https://doi.org/10.1016/0014-5793(94)00773-x

Roth MD, Marques-Magallanes JA, Yuan M et al (2001) Induction and regulation of the carcinogen-metabolizing enzyme CYP1A1 by marijuana smoke and delta (9)-tetrahydrocannabinol. Am J Respir Cell Mol Biol 24:339–344. https://doi.org/10.1165/ajrcmb.24.3.4252

Russo E (1998) Cannabis for migraine treatment: the once and future prescription? A historical and scientific review. Pain 76:3–8. https://doi.org/10.1016/s0304-3959(98)00033-5

Russo EB (2001) Hemp for headache: an in-depth historical and scientific review of cannabis in migraine treatment. J Cannabis Ther 1:21–92. https://doi.org/10.1300/J175v01n02_04

Russo EB (2002) Cannabis treatments in obstetrics and gynecology: a historical review. J Cannabis Ther 2:5–35. https://doi.org/10.1300/J175v02n03_02

Russo EB (2003) Cannabis and cannabis-based medicine extracts: additional results. J Cannabis Ther 3:153–162. https://doi.org/10.1300/J175v03n04_03

Russo EB (2004) Clinical endocannabinoid deficiency (CECD): can this concept explain therapeutic benefits of cannabis in migraine, fibromyalgia, irritable bowel syndrome and other treatment-resistant conditions? Neuro Endocrinol Lett 25:31–39

Russo EB (2005) Cannabis in India: ancient lore and modern medicine. In: Mechoulam R (ed) Cannabinoids as therapeutics, 1st edn. Birkhäuser Verlag, Switzerland, pp 1–23

Sam AH, Salem V, Ghatei MA (2011) Rimonabant: from RIO to ban. J Obes 432607:1. https://doi.org/10.1155/2011/432607

Sharma P, Murthy P, Bharath MM (2012) Chemistry, metabolism and toxicology of cannabis: clinical implications. Iran J Psychiatr 7:149–156

Solowij N, Battisti R (2008) The chronic effects of cannabis on memory in humans: a review. Curr Drug Abuse Rev 1(1):81–98. https://doi.org/10.2174/1874473710801010081

Stella N, Schweitzer P, Piomelli D (1997) A second endogenous cannabinoid that modulates long-term potentiation. Nature 388:773–778. https://doi.org/10.1038/42015

Sulak D (2018) A physician's perspective on optimal cannabis dosing. https://www.leafly.com/news/health/a-physicians-perspective-on-optimal-cannabis-dosing, Accessed 14 August 2020

Sun X, Dey SK (2002) Endocannabinoid signaling in female reproduction. ACS Chem Neurosci 3:349–355. https://doi.org/10.1021/cn300014e

Tambe Y, Tsujiuchi H, Honda G, Ikeshiro Y, Tanaka S (1996) Gastric cytoprotection of the non-steroidal anti-inflammatory sesquiterpene, beta-caryophyllene. Planta Med 62:469–470. https://doi.org/10.1055/s-2006-957942

Tikka SK, D'Souza DC (2019) The association between cannabinoids and psychosis. In: Montoya, Weiss (eds) Cannabis use disorders, 1st edn. Springer Nature, Switzerland, AG, pp 127–155. https://doi.org/10.1007/978-3-319-90365-1_14

Treister-Goltzman Y, Freud T, Press Y, Peleg R (2019) Trends in publications on medical cannabis from the year 2000. Popul Health Manag 22:362–268. https://doi.org/10.1089/pop.2018.0113

UN News (2020) UN commission reclassifies cannabis, no longer considered risky narcotic. https://news.un.org/en/story/2020/12/1079132. Accessed 4 Dec 2020

Volicer L, Stelly M, Morris J et al (1997) Effects of dronabinol on anorexia and disturbed behavior in patients with Alzheimer's disease. Int J Geriatr Psychiatry 12:913–919. org/10.1002/(SICI)1099-1166(199709)12:9<913::AID-GPS663>3.0.CO;2-D

Wade DT, Robson P, House H et al (2003) A preliminary controlled study to determine whether whole-plant cannabis extracts can improve intractable neurogenic symptoms. Clin Rehabil 17:21–29. https://doi.org/10.1191/0269215503cr581oa

Wall ME, Perez-Reyes M (1981) The metabolism of delta 9-tetrahydrocannabinol and related cannabinoids in man. J Clin Pharmacol 21(S1):178S–189S. https://doi.org/10.1002/j.1552-4604.1981.tb02594.x

Wallace MJ, Blair RE, Falenski KW et al (2003) The endogenous cannabinoid system regulates seizure frequency and duration in a model of temporal lobe epilepsy. J Pharmacol Exp Ther 307 (1):129–137. https://doi.org/10.1124/jpet.103.051920

Wallace MJ, Martin BR, DeLorenzo RJ (2002) Evidence for a physiological role of endocannabinoids in the modulation of seizure threshold and severity. Eur J Pharmacol 452:295–301. https://doi.org/10.1016/s0014-2999(02)02331-2

Wing N (2018) The exhaustive list of everyone who's died of a marijuana overdose: because cannabis is among the most dangerous drugs, according to the government. https://www.huffpost.com/entry/marijuana-lethaldose_n_58f4ec07e4b0b9e9848d6297, Accessed 14 August 2020

Wright K, Rooney N, Tate J et al (2003) Functional cannabinoid receptor expression in human colonic epithelium. In: 2003 Symposium on the Cannabinoids. International Cannabinoid Research Society, Cornwall, ON, Canada, 25

Yang ZM, Paria BC, Dey SK (1996) Activation of brain-type cannabinoid receptors interferes with preimplantation mouse embryo development. *Biol Reprod* 55:756–761. https://doi.org/10.1095/biolreprod55.4.756

Zajicek J, Fox P, Sanders H et al (2003) Cannabinoids for treatment of spasticity and other symptoms related to multiple sclerosis (CAMS study): multicentre randomised placebo-controlled trial. Lancet 362:1517–1526. https://doi.org/10.1016/S0140-6736(03)14738-1

Zuardi AW, Crippa JA, Hallak JE, Moreira FA, Guimarães FS (2006) Cannabidiol, a Cannabis sativa constituent, as an antipsychotic drug. Braz J Med Biol Res 39:421–429. https://doi.org/10.1590/s0100-879x2006000400001

# Neurotoxicity of Polyherbal Formulations: Challenges and Potential Solutions

**Saraswathy Nachimuthu, Ruckmani Kandasamy, Ramalingam Ponnusamy, Muralikrishnan Dhanasekaran, and Sivasudha Thilagar**

## Contents



**Abstract** Herbal products with potential therapeutic and nutritional values are gaining importance among people around the world. Herbal products are generally considered safe for human applications. Extracts from various herbal products or purified bioactive compounds are prepared and marketed in various forms.

S. Nachimuthu (✉)
Department of Biotechnology, Kumaraguru College of Technology, Coimbatore, India
e-mail: saraswathy.n.bt@kct.ac.in

R. Kandasamy
Department of Pharmaceutical Technology, University College of Engineering, Bharathidasan Institute of Technology Campus, Anna University, Tiruchirappalli, India

R. Ponnusamy · M. Dhanasekaran
Department of Drug Discovery and Development, Harrison School of Pharmacy, Auburn University, Auburn, AL, USA
e-mail: dhanamu@auburn.edu

S. Thilagar
Department of Environmental Biotechnology, School of Environmental Sciences, Bharathidasan University, Tiruchirappalli, India

 
D. C. Agrawal, M. Dhanasekaran (eds.), *Medicinal Herbs and Fungi*,
https://doi.org/10.1007/978-981-33-4141-8_7

Polyherbal products are formulations with more than two herbal extracts. They are considered prophylactically or therapeutically effective on many occasions due to their complementary and/or potentiation activities due to each other's benefits. In most of the incidents, these herbal products have not been scientifically validated. Hence, patients and/or consumers are at risk for various adverse effects, leading to acute and chronic toxicity. General toxicity tests are generally conducted for herbal products; however, the neurotoxicity tests are not routinely conducted. Heavy metals such as lead, mercury, and arsenic present in the polyherbal products often cause severe neurotoxicity. The major challenge of using polyherbal formulations products for prophylactic and/or therapeutic purposes is the nonavailability of valid scientific information on the complete metabolite profile, human equivalent dose, potential adverse effects, and possible antidotes. In this chapter, we look into the potentially toxic substances present in the polyherbal products, and various general toxicity and neurotoxicity tests of the herbal products are discussed.

**Keywords** Herbal products · Neurotoxicity · Neurotoxin · Polyherbal formulation

## Abbreviations

| | |
|---|---|
| AAS | Atomic absorption spectroscopy |
| CNS | Central nervous system |
| GABA | γ-aminobutyric acid |
| GMP | Good manufacturing practice |
| HED | Human equivalent dose |
| JNK | C-jun-N-terminal kinase |
| MAPK | matrix metalloproteinase kinase |
| RNS | Reactive nitrogen species |
| ROS | Reactive oxygen species |
| RT-PCR | Reverse transcription-polymerase chain reaction |

## 1 Introduction

World Health Organization (WHO) estimated that approximately 80% of the global population relies on herbal-based medicines to treat various ailments. The current global market for the herbal formulation is UD$1.5 billion, and it is expected to grow exponentially due to the ever-increasing demand for natural remedies for preventing and treating various diseases. Herbal medicines are accepted as complementary alternative medicine worldwide and are generally considered safe (Yuan et al. 2016). Herbs/plants with ethnomedicinal values are continuously explored and well documented in the literature. Either single herbal or polyherbal formulations are used for the treatment of various diseases in various traditional treatment systems such as Ayurveda, Unani, Siddha, and other traditional systems of medicine around the world (Pole 2006). Herbs/plants produce various secondary metabolites such as alkaloids, phenolic compounds, terpenes, catechols, flavonoids, stilbenoids, puerarin, baicalein, ginsenosides, saponin, phytosterols, tannin, and so on, to protect themselves from external adversaries/stimuli such as insects, microbes, and other living creatures. Medicinal plants contain significant quantities of some of these compounds belonging to flavonoids, alkaloids, isoflavanols, catechols, polyphenols, lignans, monoterpenes, triterpenes, quinones, coumarins, and lectins are responsible for exerting a beneficial role in herbal extracts (Song et al. 2012; Aslam et al. 2016; Konieczynski et al. 2018; Anand et al. 2019). The therapeutic efficacy of the herbal formulations depends on the quality and quantity of the bioactive compounds. Various advanced analytical methods like chromatography, spectrophotometry, and molecular fingerprinting are employed to determine these phytoconstituents. Apart from prophylactic and/or therapeutic properties, compounds present in certain herbs/plants also possess toxic substances that affect the various cells, tissues, or organs of the human body. The brain, spinal cord, and peripheral nervous system control the whole body. Some of the phytochemicals from herbal extracts cause potential structural and/or functional nerve/neuronal damages and are called neurotoxins. If the herbal extracts are not purified to remove the toxins and tested for its presence through valid and sensitive tests, there is a great possibility for the production of herbal medicines with potential toxins, including neurotoxins. Although regulations with respect to toxicity testing of herbal formulations are currently present, they are not strictly enforced by regulatory authorities and practiced by the manufactures. Hence, most of the currently available herbal formulations in the market are sold without complete testing details. Although a polyherbal formulation is therapeutically efficient unless scientific data on its safety is evaluated, it cannot be recommended for human or animal applications. Most of the polyherbal formulations are nontoxic at low concentration but at higher concentrations, there might be potential neurotoxicity.

## 2 Polyherbal Formulations

Herbal-based therapy is one of the oldest treatments for many diseases across the world. Formulations with more than two herbal ingredients are called polyherbal formulations (Parasuraman et al. 2014). More than one herbal bioactive compounds are used to prepare herbal formulation to increase the prophylactic and/or therapeutic efficiency. Most of the time, it has been noted that bioactive compounds from a single herb do not successfully show an effective therapeutic index (Petchi et al. 2014). When more than two herbal ingredients are added, better therapeutic efficacy is recorded due to the synergistic effect for most of the diseases (Spinella 2002). A combination of herbal extracts in a particular ratio provides an enhanced therapeutic effect than the individual herbal extracts (Sachdeva et al. 2013; Ghelani et al. 2014). Furthermore, when more than two herbal extracts are added, either therapeutic efficiency is increased, or toxicity due to any one of the herb is decreased

**Table 1** Advantages and disadvantages of the polyherbal formulations

| Polyherbal Formulations | |
|---|---|
| Advantages | Disadvantages |
| ✓ Combination of herbal active ingredients may bind to more than one target to produce faster relief.<br>✓ Enhanced pharmacology actions<br>✓ The presence of multiple active compounds, hence higher therapeutic efficacy<br>✓ Reduce adverse effects due to a lower dose of the polyherbal formulation<br>✓ Helping patients to take a single dose of polyherbal formulation rather than taking many single herbal formulations to treat various ailments | • Incompatibility between herbal extracts in a polyherbal formulation may lead to precipitation of ingredients and instability during storage<br>• Toxicity due to ingredients present in the herbal plant extracts affect the polyherbal formulation as a whole |

(Parasuraman et al. 2010). Some of the herbal extracts used for treating central nervous system disorders are *Convolvulus pluricaulis, Ginkgo biloba, Bacopa monniera* (Kumar 2006). Neurodegenerative diseases are caused due to damage to the neurons in the central nervous system.

"Sarangdhar Samhita" is the concept of using two or more herbal extracts in an appropriate ratio for treatment. It has been well documented in the ancient Ayurvedic literature way back to 1300 A.D. When the polyherbal formulations are used for treatment due to their synergistic activity, pharmacokinetic (absorption, distribution, metabolism, and elimination) and/or pharmacodynamic (possible interaction between herbal drug) parameters are altered (Bhope et al. 2011; Karole et al. 2019; Srivastava et al. 2012). Many studies were conducted to study the herb–drug interactions and reported the safety of polyherbal formulation (Undale et al. 2014; Pandit et al. 2017); however, still, on certain occasions, the interaction between drug and herbal extracts showed adverse effects (Chavez et al. 2006). For example, turmeric (active ingredient is curcumin) is taken with black pepper and asafoetida to get the maximum immunity-boosting effect. Major advantages and disadvantages of polyherbal formulations over single herbal formulations are listed in Table 1, and some of the polyherbal formulations tested for various diseases are given in Table 2.

## 3 Toxicity of Polyherbal Formulation

Plants/herbs produce secondary metabolites as a natural defense mechanism (Wink 2003). When herbal extracts are used for prophylactic/therapeutic purposes with partial purification, these toxic metabolites cause adverse effects (Fatima and Nayeem 2016). Some of the secondary metabolites are synthesized by plants to kill the insects since few receptors are conserved across animals; the toxic effect is also shown in humans. Apart from bioactive compounds responsible for therapeutic potential, herbal extracts used for formulation contains other constituents, among

**Table 2** Polyherbal formulations currently used for the treatment of various diseases

| Plant species | Disease | In vitro tests | References |
|---|---|---|---|
| *Glycosmis pentaphylla, Tridax procumbens, Mangifera indica* | Diabetic mellitus | Acute toxicity Histopathological | Petchi et al. (2014) |
| *Aegle marmelos, Benincasa hispida, Garcinia indica, Musa paradiasica, Rosa indica, Hibiscus rosa sinensis* | Hypertension | Oral toxicity tests to the whole animal | Ghelani et al. (2014) |
| *Adhatoda vasica, Glycyrrhiza glabra, Piper longum, Ocium sanctum, Zingiber officinale, Azadirachta indica* | Anxiety | Oral toxicity and observations on elevated pulse maze, light/ dark exploration test | Mathew and Babu (2011) |
| *Curcuma longa, Ocimum sanctum, Murraya koenigii, Nyctanthes arbortristis* | Hepatotoxicity | Acute, subacute, subchronic oral toxicity tests | Sachdeva et al. (2013) |
| *Cicer arietinum, Phaseolus mungo, Mucuna pruriens, Triticum sativum, Allium cepa* | Methionine supplement | Acute toxicity | Rajurker et al. (2009) |
| *Bacopa monneiri, Evolvulus alsinoids, Acorus calamus, Saussurea lappa* | Epilepsy | Acute toxicity, behavioral tests | Achliya et al. (2005) |
| *Ginkgo biloba, Beta vulgaris, Emblica officinalis* | Diabetic mellitus Cataract | Eye test (eye opacity) | Mahajan et al. (2018) |
| *Withania somnifera, Nardostachys jatamansi, Rauwolfia serpentina, Evolvulus alsinoides, Asparagus racemosus, Emblica officinalis, Mucuna pruriens, Hyoscyamus niger* | Schizophrenia/ psychosis/ bipolar disorder | Passive avoidance learning and elevated plus-maze model (EPM) | Shah and Goyal (2011) |
| *Andrographis paniculata, Picrorhiza kurroa* | Hepatotoxicity | Biochemical and histopathological tests | Okhuarobo et al. (2014) |
| *Prunus amygdalus, Arachis hypogaea, Citrullus lanatus* | Parkinson's disease | Assay for dopamine, lipid peroxidation, glutathione, and superoxide dismutase, histopathology of the brain tissue | Nandagopal and Ali Khan (2020) |
| *Cichorium intybus, Gymnema sylvestre, Nigella sativa, Trigonella foenum graecum* | Cardio-metabolic disorders | Measuring vasomodulatory effect | Malik et al. (2017) |

which neurotoxins are potentially dangerous (Williamson 2017). In most cases, these neurotoxins are not correctly identified, and inadequate processing of the herbal extracts from various herbs for preparing polyherbal formulations and, at times, adulteration with other herbal extracts also introduce neurotoxins in the polyherbal formulations. Usually, more than one biological compound is present in the herbal extracts. Often, the concentration of the extract varies from one batch to the other depending on environmental factors under which the herbs are grown, techniques of extraction, and storage conditions, which also contribute to the presence of toxic substances in the herbal formulation. Many polyherbal products are reported to possess side effects. The awareness and scientific analysis of their side effects are not studied yet.

There is a limited understanding of the interaction of herbal formulations with the drugs. Pharmacodynamics studies need to be conducted to understand the mechanism of herb-drug interactions (Ifeoma and Oluwakanyinsola 2013). WHO has formulated guidelines to be followed for herbal medicines, and this has been circulated to the manufacturers of herbal formulations. Noncompliance with these guidelines often leads to a problem of low-quality herbal formulation and subsequent adverse effects to the consumers and patients. Furthermore, WHO also formulated guidelines for the appropriate use of herbal medicines as a prophylactic and therapeutic agent. However, the Food and Drug Administration (FDA) has given strict guidelines to be followed for medications, which are not directly applicable to herbal formulations.

The general perception is that herbal formulations are safe and do not cause any adverse effects compared to allopathic chemicals. But many clinical reports revealed the toxic effects of herbal products due to the presence of various phytochemicals and heavy metals, which are present in the herbal formulations (Kabelitz 1998). Sometimes, the natural compounds from the herbal extracts are metabolized into toxic molecules after human consumption. Hence intermediate metabolic profiling is needed to identify the intermediate toxicity. The safety of herbal formulations, in general, is a big debate as continuous intake of herbal formulation has been reported with various toxicity issues. Some of the neurotoxic effects are convulsion, psychosis, encephalopathy, and neurovascular damages. Other in vitro and in vivo toxicity tests are to be conducted to identify and validate the safety of the polyherbal formulations. The testing of polyherbal formulations is challenging because of the presence of many compounds whose structure and assay procedure are not known. When the herbal extract is used as a whole, the neurotoxicity effect is not the same as purified compounds. As per the regulations, even a trace amount of heavy metals is not permitted as they are well known to cause adverse health complications. But in certain polyherbal preparations, metals are added intentionally to improve the therapeutic efficacy of the preparation (Dwivedi and Dey 2002). However, according to herbal experts, the presence of metals causes toxicities. Some of the heavy metals present in the polyherbal formulation are mercury, lead, arsenic, and cadmium, which causes neurological disturbances (Dargan et al. 2008). Certain herbal extracts contain a higher level of neurotoxic heavy metals (Ernst 2002). Therefore, polyherbal formulations are to be systematically tested for the presence of heavy metals, particularly lead and mercury (Choi 2005). Lead is assessed using sensitive techniques like AAS to detect the lowest possible level. The heavy metals cause neurotoxicity by altering the free radical production pathways to induce oxidative stress, which can induce neuronal damage. Reactive nitrogen species (RNS) and reactive species (ROS) are produced in excess and cause damage to the nervous system. Various types of toxicities caused by herbal formulations (Fig. 1).

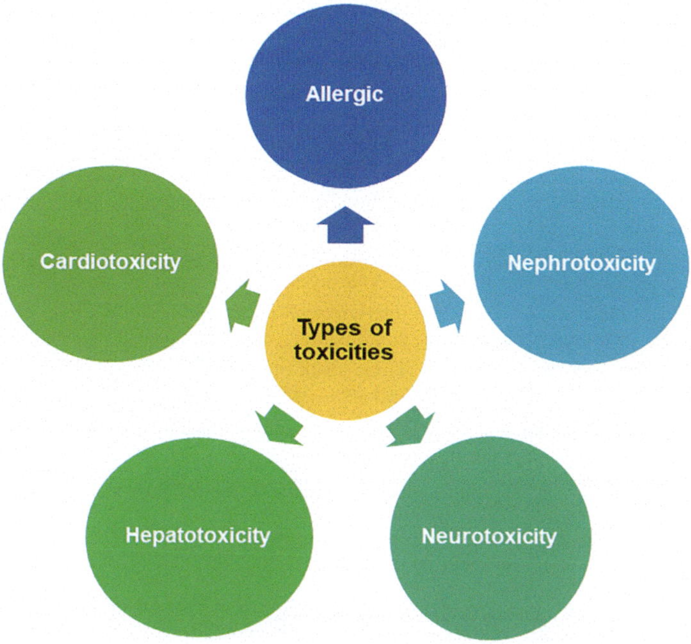


**Fig. 1** Toxicities associated with medicinal herbal formulations

## 4 Neurotoxicity of the Polyherbal Formulations

Neurotoxicity refers to the study of anatomical (structural) and/or physiological (functional) alterations of the nerves/neurons in the central and peripheral nervous system. Due to the structural and complex network of the nervous system, toxic substances affect the various sites and cause different neurotoxicity. The various sites are the axons, dendrites, neuronal membranes, mitochondria, nucleus, and endoplasmic reticulum. The nervous system is one of the complex organs of the human body. It consists of different cell types such as neuron and gila, which makes two important parts, the central nervous system (CNS) and the peripheral nervous system (PNS). The functional molecules, neurotransmitters, such as acetylcholine, monoamines, glutamate, and GABA, play major roles in controlling nervous system activity. The cholinergic system is one of the most important neuromodulators which control the movement, cognitive, and physiological functions of the body (Kawashima et al. 2007). Any compound which binds to the acetylcholine receptor (nicotinic and muscarinic) affects the metabolism (synthesis/degradation), increases release, which can lead to an increase in the neurotransmission, which is referred to as cholinergic substances. The parasympathetic nervous system is part of the autonomous nervous system which is present in the peripheral nervous system. In the brain, there is a specific cholinergic tract, starting from the basalis meynert and extending to the cortex and hippocampus. Some of the neurotoxins present in herbal formulations are given in Table 3 and the corresponding structure in Fig. 2.

**Table 3** Common neurotoxins present in medicinal herbal plants

| Scientific name | Toxin | Mode of action | References |
|---|---|---|---|
| *Coscinium fenestratum* | Berberine | Deceases neuron density | Wattanathorn et al. (2006) |
| *Vinca rosea* | Vincristine | Inhibits anterograde and retrograde axonal transport | Nicolini et al. (2015) |
| *Datura stramonium* | Tropane alkaloid | Affects the central nervous system | Devi et al. (2011) |
| *Atropa belladonna* | Atropine (tropane alkaloid) | Affects central and peripheral nervous system | Kwakye et al. (2018) |
| *Cannabis sativa* | Delta-9-tetrahydrocannabinol | Interfere with psychomotor coordination | Rocchetti et al. (2013) |
| *Veratrum nigrum L.* | Veratridine | Toxic insult of peripheral nerves and affects MAPK pathway | Zhang et al. (2018) |
| *Annona squamosa* | Acetogenin | Causes neuronal cell death | Bonneau et al. (2017) |
| *Ginkgo biloba* | Ginkgo toxin | Inhibits glutamate decarboxylase and reduces GABA | Arenz et al. (1996) |
| *Tanacetum vulgare* | Thujone | GABA-A receptor antagonist | Zámboriné and Thi Nguyen (2020) |
| *Digitalis purpurea* | Cardiac glycosides | Inhibits $Na^+/K^+$-ATPase | Hauptman and Kelly (1999) |
| *Rhododendron maximum* | Andromedotoxin | Interfere with voltage-gated sodium channels | Moran et al. (1954) |

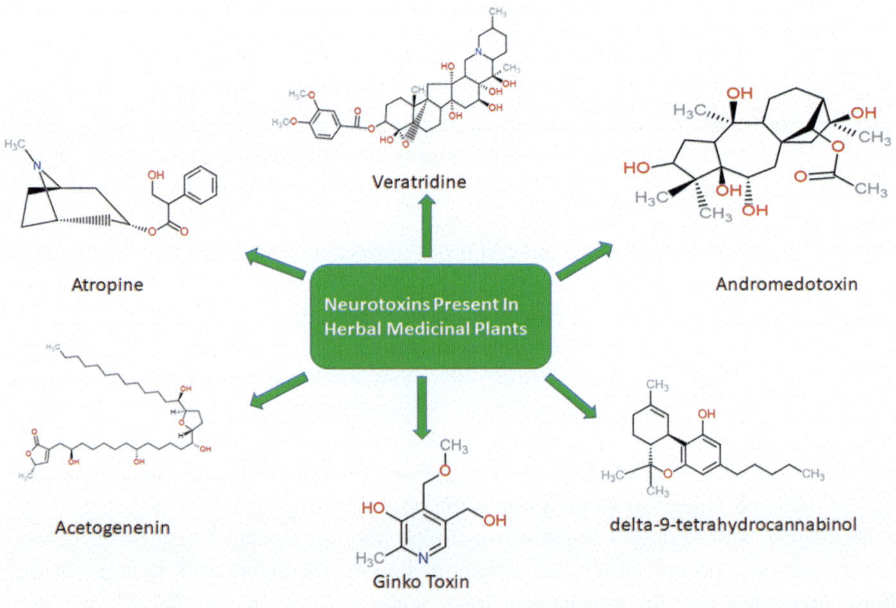


**Fig. 2** Neurotoxins present in medicinal plants

Depending on the nature of the neurotoxin, it might cause acute or chronic neurotoxicity. When the polyherbal formulation is prepared with extracts having neurotoxin is consumed, typical symptoms of neurotoxicity can occur that include headache, impaired vision, fatigue, memory loss, sexual dysfunction, and numbness of limbs. Neurotoxins are those compounds that affect or damage the neurons and render them inactive. Otherwise, neurotoxins affect the brain functions and result in many psychological symptoms like anxiety, depression, headache, impaired vision, and limb weakness that occur to the persons who have taken the polyherbal formulations. Neuropeptide and neurotransmitters are inactivated because of the binding to the neurotoxins. Some of the alkaloids produced by plants are known to interfere with neurotransmitters in humans. The conserved nature of acetylcholine, an important neurotransmitter, is found in all life forms. The parameters to be measured for neurological tests for the central nervous system include motor activity, behavioral coordination, electrophysiological examinations, and sensory reflex responses.

## 5 Neurotoxicity Tests for the Polyherbal Formulations

### *5.1 In Vitro Neurotoxicity Tests*

Toxicity testing of a polyherbal formulation is essential to know the presence of potentially toxic molecules and also determine the concentration of the herbal extracts that can be permitted to use as therapeutic substances. The results of the toxicity analysis at preclinical and clinical stages will help to take appropriate steps to either define the maximum limit or concentration that can be used to prepare the formulation and also chemical modification, or addition of suitable additives may be done to reduce the adverse effects. Commonly followed in vitro toxicity tests for herbal formulations are:

- Acute high-dose toxicity tests.
- Chronic low-dose toxicity tests.
- Histopathological studies.
- Cellular or organ-specific testing.
- Cytotoxicity tests.

The various in vitro toxicity tests are mentioned in Fig. 3.

Various factors that need to be considered while carrying the neurotoxicity tests of the polyherbal extracts are as follows: the age of the animal, relevant neurotoxicity biomarkers, a dose of the extract, and duration of the exposure. Acute toxicity effects of herbal extracts showed convulsions, whereas the subacute and chronic effects of herbal extracts showed encephalopathy and psychosis.

In vivo toxicity studies are carried out using whole animals. Four major tests under in vivo toxicity studies: (a) acute (b) subacute (c) chronic and (d) subchronic.

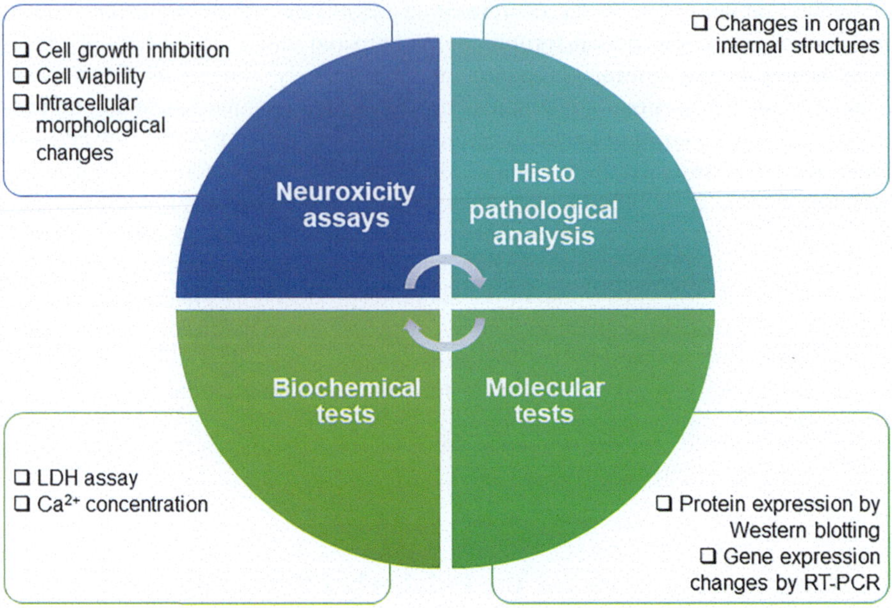


**Fig. 3** Various in vitro and in vivo toxicity studies for polyherbal formulations

All these studies are performed to test the safety level of herbal formulations. At present, well-established experimental protocols are available, which are carried out in compliance with the regulatory guidelines. The findings of the in vivo experiments help to predict the possible adverse effects of herbal formulations in addition to information on safety dose for human use. Commonly used animal models are rat, mouse, and to some extent, hen and dog. The morbidity and mortality effects of herbal formulations were also investigated using in vivo studies. At the end of the experimental study period, the surviving animals are sacrificed. The vital organs such as the liver, kidney, brain, thymus, spleen, and lung are carefully removed and subjected to histopathological examination to assess the damages caused due to the herbal products (Saiyed et al. 2015; Aydın et al. 2016; Fonseca et al. 2018).

## 5.2 *Cytotoxicity Tests*

Cytotoxic tests are the first stage of testing the polyherbal formulations for any possible toxicity before proceeding to other tests such as the safety of the herbal formulation in whole animal models. Cytotoxic tests are performed with various cell types, which include regular as well as transformed cell types. The cytotoxic effect of herbal compounds is measured in terms of the viability of the cells, inhibition of cell growth detected through a change of cell membrane and intracellular structural deformities (Aslantürk 2018). The cytotoxic tests are an indispensable part of the toxicity evaluation of any herbal formulation. Cytotoxicity analysis provides first-hand information on the safety of the polyherbal formulation and reduces animal usage, and it is easy to obtain results with high throughput screening. The most commonly used cell lines for toxicity analysis are mouse fibroblast cell lines and Syrian Hamster embryo cell lines. Some of the cell lines commonly used for neurotoxicity experiments are SH-SY5Y, PC12, and N27 (Heusinkveld and Westerink 2017).

## 5.3 *Histopathological Tests*

Neuropathology of neuronal cells provides evidence for the damage of nerve cells and their associated disruption of neural communication. Most of the time, it becomes difficult to directly associate behavioral changes with corresponding neural system damage that occur either in the CNS or PNS. Histopathological image analysis plays a major role in identifying the neurotoxicity of the toxins. To understand the neurotoxicity of the polyherbal formulation, brain tissue from the test animal after treating the formulation is collected, and hematoxylin–eosin staining is used to stain the brain samples. Different regions of the brains (nucleus accumbens, cortex, and hippocampus) are mainly analyzed as they control the memory and learning process of animals. The number of neurons is counted to know the extent of brain damage. Any change in the nerve cell membrane or other structures in these regions is also an indication of neurotoxicity. The behavioral changes are also associated with histopathological and neurochemical changes in the brain tissue. Accumulation of excessive protein and lysosomal bodies in the nerve cells indicates the neurodegenerative effect. The common neurotoxicity tests are qualitative and semiquantitative, and the results are recorded based on the extent of damage of the brain tissue and represented on a scale from 0 to 5; 0 for no damage, 1 for mild, 2 for moderate, and 3 for severe. Recently argyrophilic staining method was employed to carry out a quantitative analysis of the brain tissue damage with the help of automated computer software-based image analysis, and these types of advanced quantitative studies can be unbiased and are high-throughput to obtain neurotoxic effects quickly.

## 5.4 *Biochemical Tests*

Biochemical assays of some of the vital enzymes such as LDH and ions such as $Ca^{2+}$ are measured and represented as indicators of toxicity. Blood samples are collected from the animals fed with polyherbal formulations to quantify biochemical parameters like aspartate transaminase (AST), alanine transaminase (ALT), creatinine level, bilirubin, and urea.

### 5.5 *Molecular Biology Tests*

Change in gene expression can be studied using the RT-PCR technique and western blotting techniques to know about the upregulation of genes and proteins in the test samples obtained from animals treated with polyherbal formulations. It has been proven that changes in gene expression occur upon neurotoxicity. Some of the most commonly recorded changes in protein expression are overexpression of matrix metalloproteinases (MMPs) and induction of mitogen-activated protein kinases (MAPK) pathway-related proteins. C-jun-N-terminal kinase (JNK) pathway, which is associated with the neurodegenerative disease, can also be studied, which provides more information on the neurotoxic effects of polyherbal formulations.

## 6 Animal Models for Neurotoxicity Tests

Animal behavioral studies provide scientific data on neurotoxicity and the safety level of polyherbal formulations (Kulig et al. 1996). As per regulatory guidelines, "animals" studies need to be conducted before recommending any polyherbal formulation for human applications. After obtaining toxicological and safety level data of the polyherbal formulation from animal models, the safe concentration of the herbal formulations has to be calculated according to the human body. Most of the time, body weight and body surface are taken into account for scaling the safety level of polyherbal formulation from animal studies. Various factors that affect the therapeutic and toxicity dose of polyherbal formulations are life span, body size, metabolic function, anatomical factors, physiological and pharmacological responses of animals. Hence, it is recommended to calculate the Human Equivalent Dose (HED) of herbal formulations through animal models such as rat, pig, rabbit, dog, and so on (Bae et al. 2015). Interpretation of animal dose and the human equivalent dose is an important criterion to be considered while formulating polyherbal formulations. It has been recommended that bovine serum albumin (BSA) can be taken as a factor to calculate HED.

## 7 Future Perspective and Conclusion

The use of herbal-based medicines and formulations is increasing in the form of alternative medicine to treat many human diseases. But considering the safety of the polyherbal formulations, further investigations following advanced techniques with sensitive equipment to detect the very low level of the herbal toxins are required. Purification methods need to be employed to remove the potential neurotoxins from the herbal extracts before it is used for preparing formulations. To increase the sensitivity of toxicity testing, a noninvasive method based on in vivo neurotoxicity animal model is reported by genetically engineered mice with a molecular reporter gene. It may help to test the neurotoxicity of the samples through time-course analysis and also dose-dependent study using a lesser number of test animals. Quantification of heavy metals in traditional herbal preparation should be made compulsory, and the general public should be educated on the severity of heavy metal toxicity on the nervous system. Stringent regulations must be enforced to grant a patent on polyherbal formulation and issuing manufacturing licenses, including Good Manufacturing Practices (GMP).

## References

Achliya GS, Wadodkar SG, Dorle AK (2005) Evaluation of CNS activity of *Bramhi Ghrita*. Indian J Pharmacol 6(1):33–36. https://doi.org/10.4103/0253-7613.13853

Anand U, Jacobo-Herrera N, Altemimi A et al (2019) A comprehensive review on medicinal plants as antimicrobial therapeutics: potential avenues of biocompatible drug discovery. Meta 9 (11):258. https://doi.org/10.3390/metabo9110258

Arenz A, Klein M, Fiehe K et al (1996) Occurrence of neurotoxic 4'-O-methylpyridoxine in *Ginkgo biloba* leaves, Ginkgo medications and Japanese Ginkgo food. Planta Med 62(6):548–551. https://doi.org/10.1055/s-2006-957967

Aslam MS, Ahmad MS, Mamat AS (2016) Phytochemical evaluation of polyherbal formulation of *Clinacanthus nutans* and *Elephantopus scaber* to identify flavonoids. Pharm J 8(6):534–541

Aslantürk ÖS (2018) *In vitro* cytotoxicity and cell viability assays: principles, advantages, and disadvantages. In: Macelo et al (Eds) Genotoxicity: a predictable risk to our actual world, vol. 2. InTech, p 64

Aydın A, Aktay G, Yesilada E (2016) A guidance manual for the toxicity assessment of traditional herbal medicines. Nat Prod Commun 11(11):1763–1773

Bae W, Kim DH, Lee WW et al (2015) Characterizing the human equivalent dose of herbal medicines in animal toxicity studies. J Ethnopharmacol 162:1–6. https://doi.org/10.1016/j.jep.2014.12.023

Bhope SG, Nagore DH, Kuber VV et al (2011) Design and development of a stable polyherbal formulation based on the results of compatibility studies. Pharm Res 3(2):122–129. https://doi.org/10.4103/0974-8490.81960

Bonneau N, Baloul L, ba Ndob IB et al (2017) The fruit of *Annona squamosa L.* as a source of environmental neurotoxins: from quantification of squamocin to annotation of Annonaceous acetogenins by LC–MS/MS analysis. Food Chem 226:32–40. https://doi.org/10.1016/j.foodchem.2017.01.042

Chavez ML, Jordan MA, Chavez PI (2006) Evidence-based drug--herbal interactions. Life Sci 78 (18):2146–2157. https://doi.org/10.1016/j.lfs.2005.12.009

Choi KG (2005) Neurotoxicity of herbal medicine. J Korean Med Assoc 48(4):308–317. https://doi.org/10.5124/jkma.2005.48.4.308

Dargan PI, Gawarammana IB, Archer JR et al (2008) Heavy metal poisoning from Ayurvedic traditional medicines: an emerging problem? Int J Environ Health 2(3–4):463–474

Devi MR, Bawari M, Paul SB et al (2011) Neurotoxic and medicinal properties of *Datura stramonium* L.–review. Assam Univ J Sci Tech 7(1):139–144

Dwivedi SK, Dey S (2002) Medicinal herbs: a potential source of toxic metal exposure for man and animals in India. Arch Environ Health 57(3):229–231. https://doi.org/10.1080/00039890209602941

Ernst E (2002) Heavy metals in traditional Indian remedies. Eur J Clin Pharmacol 57(12):891–896. https://doi.org/10.1007/s00228-001-0400-y

Fatima N, Nayeem N (2016) Toxic effects as a result of herbal medicine intake. In: Toxicology—new aspects to this scientific conundrum. InTech, UK, pp 193–204

Fonseca AG, Ribeiro Dantas LL, Fernandes JM et al (2018) *In vivo* and *in vitro* toxicity evaluation of hydroethanolic extract of *Kalanchoe brasiliensis (Crassulaceae)* leaves. J Toxicol. https://doi.org/10.1155/2018/6849765

Ghelani HS, Patel BM, Gokani RH et al (2014) Evaluation of polyherbal formulation (SJT-HT-03) for antihypertensive activity in albino rat. Ayu 35:452–457. https://doi.org/10.4103/0974-8520.159034

Hauptman PJ, Kelly RA (1999) Digitalis. Circulation 99(9):1265

Heusinkveld HJ, Westerink RH (2017) Comparison of different *in vitro* cell models for the assessment of pesticide-induced dopaminergic neurotoxicity. Toxicol In Vitro 45:81–88. https://doi.org/10.1016/j.tiv.2017.07.030

Ifeoma O, Oluwakanyinsola S (2013) In: Gowder (ed) Screening of herbal medicines for potential toxicities, New insights into toxicity and drug testing. InTech, pp 63–88

Kabelitz L (1998) Heavy metals in herbal drugs. Eur J Herb Med 4:25–29

Karole S, Shrivastava S, Thomas S et al (2019) Polyherbal formulation concept for synergic action: a review. J Drug Deliv Ther 9(1S):453–466. https://doi.org/10.22270/jddt.v9i1-s.2339

Kawashima K, Misawa H, Moriwaki Y et al (2007) Ubiquitous expression of acetylcholine and its biological functions in life forms without nervous systems. Life Sci 80(24–25):2206–2209. https://doi.org/10.1016/j.lfs.2007.01.059

Konieczynski P, Viapiana A, Lysiuk R et al (2018) Chemical composition of selected commercial herbal remedies in relation to geographical origin and inter-species diversity. Biol Trace Elem Res 182(1):169–177. https://doi.org/10.1007/s12011-017-1078-z

Kulig B, Alleva E, Bignami G et al (1996) Animal behavioural methods in neurotoxicity assessment: SGOMSEC joint report. Environ Health Perspect 104(Suppl 2):193–204. https://doi.org/10.1289/ehp.96104s2193

Kumar V (2006) Potential medicinal plants for CNS disorders: an overview. Phytother Res 20 (12):1023–1035. https://doi.org/10.1002/ptr.1970

Kwakye GF, Jiménez J, Jiménez JA et al (2018) *Atropa belladonna* neurotoxicity: implications to neurological disorders. Food Chem Toxicol 116:346–353. https://doi.org/10.1016/j.fct.2018.04.022

Mahajan NM, Lokhande BB, Thenge RR et al (2018) Polyherbal formulation containing antioxidants may serve as a prophylactic measure to diabetic cataract: preclinical investigations in rat model. Phcog Mag 14(58):572–577

Malik A, Mehmood MH, Channa H et al (2017) Pharmacological basis for the medicinal use of polyherbal formulation and its ingredients in cardiovascular disorders using rodents. BMC Complement Altern Med 17(1):142. https://doi.org/10.1186/s12906-017-1644-0

Mathew L, Babu S (2011) Phytotherapy in India: transition of tradition to technology. Curr Bot 2:17–22

Moran NC, Dresel PE, Perkins ME et al (1954) The pharmacological actions of andromedotoxin, an active principle from *Rhododendron maximum*. J Pharmacol Exp Ther 4:415–432

Nandagopal A, Ali Khan MA (2020) Neuroprotective effect of polyherbal formulation in 'Parkinson's animal model. Asian J Pharm Clin Res 13(3):121–125. https://doi.org/10.22159/ajpcr.2020.v13i3.36549

Nicolini G, Monfrini M, Scuteri A (2015) Axonal transport impairment in chemotherapy-induced peripheral neuropathy. Toxics 3(3):322–341. https://doi.org/10.3390/toxics3030322

Okhuarobo A, Falodun JE, Erharuyi O et al (2014) Harnessing the medicinal properties of *Andrographis paniculata* for diseases and beyond: a review of its phytochemistry and pharmacology. Asian Pacific J Tropic Dis 4(3):213–222. https://doi.org/10.1016/S2222-1808(14)60509-0

Pandit S, Kanjilal S, Awasthi A et al (2017) Evaluation of herb-drug interaction of a polyherbal Ayurvedic formulation through high throughput cytochrome P450 enzyme inhibition assay. J Ethnopharmacol 197:165–172. https://doi.org/10.1016/j.jep.2016.07.061

Parasuraman S, Kumar EP, Kumar A et al (2010) Anti-hyperlipidemic effect of triglize, a polyherbal formulation. Int J Pharm Pharm Sci 2(3):118–122

Parasuraman S, Thing GS, Dhanaraj SA (2014) Polyherbal formulation: concept of Ayurveda. Pharmacogn Rev 8(16):73–80

Petchi RR, Vijaya C, Parasuraman S (2014) Antidiabetic activity of polyherbal formulation in streptozotocin - nicotinamide induced diabetic wistar rats. J Tradit Complement Med 4 (2):108–117. https://doi.org/10.4103/2225-4110.126174

Pole S (2006) Ayurvedic medicine: the principles of traditional practice. Elsevier, UK

Rajurker S, Rekhe D, Maini S et al (2009) Acute toxicity studies of polyherbal formulation (Methiorep Premix). Vet World 2:58–59

Rocchetti M, Crescini A, Borgwardt S et al (2013) Is cannabis neurotoxic for the healthy brain? A meta-analytical review of structural brain alterations in non-psychotic users. Psychiatry Clin Neurosci 67(7):483–492

Sachdeva M, Bajpai M, Razdan B (2013) Toxicity studies of a developed hepatoprotective polyherbal formulation in experimental rats. Asian J Pharm Clin Res 6(4):47–50

Saiyed ZM, Sengupta K, Krishnaraju A et al (2015) Safety and toxicological evaluation of Meratrim®: an herbal formulation for weight management. Food Chem Toxicol 78:122–129

Shah JS, Goyal RK (2011) Investigation of neuropsychopharmacological effects of a polyherbal formulation on the learning and memory process in rats. J Young Pharm 3(2):119–124

Song JX, Sze SC, Ng TB et al (2012) Anti-Parkinsonian drug discovery from herbal medicines: what have we got from neurotoxic models? J Ethnopharmacol 139(3):698–711. https://doi.org/10.1016/j.jep.2011.12.030

Spinella M (2002) The importance of pharmacological synergy in psychoactive herbal medicines. Altern Med Rev 7(2):130–137

Srivastava S, Lal VK, Pant KK (2012) Polyherbal formulations based on Indian medicinal plants as antidiabetic phytotherapeutics. Phytopharmacology 2(1):1–15

Undale R, Bhosale AD, Upasani C (2014) Study of Pharmacodynamic interaction between a Polyherbal formulation BSL-150 and metformin. Pharma Crops 5(1):67–76

Wattanathorn J, Uabundit N, Itarat W et al (2006) Neurotoxicity of *Coscinium fenestratum* stem, a medicinal plant used in traditional medicine. Food Chem Toxicol 44(8):1327–1333. https://doi.org/10.1016/j.fct.2006.02.012

Williamson EM (2017) Herbal neurotoxicity: an introduction to its occurrence and causes. In: Toxicology of herbal products. Springer, pp 345–362

Wink M (2003) Evolution of secondary metabolites from an ecological and molecular phylogenetic perspective. Phytochemistry 64(1):3–19. https://doi.org/10.1016/s0031-9422(03)00300-5

Yuan H, Ma Q, Ye L et al (2016) The traditional medicine and modern medicine from natural products. Molecules 21(5):559. https://www.mdpi.com/1420-3049/21/5/559

Zámboriné NÉ, Thi Nguyen H (2020) Thujone, a widely debated volatile compound: what do we know about it? Phytochem Rev 19:405–423. https://doi.org/10.1007/s11101-020-09671-y

Zhang X, Wang Y, Shao S et al (2018) Neurotoxicity of *Veratrum nigrum L.* and the molecular mechanism of veratridine toxicity. Int J Clin Exp Med 11(7):6547–6559

# Balancing the Neuroprotective Versus Neurotoxic Effects of Cannabis

**Shravanthi Mouli, Sindhu Ramesh, Manoj Govindarajulu, Mohammed Almaghrabi, Ayaka Fujihashi, Rishi M. Nadar, Julia M. Salamat, Jack Deruiter, Randall C. Clark, Timothy Moore, Satyanarayana R. Pondugula, Dinesh Chandra Agrawal, and Muralikrishnan Dhanasekaran**

## Contents




S. Mouli · S. Ramesh · M. Govindarajulu · M. Almaghrabi · A. Fujihashi · R. M. Nadar · J. Deruiter · R. C. Clark · T. Moore · M. Dhanasekaran (✉)
Department of Drug Development and Discovery, Harrison School of Pharmacy, Auburn University, Auburn, AL, USA
e-mail: dhanamu@auburn.edu

J. M. Salamat · S. R. Pondugula
Department of Anatomy, Physiology and Pharmacology, Auburn University, Auburn, AL, USA

D. C. Agrawal (✉)
Department of Applied Chemistry, Chaoyang University of Technology, Taichung, Taiwan
e-mail: dcagrawal@gm.cyut.edu.tw





D. C. Agrawal, M. Dhanasekaran (eds.), *Medicinal Herbs and Fungi*,
https://doi.org/10.1007/978-981-33-4141-8_8

**Abstract** Cannabinoids are chemicals that can be endogenous, natural, and synthetic substances which has the ability to bind to the cannabinoid receptors and exhibit a plethora of pharmacological and toxicological effects. Cannabinoids, the multipotent substance has a well-defined cellular signaling pathway that leads to a diverse physiological action in the body. Based on their chemical structural attributes, these substances can cross the blood–brain barrier, and hence, can exert an effect both in the central nervous system and the peripheral system in the body. Manipulation of the cannabinoid signaling with the natural and/or synthetic ligands can result in a plethora of pharmacological effects that can be used for various prophylactic and therapeutic treatment strategies in animals and humans. However, cannabinoids also induce severe adverse effects in the central and peripheral nervous systems. Hence, in this chapter, we review the current neuropharmacological and neurotoxicological properties of cannabinoids.

**Keywords** Cannabinoids · Central nervous system · Neuroprotection · Neurotoxicity · Pharmacology

## Abbreviations

| | |
|---|---|
| AAN | American Academy of Neurology |
| ABCD | Adolescent Brain Cognitive Development |
| AD | Alzheimer's disease |
| AD | Anno Domini |
| AEA | Arachidonoyethanolamine (Anandamide) |
| AIDS | Acquired immunodeficiency syndrome |
| ALD | Adrenoleukodystrophy |
| ALS | Amyotrophic lateral sclerosis |
| AMA | American Medical Association |
| AMPAR | α-Amino-3-hydroxy-5-methyl-4-isoxazolepropionic acid receptor |
| Aβ | Amyloid-β |
| BC | Before Christ |
| BDNF | Brain-derived neurotrophic factor |
| cAMP | 3′,5′-cyclic adenosine monophosphate |
| CB1 | Cannabinoid type 1 receptors |
| CB2 | Cannabinoid type 2 receptors |
| CBD | Cannabidiol |
| CBG | Cannabigerol |

| | |
|---|---|
| CBN | Cannabinol |
| CINV | Chemotherapy-induced nausea and vomiting |
| CNS | Central nervous system |
| DGL | Diacylglycerol lipase |
| DSE | Depolarization-induced suppression of excitation |
| DSI | Depolarization-induced suppression of inhibition |
| eCB | Endocannabinoids |
| FAAH | Fatty-acid amide hydrolase |
| GABA | Gamma-aminobutyric acid |
| GI | Gastrointestinal |
| *GM-CSF* | *G*ranulocyte–macrophage colony-stimulating factor |
| GPCRs | G-protein coupled receptors |
| HD | Huntington disease |
| HIV | Human immunodeficiency virus |
| IBD | Inflammatory bowel disease |
| *IGF-1* | *Insulin-like growth factor 1* |
| *IL*-6 | I*nterleukin*-6 |
| LTD | Long-term depression |
| LTP | Long-term potentiation |
| mAChRs | Muscarinic acetylcholine receptors |
| MS | Multiple sclerosis |
| NADA | *N*-Arachidonoyl-dopamine |
| NAPEPLD | N-acyl phosphatidylethanolamine phospholipase D |
| NF-Kb | N*uclear factor* kappa-light-chain-enhancer of activated B cells |
| NIH | National Institutes of Health |
| NMDA | *N*-Methyl-D-aspartate |
| OCE | Oral *cannabis* extract |
| OMNI-CAN | Outcomes Mandate National Integration with Cannabis as Medicine |
| PCL-β | Phospholipase C-β |
| PD | Parkinson's disease |
| PKA | Protein kinase A |
| PTSD | P*osttraumatic stress disorder* |
| SN | Substantia nigra |
| THC | Tetrahydrocannabinol |
| TNF-α | Tumor necrosis factor-α |
| TRPV1 | Transient receptor potential vanilloid 1 |
| VGCC | Voltage-gated calcium channel |

## 1 Introduction

*Cannabis sativa L.* (cannabis) is one of the earliest cultivated plants and is indigenous to central Asia and the Indian subcontinent. The earliest usage dates were reported in China, where archaeological findings indicate that the plant was cultivated for fibers since 4000 BC (Campos 2012). Furthermore, in western Asia and Egypt, *Cannabis sativa* was used for the cultivation of textile fiber, making it one of the oldest plants to be used as sources of food and fiber. In North America, cannabis was used in the form of hemp and was grown on several plantations for use in rope, cloth, and paper. The seeds of cannabis have been used for animal feed, its fiber for hemp rope, and its oil as a vehicle for paint (Small et al. 2002). In 1851, cannabis found its way into the third edition of Pharmacopeia of the United States, where it was used as a treatment for opioid withdrawal, pain, appetite stimulation, nausea, and vomiting (Pacula and Smart 2017). Growing concerns regarding its usage led to the prohibition of cannabis in several states in the early 1900s. Finally, in 1937, with the passage of the Marijuana Tax Act, the use and sale of the drug were prohibited by federal law. In 1942, Marijuana was removed from the 12th edition of *U.S Pharmacopeia* by American Medical Association (AMA) and was discredited for not having any medical use (Pacula and Smart 2017). In 2014, cannabis was decriminalized by Colorado, followed shortly by six other states that have legalized it for recreational use (Groce 2018).

## *1.1 Components of Cannabis*

More than 100 different compounds have been isolated from cannabis sativa and have been given the term cannabinoids. The three classes of cannabinoids are (a) The plant-based cannabinoids (THC, cannabidiol, cannabinol) (Fig. 1), (b) Endogenous cannabinoids, anandamide (AEA) and 2-arachidonoylglycerol (2-AG) (Fig. 2), and (c) The synthetic cannabinoids (e.g., WIN55212–2, CP-55940, JWH-015) (Fig. 3) (Maroon and Bost 2018). Representative members of each of these classes are shown in Fig. 1–3.

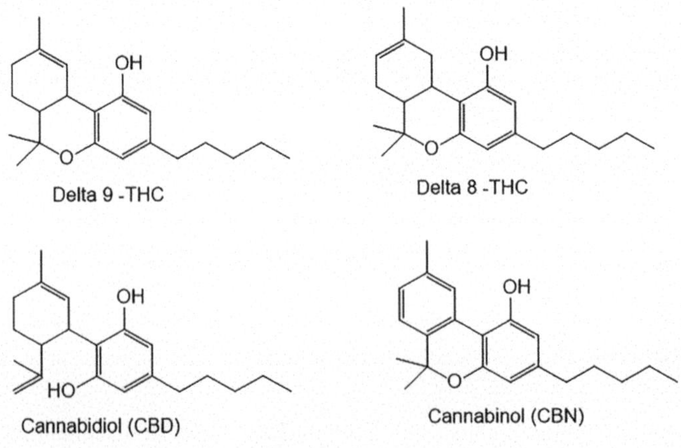


**Fig. 1** Plant-based cannabinoids

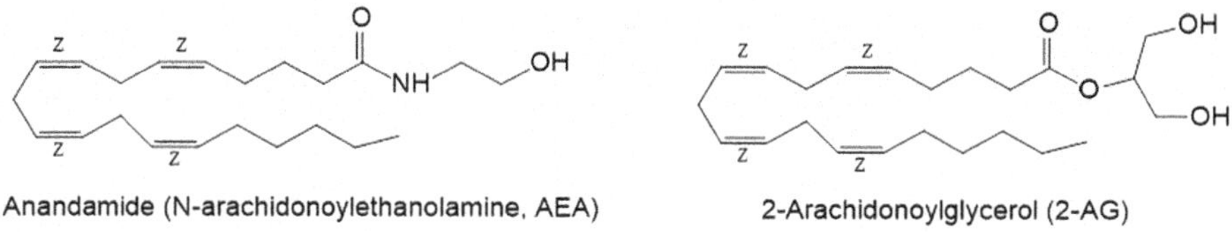


**Fig. 2** Endocannabinoids

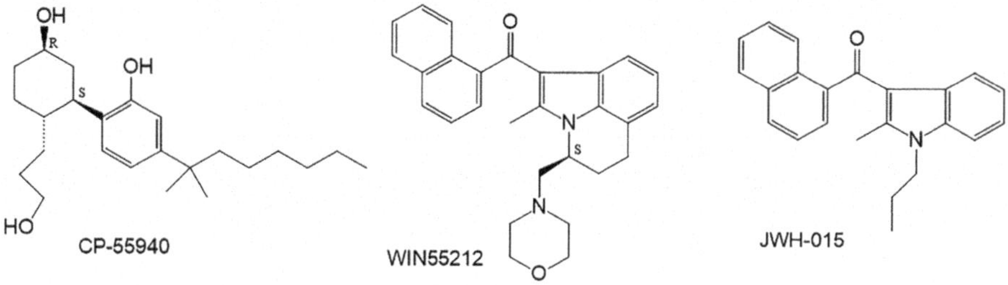


**Fig. 3** Synthetic cannabinoids

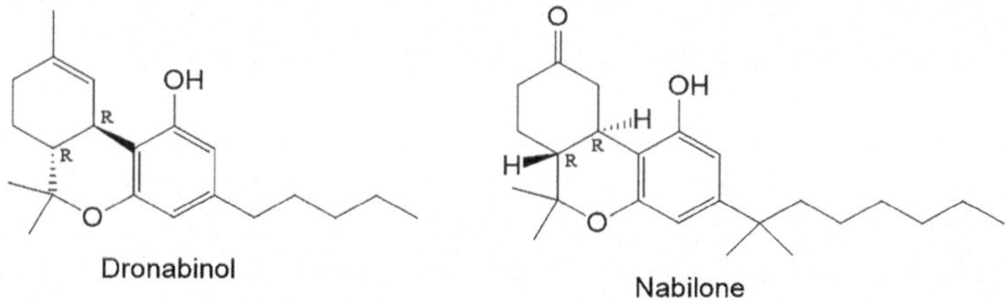


**Fig. 4** THC drug derivatives

The most abundant and the primary psychoactive plant-based cannabinoid is tetrahydrocannabinol (THC) or Δ-9-tetrahydrocannabinol (Δ-9–THC). It exerts its effects on the central nervous system (CNS) through partial agonism of G-protein coupled cannabinoid type 1 receptors (CB1) and type 2 receptors (CB2). Other natural cannabinoids include cannabidiol (CBD), a nonintoxicating anti-inflammatory and antianxiety agent, and Δ-8-THC. In 1985, THC derivatives—dronabinol and nabilone (Fig. 4) were approved for therapeutic use (Maroon and Bost 2018). The pharmacological and therapeutic properties of THC, CBD, and other natural cannabinoids are being extensively studied today. Cannabinol (CBN), produced by the oxidation of THC, could be a predictor of cannabis exposure to heat during storage. They possess mild psychoactive effects and have a high affinity for CB2 receptors. CBN is a potential antiemetic, antioxidant, anticonvulsant, and helps in the management of insomnia and possesses analgesic effects like THC (Maroon and Bost 2018).

## 1.2 Medicinal Uses of Cannabis

The medicinal use of cannabis dates back 5000 years ago when the dried flowers of cannabis, also known as marijuana was used as an important ingredient in holy anointing oil referenced in the original Hebrew version of Exodus. The Egyptians have reportedly used marijuana to treat glaucoma as well as inflammation, and by 100 AD, the Chinese had identified more than 100 medicinal uses for the plant. The roots of the plants were used by the Romans in 70 AD to treat gout, arthritis, and generalized pain and by Arabians in 800–900 AD for migraines, pain, and syphilis. The British discovered several other uses of marijuana, such as for menstrual cramps, convulsions, rheumatism, gout, and joint pain, insomnia, and muscle spasms. Studies have found medical cannabis and cannabinoids to be more effective than standard antiemetics in controlling chemotherapy-induced nausea and vomiting (CINV) although high side effects were equally observed (Tramer et al. 2001). The analgesic properties of cannabis have been clinically proven in conditions like neuropathic pain from multiple sclerosis (MS), human immunodeficiency virus (HIV) infection, acquired immune deficiency syndrome (AIDS), rheumatoid arthritis, cancer, migraine, muscle spasticity, Crohn's disease, and ulcerative colitis (inflammatory bowel disease) (IBD) (Pacula and Smart 2017). In addition, cannabis has been legalized for use for a variety of medical conditions, including Alzheimer's disease, epilepsy, amyotrophic lateral sclerosis, Parkinson's disease, post-traumatic stress disorder, Tourette syndrome, and cachexia. Despite its several uses, decriminalizing cannabis for medicinal purposes is still controversial and is expected to be a major topic of discussion in the forthcoming elections. Similarly, while cannabinoids have demonstrated a significant clinical reduction in symptoms in patients with chronic pain and CINV, further studies are required to determine adverse effects, optimum drug, and dosage choices, long-term effects, the risk–benefit ratio of combinatorial drugs.

CBD is considered the therapeutic champion of cannabinoids. CBD is a 5-HT1A receptor agonist and accounts for neuroprotective and antidepressive effects of the group and also plays a major role in the management of epilepsy, inflammation, and nausea. Furthermore, CBD has shown to stabilize the *N*-methyl-D-aspartate (NMDA) receptor pathway, thus resulting in the reduction of schizophrenia equivalent to amisulpride. The combination of THC with the non-psychotropic CBD (buccal spray, Sativex) was effective in treating neuropathic pain and in reducing sleep disturbances in patients with MS (Bridgeman and Abazia 2017). Furthermore, low doses of THC have been proven to reduce atherosclerosis by its pleiotropic, immunomodulatory effects on myeloid and lymphoid cells (Steffens et al. 2005).

## 1.3 Adverse Effects of Cannabinoids

Cognitive impairment, such as memory loss, psychosis, and paranoia, are some of the potential short-term adverse effects of cannabinoids (Wilkinson et al. 2014). Long-term adverse effects in adolescents include cognitive impairment, altered brain development, and negative impact on learning. Adverse effects of CBD usage include somnolence, reduced appetite, diarrhea, and fatigue (Wilkinson et al. 2014; Huestis et al. 2019).

# 2 Endocannabinoid Signaling System

The endocannabinoid signaling system is comprised of the CB receptors, their endogenous ligands or endocannabinoids, synthetic and degradative enzymes that inactivate the endocannabinoids, and the transporters of the endocannabinoids.

## 2.1 Cannabinoid Receptors in Brain

### 2.1.1 CB1

One of the most abundant G-protein coupled receptors (GPCRs) found in the brain is the CB1 receptor and is present in high levels in the hippocampus, basal ganglia, cerebellum, and the brain stem, where it may restrict their function to presynaptic and axonal compartments (Marsicano and Kuner 2008). It has attracted considerable attention by exhibiting a variety of brain functions such as execution, emotion, reward, and memory processing by indirect interaction with dopaminergic, glutamatergic, and GABAergic systems. CB1 receptors bind synthetic cannabinoids such as CP55940, JWH-015, WIN55212–2, and endocannabinoids, including arachidonic acid derivatives arachidonylethanolamine (AEA) and 2-AG. Upon activation, CB1 receptors bind to pertussis toxin $G_i$/o-proteins leading to a decrease in cAMP levels (Kendall and Yudowski 2017). Further, its presynaptic localization and inhibition of adenylate cyclase and voltage-dependent calcium channels suggest an important role of CB1 receptors in neurotransmitter release. The neurotransmitters controlled by the CB1 receptors include GABA, glutamate, glycine, acetylcholine, norepinephrine, dopamine, serotonin, and cholecystokin.CB1 receptors are involved in several neural activities such as appetite, learning, memory, and cognition, as well as several disease states, including anxiety, depression, schizophrenia, stroke, MS, epilepsy, and addiction (Di Marzo et al. 2014). CB1 receptors control appetite from the hypothalamus in the CNS and regulate the energy and food intake from the GI tract. One of the plausible mechanisms by which CB1 mediates neuroprotection against neurological disorders like epilepsy and neurodegenerative diseases is by inhibiting GABA and glutamate release from the presynaptic terminals of the hippocampal interneurons and the dorsal striatum, respectively (Katona et al. 1999; Di Marzo et al. 2014). CB1 receptors have been indicated in several neurodegenerative diseases such as Huntington's disease, multiple sclerosis, and Alzheimer's disease (Scotter et al. 2010).

#### 2.1.2 CB2

Unlike CB1, CB2 has been referred to as the "peripheral cannabinoid receptor" because it is predominantly found in the cells and tissues of the immune system and is responsible for the anti-inflammatory effects of cannabis (Buckley et al. 2000). The receptor is localized to microglia, and this is of great relevance since microglial cells play a significant role in Alzheimer's disease (Roche and Finn 2010). However, contradicting studies by Benito et al. have recently identified functional CB2 throughout the CNS (Benito et al. 2003). CB2 expression can be induced in the brain, suggesting its potential role in neurological diseases. While CB1 is expressed in the presynaptic terminals, CB2 is expressed in postsynaptic areas (somatodendritic), thus, suggesting contrasting roles neuronal firing and neurotransmitter release (Roche and Finn 2010). Thus, CB2 equally plays an important role in neuroprotection, and targeting CB2 offers a novel therapeutic approach in treating neurological diseases.

### *2.2 Endocannabinoid-Mediated Synaptic Transmission—Mechanism of Signaling*

For nearly a decade, neurophysiologists were looking for candidate molecules that mediated retrograde synapsis until when in 2001; cannabinoids were discovered to produce such an effect in the central synapse (Ohno-Shosaku et al. 2001; Maejima et al. 2001). The endocannabinoids serve as retrograde messengers that mediate feedback inhibition modulating synaptic plasticity (Katona and Freund 2012). Endocannabinoids are released from postsynaptic neurons due to receptor activation and/or postsynaptic depolarization. The released endocannabinoids activate CB1 receptors at the presynaptic terminal. The activation of the CB1 receptor leads to activation of inwardly rectifying K+ channel, inhibition in N-type, and P/Q-type voltage-dependent $Ca^{2+}$ channel and endocannabinoid production (Mackie et al. 1995; Twitchell et al. 1997). This leads to suppression of inhibitory neurotransmitter GABA (a.k.a depolarization-induced suppression of inhibition or DSI) or excitatory neurotransmitter glutamate (a.k.a depolarization-induced suppression of excitation or DSE) or long-term depression or potentiation (LTD/LTP) during synaptic plasticity. Also, the presence of CB1 receptors at the mitochondrial membrane reduces mitochondrial respiration and contributes to DSI. The DSI and DSE have been characterized in the cerebellum and hippocampus parts of the brain (Heifets and Castillo 2009).

## *2.3 Endocannabinoid Mediated Long-Term Plasticity*

The involvement of endocannabinoids in long-term plasticity or long-term depression was observed in 2002 at excitatory synapses in the dorsal striatum and has since been observed in the accumbens, amygdala, hippocampus, visual cortex, somatosensory cortex, prefrontal cortex, cerebellum, ventral tegmental area, and brain stem. The induction of LTD begins with the activation of postsynaptic metabotropic glutamate receptors, following presynaptic glutamate activation. The receptors are coupled to Phospholipase C-β (PCLβ) and $G\alpha_{q11}$ leading to diacylglycerol formation, which is converted to 2-AG by diacylglycerol lipase (DGL). 2-AG is released from postsynaptic by endocannabinoid membrane transporters and binds the CB1 receptors in the excitatory terminal. In some cases, $Ca^{2+}$ in the postsynaptic terminal also contributes to the endocannabinoid mobilization in a PLC dependent (activating it) or independent manner. In others, LTD occurs independently of $Ca^{2+}$ release. The CB1 receptors at the presynaptic terminal inhibit adenylate cyclase and protein kinase A via $G\alpha_{q11}$, suppressing the release of neurotransmitters. Reduced protein kinase A (PKA) activity, together with $Ca^{2+}$-sensitive phosphatase calcineurin, shifts the phosphorylation/kinase balance leading to dephosphorylation of a presynaptic target that is involved in the reduction of neurotransmitter release. Other players include active zone protein RIM1α, and vesicle-associated rab3b is required for PKA-dependent inhibitory LTP (Heifets and Castillo 2009; Castillo et al. 2012).

## *2.4 Endocannabinoid-Mediated Short-Term Plasticity*

The release of endocannabinoids characterized by DSI or DSE triggers $Ca^{2+}$ influx into postsynaptic neurons. Alternatively, other methods of endocannabinoid release from hippocampal cells have been proposed, which include activation of metabotropic postsynaptic glutamate receptors or muscarinic acetylcholine receptors (mAChRs). Activation of metabotropic glutamate receptors contributes to the endocannabinoid mobilization by triggering PLC. Furthermore, DGL promotes the synthesis and release of 2-AG release, targeting presynaptic CB1 receptors. The beta and gamma subunits couple with the presynaptic voltage-gated calcium channel (VGCC) to reduce neurotransmitter release (Castillo et al. 2012; Diana and Marty 2004).

## 2.5 *Endocannabinoid Signaling in Astrocytes*

Growing evidence supports the endocannabinoid signaling in astrocyte–neuron communication and synaptic transmission in hippocampal and cortical astrocytes. CB1 receptor activation by endocannabinoids in hippocampal astrocytes has been reported, resulting in increased intracellular $Ca^{2+}$ levels. The increase in $Ca^{2+}$ levels from pyramidal neurons activated by endocannabinoids stimulates glutamate release from astrocytes. This evokes the release of slow inward currents in an adjacent CA1 pyramidal neuron, thus indicating that endocannabinoids play a major role in the communication between neurons and astrocytes via eCB glutamate signaling (Navarrete et al. 2014).

## 2.6 *Endocannabinoid-Mediated Non-Retrograde Signaling*

There is considerable evidence that endocannabinoids function to regulate the neural system and synaptic transmission in a non-retrograde or autocrine fashion using transient receptor potential vanilloid 1 (TRVP1) (expressed predominantly in the afferent sensory neurons and activates synaptic transmission associated with pain) and the postsynaptic CB1 receptors. The AEA-LTD is observed in nucleus accumbens, dentate granule cells, bed nucleus of stria terminalis, and induces LTD at the postsynaptic terminal. Activation of metabotropic glutamate receptors, via PLC activation of intracellular $Ca^{2+}$ release, leads to the synthesis of AEA, which in turn activates TRPV1. $Ca^{2+}$ release triggers α-amino-3-hydroxy-5-methyl-4-isoxazole propionic acid receptor (AMPAR) endocytosis and LTD via calcineurin/dynamin. In another signaling mechanism proposed by Bacci et al., neocortical GABA-containing interneurons induce CB1-dependent postsynaptic hyperpolarization reducing its excitability (Bacci et al. 2004). The self-induced inhibition is activated by intracellular $Ca^{2+}$ endocannabinoid mobilization and CB1 receptor activation leading to activation of inwardly rectifying $K^+$ channel (Castillo et al. 2012).

# 3 Endogenous Ligands, Natural and Synthetic Compounds Acting on Cb Receptors

## 3.1 *Endogenous Ligands that Act upon Endocannabinoids*

### 3.1.1 Arachidonoylethanolamine (Anandamide, Aea) and 2-AG

Cannabinoids receptors have neutral and lipophilic ligands that are derived from arachidonic fatty acids. The most common types are AEA and 2-AG. AEA was first

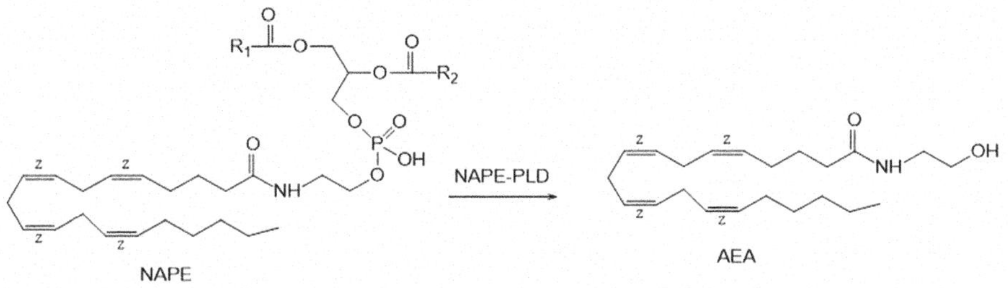


**Fig. 5** Formation of AEA

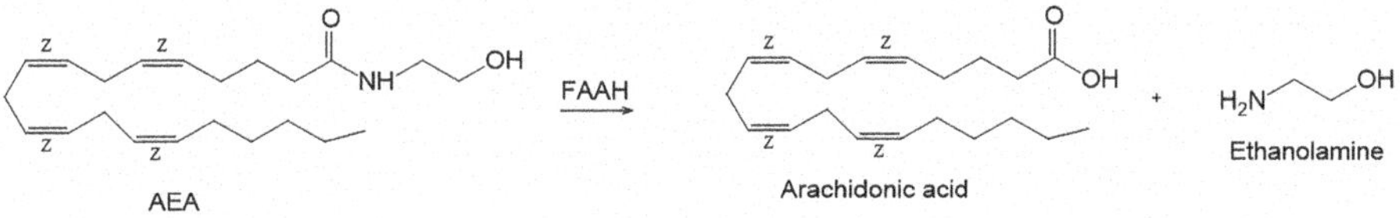


**Fig. 6** Metabolism of AEA

isolated from the porcine brain (Devane et al. 1992). AEA has high affinity and serves as a partial agonist for the CB1 receptor and has a low affinity for CB2 receptors. 2-AG has a moderate-to-low affinity and serves as a full agonist at both CB1 and CB2 receptors (Di Marzo and De Petrocellis 2012). Nyilas et al. have demonstrated that N–acyl phosphatidylethanolamine hydrolyzing phospholipase D (NAPE-PLD), an enzyme involved in the formation of AEA, is present at the hippocampal excitatory presynaptic terminal and is associated with intracellular $Ca^{2+}$ reserves (Nyilas et al. 2008), leading to the belief that unlike 2-AG, AEA has a presynaptic origin (Fig. 5). Following CB1 receptor activation, AEA is taken up by postsynaptic cells by transport proteins on both neurons and glia. Furthermore, inside the postsynaptic cell, AEA is metabolized to arachidonic acid and ethanolamine by an enzyme called fatty-acid amide hydrolase (FAAH) that is located within the endoplasmic reticulum (Ahn et al. 2009) (Fig. 6). A study by Pertwee has shown that AEA, along with other CB1/CB2 receptor agonists such as Δ-9–THC, HU-210, and WIN55,212, induces signs of antinociception in mouse, rat, and monkey models of acute thermal, inflammatory, and neuropathic pain (Pertwee 2012).

Other endogenous endocannabinoids identified are 2-eicosa-5′,8′,H′,14′-tetraenylglycerol (2-AG ether, noladin ether), and virodhamine and *N*-arachidonoyl-dopamine (NADA) (Fig. 7). Virodhamine is arachidonic acid, and ethanolamine conjugated by ester and has antagonistic properties against the CB1 receptor. NADA is a "capsaicin" like substance observed in mammalian nervous tissue and activates CB1 receptors and is an agonist for vanilloid receptors (Console-Bram et al. 2012). Activation of CB1 modulates adenylate cyclase to inhibit 3′,5′-cyclic adenosine monophosphate (cAMP) accumulation, VGCC, K+ channels, and 6 release. The basal levels of 2-AG have been found to be 1000 times higher than AEA in the brain. 2-AG also has been shown to play a significant role in

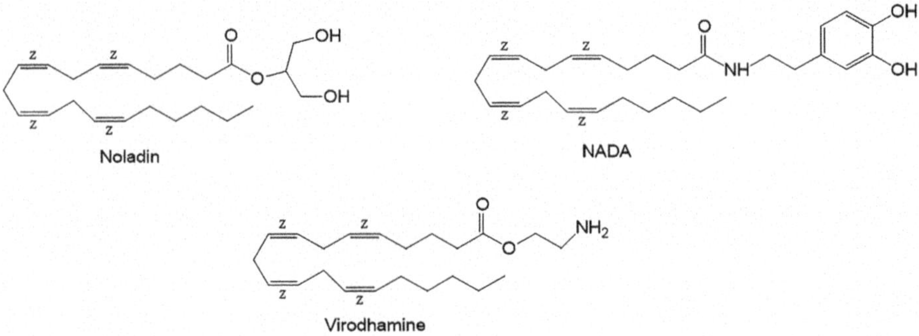


**Fig. 7** Other endogenous cannabinoids

retrograde signaling. In light of all these findings, it is established that 2-AG is the primary endogenous ligand for CB receptors in the CNS (Zou and Kumar 2018).

## 3.2 *Natural and Synthetic Cannabinoids and their Therapeutic Applications*

### 3.2.1 Synthetic Cannabinoids

Synthetic compounds involved in stimulating the cannabinoid receptors mimic the psychoactive effects of cannabis variably. Research on these natural compounds, especially THC, has led to several modifications in basic molecular structures, chemistry, and pharmacology. A number of synthetic cannabinoids have been tested for their clinical effects. Three cannabinoids that are currently being used in the clinic are Cesamet (nabilone), Marinol (dronabinol), and Sativex (cannabidiol)—all three activate CB1/CB2 receptors.

### 3.2.2 Natural Cannabinoids

CBN is one of the nonpsychoactive cannabinoids present in the cannabis plant at low levels and occurs as a degradant of THC. CBN is produced in trace amounts when THC is exposed to air, light, and heat, and when the plant dies. With potential anti-inflammatory and immunosuppressive effects, CBN has a low affinity for CB1 but binds the CB2 receptor, triggers apoptosis, and inhibits the production of cytokines. CBN has also been proposed as an antiemetic, anticonvulsant, and in the management of insomnia (Holt 2016).

CBD is a nonpsychoactive phytocannabinoid derived from cannabis and has shown promise with anti-inflammatory, analgesic, anticonvulsant, muscle relaxant, neuroprotective, antioxidant, and antineoplastic properties. CBD is known to act as a negative allosteric modulator of the CB1 receptor. In addition to its effects on CB1

**Table 1** A selective list of medications composed of natural and synthetic cannabinoids (Hruba and McMahon 2014)

| Medications | Comments |
|---|---|
| Dronabinol | • THC that is used to treat anorexia in HIV patients and CINV |
| Nabilone | • Synthetic cannabinoid and analog of Marinol<br>• Approved for CINV |
| Sativex | • Contains THC, CBD, and other cannabinoids<br>• Licensed in Canada in 2005 for relief of neuropathic pain and spasticity in MS<br>• Approved as an analgesic for patients with cancer |
| Rimonabant | • Trade name: Acomplia and Zimulti<br>• CB1 receptor inverse agonist<br>• Approved in the EU market for use as an anorectic anti-obesity drug and withdrawn in 2008 due to causation of depression and suicide |
| JWH-018 | • A synthetic cannabinoid agonist included in herbal incense mixtures<br>• Named "spice," "K2"<br>• Increasingly sold in "legal" smoke blends that are illegal |
| CP55,940 | • A synthetic cannabinoid that mimics THC but has been identified to be more potent than THC |
| HU-210 | • Synthetic cannabinoid<br>• At least 100 times more potent than THC |
| HU-331 | • Anticarcinogenic drug synthesized from CBD<br>• Inhibits DNA topoisomerase II |

and CB2 receptors, it is known to activate 5-HT1A/2A/3A serotonergic and TRPV1–2 vanilloid receptors and antagonizes alpha-1 adrenergic receptors and μ-opioid receptors. It is also reported to trigger the synaptosomal uptake inhibition of noradrenaline, dopamine, serotonin, and gamma-aminobutyric acid (GABA). CBD acts on mitochondrial $Ca^{2+}$ stores, blocks T-type $Ca^{2+}$ channel, and stimulates inhibitory glycine and FAAH. Nabiximol contains CBD and THC in equal proportions. Epidiolex, an oral CBD solution, has been approved for the treatment of refractory epilepsy due to Lennox–Gastaut syndrome and Dravet syndrome. THC and CBD account for 95% of marijuana's active ingredient.

Cannabigerol (CBG)—Both THC and CBD are produced from precursor CBG, which is subject to enzymatic conversion to CBD, THC, and other cannabinoids. Studies have suggested that CBG is nonpsychoactive and has anti-inflammatory and antioxidant effects and has been used for treatment in IBD. CBG also possesses COX-2 inhibitor properties. CBG has been investigated for use in prostate cancer. Other natural cannabinoids, such as cannabidivarin, are being investigated for epilepsy. The cannabinoid tetrahydrocannabivarin is currently being investigated for its role in diabetes mellitus and metabolic syndrome (Jadoon et al. 2016).

A selective list of medications composed of natural and synthetic cannabinoids has been given in Table 1 and the chemical structures are shown in Fig. 8.

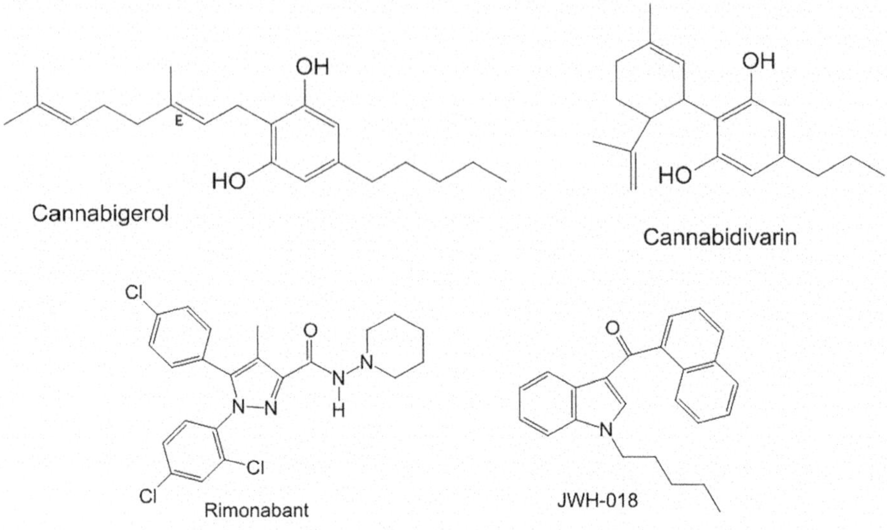


**Fig. 8** Other natural and synthetic cannabinoids

# 4 Neuroprotective Effects of Cannabis

In 2600 BC, the Chinese emperor Huang Ti prescribed taking cannabis for the relief of cramps in rheumatic and menstrual pain. THC has been used to improve symptoms of neurodegeneration in multiple sclerosis (MS), adrenoleukodystrophy (ALD), Huntington disease (HD), amyotrophic lateral sclerosis (ALS), and Alzheimer's disease (AD). Furthermore, THC offers neuroprotection on neuronal and nonneuronal elements in the brain via modulation of excitatory and glutamatergic and synaptic transmission, immune modulation, and release of anti-inflammatory mediators (Maroon and Bost 2018).

## *4.1 Alzheimer's Disease*

Alzheimer's disease (AD) is characterized by enhanced deposition of β-amyloid peptide A (βA) and glial activation in senile plaques, neuronal loss, and cognitive impairment. Cannabinoids have been found to be neuroprotective against excitotoxicity and brain damage. In AD patients, studies on senile plaques have demonstrated expressions of CB1 and CB2 receptors as well as markers of microglial activation. CB1-positive neurons were high in controls compared to areas of microglial activation. Furthermore, AD brains have decreased GP-receptor coupling and CB1 receptor expression. Cannabinoids HU-210, WIN55,212–2, and JWH-133 have blocked β-amyloid-induced activation of microglial cells, and administration of WIN55,212–2 to rats have known to prevent βA-induced microglial activation and cognitive impairment (Ramirez et al. 2005).

### 4.2 *Parkinson's Disease (PD)*

PD is the second common neurodegenerative disorder characterized by the degeneration of "dopaminergic" neurons in substantia nigra. The center for disease control and prevention (CDC) has rated the complications from PD as the 14th leading cause of death in the United States. CBD acts via antioxidant properties independent of cannabinoid receptors and has been found to be neuroprotective in an experimental model of Parkinsonism and is effective in attenuating PD-related dystonic movements (Maroon and Bost 2018). CBD was found to upregulate the mRNA levels of Cu/Zn-superoxide dismutase, a key enzyme involved in the defense mechanism against oxidative stress, indicating its neuroprotection against the degeneration of nigrostriatal dopaminergic neurons in PD. Activation of CB2 and not CB1 has also been implicated in the contribution of cannabinoids in PD (Milano and Capasso 2018).

### 4.3 *Amyotrophic Lateral Sclerosis (ALS)*

ALS is a progressive neurodegenerative disease affecting the motor neurons in the brain and spinal cord, leading to their destruction. Areas of degeneration are usually characterized by the presence of cytokines, immune cells like T-cells, microglia, and astrocytes. Currently, no effective pharmacological treatments have been established for this debilitating disease; however, cannabinoids have been found to reduce the progression of ALS through interactions with CB1 and CB2 receptors, delaying the progression of the disease (Milano and Capasso 2018). In an animal model of ALS, endogenous cannabinoids were found to be elevated in the spinal cords of mice, and treatment with nonselective cannabinoids delayed disease progression and prolonged survival (Raman et al. 2009). Furthermore, CB2 receptors, which primarily exist in the periphery, are upregulated in the inflamed neural tissues in ALS models. The administration of selective CB2 agonist AM-1241 increased the survival interval by 56%, suggesting that CB2 agonists may slow mortar degeneration and preserve mortar function, and may represent a novel therapeutic target for the treatment of ALS (Shoemaker et al. 2007). A phase 2 trial in Italy called the CANALS, or Safety and efficacy of nabiximols on spasticity symptoms in patients with motor neuron disease, studied the safety, efficacy, and tolerability of Sativex in ALS patients demonstrated that Sativex was well tolerated with no serious side effects (Riva et al. 2019). An ongoing phase 3 study, Efficacy of Cannabinoids in ALS or Motor Neuron Disease, is currently evaluating the effects of CBD oil capsules on slowing disease progression in ALS (Urbi et al. 2019). Another large observational study, OMNI-CAN (Outcomes Mandate National Integration With Cannabis as Medicine), which included people with several disease conditions, used an anonymous online questionnaire to determine the potential benefits and side effects of cannabis.

## 4.4 Huntington Disease (HD)

Despite its clinical utility in many neurodegenerative diseases, cannabis has not received a lot of attention from the HD scientific community. It is significant to note that, in cases like HD and MS, the clinical effects of CBD reported to date are observed only when combined with THC in a ratio of 1:1 (Sativex). In patients who were administered Sativex and THC in a ratio of 1:1, there was attenuated GABA deficiency, decreased loss of Nissl-stained neurons, downregulation of CB1 receptors, i*nsulin-like growth factor 1* (*IGF-1*) expression, and upregulation of calpain expression, and reversed superoxide dismutase expression (Sagredo et al. 2011). Preclinical and postmortem studies in HD patients suggested an inverse relationship between CB1 receptor decrease and CB2 receptor increase in glial elements, astrocytes, and reactive microglial cells (Milano and Capasso 2018). Another study confirmed that CB1 receptor reduction in basal ganglia is linked to increases in chances of Huntington-like symptoms and associated lower levels of brain-derived neurotrophic factor (BDNF) with a lack of CB1 receptors, thereby leading to the development of HD (Blázquez et al. 2011).

## 4.5 Multiple Sclerosis (MS)

MS is a chronic illness where the immune system attacks the myelin sheath causing demyelination of neurons, compromised oligodendrocytes, and aberrant neuronal firing leading to spasticity and neuropathic pain. The pathological changes in MS include inflammation, excitotoxicity, neurodegeneration including AD, and cerebral ischemia. The use of cannabis in the treatment of MS has a long history. It is reported that healers in Greece, Rome, China, and India have used cannabis to treat muscle cramps, spasms, and pain related to MS. In an experimental model consisting of encephalomyelitis rats, treatment with THC reduced neuroinflammation and improved neurological outcomes. A combination of Δ-8-tetrahydrocannabinol (a more stable analog of THC) and non-psychotropic Dexanabinol was found to reduce the neurological deficits, thereby suggesting a tonic control in muscle tone by the endocannabinoid signaling system. In another murine model of MS, treatment with cannabinoids reduced the progression of symptoms, leading to downregulation of delayed-type hypersensitivity, interferon production, and inhibited proinflammatory cytokine production. In experimental models of MS, activation of CB1 and CB2 receptors appears to be beneficial against inflammation, thus

**Table 2** Effects of medical marijuana on select neurological diseases

| Condition | Effects |
|---|---|
| Spasticity in MS | Oral *cannabis* extract (OCE) is effective. Nabiximol (Sativex) and THC were probably effective for objective measures and patient-related symptoms |
| Central pain and painful spasms in MS | OCE is effective<br>THC or nabiximol may be effective |
| Urinary dysfunction in MS | Nabiximol is probably effective for reducing the bladder voids after 10 weeks<br>THC and OCE are ineffective |
| Tremor in MS | THC and OCE are probably ineffective. Nabiximol is probably ineffective |
| Other neurological conditions | OCE is ineffective in treating levodopa-induced dyskinesia in Parkinson's disease |

suggesting a reduction in frequency in relapses in patients with MS when smoking marijuana. In another multicenter study in the United States and United Kingdom, treatment with Marinol (synthetically produced dronabinol) or cannador (plant extract, consisting of THC and CBD in the ratio of 2:1), significant improvement was observed in spasticity, pain, sleep quality, and reduction in admissions for relapse (Milano and Capasso 2018).

In 2014, the American Academy of Neurology (AAN) published a systematic review of the effects of medical marijuana on select neurological diseases. Conclusions from the study are summarized in Table 2.

## 4.6 Post-Traumatic Stress Disorder

Treatment with CBD has been reported to reduce the symptoms such as chronic anxiety associated with p*ost-traumatic stress disorder* (PTSD). Also, treatment with cannabis resin containing CBD reduced the depression and anxiety in patients with PTSD. In rodent models reexposed to stressful situations, CBD treatment led to fear of memory extinction by causing a significant decrease in freezing time and attenuated fearful memories associated with past experiences. CBD has also demonstrated a decrease in the production of inflammatory cytokines, activation of glial cells, and brain leukocyte infiltration (Maroon and Bost 2018; Mecha et al. 2012). CBD has been involved in the modulation of receptors outside the endocannabinoid signaling system as well. Adenosine receptor activation by CBD leads to the enhancement of adenosine signaling activation and mediates anti-inflammatory and immunosuppressive effects. Furthermore, CBD is known to improve hippocampal neurogenesis. Animal studies have indicated that treatment with CBD improves AEA, which reduces short-term brain damage, improving the brain's metabolic activity and causing a reduction in the cerebral hemodynamic impairment brain edema and seizure. In addition to helping reducing seizures, CBD is also proven to produce anti-epileptiform and anticonvulsant effects in vitro and in vivo models (Maroon and Bost 2018).

In conclusion, there exists a strong link between neurodegenerative diseases and cannabinoids. Furthermore, CB1 receptors play a significant role in the neuroprotective role of cannabis and thus appear to be a promising therapeutic target. Increased research on cannabis and cannabinoid signaling systems may lead to novel therapies to delay the progression and onset of these diseases and their symptoms.

## 5 Neurotoxic Effects of Cannabis

Evidence of cannabis associated neurotoxicity suggests long-term changes in the human brain such as (a) reduced orbitofrontal cortex gray matter volume (enriched in CB1 receptors, implicated in addictive behavior under the possible mechanism neuronal loss, changes in cell size, and reduction in CBI density), (b) increased structural and functional connectivity, and (c) changes in neural architecture impacted by the age of onset of disease and its duration. Several studies have demonstrated poorer cognitive performance upon early-onset chronic cannabis exposure (Filbey et al. 2014). A study by Rocchetti et al. also concluded a significant reduction in brain volume, and that the frequency of use significantly impacted hippocampus functioning (Rocchetti et al. 2013). At a neurophysiological level, the effect of cannabinoids in the hippocampus was envisaged by neurocognitive impairments. Prolonged use of cannabis (almost daily use) leads to the greatest risk of adverse health outcomes. Thus, chronic administration of cannabis has been determined to decrease hippocampal volume causing memory dysfunction (Rocchetti et al. 2013). Impaired coding, storage, manipulation, and retrieval mechanisms have been observed in long-term and/or heavy cannabis users (Solowij and Battisti 2008). In cultured neurons, THC induced shrinkage of neuronal cell bodies, nuclei, and genomic DNA strands, which includes hallmarks of apoptosis (Chan et al. 1998).

The effects of cannabinoids are characterized by several responses, including a reduction in memory, antinociception, hypothermia, catalepsy, and diminished motion. Lesions in the hippocampus reduce the acquisition and storage of new information or reduce the capacity to recover previous knowledge. Further cannabinoids within the intrahippocampal impair spatial memory, suggesting the primary role of cannabinoid receptors in meditating memory deficits with the gradual development in alteration of brain function. The activation of cannabinoid receptors induces the expression of early genes Krox24, Krox20, and Jun B increases NF-Kb (*nuclear factor* kappa-light-chain-enhancer of activated B cells) activity, which further activates the expression of tumor necrosis factor- α (TNF-α), which induces apoptosis. Furthermore, activation of these pathways with the release of arachidonic acid and ROS may be possible in neurotoxicity. Therefore, it can be concluded that THC mediates a transcription-mediated cell death on hippocampal neurons. Cannabinoids mediate cell death in certain populations of neurons in response to the release of anandamide. Several processes during neurogenesis and neuroplasticity and dysregulation of the endocannabinoid system interact to cause neuroanatomical interactions. Chronic cannabis exposure leads to profound morphological changes in hippocamps. Three months of exposure decreases the volume of neurons, nuclei, synaptic density, and the dendritic length of CA3 pyramidal neurons. Cannabinoids and marijuana have been said to exert deleterious effects on the immune system (Chan et al. 1998). According to Baldwin et al. alveolar macrophages were unable to phagocytose *Staphylococcus aureus* and were unable to kill bacteria and tumor cells. Furthermore, these alveolar macrophages produced fewer amounts of TNF-α, granulocyte–macrophage colony-stimulating factor (*GM-CSF), and interleukin*-6 (*IL*-6) upon lipopolysaccharide stimulation (activator of macrophage) (Baldwin et al. 1997). Further THC reduced the TNF-α production when exposed to large granulocyte lymphocyte and cytolytic activity against K562 leukemia cells (Kusher et al. 1994). An anecdotal report by Juel-Jensen indicated that marijuana exposure altered resistance to infections in humans (Juel-Jensen 1972). Marijuana smokers with herpes simplex virus had an increased recurrence of lesions. Studies by Nadia Solowij determined the neuropsychiatric and psychobiological aspects of taking cannabis. Acute intoxication of cannabis exposure is said to lead to mild hallucinations, perpetual distortions, and delusions (Parrott et al. 2016). There has also been a considerable amount of evidence suggesting a causal relationship between usage of cannabis during adolescence and the development of depression, anxiety, and schizophrenia. Besides, the neurobiological effects of cannabis can lead to psychosis and/or may affect a preexisting condition to cause psychosis. Cannabis is known to exacerbate symptoms of schizophrenia by complex interactions between dopaminergic, serotonergic, and glutamatergic, GABAergic, cholinergic, opioid, and endogenous cannabinoid systems. Chronic cannabis exposure has been found to increase cannabinoids leading to schizophrenia-like neurotransmitter conditions in the prefrontal cortex with the desynchronized cortex, resulting in cognitive dysfunction (Parrott et al. 2016). Another study demonstrated that chronic cannabis exposure was associated with diminished neuronal integrity in the dorsolateral prefrontal cortex (Jacobus and Tapert 2014). In addition, THC, Δ-8-THC, tetrahydrocannabivarin, and cannabinol that are known to have different effects still have undefined roles.

Several studies have investigated the potential causes of neurotoxicity. According to Shamarka et al. the neurotoxicity of THC was a result of the prostanoid synthesis pathway and generation of free radicals by cyclooxygenase. El-Shamarka et al. observed that coadministration of cannabis in adolescent rats potentiated the neurotoxic effects of nandrolone decanoate, a commonly used anabolic–androgenic steroid, by mediating oxidative stress and inflammation, intrinsic and extrinsic apoptosis in the hippocampus and prefrontal cortex of rats (El-Shamarka et al. 2020). Studies by Chan et al. suggested that THC-mediated apoptosis was attributed to the activation of phospholipase $A_2$—cyclooxygenase pathway (Chan et al. 1998).

While researchers do not fully understand the long-term risks and effects associated with cannabis use, research is currently ongoing. However, it can be said with confidence that long-term exposure to cannabis increases the risk of substance abuse, thereby leading to memory and concentration deficits. Currently, the *National Institutes of Health* (NIH), USA, has enrolled more than 11,000 children between ages 9 and 10 for the ongoing "Adolescent Brain Cognitive Development" (ABCD) study. Researchers plan to utilize functional and structural neuroimaging tools to track them into late adulthood.

## 6 Current Treatments and Future of Cannabis Pharmacology

The use of medical cannabis in the legal system faces a few challenges. Effective delivery systems are quintessential and are being developed. Since the cannabinoids are volatile and vaporize at a low temperature, when the heated air is drawn through marijuana, the active compounds can be vaporized into a mist, which is then dosed and inhaled without the generation of smoke. Patient individuality, including the severity of the condition, lung capacity, inhalation and exhalation habits, GI absorption, greatly varies. Other concerns with the use of medical cannabis include the constant need for monitoring and prevention of addiction. The future is likely to see extensive research that digs into the therapeutic effects of cannabinoid receptor agonists and determine its side effects. Despite good therapeutic efficacy, the production of potent medications can be hindered by the fact that these natural compounds in cannabis work at its best only when all the natural cannabinoids are allowed to work concurrently. This "orchestration" of effects can be seen best in the combination of THC and CBD, which is known to produce anxiolytic effects (Bridgeman and Abazia 2017). Refinement of cannabinoids is expected to facilitate cleaner and highly efficacious drugs; however, this may also lead to dire consequences. For example, in patients with ALS, dronabinol has been found to be too sedating and less effective compared to natural cannabis (Group AL 2012). The legalization of cannabis for medicinal use of cannabis warrants a lot of research into the pharmacokinetics and pharmacodynamics of cannabis. Oral THCs nabiximol and nabilone have been shown to improve cannabis withdrawal symptoms. The effectiveness of nabilone in preventing relapse has yet to be tested in a clinical setting. In contrast, nabiximol, although, has been approved by 15 countries, requires further studies to determine its longitudinal effects (Copeland and Pokorski 2016). Despite the therapeutic effects of THC and CBD, a lack of well-controlled, double-blind, randomized, clinical trial fails to establish its long-term and short-term efficacy. In addition, safety issues impede their use in the clinic. THC and CBD can have undesirable effects on the CNS, and dose optimization may not be an option before the onset of side effects. However, CBD receptor agonists that modulate the endocannabinoid system are promising agents in the treatment of unmanageable disorders. Further research is required to completely understand the endocannabinoid system in antinociception/pain and antispasticity. Improved methodologies, a larger study population, standard formulations are required to establish the use of medical cannabis. Although medical cannabis has been approved for muscle spasms and pain, especially when standard therapy failed, the adverse events involving CNS and gastrointestinal (GI) system may preclude its usage. Extreme caution is advised in patients with a history of cardiovascular or mental disorders.

Cannabis can be regarded as neither a miracle drug nor be regarded as being responsible for people's illness. Several aspects must be considered in the risk to benefit ratio. The United States has spent a tremendous amount of money on curbing the illicit use of cannabis with limited success. Regarding medical use, laws must consider scientific research and logic during policy-making decisions rather than political views that debate over the harmful effects of the illicit use of recreational marijuana. Clinicians and students must be encouraged to improve their knowledge base and educate themselves in the science of cannabis pharmacology and medicine. Although nabilone and dronabinol are well-regulated drugs that have demonstrated safety and efficacy, there is a dearth of evidence of medical use of cannabis, which requires the drugs to be used with the greatest caution.

## References


Ahn K, Johnson DS, Cravatt BF (2009) Fatty acid amide hydrolase as a potential therapeutic target for the treatment of pain and CNS disorders. Expert Opin Drug Discov 4(7):763–784. https://doi.org/10.1517/17460440903018857

Bacci A, Huguenard JR, Prince DA (2004) Long-lasting self-inhibition of neocortical interneurons mediated by endocannabinoids. Nature 431(7006):312–316. https://doi.org/10.1038/nature02913

Baldwin GC, Tashkin DP, Buckley DM et al (1997) Marijuana and cocaine impair alveolar macrophage function and cytokine production. Am J Respir Crit Care Med 156 (5):1606–1613. https://doi.org/10.1164/ajrccm.156.5.9704146

Benito C, Nunez E, Tolon RM et al (2003) Cannabinoid CB2 receptors and fatty acid amide hydrolase are selectively overexpressed in neuritic plaque-associated glia in Alzheimer's disease brains. J Neurosci 23(35):11136–11141

Blázquez C, Chiarlone A, Sagredo O et al (2011) Loss of striatal type 1 cannabinoid receptors is a key pathogenic factor in Huntington's disease. Brain 134(1):119–136. https://doi.org/10.1093/brain/awq278

Bridgeman MB, Abazia DT (2017) Medicinal cannabis: history, pharmacology, and implications for the acute care setting. P T 42(3):180–188

Buckley NE, McCoy KL, Mezey E et al (2000) Immunomodulation by cannabinoids is absent in mice deficient for the cannabinoid CB(2) receptor. Eur J Pharmacol 396(2–3):141–149. https://doi.org/10.1016/s0014-2999(00)00211-9

Campos I (2012) Home grown marijuana and the origins of Mexico's war on drugs. University of North Carolina Press. https://doi.org/10.5149/9780807882689_campos

Castillo PE, Younts TJ, Chavez AE et al (2012) Endocannabinoid signaling and synaptic function. Neuron 76(1):70–81. https://doi.org/10.1016/j.neuron.2012.09.020

Chan GC-K, Hinds TR, Impey S et al (1998) Hippocampal neurotoxicity of δ9-tetrahydrocannabinol. J Neurosci 18(14):5322–5332. https://doi.org/10.1523/jneurosci.18-14-05322.1998

Console-Bram L, Marcu J, Abood ME (2012) Cannabinoid receptors: nomenclature and pharmacological principles. Prog Neuro-Psychopharmacol Biol Psychiatry 38(1):4–15. https://doi.org/10.1016/j.pnpbp.2012.02.009

Copeland J, Pokorski I (2016) Progress toward pharmacotherapies for cannabis-use disorder: an evidence-based review. Subst Abus Rehabil 7:41–53. https://doi.org/10.2147/sar.S89857

Devane WA, Hanus L, Breuer A et al (1992) Isolation and structure of a brain constituent that binds to the cannabinoid receptor. Science 258(5090):1946–1949. https://doi.org/10.1126/science.1470919

Di Marzo V, De Petrocellis L (2012) Why do cannabinoid receptors have more than one endogenous ligand? Philos Trans R Soc Lond Ser B Biol Sci 367(1607):3216–3228. https://doi.org/10.1098/rstb.2011.0382

Di Marzo V, Stella N, Zimmer A (2014) Endocannabinoid signalling and the deteriorating brain. Nat Rev Neurosci 16(1):30–42. https://doi.org/10.1038/nrn3876

Diana MA, Marty A (2004) Endocannabinoid-mediated short-term synaptic plasticity: depolarization-induced suppression of inhibition (DSI) and depolarization-induced suppression of excitation (DSE). Br J Pharmacol 142(1):9–19. https://doi.org/10.1038/sj.bjp.0705726

El-Shamarka ME-S, Sayed RH, Assaf N et al (2020) Combined neurotoxic effects of cannabis and nandrolone decanoate in adolescent male rats. Neurotoxicology 76:114–125. https://doi.org/10.1016/j.neuro.2019.11.001

Filbey FM, Aslan S, Calhoun VD et al (2014) Long-term effects of marijuana use on the brain. Proc Natl Acad Sci U S A 111(47):16913–16918. https://doi.org/10.1073/pnas.1415297111

Groce E (2018) The health effects of cannabis and cannabinoids: the current state of evidence and recommendations for research. J Med Regul 104(4):32–32. https://doi.org/10.30770/2572-1852-104.4.32

Group AL (2012) ALSUntangled no. 16: cannabis. Amyotroph Lateral Scler 13(4):400–404. https://doi.org/10.3109/17482968.2012.687264

Heifets BD, Castillo PE (2009) Endocannabinoid signaling and long-term synaptic plasticity. Annu Rev Physiol 71:283–306. https://doi.org/10.1146/annurev.physiol.010908.163149

Holt S (2016) The Cannabis revolution: what you need to know. iUniverse, Holt Institute of Medicine (www.hiom.org)

Hruba L, McMahon LR (2014) The cannabinoid agonist HU-210: pseudo-irreversible discriminative stimulus effects in rhesus monkeys. Eur J Pharmacol 727:35–42. https://doi.org/10.1016/j.ejphar.2014.01.041

Huestis MA, Solimini R, Pichini S et al (2019) Cannabidiol adverse effects and toxicity. Curr Neuropharmacol 17(10):974–989. https://doi.org/10.2174/1570159X17666190603171901

Jacobus J, Tapert S (2014) Effects of cannabis on the adolescent brain. Curr Pharm Des 20 (13):2186–2193. https://doi.org/10.2174/13816128113199990426

Jadoon KA, Ratcliffe SH, Barrett DA et al (2016) Efficacy and safety of cannabidiol and tetrahydrocannabivarin on glycemic and lipid parameters in patients with type 2 diabetes: a randomized, double-blind, placebo-controlled, parallel group pilot study. Diabetes Care 39 (10):1777. https://doi.org/10.2337/dc16-0650

Juel-Jensen BE (1972) Cannabis and recurrent herpes simplex. Br Med J 4(5835):296. https://doi.org/10.1136/bmj.4.5835.296-a

Katona I, Freund TF (2012) Multiple functions of endocannabinoid signaling in the brain. Annu Rev Neurosci 35:529–558. https://doi.org/10.1146/annurev-neuro-062111-150420

Katona I, Sperlagh B, Sik A et al (1999) Presynaptically located cb1 cannabinoid receptors regulate GABA release from axon terminals of specific hippocampal interneurons. J Neurosci 19 (11):4544–4558

Kendall DA, Yudowski GA (2017) Cannabinoid receptors in the central nervous system: their signaling and roles in disease. Front Cell Neurosci 10:294. https://doi.org/10.3389/fncel.2016.00294

Kusher DI, Dawson LO, Taylor AC et al (1994) Effect of the psychoactive metabolite of marijuana, δ9-tetrahydrocannabinol (THC), on the synthesis of tumor necrosis factor by human large granular lymphocytes. Cell Immunol 154(1):99–108. https://doi.org/10.1006/cimm.1994.1060

Mackie K, Lai Y, Westenbroek R et al (1995) Cannabinoids activate an inwardly rectifying potassium conductance and inhibit Q-type calcium currents in AtT20 cells transfected with rat brain cannabinoid receptor. J Neurosci 15(10):6552–6561

Maejima T, Hashimoto K, Yoshida T et al (2001) Presynaptic inhibition caused by retrograde signal from metabotropic glutamate to cannabinoid receptors. Neuron 31(3):463–475. https://doi.org/10.1016/s0896-6273(01)00375-0

Maroon J, Bost J (2018) Review of the neurological benefits of phytocannabinoids. Surg Neurol Int 9:91. https://doi.org/10.4103/sni.sni_45_18

Marsicano G, Kuner R (2008) Anatomical distribution of receptors, ligands and enzymes in the brain and in the spinal cord: circuitries and neurochemistry. In: Köfalvi A (ed) Cannabinoids and the brain. Springer, Boston, MA. https://doi.org/10.1007/978-0-387-74349-3_10

Mecha M, Torrao AS, Mestre L et al (2012) Cannabidiol protects oligodendrocyte progenitor cells from inflammation-induced apoptosis by attenuating endoplasmic reticulum stress. Cell Death Dis 3(6):e331–e331. https://doi.org/10.1038/cddis.2012.71

Milano W, Capasso A (2018) Neuroprotection by cannabinoids in neurodegenerative diseases. Alzheimer's Dement Cogn Neurol 2(1). https://doi.org/10.15761/adcn.1000120

Navarrete M, Diez A, Araque A (2014) Astrocytes in endocannabinoid signalling. Philos Trans R Soc Lond Ser B Biol Sci 369(1654):20130599. https://doi.org/10.1098/rstb.2013.0599

Nyilas R, Dudok B, Urbán GM et al (2008) Enzymatic machinery for endocannabinoid biosynthesis associated with calcium stores in glutamatergic axon terminals. J Neurosci 28(5):1058–1063. https://doi.org/10.1523/jneurosci.5102-07.2008

Ohno-Shosaku T, Maejima T, Kano M (2001) Endogenous cannabinoids mediate retrograde signals from depolarized post-synaptic neurons to presynaptic terminals. Neuron 29(3):729–738. https://doi.org/10.1016/s0896-6273(01)00247-1

Pacula RL, Smart R (2017) Medical marijuana and marijuana legalization. Annu Rev Clin Psychol 13:397–419. https://doi.org/10.1146/annurev-clinpsy-032816-045128

Parrott AC, Gouzoulis-Meyfrank E, Rodgers J et al (2016) Ecstasy/MDMA and cannabis: the complexities of their interactive neuropsychobiological effects. J Psychopharmacol 18 (4):572–575. https://doi.org/10.1177/026988110401800416

Pertwee RG (2012) Targeting the endocannabinoid system with cannabinoid receptor agonists: pharmacological strategies and therapeutic possibilities. Philos Trans R Soc Lond Ser B Biol Sci 367(1607):3353–3363. https://doi.org/10.1098/rstb.2011.0381

Raman C, McAllister SD, Rizvi G et al (2009) Amyotrophic lateral sclerosis: delayed disease progression in mice by treatment with a cannabinoid. Amyotroph Lateral Scler 5(1):33–39. https://doi.org/10.1080/14660820310016813

Ramirez BG, Blazquez C, Gomez del Pulgar T et al (2005) Prevention of Alzheimer's disease pathology by cannabinoids: neuroprotection mediated by blockade of microglial activation. J Neurosci 25(8):1904–1913. https://doi.org/10.1523/JNEUROSCI.4540-04.2005

Riva N, Mora G, Sorarù G et al (2019) Safety and efficacy of nabiximols on spasticity symptoms in patients with motor neuron disease (CANALS): a multicentre, double-blind, randomised, placebo-controlled, phase 2 trial. Lancet Neurol 18(2):155–164. https://doi.org/10.1016/s1474-4422(18)30406-x

Rocchetti M, Crescini A, Borgwardt S et al (2013) Is cannabis neurotoxic for the healthy brain? A meta-analytical review of structural brain alterations in non-psychotic users. Psychiatry Clin Neurosci 67(7):483–492. https://doi.org/10.1111/pcn.12085

Roche M, Finn DP (2010) Brain CB(2) receptors: implications for neuropsychiatric disorders. Pharmaceuticals (Basel) 3(8):2517–2553. https://doi.org/10.3390/ph3082517

Sagredo O, Pazos MR, Satta V et al (2011) Neuroprotective effects of phytocannabinoid-based medicines in experimental models of Huntington's disease. J Neurosci Res 89(9):1509–1518. https://doi.org/10.1002/jnr.22682

Scotter EL, Abood ME, Glass M (2010) The endocannabinoid system as a target for the treatment of neurodegenerative disease. Br J Pharmacol 160(3):480–498. https://doi.org/10.1111/j.1476-5381.2010.00735.x

Shoemaker JL, Seely KA, Reed RL et al (2007) The CB2 cannabinoid agonist AM-1241 prolongs survival in a transgenic mouse model of amyotrophic lateral sclerosis when initiated at symptom onset. J Neurochem 101(1):87–98. https://doi.org/10.1111/j.1471-4159.2006.04346.x

Small E, Marcus D et al (2002) Hemp: a new crop with new uses for North America. In: Janick, Whipkey (eds) Trends in new crops and new uses. ASHS Press, Alexandria, VA

Solowij N, Battisti R (2008) The chronic effects of cannabis on memory in humans: a review. Curr Drug Abuse Rev 1(1):81–98. https://doi.org/10.2174/1874473710801010081

Steffens S, Veillard NR, Arnaud C et al (2005) Low dose oral cannabinoid therapy reduces progression of atherosclerosis in mice. Nature 434(7034):782–786. https://doi.org/10.1038/nature03389

Tramer MR, Carroll D, Campbell FA et al (2001) Cannabinoids for control of chemotherapy induced nausea and vomiting: quantitative systematic review. BMJ 323(7303):16–21. https://doi.org/10.1136/bmj.323.7303.16

Twitchell W, Brown S, Mackie K (1997) Cannabinoids inhibit N- and P/Q-type calcium channels in cultured rat hippocampal neurons. J Neurophysiol 78(1):43–50. https://doi.org/10.1152/jn.1997.78.1.43

Urbi B, Broadley S, Bedlack R et al (2019) Study protocol for a randomised, double-blind, placebo-controlled study evaluating the efficacy of cannabis-based medicine extract in slowing the disease progression of amyotrophic lateral sclerosis or motor neurone disease: the emerald trial. Br Med J Open 9(11):e029449. https://doi.org/10.1136/bmjopen-2019-029449

Wilkinson ST, Radhakrishnan R, D'Souza DC (2014) Impact of cannabis use on the development of psychotic disorders. Curr Addict Rep 1(2):115–128. https://doi.org/10.1007/s40429-014-0018-7

Zou S, Kumar U (2018) Cannabinoid receptors and the endocannabinoid system: signaling and function in the central nervous system. Int J Mol Sci 19(3):833. https://doi.org/10.3390/ijms19030833

# Alpha-Synuclein: Biomarker for Parkinson's Disease, It's Estimation Methods, and Targeted Medicinal Therapies

**Shivani V. Dhokne, Vaishali R. Undale, Dinesh Chandra Agrawal, and Sharad D. Pawar**

## Contents




S. V. Dhokne
Dr. D.Y. Patil Institute of Pharmaceutical Sciences and Research, Pune, Maharashtra, India

V. R. Undale
Pharmacology, Dr. D.Y. Patil Institute of Pharmaceutical Sciences and Research, Pune, Maharashtra, India

D. C. Agrawal (✉)
Department of Applied Chemistry, Chaoyang University of Technology, Taichung, Taiwan
e-mail: dcagrawal@gm.cyut.edu.tw

S. D. Pawar (✉)
Regional Ayurveda Institute for Fundamental Research, Pune, Maharashtra, India
e-mail: rophraifr.pu-ayush@gov.in


 
D. C. Agrawal, M. Dhanasekaran (eds.), *Medicinal Herbs and Fungi*,
https://doi.org/10.1007/978-981-33-4141-8_9

**Abstract** Parkinson's disease (PD) is the second most common neurodegenerative disease after Alzheimer's, affecting more than ten million people in the world. The highest priority in PD research has been for the development of a reliable biomarker to detect the disease at an early stage. The presence of alpha-synuclein (α-synuclein) is one of the most significant pathological hallmarks of the PD. Accumulation of cellular levels of α-synuclein and increased release of α-synuclein from the synaptic vesicles results in its toxicity by minimizing dopaminergic neurotransmission in the brain. Therefore, it is imperative to devise appropriate techniques for measurement of total α-synuclein, oligomeric α-synuclein, phosphorylated α-synuclein, and seeding activity of α-synuclein in peripheral tissues and body fluids. Inhibition of α-synuclein aggregation, oligomerization, and fibrillation are important approaches for PD treatment. Also, for the development of new drugs for PD, specific plants and phytochemicals must be identified to provide selective molecules. The chapter provides an overview of the role of α-synuclein in Parkinson's disease, its estimation methods, and the use of phytochemicals in targetting α-synuclein for preventing neurotoxicity.

**Keywords** Alpha-synuclein · Biomarker · Neurodegeneration · Neuroprotection · Neurotoxicity · Parkinson's disease

## Abbreviations

| | |
|---|---|
| ATP | Adenosine triphosphate |
| CNS | Central nervous system |
| CSF | Cerebrospinal fluid |
| CSP-α | Cysteine string protein-alpha |
| CT | Computed tomography |
| DAT | Dopamine transporter |
| DNA | Deoxyribonucleic acid |
| ECL | Electrochemiluminescence |
| ELISA | Enzyme-linked immune sorbent assay |
| ER | Endoplasmic reticulum |
| GI | Gastrointestinal |
| HGB | Haemoglobin |
| HPLC | High-performance liquid chromatography |
| HRP | Horseradish peroxidase |
| KCl | Potassium chloride |

| | |
|---|---|
| $KH_2PO_4$ | Potassium phosphate monobasic |
| LB | Lewy bodies |
| LN | Lewy neurites |
| MgCl | Magnesium chloride |
| MRI | Magnetic resonance imaging |
| MRM | Multiple reaction monitoring |
| MSA | Multiple system atrophy |
| $Na_2CO_3$ | Sodium carbonate |
| NAC | Non-Aβ component |
| NaCl | Sodium chloride |
| $NaHPO_4$ | Sodium phosphate |
| NaOH | Sodium hydroxide |
| $NH_4HCO_3$ | Ammonium bicarbonate |
| NMDA | N-methyl-D-aspartate receptor |
| PAGE | Polyacrylamide gel electrophoresis |
| PBS | Phosphate buffer saline |
| PD | Parkinson's disease |
| PET | Positron emission tomography |
| PNS | Peripheral nervous system |
| PUFA | Peroxidation prone polyunsaturated fatty acids |
| PVDF | Polyvinylidene fluoride membrane |
| RBCs | Red blood corpuscles |
| ROS | Reactive oxygen species |
| RT | Room temperature |
| SDS | Sodium dodecyl sulfate |
| SiRNA | Small interfering RNA (ribonucleic acid) |
| SNARE | Soluble N-ethylmaleimide sensitive factor attachment protein receptor |
| SNc | Substantia nigra pars compacta |
| SNCA | Synuclein alpha |
| TRIS | Tris(hydroxymethyl) aminomethane |
| UPS | Ubiquitin–proteasome system |

## 1 Introduction

After Alzheimer's, Parkinson's disease (PD) is the second most common neurodegenerative disease affecting more than ten million people in the world (Kalia 2018). However, the precise causes of PD are not yet known. Several factors, including genetic, oxidative stress, environment, disturbance in synaptic function, and Lewy bodies formation, have been reported (Zheng et al. 2018). The characteristic motor features of PD include micrographia (abnormally small or cramped handwriting), dystonia (a movement disorder in which muscles contract involuntarily, causing repetitive or twisting movements), gait disturbance, hypotonia (decreased muscle tone), and sialorrhea; (excessive salivation), while nonmotor features include depression, anxiety, cognitive impairment, sensory abnormalities, anosmia, sleep disturbance, and neuropsychiatric features like changes in mood, cognition, and so on (Ahsan et al. 2018).

In PD, there is a selective loss of dopaminergic neurons in the substantia nigra of the midbrain caused by a reduction in dopamine levels in the striatum (Kalia 2018). It leads to an imbalance in neurotransmitters, acetylcholine (Ach), and dopamine, and thus results in the PD symptoms (Kalia 2018; Farooqui and Farooqui 2009). Also, PD is characterized by the accumulation of presynaptic protein α-synuclein. At a molecular level, PD is characterized by proteinopathy (a neurodegenerative disease caused by a malformed protein), synucleinopathy (a neurodegenerative disorder in which the protein α-synuclein accumulates abnormally to form inclusions in the cell bodies or axons of neurons or oligodendrocytes), and oxidative stress (Kalia 2018; Singh et al. 2019; Farooqui and Farooqui 2009).

Presently, there are hardly any reliable specific tests for PD diagnosis, especially in the early stages. While examining patients, a neurologist evaluates the person's motor function or movement, such as walking, coordination and balance, some simple hand tasks, the agility of the legs and arms, muscle tone, and so on. Other diagnostic tests include a blood test, MRI/CT scan, DAT scan, and PET scan (Kalia 2018; Farooqui and Farooqui 2009).

The treatment for PD includes dopamine precursors, peripheral decarboxylase inhibitors, and dopamine agonists while, the surgical treatment includes deep brain stimulation, immunotherapy, and cell transplantation. Limitations and bottleneck in the treatment of PD are the blood–brain barriers and restricted access across barrier layers. Alpha-synuclein is considered the major pathological hallmark for diagnosing the PD. In the present chapter, the authors provide an overview of the role of α-synuclein in Parkinson's disease, its estimation methods, and the use of phytochemicals in targetting α-synuclein for preventing the neurotoxicity.

## 2 Structure and Functions of α-Synuclein

Alpha-synuclein is a small protein consisting of 140 amino acids found in the presynaptic terminals of neurons that contain three functional domains: the N-terminal domain (amino acid residues (AA) 1–60), the NAC region (AA 1–95), and the C-terminal domain (AA 96140) (Kalia 2018). It is a highly acidic and unfolded protein encoded by the synuclein alpha (SNCA) gene and is abundantly present in the brain (Fig. 1).

It is a synaptic protein that is a major component of Lewy bodies (Javed et al. 2019). It is mainly a neuronal protein found in the neuroglial cells (Tokuda et al. 2010) and expresses in the neocortex, hippocampus, substantia nigra, thalamus, and cerebellum. It is highly expressed in the mitochondria in the olfactory bulb, hippocampus, striatum, and thalamus, where the cytosolic α-synuclein content is high (Schaser et al. 2019). Alpha-synuclein interacts with tubulin and acts as a potential microtubule-associated protein-like tau. In the formation of soluble

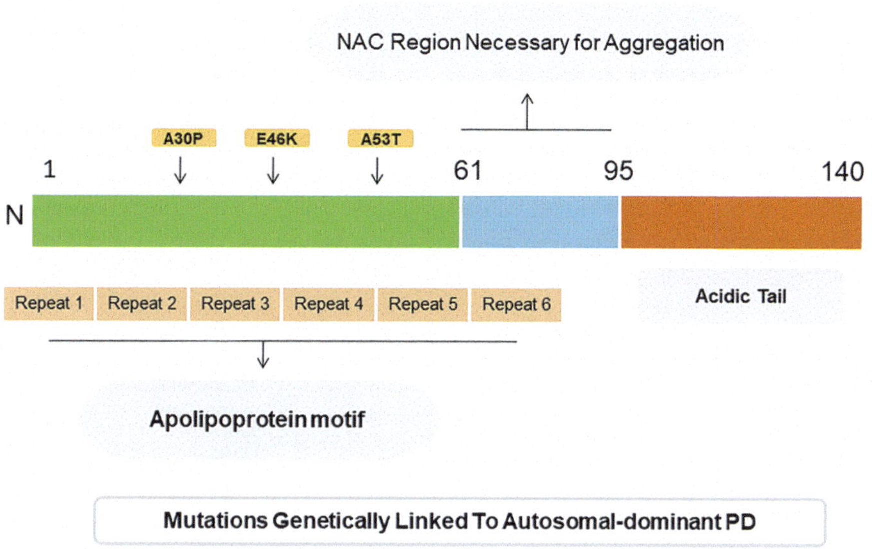


**Fig. 1** Structure of α-synuclein

N-ethylmaleimide sensitive factor attachment protein receptor (SNARE) complexes, it functions as a molecular chaperone (Ghosh et al. 2017). Alpha-synuclein is vital for the normal expansion of a cognitive function. It has antioxidant activity in the membrane and shows autoproteolytic action. Membrane interaction of α-synuclein affects its rate of aggregation, where it perfuses into the lipid vesicles. Thus, aggregated states of α-synuclein are formed upon interaction with PUFA (peroxidation prone polyunsaturated fatty acids) (Schaser et al. 2019).

Alpha-synuclein is an intracellular protein found in biological fluids like cerebrospinal fluid (CSF), plasma, blood, and saliva. It is the chief component of Lewy bodies, which on mutation leads to familial PD (Kalia 2019). Thus, α-synuclein is considered a major pathological hallmark for diagnosing the PD. There are four types of α-synuclein isoforms with varying amino acids and cluster potential, which are encoded by the synuclein alpha (SCNA) gene (Kalia 2019; Gautam et al. 2014). Dimer formation is the first step for the cascade of α-synuclein aggregation, followed by tiny oligomers/protofibrils and leading to the development of α-synuclein fibrils, which is rich in β-sheet (Mollenhauer et al. 2008; Gautam et al. 2014). These eventually lead to the formation of a major component of Lewy bodies that are end-stage fibrils and aggregated α-synuclein. The pathway to fibril formation includes numerous intermediates. Initially, soluble oligomeric forms of α-synuclein are seen as spherical rings under an electron microscope (Mollenhauer et al. 2008). These are collectively termed as protofibrils, which are SDS-stable and can be detected on SDS/polyacrylamide gel electrophoresis (PAGE) (Rabenstein et al. 2019). In contrast, other forms are only captured on native gels or by size-exclusion chromatography. The cascade of events, starting from the natively

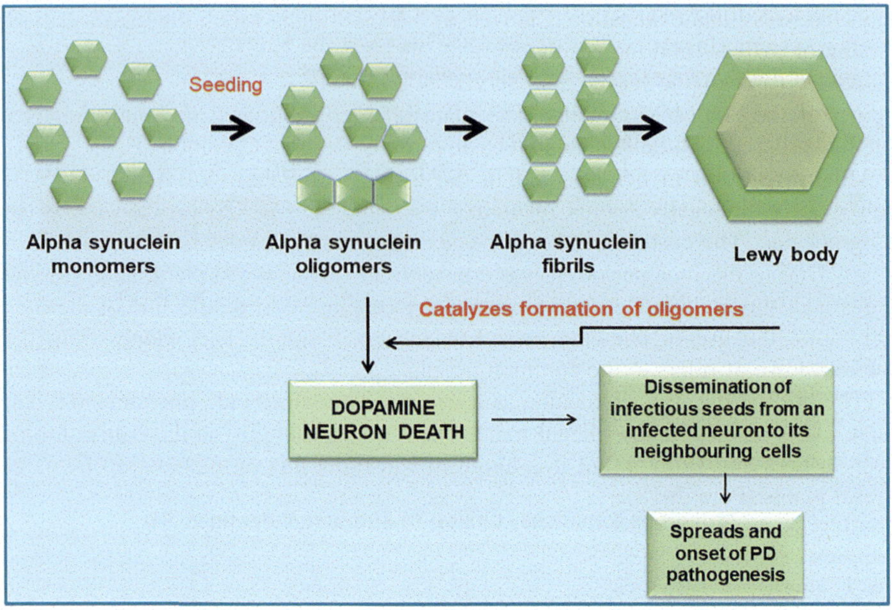


**Fig. 2** Different stages of α-synuclein

unfolded protein and culminating into mature fibril formation, is termed as α-synuclein aggregation. Misfolding of protein leads to aggregation and proliferation (Beach et al. 2008).

Factors such as oligomers and fibrils, oxidative stress, proteolysis, fatty acid concentration, phospholipids, and metal ions, which can lead to altered protein formation, modulate α-synuclein structure (Ardah et al. 2015). Additionally, phosphorylation, ubiquitination, nitration, and truncation also result in altered protein size, structure, or charge (Breydo et al. 2012). PD and other synucleinopathies are characterized by phosphorylation at Ser-129, a major phosphorylation site (Kalia 2018; Ingelsson 2016).

Initially, α-synuclein monomers interact to form unstable dimers, which slowly expands to initiate oligomer's genesis that polymerizes to amyloid fibrils, terms as "nucleated polymerization mechanism" (Fig. 2).

Alpha-synuclein is transmitted via direct penetration, diffusion, pore formation, endocytosis, micropinocytosis, and nanotube tunneling. It is the main component of Lewy bodies (LB) and Lewy neurites (LN) inclusions (Breydo et al. 2012), where it regulates vesicular transport between brain cells. A potential therapeutic target during the aggregation process and the etiopathogenesis of PD is an upstream form of the protein known as monomeric α-synuclein (Ardah et al. 2015; Breydo et al. 2012; Stefanis 2012).

It has been reported that α-synuclein robustly expresses in Lewy bodies. Currently, in the neuropathological evaluation of PD, α-synuclein immunohistochemistry is the most accepted method. The pathology begins from the olfactory bulb and vagal nucleus, followed into cortical regions. Only substantia nigra is affected in this staging system (Ingelsson 2016). Finding the α-synuclein in Lewy bodies is a significant finding. It is somewhat clear that abnormal α-synuclein deposition occurs in the early stages of PD (Tokuda et al. 2010). As it does not have a defined structure, it is termed as "natively unfolded protein." On binding to negatively charged lipids like phospholipids present in cellular membranes and β-sheets, rich structures form α-helical structures (Gautam et al. 2014), and where the protein consists of three regions: (1) amino terminus; (2) hydrophobic region, so-called NAC (non-Aβ component); and (3) carboxyl terminus. This protein comprises 1% total cytosolic proteins (Mollenhauer et al. 2008; Gautam et al. 2014). Alpha-synuclein antibodies endow with neuropil staining pattern in normal brains immunohistochemistry. Alpha-synuclein expression levels are harmonized in conditions that alter plasticity or prevent injury. Thus, α-synuclein is a modulator of synaptic transmission (Beach et al. 2008; Zheng et al. 2018).

It is a chaperone that acts in unity with CSP-α in the synod of the SNARE complex, and this involves binding to synaptotagmin-2. Thus, the main function of α-synuclein is control of neurotransmitter release through effects on the SNARE complex (Ardah et al. 2015; Ghosh et al. 2017). It appears to induce mitochondrial fragmentation, which is responsible for mitochondrial dysfunction and death (Rabenstein et al. 2019). The same mechanism of fragmentation also induces inappropriate, excessive mitophagy, which leads to mitochondrial removal and

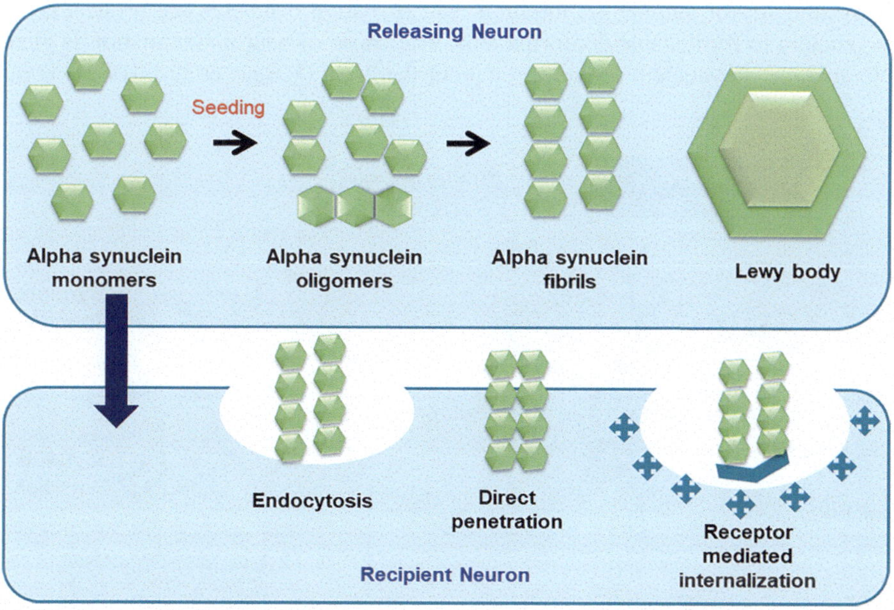


**Fig. 3** Aggregation and spreading mechanism of α-synuclein: Via endocytosis, direct penetration and receptor-mediated internalization, aggregation, and spreading of α-synuclein occur. For amplification of α-synuclein cascade, amyloid α-synuclein seeds activate aggregation and fibrils fragmentation

neuronal death. Alpha-synuclein potentially generates reactive oxygen species (ROS) production, oxidative stress, and resultant neuronal death (Fig. 3) (Zheng et al. 2018).

Rare point mutations occur in the N-terminal of α-synuclein. In the process of degradation in neurodegeneration, phosphorylation of α-synuclein is important (Fjorback et al. 2007). The ubiquitin–proteasome system (UPS) and the autophagy–lysosomal pathway degrade α-synuclein. Findings show α-synuclein in monomeric and oligomeric forms can be detected in solid and liquid tissues (Beach et al. 2008; Appel-Cresswell et al. 2013).

## 3 Pathways Implicated in α-Synuclein Neurotoxicity

An increased release of α-synuclein and increased cell-to-cell transfer or via accumulated cellular levels of the protein results in aggregation of α-synuclein. The physiological function and pathological dysfunction of α-synuclein trigger its toxicity (Rabenstein et al. 2019; Naoi et al. 2019). Factors responsible for neuronal toxicities of α-synuclein are genetic mutations or epigenetic mechanisms, cellular functions including endocytosis, Golgi homeostasis, ER-to-Golgi transport, presynaptic trafficking, UPS, autophagy, ER, and oxidative and nitration stress (Rabenstein et al. 2019; Ahsan et al. 2018; Ghosh et al. 2017) (Fig. 4).

Dysfunction of autophagy, nitration, and oxidation decreases the proneness of α-synuclein to form stable conformations. The cause of neuroinflammation is glial cells making α-synuclein spread throughout the brain (Vargas et al. 2019). One of

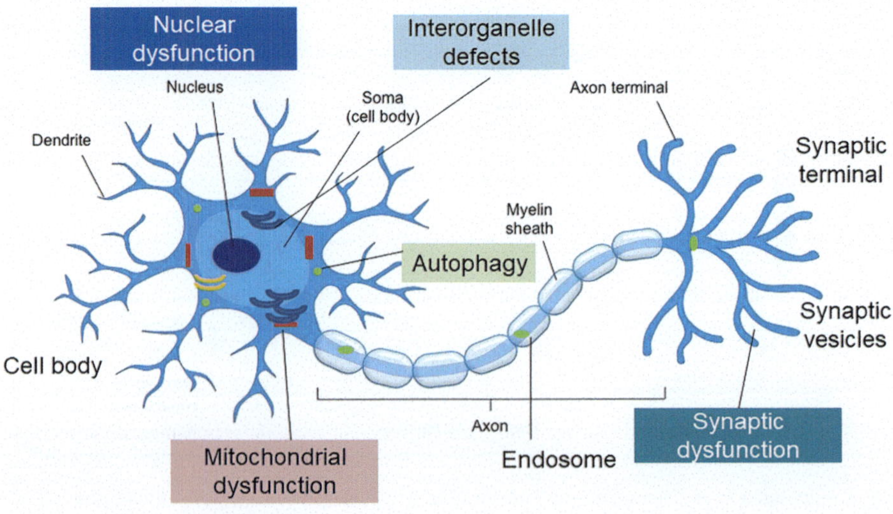


**Fig. 4** Implicated toxicities of *α*-synuclein

the toxicity is mitochondrial dysfunction, as mitochondria are vital for ATP synthesis, lipid metabolism, calcium uptake, and storage and survival of neurons (Atik et al. 2016). Protein toxicity disrupts mitochondrial hemostasis, increases DNA damage, and mitophagy (Ghosh et al. 2017).

Malfunctioning of dopamine and reduced reuptake of dopamine is due to the overexpression of α-synuclein (Vargas et al. 2019). Raised α-synuclein expression obstructs endosomal transport as it is necessary for protein folding, Golgi trafficking, calcium buffering, and the unfolded protein response (Rosborough et al. 2017). An increase in the α-synuclein level also upraises cytoplasmic calcium levels, which is responsible for activation of toxic calmodulin–calcineurin cascades (Ghosh et al. 2017; Rosborough et al. 2017).

Alpha-synuclein-associated mutations in PD (A30P, A53T) increase nuclear localization. Axonal organelles become highly dynamic and undergo axonal transport, which is regulated by phosphorylation and calcium signaling. Alpha-synuclein fibrils impair the axonal transport of autophagosomes and result in the misregulation of organelle dynamics (Vargas et al. 2019; Mollenhauer et al. 2008). Extracellular α-synuclein disrupts actin turnover and waves along axons on neuronal exposure (Atik et al. 2016). Oxidative stress increases with age and contributes to modification of α-synuclein such as nitration of tyrosine residues in PD patients (Rosborough et al. 2017). Alpha-synuclein nitration facilitates its aggregation, and its degradation increases its accumulation. With the increasing age of human substantia nigra, the α-synuclein level increases (Mollenhauer et al. 2008; Gautam et al. 2014).

## 4 Alpha-Synuclein as a Diagnostic Biomarker in Parkinson's Disease

The diagnosis of PD is mostly relying on clinical facets because the neuropathological examination mainly depends on autopsy. Due to unique peculiarity in incident and generation of synucleinopathies not only in the CNS but also in the PNS, salivary gland, plasma, skin, and submandibular gland, makes α-synuclein a good candidate as a biomarker for PD (Kalia 2018).

Previous studies provide affirmations that α-synuclein is present in CSF of PD patients. Devic et al. (2011) inspected human saliva and propounded saliva α-synuclein as a potential biomarker for detection and advancement. The presence of α-synuclein reactive antibodies in the serum of PD patients became an important topic. Plasma and urine are other sources of easily accessible biofluids (Ren et al. 2019). From a few patients with multiple system atrophy (MSA) and PD, the skin was tested by Zange et al. (2015), where phosphorylated α-synuclein in the specimens was investigated by immunohistochemistry. These findings supported that phosphorylated α-synuclein deposition causes nerve degeneration in PD (Kalia 2018). Several studies have detected the presence of α-synuclein in gastric, colonic specimens, and salivary glands. Total α-synuclein, as well as its oligomeric and phosphorylated forms, can be measured in body fluids and solid tissues using several methods such as ELISA, Western Blot, Luminex assay/mass spectroscopy, size-exclusion chromatography, and immunohistochemistry (Vivacqua et al. 2019; Atik et al. 2016).

### *4.1 Measurement of α-Synuclein in Peripheral Tissues and Body Fluids*

The aim is to seek α-synuclein inclusions within specific tissues, which also may be a possible target for PD diagnosis (Kalia 2018). The tissue samples (cardiac plexus, sympathetic ganglia, gastric myenteric plexus, colonic tissue, GI tract, cardiac sympathetic nervous system, heart, salivary gland, and vagus nerve), which were collected from postmortem and antemortem had increased α-synuclein staining (Vivacqua et al. 2019). Detection of phosphorylated α-synuclein in skin nerves will be a sensitive marker for PD according to recent studies. Alpha-synuclein has been detected in dystrophic olfactory epithelium neurites, receptor neurons, and epithelium basal cells in all PD patients in several postmortem studies (Fjorback et al. 2007). In another study, Lewy pathology has been noted in 100% of PD patients but not in controls when submandibular gland biopsies have been stained for α-synuclein detection. In the submucosa of the gut, the presence of phosphorylated α-synuclein was shown in 80% of PD patients. Collection of CSF and solid tissues would be comparatively difficult to assess, therefore alternative to it could be plasma and serum, which requires relatively less invasive technique to obtain the sample. One of the studies showed, the increased total plasma α-synuclein levels in PD patients in two studies using ELISA, but decreased levels in on the use of Western blot (Tokuda et al. 2010). Contamination can happen in plasma/serum samples, which may affect α-synuclein levels. Measurement of total α-synuclein in RBCs may be another source for PD diagnosis. Also, salivary α-synuclein is more readily accessible compared to other biofluids (Tokuda et al. 2010).

## 5 Methods for Estimation of α-Synuclein

### *5.1 Measurement of α-Synuclein in Human Cerebrospinal Fluid*

Estimation of α-synuclein in human CSF can be attained by ELISA, Western blot, and Luminex assay. ELISA is used to measure the concentration of an analyte (usually antibodies or antigens) in solution and involves at least one antibody with specificity for a particular antigen (Si et al. 2017). Enzyme-conjugated substrates bound to a secondary antibody bind with the antibody–antigen complex. Four different immunoassays show correlation and analytical precision between each other (Wong and Krainc 2017) (Table 1).

### 5.1.1 Assay Procedure

Precipitate human CSF samples overnight using a rabbit polyclonal anti-α-synuclein antibody. On the following day, elute beads and antibody-bound α-synuclein complexes and spike with an accurately quantified heavy labeled, recombinant human α-synuclein (acts as calibrator). Expose immunoprecipitated samples to trypsin and Glu-C proteolysis (Si et al. 2017). Maximal α-synuclein can be procured through combined proteolysis, where overnight digestion of immunoprecipitated CSF samples transpires with trypsin, followed by 8 hours digestion with Glu-C protease. The processed CSF samples can be analyzed by nano-liquid chromatography coupled with multiple reaction monitoring (MRM) using mass spectrometry (Wong and Krainc 2017).

### 5.1.2 Dot Blot

For formic acid denaturation of ASO mouse brain synaptosomal extracts, Dot blot could be used for validation of the binding of the conformational specific antibody to oligomers (Ren et al. 2019). It can be spotted directly onto a nitrocellulose membrane using vacuum samples (Ahsan et al. 2018). Incubate membrane with primary (anti-α-synuclein (BD), ASY-1, or FILA-1) and secondary horseradish peroxidase (HRP) conjugated antibodies. Visualize proteins using ECL (Huang et al. 2019).

**Table 1** Different types of immunoassays for the measurement of α-synuclein in human CSF

| Total α-synuclein prototype assay | Human α-synuclein kit | Biolegend α-synuclein immunoassay | α-synuclein prototype immunoassay |
|---|---|---|---|
| ELECSYS | MSD kit | Biolegend | ADX |
| This ELECSYS prototype assay is based on two anti-α-synuclein antibodies and utilizes electrochemiluminescence detection technology. | This kit is a sandwich immunoassay with the biotinylated capture antibody, and SULFO-TAG labeled detection antibody specific for α-synuclein by electrochemiluminescence | It is a sandwich-type immunoassay with luminescence readout developed using the same α-synuclein antibodies | Absorbance based sandwich ELISA with readout at $OD_{450-600\ nm}$ |

## 5.2 *Measurement of α-Synuclein in Human Plasma*

### 5.2.1 Preparation Method

Use alkaline phosphatase-conjugated ASY-1 antibody. Adjust the antibody to a concentration of 2 mg/mL protein in 0.136 M NaCl, 0.002 M KCl. 0.03 M $NaHPO_4$, 0.02 M $KH_2PO_4$ (PBS) pH 7.2 (Vivacqua et al. 2019). Add alkaline phosphate enzyme to a final concentration of 2 mg/mL. Dialyze enzyme and antibody mixture in PBS pH 7.2 overnight. Add glutaraldehyde to a final concentration of 0.2% and incubate it at room temperature (RT) for 2 hours followed by second round of dialysis against 0.05 M TRIS, 1 mM $MgCl_2$, pH 8.0 overnight (Vivacqua et al. 2019; Rabenstein et al. 2019). Supplement conjugated antibody with bovine albumin and $NaNO_3$ to a final concentration of 1% and 0.02%. Store alkaline phosphates conjugated antibody in the dark at 4 °C until further use (Rabenstein et al. 2019).

### 5.2.2 Detection Method

Coat the ELISA plates with 100 mL/well of ASY-1 IgG (1 mg/mL) in a 400 mM $NaHCO_3$, $Na_2CO_3$, pH 9.6 overnight at 4 °C. Wash three times in PBS, 0.05% Tween, followed by 1 hour incubation in PBS, 1% bovine serum albumin at RT (Vivacqua et al. 2019; Rabenstein et al. 2019; Malek et al. 2014). For measurement of α-synuclein concentration in human plasma, add the plasma (100 mL) to each of the wells and incubate for 2 hours at 37 °C and obtain a standard curve. Detect bound α-synuclein using 100 mL/well of alkaline phosphate-conjugated ASY1 dilute it in PBS, incubate in the dark for 30 min. Subject the plates to colorimetric analysis with the addition of 100 mL 1 mg/mL para-nitro-phenyl-phosphate, 400 mM $NaHCO_3$, $Na2CO_3$ and 2 mM $MgCl_2$, pH 9.6. Incubate for 30 min in the dark, stop the reaction by 100 mL 2.5 M NaOH, and read it in a microplate reader at 405 nm. Test the sample in triplicates (Malek et al. 2014; Fares et al. 2016).

## 5.3 *Measurement of α-Synuclein in Saliva*

Refrain patients from eating prior 60 min of sample collection. To control the potential effects of circadian rhythm, saliva should be collected in an unstimulated state (9–11 a.m.). Collect saliva samples by drooling and then place it on ice (Pujols et al. 2017). Then preclear it by low spin 2600 × g for 15 min at 4 °C followed by high spin at 1500 × g for 15 min at 4 °C. Divide supernatant into 0.5 mL aliquots and store at −80 °C (Fjorback et al. 2007).

### 5.3.1 Saliva Sample Collection and Preparation

Collect 1 mL saliva and concentrate it to 0.5 mL to increase protein concentration by using a low-temperature vacuum dryer (Huang et al. 2019; Zheng et al. 2018). Integrate 0.2 mL saliva sample and interpret by HPLC on the column, equipoise it with 50 Mm $NH_4HCO_3$ (pH 7.4). Elute protein from the column at a flow rate of 0.8 mL/min. Collect 0.5 mL fractions and concentrate by centrifugation under vacuum (Huang et al. 2019).

### 5.3.2 Western Blot

Elute salivary proteins from the column with equipoise solution and boil concentrated fractions in 4× SDS loading buffer followed by samples. Separate by SDS–polyacrylamide gel electrophoresis (Ren et al. 2019; Ahsan et al. 2018; Kramer et al. 2008). Then, transfer it onto (PVDF) polyvinylidene fluoride membrane. Detect it by the immunoblotting process. Assess band intensities by densitometry. Perform the experiment in triplicate (Huang et al. 2019; Kramer et al. 2008; Xu and Pu 2016; Fjorback et al. 2007).

### 5.3.3 Luminex Assay

Couple a mouse antihuman-α-synuclein antibody with magnetic COOH beads with an amine coupling kit. Activate 100 μL beads with 10 μL EDC (1-ethyl-3-[3-dimethylaminopropyl] carbodiimide, 50 mg/mL) and Sulfo-NHS (N-hydroxysulfo succinimide, 50 mg/mL). Add 20 μg of primary antibody to activate beads and incubate for 2 hours. Resuspend the coupled beads in a 150 μL buffer. Perform determination of bead concentration using Coulter 7.2 counter/hemocytometer (Stefanis 2012; Fjorback et al. 2007).

### 5.3.4 Electron Microscopy

Collect 1 mL saliva supernatant through centrifugation and precipitation. Extract EVs in samples using the XYCQEV enrichment kit. Drop 500 mL PBS on the copper grid for 6 min. Stain the grid with uranyl acetate for 1 min and leave it at room temperature to dry. Perform imaging on an electron microscope (Giehm et al. 2011; Huang et al. 2019).

## 5.4 Immunohistochemistry

Remove tissue samples on autopsy from zones representing all the prime neuroanatomical structures within the brain and brainstem from PD convalescent (Ren et al. 2019). Fix each sample by immersing it in 70% ethanol with 150 Mm/L NaCl or 10% buffered formalin for 24–36 hours. Dehydrate samples through a chain of graded ethanols to xylene at room temperature. Then intrude each sample on paraffin at 60 °C. Cut the blocks into manifolds about 6 μm sections for immunohistochemical staining (Giehm et al. 2011; Xu and Pu 2016; Ingelsson 2016; Lin et al. 2017; Gautam et al. 2014).

## 5.5 Thioflavin T Fluorescence Assay

### 5.5.1 Purification of Acetylated α-Synuclein

The pT7–7 vector, together with the pNatB vector, contains the N-terminal acetylated enzyme NatB, coexposes the codon-optimized α-synuclein. Conduct the expression in *Escherichia coli* BL21DE3 grown in 2YT medium supplemented with 50 mM K/Na-phosphate buffer pH 7.2, 2 mM $MgSO_4$ and 0.4% glycerol (v/v), with 100 μg/mL ampicillin and 35 μg/mL chloramphenicol (Pujols et al. 2017). From a glycerol stock (prepare by overnight growth of a single colony in 2YT medium, mix with 50% glycerol (sterile) and store at −80 °C) inoculate the preculture (Xu et al. 2019). Induce the expression by adding 1 mM isopropyl β-D-1-thiogalactopyranoside (IPTG) and endorse while shaking at 120 rpm for 4 hours at 37 °C. Harvest cells by centrifugation at 6000 × g for 15 min at 4 °C (Vivacqua et al. 2019; Ahsan et al. 2018). Resuspend individual cell pellet in 15 mL $H_2O$ and freeze at −20 °C. Transfer the cell suspension to preheated block and incubate for 30 min to precipitate the proteins by denaturation (Caruana et al. 2012). Cooldown the lysed cells on ice for 10 min, then make the pellet of cell debris and precipitated proteins at 15,000 × g for 30 min at 4 °C. Filter the supernatant through a 0.2 μm filter. To saturate, saturated ammonium sulfate solution is added, thereby precipitating α-synuclein (Caruana et al. 2012; Paul et al. 2019). Incubate the turbid solution on ice for 20 min and make a pellet at 15,000 × g for 30 min at 4 °C. Discard the supernatant and resuspend the pellet into the IEC buffer. Load the solution onto the HiTrap Q HP ion-exchange chromatography column (Huang et al. 2019; Paul et al. 2019). Around 250 mM NaCl on a 0–500 mM NaCl gradient to IEC buffer alpha-synuclein elutes out. Resuspend the pellet into size exclusion chromatography buffer for purification. Using a 0.2 μm filter, sterile filter α-synuclein. Measure the concentration on the spectrophotometer. In liquid nitrogen, flash freeze the aliquot of α-synuclein and store at −80 °C (Caruana et al. 2011; Stefanis 2012; Karpinar et al. 2009).

### 5.5.2 Sample Preparation Method

Stock solution: 1 mM Thioflavin T in $H_2O$, 10% (w/v) sodium azide in H2O. Sodium azide, Thioflavin T, and protein solution into the 1.5 mL reaction tubes. Vortex the samples to mix them for a short time and put them back on ice. Distribute samples per plate and seal the plate and transfer it to plate reader (Paul et al. 2019; Stefanis 2012; Karpinar et al. 2009).

## 6 Alpha-Synuclein as a Therapeutic Target

There are four common ways to curb the neurotoxicity triggered by α-synuclein; by decreasing α-synuclein aggregation, by controlling its propagation, by increasing α-synuclein clearance, and by stabilizing its existing forms (Foltynie and Langston 2018). Intercessors are responsible for inducing neurotoxicity α-synuclein namely, inflammatory protease caspase-1 mediates the C-terminal truncation (Javed et al. 2019). A caspase-1 inhibitor potentially showed neuroprotective activity by reducing α-synuclein splitting. Passive and active immunization approaches namely, gene silencing or active protein immunization deters the congeries (Javed et al. 2019; Ardah et al. 2015). The pooling of oxidized proteins and aging is the most prominent players to increase the risk of neurodegeneration and the promotion of synucleinopathies. By activating and suppressing, the immune system accumulation of truncated α-synuclein can be prevented. The use of interleukin, interferon, and growth factors has therapeutic insights for treating the PD (Foltynie and Langston 2018; Ardah et al. 2015; Vargas et al. 2019).

In several recent reports, plant extracts and its phytochemicals have been highlighted for the neuroprotective potential in PD due to their antioxidant and anti-inflammatory activities. In phase 1 studies, α-synuclein antibodies (PRX002) were found to be safe and passed for the next clinical phase (Javed et al. 2019). The neuroprotective activity of formulations based on a plant extract, their mechanism of action, and phytochemicals which targets the fibrillation, oligomerization, aggregation, and toxicity of α-synuclein in different experimental models of PD have been discussed, and depicted in Table 2 (Ardah et al. 2015).

### *6.1 Medicinal Plants Targeting α-Synuclein Cascade and Neurotoxicity*

Several plants have been reported to suppress oligomerization and fibrillization of α-synuclein and may be used as a therapeutic target in PD (Ghosh et al. 2017). The plant extracts providing neuroprotective action by targeting α-synuclein in PD are listed in Table 2. The specific biochemical interaction of the plant or its

**Table 2** Plants providing neuroprotective action in PD by targeting α-synuclein

| Plant species/common name | Family | Neuroprotective action by targeting α-Synuclein |
|---|---|---|
| *Acanthopanax senticosus* (Siberian Ginseng) | Araliaceae | Inhibits caspase-3, α-synuclein |
| *Alaria esculenta* (Winged kelp) | Araliaceae | Decreases melting point of α-synuclein; Curbs aggregation and fibrils formation by associating with an unfolded form of α-synuclein |
| *Carthamus tinctorius* (Safflower) | Asteraceae | Reduces α-synuclein aggregation and astrogliosis |
| *Crocus sativus* L. (Saffron) | Iridaceae | Prevents dissociation of fibrils and inhibit α-synuclein aggregation |
| *Centella asiatica* (Asiatic pennywort) | Apiaceae | Inhibits α-synuclein aggregation and stabilizes oligomer; disintegrates preformed fibrils |
| *Cinnamon zeylanicum* (Cinnamon) | Lauraceae | Inhibits α-synuclein aggregation, stabilizes soluble oligomers of α-synuclein |
| *Panax ginseng* (Asian ginseng) | Araliaceae | Prevents dopaminergic loss and locomotor deficits; attenuates α-synuclein aggregation |
| *Polygala tenuifolia* (Yuan Zhi) (In Chinese) | Polygalaceae | Reduces α-synuclein phosphorylation |
| *Scutellaria pinnatifida* (Skullcap) | Lamiaceae | Reduces α- synuclein aggregation and scavenges free radicals |
| *Scutellaria baicalensis* (Baikal skullcap) | Lamiaceae | Attenuates astrocytosis/microgliosis, improve dopamine |
| *Bupleurum scorzonerifolfium* (Hong Chai Hu) (In Chinese) | Apiaceae | Inhibits α-synuclein aggregation in substantia nigra pars compacta (SNc) |

phytochemical on oligomerization, fibrillation, and aggregation of α-synuclein is illustrated in Fig. 5 (Vargas et al. 2019; Rosborough et al. 2017; Gautam et al. 2014).

There are several reports on certain plant species targeting α-synuclein in experimental models of PD. Some of these plant species include *Eleutherococcus senticosus* (Rupr. and Maxim.); *Bacopa monnieri* (L.) Wettst; *Polygala tenuifolia* Wild; *Carthamus tinctorius* L.; *Cinnamomum verum* J.Presl; *Centella asiatica* (L.) Urb.; *Rehmannia glutinosa* (Gaertn.) DC*;* *Trema micranthum* (L.) Blume; *Panax ginseng* C.A.Mey.; *Opuntia ficus-indica* (L.) Mill.*;* *Sagina japonica* (Sw. ex Steud.) Ohwi; and *Crocus sativus* L. (Javed et al. 2019; Ardah et al. 2015; Vargas et al. 2019; Atik et al. 2016). In vitro, α-synuclein aggregation arrest was demonstrated by *Centella asiatica* (L.) Urb. In vitro and in vivo studies in flies with A53T α-synuclein-induced PD exhibited α-synuclein aggregation inhibition and stabilization of oligomers on treatment with *Cinnamon* extract. However, its prime constituent, cinnamaldehyde, is not so far investigated (Atik et al. 2016). *Crocus sativus* L. (saffron) and its constituents crocetin, crocin-1, crocin-2, safranal, norbixin, hexadecanedioic acid, and trans-muconic acid are identified to modify α-synuclein fibrillation and aggregation (Rosborough et al. 2017; Gautam et al. 2014). In experimental PD models, Tianma Gouteng Yin, a traditional Chinese medicine decoction containing extracts of *Bupleurum scorzonerifolfium* and *Scutellaria baicalensis,* is found to decrease α-synuclein aggregation and accumulation (Javed et al. 2019; Ardah et al. 2015; Vargas et al. 2019).

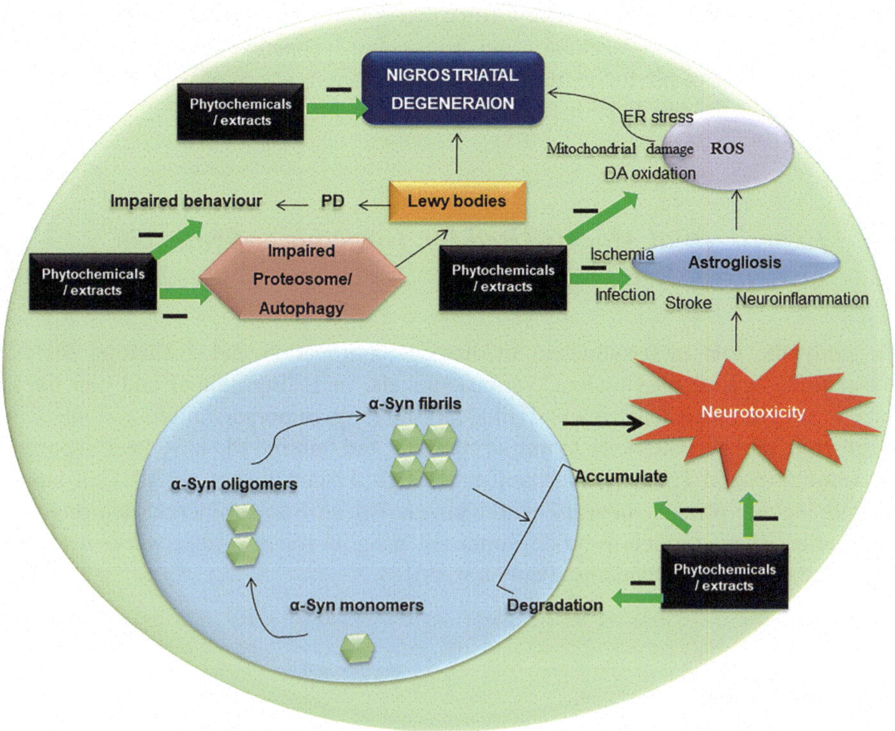


**Fig. 5** Target sites of plant extracts and their phytochemicals for their action on α-synuclein

## *6.2 Phytochemicals Targeting α-Synuclein Assembly and Neurotoxicity*

Phytochemical is a chemical substance, biologically active, but nonnutritive metabolite produced by plants, and which plays a significant role in drug discovery and development (Ramezani et al. 2019). Several classes of phytochemicals such as lignans, alkaloids, glycosides, saponins, and carotenoids have been found to act at different stages of pathogenicity and targeting α-synuclein. Reduction in neurotoxicity has been reported with inhibition of fibril formation or α-synuclein aggregation by alkaloids (3α-acetoxyeudesma-1,4 (15), acetylcorynoline, 11 (13)-trien-12, corynoxine B, 6α-olide, dl-3-n-butylphthalide, squamosamide and isorhynchophylline) and glucosides (paeoniflorin, jatamanin 11, 2,3,5,4-′-tetrahydroxy stilbene-2-O-β-D-glucoside, 3-O-demethylswertipunicoside, strophanthidine, and 10-O-trans-p-coumaroylcatalpol) (Foltynie and Langston 2018; Pronchik et al. 2010). Terpenoids such as celastrol, reynosin, 12 dioxo-oleana-1, 9-dien-28-oic acid, 2-cyano-3, geraniol, ginkgolide A, B, and C and thymoquinone also reduce α-synuclein aggregation. Recently, it has been reported that phytochemicals present in our daily food-restricted α-synuclein aggregation, fibrillation, and formation of amyloid protofilaments and fibrils, for example, curcumin (turmeric), oleuropein (olive oil), resveratrol (grapes), catechins (black and green tea), astaxanthin (vegetables and fruits), and lycopene (tomato) (Javed et al. 2019). Categories of phytochemicals that act at different stages of pathogenicity and target α-synuclein have been listed (Table 3) (Pronchik et al. 2010).

### *6.3 Experimental Models Used for Targeting α-Synuclein Toxicity in Drug Screening*

To study the role of α-synuclein in disease modification and evaluation of test compounds against α-synuclein, various models, including animal and cell lines, have emerged. The experimental models showing oligomerization, α-synuclein fibrillation, and neurotoxicity to mimic sporadic and familial PD have been reported are summarized in Table 4 (Ramezani et al. 2019; Pronchik et al. 2010). Cell-based models include primary neurons with characteristics of dopaminergic neurons and stem cells. These models provide important insights for drug discovery in PD to know the cellular mechanisms (Paul et al. 2019).

## 7 Conclusions

Despite extensive research, there is no reliable diagnostic test for Parkinson's disease. Thus, there is an urgent need to verify readily available biomarkers in biological samples, which may help in detecting PD. Exploration of a combination of biomarkers is essential to understand their accuracy in diagnostics. Alpha-synuclein aggregation and amyloid formation are the major hallmark for the pathogenesis of PD. It is the prime component of Lewy bodies. It is understood that oligomeric intermediates formed during the fibrillation pathway of α-synuclein are potential toxic entities responsible for neurodegeneration in PD. Thus, α-synuclein represents a valid therapeutic target in PD and relates to synucleinopathies. Several studies have been reported on the implication of α-synuclein toxicities and its contributing factors. Also, several estimation methods are available to detect the presence of α-synuclein in solid tissues and body fluids. These findings will help target the protein specifically. Gene targeting therapy and biotherapy are the current hot topics for the treatment of PD. The review also presents the above-mentioned phytochemicals and their underlying mechanisms in targeting α-synuclein toxicity. It is observed that due to physiological, biochemical, and pharmacological relevancy, the screening of plant extracts and their phytochemicals in cell lines shortfalls clinical applicability. Future perspectives could be inhibiting the α-synuclein production by siRNA, ASO (antisense oligonucleotide), increasing anti-apoptotic proteins, increasing antioxidant enzymes. It can be concluded based on the available preclinical studies that medicinal plants and their phytochemicals may be novel drug candidates for PD.

**Table 3** Phytochemicals that act at different stages of pathogenicity and target α-synuclein

| Phytoconstituents | Mechanism of action |
|---|---|
| Stimoval, 7,8-dihydroxy coumarin, etorphine, propoxyphene, pentazocine | Inhibition of α-synuclein in in silico studies |
| Dihydromyricetin (DHM), salvianolic acid B | Inhibit α-synuclein aggregation and enhance chaperone-mediated autophagy |
| Isoquinoline, Indole, Vinca, Pyrrolindole, Oxindole, β-carboline, Piperidine, pyridine, Aporphine, Methylxanthene, Lycopodium, Erythrine by-products | Agonist of muscarinic and adenosine receptors, an inhibitor of antioxidant monoamine oxidase (MAO), an inhibitor of AchE. Butyrylcholinesterase, inhibition of α-synuclein aggregation, dopaminergic and nicotine agonist and N-methyl-D-aspartate (NMDA) antagonist |
| Ginsenoside Rb1 | Inhibits α-synuclein toxicity and fibril formation |
| Oleuropein derivative—Olive fruit extract | Reduces α-synuclein fibrillation and oligomer toxicity |
| Polyphenols with B-cyclodextrin | α-Synuclein amyloids inhibition and disaggregation |
| *Aegeline (Aegle marmelos)* | Mimics yeast SNARE protein, suppresses α-synuclein and Bax toxicities |
| Ferulic 3,4-dimethoxycinnamic acid | Present in human blood, inhibits fibrillation, aggregation, and oligomerization of α-synuclein |

**Table 4** Experimental models used for targeting α-synuclein toxicity in drug screening

| Model | Effect on α-synuclein |
|---|---|
| • MPTP-intoxicated monkeys showing PD features | Accumulation of α-synuclein oligomers in the striatum |
| • MPP (+)-induced toxicity in SH-SY5Y cells.<br>• Cytotoxicity in catecholaminergic PC12 cells.<br>• MPTP/MPP+ induced neurotoxicity in PC12 cells.<br>• Dopamine induction in SH-SY5Y cells | Increases α-synuclein level and expression |
| • Rotenone neurotoxicity in SH-SY5Y and PC12 cells | Enhances the degradation of α-synuclein. |
| • Rotenone-induced neurotoxicity in cell lines | Increases α-synuclein aggregation and synphilin-1 deposits |

## References

Ahsan N, Siddique IA, Gupta S, Surolia A (2018) A routinely used protein staining dye acts as an inhibitor of wild type and mutant alpha-synuclein aggregation and modulator of neurotoxicity. Eur J Med Chem 143:1174–1184. https://doi.org/10.1016/j.ejmech.2017.10.002

Appel-Cresswell S, Vilarino-Guell C, Encarnacion M et al (2013) Alpha-synuclein p.H50Q, a novel pathogenic mutation for Parkinson's disease. Mov Disord 28(6):811–813. https://doi.org/10.1002/mds.25421

Ardah MT, Paleologou KE, Lv G, Menon SA, Abul Khair SB, Lu JH (2015) Ginsenoside Rb1 inhibits fibrillation and toxicity of alpha-synuclein and disaggregates preformed fibrils. Neurobiol Dis 74:89–101. https://doi.org/10.1016/j.nbd.2014.11.007

Atik A, Stewart T, Zhang J (2016) Alpha-Synuclein as a biomarker for Parkinson's disease. Brain Pathol 26(3):410–418. https://doi.org/10.1111/bpa.12370

Beach TG, White CL, Hamilton RL, Duda JE, Iwatsubo T, Dickson DW (2008) Evaluation of α-synuclein immunohistochemical methods used by invited experts. Acta Neuropathol 116 (3):277–288. https://doi.org/10.1007/s00401-008-0409-8

Breydo L, Wu JW, Uversky VN (2012) Alpha-synuclein misfolding and Parkinson's disease. Biochim Biophys Acta 1822(2):261–285. https://doi.org/10.1016/j.bbadis.2011.10.002

Caruana M, Högen T, Levin J, Hillmer A, Giese A, Vassallo N (2011) Inhibition and disaggregation of α-synuclein oligomers by natural polyphenolic compounds. FEBS Lett 585(8):1113–1120. https://doi.org/10.1016/j.febslet.2011.03.046

Caruana M, Neuner J, Högen T, Schmidt F, Kamp F, Scerri C (2012) Polyphenolic compounds are novel protective agents against lipid membrane damage by α-synuclein aggregates in vitro. Biochim Biophys Acta 1818(11):2502–2510. https://doi.org/10.1016/j.bbamem.2012.05.019

Devic I, Hwang H, Edgar JS et al (2011) Salivary α-synuclein and DJ-1: potential biomarkers for Parkinson's disease. Brain 134:e178. https://doi.org/10.1093/brain/awr015

Fares MB, Maco B, Oueslati A, Rockenstein E, Ninkina N, Buchman VL (2016) Induction of de novo α-synuclein fibrillization in a neuronal model for Parkinson's disease. Proc Natl Acad Sci U S A 113(7):E912–E921. https://doi.org/10.1073/pnas.1512876113

Farooqui T, Farooqui AA (2009) Aging:an important factor for the pathogenesis of neurodegenerative diseases. Mech Ageing Dev 130(4):203–215. https://doi.org/10.1016/j.mad.2008.11.006

Fjorback WA, Varming K, Jensen HP (2007) Determination of α-synuclein concentration in human plasma using ELISA. Scand J Clin Lab Invest 67(4):431–435. https://doi.org/10.1080/00365510601161149

Foltynie T, Langston JW (2018) Therapies to slow, stop, or reverse Parkinson's disease. J Parkinsons Dis 8:S115–S121. https://doi.org/10.3233/JPD-181481

Gautam S, Karmakar S, Bose A, Chowdhury PK (2014) β-Cyclodextrin and curcumin, a potent cocktail for disaggregating and/or inhibiting amyloids: a case study with α-synuclein. Biochemistry 53(25):4081–4083. https://doi.org/10.1021/bi500642f

Ghosh D, Mehra S, Sahay S, Singh PK, Maji SK (2017) Alpha-synuclein aggregation and its modulation. Int J Biol Macromol 100:37–54. https://doi.org/10.1016/j.ijbiomac.2016.10.021

Giehm L, Lorenzen N, Otzen DE (2011) Assays for α-synuclein aggregation. Methods 53 (3):295–305. https://doi.org/10.1016/j.ymeth.2010.12.008

Huang YR, Xie XX, Ji M et al (2019) Naturally occurring autoantibodies against α-synuclein rescues memory and motor deficits and attenuates α-synuclein pathology in mouse model of Parkinson's disease. Neurobiol Dis 124:202–217. https://doi.org/10.1016/j.nbd.2018.11.024

Ingelsson M (2016) Alpha-synuclein oligomers-neurotoxic molecules in Parkinson's disease and other Lewy body disorders. Front Neurosci 10:1–10. https://doi.org/10.3389/fnins.2016.00408

Javed H, Nagoor Meeran MF, Azimullah S et al (2019) Plant extracts and phytochemicals targeting α-synuclein aggregation in Parkinson's disease models. Front Pharmacol 9:1555. https://doi.org/10.3389/fphar.2018.01555

Kalia LV (2018) Biomarkers for cognitive dysfunction in Parkinson's disease. Parkinsonism Relat Disord 46(Suppl 1):S19–S23. https://doi.org/10.1016/j.parkreldis.2017.07.023

Kalia LV (2019) Diagnostic biomarkers for Parkinson's disease: focus on α-synuclein in cerebrospinal fluid. Parkinsonism Relat Disord 59:21–25. https://doi.org/10.1016/j.parkreldis.2018.11.016

Karpinar DP, Balija MBG, Kügler S, Opazo F, Rezaei-Ghaleh N, Wender N (2009) Pre-fibrillar α-synuclein variants with impaired B-structure increase neurotoxicity in Parkinson's disease models. EMBO J 28(20):3256–3268. https://doi.org/10.1038/emboj.2009.257

Kramer ML, Behrens C, Schulz-Schaeffer WJ (2008) Selective detection, quantification, and subcellular location of α-synuclein aggregates with a protein aggregate filtration assay. BioTechniques 44(3):403–411. https://doi.org/10.2144/000112691

Lin CH, Yang SY, Horng HE et al (2017) Plasma α-synuclein predicts cognitive decline in Parkinson's disease. J Neurol Neurosurg Psychiatry 88(10):818–824. https://doi.org/10.1136/jnnp-2016-314857

Malek N, Swallow D, Grosset KA, Anichtchik O, Spillantini M, Grosset DG (2014) Alpha-synuclein in peripheral tissues and body fluids as a biomarker for Parkinson's disease - a systematic review. Acta Neurol Scand 130(2):59–72. https://doi.org/10.1111/ane.12247

Mollenhauer B, Cullen V, Kahn I, Krastins B, Outeiro TF, Pepivani I (2008) Direct quantification of CSF α-synuclein by ELISA and first cross-sectional study in patients with neurodegeneration. Exp Neurol 213(2):315–325. https://doi.org/10.1016/j.expneurol.2008.06.004

Naoi M, Shamoto-Nagai M, Maruyama W (2019) Neuroprotection of multifunctional phytochemicals as novel therapeutic strategy for neurodegenerative disorders: anti-apoptotic and antiamyloidogenic activities by modulation of cellular signal pathways. Future Neurol 14(1). https://doi.org/10.2217/fnl-2018-0028

Paul A, Zhang BD, Mohapatra S, Li G, Li YM, Gazit E (2019) Novel mannitol-based small molecules for inhibiting aggregation of α-synuclein amyloids in Parkinson's disease. Front Mol Biosci 6:1–10. https://doi.org/10.3389/fmolb.2019.00016

Pronchik J, He X, Giurleo JT, Talaga DS (2010) In vitro formation of amyloid from α-synuclein is dominated by reactions at hydrophobic interfaces. J Am Chem Soc 132(28):9797–9803. https://doi.org/10.1021/ja102896h

Pujols J, Peña-Díaz S, Conde-Giménez M, Pinheiro F, Navarro S, Sancho J (2017) High-throughput screening methodology to identify alpha-synuclein aggregation inhibitors. Int J Mol Sci 18 (3):478. https://doi.org/10.3390/ijms18030478

Rabenstein M, Besong Agbo D, Wolf E, Dams J, Nicolai M, Roeder A (2019) Effect of naturally occurring α-synuclein-antibodies on toxic α-synuclein-fragments. Neurosci Lett 704:181–188. https://doi.org/10.1016/j.neulet.2019.04.004

Ramezani M, Wilkes MM, Das T, Holowka D, Eliezer D, Baird B (2019) Regulation of exocytosis and mitochondrial relocalization by alpha-synuclein in a mammalian cell model. NPJ Parkinsons Dis 5(1):12. https://doi.org/10.1038/s41531-019-0084-6

Ren X, Zhao Y, Xue F, Zheng Y, Huang H, Wang W (2019) Exosomal DNA Aptamer targeting α-synuclein aggregates reduced neuropathological deficits in a mouse Parkinson's disease model. Mol Ther Nucleic Acids 17:726–740. https://doi.org/10.1016/j.omtn.2019.07.008

Rosborough K, Patel N, Kalia LV (2017) Alpha-synuclein and parkinsonism: updates and future perspectives. Curr Neurol Neurosci Rep 17(4):31. https://doi.org/10.1007/s11910-017-0737-y

Schaser AJ, Osterberg VR, Dent SE, Stackhouse TL, Wakeham CM, Boutros SW (2019) Alpha-synuclein is a DNA binding protein that modulates DNA repair with implications for Lewy body disorders. Sci Rep 9(1):1–19. https://doi.org/10.1038/s41598-019-47227-z

Si X, Pu J, Zhang B (2017) Structure, distribution, and genetic profile of α-synuclein and their potential clinical application in Parkinson's disease. J Mov Disord 10(2):69–79. https://doi.org/10.14802/jmd.16061

Singh A, Zhi L, Zhang H (2019) LRRK2 and mitochondria: recent advances and current views. Brain Res 1702:96–104. https://doi.org/10.1016/j.brainres.2018.06.010

Stefanis L (2012) Alpha-synuclein in Parkinson's disease. Cold Spring Harb Perspect Med 2(2): a009399. https://doi.org/10.1101/cshperspect.a009399

Tokuda T, Qureshi MM, Ardah MT et al (2010) Detection of elevated levels of α-synuclein oligomers in CSF from patients with Parkinson disease. Neurology 75(20):1766–1772. https://doi.org/10.1212/WNL.0b013e3181fd613b

Vargas JY, Grudina C, Zurzolo C (2019) The prion-like spreading of α-synuclein: from in vitro to in vivo models of Parkinson's disease. Ageing Res Rev 50:89–101. https://doi.org/10.1016/j.arr.2019.01.012

Vivacqua G, Suppa A, Mancinelli R et al (2019) Salivary alpha-synuclein in the diagnosis of Parkinson's disease and progressive supranuclear palsy. Parkinsonism Relat Disord 63:143–148. https://doi.org/10.1016/j.parkreldis.2019.02.014

Wong YC, Krainc D (2017) Alpha-synuclein toxicity in neurodegeneration: mechanism and therapeutic strategies. Nat Med 23(2):1–13. https://doi.org/10.1038/nm.4269

Xu L, Pu J (2016) Alpha-Synuclein in Parkinson's disease : from Pathogenetic dysfunction to potential clinical application. Parkinson's Dis 1720621:1–10. https://doi.org/10.1155/2016/1720621

Xu M, Loa Kum-Cheung W, Zhang H, Quinn RJ, Mellick GD (2019) Identification of a new α-synuclein aggregation inhibitor via mass spectrometry based screening. ACS Chem Neurosci 10(6):2683–2691. https://doi.org/10.1021/acschemneuro.9b00092

Zange L, Noack C, Hahn K et al (2015) Phosphorylated α-synuclein in skin nerve fibres differentiates Parkinson's disease from multiple system atrophy. Brain 138:2310–2321. https://doi.org/10.1093/brain/awv138

Zheng Y, Qu J, Xue F, Zheng Y, Yang B, Chang Y (2018) Novel DNA aptamers for Parkinson's disease treatment inhibit α-synuclein aggregation and facilitate its degradation. Mol Ther Nucleic Acids 11:228–242. https://doi.org/10.1016/j.omtn.2018.02.011

# Screening of Herbal Medicines for Neurotoxicity: Principles and Methods

**Alshaimaa M. Almehmady, Nabil A. Alhakamy, and Waleed S. Alharbi**

## Contents



**Abstract** Although there are some benefits derived from the use of medicinal herbs as remedies, some of them are associated with neurological adverse effects. Such

A. M. Almehmady · N. A. Alhakamy · W. S. Alharbi (✉)
Department of Pharmaceutics, Faculty of Pharmacy, King Abdulaziz University, Jeddah, Saudi Arabia
e-mail: wsmalharbi@kau.edu.sa

 
D. C. Agrawal, M. Dhanasekaran (eds.), *Medicinal Herbs and Fungi*,
https://doi.org/10.1007/978-981-33-4141-8_10

potential hazards could be caused by several factors, including intrinsic toxicity of herbal products, contamination, adulteration, and so on. To assess the toxicity of the herbal products, regulatory institutes usually require a list of tests to identify the cause of herbal neurotoxicity. This chapter aims to review screening methods of herbal medicines for neurotoxicity. Despite advanced procedures used to assess the toxicity of herbal products, considerable efforts are required to provide information on their safety aspects. There are many problems with neurological assessments, which are not only limited to the herbal aspects but also apply to neurotoxicity tests. Concerted efforts are needed to establish databases that can be used in the assessment tools to predict the toxicity of herbal products.

**Keywords** Herbal extracts · Herbal neurotoxicity · Medicinal herbs · Neurotoxicity assessment · Safety

## Abbreviations

| | |
|---|---|
| ALS | Amyotrophic lateral sclerosis |
| CNS/ANS | Central and/or autonomic nervous systems |
| DDD | Dichlorodiphenyldichloroethane |
| DDT | Dichlorodiphenyltrichloroethane |
| DNT | Developmental neurotoxicity |
| FAAS | Flame atomic absorption spectrometry |
| FOB | Functional observation battery |
| HPLC | High-performance liquid chromatography |
| ICP-MS | Inductively coupled plasma mass spectrometry |
| ICP-OES | Inductively coupled plasma optical emission spectrometry |
| LUHMES | Lund human mesencephalic |
| MDNs | Midbrain dopaminergic neurons |
| PD | Parkinson's disease |

## 1 Introduction

The nervous system comprises different cells having different tasks and vulnerabilities. Several distinctive characteristics of the nervous system have an impact on its response to toxic agents, such as its inadequate regenerative ability, extraordinary anatomy (long axons), various functions, expansive cell connections, energy supply reliant on glucose, and presence of extremely specific cell supplements. Furthermore, the complicated relationships between the nervous system and other organs can be altered by environmental toxicants (National Research Council (US) Committee on Neurotoxicology and Models for Assessing Risk, Committee NRC (US) 1992). The blood–brain barrier, the tight connections between the capillary endothelial cells, which provides brain tissue and cell–astrocyte relationships with endothelium, prevents harmful chemicals in the blood from reaching the brain. However, lipophilic substances, such as nutrients, hormones, amino acids, peptides, proteins, and fatty acids required for the brain functions, pass through this barrier through various transport processes (Pardridge 1988). Thus, lipid-soluble toxicants and substances similar to those typically absorbed by the nervous system could also gain entry into the brain. Moreover, the blood–brain barrier in immature animals is less efficient. Besides, certain areas of the brain, including the circumventricular region and around the olfactory nerve running straight from the nose to the front, lack the blood–brain barrier (Broadwell 1989).

Neurotoxicity is the ability of certain chemical, biological, or physical substances to trigger unfavorable functional or structural changes in the nervous system. The impact of toxic substances on the nervous system is diverse. Neurotoxicity could occur at any time in the life cycle, from gestation to senescence, can be affected by neurotoxins at various time points and the manifestation of neurotoxic effects is age-related (Cushner 1981; National Research Council (US) Committee on Neurotoxicology and Models for Assessing Risk, Committee NRC (US) 1992; Blair et al. 2008). The study of the natural source of several neurological and psychiatric disorders is not well understood. Since Roman times, the association of high-dose lead exposure with amnesia, convulsion, and colic was acknowledged only as a striking manifestation of neurotoxicity (Waldron 1973; Kjellstrom 1979). Other long-recognized associations include increasing incidents of depression and suicide among workers exposed to carbon disulfide and paralysis following the consumption of tri-o-cresyl phosphate-containing products (US Ginger Jake and Morocco cooking oil). Furthermore, exposure to tetraethyl lead inhalers and elemental mercury can lead to toxic psychoses, such as erethism (Vigliani 1954; Smith and Spalding 1959; Waldron 1973; Kjellstrom 1979; Greenberg 2006). Exposure to toxic chemicals and substances are also known to cause neurodegenerative illnesses in later life. Calne et al. reported that multiple environmental factors contribute to the development of Alzheimer's disease, Parkinson's disease, or amyotrophic lateral sclerosis (ALS) that causes rapid degeneration of neuronal resources (Calne et al. 1986). For instance, ALS/Parkinsonism dementia complex, a neurodegenerative disorder common among people of Guam, was proposed to be caused by the previous intake of chemicals present in seeds of *Cycas circinalis* L., the fake sago palm (Spencer et al. 1978).

Plant toxins possessing various poisonous constituents can affect most of the organs. The dominant adverse effect depends on the situation, the developmental phase, and the section of the plant used, the intake dosage, and the vulnerability of the people around the world (Botha and Penrith 2008).

Physostigmine, harmaline, and bicuculine (plant-derived drugs) can have significant effects on the central nervous system (CNS) and/or autonomic nervous system (ANS). Plants comprising these neurotoxins are typically not regarded as medicinal products, and several natural substances, such as *Annona muricata* (soursop), have been used to induce neurotoxicity in animal models. The use of extremely toxic plant species is common in traditional medicines, such as Aconitum, Rhododendron, and Strychnos in Chinese medicine (Liu et al. 2014). Poisonous and psychoactive plants are generally illegal in marketed American and European herbal products (except errors or fraud) (Pelkonen et al. 2017). CNS/ANS stimulants derived from some plants are used clinically, such as atropine derived from *Atropa belladonna* and morphine and codeine derived from *Papaver somniferum* (Greenberg 2006; Pelkonen et al. 2017).

Despite sanctions on herbal neurotoxins, there are increasing reports of clinical CNS toxicity needing immediate hospitalization and mostly causes patient death in severe cases. *Aconitum* plant commonly causes acute toxicity, which depends on the dose, misidentification, adulteration, wrongful preparation method, and handling of the plant material (Chan 2015). CNS herbal toxicity is the consequence of complex variables, such as the ingestion in this sense of non-herbal medicinal plants and the unlawful use of recreational products. Intake of herbal drugs such as *Strychnos ignatia*, *Salvia*, *Rosmarinus*, *Atropa belladonna* leaves, *Erythroxylum coca*, *Illicium anisatum*, *Gingko biloba*, *Aconitum* species, *Annona muricata*, and *Hypericum perforatum* could lead to severe CNS toxicity.

Remarkably, few scientific articles have reviewed herbal neurotoxins as a category: the "plant and neurotoxin" database search reports almost solely relied upon the use, among other agents, of plant extracts to reduce neurotoxicity. In the next section, the causes of neurotoxicity will be reviewed.

## 2 Causes of Neurotoxicity

CNS toxicity occurs when lipophilic toxic compounds pass through or damage the blood–brain barrier permeability to blood, enabling the entry of otherwise impermeable hydrophilic molecules (Wager et al. 2012). CNS toxicity manifests as restlessness, confusion, tremors, convulsions, psychosis, or low consciousness rates, and all these symptoms could also coexist (Williamson 2017). Excessive CNS stimulation by an overdose of caffeine or cocaine results in seizures. Moreover, vision impairment, hearing loss, discomfort, or parenthesis can also occur if sensory cells or the neurons are damaged.

Concerning sensory awareness gaps, there may be a deficiency of motor functions, which may lead to muscle weakness and paralysis. At the same time, excessive neuronal arousal or rhythmic firing may trigger movement disorders. Figure 1 summarizes the main reasons for herbal neurotoxicity: intrinsic neurotoxicity of herbal medicines, heavy metal contaminants, and organic contaminants.

## 3 Intrinsic Neurotoxicity of Herbal Medicines

Herbs produce several well-known neurotoxins. Graviola, also known as *Annona muricata* (from the Annonaceae family), has been used for its anticancer activity but can cause neurodegeneration and atypical Parkinson's disease (Moghadamtousi et al. 2015; Rady et al. 2018). In some cases, medicinal herbs used to treat

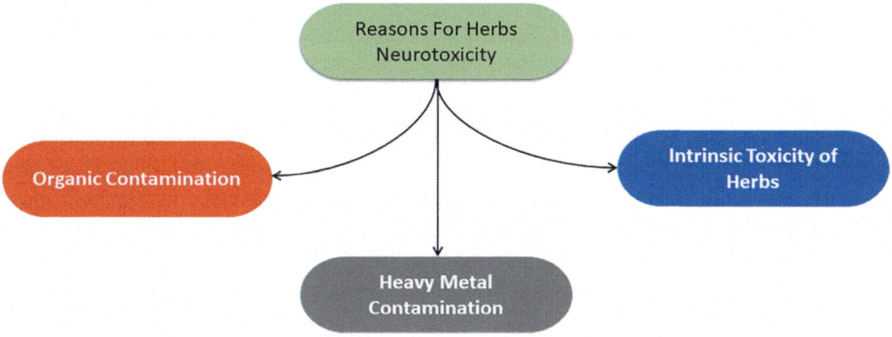


**Fig. 1** Reasons for herbal neurotoxicity

neurological conditions contain neurotoxins. For example, Siberian Ginseng is used for its neuroprotective effect, although Zhang et al. reported that it produces several metabolites that can cause neurotoxic effects (Zhang et al. 2014).

Overdosing on a CNS stimulant must be differentiated from overdosing on a CNS toxin that does not have a therapeutic CNS effect. Cyanide toxicity results from taking excessive amounts of cyanogenic glycosides, for example, when apricot kernels are used as a proven, damaging alternative cancer treatment compared with when apricot kernels are used as a nonproven and harmful alternative therapy for cancer (Williamson 2017). Soursops are another nonproven agent used in cancer therapy. Although the use of fruit and juice is common and relatively secure, poisonous acetogenins present in leaves and barks can trigger atypical Parkinson's disease (PD), as shown in animal models, and can destroy dopaminergic cells. Other neurotoxins in this plant are reticularine and *N-Methylcoclaurine* alkaloids (Kotake et al. 2004; Champy et al. 2009). Considerably higher incidences of atypical PD are reported in areas, such as Guadeloupe, where the leaf is used as an herbal medication. Other potent toxins, such as the cicutoxin discovered in hemlock water dropwort, are not used for medicinal purposes, but they trigger toxicity through misidentification. The water dropworts belong to the family Apiaceae and are readily confused for other safe species. *Conium maculatum*, which also belongs to the family Apiaceae and includes the deadly alkaloid coniine, could also be easily misidentified. It has been used as a rather dangerous sedative for years (Williamson 2017).

In many instances, plants are used for their CNS effects, and their components are not referred to as toxins. Valerian and skullcap (genus *Scutellaria*) have a moderate amount of GABA agonists. They are used to treat mild anxiety, and no harmful effects have ever been recorded for these plants (Williamson 2017). CNS toxicity has continuously been caused by muscimol, also a GABA agonist. The source of this fly-agaric fungus is not used without permission. Discovered in St. John's wort, hyperforin is an inhibitor of GABA regeneration, but its mild, calming influence differs from that reported for serotonergic results offered by hypericum (Russo et al. 2014). Ergot alkaloids are widely used in alternative medicine, but many of them are psychoactive agents, and can be regarded as neurotoxins, and can even cause epileptiform seizures ("St. Anthony's Fire"). Ergot was previously utilized as a medicinal plant in facilitating labor but is now deemed highly poisonous, with the preference given to personal alkaloids and their derivatives.

Caffeine is a prevalent CNS stimulant and an adenosine inhibitor that is often ingested through tea and coffee by the majority of the population. Among herbal products (e.g., guarana, in which "guaranine" was first referred to as caffeine), chocolate and maté tea also contain caffeine. Caffeine is also used as an herbal supplement. Other methylxanthines (e.g., theobromine and theophylline) demonstrate toxic CNS effects, primarily insomnia, tremor, and significant palpitations (Harborne and Baxter 1996; Eastwood 2001; Hussain et al. 2014). These are examples that illustrate the inextricable link to other toxicity symptoms in the ANS and CNS and cannot be regarded alone. Medicines that instantly affect neurotransmission can affect the cardiovascular system. Ephedrine, a component of the Chinese medicine "Ma Huang" used to treat bronchitis and asthma, is an $\alpha 1$-adrenergic mimetic that is used in decongestive medication, similar to its racemic mixture, pseudoephedrine (Harborne and Baxter 1996; Eastwood 2001). Concerning cardiovascular and CNS/ANS effects, the use of ephedrine products for weight loss has recently created concern about their impacts on the immune response (Senchina et al. 2014). P-synephrine, a bitter orange component that is used to help weight loss, is another sympathomimetic amine. Issues have also been raised regarding the use of P-synephrine, centered on the chemical structural resemblance with f-phenylephrine (also called M-synephrine), but those have been contested. P-synephrine was administered together with caffeine in several research studies, and it appears to significantly improve their toxicity (Stohs et al. 2011; Hansen et al. 2012).

Camphor suppresses nicotinic enzymes and is a multichannel TRVP agonist. Its CNS stimulant impacts can trigger a wide range of severe side effects, such as nausea, pain, headache, twisting, seizures, and delirium (dose-dependent). In an acute overdose, a condition that lasts for several hours, coma, and death by asphyxia or fatigue can occur. Because of its impacts on TRPV1 desensitizing and TRPA1 blockage, camphor is used as a local analgesic. One of the primary problems with its toxicity is that it is carried very readily by the body. Camphor has an influence on desensitization and blockage of TRPV1 as a local analgesic. It is used as an analgesic. The fact that it is very readily absorbed through the skin is a significant problem when discussing its toxicity (Harborne and Baxter 1996; Chen et al. 2013; Williamson 2017).

Taxol, a complex molecule belonging to the taxine alkaloid category, is a plant neurotoxin that is widely known for its efficacy in chemotherapy treatment for at least one type of cancer. Taxol exists in most Western yew plants and is separated from bark (formerly regarded to be a molester tree in forestry; however, it was not available until additional sources were created following the finding of the medicinal significance of taxol). Taxol ingestion produces a variety of neurotoxic consequences, such as sensory neuropathy, nausea, gastrointestinal upsets, and respiration impairment (Manahan 2002). The cause of taxol neurotoxicity is tubulin binding, a microtubule-built protein that is a component of the cell function of *Taxus brevifolia*

(Manahan 2002). This binding of tubulin in the microtubules of nervous cells helps to stabilize microtubules and to avoid their dissociations. The alkaloid–nerve toxin coniine is found in spotted hemlock. Symptoms of nervousness, tremors, arrhythmia, and bradycardia occur 15 min following ingestion. The temperature of the body may drop, and deadly paralysis may occur (Manahan 2002; Tamokou and Kuete 2014). In general, the CNS reacts to toxicity with signs such as seizures, hallucinations, ataxia, and coma (absence of control of voluntary body motions). Other symptoms of cognitive toxicity include disturbance and hyperactivity. *Boophane disticha* contains multiple alkaloids, such as buphanidrine, buphanisine, and buphanamin, including bushman poison bulbs. Poisoning generally occurs in people who use these bulbs for medical reasons. Acute toxicity leads to nausea, weakness, coma, and death (Steenkamp 2005). Cerpegin, separated from chloroform extract of *Ceropegia junceae* (Apocynaceae) (Harry et al. 1998), was found to be poisonous at concentrations above 400 mg/kg, with mice shown to be excited and irritable and caused seizures and respiratory paralysis. Two plants, *Datura stramonium*, and *Datura ferox* belonging to the family Solanaceae are endemic weeds containing parasympathetic alkaloids such as hyoscine and atropine and produce several other sympathetic alkaloids.

## 4 Heavy Metal Contaminants

The most common heavy metals contaminants that can cause neurotoxic effects are lead (Das et al. 2011), arsenic (Bolan et al. 2017), cadmium, and mercury. Origins of heavy metal contamination can come from different sources (Bolan et al. 2017; de Oliveira et al. 2018; Wang et al. 2019), most often, environmental contamination with heavy metals could result in contamination of plants. For example, many herbs cultivated in contaminated soil uptake toxic concentrations of metals. Another source of heavy metal contamination is the irrigation of herbs with wastewater contaminated with heavy metals (Bolan et al. 2017). The quantity of heavy metal contamination can be detected by various sensitive instruments such as flame atomic absorption spectrometry (FAAS) (Cheremisinoff 1996), gas chromatograph–mass spectrometry (GC-MS) (Das et al. 2011), inductively coupled plasma optical emission spectrometry (ICP-OES) or mass spectrometry (ICP-MS) (Mulaudzi et al. 2017), high-performance liquid chromatography (HPLC) (Yang et al. 2004), and X-ray fluorescence spectroscopy (Shaban et al. 2016).

## 5 Organic Contaminants

Pesticides are the most common organic contaminant of herbs (Shaban et al. 2016). Examples of pesticides that cause neurotoxic symptoms are those derived from organochlorines such as dichlorodiphenyltrichloroethane (DDT) and dichlorodiphenyldichloroethane (DDD). Safe levels of pesticides are usually controlled by regulatory institutes and are reviewed (World Health Organization 1993; Boyes 1997; Selvi and Paramasivam 2017). The quantity of pesticides available in herbs can be detected by chromatographic methods such as GC and HPLC (Selvi and Paramasivam 2017). Khan et al. have reported the determination of the level of pesticides such as DDT and DDD (Khan et al. 2001).

# 6 Goals of Neurotoxicity Assessments

Neurotoxicology is known as "the study of the adverse effects of chemical, biological and certain physical agents on the nervous system and/or behavior during development and in maturity" (Harry et al. 1998). Herbs can be neurotoxic in different ways (Williamson 2017). Therefore, the main aim of neurotoxicity assessments is first to rule out the cause of herbal neurotoxicity, determine adverse neurotoxic effects (Ifeoma and Oluwakanyinsol 2013) and determine the exposure level that would cause toxicity (Ifeoma and Oluwakanyinsol 2013; Williamson 2017). Three main factors make a herb neurotoxic; intrinsic toxicity of herbs, heavy metal contamination, and organic contamination (Fig. 1) (Bogusz and Al-Tufail 2008). Neurotoxicity is considered a major toxicity endpoint. Regulatory institutes usually require a list of tests for detailing the determination of the toxicity of substances. These tests are completed with in vivo tests that involve observation of the effect of a particular toxin on animals in cage settings (OECD 1995, 1997). Understanding the mechanistic toxicity of a substance is essential to apply the correct screening test. Neurotoxicity of substances usually does not occur alone but is associated with other symptoms in body systems. Many general screening tests exist and are performed for toxic substances, which are mentioned in other reviews and books (Harry et al. 1998; Bogusz and Al-Tufail 2008; Ifeoma and Oluwakanyinsol 2013).

# 7 Neurotoxicity Assessment Methods

## *7.1 Flame Atomic Absorption Spectrometry (FAAS)*

FAAS works by illuminating light at a specific wavelength. A sample substance is taken by a capillary and heated by a flame inside the machine. The light beam goes through the heated sample. The sample absorbs some of the light according to its atomic number of elements. The light then goes through a series of mirrors until it reaches a detector that can quantify light beams according to sample concentration. The detector is connected to a computer to produce data, which is then analyzed by specific software (Fig. 2) (Cheremisinoff 1996).

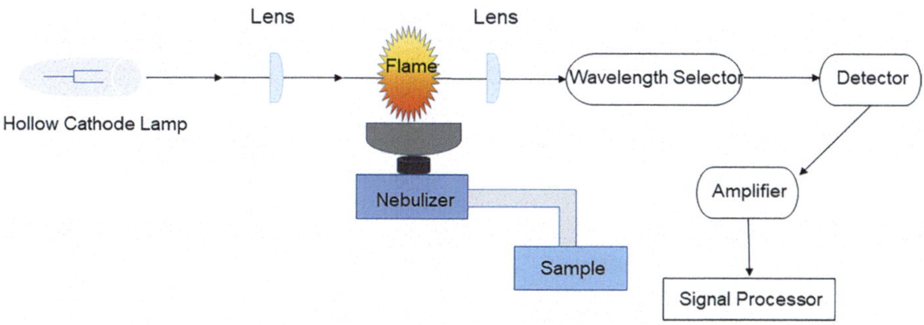


**Fig. 2** Flow chart showing parts of flame atomic absorption spectrometry

### 7.2 *High-Performance Liquid Chromatography (HPLC)*

HPLC is a well-known method of analyzing and separating chemicals based on their molecular weight or polarity of materials. HPLC is not generally used as an analytical method for the determination of heavy metal concentration; however, scientists have developed a method to measure the metal concentration in Chinese herbs by complexing with a chelating agent and then measuring the concentration of the chelating agent (Yang et al. 2004).

### 7.3 *Inductively Coupled Plasma Optical Emission Spectrometry (ICP-OES) or Mass Spectrometry (ICP-MS)*

This technique is used to determine the trace heavy metal quantity by bombarding light into the sample and measuring the specific wavelength emitted from the metal (Mulaudzi et al. 2017).

## 8 In Vitro Models of Neurotoxicity Assessments

In vitro models are cell-based assays, based on the exposure of toxins for a period onto cultured cells on plastic plates. The goal of testing neuroactive materials in in vitro models is to generally screen the neuroprotective and neurotoxic effects of materials before in vivo testing (i.e., animal testing). This stage is crucial as it plays a vital role in reducing the number of tests directly performed on animals, thus reducing cost and time spent in the primary stage of chemical screening.

There are numerous cell lines derived from cancerous neuronal cells available in the market, such as PC12, SH-SY5Y, LUHMES, and SK-N-SH (Table 1). The

**Table 1** Most common nerve-derived cell lines used to test neuroactive herbal extracts

| Cells line | Origin | Application | References |
|---|---|---|---|
| PC12 | Neuroendocrine cells, rat adrenal medulla (rat pheochromocytoma) | Neurosecretion model, exocytosis model | Lee et al. (2003), Westerink and Ewing (2007), Badisa et al. (2018) |
| SH-SY5Y | Neuroblastoma cells, human | Neuropharmacology and neurotoxicology models | Xie et al. (2010), Kovalevich and Langford (2013), Ko et al. (2018) |
| MES23.5 | Dopaminergic neuroblastoma, cell fusion of rat Embryonic mesencephalon cells and the murine Neuroblastoma–glioma cell line | Neuropharmacology and neurotoxicology models | Shen et al. (2014), Ye et al. (2017) |
| SK-N-SH | Neuroblastoma cell line, human | Neuropharmacology and neurotoxicology models | Preis et al. (1988), Gao et al. (2009) |
| LUHMES | Immortalized human dopaminergic neuronal precursor cells, human | Neuropharmacology and neurotoxicology models | Höllerhage et al. (2015) |

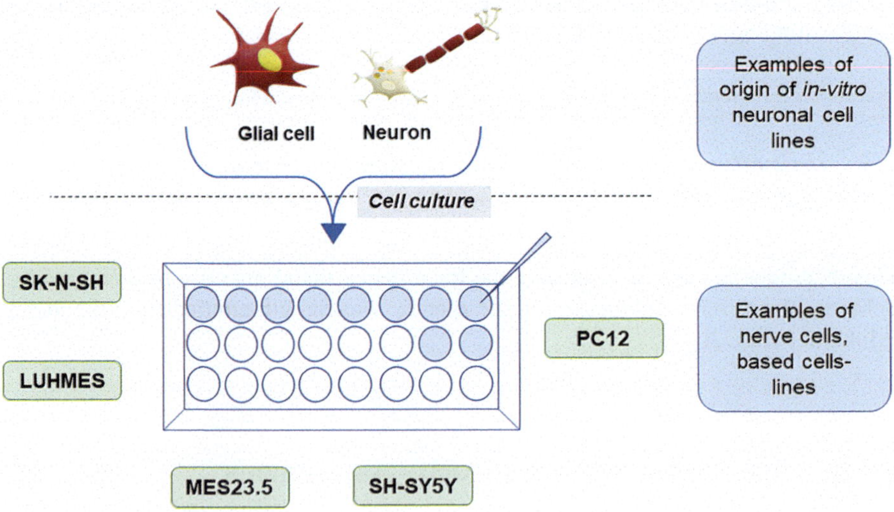


**Fig. 3** Examples of nerve cell-based cell lines

following are examples of cell lines that have been used to test the neurotoxicity of various herbal medicines (Fig. 3).

### 8.1 *Lund Human Mesencephalic (LUHMES) Cell Line*

Immortalized human dopaminergic neuronal precursor cells, known as the LUHMES cell line, are human cell line derived from mesencephalic tissue. The LUHMES cell line has neuronal morphology. In a study by Höllerhage et al., dietary supplements that contained Graviola stems and leaves or pawpaw twigs, which are known to have neurotoxins, were tested using LC-MS tests and were found to have a toxic concentration of Graviola (Höllerhage et al. 2015).

### 8.2 *Primary Neurons Cell Lines*

The culture of mouse cortical cells, containing glutaminergic and GABAergic neurons and glia to test the seizure effects of materials, was conducted using rat cortical cells that contained glutamatergic and GABAergic neurons and glia. The primary neuronal cell lines used were derived from mouse embryos (Galofré et al. 2010).

### 8.3 *Primary Cultured Midbrain Dopaminergic Neurons (MDNs)*

MDNs are cells highly enriched with dopaminergic receptors. However, their application is limited due to ethical issues and technical difficulties in handling the cells (Leung et al. 2011).

### 8.4 *Neuroblastoma Cell Line*

Alternatively, neuroblastoma cell line derivatives can be used as they are enriched with dopaminergic neurons. Several cell lines have been used to test the neurotoxicity of well-known neurotoxic chemicals such as SK-N-SH, MES23.5, SH-SY5Y, and MN9D.

### 8.5 *Rat Pheochromocytoma Cell Line (PC12)*

The rat pheochromocytoma cell line (PC12) is another cell line that can be used as an alternative to dopaminergic neurons as they have similar properties. For example, they both express tyrosine hydroxylase and synthesize dopamine.

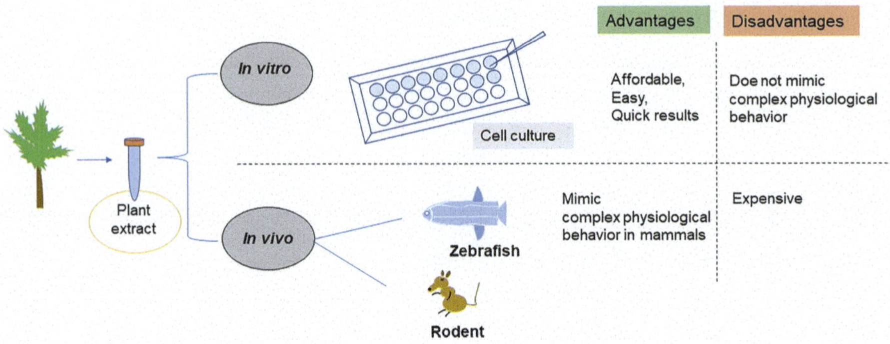


**Fig. 4** Comparison between in vitro and in vivo based tests

The 3Rs organization has done great work in supporting research into developing in vitro neurotoxicity assays (Harry et al. 1998). One currently trending technique is 3D cell culture, which is a cell culture system that can mimic a more complex environment for cell growth rather than growing cells on plastic plates.

To use in vitro models, an understanding of cellular mechanistic responses is needed. One of the disadvantages of in vitro models is that the origin of cells is commonly cancerous cells. This fact possibly makes the exclusion of neurotoxicity more complex. Moreover, in vitro models cannot mimic the complex physiological process that occurs in animals; therefore, they still cannot be thought of as a replacement for in vivo models. There are tremendous efforts ongoing to invent more complex cell culture in vitro methods, which may mimic biological processes in living beings.

While in vitro models are not considered sufficient for neurotoxicity assessment, they can still be used for high throughput screening methods to assess the neurotoxicity of particular substances. They can also be regarded as a tiered system for the screening of chemicals. In vitro assays are very useful to speed up assessment processes and to rule out/help choose chemicals that urgently require in vivo testing (Fig. 4).

## 9 In Vivo Procedures of Neurotoxicity Assessments

The aim of in vivo procedures is to study the integrative and complex functions that are controlled by the nervous system such as acuity, motor, and sensory functions. Animal species that are mainly involved in in vivo studies are rats, hens, drosophila, zebrafish, and mice. In vivo assessments are mostly behavioral assessments, functional observations, motor activity testing, and specialized tests of neurologic function (National Research Council (US) Committee on Neurotoxicology and Models for Assessing Risk, Committee NRC (US) 1992).

## 10 Behavioral Assessment

Behavioral assessments commonly involve testing the effects of neurotoxins that may alter the nervous system's structure and functionality. While behavior is one of the significant toxicity endpoints in neurotoxicology, it is one of the most challenging parameters to assess compared with other endpoints such as morphology, neurophysiology, and neurochemistry. A change in behavior may indicate a disturbance in autonomic, sensory, motor, or cognitive function.

Behavioral assessment methods vary in their specificity, complexity, and their usage. For example, some behavioral assessments can be used for screening for pharmacological effects and toxicological identification. In contrast, other specific and more complex tests involve training of animals before exposure to neurotoxins and the use of sophisticated instrumentation.

## 11 Functional Observation Battery (FOB) Test

FOBs are a group of tests that evaluate visible changes in neurological functions (Cha et al. 2015), and they can be assessed by observation of animal behavior either in a home cage or open field. In the neurotoxicology field of science, the apparent change in neurological function is called an endpoint. Examples of assessed parameters are posture, convulsions and tremors, lacrimation, piloerection, salivation, vocalizations, rearing, mobility, and urination. FOBS are well-established tests to screen neurotoxicity induced by neurotoxins, drugs, and pesticides. In 1968, Irwin described several assessments to examine the effect of neuroactive drugs on the sensorimotor function of a group of mice (Irwin 1968). In another critical report, Gad listed several tests to evaluate the neuromuscular effects of synthetic chemicals in a group of rats (Gad 1982). A study suggested that neurologic components of endpoints to assess various neurological functions (e.g., autonomic, motor, and sensorimotor functions) can overlap with the function they are supposed to assess (Moser 1989).

## 12 Motor Activity Tests

Motor activity tests usually involve testing a wide range of behaviors that are controlled by sensory, motor, and integrative processes. This type of testing is popular and well-accepted as it is fully automated and does not involve a high level of interaction between the examiner and tested animal. Moreover, it is a noninvasive test. However, motor activity measures have been criticized due to limited specificity for adverse effects on the nervous system (National Research

Council (US) Committee on Neurotoxicology and Models for Assessing Risk, Committee NRC (US) 1992).

## 13 Herbal Toxicokinetics

Herbal toxicokinetics is the science that deals with the toxicity of herbs related to disposition and absorption, distribution, metabolism, and excretion or herb–drug interaction. Studies of herbal toxicity that have resulted in the disposition of herbal components inside the body have raised the concern to conduct more toxicity testing (Pelkonen et al. 2017). This type of study has become popular in herbal toxicology to exclude the effect of extract or whole plant on the pharmacokinetics of the human body.

## 14 Developmental Neurotoxicity (DNT) Testing

The developing brain is more vulnerable to toxic substances than the adult brain. This vulnerability may be due to the complex growth and developing physiological processes going on in the living body. Thus, there has been an increasing number of reports which focus on developmental neurotoxicity (DNT) testing (Spyker 1975; Rodier 1995; Grandjean and Landrigan 2006).

Due to the increase in exposure to environmental pollution (herbicides, pesticides, adulterated botanicals, and synthetic chemicals), the number of cases associated with the damage of the number of toxic substances in the environment, in particular, those that could affect the nervous systems in children such as pesticides and synthetic chemicals have significantly increased. A report released by the National Press Academy in 1993 has highlighted the significant toxic traces of pesticides in the food and diet of children (National Research Council (US) Committee on Pesticides in the Diets of Infants and Children 1993). On the other hand, to reach a neurotoxicity endpoint, the battery of toxicity tests is expensive and time-consuming. Therefore, improving strategies to screen toxic substances that can potentially cause DNT has been an area of extensive research (Bal-Price et al. 2008; Bal-Price and Meek 2017). Aoyama et al. have reviewed the guidelines of DNT studies (Aoyama et al. 2015). Besides, an excellent report by Bal-Price et al. has reviewed various aspects of screening of toxic chemicals using in vitro models for developmental neurotoxicity (DNT) (Bal-Price et al. 2018).

## 15 Challenges in Assessment of Herbal Medicines

Despite advanced techniques employed to evaluate the safety of herbal products, much effort is still needed to provide details on the toxicity of medicinal herbs. The challenges associated with neurotoxicological assessment come up from either the complexity of the herbal constitutes or the methods related to the neurotoxicity testing (Gomase et al. 2008; Giordano and Costa 2012).

The net damaging or therapeutic effects of plants are primarily based on a specific component or the whole plant extracts. At the same time, it is difficult to conclude that the entire ingredients of a given herb are toxic based on individual components (Wagner and Ulrich-Merzenich 2009; Rasoanaivo et al. 2011). Basil and licorice are good examples showing that the toxicity of a specific herbal product does not necessarily indicate the same toxicity level of the whole plant (Cantelli-Forti et al. 1994; Jeurissen et al. 2008). Many authors have highlighted different conventional methods of evaluating the toxicity of herbal mixtures either by comparing the toxicity of a defined blend of the herbs of interest or by taking into consideration the toxicity of the specific components to predict any possible future interaction (McCarty and Borgert 2006; Borgert 2007). Without having a proper toxicological characterization of a given plant and its components, the use of data on isolated phytochemicals must be carefully considered (Schilter et al. 2003). Using omics technologies to identify the possible effect of the interaction of certain compounds in plant matrixes would be one of the potential solutions to solve this lack of information (Lord et al. 2006; Knasmüller et al. 2008; Amir-Aslani 2008).

Neurotoxicity testing has another difficulty associated with the evaluation of medicinal plants. It is essential to take into consideration that four elements should be considered independently during the assessment phase, including the potential, potency, hazard, and risk of the neurotoxic substances (Boyes 1997). Given that the brain has different kinds of cells with particular functions, distinct chemical mediators as well as the network connections between the brain and other systems, both the cumulative and the individual impact of herbal compounds could be challenging to explore. Besides, some neurotoxic agents have an age-dependent effect on the nervous system, which could not be assessed in vitro. Also, there is no single screening test that can assess all neurological dysfunctions. Therefore, in the initial stages of neurotoxicological testing, there is a need to incorporate different methods of evaluation, which may measure the effect of chemicals on functional and structural changes in the nervous system. Such tests should be conducted in all phases of CNS development.

Several issues are also recognized in the assessment of herbal products using animal tests such as animal cost, species variations, which may influence the type of observed effects and length of the experiment. It is necessary to have a limited number of animals and refined evaluation techniques to reduce the suffering of test animals. Also, appropriate use of validated alternative methods of screening will minimize the need to use experimental animals.

## 16 Conclusions

The crucial capability to recognize poisonous plants or toxic medicinal herbal contaminants is an essential task in protecting human health. There are many ways in which plants/medicinal herbs can cause neurotoxicity. Neurotoxicity testing of plants/medicinal herbs continues to be an extremely complicated task. It is essential to create databases that can be used to fully exploit the assessment tools to predict the safety of herbal extracts. It may be accomplished by generating genomic signatures for herbal phytochemicals to be used as data libraries. Furthermore, the use of plants/ medicinal herbs as remedies in the future could have a significant impact on treating the toxicity induced by heavy metals, insecticides, pesticides, and other infections.

## References

Amir-Aslani A (2008) Toxicogenomic predictive modeling: emerging opportunities for more efficient drug discovery and development. Technol Forecast Soc Change 75:905–932. https://doi.org/10.1016/j.techfore.2007.10.002

Aoyama H, Takahashi N, Shutoh Y et al (2015) Developmental neurotoxicology: history and outline of developmental neurotoxicity study guidelines. Food Saf 3:48–61. https://doi.org/10.14252/foodsafetyfscj.2015012

Badisa RB, Batton CS, Mazzio E et al (2018) Identification of biochemical and cytotoxic markers in cocaine treated PC12 cells. Sci Rep 8:2710. https://doi.org/10.1038/s41598-018-21182-7

Bal-Price A, Meek MEB (2017) Adverse outcome pathways: application to enhance mechanistic understanding of neurotoxicity. Pharmacol Ther 179:84–95. https://doi.org/10.1016/j.pharmthera.2017.05.006

Bal-Price A, Pistollato F, Sachana M et al (2018) Strategies to improve the regulatory assessment of developmental neurotoxicity (DNT) using in vitro methods. Toxicol Appl Pharmacol 354:7–18. https://doi.org/10.1016/j.taap.2018.02.008

Bal-Price AK, Suñol C, Weiss DG et al (2008) Application of in vitro neurotoxicity testing for regulatory purposes: symposium III summary and research needs. Neurotoxicology 29:520–531. https://doi.org/10.1016/j.neuro.2008.02.008

Blair VW, Hollenbeck AR, Smith RF, Scanlon JW (2008) Neonatal preference for visual patterns: modification by prenatal anesthetic exposure? Dev Med Child Neurol 26:476–483. https://doi.org/10.1111/j.1469-8749.1984.tb04474.x

Bogusz MJ, Al-Tufail M (2008) Toxicological aspects of herbal remedies. In: Handbook of analytical separations, 2nd edn. Elsevier B.V., pp 589–610

Bolan S, Kunhikrishnan A, Seshadri B et al (2017) Sources, distribution, bioavailability, toxicity, and risk assessment of heavy metal(loid)s in complementary medicines. Environ Int 108:103–118. https://doi.org/10.1016/j.envint.2017.08.005

Borgert CJ (2007) Predicting interactions from mechanistic information: can omic data validate theories? Toxicol Appl Pharmacol 223:114–120. https://doi.org/10.1016/j.taap.2007.01.002

Botha C, Penrith M-L (2008) Poisonous plants of veterinary and human importance in southern Africa. J Ethnopharmacol 119:549–558. https://doi.org/10.1016/j.jep.2008.07.022

Boyes W (1997) EPA's neurotoxicity risk assessment guidelines. Fundam Appl Toxicol 40:175–184. https://doi.org/10.1006/faat.1997.2388

Broadwell RD (1989) Transcytosis of macromolecules through the blood-brain barrier: a cell biological perspective and critical appraisal. Acta Neuropathol 79:117–128. https://doi.org/10.1007/BF00294368

Calne D, Mcgeer E, Eisen A, Spencer P (1986) Alzheimer's disease, Parkinson's disease, and motoneurone disease: abiotropic interaction between ageing and environment? Lancet 328:1067–1070. https://doi.org/10.1016/S0140-6736(86)90469-1

Cantelli-Forti G, Maffei F, Hrelia P et al (1994) Interaction of licorice on glycyrrhizin pharmacokinetics. Environ Health Perspect 102:65–68. https://doi.org/10.1289/ehp.94102s965

Cha HJ, Seong YH, Song MJ et al (2015) Neurotoxicity of synthetic cannabinoids JWH-081 and JWH-210. Biomol Ther (Seoul) 23:597–603. https://doi.org/10.4062/biomolther.2015.057

Champy P, Guérineau V, Laprévote O (2009) MALDI-TOF MS profiling of annonaceous acetogenins in *Annona muricata* products for human consumption. Molecules 14:5235–5246. https://doi.org/10.3390/molecules14125235

Chan TYK (2015) Incidence and causes of *Aconitum* alkaloid poisoning in Hong Kong from 1989 to 2010. Phytother Res 29:1107–1111. https://doi.org/10.1002/ptr.5370

Chen W, Vermaak I, Viljoen A (2013) Camphor—a fumigant during the black death and a coveted fragrant wood in ancient Egypt and Babylon—a review. Molecules 18:5434–5454. https://doi.org/10.3390/molecules18055434

Cheremisinoff NP (1996) Elemental and structural characterization tests. In: Polymer characterization, 1st edn. Elsevier, pp 43–81

Cushner IM (1981) Maternal behavior and perinatal risks: alcohol, smoking, and drugs. Annu Rev Public Health 2:201–218. https://doi.org/10.1146/annurev.pu.02.050181.001221

Das S, Grewal A, Banerjee M (2011) A brief review : heavy metal and their analysis. Int J Pharm Sci Rev Res 11:13–18

de Oliveira LM, Das S, da Silva EB et al (2018) Metal concentrations in traditional and herbal teas and their potential risks to human health. Sci Total Environ 633:649–657. https://doi.org/10.1016/j.scitotenv.2018.03.215

Eastwood MA (2001) A molecular biological basis for the nutritional and pharmacological benefits of dietary plants. QJM 94:45–48. https://doi.org/10.1093/qjmed/94.1.45

Gad SC (1982) A neuromuscular screen for use in industrial toxicology. J Toxicol Environ Health 9:691–704. https://doi.org/10.1080/15287398209530197

Galofré M, Babot Z, García DA et al (2010) GABAA receptor and cell membrane potential as functional endpoints in cultured neurons to evaluate chemicals for human acute toxicity. Neurotoxicol Teratol 32:52–61. https://doi.org/10.1016/j.ntt.2009.01.010

Gao QG, Chen WF, Xie JX, Wong MS (2009) Ginsenoside Rg1 protects against 6-OHDA-induced neurotoxicity in neuroblastoma SK-N-SH cells via IGF-I receptor and estrogen receptor pathways. J Neurochem 109:1338–1347. https://doi.org/10.1111/j.1471-4159.2009.06051.x

Giordano G, Costa LG (2012) Developmental neurotoxicity: some old and new issues. ISRN Toxicol 2012:1–12. https://doi.org/10.5402/2012/814795

Gomase V, Changbhale S, Patil S, Kale K (2008) Metabolomics. Curr Drug Metab 9:89–98. https://doi.org/10.2174/138920008783331149

Grandjean P, Landrigan P (2006) Developmental neurotoxicity of industrial chemicals. Lancet 368:2167–2178. https://doi.org/10.1016/S0140-6736(06)69665-7

Greenberg M (2006) Encyclopedia of terrorist, natural, and man-made disasters, 1st edn. Jones & Bartlett Learning Publishers

Hansen DK, George NI, White GE et al (2012) Physiological effects following administration of *Citrus aurantium* for 28 days in rats. Toxicol Appl Pharmacol 261:236–247. https://doi.org/10.1016/j.taap.2012.04.006

Harborne JB, Baxter H (eds) (1996) Dictionary of plant toxins. John Wiley & Sons, Chichester

Harry GJ, Billingsley M, Bruinink A et al (1998) *In vitro* techniques for the assessment of neurotoxicity. Environ Health Perspect 106:131–158. https://doi.org/10.1289/ehp.98106s1131

Höllerhage M, Rösler TW, Berjas M et al (2015) Neurotoxicity of dietary supplements from Annonaceae species. Int J Toxicol 34:543–550. https://doi.org/10.1177/1091581815602252

Hussain K, Parvaiz M, Khalid S et al (2014) A review: medicinal importance of *Glycyrrhiza glabra* L. (Fabaceae Family). Glob. J Pharmacol 8:8–13. https://doi.org/10.5829/idosi.gjp.2014.8.1.81179

Ifeoma O, Oluwakanyinsol S (2013) Screening of herbal medicines for potential toxicities. In: Gowder SJT (ed) New insights into toxicity and drug testing. InTech Open

Irwin S (1968) Comprehensive observational assessment: Ia. A systematic, quantitative procedure for assessing the behavioral and physiologic state of the mouse. Psychopharmacologia 13:222–257. https://doi.org/10.1007/BF00401402

Jeurissen SMF, Punt A, Delatour T, Rietjens IMCM (2008) Basil extract inhibits the sulfotransferase mediated formation of DNA adducts of the procarcinogen 1′-hydroxyestragole by rat and human liver S9 homogenates and in HepG2 human hepatoma cells. Food Chem Toxicol 46:2296–2302. https://doi.org/10.1016/j.fct.2008.03.010

Khan IA, Allgood J, Walker LA et al (2001) Determination of heavy metals and pesticides in ginseng products. J AOAC Int 84:936–939. https://doi.org/10.1093/jaoac/84.3.936

Kjellstrom T (1979) The biogeochemistry of lead in the environment. Part B: biological effects. Topics in environmental health, volume 1B. J. O. Nriagu. Q Rev Biol 54:471–471. https://doi.org/10.1086/411541

Knasmüller S, Nersesyan A, Mišík M et al (2008) Use of conventional and -omics based methods for health claims of dietary antioxidants: a critical overview. Br J Nutr 99:ES3–ES52. https://doi.org/10.1017/S0007114508965752

Ko JH, Lee J-H, Choi B et al (2018) Neuroprotective effects of Gagam-Sipjeondaebo-tang, a novel herbal formula, against MPTP-induced Parkinsonian mice and MPP + −induced cell death in SH-SY5Y cells. Evid Based Complement Alternat Med 2018:1–9. https://doi.org/10.1155/2018/2420809

Kotake Y, Okuda K, Kamizono M et al (2004) Detection and determination of reticuline and N-methylcoclaurine in the Annonaceae family using liquid chromatography-tandem mass spectrometry. J Chromatogr B 806:75–78. https://doi.org/10.1016/j.jchromb.2004.03.017

Kovalevich J, Langford D (2013) Considerations for the use of SH-SY5Y Neuroblastoma cells in neurobiology. Humana Press, Totowa, NJ, pp 9–21

Lee H, KIM YO, KIM H et al (2003) Flavonoid wogonin from medicinal herb is neuroprotective by inhibiting inflammatory activation of microglia. FASEB J 17:1943–1944. https://doi.org/10.1096/fj.03-0057fje

Leung GPH, Shaw PC, Ng TB et al (2011) Anti-Parkinsonian drug discovery from herbal medicines: what have we got from neurotoxic models? J Ethnopharmacol 139:698–711. https://doi.org/10.1016/j.jep.2011.12.030

Liu X, Wang Q, Song G et al (2014) The classification and application of toxic Chinese Materia Medica. Phytother Res 28:334–347. https://doi.org/10.1002/ptr.5006

Lord PG, Nie A, McMillian M (2006) Application of genomics in preclinical drug safety evaluation. Basic Clin Pharmacol Toxicol 98:537–546. https://doi.org/10.1111/j.1742-7843.2006.pto_444.x

Manahan SE (2002) Toxicological chemistry and biochemistry, 3rd edn. CRC Press, pp 1–425

McCarty LS, Borgert CJ (2006) Review of the toxicity of chemical mixtures: theory, policy, and regulatory practice. Regul Toxicol Pharmacol 45:119–143. https://doi.org/10.1016/j.yrtph.2006.03.004

Moghadamtousi SZ, Fadaeinasab M, Nikzad S et al (2015) *Annona muricata* (Annonaceae): a review of its traditional uses, isolated acetogenins and biological activities. Int J Mol Sci 16:15625–15658. https://doi.org/10.3390/ijms160715625

Moser VC (1989) Screening approaches to neurotoxicity: a functional observational battery. J Am Coll Toxicol 8:85–93. https://doi.org/10.3109/10915818909009095

Mulaudzi RB, Tshikalange TE, Olowoyo JO et al (2017) Antimicrobial activity, cytotoxicity evaluation and heavy metal content of five commonly used south African herbal mixtures. South African J Bot 112:314–318. https://doi.org/10.1016/j.sajb.2017.06.024

National Research Council (US) Committee on Neurotoxicology and Models for Assessing Risk, Committee NRC (US) (1992) Environmental neurotoxicology. National Academy Press

National Research Council (US) Committee on Pesticides in the Diets of Infants and Children (1993) Pesticides in the diets of infants and children. National Academies Press, Washington, DC

OECD (1995) Test No. 418: Delayed neurotoxicity of organophosphorus substances following acute exposure. OECD

OECD (1997) Test No. 424: Neurotoxicity study in rodents. OECD

Pardridge WM (1988) Recent advances in blood-brain barrier transport. Annu Rev Pharmacol Toxicol 28:25–39. https://doi.org/10.1146/annurev.pa.28.040188.000325

Pelkonen O, Duez P, Vuorela PM, Vuorela H (eds) (2017) Toxicology of herbal products. 1st edn, Springer International Publishing, p 1–479

Preis PN, Saya H, Nádasdi L et al (1988) Neuronal cell differentiation of human neuroblastoma cells by retinoic acid plus herbimycin A. Cancer Res 48:6530–6534

Rady I, Bloch MB, Chamcheu RCN et al (2018) Anticancer properties of graviola (*Annona muricata*): a comprehensive mechanistic review. Oxidative Med Cell Longev 2018:1826170. https://doi.org/10.1155/2018/1826170

Rasoanaivo P, Wright CW, Willcox ML, Gilbert B (2011) Whole plant extracts versus single compounds for the treatment of malaria: synergy and positive interactions. Malar J 10:S4. https://doi.org/10.1186/1475-2875-10-S1-S4

Rodier PM (1995) Developing brain as a target of toxicity. Environ Health Perspect 103:73–76. https://doi.org/10.1289/ehp.95103s673

Russo E, Scicchitano F, Whalley BJ et al (2014) Hypericum perforatum:pharmacokinetic, mechanism of action, tolerability, and clinical drug-drug interactions. Phytother Res 28:643–655. https://doi.org/10.1002/ptr.5050

Schilter B, Andersson C, Anton R et al (2003) Guidance for the safety assessment of botanicals and botanical preparations for use in food and food supplements. Food Chem Toxicol 41:1625–1649. https://doi.org/10.1016/S0278-6915(03)00221-7

Selvi C, Paramasivam M (2017) Review on pesticide residue analytical methods and residue status in medicinal plants. J Entomol Zool Stud 5:945–950

Senchina DS, Hallam JE, Kohut ML et al (2014) Alkaloids and athlete immune function: caffeine, theophylline, gingerol, ephedrine, and their congeners. Exerc Immunol Rev 20:68–93

Shaban NS, Abdou KA, Hassan NE-HY (2016) Impact of toxic heavy metals and pesticide residues in herbal products. Beni-Suef Univ J Basic Appl Sci 5:102–106. https://doi.org/10.1016/j.bjbas.2015.10.001

Shen XL, Jia FJ, Song N et al (2014) Protection of MES23.5 dopaminergic cells by obestatin is mediated by proliferative rather than anti-apoptotic action. Neurosci Bull 30:118–124. https://doi.org/10.1007/s12264-013-1405-0

Smith H, Spalding J (1959) Outbreak of paralysis in Morocco due to ortho-cresyl phosphate poisoning. Lancet 274:1019–1021. https://doi.org/10.1016/S0140-6736(59)91486-2

Spencer PS, Bischoff MC, Schaumburg HH (1978) On the specific molecular configuration of neurotoxic aliphatic hexacarbon compounds causing central-peripheral distal axonopathy. Toxicol Appl Pharmacol 44:17–28. https://doi.org/10.1016/0041-008X(78)90280-6

Spyker JM (1975) Assessing the impact of low level chemicals on development: behavioral and latent effects. In: Persaud TVN (ed) Problems of birth defects, 1st edn. Springer Netherlands, Dordrecht, pp 242–251

Steenkamp P (2005) Chemical analysis of medicinal and poisonous plants of forensic importance in South Africa. University of Johannesburg

Stohs SJ, Preuss HG, Shara M (2011) The safety of *Citrus aurantium* (bitter orange) and its primary protoalkaloid p-synephrine. Phytother Res 25:1421–1428. https://doi.org/10.1002/ptr.3490

Tamokou JD, Kuete V (2014) Toxic plants used in African traditional medicine. In: Kuete V (ed) Toxicological survey of African medicinal plants, 1st edn. Elsevier, pp 135–180. https://doi.org/10.1016/C2013-0-15406-2

Vigliani EC (1954) Carbon disulphide poisoning in viscose rayon factories. Occup Environ Med 11:235–244. https://doi.org/10.1136/oem.11.4.235

Wager TT, Liras JL, Mente S, Trapa P (2012) Strategies to minimize CNS toxicity: *in vitro* high-throughput assays and computational modeling. Expert Opin Drug Metab Toxicol 8:531–542. https://doi.org/10.1517/17425255.2012.677028

Wagner H, Ulrich-Merzenich G (2009) Synergy research: approaching a new generation of phytopharmaceuticals. Phytomedicine 16:97–110. https://doi.org/10.1016/j.phymed.2008.12.018

Waldron HA (1973) Lead poisoning in the ancient world. Med Hist 17:391–399. https://doi.org/10.1017/s0025727300019013

Wang Z, Wang H, Wang H et al (2019) Heavy metal pollution and potential health risks of commercially available Chinese herbal medicines. Sci Total Environ 653:748–757. https://doi.org/10.1016/j.scitotenv.2018.10.388

Westerink RHS, Ewing AG (2007) The PC12 cell as model for neurosecretion. Acta Physiol 192:273–285. https://doi.org/10.1111/j.1748-1716.2007.01805.x

Williamson EM (2017) Herbal neurotoxicity: an introduction to its occurrence and causes. In: Pelkonen et al (eds) Toxicology of herbal products, 1st edn. Springer International Publishing, pp 345–362

World Health Organization (1993) Research guidelines for evaluating the safety and efficacy of herbal medicines. WHO, Regional Office for the Western Pacific, Manila, Philippines

Xie H, Hu L, Li G (2010) SH-SY5Y human neuroblastoma cell line: *in vitro* cell model of dopaminergic neurons in Parkinson's disease. Chin Med J 123:1086–1092. https://doi.org/10.3760/cma.j.issn.0366-6999.2010.08.021

Yang Y, Guangyu Y, Lin Q (2004) Determination of heavy metal ions in Chinese herbal medicine by microwave digestion and RP-HPLC with UV-Vis detection. Microchim Acta 144:297–302. https://doi.org/10.1007/s00604-003-0020-6

Ye S, Koon HK, Fan W et al (2017) Effect of a traditional Chinese herbal medicine formulation on cell survival and apoptosis of MPP + −treated MES 23.5 dopaminergic cells. Parkinson's Dis 2017:1–10. https://doi.org/10.1155/2017/4764212

Zhang SN, Li XZ, Wang Y et al (2014) Neuroprotection or neurotoxicity? New insights into the effects of *Acanthopanax senticosus* harms on nervous system through cerebral metabolomics analysis. J Ethnopharmacol 156:290–300. https://doi.org/10.1016/j.jep.2014.08.037

# Plants with Phytomolecules Recognized by Receptors in the Central Nervous System

**Hinanit Koltai and Zohara Yaniv**

## Contents



**Abstract** Multiple plants were used traditionally, throughout history, for a variety of purposes. Many of these traditionally used plants produce various molecules,

H. Koltai · Z. Yaniv (✉)
Institute of Plant Science, Agriculture Research Organization, Rishon LeZion, Israel
e-mail: zohara@zyb.onmicrosoft.com

 
D. C. Agrawal, M. Dhanasekaran (eds.), *Medicinal Herbs and Fungi*,
https://doi.org/10.1007/978-981-33-4141-8_11

which directly, and sometimes with high affinity, bind and activate or block from activation human receptors. Among the most effective molecules are those that serve as ligands to receptors found in the central nervous system (CNS) of the human brain. Here we present some of these plants, their traditional uses, the molecules they produce, and some of their modes of action. In accordance, the first part of the chapter describes the traditional, medicinal, and recreational uses of the plants. The second part of the chapter reviews the compounds present in these different plants and their binding properties to CNS receptors. Lastly, insight and suggestions will be provided into the man-made evolution of these plants, which might have led to their high potency.

**Keywords** Central nervous system · Neurotransmitters · Phytomolecules · Receptors · Traditional medicines

## Abbreviations

| | |
|---|---|
| 2-AG | 2-Arachidonoylglycerol |
| 5-HT | 5-Hydroxytryptamine or (serotonin) |
| Anandamide | N-Arachidonoylethanolamine |
| BCE | Before the Common Era |
| $CB_1$ | Endocannabinoid receptor type 1 |
| $CB_2$ | Endocannabinoid receptor type 2 |
| CBCA | Cannabichromenic acid |
| CBD | Cannabidiol |
| CBDA | Cannabidiolic acid |
| CBGA | Cannabigerolic acid |
| CNS | Central nervous system |
| DA | Dopamine |
| EGCG | (−)-Epigallocatechin gallate |
| FDA | Food and Drug Administration |
| $GABA_ARs$ | γ-Aminobutyric acid type A receptor |
| NA | Noradrenaline |
| nAChRs | Acetylcholine receptors |
| THC | $\Delta^9$-Tetrahydrocannabinol |
| THCA | Tetrahydrocannabinolic acid |
| TRPV1 | Transient receptor potential vanilloid 1 |
| σ1R | Sigma-1 receptor |

# 1 History of the Uses of Plants with Phytomolecules Recognized by Receptors in the Central Nervous System

Plants were recognized as beneficial to human culture since ancient times. Some of them were used as stimulating drinks (coffee, tea, cacao, guarana, cola), some for smoking (tobacco, lobelia), for chewing (areca nut), as sedatives and tranquilizers (Valeriana, kava kava), as stimulants (ephedra, khat, coca leaves), and some as painkillers (the opium poppy, cannabis). Their desired effects were probably discovered accidentally and further refined by trial and error.

## *1.1 Plants Having Stimulating Effects*

### 1.1.1 *Camellia sinensis*

According to Chinese tradition, in 2723 BCE, some leaves of the plant *Camellia sinensis* (Fig. 1a) fell into a warm cup of water of the famous Chinese emperor Shen-Nung. He recommended adding this yellow drink to his people as an important part

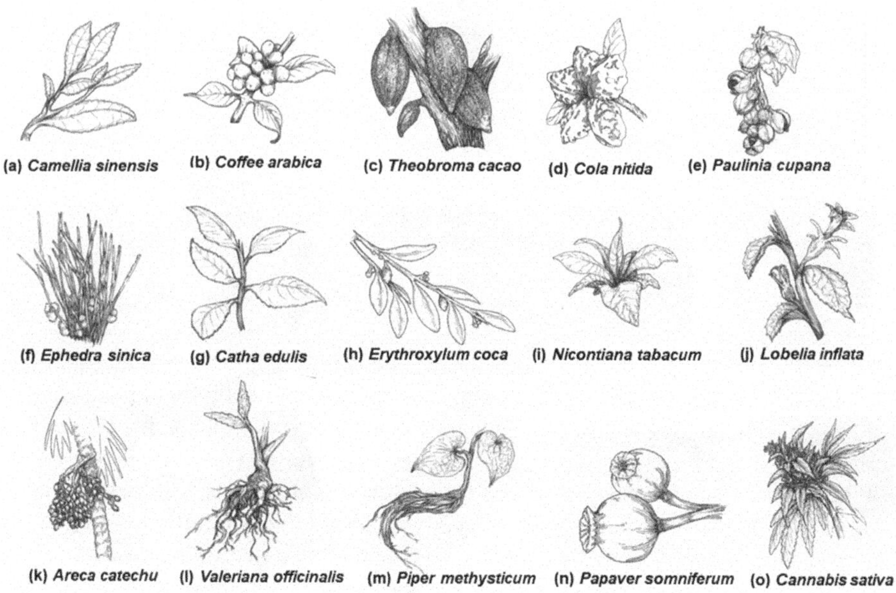


**Fig. 1** (**a–o**) Drawings of plant species that produce compounds with activity on the human central nervous system (illustrated by Omer Koltai). (**a**) *Camellia sinensis*; (**b**) *Coffee Arabica*; (**c**) *Theobroma cacao*; (**d**) *Cola nitida*; (**e**) *Paulinia cupana*; (**f**) *Ephedra sinica*; (**g**) *Catha edulis*; (**h**) *Erythroxylum coca*; (**i**) *Nicontiana tabacum*; (**j**) *Lobelia inflate*; (**k**) *Areca catechu*; (**l**) *Valeriana officinalis*; (**m**) *Piper methysticum*; (**n**) *Papaver somniferum*; (**o**) *Cannabis sativa*

of their diet. This was the beginning of a very long tradition of green tea drinking, first in China, and later in different parts of the world (Yaniv Bachrach, unpublished). The tea plant, *C. sinensis* is native to South East Asia, yet now it is cultivated in more than 30 countries. It is generally accepted that next to the water, tea is the most consumed beverage in the world, with a per capita consumption of 120 mL/day.

Among the types of tea produced in the world, 78% is black and 20% is green. Black tea is consumed mainly in Western countries, whereas green tea is consumed primarily in Asian countries. Black tea is produced by allowing the fermentation of the freshly harvested leaves, whereas green tea is produced by steaming the freshly harvested tea leaves. Steaming causes the inactivation of the fermentation enzymes, so that the green color of the leaves is maintained (Graham 1992).

Green tea has been associated with the prevention and reduction of a wide range of severe health conditions such as cancer, cardiovascular diseases, malaria, and Alzheimer's disease. The health benefits associated with green tea consumption have been predominantly attributed to green tea polyphenol—EGCG, (−)-epigallocatechin gallate. In the past, leaves of *C. sinensis* were inserted directly into boiling water, and the subsequent tea was consumed. In the present time, extracts are prepared from the leaves, added to water, and consumed (Bachrach and Yaniv 2016).

### 1.1.2 *Coffee arabica*

The coffee plant *Coffee arabica* (Fig. 1b) is a small tree with shiny leaves. It bears white and fragrant flowers. The fruit, berry, is green and turns red when ripe. Normally, the fruit has two seeds (beans) in each berry. The coffee bean contains about 5% phenolic acids (quinic acid, caffeic acid, and chlorogenic acid), and their caffeine concentration ranges from 0.6 to 2%, yet it could be higher in some varieties.

The origin of *C. arabica* is in the mountain regions of southern Ethiopia. The Indigenous peoples of Africa were the first users of coffee while they used various parts of the coffee plant: leaves, fruits, and seeds for food. The pulp containing the stimulating caffeine was discovered in the ripe fruit, and people began chewing the seeds. According to historians, the use of coffee beans was introduced from Africa to Arabia in pre-Islamic days, where it was eaten as a compressed product (Teketay 1999).

Coffee is consumed globally in various ways. Initially cultivated by the Arabs, it made its way quickly to the world of Islam. It was introduced to Europe by the Venetians in 1615 AD and to France in 1644. Despite specific precautions led by the Arabs to prevent live seeds from crossing the Atlantic by preventing the export of unheated seeds, several plants were introduced to Brazil, India, and Sri Lanka. Brazil later became the world leader in coffee production (Bruneton 1995).

People most commonly drink coffee to relieve mental and physical fatigue and to increase mental alertness. Coffee is also used traditionally to prevent Parkinson's disease, Alzheimer's disease, dementia, and cognitive decline, gallstones, gout, type 2 diabetes mellitus, and rheumatic pains. It stimulates the kidneys and is even believed to treat certain types of cancer. Ethnobotanical uses of coffee in the Middle-East include the treatment of headaches, colds, and diarrhea. The Bedouins, for example, use the beans to prevent bleeding nose and as a component for tooth remedy and skin burns (Danino and Amar 2013).

### 1.1.3 *Theobroma cacao*

*Theobroma cacao* (Fig. 1c) is a small tree, indigenous to the tropical regions, mainly to forests of Central America, and is native in the equator of South America. It is cultivated in similar geographic regions in West Africa. The flowers stem directly from the trunk and on the main branches. The fruits are the cacao pods, which have a rough, leathery rind about 2–3 cm thick (this varies with the origin and variety of pod) that encloses 30–50 large seeds. The rind is filled with sweet lemonade-like tasting, mucilaginous pulp called "baba de cacao" in South America. The fresh seeds are odorless and bitter. During harvest, the pods are opened, the seeds are kept, the empty pods are discarded, and the pulp is made into juice. This sweet drink was consumed by the early settlers of the Americas (Schnepel 2002).

The seeds are then fermented and become brown only after lengthy fermentation. The kernel of the seeds contains about 50% lipids (cacao butter), polyphenols that are oxidized during fermentation, alkaloids including theobromine (1–3%), and a small amount of caffeine (0.05–0.3%; Bruneton 1995).

The cacao tree was domesticated in Central America more than 4000 years ago. The fruit and beans were used as food and in spiritual ceremonies by the Mayan and Olmec civilizations. Interestingly, the cocoa bean was a common currency throughout Mesoamerica before the Spanish conquest (Wood and Lass 2001).

Cacao beans and chocolate were first used by the ancient Maya and Aztec civilizations of South America to treat a variety of medical ailments involving the cardiovascular, gastrointestinal, and nervous systems. Also today, diets rich in foods containing abundant polyphenols, as found in cocoa, are considered as strong antioxidants in nature and are recommended by dietitians worldwide, despite their high caloric value (Hamed et al. 2008).

### 1.1.4 *Cola nitida*

The kola tree *Cola nitida* (Fig. 1d) is an evergreen tree, medium in size. It grows in West Africa, mainly in Nigeria, Gabon, and Sierra Leone. *C. nitida* is the main species to be cultivated among other *Cola* species. The fruits contain about 5–10 seeds, and the seeds contain about 1.5% caffeine. When the seeds are collected in the wild, the kola seed, or nut, is consumed fresh for chewing.

The use of kola nut by humans appears to be of ancient origin. Kola nuts are used for mental stimulation in West African cultures, both in private and a social gathering. Also, today, kola nuts play an important role in spiritual practices in West African countries, including Nigeria, Niger, Liberia, and Sierra Leone. Kola nuts are used as a religious object and sacred offering during prayers and important life events such as naming ceremonies, weddings, and funerals, and in Obi, a traditional divination act. In Western cultures, kola nuts were used as a flavoring ingredient and as a source of caffeine in cola and other flavored beverages. However, kola nut extract or flavoring is no longer used in major commercial cola drinks, such as Coca-Cola (Epega 2003; Greenwood 2016). In traditional medicine, kola nuts are used for improved digestion. Ground kola nuts mixed with honey are used to treat coughs (Odebunmi et al. 2009).

### 1.1.5 *Paulinia cupana*

Guarana, *Paullinia cupana* (Fig. 1e) is a creeping plant native to the lower Amazon region. It has small bright red fruits. According to popular tradition, the seeds are roasted, the external layer is removed, and then the seeds are crushed with water to a paste. This paste is rolled into sticks and then smoked (Bruneton 1995). Guarana is valued mainly for its stimulant property because of its high content of caffeine, which can be up to 6% in the seeds (Schimpl et al. 2013). In addition to caffeine, the seeds contain saponins and tannins. The ethnobotanical uses of Guarana seeds include the treatment of mild diarrhea, functional asthenia, and as a beverage for weight loss (Bruneton 1995). Besides, the seeds of Guarana are known as antimicrobial, antioxidant, anticarcinogenic, and stimulating (Marques et al. 2019).

### 1.1.6 *Ephedra sinica*

*Ephedra sinica* (Fig. 1f) or Ma-huang is an evergreen shrub native to central Asia. Ephedrine is the primary active constituent and affects the central nervous system (Bruneton 1995). Ephedra has been used for thousands of years in the treatment of asthma, sinusitis, and as a stimulant. Traditionally, the aerial parts were dried and brewed into tea. Beverages made from the plant have been referred to by many names, e.g., yellow river, Mormon tea, and whorehouse tea. Ma-huang was used as a diaphoretic, circulatory stimulant, antipyretic, and sedative for cough.

Furthermore, today, ephedra extracts are available in many prepared dosage forms such as tablets, capsules, extracts, and tinctures and are used mainly in Chinese medicine. Active phytochemicals in ephedra for the treatment of asthma were identified as ephedrine and pseudoephedrine (Chan et al. 2016). Until the early twentieth century, ephedra was sold as an herbal supplement and was increasingly used by athletes for its stimulant properties. It was popular as a dietary supplement (appetite suppressant) for weight loss. However, the US Food and Drug Administration (FDA) found it too risky to be used freely. Abuse of ephedra led to many incidents of adverse cardiovascular outcomes such as heart attack, stroke, and even death. In 2004, the FDA banned the sale of dietary supplements containing ephedrine alkaloids (Bruneton 1995).

### 1.1.7 *Catha edulis*

Khat is a small evergreen tree, *Catha edulis* (Fig. 1g), native to the horn of Africa and the Arabian Peninsula. The bitter-tasting leaves and young buds are chewed for the stimulant components cathinone and cathine, which produce mild euphoria. Chewing khat reduces fatigue, suppress the appetite, and enhances mental power and skills of communication. Excessive chewing can cause hyperactivity and insomnia. The compulsive user can become aggressive and paranoiac (Bruneton 1995). Khat is an important cash crop in Yemen, Somalia, and Ethiopia and is often cultivated in areas where no other crops can grow. Although the drug is central to social life in some countries, the plant and cathinone are considered controlled substances in much of Europe, the United States, and China (Getasetegn 2016).

### 1.1.8 *Erythroxylum coca*

*Erythroxylum coca* (Coca) (Fig. 1h) is a shrub or a small tree with reddish branches and oval, entire and petiolate leaves. It grows wild in the Peruvian and Bolivian Andes (Bruneton 1995). Widely used as the source of the drug cocaine, the leaves have long been used by local people as a stimulant to overcome hunger, thirst, physical stress, and exhaustion. Chewed with lime or plant ashes, the leaves cause a feeling of easiness and increasing energy.

An infusion of the leaves also serves as a remedy for altitude sickness in the height of the Andes Mountains. The plant is often cultivated in South America (in the borderland of the Amazonian basin of Peru, Bolivia, Colombia, Ecuador, and Brazil up to 2000 m height), and in the tropical mountains of Africa. Coca is cultivated both legally for its medicinal uses and illegally to provide the street drug cocaine. The custom of chewing coca leaves in Peru goes back thousands of years. During the Inca period, it was considered to be a sacred plant, and only the kings and important leaders were permitted to use it. With the decline of the Inca imperium, the use of chewing coca leaves passed rapidly and widely from the exclusive use of rulers to the common people.

The present-day use of cocaine as a narcotic is a worldwide phenomenon, and there exists a considerable illegal trade. There are restrictions on the cultivation of coca in Peru and Bolivia, but it is tough to substitute this profitable crop by other useful plants (Fern 2018).

### 1.1.9 *Nicontiana tabacum*

The Tobacco plant *Nicontiana tabacum* (Fig. 1i) is an annually grown herbaceous plant, and its leaves are commercially grown in many countries to be processed into tobacco. The leaves are large, and the flowers are characterized by a tubulous flared corolla with colors ranging from pink to yellow to red (Bruneton 1995).

*N. tabacum* is a native of tropical and subtropical America, but it is now commercially cultivated worldwide. Other varieties are cultivated as ornamental plants or grown like a weed. Tobacco cultivation began in the Americas circa 6000 BC. Native American Indians began using tobacco in various ways, such as in religious and medicinal practices. Tobacco was believed to be a cure-all and was used to dress wounds, as well as a pain killer. Chewing tobacco was thought to relieve the pain of a toothache (Musk and De Klerk 2003).

On October 15, 1492, Christopher Columbus was offered dried tobacco leaves as a gift from the American Indians. Soon after, sailors brought tobacco back to Europe, and the cultivation of Tobacco in Europe began. The Europeans considered it as a healing plant and used it to cure all: from wounds to cancer. In 1571, in a book about the history of medicinal plants of the new world written by a Spanish doctor, Nicolas Monardes, it was claimed that tobacco could cure 36 health problems. During the 1600s, tobacco was so popular that it was frequently used as money, and it was considered as valuable as gold.

### 1.1.10 *Lobelia inflata*

*Lobelia inflata* (Indian tobacco) (Fig. 1j) is a small annual herb. The stems are erect. Blue flowers appear in terminal inflorescence. *L. inflata* is a native of North America, especially abundant in the Appalachian Mountains. The native Indians of North America used Indian tobacco as a substitute for tobacco for centuries. It made its way to Europe at the beginning of the nineteenth century and gained a reputation as a remedy for asthma, named under "the asthma weed." Other popular names included "emetic weed" (as it can cause vomiting) and "wild tobacco."

*L. inflata* has a prolonged use as medicinally. It is used as a psychoactive agent, for respiratory aid, and as an emetic and dermatological agent. Native Americans, including the Cherokee, Iroquois, Penobscot, and other indigenous people, used it for ceremonial medicine. To treat respiratory and muscle disorders, leaves were chewed and smoked. The Cherokee burned the foliage of the plant as an insecticide against gnats (Sanders 2007). The alkaloid lobeline, one of the main components of the herbal extract, is used as a stimulant, antiasthmatic, and expectorant in cases of bronchitis (Krochmal et al. 1970). Lobelia is still used medicinally nowadays; however, there are limitations to its use. Adverse effects can include nausea, vomiting, sweating, hypothermia, convulsions, tremors, rapid heartbeat, diarrhea, mental confusion, coma, and possibly death.

### 1.1.11 *Areca catechu*

The seed of the areca palm, *Areca catechu,* is the areca nut, commonly referred to as "betel nut" (Fig. 1k). It grows in parts of East Africa, most of the tropical Pacific, Southeast, and South Asia and India. The term areca originates from the name "Kerala" (India), since Portuguese and Dutch sailors took the nut in the sixteenth century from India to Europe. It is consumed widely in all of the territories, as mentioned above.

Consumption of areca nut was found to be harmful and even carcinogenic. It was found that various compounds, including arecoline that is similar to nicotine and is the primary psychoactive ingredient, contribute to histologic changes in oral mucosa and the development of squamous cell carcinoma of the mouth and esophagus. Despite efforts to discourage its use by preventive efforts, it is still being consumed worldwide by hundreds of millions of people, mainly in southern and eastern Asia. This continuing use was described as a "neglected global public health emergency" (Gupta and Ray 2004).

Folkloristic use includes mixing cut up areca nut with lime (the origin of lime could be crushed shells, crushed corals, or ashes). The mix is rolled in a betel leaf, with the addition of spices. The selection of spices depends on the location: cardamom, nutmeg, cloves, saffron, or tobacco. During chewing, the user frequently spits red spitted saliva; the red color is due to the oxidation of polyphenols in the mix (Bruneton 1995).

At least 4000 years ago, in Thailand, Indonesia, and the Philippines, the areca nut and the betel leaf were combined into a psychoactive drug. Chewing the mixture of betel leaf and areca nut became a tradition and custom and was practiced as a part of rituals in South Asia and the Pacific. It is still present as a popular cultural activity in Asian and Oceanic countries (Strickland 2002).

## 1.2 Plants Used as Tranquilizers and Sedatives

### 1.2.1 *Valeriana officinalis*

*Valeriana officinalis* (Fig. 1l) is a perennial herb that belongs to the Valerianaceae family. It has an aromatic smell due to the essential oils in the roots and flowers. It is a well-known medicinal plant, widely used globally but especially in Europe, North America, China, and the Middle East. Roots are the source of active compounds. They have been used medicinally for ~2000 years. In traditional herbal medicine, the roots of *V. officinalis* have long been utilized for sedative and antispasmodic purposes, for curing cardiac arrhythmia, to treat digestive problems, and urinary tract infections (Pilerood and Prakash 2013Pilerood and Prakash 2013). As a treatment for neuronal disorders, Valeriana was firstly used within the late sixteenth century. It is still being used today for the treatment of various nervous system disorders, as a sedative and to induce sleep (Nandhini et al. 2018). Valeriana is also used to treat hysteria (psychosis), hypochondriasis, nervous unrest (anxiety), and similar emotional states (Nandhini et al. 2018).

### 1.2.2 *Piper methysticum*

Kava Kava, *P. methysticum* (Fig. 1m), is a pepper tree that grows in the islands of Western Polynesia, such as Fiji, Tonga, and Samosa, and can be found as far as Haiti. The famous part of the plant, which gave its reputation, is the rhizome. Rhizomes are horizontal underground stems capable of producing lateral shoots and adventitious roots. The world-renowned Kava beverage is made from the rhizomes in the following manner: The hard rhizome is ground with a pestle or chewed and spitted. Then the ground parts are soaked in water. The resulting drink has been consumed for centuries following a traditional ritual (Bruneton 1995).

Kava is greatly respected throughout the Pacific, being used for medicinal and social purposes. In Fiji, a formal ceremony called yaqona (kava) often accompanies important social, political, or religious functions. This ceremony often involves a ritual presentation of the bundled roots as a gift. Also, the yaqona beverage that has a sedative and tranquilizing activity is being consumed in this ceremony (Tomlinson 2007).

In the Southern Pacific Islands, kava was used traditionally to treat the CNS disorder for relaxation and additional peripheral organs. Significant subtlety and nuance to the effect were obtained based on the precise strain, plant part (i.e., leaf, stem, root), and preparative methods that were used. Kava is indicated to treat reproductive and women's health for improving urogenital conditions (e.g., gonorrhea infections, chronic cystitis, difficulty urinating), gastrointestinal upsets, and respiratory ailments such as coughs, asthma, and tuberculosis. Kava is also used as an analgesic and to treat skin diseases and topical wounds (Showman et al. 2015).

## *1.3 Plants as Pain Killers*

### 1.3.1 *Papaver somniferum*

The opium poppy *Papaver somniferum* (Fig. 1n) is an annual plant of 1–1.5 m in height. The stem bears a single flower. The color of the flower can be red, white, or purple, depending on the cultivar. The fruit is a capsule, usually indehiscent, containing a significant number of small seeds. All parts of the plant, upon incision, or when wounded, exude a white latex. This latex contains the opiate alkaloids (Bruneton 1995).

The opium poppy is used by humanity long before written history. Images of opium poppies are found in ancient Sumerian artifacts (circa 4000 BC). The ancient Minoans knew the making of opium and its use. The ancient Greeks named its sap "opion"; this is the origin of the modern name "opium." Opium was used in ancient times for treating stomach illnesses, asthma, and bad eyesight. Opium poppies (flower and fruit) appear as images in the coat of arms of the Royal College of Anesthetists in Great Britain (Askitopoulou et al. 2002).

Poppy is grown as an agricultural crop on a large scale. This is for the production of (1) poppy seeds that are for human consumption, (2) opium for pharmaceutical industry use, (3) other alkaloids, mainly thebaine and oripavine. These compounds are processed into drugs such as hydrocodone and oxycodone in the pharmaceutical industry. Special breeds are cultivated for each of these purposes, and breeding efforts are still ongoing. *P. somniferum* is also produced commercially at a smaller scale for ornamental purposes ("Shake-Up on Opium Island." The New York Times. Retrieved January 8, 2018).

### 1.3.2 *Cannabis sativa*

Cannabis, *Cannabis sativa* (Fig. 1o), was used by humanity for thousands of years. Evidence for the first domestication and initial cultivation date to the third millennium BCE in China and Japan (Stevens et al. 2016). However, there is evidence for independent cultivation of *C. sativa* in Europe dating to the Copper Age, approximately 7000 years ago (McPartland and Hegman 2018; Koltai and Namdar 2020). Ancient medicinal texts of India and China (Touw 1981), and Greece and Rome (Butrica 2002) list different specific uses of cannabis. Cannabis was used to treat approximately 80 different medical conditions during ancient times (Russo 2014). The use of cannabis in the Americas was dated to only after 1492 CE (Merlin 2003). Residues of cannabis used for medical and ceremonial purposes can be found in vessels from archeological sites in different parts of the world (Thackeray et al. 2001; Koltai and Namdar 2020).

Today, cannabis is used for the treatment of a wide variety of medical conditions. It is being used for pain relief in patients with chronic (non-cancer-related) pain, neuropathy pain in diabetic patients, cancer patients, or in patients suffering from multiple sclerosis-related pain. Cannabis is also used for reducing symptoms related to epilepsy and neurodegenerative disorders, including Alzheimer's disease and Parkinson's disease. CBD-based medical cannabis was shown to be a beneficial treatment for refractory behavioral problems in children, and cannabis has also been reported for the treatment of cachexia–anorexia eating disorder (reviewed by Koltai and Namdar 2020).

## 2 Mode of Action of Phytomolecules in the Nervous System

Many of the phytomolecules produced in these plants (as described above) are effective due to their ability to bind human receptors, mainly those present on cells (neurons/nerves) of the nervous system. Below are some examples of the phytomolecules in the different ethnobotanical plants and their (at least) partial mode of action.

## 2.1 Plants Containing Methylxanthines

Many plant species that have a simulative effect contain caffeine or similar compounds. *Coffee* spp. seeds, Tea (*C. sinensis*) leave, and *Cola* spp. leaf extract contains mainly purinergic compounds designated as methylxanthines, and include mainly caffeine (Fig. 2a; 1,3,7-trimethylxanthine). These extracts contain additional purine metabolites, theobromine, 3,7-dimethylxanthine, theophylline, 1,3-dimethylxanthine, and paraxanthine at lower concentrations. Theophylline and theobromine are present in high concentrations in cacao (*T. cacao*) (Camandola et al. 2019). Methylxanthines (mainly caffeine, theobromine, and theophylline) and tannins are present in high levels in seeds of Guarana (*Paullinia cupana* Kunth, Sapindaceae) (Marques et al. 2019).

The stomach and the small intestine easily absorb consumed caffeine. Caffeine goes through cytochrome P450 enzymes metabolism in the liver to form paraxanthine (Camandola et al. 2019). Caffeine, paraxanthine, and theophylline act as nonselective antagonists to the adenosine receptor (Snyder et al. 1981). Adenosine is an endogenous neuromodulator that affects many functions in the CNS. Adenosine receptors are G protein-coupled receptors expressed in different body tissues and in the brain (Fredholm et al. 2005). In the CNS, adenosine binding to its receptors regulates multiple physiological functions, including synaptic plasticity (De Mendonça and Ribeiro 2001), sleep or wakefulness (Reichert et al. 2016), and sensorimotor integration (Fredholm et al. 2005).

Methylxanthines bind to the brain adenosine-receptor subtypes A1 and A2A (Costenla et al. 2010). Adenosine receptors A1 and A2A affect heart rate, oxygen consumption, and body temperature, and A2A regulates sleep and motor activity (Yang et al. 2009). Methylxanthines are competitive antagonists of A1 and A2A adenosine receptors with affinities at the low micromolar range; theophylline and paraxanthine have higher binding affinity than caffeine, while theobromine is a lower affinity adenosine-receptor antagonist (Snyder et al. 1981).

Interestingly, an increase in synaptic responses in the hippocampal CA2 pyramidal neurons was recorded following oral administration of caffeine in vivo, or direct application of caffeine to hippocampal slices (Simons et al. 2012). Indeed, the CA2 region of the hippocampus highly expresses A1 receptors (Ochiishi et al. 1999). It was suggested that the antagonistic effect on these receptors by caffeine induce and stabilize activity-dependent long-term potentiation strongly associated with learning and memory (Dunwiddie and Masino 2001). However, while the antagonistic activity of the adenosine receptor can clarify most of the effects of methylxanthines on the CNS, additional unknown mechanisms may be active as well (Camandola et al. 2019).

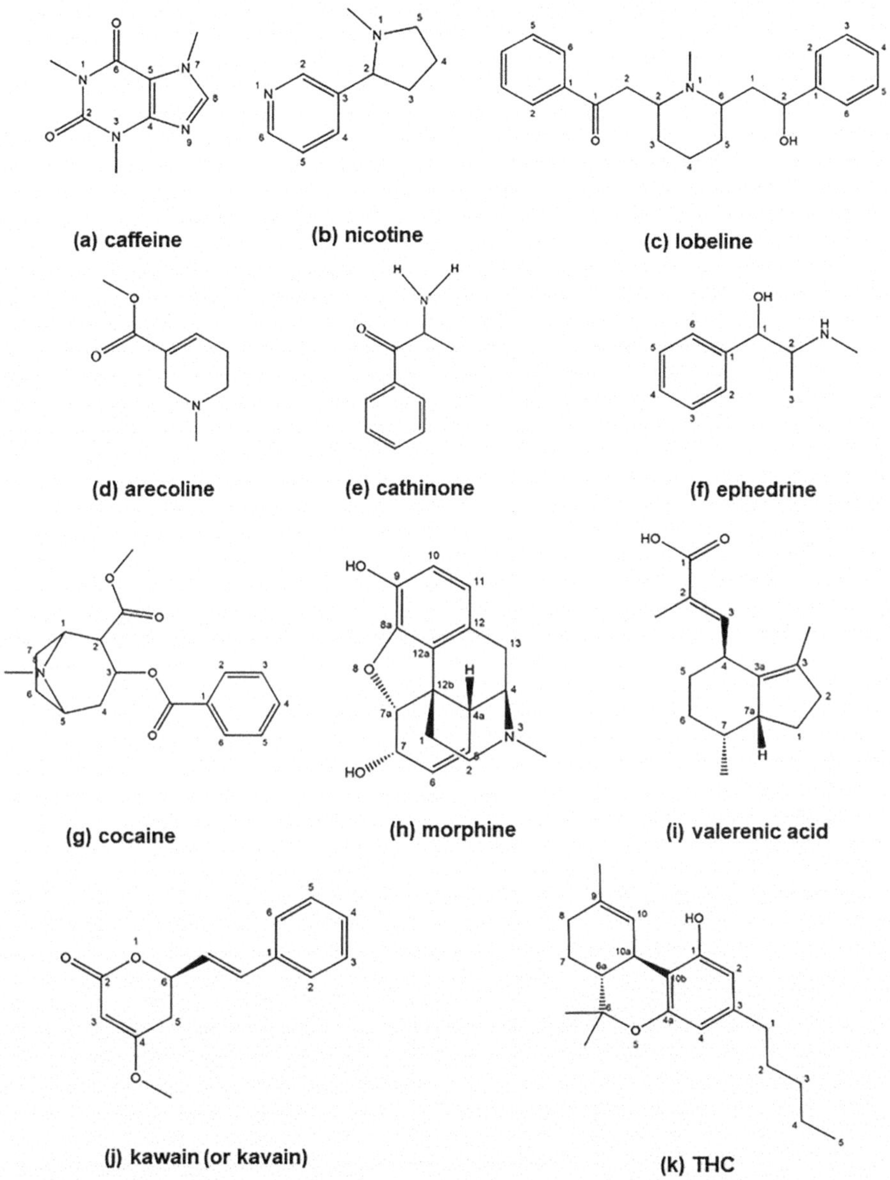


**Fig. 2** (**a–k**) Drawing of molecules produced by plants that bind to receptors present in the human central nervous system (illustrated by Dr. Dvora Namdar using by ChemDraw Professional version 18.2.). Of the main compounds listed: (**a**) caffeine; (**b**) nicotine; (**c**) lobeline; (**d**) arecoline (or arecoline); (**e**) cathinone; (**f**) ephedrine; (**g**) cocaine; (**h**) morphine; (**i**) valerenic acid; (**j**) kawain (or kavain); (**k**) THC

## 2.2 Plants Containing Nicotine and Nicotine-Like Compounds

Another group of plants with stimulating effects is the group containing nicotine. Nicotine (Fig. 2b) is the primary alkaloid in tobacco (*N. tabacum*). It is mostly responsible for the addiction and dependence of chronic relapsing disorder associated with tobacco use (Le Foll and Goldberg 2009). Nicotine binds to the ionotropic nicotinic acetylcholine receptors (nAChRs), where the endogenous agonist is acetylcholine. nAChRs are ligand-gated ion channels members of a cys-loop superfamily whose activity is induced in the micro- to milliseconds range (Albuquerque et al. 2009). These ligand-gated ion channels are formed by the assembly of a diverse family of distinct nAChR subunits around a central hydrophilic pore (Albuquerque et al. 2009). In the CNS, nAChRs primarily mediate fast synaptic transmission, neurotransmitter release, and neuronal excitability. In the peripheral nervous system, they modulate other neurotransmitter functions. They have roles in development and synaptic plasticity and participate in learning, attention, and memory (Dani and Bertrand 2007).

By binding to nAChR, nicotine as an agonist leads to functional response mediated via cation-selective flux through the nAChR ion channel. It was found that chronic exposure to nicotinic drugs causes a gradual decrease in the rate of this ionic response (milliseconds to minutes), leading to a desensitized, closed state of the receptor. Nicotine and its primary metabolite cotinine, the latter a nAChR partial agonist, also upregulate the number of nAChR receptors on the surface of neuronal cells. Yet, nicotine addiction is associated with a series of transitions in different brain parts and neuronal networks, which is characterized by a compulsory need to voluntarily consume the drug due to its reinforcing effects (Fox-Loe et al. 2016; Changeux 2010).

Disruption of nAChRs activity was found to contribute to medical conditions, including epilepsy, Parkinson's disease, schizophrenia, dementia, autism, Alzheimer's disease, and addiction (Dani and Bertrand 2007). nAChRs are dysregulated in schizophrenia leading to positive, negative, and cognitive symptoms, and it was suggested that nicotine intake might be a strategy to alleviate these symptoms by normalizing nAChR deficit. However, nicotine is highly addictive, and this property substantially limits its use as a drug candidate. A viable therapeutic option is to develop nonaddictive chemical ligands that target specific nAChRs in the brain (Cole and Parikh 2019).

Other plant species produce nicotine-like compounds. *L. tupa* leaves produce lobeline (Fig. 2c), lobelanidine, and piperidine alkaloid called (−)-pentylsedinine that act as partial agonists at nAChRs (Paz et al. 2015). While nicotine interacts with both high- and low-affinity nAChRs, lobeline selectively interacts with high-affinity nAChRs and thereby may have a different effect than nicotine on the brain. For example, the application of nicotine, but not of lobeline, reduced dopamine (DA) release in freely moving rats (Lecca et al. 2000).

Interestingly, due to the partial nicotine agonist effect of lobeline at nAChRs, it may be used for the treatment of nicotine addiction; however, there is no reliable evidence from long-term trials to indicate that lobeline can aid smoking cessation (Stead and Hughes 2012). Nevertheless, some studies suggested lobeline to have antidepressant-like effects in vivo by targeting brain nAChRs (Roni and Rahman 2013; Roni and Rahman 2015) and to be potentially effective, at least in vivo in attenuating nicotine withdrawal-induced depression-like behavior (Roni and Rahman 2014).

Another plant producing a nicotine-like molecule is Areca Nut (*A. catechu*). Nuts contain arecoline (Fig. 2d), its primary psychoactive ingredient. Similar to nicotine, arecoline has addictive properties and activity on selected nAChR subtypes, including the two classes of nAChR related to the addictive properties of nicotine (Papke et al. 2015).

Arecoline also binds as a relatively nonselective muscarinic partial agonist, accounting for its many effects on the peripheral and central nervous system. Muscarinic receptors are G-protein-coupled receptors of Class A, rhodopsin-like family with the endogenous ligand agonist being acetylcholine. These receptors mediate the pro-cognitive effects of acetylcholine released within hippocampal and cortical circuits. They are thought to be involved in psychiatric conditions such as schizophrenia and depression and neurodegenerative disease such as Alzheimer's and Parkinson's diseases (Broad et al. 2019; Papke et al. 2015).

## 2.3 *Plants Containing Alkaloids*

Amphetamine-type alkaloids compounds exert psychoactive effects and have a clear addictive potential. Cathinone (Fig. 2e) may be found as an intermediate of a plant biosynthetic pathway in several plant species, including *C. edulis* Forsk., *Ephedra gerardiana* sikkimensis, and *E. sinica*. Cathinone is converted to norephedrine (−)-norephedrine (1R,2S-(−)-norephedrine) and norpseudoephedrine (+)-norpseudoephedrine 1S,2S-(+)-norpseudoephedrine; cathine) in *E. gerardiana* sikkimensis and *C. edulis*, and to pseudoephedrine and ephedrine (Fig. 2f) in *E. sinica* (Patel 2018; Zhang et al. 2018).

Cathinone increased DA levels by inhibiting the reuptake of this neurotransmitter. It increased the release of serotonin (5-HT), although no change in whole-brain levels was found. It also increased the release of noradrenaline (NA) (Patel 2018). Cathinone was shown to bind to α1-adrenoceptors and with a low affinity to the trace-amine associated receptor (Patel 2018). The latter has an important role in the modulation of serotoninergic and dopaminergic activity (Revel et al. 2011).

Ephedra herbal extract was shown to be an activator of transient receptor potential vanilloid 1 (TRPV1), a nonselective ligand-gated cation channel, which plays an essential role in nociception on sensory neurons and thereby has analgesic effects. Nevertheless, ephedrine does not show such effects (Zhang et al. 2018; Hyuga et al. 2016).

But perhaps two of the most effective and well-known alkaloid compounds are cocaine and morphine.

Cocaine (Fig. 2g) is extracted from coca (*Erythroxylum coca*) leaves in a chemical process resulting in cocaine hydrochloride. Cocaine's main activity is inhibition of monoaminergic neurotransmitter transporters, thereby blocking uptake of the monoamines such as DA, 5-HT, and norepinephrine back into the neurons that released these neurotransmitters. As a result, the level of monoamines in the extracellular space is quickly increased. This effect underlies the positively reinforcing effects that cocaine users experience. In turn, it results in a number of severe adverse effects and underlies the development of addiction (Shorter and Kosten 2011).

In addition, evidence suggests that cocaine interacts with the sigma-1 receptor (σ1R), associated with the "hunger" hormone, ghrelin (Aguinaga et al. 2019), and that it induces pathological A2AR-DA D2 receptor—σ1R complexes. These complexes may form a strong and permanent DA D2 receptor brake and result in a long-term memory leading to cocaine addiction (Borroto-Escuela et al. 2018).

Morphine (Fig. 2h) and additional opioid alkaloids, including codeine, papaverine, noscapine, thebaine, oripavine, reticuline, and narceine, are natural derivatives of the opium plant, *P. somniferum* (Labanca et al. 2018). Morphine is considered as the major psychoactive and analgesic opiates, and its usage may lead to opiate dependence. Morphine acts as an agonist of opioid receptors. The opioid system consists of three G protein-coupled receptors, mu, delta, and kappa, which are stimulated by a family of endogenous opioid peptides (Contet et al. 2004).

Morphine activates mainly mu (μ) opioid receptors and thereby mediates both the therapeutic and the adverse activities of this compound. Multiple μ opioid receptors were identified. The variants, once activated by an agonist, act (synergistically in some cases), to block the transmission of signaling from pain-modulating neurons to the spinal cord. Agonists to morphine are in intensive medical use, and cause, in addition to potent analgesia, a number of adverse effects including constipation, sedation, euphoria, physical dependence, respiratory depression, and addiction (Pacifici 2016; Pasternak 1993).

Interestingly, reinforcing properties of alcohol, cannabinoids, and nicotine are sharply diminished in mice lacking the μ opioid receptor, suggesting μ receptors to act as convergent molecular switches for reinforcement of these drugs' effects (Contet et al. 2004).

## 2.4 Plants Producing Lactone and Sesquiterpenoid Compounds

Valerenic acid (Fig. 2i), a sesquiterpenoid constituent of the essential oil of *V. officinalis*, was shown to bind to a specific binding site on γ-Aminobutyric acid type A receptors ($GABA_ARs$), a class of physiologically and therapeutically important ligand-gated ion channels. Valerenic acid binds $GABA_ARs$ with nM affinity in vivo, it acts as a partial agonist and thereby exerts an anxiolytic effect (Benke et al. 2009; Trauner et al. 2008; Yuan et al. 2004). Valerian and valerenic acid were also suggested to act as partial agonists of the 5-$HT_{5a}$ serotonin receptor in vivo (Dietz et al. 2005).

Kavalactones, including kawain (Fig. 2j), dihydrokavain, methysticin, dihydromethysticin, yangonin, and desmethoxyyangonin, are the pharmacologically active constituents in extracts of the pepper plant kava (*P. methysticum*) and were shown to be effective in alleviating anxiety in clinical trials (Sarris et al. 2013). The in vivo molecular target of kavalactones is $GABA_A$Rs: Kavain was found to positively modulate all $GABA_A$Rs receptors regardless of their composition of subunits (Chua et al. 2016).

## 2.5 *Plants Producing Phytocannabinoids*

*C. sativa* L. strains produce around 150 different phytocannabinoids, structurally homogenous class of monoterpenoids indicative of the plant. Phytocannabinoids can be divided into 11 main subfamilies (Hanuš et al. 2016; Berman et al. 2018). The main phytocannabinoids are synthesized in the plant in their acid form (e.g., $\Delta^9$-tetrahydrocannabinolic acid [THCA], cannabidiolic acid [CBDA], cannabichromenic acid [CBCA] from a single cannabigerolic acid (CBGA) precursor (Hanuš et al. 2016). Heat exposure of these acid molecules is needed for decarboxylation, i.e., the release of the acidic residue from the molecules. Then, the decarboxylated molecule of THCA, i.e., $\Delta^9$-tetrahydrocannabinol (THC; Fig. 2k), is the psychoactive cannabinoid (Hanuš et al. 2016). However, both the acid forms of the phytocannabinoids and their decarboxylated forms are active for therapeutic purposes (as described above).

THC, $\Delta^8$-THC its isomer, and, to a minor extent, their degradation derivative cannabinol (CBN), bind to endocannabinoid receptor type 1 ($CB_1$) and type 2 ($CB_2$) (Pertwee et al. 2010), whereas cannabidiol (CBD) is an allosteric modulator of $CB_1$ signaling (Laprairie et al. 2015). The endocannabinoid receptors are part of the endocannabinoid system that consists, in addition to the endocannabinoids receptors, also the endogenous ligands (endocannabinoids), their biosynthetic and catabolic enzymes. The endocannabinoids receptors are composed of two 7-transmembrane-domain and G protein-coupled receptors, and their endogenous endocannabinoid ligands include N-arachidonoylethanolamine (anandamide) and 2-arachidonoylglycerol (2-AG) (Di Marzo and Piscitelli 2015). The endocannabinoid receptors are involved in the regulation of multiple neuronal CNS processes, including activity, proliferation, survival, and migration of neural cells and stem cells in the young and adult brain (Oddi et al. 2020; Morena et al. 2016).

Other plant species were identified to produce "phytocannabinoids" that may bind the CB receptors (Gertsch et al. 2010) among them *P. methysticum*; Yangonin, one of the kavalactones, was suggested to an exhibited affinity for the human recombinant CB1 receptor with a $K_i = 0.72$ μM (Ligresti et al. 2012).

## 3 Conclusions and Prospects

Many plants produce molecules that specifically interact with receptors in the CNS, some with high potency. Many of these plants were substantially used throughout history. However, why would plants produce molecules that interact with human CNS receptors? It might be that some of the phytomolecules were initially utilized in plants for protection against insect pests, herbivores, or other plant species. However, the fecundity of CNS-active phytomolecules in plants could represent a result of domestication. Human-made evolution is at the base of domestication, where human interaction and/or use lead to the evolvement of plant and animal species. During this process, due to human activity based on a conscious choice and agricultural and/or medical knowledge, plant and animal species result in morphological and physiological characteristics distinct from wild ancestors (Pisanti and Bifulco 2019; Kerem et al. 2007). Selective cultivation toward specific beneficial activity was shown for plant species such as legumes and cannabis (Kerem et al. 2007; Small 2015; Clarke and Merlin 2016; Koltai and Namdar 2020). Interestingly, in the case of cannabis, breeding and selection were in favor of strains with aromatic properties defined by their monoterpene assemblages or with high levels of the psychoactive THC molecule (Mudge et al. 2018; Small and Beckstead 1973; Koltai and Namdar 2020). It was found that domestication led to the loss of some phytocannabinoid pathways in some strains and the reallocation of resources between pathways in others. It also led to a loss of biodiversity in modern cannabis strains (Mudge et al. 2018).

Taken together, it is likely that throughout history, humans identified the potentiating activity of plants producing CNS-active phytomolecules. Then, humans domesticated these plants by enrichment and/or breeding and selection to contain a higher amount and possibly wider variety or even higher potency of active phytomolecules, perhaps in some cases on the expanse of biodiversity of other less or non-active-on-human phytomolecules. Thereby, in what might be considered as a positive feedback loop, the use of these plants for ritual, social, or medical purposes was further enhanced.

**Acknowledgments** We thank Dr. Dvora Namdar and Omer Koltai for their drawings and comments on the manuscript.

## References

Aguinaga D, Casanovas M, Rivas-Santisteban R, Reyes-Resina I, Navarro G, Franco R (2019) The sigma-1 receptor as key common factor in cocaine and food-seeking behaviors. J Mol Endocrinol 63(4):R81–R92. https://doi.org/10.1530/JME-19-0138

Albuquerque EX, Pereira EF, Alkondon M, Rogers SW (2009) Mammalian nicotinic acetylcholine receptors: from structure to function. Physiol Rev 89(1):73–120. https://doi.org/10.1152/physrev.00015.2008

Askitopoulou H, Ramoutsaki IA, Konsolaki E (2002) Archaeological evidence on the use of opium in the Minoan world. Int Congr Ser 1242:23–29. https://doi.org/10.1016/S0531-5131(02)00769-0, Elsevier

Bachrach U, Yaniv Z (2016) Green tea and its role in cancer prevention and therapy. In: Tsay et al (eds) Medicinal plants-recent advances in research and development, 1st edn. Springer, Singapore, pp 309–326. https://doi.org/10.1007/978-981-10-1085-9_13

Benke D, Barberis A, Kopp S et al (2009) GABAA receptors as in vivo substrate for the anxiolytic action of valerenic acid, a major constituent of valerian root extracts. Neuropharmacology 56 (1):174–181. https://doi.org/10.1016/j.neuropharm.2008.06.013

Berman P, Futoran K, Lewitus GM et al (2018) A new ESI-LC/MS approach for comprehensive metabolic profiling of phytocannabinoids in Cannabis. Sci Rep 8(1):1–15. https://doi.org/10.1038/s41598-018-32651-4

Borroto-Escuela DO, Wydra K, Filip M, Fuxe K (2018) A2AR-D2R heteroreceptor complexes in cocaine reward and addiction. Trends Pharmacol Sci 39(12):1008–1020. https://doi.org/10.1016/j.tips.2018.10.007

Broad LM, Sanger HE, Mogg AJ et al (2019) Identification and pharmacological profile of SPP1, a potent, functionally selective and brain penetrant agonist at muscarinic M1 receptors. Br J Pharmacol 176(1):110–126. https://doi.org/10.1111/bph.14510

Bruneton J (1995) Pharmacognosy, phytochemistry, medicinal plants. Lavoisier Publishing, Paris

Butrica JL (2002) The medical use of cannabis among the Greeks and Romans. J Cannabis Therapeutics 2(2):51–70. https://doi.org/10.1300/J175v02n02_04

Camandola S, Plick N, Mattson MP (2019) Impact of coffee and cacao purine metabolites on neuroplasticity and neurodegenerative disease. Neurochem Res 44(1):214–227. https://doi.org/10.1007/s11064-018-2492-0

Chan N, Li S, Perez E (2016) Interactions between Chinese nutraceuticals and Western medicines. In: Grumezescu A (ed) Nutraceuticals. Elsevier, Amsterdam, pp 875–882. https://doi.org/10.1016/B978-0-12-802147-7.00061-9

Changeux JP (2010) Nicotine addiction and nicotinic receptors: lessons from genetically modified mice. Nat Rev Neurosci 11(6):389–401. https://doi.org/10.1038/nrn2849

Chua HC, Christensen ET, Hoestgaard-Jensen K et al (2016) Kavain, the major constituent of the anxiolytic kava extract, potentiates GABAA receptors: functional characteristics and molecular mechanism. PLoS One 11(6). https://doi.org/10.1371/journal.pone.0157700

Clarke RC, Merlin MD (2016) Cannabis domestication, breeding history, present-day genetic diversity, and future prospects. CRC Crit Rev Plant Sci 35(5–6):293–327. https://doi.org/10.1080/07352689.2016.1267498

Cole RD, Parikh V (2019) Nicotine dependence in schizophrenia: contributions of nicotinic acetylcholine receptors. In: Preedy V (ed) Neuroscience of nicotine, 1st edn. Academic Press, pp 135–143. https://doi.org/10.1016/B978-0-12-813035-3.00017-4

Contet C, Kieffer BL, Befort K (2004) Mu opioid receptor: a gateway to drug addiction. Curr Opin Neurobiol 14(3):370–378. https://doi.org/10.1016/j.conb.2004.05.005

Costenla AR, Cunha RA, De Mendonça A (2010) Caffeine, adenosine receptors, and synaptic plasticity. J Alzheimers Dis 20(s1):S25–S34. https://doi.org/10.3233/JAD-2010-091384

Dani JA, Bertrand D (2007) Nicotinic acetylcholine receptors and nicotinic cholinergic mechanisms of the central nervous system. Annu Rev Pharmacol Toxicol 47:699–729. https://doi.org/10.1146/annurev.pharmtox.47.120505.105214

Danino D, Amar Z (2013) Little Ethiopia: an ethnopharmacological study of the Ethiopian community in Israel. Int J Ethiopian Stud 4:105–134. https://www.jstor.org/stable/27828907
De Mendonça A, Ribeiro JA (2001) Adenosine and synaptic plasticity. Drug Dev Res 52 (1–2):283–290. https://doi.org/10.1002/ddr.1125
Di Marzo V, Piscitelli F (2015) The endocannabinoid system and its modulation by phytocannabinoids. Neurotherapeutics 12(4):692–698. https://doi.org/10.1007/s13311-015-0374-6
Dietz BM, Mahady GB, Pauli GF, Farnsworth NR (2005) Valerian extract and valerenic acid are partial agonists of the 5-HT5a receptor in vitro. Mol Brain Res 138(2):191–197. https://doi.org/10.1016/j.molbrainres.2005.04.009
Dunwiddie TV, Masino SA (2001) The role and regulation of adenosine in the central nervous system. Annu Rev Neurosci 24(1):31–55. https://doi.org/10.1146/annurev.neuro.24.1.31
Epega AA (2003) Obi Divination. Athelia Henrietta Press, Brooklyn, pp 1–2. ISBN: 978-1890157340
Fern K (2018) Tropical plants database. http://tropical.theferns.info/
Fox-Loe AM, Dwoskin LP, Richards CI (2016) Nicotinic acetylcholine receptors as targets for tobacco cessation therapeutics: cutting-edge methodologies to understand receptor assembly and trafficking. In: Li MD (ed) Nicotinic acetylcholine receptor technologies, 1st edn. Springer, New York, pp 119–132. https://doi.org/10.1007/978-1-4939-3768-4_7
Fredholm BB, Chen JF, Cunha RA, Svenningsson P, Vaugeois JM (2005) Adenosine and brain function. Int Rev Neurobiol 63(1):191–270
Gertsch J, Pertwee RG, Di Marzo V (2010) Phytocannabinoids beyond the Cannabis plant–do they exist? Br J Pharmacol 160(3):523–529. https://doi.org/10.1111/j.1476-5381.2010.00745.x
Getasetegn M (2016) Chemical composition of *Catha edulis* (khat): a review. Phytochem Rev 15 (5):907–920. https://doi.org/10.1007/s11101-015-9435-z
Graham HN (1992) Green tea composition, consumption, and polyphenol chemistry. Prev Med 21 (3):334–350. https://doi.org/10.1016/0091-7435(92)90041-F
Greenwood V (2016). The little-known nut that gave Coca-Cola its name. BBC News—Future. Accessed 23 Dec 2019
Gupta P, Ray C (2004) Epidemiology of betel quid usage. Ann Acad Med Singap 33:31–36
Hamed MS, Gambert S, Bliden KP, Bailon O, Singla A, Antonino MJ, Hamed F, Tantry US, Gurbel PA (2008) Dark chocolate effect on platelet activity, C-reactive protein and lipid profile: a pilot study. South Med J 101(12):1203–1208. https://doi.org/10.1097/smj.0b013e31818859eb
Hanuš LO, Meyer SM, Muñoz E et al (2016) Phytocannabinoids: a unified critical inventory. Nat Prod Rep 33(12):1357–1392. https://doi.org/10.1039/C6NP00074F
Hyuga S, Hyuga M, Oshima N et al (2016) Ephedrine alkaloids-free Ephedra herb extract: a safer alternative to ephedra with comparable analgesic, anticancer, and anti-influenza activities. J Nat Med 70(3):571–583. https://doi.org/10.1007/s11418-016-0979-z
Kerem Z, Lev-Yadun S, Gopher A, Weinberg P, Abbo S (2007) Chickpea domestication in the Neolithic Levant through the nutritional perspective. J Archaeol Sci 34(8):1289–1293. https://doi.org/10.1016/j.jas.2006.10.025
Koltai H, Namdar D (2020) Cannabis phytomolecule 'entourage': from domestication to medical use. Trends Plant Sci 25:976–984. https://doi.org/10.1016/j.tplants.2020.04.007
Krochmal A, Wilken L, Millie C (1970) Lobeline content of Lobelia inflata: structural, environmental and developmental effects. Res Pap NE-178 Upper Darby, PA: US Department of Agriculture, Forest Service, Northeastern Forest Experiment Station 13, p 178
Labanca F, Ovesna J, Milella L (2018) *Papaver somniferum* L. taxonomy, uses and new insight in poppy alkaloid pathways. Phytochem Rev 17(4):853–871. https://doi.org/10.1007/s11101-018-9563-3
Laprairie R, Bagher A, Kelly M, Denovan-Wright E (2015) Cannabidiol is a negative allosteric modulator of the cannabinoid CB1 receptor. Br J Pharmacol 172(20):4790–4805. https://doi.org/10.1111/bph.13250

Le Foll B, Goldberg SR (2009) Effects of nicotine in experimental animals and humans: an update on addictive properties. In: Henningfield et al (eds) Nicotine psychopharmacology, 1st edn. Springer, Berlin, pp 335–367

Lecca D, Shim I, Costa E, Javaid JI (2000) Striatal application of nicotine, but not of lobeline, attenuates dopamine release in freely moving rats. Neuropharmacology 39(1):88–98. https://doi.org/10.1016/S0028-3908(99)00085-4

Ligresti A, Villano R, Allarà M, Ujváry I, Di Marzo V (2012) Kavalactones and the endocannabinoid system: the plant-derived yangonin is a novel CB1 receptor ligand. Pharmacol Res 66(2):163–169. https://doi.org/10.1016/j.phrs.2012.04.003

Marques LLM, Ferreira EDF, de Paula MN, Klein T, de Mello JCP (2019) *Paullinia cupana*: a multipurpose plant–a review. Rev Bras Farmacogn 29(1):77–110. https://doi.org/10.1016/j.bjp.2018.08.007

McPartland JM, Hegman W (2018) Cannabis utilization and diffusion patterns in prehistoric Europe: a critical analysis of archaeological evidence. Veg Hist Archaeobot 27(4):627–634. https://doi.org/10.1007/s00334-017-0646-7

Merlin MD (2003) Archaeological evidence for the tradition of psychoactive plant use in the old world. Econ Bot 57(3):295–323. https://doi.org/10.1663/0013-0001(2003)057[0295:AEFTTO]2.0.CO;2

Morena M, Patel S, Bains JS, Hill MN (2016) Neurobiological interactions between stress and the endocannabinoid system. Neuropsychopharmacology 41(1):80–102. https://doi.org/10.1038/npp.2015.166

Mudge EM, Murch SJ, Brown PN (2018) Chemometric analysis of cannabinoids: chemotaxonomy and domestication syndrome. Sci Rep 8(1):1–9

Musk AW, De Klerk NH (2003) History of tobacco and health. Respirology 8(3):286–290. https://doi.org/10.1038/s41598-018-31120-2

Nandhini S, Narayanan K, Ilango K (2018) *Valeriana officinalis*: a review of its traditional uses, phytochemistry and pharmacology. Asian J Pharm Clin Res 11(1):36–41. https://doi.org/10.22159/ajpcr.2018.v11i1.22588

Ochiishi T, Saitoh Y, Yukawa A et al (1999) High level of adenosine A1 receptor-like immunoreactivity in the CA2/CA3a region of the adult rat hippocampus. Neurosci 93(3):955–967. https://doi.org/10.1016/S0306-4522(99)00179-7

Oddi S, Scipioni L, Maccarrone M (2020) Endocannabinoid system and adult neurogenesis: a focused review. Curr Opin Pharmacol 50:25–32. https://doi.org/10.1016/j.coph.2019.11.002

Odebunmi E, Oluwaniyi O, Awolola G, Adediji O (2009) Proximate and nutritional composition of kola nut (*Cola nitida*), bitter cola (*Garcinia cola*) and alligator pepper (*Afromomum melegueta*). Afr J Biotechnol 8(2):1684–5315

Pacifici GM (2016) Metabolism and pharmacokinetics of morphine in neonates: a review. Clinics 71(8):474–480. https://doi.org/10.6061/clinics/2016(08)11

Papke RL, Horenstein NA, Stokes C (2015) Nicotinic activity of arecoline, the psychoactive element of "betel nuts", suggests a basis for habitual use and anti-inflammatory activity. PLoS One 10(10). https://doi.org/10.1371/journal.pone.0140907

Pasternak GW (1993) Pharmacological mechanisms of opioid analgesics. Clin Neuropharmacol 16(1):1–18. https://doi.org/10.1097/00002826-199302000-00001

Patel NB (2018) Khat—a natural source of cathinone. In: Jolanta BZ (ed) Synthetic cathinones. Springer, Cham, pp 25–40. https://doi.org/10.1007/978-3-319-78707-7_3

Paz C, Becerra J, Silva M et al (2015) (−)-Pentylsedinine, a new alkaloid from the leaves of *Lobelia tupa* with agonist activity at nicotinic acetylcholine receptor. Nat Prod Commun 10(8). https://doi.org/10.1177/1934578X1501000810

Pertwee RG, Howlett A, Abood ME et al (2010) International union of basic and clinical pharmacology. LXXIX. Cannabinoid receptors and their ligands: beyond CB1 and CB2. Pharmacol Rev 62(4):588–631. https://doi.org/10.1124/pr.110.003004

Pilerood SA, Prakash J (2013) Nutritional and medicinal properties of valerian (*Valeriana officinalis*) herb: a review. Int J Food Sci Nutr Diet 1:25–33

Pisanti S, Bifulco M (2019) Medical cannabis: a plurimillennial history of an evergreen. J Cell Physiol 234(6):8342–8351. https://doi.org/10.1002/jcp.27725
Reichert CF, Maire M, Schmidt C, Cajochen C (2016) Sleep-wake regulation and its impact on working memory performance: the role of adenosine. Biology 5(1):11. https://doi.org/10.3390/biology5010011
Revel FG, Moreau JL, Gainetdinov RR et al (2011) TAAR1 activation modulates monoaminergic neurotransmission, preventing hyperdopaminergic and hypoglutamatergic activity. Proc Natl Acad Sci U S A 108(20):8485–8490. https://doi.org/10.1073/pnas.1103029108
Roni MA, Rahman S (2013) Antidepressant-like effects of lobeline in mice: behavioral, neurochemical, and neuroendocrine evidence. Prog Neuro-Psychopharmacol Biol Psychiatry 41:44–51. https://doi.org/10.1016/j.pnpbp.2012.11.011
Roni MA, Rahman S (2014) The effects of lobeline on nicotine withdrawal-induced depression-like behavior in mice. Psychopharmacol 231(15):2989–2998. https://doi.org/10.1007/s00213-014-3472-y
Roni MA, Rahman S (2015) Effects of lobeline and reboxetine, fluoxetine, or bupropion combination on depression-like behaviors in mice. Pharmacol Biochem Behav 139:1–6. https://doi.org/10.1016/j.pbb.2015.10.006
Russo EB (2014) The pharmacological history of cannabis. In: Handbook of Cannabis, pp 23–43
Sanders M (2007) Inhalation therapy: an historical review. Prim Care Respir J 16(2):71–81. https://doi.org/10.3132/pcrj.2007.00017
Sarris J, Stough C, Teschke R et al (2013) Kava for the treatment of generalized anxiety disorder RCT: analysis of adverse reactions, liver function, addiction, and sexual effects. Phytother Res 27(11):1723–1728. https://doi.org/10.1002/ptr.4916
Schimpl FC, da Silva JF, de Carvalho Gonçalves JF, Mazzafera P (2013) Guarana: revisiting a highly caffeinated plant from the Amazon. J Ethnopharmacol 150(1):14–31. https://doi.org/10.1016/j.jep.2013.08.023
Schnepel E (2002) Chocolate: from bean to bar. Gastronomica 2(4):98–100. https://www.jstor.org/stable/10.1525/gfc.2002.2.4.98
Shorter D, Kosten TR (2011) Novel pharmacotherapeutic treatments for cocaine addiction. BMC Med 9(1):119. https://doi.org/10.1186/1741-7015-9-119
Showman AF, Baker JD, Linares C et al (2015) Contemporary Pacific and Western perspectives onawa (*Piper methysticum*) toxicology. Fitoterapia 100:56–67. https://doi.org/10.1016/j.fitote.2014.11.012
Simons SB, Caruana DA, Zhao M, Dudek SM (2012) Caffeine-induced synaptic potentiation in hippocampal CA2 neurons. Nat Neurosci 15(1):23–25. https://doi.org/10.1038/nn.2962
Small E (2015) Evolution and classification of *Cannabis sativa* (marijuana, hemp) in relation to human utilization. Bot Rev 81(3):189–294. https://doi.org/10.1007/s12229-015-9157-3
Small E, Beckstead H (1973) Cannabinoid phenotypes in *Cannabis sativa*. Nature 245 (5421):147–148
Snyder SH, Katims JJ, Annau Z, Bruns RF, Daly JW (1981) Adenosine receptors and behavioral actions of methylxanthines. Proc Natl Acad Sci U S A 78(5):3260–3264. https://doi.org/10.1073/pnas.78.5.3260
Stead LF, Hughes JR (2012) Lobeline for smoking cessation. Cochrane Database Syst Rev 2. https://doi.org/10.1002/14651858.CD000124.pub2
Stevens CJ, Murphy C, Roberts R et al (2016) Between China and South Asia: a middle Asian corridor of crop dispersal and agricultural innovation in the Bronze Age. The Holocene 26 (10):1541–1555. https://doi.org/10.1177/0959683616650268
Strickland S (2002) Anthropological perspectives on use of the areca nut. Addict Biol 7(1):85–97. https://doi.org/10.1080/13556210120091446
Teketay D (1999) History, botany and ecological requirements of coffee. Walia 20:28–50
Thackeray J, Van der Merwe N, Van der Merwe T (2001) Chemical analysis of residues from seventeenth-century clay pipes from Stratford-upon-Avon and environs: research in action. S Afr J Sci 97(1–2):19–21. https://hdl.handle.net/10520/EJC97282

Tomlinson M (2007) Everything and its opposite: kava drinking in Fiji. Anthropol Q 80 (4):1065–1081. https://www.jstor.org/stable/30052773

Touw M (1981) The religious and medicinal uses of Cannabis in China, India and Tibet. J Psychoactive Drugs 13(1):23–34. https://doi.org/10.1080/02791072.1981.10471447

Trauner G, Khom S, Baburin I, Benedek B, Hering S, Kopp B (2008) Modulation of GABAA receptors by valerian extracts is related to the content of valerenic acid. Planta Med 74 (01):19–24

Wood GAR, Lass RA (2001) Cocoa, 4th edn. Wiley-Blackwell Science, p 620

Yang JN, Chen JF, Fredholm BB (2009) Physiological roles of A1 and A2A adenosine receptors in regulating heart rate, body temperature, and locomotion as revealed using knockout mice and caffeine. Am J Physiol Heart Circ Physiol 296(4):H1141–H1149. https://doi.org/10.1152/ajpheart.00754.2008

Yuan CS, Mehendale S, Xiao Y et al (2004) The gamma-aminobutyric acidergic effects of valerian and valerenic acid on rat brainstem neuronal activity. Anesth Analg 98(2):353–358. https://doi.org/10.1213/01.ANE.0000096189.70405.A5

Zhang BM, Wang ZB, Xin P et al (2018) Phytochemistry and pharmacology of genus Ephedra. Chin J Nat Med 16(11):811–828. https://doi.org/10.1016/S1875-5364(18)30123-7

# Reserpine-Induced Depression and Other Neurotoxicity: A Monoaminergic Hypothesis

**Manoj Govindarajulu, Tharanath Shankar, Shriya Patel, Mary Fabbrini, Amulya Manohar, Sindhu Ramesh, Prashanth Boralingaiah, Sreelakshmi Sharma, Randall C. Clark, Jack Deruiter, Timothy Moore, Dinesh Chandra Agrawal, and Muralikrishnan Dhanasekaran**

## Contents



---

Manoj Govindarajulu, Tharanath Shankar, Shriya Patel, and Mary Fabbrini contributed equally to this work.

---

M. Govindarajulu · S. Patel · M. Fabbrini · S. Ramesh · R. C. Clark · J. Deruiter · T. Moore · M. Dhanasekaran (✉)
Department of Drug Discovery and Development, Harrison School of Pharmacy, Auburn University, Auburn, AL, USA
e-mail: dhanamu@auburn.edu

T. Shankar
Department of Internal Medicine, Ramaiah Medical College, and Hospital, Bengaluru, Karnataka, India

A. Manohar
Department of Pediatrics, Ramaiah Medical College and Hospital, Bengaluru, Karnataka, India

P. Boralingaiah · S. Sharma
ESIC Medical College, Bengaluru, Karnataka, India

D. C. Agrawal (✉)
Department of Applied Chemistry, Chaoyang University of Technology, Taichung, Taiwan
e-mail: dcagrawal@gm.cyut.edu.tw

 
D. C. Agrawal, M. Dhanasekaran (eds.), *Medicinal Herbs and Fungi*,
https://doi.org/10.1007/978-981-33-4141-8_12

**Abstract** Depression is a disease that many suffer at some point in their lifetime. Depression can present as an array of symptoms spanning from tiredness to suicidal thoughts. The cause of depression can be either due to internal or external factors, including genetics, stress, and drug use. At a molecular level, depression can alter the synaptic plasticity of the brain and cause atrophic changes in the cortical and hippocampal regions. Some scientists proposed that a large decline of monoamine neurotransmitters induces depression.

Furthermore, the antihypertensive drug reserpine was considered a model for the monoamine depletion hypothesis. After multiple reserpine users began experiencing depression, scientists discovered that it caused a significant decrease in the amount of monoamine neurotransmitters produced. Following this discovery, the monoamine depletion hypothesis came to light. Antidepressants became focused on increasing monoamine neurotransmitter levels as their success grew, so did the support of the hypothesis. Criticisms and counterarguments arose against the monoamine depletion hypothesis, the main argument being the time lag between the intake of antidepressants and the time they begin working. Other counterarguments included why antidepressants worked just as well on patients not diagnosed with depression, like patients with obsessive–compulsive disorder or anxiety. This review summarizes depression and the monoamine depletion hypothesis, with a focus on the drug reserpine and its role in establishing the hypothesis.

**Keywords** Cholinergic neurotransmission · Depression · Dopamine · Monoamine neurotransmission · Reserpine · Serotonergic neurotransmission

## Abbreviations

| | |
|---|---|
| 5-HIAA | 5-Hydroxyindoleacetic acid |
| 5-HT | 5-Hydroxytryptamine |
| BACE-1 | Beta-secretase-1 |
| CNS | Central nervous system |
| CYP | Cytochrome P450 |
| DAT | Dopamine transporters |
| FDA | Food and Drug Administration |
| HPA | Hypothalamic–pituitary axis |
| MAOIs | Monoamine oxidase inhibitors |
| MDD | Major depressive disorder |
| NAc | Nucleus accumbens |
| NDRIs | Norepinephrine–dopamine reuptake inhibitors |
| NE | Norepinephrine |

| | |
|---|---|
| NET | Norepinephrine transporter |
| SNRI | Serotonin norepinephrine reuptake inhibitors |
| SSRI | Selective serotonin reuptake inhibitors |
| TCAs | Tricyclic antidepressants |
| TNF-α | Tumor necrosis factor-alpha |
| VMAT | Vesicular monoamine transporters |
| VTA | Ventral tegmental area |

# 1 Introduction

Depression is a chronic mental illness that has been recognized as a syndrome for over 2000 years (Beck and Alford 2009). Depression can affect thoughts, mood, and physical health. It is typical to note a lack of energy, insomnia, lack of livelihood, and sadness when experiencing depression (Kraus et al. 2019). One of the serious consequences of this mental illness is the increased thought of and risk for suicide. Multiple internal and external factors can induce depression in individuals, including genetics, stress, alcohol and drug use, severe life events, cardiovascular disease, chronic insomnia, and cerebrovascular disease (Nabeshima and Kim 2013).

Depression is almost twice more common in females than males, and females tend to have more symptoms than males (Albert 2015). Also, females tend to have more identifiable symptoms such as loss of appetite and sleep, fatigue, somatic anxiety, and hypochondriasis (Sloan and Sandt 2006). Women might experience more preexisting anxiety, while males may experience more materialized disorders like drug use, alcoholism, and antisocial personality (Sloan and Sandt 2006). Another possible explanation is that women are exposed to more risk factors that could increase their chances of developing a mental illness like depression. Females are at greater risk of sexual abuse, they have to cope with role overload and competing social roles and tend to be more sensitive regarding adverse experiences in childhood and life (Sloan and Sandt 2006). Interestingly, the incidence increases with age in both sexes.

The American Psychiatric Association's *Diagnostic Statistical Manual of Mental Disorders*, Fifth Edition (DSM-5) classifies the depressive disorders as disruptive mood dysregulation disorder, major depressive disorder (including major depressive episode), persistent depressive disorder (dysthymia), premenstrual dysphoric disorder, and depressive disorder due to another medical condition. Almost 15% of the population experiences a major depressive episode at some point in life, and 6–8% of all outpatients in primary care settings satisfy diagnostic criteria for the disorder (Wang et al. 2017). Depression is often undiagnosed and inadequately treated. An estimated 10–15% of cases are secondary to a medical illness or substance abuse. Because depression is a crippling and chronic illness, major depressive disorder (MDD) is considered as one of the greatest medical burdens in the world. In the United States, an estimated 16.1 million adults aged 18 or older had at least one major depressive episode in the year 2015.

Patients may exhibit the signs and symptoms of major depression during grief, although the emphasis is usually on feelings of emptiness and loss, rather than anhedonia and loss of self-esteem, and the duration is generally limited. In some instances, however, the diagnosis of major depression may be warranted even in the context of a significant loss. Twin studies indicate that major depression of early onset is mainly genetic in origin (Dunn et al. 2015). Adverse life events can either cause or aggravate depression, although genetic factors influence the sensitivity of individuals to these stressful events. In most cases, both biologic and psychosocial factors are involved in the precipitation and unfolding of depressive episodes. The most influential stressors appear to involve the death of a relative, assault, or severe marital or relationship problems.

With advances in modern medicine, medical experts now know that depression can be induced by either a reduction or an excessive amount of certain neurotransmitters in the central nervous system. The monoamine neurotransmitters dopamine, norepinephrine, and especially serotonin, are the focused neurotransmitters implicated in depression (Nutt 2008). On a molecular level, researchers have focused on the cortical and hippocampal regions and noted many atrophic changes (Krishnan and Nestler 2010). These atrophic manifestations include reduced neuronal size, decreased glial cell number, local reductions in dendrite densities, and trophic factors (Krishnan and Nestler 2010). These subtle but intense adjustments can affect synaptic plasticity, which in turn affects the neurotransmitters being regulated by these synapses (Duman et al. 2016).

Altered neurotransmission, hypothalamic–pituitary axis (HPA) abnormalities, involved in chronic stress, inflammation, reduced neuroplasticity, and network dysfunction, are some of the possible pathophysiological mechanisms reported for depression (Keller et al. 2017). All these mechanisms are integrally related and interconnected, which may explain the heterogeneity found in depressed patients. Furthermore, psychological factors affect the neurodevelopmental process, predisposing to depression, while biological factors predispose to psychological pathology as well (Hasler 2010). Depression not only can alter synaptic plasticity on the molecular level but also influence the levels of other regulatory factors including inflammatory enzymes, hormones, metabolic levels, and many more. One study discovered that inflammatory enzymes such as tumor necrosis factor-alpha (TNF-α) and interleukin-6 were significantly higher in depressed subjects versus the healthy controls (Lang and Borgwardt 2013). The same study found that increased leptin levels, a hormone commonly found in relation to the gut, are associated with depressed mood and sleep disturbances (Lang and Borgwardt 2013).

# 2 Depression and the Monoamine Depletion Hypothesis

## *2.1 Monoamine Depletion Hypothesis of Depression*

Major depressive disorder (MDD) is a psychiatric brain disorder with symptoms such as impaired memory, attention, and learning (Boku et al. 2018). MDD often causes individuals to have a low mood and feel a sudden disinterest in their lives. In an attempt to find the physical cause underlying the depressive state of mind, the monoamine hypothesis of depression was developed. This hypothesis attempts to correlate neurotransmitter availability and mood in the central nervous system. This hypothesis was first formulated four decades ago and proposed that depletion or deficiency of monoamine neurotransmitters such as serotonin and norepinephrine underlies the basis for depressive manifestations. This formed the basis to explain the mechanism of action of several classes of antidepressant medications, which increase the levels of monoamines in the brain. First, researchers made the connection that people who died by suicide tended to have low levels of monoamines, specifically serotonin (Stockmeier 2003). Researchers then took this information and developed antidepressants drugs that enhanced the concentrations of these monoamines, and these treatments began to see success.

The therapeutic use of antidepressant drugs has significantly increased over the last 20 years, especially in Western countries, and this has stimulated renewed interest in understanding the neurochemical mechanisms by which these drugs ameliorate symptoms of depression (Cipriani et al. 2009). An antidepressant will target the termination process of neurotransmitters to increase the concentration of monoamines in the synaptic cleft, which can be mediated by either blocking the reuptake process of the neurotransmitter or by blocking the enzymes that would normally degrade the neurotransmitter. These approaches have led to different mechanistic classes of antidepressants over the years. The major classifications of antidepressants include monoamine oxidase inhibitors (MAOIs), tricyclics tricyclic antidepressants (TCAs), and selective serotonin reuptake inhibitors (SSRIs), SNRIs, NDRIs, TCAs, NaSSAs, and mixed 5HTs. Each of them has a unique way of increasing monoamines in the brain (Hirschfeld 2000). As success grew with the development of drugs, the support of the monoamine depletion hypothesis began to grow.

## *2.2 Serotonergic System in Depression*

The first evidence that the serotonergic system may play a role in depression came from observations that reserpine, when used to treat hypertension, was capable of precipitating depression. Furthermore, studies showed that reserpine depleted brain serotonin stores (Brodie et al. 1966) and increased the concentration of its metabolites such as 5-hydroxyindoleacetic acid (5-HIAA) in the urine (Sreemantula et al.

2004). From these early studies, it was discovered that reserpine interferes with vesicular storage of serotonin, thereby depleting presynaptic levels of monoamines available for release from the synapse.

Further evidence for the serotonergic theory includes the finding that serotonin metabolites are reduced in patients diagnosed with MDD (Pan et al. 2018), and that serotonin uptake inhibitors (SSRIs) and serotonin–norepinephrine reuptake inhibitors (SNRIs) help increase the amount of serotonin produced in the brain (Machado and Einarson 2010). Also, chronic antidepressant treatment has been consistent with downregulating inhibitory presynaptic autoreceptors (Albert 2012; Taylor et al. 2005). This is helpful because presynaptic autoreceptors inhibit 5-HT release. Downregulating the autoreceptors helps increase the likelihood of 5-HT release, which is linked to the proper antidepressant response (Richelson 2001). Correspondingly, it has been shown that the depletion of tryptophan, an essential amino acid required for the synthesis of 5-HT, prompts depression-like symptoms in patients who previously were treated for their depression via an antidepressant (Stewart et al. 2018). True to the findings, it is evident that increased serotonin levels are required for antidepressant medication effects, although the depletion of serotonin levels alone might not be enough cause to induce depressive symptoms.

## 2.3 *Adrenergic System in Depression*

In addition to serotonin, norepinephrine (NE) has also been implicated in the etiology of major depressive disorders (Moret and Briley 2011). It is critical to note the relationship between NE pathways and specific symptoms associated with depression and its pathophysiology. In support of the role of NE in depression was the finding that NE projections innervate the limbic system, which is involved in the modulation of emotions and cognition (Moret and Briley 2011). Previous studies have discovered increased β-adrenergic receptor binding in the frontal cortex of suicide victims (Stockmeier and Meltzer 1991). Also, postmortem findings of such victim revealed altered density and sensitivity of α2A-adrenoceptors, which modulate NE release, decreased NE, transporter binding, and alteration in peripheral markers of NE (decreased α2-adrenoceptor density in platelets). These alterations in the neurotransmitter function are considered to contribute to the pathophysiology of depression. Genetic studies of NE function in depression have investigated the functional deletion (knockout) of the NE transporter in mice (Fentress et al. 2013). These knockouts have increased extracellular levels of NE because these mice are resistant to the stress-induced depressive-like changes in behavior and brain neurotrophins expression. Human genetic studies have suggested that variations in the gene coding for NE transporter, which alter neurotransmitter release, are related to individual differences in behavior and susceptibility to depression (Lin and Madras 2006). For instance, NET-T182C polymorphism in the NE transporter is linked with increased susceptibility to depression. NE depletion is seen in depressed patients in remission, and discontinued antidepressant medication results in the prompt reappearance of depressive symptoms (Ono et al. 2004). Therefore, antidepressants that enhance NE activity offer a therapeutic benefit in the treatment of symptoms associated with reduced symptoms of depression.

## 2.4 Dopaminergic System in Depression

Dopamine is one of the major neurotransmitters in the central nervous system. Four dopaminergic pathways have been discovered in the brain, including the mesolimbic, mesocortical, nigrostriatal pathway, and tuberoinfundibular dopaminergic pathway (Leggio et al. 2013). Dopaminergic mesocortical and mesolimbic systems are directly related to two crux symptoms of depression—the inability to feel pleasure and loss of motivation (Yadid and Friedman 2008). The mesolimbic dopaminergic system constitutes the nucleus accumbens (NAc) and its dopaminergic input from the ventral tegmental area (VTA). This pathway has been implicated in the pathophysiology and etiology of depression (Nestler and Carlezon Jr 2006). Alterations in the above-mentioned pathways, namely NAc and VTA, predispose to the development of symptoms of depression, notably anhedonia, decreased energy level, and reduced motivation (Nestler and Carlezon Jr 2006). Circuits in the brain associated with dopamine have an act in the regulation of motivation, psychomotor speed, concentration, and the ability to experience pleasure and disturbances. Foundational features of depressive symptoms can be noted in these functions (Dunlop and Nemeroff 2007). The strongest finding supporting a role for dopamine in depression is derived from the observation that concentrations of homovanillic acid, a metabolite of dopamine, in the cerebrospinal fluid of depressed patients are low (Mendels et al. 1972). Specifically, it has been reported that the anhedonic symptoms in depressed subjects are associated with altered mesolimbic dopamine transmission and may be improved by dopaminergic therapies (Mann and Kapur 1995; Cassano et al. 2004). In addition, a decreased density of dopamine transporters (DATs) and increased binding of D2/D3 dopamine receptors have been reported in the central and basal nuclei of the postmortem amygdala of patients with depression as compared to healthy subjects (Pare et al. 1969). Furthermore, multiple studies conducted using neuroimaging have shown elevated levels of striatal D2 receptor and increased temporal cortex D2/3 binding. Important to note, however, some studies were unable to produce images that showed D2 density alterations in depressed patients (Yadid and Friedman 2008; Lehto et al. 2008).

Alongside previous results, patients experiencing depression have shown a sensitivity to upregulated binding of dopamine D2/D3 receptors and a decreased activity of DAT, insinuating possible compensatory adjustments following a shortage of the mesolimbic dopaminergic system in patients with depression (Meyer et al. 2001; Klimek et al. 2002). The nucleus accumbens accepts input from the mesolimbic dopaminergic neurons, most likely via enhancing dopaminergic function in the mesolimbic system by upregulating expression of dopamine D2 and D3 receptors and dopamine release. It is possibly also encompassed in the pathophysiology of depression and mechanisms of some antidepressant drugs (Willner et al. 1991; Ainsworth et al. 1998; Lammers et al. 2000). Decreased dopaminergic neurotransmission within the mesolimbic system has been propositioned in the pathogenesis of the forefront symptoms of severe depression, including anhedonia, deficiency of motivation, and psychomotor obstruction (Leggio et al. 2013). Differing results have testified about the link between the D4 receptor alleles and the symptomatology of depression (Lammers et al. 2000; Serretti et al. 2002).

## 2.5 *Limitations of Monoamine Depletion Hypothesis*

Despite a wealth of supporting scientific evidence, the monoamine hypothesis has shown over time to have limited application as a foundational rationale to explain depression and its symptoms (Hirschfeld 2000). One major limitation of this hypothesis is that there is a delay between the start of the antidepressant treatment and recognized improvement in symptoms (Hinz et al. 2012). This means that even though most antidepressant drugs produce neurochemical and physiological effects in the brain, the clinical effects develop only upon chronic use of the antidepressant. For example, the initial administration of SSRIs results in a presynaptic reuptake blockade to increase 5-HT in the synapse and increase the activity at both postsynaptic and autoreceptors. Autoreceptor activation reduces the rate of firing of the cell as well as the rate of synthesis that releases 5-HT. It is likely that, with chronic SSRI use, clinical effects can develop because these autoreceptors become less sensitive, and inhibition to serotonergic neurons eventually decrease. This mechanism allows the increased 5-HT to have prolonged effects. When enough of these receptors are activated in other parts of the brain, such as the cortex and hippocampus, even more, clinical effects can be seen because there is a higher probability of neuronal excitation in these regions compared to acute SSRI intake.

Another minor limitation of the monoamine hypothesis, yet still important, is that there is not enough evidence that supports the theory that a deficit in NT directly and solely causes a decrease in mood in patients with depression. Although these limitations are significant to say that the monoamine hypothesis is not completely sufficient in explaining MDD, they detract from the current supporting evidence for this hypothesis. Still, they may imply the need for further research on the pathophysiological causes behind MDD and its symptoms.

The monoamine hypothesis also does not address other key issues related to antidepressant drug activity and therapeutic utility. For example, this hypothesis does not explain why antidepressants are effective in patients experiencing other mental illnesses such as obsessive–compulsive disorder, panic disorder, or bulimia (Hirschfeld 2000). It also does not provide a uniform explanation of why drugs of different mechanistic classes have variable efficacy in depressed patients, and even have little effect on some depressed patients (Hirschfeld 2000). Some studies have shown that reducing monoamines in unaffected patients or reducing levels in patients who have recovered does not induce a state of depression or worsen present symptoms, which conflicts the basic tenets of this hypothesis. On the contrary, there have been studies conducted that have reported depletion of monoamines affects some basic emotions, like happy and sad (Harmer et al. 2009). These basic emotions are thought to be the "building blocks" of other emotions, which would support the monoamine depletion hypothesis (Harmer et al. 2009).

Other theories have been advanced in an attempt to characterize the etiology of depression, including the neuroplasticity hypothesis and the neurogenesis hypotheses. Both of these hypotheses focus on stress as a significant factor in the negative feedback loop of the hypothalamic–pituitary–adrenal (HPA) axis that facilitates responses of stress in mammals (Hirschfeld 2000). The two hypotheses are not nearly as simple and common as the monoamine hypothesis of depression for the production of antidepressant drugs. Still, they are better able to explain the latency of clinical effects (Boku et al. 2018). Furthermore, understanding how stress can lead to MDD is important in acknowledging a prior factor in decreased hippocampal volume often seen in depressed patients (Xue et al. 2020). In general, prolonged stress can lead to an increase in stress hormones, such as cortisol. Cortisol is produced by the adrenal glands and is involved in glucose metabolism, and regulation of blood pressure as well as many other physiologic functions. However, increased cortisol for prolonged periods initiates a significant decrease in brain-derived neurotropic factors that would generally support the survival of sounding cells. It is proposed that yet, due to high-stress levels, cell atrophy occurs, causing the onset of depression and suggesting a neurodegenerative cause.

The neuroplasticity hypothesis is focused on decreased hippocampal volume as a result of shortening the dendritic spine. In contrast, the neurogenesis hypothesis states that the decrease in hippocampal volume is caused instead by fewer cell genesis, especially in the dentate gyrus of the hippocampus (Boku et al. 2018). For many years, the monoamine hypothesis of depression has been more common than any other hypothesis in explaining the basic factor leading to depression. Still, over time, the hypothesis has been shown to have its limitations that demonstrate the need for more research about other potential biological causes that lead to depression in humans. Nonetheless, this does not mean there is not sufficient evidence to support a significant contribution by the original monoamine hypothesis of depletion in depression.

# 3 Reserpine

## *3.1 Chemical Composition*

Rauwolfia encompasses diverse phytochemicals, including alcohols, sugars and glycosides, fatty acids, flavonoids, phytosterols, oleoresins, steroids, tannins, and alkaloids. The vital alkaloids located in the plant are indole alkaloids, with more than 50 of those alkaloids having been detected in the plant (Paula et al. 2006). Indole alkaloids are a group of nitrogenous compounds that are taken from the amino acid

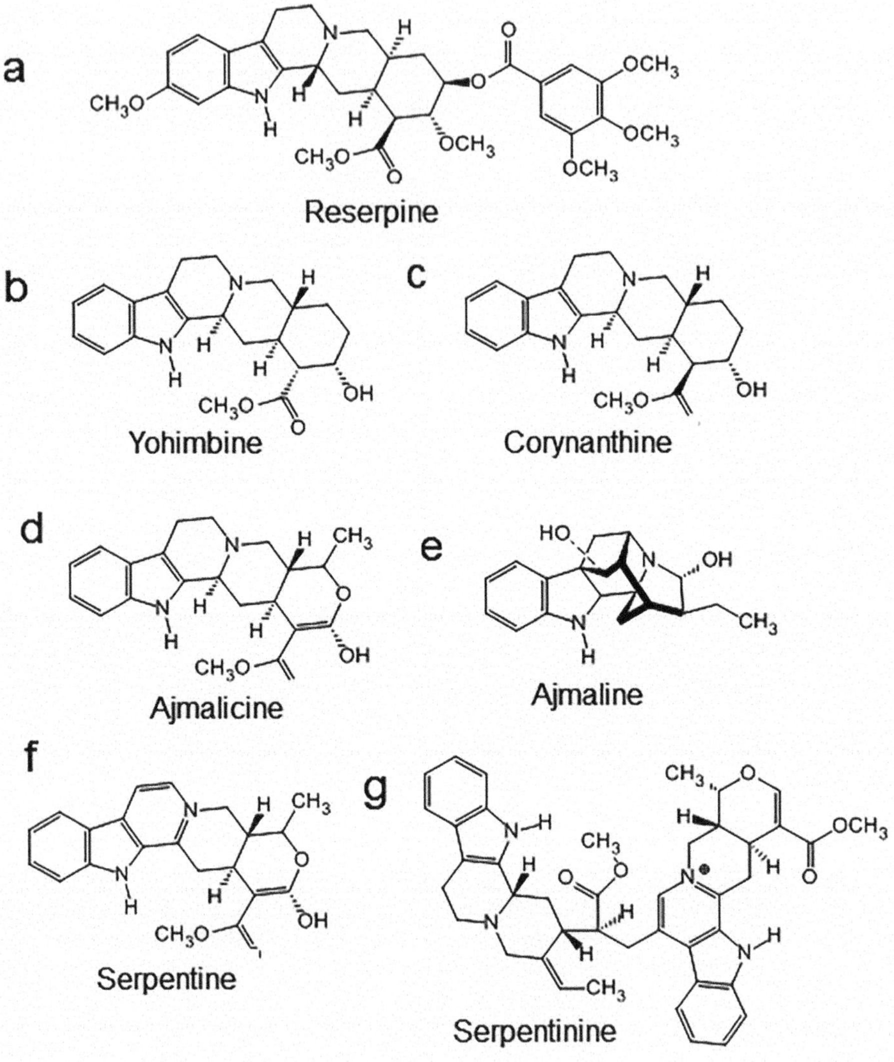


**Fig. 1** (**a–g**) Various indole alkaloids extracted from *Rauwolfia* plant

tryptophan. They share a mutual five and six carbon heterocyclic ring structure with one nitrogen molecule (Ruyter et al. 1991) (Fig. 1a–h). All components of the plant, including the stem and leaves, hold indole alkaloids, but they are located in the highest concentration in the bark of the root (Ruyter et al. 1991). The known indole and indole alkaloids include ajmalidine, ajmaline, ajmalinine, ajmalicine, aricine, canescine, coryanthine, deserpidine, isoajmaline, isoserine, isoserpiline, lankanescine, neoajmaline, papaverine, raubasine, raucaffricine, rauhimbine, rauwolfinine, recanescine, rescinnamine, reserpiline, reserpine, reserpinine, sarpagine, serpentine, serpentinine, thebaine, yohimbine, and yohimbinine

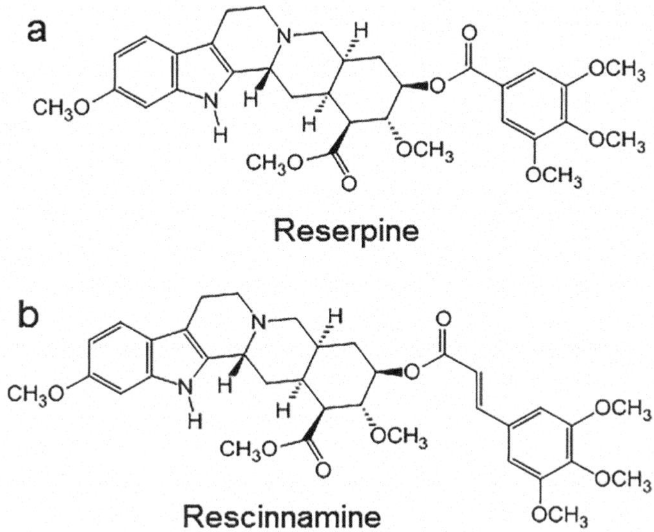


**Fig. 2** Structural comparison of (**a**) reserpine with (**b**) rescinnamine

(Woodson et al. 1957; Panwar and Guru 2011) (Fig. 1a–h). The precise concentration of alkaloids varies. One study discovered the amount of total alkaloids ranged from 0.8% to 1.3% of the dry weight of the plant (Woodson et al. 1957). Another study pinned the total amount of alkaloids, between 0.7% and 3.0% of the root content (Biradar et al. 2016). The largest alkaloid content detected in regenerated roots was 3.3% (Panwar and Guru 2011). Other species in the Rauwolfia genus have been sourced in place of *Rauwolfia serpentina*, including *Rauwolfia vomitoria* and *Rauwolfia caffra* from Africa and *Rauwolfia heterophylla* and *Rauwolfia tetraphylla* from Central and South America. The species of the same genus comprised varying amounts of indole and indole alkaloids, which could be used as potential alternatives to *R. serpentina*.

Reserpine is one of the key alkaloids of the Rauwolfia plant. The reserpine matter is found at its peak in the root and at lower levels in the stems and leaves. It is the most common indole alkaloid in the plant; however, different assays have contested that finding. In one study, the concentration of reserpine in the plant was found to differ from 0.03% to 0.14% of the dry weight of the plant (Kumar et al. 2010). Furthermore, the reserpine content of the root differed from 0.038% to 0.14% in different plants (Kumar et al. 2010). In another study, the reserpine amount was 33 mg of 496 mg of total alkaloids per gram of root (Leete 1960). In another study, reserpine matter was 0.955 mg/g in the Rauwolfia root (Deshmukh et al. 2012). Other alkaloids in the plant that are noted to have biochemical medicinal properties, including canescine, deserpidine, recanescine, and rescinnamine, are shown in Fig. 2a, b.

## 3.2 Pharmacology

Reserpine is the most prevalent alkaloid found in *R. serpentina*. The original paper on reserpine was published in 1931 in the Indian Medical Journal by Sen and Bose. Reserpine's first isolation and use were reported by Robert Wallace Wiggins in 1950 (Jerie 2007). CIBA Labs in Switzerland printed the first full report in 1952, on the chemistry and pharmacology of reserpine (Jerie 2007). Also, in 1952, isolated reserpine was announced as the drug Serpasil used as a medication for hypertension, tachycardia, and thyrotoxicosis. Reserpine is classed as an indole alkaloid. Normally white to yellow, it is in the powered form and becomes darker when exposed to light. Reserpine is fragrance-free, impenetrable in water, slightly soluble in alcohol, and spontaneously soluble in acetic acid. It has a chemical formula of $C_{33}H\ _0N_{209}$, a molecular mass of 609 g, and is said to have a bitter flavor. After being ingested orally and breaking down, the bioavailability of reserpine is said to be between 50% and 70%, although most studies have shown it to be approximately 50%. Absorption is relatively fast, occurring 1–2 h after oral ingestion, although slower absorption of between 2 and 4 h has been reported. Reserpine is distributed throughout the body, including to organs such as the brain, liver, spleen, kidney, and as well to adipose tissue. Traces of reserpine have been noted in the breast milk of pregnant women, and it can cross the placenta and blood–brain barrier. The half-life in the blood has been stated to be between 4 and 5 h, with an elimination half-life between 45 and 168 h in plasma. Its prolonged elimination half-life is believed to be caused by its ability to bind to proteins and red blood cells. Reserpine is metabolized hepatically by cytochrome P450 system; however, it neither induces nor inhibits cytochrome P450 (CYP) activity. Renal elimination accounts for less than 8%. Fecal excretion is the main form of elimination for reserpine (Shore and Giachetti 1978). We confirmed the previous pharmacokinetic and pharmacodynamic properties of reserpine by probing reserpine and tetrabenazine, vesicular monoamine transporter 2 inhibitors like reserpine. We utilized QikProp filter from Schrödinger to predict many aspects of reserpine pharmacokinetics and pharmacodynamics properties such as (SASA, FOSA, FISA, PISA, #metabolites, CNS distribution, QPlog BB, Donor H.B., Accept H.B., logP, % human oral absorption, and Rule of 5) as shown in Table 1. The QikProp filter shows that (SASA) is a solvent-accessible surface area, which predicts the contacting of the surface area of a molecule with a biological solvent. The software illustrates that SASA results occur within the normal range (300–1000) for reserpine and Tetrabenazine, but the reserpine has higher SASA value. Other parameters like hydrophobic components of the SASA (FOSA), hydrophilic components of the SASA (FISA), and $\pi$ (carbon and attached hydrogen) components of the SASA (PISA) values for reserpine and tetrabenazine are all within the acceptable range. Number of hydrogen atoms that could be donated by reserpine to the solvent, or the number of hydrogen atoms that would be accepted by reserpine from the solvent also in the normal range, reserpine has a double number of hydrogen acceptors than tetrabenazine. Predicted solubility of reserpine (QPlogPo/w) and predicted brain/blood partition coefficient (QPlogBB) and other parameters show

**Table 1** QikProp analysis comparing reserpine and tetrabenazine

| Drugs | CNS | SASA | FOSA | FISA | PISA | Donor hub | Accept HB | QPlog Po/w | QPlog BB | Rule of five | #metab | % Human oral absorption (%) |
|---|---|---|---|---|---|---|---|---|---|---|---|---|
| Reserpine | 1 | 903.546 | 676.367 | 75.938 | 151.241 | 1 | 10.7 | 4.914 | −0.352 | 2 | 7 | 77.635 |
| Tetrabenazine | 2 | 578.094 | 504.492 | 23.498 | 50.103 | 0 | 5.5 | 2.85 | 0.512 | 0 | 6 | 100 |

**Table 2** Normal values for QikProp analysis

| Property | Range or recommended values |
|---|---|
| CNS | −2 to +2 |
| SASA | 300.0–1000.0 |
| FOSA | 0.0–750.0 |
| FISA | 7.0–330.0 |
| PISA | 0.0–450.0 |
| donorHB | 0.0–6.0 |
| accptHB | 2.0–20.0 |
| QPlogPo/w | −2.0 to 6.5 |
| QPlogBB | −3.0 to 1.2 |
| Rule of five | Maximum is 4 |
| #metab | 1–8 |
| Percent human oral absorption | >80% is high<br><25% is poor |

that all occur within the acceptable range. Tetrabenazine has a higher percentage of human oral absorption prediction than reserpine. The properties and their corresponding normal recommended values are depicted in Table 2.

## 3.3 *Mechanism of Action*

Reserpine's mechanism of action is well studied and documented. Reserpine attaches to protein receptors called vesicular monoamine transporters (VMATs) in the organelle membranes of specialized secretory vesicles of presynaptic neurons. Reserpine inhibits intracellular neurotransmitters from binding to VMAT proteins and halts secretory vesicles from taking up neurotransmitters (Dahlström et al. 1965). Eventually, the use of reserpine results in neurotransmitter deficiency, so that little or no neurotransmitters are let go from the presynaptic neuron. The result is no or only slight transmission of the nerve impulse occurring in the postsynaptic neuron. Two isoforms of vesicular transport proteins have been branded as VMAT1 and VMAT2. VMAT1 is commonly found in the neuroendocrine cells of the peripheral nervous system, specifically in the chromaffin granules in the adrenal medulla, sympathetic neurons, and platelets (Wimalasena 2011). VMAT2 is largely found in the brain, sympathetic nervous system, mast cells, and cells containing histamine in the gut and pancreas (Yaffe et al. 2018). Reserpine's affinity for VMAT2 is three times greater than its affinity for VMAT1. It binds to specific receptors on VMAT, particularly VMAT2, with nearly irreversible affinity.

### 3.4 *Indications*

The only FDA approved indication of reserpine is to treat hypertension. Its use is limited as a second-line treatment due to the availability of better and favorable antihypertensive medications with better side effects (Cheung and Parmar 2020). The antihypertensive effects are due to decreased cardiac output and decreased peripheral vascular resistance attributed to monoamine depletion. It is usually prescribed alone or in combination with other antihypertensives such as vasodilator or thiazide diuretic (Cheung and Parmar 2020). It is also useful in treating hypertensive emergencies, although its clinical utility has decreased due to the availability of better medications.

Reserpine has historically had other uses. Reserpine was previously utilized for treating schizophrenia and tardive dyskinesia; however, it is not currently in use in clinical practice for these indications (Cheung and Parmar 2020).

## 4 Adverse Drug Reactions and Contraindications

Adverse side effects of reserpine are categorized by organ system and are listed in decreasing order of severity.

- Neurologic: depression and suicidal tendencies, parkinsonian syndrome, and other extrapyramidal tract symptoms, dizziness, headache, paradoxical anxiety, nervousness, nightmares, drowsiness.
- Cardiovascular: arrhythmias (particularly when used concurrently with digitalis or quinidine), postural hypotension, syncope, angina-like symptoms, bradycardia, and edema.
- Respiratory: dyspnea, epistaxis, nasal congestion.
- Gastrointestinal: vomiting, diarrhea, nausea, anorexia, dryness of mouth, hypersecretion.
- Musculoskeletal: muscular aches.
- Genitourinary: pseudolactation, impotence, dysuria, gynecomastia, decreased libido, breast engorgement.
- Metabolic: weight gain.
- Hypersensitive reactions: purpura, rash, pruritus.
- Special senses: deafness, optic atrophy, glaucoma, uveitis, conjunctival injection.
- Contraindications: peptic ulcer disease, Parkinson's disease, depression, ulcerative colitis, cardiac arrhythmias, myocardial infarction, asthma.

## 5 Drug Interaction

In patients taking reserpine, MAO inhibitors should generally be avoided altogether or used with extreme caution. The coexisting use of TCAs with reserpine may diminish the antihypertensive effect of the drug. The use of reserpine and direct-or-indirect acting sympathomimetics contemporaneously should be monitored constantly and closely. Direct-acting amines (epinephrine, isoproterenol, phenylephrine, metaraminol), specifically their actions, may be lengthened when taken at the same time at reserpine. The action of indirect-acting amines (ephedrine, tyramine, amphetamines) is obstructed completely. Because cardiac arrhythmias have occurred with Rauwolfia preparations, reserpine should be used warily with digitalis and quinidine. The parallel use of reserpine with other antihypertensive agents requires meticulous titration of dosage with each occurring agent.

Possible other drug interactions include cardiac glycosides, ephedra, alcohol, antipsychotic drugs, barbiturates, digoxin, diuretics, ephedrine, levodopa, monoamine oxidase inhibitors, propranolol, stimulant drugs, and tricyclic antidepressants. Rauwolfia could affect the following lab tests, tests for corticosteroids, bilirubin, catecholamines, gastric acidity, norepinephrine, prolactin, thyroxine, and vanillylmandelic acid.

## 6 Other Biological Properties

In vitro studies utilizing *Rauvolfia serpentina* have demonstrated antioxidant, antihypertensive, and antibacterial activity (Ahmed et al. 2002). A recent study showed neuroprotective activity against amyloid-beta toxicity in neuronal cell cultures in addition to exhibiting antioxidant properties. Furthermore, this study demonstrated reserpine and ajmalicine to exhibit dual cholinesterase inhibitory activity, inhibition of MAO-B, and BACE-1 (beta-secretase-1). Hence, the authors proposed that reserpine and ajmalicine could act as a disease-modifying therapeutic compound in Alzheimer's disease (Kashyap et al. 2020). Similarly, several in vivo studies have demonstrated different pharmacological properties of *R. serpentina*, which have been summarized in Table 3.

## 7 Conclusion

Depression is an illness that can affect anyone at any time during their life. Through modern medical advances, it can be noted that decreased production of the monoamine neurotransmitters plays a major role in the molecular changes linked to depression. This discovery laid the groundwork for the monoamine depletion hypothesis, which states that unstable levels of the monoamine neurotransmitters,

**Table 3** In vivo pharmacological activity of *Rauwolfia serpentine*

| Biological activity | Part used | Extract | Method | Animal | Dose | References |
|---|---|---|---|---|---|---|
| Hypolipidemic | Roots | Powder | – | Rabbit | 30 mg/kg | Qureshi and Udani (2009) |
| Hepatoprotective | Rhizome | Aqueous ethanolic | Paracetamol-induced hepatic damage | Albino rats | 425 mg/kg | Gupta (2006) |
| Hyperglycemic | Roots | Methanolic | Alloxan-induced diabetic mice | Mice | 60 mg/kg | Azmi and Qureshi (2013) |
| Antidiarrheal | Leaves | Methanolic | Castor oil-induced diarrhea | Mice | 100–400 mg/kg | Ezeigbo et al. (2012) |
| Antivenom | Whole plants | Ethanolic | – | Humans | 0.14 mg | Rajasree et al. (2013) |

dopamine, norepinephrine, and serotonin, were the cause of depression in patients. A major contributor to this theory was the hypertensive drug reserpine. Extracted from the *Rauwolfia* plant, reserpine showed great advances in lowering diastolic and systolic blood pressure and proved to have a multitude of other beneficial uses. After many patients taking reserpine began to experience depression, the mechanism of reserpine was revealed to decrease monoamine neurotransmitters. Building off the hypothesis, antidepressants were reevaluated and recreated to address the lack of the monoamine neurotransmitters and showed great progress. As the monoamine depletion hypothesis grew, soon arguments arose against it. The biggest argument to the hypothesis was the time lag between antidepressant ingestion and noticed symptom improvement. While antidepressants still rely heavily on increasing the levels of the monoamine neurotransmitters, research is still being conducted on the exact cause of "major depressive disorder" in patients.

## References

Ahmed S, Amin M, Anjum A, Haque M (2002) In vitro antibacterial activity of Rauvolfia serpentina and its tissue culture. Nigerian J Nat Prod Med 6(1):45–49. https://doi.org/10.4314/njnpm.v6i1.11693

Ainsworth K, Smith SE, Zetterström TS, Pei Q, Franklin M, Sharp T (1998) Effect of antidepressant drugs on dopamine D1 and D2 receptor expression and dopamine release in the nucleus accumbens of the rat. Psychopharmacology (Berl) 140(4):470–477. https://doi.org/10.1007/s002130050791

Albert PR (2012) Transcriptional regulation of the 5-HT1A receptor: implications for mental illness. Philos Trans R Soc Lond Ser B Biol Sci 367(1601):2402–2415. https://doi.org/10.1098/rstb.2011.0376

Albert PR (2015) Why is depression more prevalent in women? J Psychiatry Neurosci 40 (4):219–221. https://doi.org/10.1503/jpn.150205
Azmi MB, Qureshi SA (2013) Rauwolfia serpentina ameliorates hyperglycemic, haematinic and antioxidant status in alloxan-induced diabetic mice. J Appl Pharma Sci 3(7):136. https://doi.org/10.7324/JAPS.2013.3726
Beck AT, Alford BA (2009) The definition of depression. In: Beck AT, Alford BA (eds) Depression: causes and treatment, 2nd edn. University of Pennsylvania Press, Philadelphia, pp 3–11
Biradar N, Hazar I, Chandy V (2016) Current insight to the uses of *Rauwolfia*: a review. Res Rev A J Pharmacog 3(3):1–4. https://doi.org/10.12300/rrjopc.v3i3.433
Boku S, Nakagawa S, Toda H, Hishimoto A (2018) Neural basis of major depressive disorder: beyond monoamine hypothesis. Psychiatry Clin Neurosci 72(1):3–12. https://doi.org/10.1111/pcn.12604
Brodie BB, Comer MS, Costa E, Dlabac A (1966) The role of brain serotonin in the mechanism of the central action of reserpine. J Pharmacol Exp Ther 152(2):340–249
Cassano P, Lattanzi L, Soldani F et al (2004) Pramipexole in treatment-resistant depression: an extended follow-up. Depress Anxiety 20(3):131–138. https://doi.org/10.1002/da.20038
Cheung M, Parmar M (2020) Reserpine. In: StatPearls [Internet]. StatPearls Publishing, Treasure Island, FL
Cipriani A, Furukawa TA, Salanti G et al (2009) Comparative efficacy and acceptability of 12 new-generation antidepressants: a multiple-treatments meta-analysis. Lancet 373 (9665):746–758. https://doi.org/10.1016/s0140-6736(09)60046-5
Dahlström A, Fuxe K, Hillarp N (1965) Site of action of reserpine. Acta Pharmacol Toxicol 22 (3):277–292. https://doi.org/10.1111/j.1600-0773.1965.tb01823.x
Deshmukh SR, Ashrit DS, Patil BA (2012) Extraction and evaluation of indole alkaloids from *Rauwolfia serpentina* for their antimicrobial and antiproliferative activities. Int J Pharm Pharm Sci 4(Suppl 5):329–334
Duman RS, Aghajanian GK, Sanacora G, Krystal JH (2016) Synaptic plasticity and depression: new insights from stress and rapid-acting antidepressants. Nat Med 22(3):238–249. https://doi.org/10.1038/nm.4050
Dunlop BW, Nemeroff CB (2007) The role of dopamine in the pathophysiology of depression. Arch Gen Psychiatry 64(3):327–337. https://doi.org/10.1001/archpsyc.64.3.327
Dunn EC, Brown RC, Dai Y, Rosand J, Nugent NR, Amstadter AB, Smoller JW (2015) Genetic determinants of depression: recent findings and future directions. Harv Rev Psychiatry 23 (1):1–18. https://doi.org/10.1097/HRP.0000000000000054
Ezeigbo II, Ezeja MI, Madubuike KG, Ifenkwe DC, Ukweni IA, Udeh NE, Akomas SC (2012) Antidiarrhoeal activity of leaf methanolic extract of *Rauwolfia serpentina*. Asian Pac J Trop Biomed 2(6):430–432. https://doi.org/10.1016/S2221-1691(12)60070-7
Fentress HM, Klar R, Krueger JJ, Sabb T, Redmon SN, Wallace NM, Shirey-Rice JK, Hahn MK (2013) Norepinephrine transporter heterozygous knockout mice exhibit altered transport and behavior. Genes Brain Behav 12(8):749–759. https://doi.org/10.1111/gbb.12084
Gupta AK (2006) Hepatoprotective activity of Rauwolfia serpentina rhizome in paracetamol intoxicated rats. J Pharmacol Toxicol 1:82–88. https://doi.org/10.3923/jpt.2006.82.88
Harmer CJ, Goodwin GM, Cowen PJ (2009) Why do antidepressants take so long to work? A cognitive neuropsychological model of antidepressant drug action. Br J Psychiatry 195 (2):102–108. https://doi.org/10.1192/bjp.bp.108.051193
Hasler G (2010) Pathophysiology of depression: do we have any solid evidence of interest to clinicians? World Psychiatry 9(3):155–161. https://doi.org/10.1002/j.2051-5545.2010.tb00298.x
Hinz M, Stein A, Uncini T (2012) The discrediting of the monoamine hypothesis. Int J Gen Med 5:135–142. https://doi.org/10.2147/IJGM.S27824
Hirschfeld RM (2000) History and evolution of the monoamine hypothesis of depression. J Clin Psychiatry 61(Suppl 6):4–6
Jerie P (2007) Milestones of cardiovascular therapy. IV. Reserpine. Cas Lek Cesk 146(7):573–577

Kashyap P, Kalaiselvan V, Kumar R, Kumar S (2020) Ajmalicine and reserpine: indole alkaloids as multi-target directed ligands towards factors implicated in Alzheimer's disease. Molecules 25 (7):1609. https://doi.org/10.3390/molecules25071609

Keller J, Gomez R, Williams G et al (2017) HPA axis in major depression: cortisol, clinical symptomatology and genetic variation predict cognition. Mol Psychiatry 22(4):527–536. https://doi.org/10.1038/mp.2016.120

Klimek V, Schenck JE, Han H, Stockmeier CA, Ordway GA (2002) Dopaminergic abnormalities in amygdaloid nuclei in major depression: a postmortem study. Biol Psychiatry 52(7):740–748. https://doi.org/10.1016/s0006-3223(02)01383-5

Kraus C, Kadriu B, Lanzenberger R, Zarate CA Jr, Kasper S (2019) Prognosis and improved outcomes in major depression: a review. Transl Psychiatry 9(1):127–127. https://doi.org/10.1038/s41398-019-0460-3

Krishnan V, Nestler EJ (2010) Linking molecules to mood: new insight into the biology of depression. Am J Psychiatry 167(11):1305–1320. https://doi.org/10.1176/appi.ajp.2009.10030434

Kumar VH, Shashidhara S, Rajendra C (2010) Reserpine content of *Rauwolfia serpentina* in response to geographical variation. Int J Pharma BioSci 1(4):1–6

Lammers CH, Diaz J, Schwartz JC, Sokoloff P (2000) Dopamine D3 receptor gene expression in the shell of nucleus accumbens is increased by chronic antidepressant treatment. Mol Psychiatry 5(3):229–229. https://doi.org/10.1038/sj.mp.4000756

Lang UE, Borgwardt S (2013) Molecular mechanisms of depression: perspectives on new treatment strategies. Cell Physiol Biochem 31(6):761–777. https://doi.org/10.1159/000350094

Leete E (1960) The biogenesis of the Rauwolfia alkaloids. I. The incorporation of tryptophan into ajmaline. J Am Chem Soc 82:6338–6342. https://doi.org/10.1021/ja01509a037

Leggio GM, Salomone S, Bucolo C et al (2013) Dopamine D(3) receptor as a new pharmacological target for the treatment of depression. Eur J Pharmacol 719(1–3):25–33. https://doi.org/10.1016/j.ejphar.2013.07.022

Lehto SM, Kuikka J, Tolmunen T et al (2008) Temporal cortex dopamine D2/3 receptor binding in major depression. Psychiatry Clin Neurosci 62(3):345–348. https://doi.org/10.1111/j.1440-1819.2008.01814.x

Lin Z, Madras BK (2006) Human genetics and pharmacology of neurotransmitter transporters. Handb Exp Pharmacol 175:327–371. https://doi.org/10.1007/3-540-29784-7_16

Machado M, Einarson T (2010) Comparison of SSRIs and SNRIs in major depressive disorder: a meta-analysis of head-to-head randomized clinical trials. J Clin Pharm Ther 35(2):177–188. https://doi.org/10.1111/j.1365-2710.2009.01050.x

Mann JJ, Kapur S (1995) A dopaminergic hypothesis of major depression. Clin Neuropharmacol 18:S57–S65

Mendels J, Frazer A, Fitzgerald RG, Ramsey TA, Stokes JW (1972) Biogenic amine metabolites in cerebrospinal fluid of depressed and manic patients. Science 175(4028):1380. https://doi.org/10.1126/science.175.4028.1380

Meyer JH, Krüger S, Wilson AA et al (2001) Lower dopamine transporter binding potential in striatum during depression. Neuroreport 12(18):4121–4125. https://doi.org/10.1097/00001756-200112210-00052

Moret C, Briley M (2011) The importance of norepinephrine in depression. Neuropsychiatr Dis Treat 7(Suppl 1):9–13. https://doi.org/10.2147/NDT.S19619

Nabeshima T, Kim H-C (2013) Involvement of genetic and environmental factors in the onset of depression. Exp Neurobiol 22(4):235–243. https://doi.org/10.5607/en.2013.22.4.235

Nestler EJ, Carlezon WA Jr (2006) The mesolimbic dopamine reward circuit in depression. Biol Psychiatry 59(12):1151–1159. https://doi.org/10.1016/j.biopsych.2005.09.018

Nutt DJ (2008) Relationship of neurotransmitters to the symptoms of major depressive disorder. J Clin Psychiatry 69(Suppl E1):4–7

Ono H, Shirakawa O, Nushida H, Ueno Y, Maeda K (2004) Association between catechol-O-methyltransferase functional polymorphism and male suicide completers. Neuropsychopharmacology 29(7):1374–1377. https://doi.org/10.1038/sj.npp.1300470

Pan JX, Xia JJ, Deng FL et al (2018) Diagnosis of major depressive disorder based on changes in multiple plasma neurotransmitters: a targeted metabolomics study. Transl Psychiatry 8 (1):130–130. https://doi.org/10.1038/s41398-018-0183-x

Panwar G, Guru S (2011) Alkaloid profiling and estimation of reserpine in Rauwolfia serpentina plant by TLC, HP-TLC and HPLC. Asian J Plant Sci 10(8):393–400. https://doi.org/10.3923/ajps.2011.393.400

Pare CM, Yeung DP, Price K, Stacey RS (1969) 5-hydroxytryptamine, noradrenaline, and dopamine in brainstem, hypothalamus, and caudate nucleus of controls and of patients committing suicide by coal-gas poisoning. Lancet 2(7612):133–135. https://doi.org/10.1016/s0140-6736(69)92442-8

Paula CR, Krebs VLJ, Auler ME et al (2006) Nosocomial infection in newborns by Pichia anomala in a Brazilian intensive care unit. Medical Mycol 44(5):479–484. https://doi.org/10.1080/13693780600561809

Qureshi SA, Udani SK (2009) Hypolipidaemic activity of Rauwolfia serpentina Benth. Pakistan J Nutr 8(7):1103–1106. https://doi.org/10.3923/pjn.2009.1103.1106

Rajasree P, Singh R, Sankar C (2013) Anti-venom activity of ethanolic extract of Rauwolfia serpentine against Naja naja (cobra) venom. Int J Drug Dis Herb Res 3:521–524

Richelson E (2001) Pharmacology of antidepressants. In: Mayo Clinic proceedings, vol 5. Elsevier, pp 511–527. https://doi.org/10.4065/76.5.511

Ruyter CM, Akram M, Illahi I, Stöckigt J (1991) Investigation of the alkaloid content of Rauwolfia serpentina roots from regenerated plants. Planta Med 57(04):328–330. https://doi.org/10.1055/s-2006-960109

Serretti A, Cristina S, Lilli R, Cusin C, Lattuada E, Lorenzi C, Corradi B, Grieco G, Costa A, Santorelli F (2002) Family-based association study of 5-HTTLPR, TPH, MAO-A, and DRD4 polymorphisms in mood disorders. Am J Med Genetics 114(4):361–369. https://doi.org/10.1002/ajmg.10356

Shore PA, Giachetti A (1978) Reserpine: basic and clinical pharmacology. In: Iversen L, Iversen S, Snyder S (eds) Handbook of psychopharmacology. Springer, New York, pp 197–219. https://doi.org/10.1007/978-1-4613-4042-3_6

Sloan DM, Sandt AR (2006) Gender differences in depression. Womens Health 2(3):425–434. https://doi.org/10.2217/17455057.2.3.425

Sreemantula S, Boini KM, Nammi S (2004) Reserpine methonitrate, a novel quaternary analogue of reserpine augments urinary excretion of VMA and 5-HIAA without affecting HVA in rats. BMC Pharmacol 4:30–30. https://doi.org/10.1186/1471-2210-4-30

Stewart RM, Hood SD, Rao P et al (2018) Using acute tryptophan depletion to investigate predictors of treatment response in adolescents with major depressive disorder: study protocol for a randomised controlled trial. Trials 19(1):434–434. https://doi.org/10.1186/s13063-018-2791-4

Stockmeier CA (2003) Involvement of serotonin in depression: evidence from postmortem and imaging studies of serotonin receptors and the serotonin transporter. J Psychiatr Res 37 (5):357–373. https://doi.org/10.1016/s0022-3956(03)00050-5

Stockmeier CA, Meltzer HY (1991) β-Adrenergic receptor binding in frontal cortex of suicide victims. Biol Psychiatry 29(2):183–191. https://doi.org/10.1016/0006-3223(91)90047-P

Taylor C, Fricker AD, Devi LA, Gomes I (2005) Mechanisms of action of antidepressants: from neurotransmitter systems to signaling pathways. Cell Signal 17(5):549–557. https://doi.org/10.1016/j.cellsig.2004.12.007

Wang J, Wu X, Lai W et al (2017) Prevalence of depression and depressive symptoms among outpatients: a systematic review and meta-analysis. BMJ Open 7(8):e017173–e017173. https://doi.org/10.1136/bmjopen-2017-017173

Willner P, Muscat R, Phillips G (1991) The role of dopamine in rewarded behavior: ability, insight, drive or incentive? Polish J Pharmacol Pharmacy 43(4):291–300

Wimalasena K (2011) Vesicular monoamine transporters: structure-function, pharmacology, and medicinal chemistry. Med Res Rev 31(4):483–519. https://doi.org/10.1002/med.20187

Woodson RE, Youngken HW, Schlittler E, Schneider JA (1957) Rauwolfia: botany, pharmacognosy, chemistry, and pharmacology. Econ Bot 11:183–184. https://doi.org/10.1007/BF02985308

Xue SS, Zhou CH, Xue F et al (2020) The impact of repetitive transcranial magnetic stimulation and fluoxetine on the brain lipidome in a rat model of chronic unpredictable stress. Progress Neuro-Psychopharmacol Biol Psychiatry 102:109946. https://doi.org/10.1016/j.pnpbp.2020.109946

Yadid G, Friedman A (2008) Dynamics of the dopaminergic system as a key component to the understanding of depression. Progress Brain Res 172:265–286. https://doi.org/10.1016/S0079-6123(08)00913-8

Yaffe D, Forrest LR, Schuldiner S (2018) The ins and outs of vesicular monoamine transporters. J Gen Physiol 150(5):671–682. https://doi.org/10.1085/jgp.201711980

# Traditional Medicinal Plants of Sri Lanka and Their Derivatives of Benefit to the Nervous System

**Pathirage Kamal Perera, Adrian Cuda Banda Meedeniya, and Nupe Hewage Akila Chamikara**

## Contents



**Abstract** Nervous system disorders are increasing in global prevalence. The prevention and treatment of nervous system disorders by modern medicines are insufficient, facilitating the rapid growth and use of traditional medicines. Plant-derived herbal preparations are used for the well-being and treatment of nervous system disorders by a large proportion of the Sri Lankan population. This ethnopharmacology may be common to the Southeast Asian region, as in Ayurvedic treatments, or endemic to Sri Lanka. While some herbs of benefits to the nervous system are currently under scientific investigation (e.g., *Centella asiatica*, *Bacopa*

P. K. Perera (✉)
Institute of Indigenous Medicine, University of Colombo, Colombo, Sri Lanka
e-mail: kamalperera@iim.cmb.ac.lk

A. C. B. Meedeniya
Griffith Health, Griffith University, Brisbane, Australia

N. H. A. Chamikara
Management and Science University, Shah Alam, Malaysia

 
D. C. Agrawal, M. Dhanasekaran (eds.), *Medicinal Herbs and Fungi*,
https://doi.org/10.1007/978-981-33-4141-8_13

*monnieri*, *Curcuma longa*, *Withania somnifera*), others are less well known (e.g., *Celastrus paniculatus*, *Evolvulus alsinoides*, *Acorus calamus*, *Sphaeranthus indicus*). Notably, many bioactive components beneficial to the nervous system have been described from "single-herb" preparations. Scientific data on the mechanisms of action of these phytochemicals from herbal plants are available. Further poly-herbal preparations are frequently used in the treatment of neurological disorders or for enhancing nervous system function. The efficacy of these treatments has been demonstrated over the centuries. The compositions of these preparations are discussed in terms of concurrent and synergistic actions, the negation of toxicity, and other features as sometimes described in their formulation. However, many of these preparations require scientific elucidation, where scientific data to determining active ingredients and mechanisms of action are needed. This, together with the growing interest in preventive health, has led to an interest in the health benefits of foods and traditional treatments of nervous system disorders.

**Keywords** Neurodegeneration · Neuroprotection · Neurotoxicity · Phytochemicals · Secondary metabolites · Traditional medicines

## Abbreviations

| | |
|---|---|
| 6-OHDA | 6-Hydroxydopamine |
| BC | Before Christ |
| CBD | Cannabidiol |
| CNS | Central nervous system |
| GABA | Gamma-aminobutyric acid |
| MPTP | 1-Methyl-4-phenyl-1,2,3,6-tetrahydropyridine |
| PD | Parkinson's disease |
| PNS | Peripheral nervous system |
| ROS | Reactive oxygen species |
| TCM | Traditional Chinese medicine |

## 1 Introduction

Plants are mostly sessile autotrophs, positioned as the foundation biotic component in ecosystems. Plants have common phytochemicals comprising the primary metabolites, supporting fundamental life functions (Hussein and El-Anssary 2018). These primary metabolites in plants, constituents of both anabolic and catabolic pathways (photosynthesis and respiration), support all life forms within an ecosystem (Griffin and Seemann 1996). The spectrum of subtle structural and physiological adaptations of plants in response to the biotic and abiotic components of habitats is driven mainly by secondary metabolites (Bennett and Wallsgrove 1994; Hartmann 2008; Hussein and El-Anssary 2018; Wink 2018).

Secondary metabolites are derived from primary metabolites (e.g., carbohydrates, amino acids, and lipids) (Bennett and Wallsgrove 1994; Hussein and El-Anssary 2018). Based on their biosynthetic pathways, these are classified as terpenoids, flavonoids, and other phenolic/polyphenolic compounds and alkaloids (nitrogen-containing or sulfur-containing) compounds (Fig. 1). These phytochemicals may be restricted to a particular taxonomic group genus, species, or family, and their biosynthesis may be highly dynamic in occurrence, being modulated by both biotic and abiotic factors. Plants have evolved a vast array of phytochemicals as secondary metabolites, estimated at over 200,000 compounds (Fiehn 2002; Hartmann 2008; Kennedy 2014), which assist in regulating their interactions with all plants and animal species in an ecosystem (Hartmann 2008; Wink 2018).

## *1.1 Phytochemicals and the Nervous System*

The acute effects of phytochemicals on the biology of animals often occur via their action on the nervous system (Bennett and Wallsgrove 1994; Hartmann 2008). An array of "neuroactive" phytochemicals predominantly regulates plant–insect interactions. Most of these neuroactive substances also impact the vertebrate nervous system as insects have similar neurotransmitters to vertebrates, e.g., acetylcholine, GABA, glutamate, histamine, tyramine, dopamine, nitric oxide, and serotonin (octopamine) (Homberg 2002; Son et al. 2008; Wink 2018). Neuroactive phytochemicals may influence the biosynthesis of neurotransmitters, release kinetics, act as agonists and antagonists, and actin postsynaptic modulation (Son et al. 2008; Venkatesan et al. 2015; Carbonel and Cecyn 2019). They may act less directly via increased synaptogenesis, improved vascularization, neurogenesis, or through reduction of oxidative stress and inflammation (Son et al. 2008; Murugaiyah and Mattson 2015; Venkatesan et al. 2015; Carbonel and Cecyn 2019).

Humans have more recently utilized phytochemicals that modulate nervous system function, e.g., the tropane alkaloids, include atropine and scopolamine, the anticholinergic drugs, and the recreational drug cocaine. These alkaloids containing a tropane ring comprise over 200 compounds that have evolved as a defense against herbivorous insects (Jirschitzka et al. 2012; Jirschitzka et al. 2013), i.e., phytochemicals such as cocaine, which cause the dysregulation of the nervous system, is primarily to kill herbivores (Jirschitzka et al. 2013; Wink 2018). Humans, however, have adopted these neuroactive phytochemicals in curative and in other cultural activities.

Human use of plants to influence brain function is documented from early history, with evidence of the use of neuroactive phytochemicals being present from the time of early human evolution dating back 35,000 years (Wiart 2006). Medicinal plants with neuroactive phytochemicals play a central ethnopharmacological role, e.g., *Papaver somniferum* (opium, Papaveraceae), *Cannabis sativa* L. (cannabis, Cannabinaceae), and *Myristica fragrans* (nutmeg), and plants in the family Erythroxylaceae (Coca). *C. sativa* was used extensively by the Chinese and Indians from 3500 years ago to induce narcosis. It was relatively recently that it

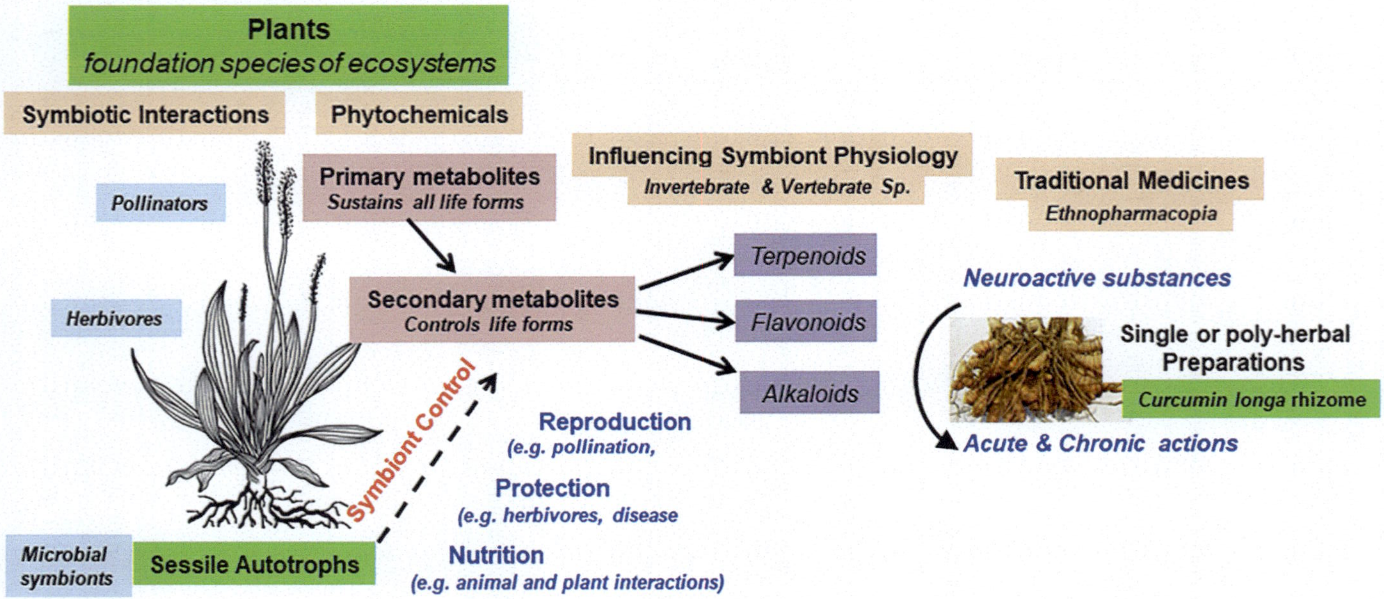


**Fig. 1** Plants phytochemicals and their actions

was introduced into Europe (in the nineteenth century), allegedly by Bonaparte's expedition to Egypt (Wiart 2006). Similarly, the leaves of *Erythroxylaceae sp.* (Coca; with its psychoactive alkaloid cocaine), were chewed with pieces of calcite to increase the potency of its alkaloid, 8000 years ago, as shown in Peruvian mummies (Dillehay et al. 2010; Valdez et al. 2015). Coca was central to South American culture, where it has been cultivated since early civilization (Dillehay et al. 2010; Valdez et al. 2015). These neuroactive phytochemicals, while deeply embedded in ethnopharmacology, are restricted in their use due to addictive properties.

The potential to acutely manipulate nervous system function in the treatment of neurological disorders, to directly enhance its function, to slow or reverse neurodegenerative diseases are a spectrum of treatment possibilities for herbal medicines (Kennedy 2014). Such approaches are currently used in the traditional medicinal systems of Sri Lanka.

## 2 Neurological Disorders and Herbal Treatments

Plant-derived herbal preparations are used for the well-being and treatment of the nervous system by a large proportion of the Sri Lankan population. The ethnopharmacology of Sri Lanka includes herbal preparations, which are common to the Southeast Asian region, such as Ayurvedic treatments, or they may be endemic to Sri Lanka. Neuroactive herbs may be used in culinary applications, e.g., spices such as nutmeg (*M. fragrans*), cloves (*Syzygium aromaticum*), chilies (*Capsicum annuum*), or as vegetables such as Gotukola (*C. asiatica*), lunuvila (*B. monnieri*). Many of these plant species commonly used as foods or spices have been investigated, and the mechanisms of action on the nervous system have been elucidated.

Many other, less commonly known plant species, which contain neuroactive phytochemicals, are used extensively as single herbs or in polyherbal preparations, for the treatment of neurological disorders or for enhancing nervous system function. The cumulative benefits from combined herbal materials include synergistic effects, augmentation of benefits, or a negation of undesired effects. The efficacy of these treatments has been demonstrated over centuries through clinical use.

There is extensive ethnopharmacological use of neuroactive plant preparations, and modern medicine, which is based solely on scientific validation, have adopted many unique phytochemicals extracted from plants. However, due to the limitations of science, to test and quantify mechanisms underlying the efficacy of polyherbal preparations, which may target multiple physiological processes to provide benefit, these traditional medicines with demonstrated efficacy through centuries of clinical use, remain to be adopted for the treatment of neurological disorders or for enhancing nervous system function.

## 2.1 Alzheimer's and Dementia Treatment

Alzheimer's disease (AD) is characterized as a progressive, multifaceted neurodegenerative disorder (Jiang et al. 2017), which primarily affects the elderly. Symptoms of AD include loss of memory, impaired cognitive function, and dementia. The specific etiological factor for AD remains uncertain, but there are some possible causative factors, including the cholinergic hypothesis, amyloid hypothesis, tau hypothesis, calcium hypothesis, and isoprenoid change (Singh et al. 2013a, b).

The amyloid precursor protein-mediated neurodegeneration is a popular hypothesis, where the amyloid precursor protein, a transmembrane glycoprotein, is implicated. Beta-amyloid is a 40–42 amino acid peptide which aggregates in the cerebrovascular system. Galantamine is a well-established acetylcholine esterase inhibitor (AChE inhibitor) used to treat Alzheimer's disease. It is found naturally in plant species belonging to the family Amaryllidaceae (Wilkinson 2001). Extracts of 26 plant species from the genus *Narcissus barring N. assoanus have shown significant AChE inhibition.*

Methanolic extraction of *Angelica sinensis* roots significantly inhibits amyloid β-peptide toxicity at lower concentrations (1–10 μg/mL), but high concentration was found to be toxic to dPC-12 cells. The active compounds are Z-ligustilide, coniferyl ferulate, 11-angeloylsenkyunolide F, and ferulic acid (Lin et al. 2008). In addition, *A. sinensis* root extracts tested on rat models lowered amyloid levels in the hippocampus, reversed behavioral impairment, inhibited apoptosis, and reduced brain inflammation. The finding suggests *A. sinensis* root extracts as an alternative treatment for Alzheimer's disease (Duan et al. 2015). Evodiamine is a quinolone alkaloid present in *Evodia rutaecarpa* (Jiang and Hu 2009). Evodiamine treatment improves spatial learning memory and enhanced cognitive function via a mechanism outside the reduction of pathological plaque formation. Evodiamine has also been shown to reduce inflammation in Alzheimer's disease models (Yuan et al. 2011). *C. asiatica,* a green leafy vegetable, and herbal medicine, frequently used in Sri Lankan households, acts as an antioxidant reducing oxidative damage in the central nervous system. *C. asiatica* significantly decreases amyloid formation in the brain via an antioxidant mechanism (Dhanasekaran et al. 2009). Furthermore, *C. asiatica* improves the central nervous system (CNS) function, e.g., cognition (Veerendra Kumar and Gupta 2003).

## 2.2 Anxiolytic and Antidepressant

Many passion fruit species (*Passiflora sp.*) exhibit anxiolytic and sedative activity and are used in traditional medicine as a mild tranquilizer (dos Santos et al. 2006). The sedative activity of the *Passiflora quadrangularis* aqueous extract was evaluated by Gazola et al. (2015). At lower doses (100 and 300 mg/kg), the extract showed diazepam-like sedative effects mediated by GABA. Based on biochemical assays, the sedative effect was attributed to apigenin, a flavonoid. da Silva et al. (2017) investigated the effects of *Passiflora incarnata* on anxiety in humans and conclude that a 500 mg dose can reduce public speaking anxiety. Garlic passion fruit (*P. tenuifila*) showed anxiolytic activity at 200 and 400 mg/kg concentrations (Holanda et al. 2020).

A decoction of the flowers and leaves of Aloysia, a well-known medicinal plant in folk medicine, is used to treat gastrointestinal disorders such as pain (Filipov 1994), nausea, vomiting, dyspepsia, and gastritis. Aloysia is also known as a sedative or tranquilizer (Hellión-Ibarrola et al. 2006). The hydroethanolic extraction of *Aloysia polystachya* has anxiolytic-like activity, but without sedative side effects (Hellión-Ibarrola et al. 2006) and antidepressant activity, when administered orally to mice (Hellión-Ibarrola et al. 2008) via a GABA-mediated mechanism. As flavonoids were absent, the antidepressant activity was attributed to phenolics and terpenes (Hellión-Ibarrola et al. 2006). Hydroalcoholic extracts of *A. polystachya* leaves containing terpenoids (thujone and carvone compounds) have an anxiolytic activity of similar efficacy to diazepam at 25 and 50 mg/kg, and antidepressant effects similar to fluoxetine at 12.5, 25, and 50 mg/kg doses (Mora et al. 2005).

The anxiolytic activity of the *Aloysia sp.* was demonstrated in a randomized double-blinded, phase-2 clinical trial on 54 adults, using 300 mg of *A. polystachya* powdered leaves, twice a day. Following 8 weeks of treatment, anxiety symptoms were significantly suppressed (Carmona et al. 2019). Importantly, an absence of or mild side effects was reported. The active phytochemicals in the leaves are acteoside, carvone, and limonene (Mora et al. 2005; Carmona et al. 2019). Similarly, an aqueous extract of *A. gratissima* was shown to mediate an antidepressant-like effect via the L-arginine-NO-cGMP pathway (Zeni et al. 2011).

Ursolic acid stearoyl glucoside (Kazmi et al. 2013) derived from methanolic extraction from *Lantana camara* has anxiolytic activity mediated via a GABA-ergic pathway (Awad et al. 2009). Similarly, in vivo studies have demonstrated anxiolytic effects and antinociceptive and anti-inflammatory effects of *Terminalia citrina* (Das et al. 2015). An anxiolytic activity like diazepam was demonstrated in aqueous extracts of *Casimiroa edulis,* using behavioral experiments with the maze test. However, a reduction of locomotion was also reported on coadministration of the extract with desipramine (Molina-Hernández et al. 2004). Remarkable sedative effects, namely anxiolytic and antidepressant-like activity mediated via binding of GABA and/or benzodiazepine receptors, have been reported in in vivo studies using ethanolic extracts of cloves (*S. aromaticum*) flowering buds (Hossain et al. 2019). Clove, a spice indigenous to Sri Lanka, is a common home remedy for chronic pain associated with a toothache and is used frequently as a topical analgesic.

## 2.3 *Anti-Cataleptic and Anti-Parkinson's Activities*

### 2.3.1 Anti-Cataleptic Activity

Catalepsy is a nervous condition characterized by muscular rigidity and fixed posture, regardless of external stimuli, as well as decreased sensitivity to pain. Catalepsy can manifest in schizophrenia, Parkinsonism, and epilepsy (Sahu et al. 2011).

The ethanolic extract of *Bauhinia racemosa* leaves are used to treat catalepsy, with efficacy in inhibiting clonidine-induced but not in haloperidol-induced catalepsy (Nirmal et al. 2011). The plant extract has a mast cell-stabilizing property. Clonidine induces mast cells to release histamine (Lakdawala et al. 1980), and haloperidol acts via dopamine receptors. Sanberg (1980) and Nirmal et al. (2011) proposed that the inhibition of catalepsy by *B. racemosa* is via the stabilization of mast cells and not via dopamine or its receptors. Similarly, the ethanolic extracts of the leaves of *Solanum xanthocarpum, C. longa, Tinospora cordifolia,* and *Piper longum* have anti-cataleptic activity mediated via the stabilization of mast cells (Dhanalakshmi et al. 2004).

The dose-dependent anti-cataleptic activity of *C. longa, Zingiber officinale,* and *Alpinia galangal* via significant inhibition of mast cell degranulation induced by clonidine has been demonstrated (Sarin 2017; Dongre et al. 2015).

A significant reduction of haloperidol-induced catalepsy, via the facilitation of dopaminergic neurotransmission by intraperitoneal administration of ethanolic and aqueous extractions of *Ocimum sanctum*, was demonstrated (Aswar and Joshi 2015). A similar anti-cataleptic activity via a dopamine-related mechanism has also been demonstrated in the seeds of *Mucuna pruriens* (Manyam et al. 2004a). Species *M. pruriens* are widely used in the treatment of dopamine-related neurological disorders as the plant contains dopamine and serotonin (5-HT) (Manyam et al. 2004b; Sahu et al. 2011). The seeds of *M. pruriens* are soaked and rinsed extensively in flowing water before their use as a food in Sri Lanka, due to potential toxicity from their high monoamine content.

### 2.3.2 Anti-Parkinson's Activity

Parkinson's disease (PD) is a complex progressive neurodegenerative disease caused by gradual loss of dopaminergic nerve cells in the substantia nigra of the midbrain (Norazit et al. 2010; Rabiei et al. 2019). Common symptoms include motor disorders such as bradykinesia, resting tremor, and rigidity, with postural instability occurring at a later stage (Dauer and Przedborski 2003). The exact etiology for the disorder is unknown, but several hypotheses are presented, including age, genetic risk factors, environmental influences such as smoking, caffeine consumption, and pesticide exposure (Norazit et al. 2011a; Kouli et al. 2018).

Dopaminergic cell death in PD is correlated to brain oxidative stress, microglial activation, and Lewy bodies formation (Adler and Ahlskog 2000; Norazit et al.

2011a; Weetman et al. 2013). Elevation of oxidative stress results in the degeneration of dopaminergic neurons in the substantia nigra in the midbrain (Norazit et al. 2010; Norazit et al. 2011a; Weetman et al. 2013). Neuronal survival is dependent on the maintenance of redox reactions, whose disruption interferes with cellular physiology leading to cell death (Norazit et al. 2011a). Oxidative stress interferes with cellular proteins, disrupts lipid membranes, and leads to mitochondrial dysfunction, which in turn produce more ROS in the brain and neuroinflammation (Norazit et al. 2011a). Therefore, the reduction of oxidative stress in the brain is a primary treatment strategy for PD.

There are several herbal treatments for reducing oxidative stress in the brain. Leaves of *Oxalis corniculate* with putative antioxidant properties are widely used in Indian and Sri Lankan traditional medicines to treat epilepsy (Senthil and Kapoor 2010). *O. corniculata* extract significantly restores peroxides and antioxidant levels to near normal in the brains of the test animals (Aruna et al. 2016).

*Gynostemma pentaphyllum* is used as an herbal tea and is widely believed to have multiple health benefits in Sri Lanka and the wider Asian region. The health benefits include protective and/or functional improvements in diabetes, depression, anxiety, fatigue, hyperlipidemia, immunity, and in tumors, putatively via reduction of oxidative stress. Experimental data on ethanol extracts of *G. pentaphyllum* leaves have shown neuroprotection against neurotoxins and oxidative stress by improving tyrosine hydroxylase activity where treatment with the extract normalized dopamine secretion in a 6-hydroxydopamine (6-OHDA)-lesioned rat model of PD (Choi et al. 2010).

*Valeriana officinalis* has anti-oxidative, neuroprotective properties, with flavonoids being the primary antioxidants (de Oliveria et al. 2008). Another *Valeriana species*, *V. moonii*, is an endemic plant of Sri Lanka, growing in the upper elevations of the island. It contains the pharmacologically important sesquiterpene jatamansone/valeranone, the principal sesquiterpene, as in *Nardostachys jatamansi* widely used in the East and Europe. It is a primary ingredient in Ayurveda for the treatment of neurological disorders (Bhatt and Kothiyal 2015). Phytochemical screening of *V. moonii* demonstrated the presence of jatamansone at significantly higher levels (25.6%) compared to *N. jatamansi* (8.9%) (Jayaratne et al. 2017). Neuronal loss in the substantia nigra results in dopamine depletion in the stratum and reduced dopaminergic signaling, causing the cardinal motor symptoms of PD (Norazit et al. 2010; Kouli et al. 2018). *Hibiscus asper* leaves are used to manage the neurological abnormalities in Parkinson's disease (Foyet et al. 2011). Methanolic extracts of *H. asper* leaves have free radical scavenging (antioxidant) and antiapoptotic activity (Hritcu et al. 2011), leading to improvement in spatial memory deficits in the 6-OHDA-lesion rodent. Similarly, *Ginkgo biloba* restores locomotor deficits in mice by increasing dopamine signaling in the striatum. *G. biloba* mediates this action as an antioxidant and inhibitor of dopamine transport and as a potent inhibiter of monoaminoxidase. It thereby increases dopamine availability in CNS by reducing dopamine reuptake into presynaptic nerves and by reducing the breakdown of dopamine (Riederer and Jellinger 1983; Sloley et al. 2000). The multiple pharmacological actions of *G. biloba* have led to its use as an anti-Alzheimer medication as well as its use to reverse dopaminergic lesions in Parkinson's disease (Ahmad et al. 2005). *Uncaria rhynchophylla* extracts can also raise midbrain dopamine levels by inhibition of mono-aminoxides, thereby increasing dopaminergic signaling (Kartika et al. 2014).

*Hyoscyamus niger* (L) of the Solanaceae family, commonly known as henbane, is widely distributed in Asia and Europe (Khatri and Juvekar 2015). *H. niger* is a potent antioxidant, protecting against oxidative stress by enhancing the levels of glutathione, reducing thiobarbituric acid reactive substance, and increasing the levels of antioxidative enzymes. *H. niger* ameliorates neurophysiological abnormalities in stereotaxically rotenone-lesioned rodent models. HPLC analysis of methanolic extracts of *H. niger* seeds demonstrated that L-DOPA is present at significant levels, putatively mediating locomotor recovery in lesioned rats through dopamine replacement in the nigra (Khatri and Juvekar 2015).

Ethanol and water extracts of *Acanthopanax senticosus* stem and bark can increase dopamine levels in the midbrain in rats via multiple pathways, resulting in a reduction of the behavioral deficits induced by MPTP (1-methyl-4-phenyl-1,2,3,6-tetrahydropyridine) lesioning. Also, it was demonstrated that protracted treatment with the extracts of *A. senticosus* showed a neuroprotective action, resulting in reduced nigral dopaminergic cell loss (Fujikawa et al. 2005).

## 2.4 Anti-Stress

Stress, a common biological phenomenon, is a response to threatened homeostasis caused by intrinsic or extrinsic adverse forces. Several endocrine and neuronal systems interact and respond to stress via activation of the "stress system." The hypothalamic–pituitary–adrenal axis, the CNS, and the peripheral nervous system (PNS) interact to control the peripheral organs in the "fight or flight" response. Monoaminergic neurotransmitters such as noradrenaline, dopamine, and 5-HT are primary mediators of the "fight or flight" response.

The response of an individual to a stressful situation varies markedly, determined by a multiplicity of factors including genetic, physiological, environmental, and developmental factors. High intensity and/or chronic exposure to stress can impair physical and mental function leading to psychosomatic disorders such as ulceration, blood pressure, stroke, diabetes, anxiety, depression, etc.

Due to the multiple activators/mediators of the stress response, engaging both the CNS and PNS, chemical agents which have a broader action are more effective in the treatment of stress. While no specific pharmacological treatments are available in Western medicine to relieve stress (excluding a hot cup of tea), many plants and their derivatives in the traditional medical system have a long history of treating stress (Prakash and Gupta 2000).

Diazepam (a benzodiazepine drug) is the standard control treatment, which mediates a nonspecific anti-stress activity via GABA-mediated neurotransmission. It is a long-acting drug that mediates the depressant activity of the CNS. It also modulates the mesocortical dopaminergic system, the noradrenergic system, and 5-HT neurotransmission in the CNS, including the hypothalamus. Opium

(*P. somniferum*) (Papaveraceae) has been used in Ayurvedic medicine in the last few centuries, with reference in the sixteenth century to its medicinal properties, in the Ayurvedic text the Yogaratnakara, in Sinhala verse (Uragoda 1983). However, human use of the opium poppy dates back to the Neolithic period (5700–5500 BC). Of the two main groups of alkaloids in opium, the phenanthrenes, e.g., morphine, codeine, and thebaine, are the main neuroactive constituents. These act via G-protein coupled opioid receptors, acting on GABAergic neurotransmission. Opiates are key sedatives of the modern pharmacopeia.

As discussed in Sect. 3.3, medicinal plants with neuroactive phytochemicals have a primary ethnopharmacological role in Sri Lanka. *C. sativa* (marijuana) present in the Chinese and Indian medical pharmacopeia for over 3500 years is used to induce narcosis. It is an herbal plant of abuse for recreational purposes. While used for centuries as an herbal mood stabilizer in Sri Lanka, cannabis is currently classified as a narcotic (an addictive drug affecting mood or behavior). Its use is, however, accepted in the traditional medicinal pharmacopeia as it contains multiple phytochemicals of therapeutic benefit, including as an antioxidant, anti-inflammatory, anticancer compound, anticonvulsant, sedative, and hypnotic properties. An extract of cannabis, Cannabidiol (CBD), has been reclassified as a narcotic drug by the World Health Organization Expert Committee on Drug Dependence. Its relatively low toxicity has led to its development for a variety of therapeutic applications of the nervous system, including schizophrenia, fragile X syndrome, encephalopathies, childhood absence seizures, neonatal hypoxic–ischemic encephalopathy, and epilepsy.

The Sri Lankan ethnopharmacopia, targeting nervous system function, includes other common plant species such as *M. fragrans* (nutmeg) and *Eugenia caryophyllus* (Clove). Clove is a popular plant used in traditional medicine to treat various localized and systemic pathological conditions. It possesses many medicinal properties, including antibacterial (Nuñez and D'Aquino 2012), antipyretic, anti-inflammatory (Taher et al. 2015), antioxidant (Ghadermazi et al. 2017), a local anesthetic (Khalilzadeh et al. 2016), and aphrodisiac activities. Ethanolic extracts of clove buds, administered orally at 100 and 200 mg/kg showed anti-stress activity. However, the presence of volatile oils such as eugenol may result in contraindications (Asad et al. 2009).

*W. somnifera* is used in India to calm the mind, nervous exhaustion, arthritis, and in helping the body to adapt to physical stress. The anti-stress activity of *W. somnifera* was reported by Kaur et al. (2001), and similarly, the anti-stress activity of *O. sanctum* has also been reported (Bhargava and Singh 1981; Prakash and Gupta 2000). A mixture of the ethanolic extracts of the roots of *W. somnifera* and the leaves of *O. sanctum* in the ratio of 1:1 exerts potent anti-stress activity (Misra et al. 2009). Extracts of *O. sanctum* per se, possess significant chronic and acute anti-stress activity mediated by ocimumosides A and B and 4-allyl-1-*O*-D-glucopyronosyl-2-hydroxybenzene (Prakash and Gupta 2000).

*Diospyros malabarica* is used in several disease conditions such as liver diseases (Mondal et al. 2005), diarrhea, diabetes, and shows antioxidant activity (Kavatagimath and Jalalpure 2016). The ethanolic extract of the root possesses anti-stress ability against anoxia stress tolerance, chemically induced stress, swimming endurance stress, and immobilization stress in rat models (Purane and Vidyadhara 2016) putatively via modulation of catecholaminergic neurotransmission.

## 2.5 CNS Stimulants and Cognition Enhancement

CSN stimulants (nootropics) are psychoactive drugs or chemicals which can temporally improve mental function by enhancing CNS synaptic activity. CNS stimulants increase alertness, attention, and physical activity. CNS stimulants are clinically used for narcolepsy, attention-deficit hyperactivity disorder, and associated disorders (Gresamma 2019).

These drugs are grouped according to their functions, i.e., analeptic stimulants, psychomotor stimulants, cognition enhancers, etc. (Mestry et al. 2016). There are five stimulant drug classes, including methylphenidate, atomoxetine, modafinil, armodafinil, and amphetamines. As these drugs are often misused and have significant adverse effects, the US Food and Drug Administration has restricted their use (Gresamma 2019). Hence, plant-based alternatives to CNS stimulants are becoming increasingly popular (Deepika et al. 2015).

The stimulant effects of caffeine have long and diverse ethnopharmacology as has nicotine, with extensive data on their mechanisms of action. Similarly, ether extracts of *Cucurbita maxima* seed have CNS stimulatory properties comparable to caffeine and are presented as a treatment for depressed CNS function (Doke et al. 2011). The ethanolic extracts of *Kigelia Africana* have greater CNS stimulatory activity than caffeine, with delayed onset and shortened duration of barbiturate induced sleeping in mice, mediated by a GABAergic mechanism (Owolabi et al. 2008). Similarly, aqueous extracts of *Boerhavia diffusa,* a plant species used in the treatment of a variety of disorders, also acts in CNS stimulation, facilitating increased motor coordination and locomotor activity (Deepika et al. 2015). Ethyl acetate and aqueous seed extracts of *Abelmoschus esculentus*, a popular and common Sri Lankan vegetable, have CNS stimulatory and antidepressant activity (Agarwal et al. 2015).

Water hyssop (*B. monnieri*; Lunuwila; Brahmi) is a creeping water plant (Rai et al. 2017). The plant is commonly available in Sri Lanka, occupying a significant position as a nootropic in the Sri Lankan and Indian pharmacopeia. *B. monnieri* influences multiple physiological systems as an analgesic, antipyretic, anti-inflammatory, antioxidant with anti-lipid peroxidase activity (Singh and Dhawan 1982). It is used in Ayurveda to treat neurological disorders such as insomnia and epilepsy (Rai et al. 2017). It possesses demonstrated cognitive and memory enhancing and neuroprotective abilities (Uabundit et al. 2010), as demonstrated for *E. alsinoides* (Sundaramurthi and Packiam 2017; Yellamma and Vanaja 2016).

*Achyranthes bidentate*, *Cnidium monnieri*, and *Ophiogon japanicus* are Chinese herbal plants that possess antiaging activity cognitive-enhancing effects (Lin et al. 2003). In traditional Chinese medicine (TCM), *Angelica gigas* roots are used to improve blood flow, treat amnesia, and as a sedative (Sowndhararajan and Kim 2017). An n-butanol soluble fraction of *A. gigas* root extract is neuroprotective

(Kang and Kim 2007). *Schizandra chinensis* fruits are used in TCM to treat cardiovascular disease (Chun et al. 2014). Furthermore, on combining *A. gigas*, *Schizandra chinensis*, and *Schizandra chinensis* extracts, at the ratio of 8:1:1, respectively, cognitive enhancement has been demonstrated in in vivo studies (Kang et al. 2005).

Ayurveda (as practiced in India and Sri Lanka) and TCM uses *Convolvulus pluricaulis* for treating CNS disorders such as insanity, epilepsy, and anxiety (Niyati et al. 2005). The pharmacologically active compounds of *C. pluricaulis* include alkaloids, flavonoids, and sumrins (Bihaqi et al. 2016). Aqueous extracts of *C. pluricaulis have shown cognitive enhancement in scopolamine-induced neurotoxic in rat models* (Balkrishna et al. 2020; Bihaqi et al. 2011). Similar aqueous extracts of *Markhamia tomentosa* bark improves cognition via putative hippocampal antioxidant activity (Ionita et al. 2017). The combination of 5:1 *Z. officinale* and *Cyperus rotundus* tuber extracts have neuroprotective and memory-enhancing properties (Sutalangka and Wattanathorn 2017).

Stimulation of the nervous system enhances synaptogenesis, dendritic growth, and axonal regeneration. This is a critical feature in maintaining a healthy nervous system or in the recovery of nervous system function.

## 2.6 Synaptogenesis, Dendritic Growth, and Axonal Regeneration

The adult nervous system is elaborately specialized (at a cellular and organ level), dynamically responding with structural and functional changes based on its activity. High levels of the spatial organization of the CNS occur during development, finely mediated by spatial and temporal expression gradients of overlapping transcription factors, which in turn are regulated by activity-driven, neurotrophic factors. In combination with the vasculature and supporting glial cells, the migration of neuroblasts, their maturation, axonal growth, dendritic arborization, and synaptogenesis are dynamically regulated until the nervous system attains its fully functional state.

The protracted time over which the nervous system develops in humans spans the embryonic, fetal, infant, and adolescent periods, extending into late puberty. Subsequent changes to the mature human CNS are limited at the cellular level, i.e., limited neurogenesis. Rewiring of the CNS through axonal regrowth occurs more readily. A high level of plasticity in the nervous system is maintained by dendritic arborization and synaptogenesis, primarily driven by neurotrophic factors.

In the aging adult human brain, this plasticity diminishes as a natural process and as a consequence of multiple factors such as vascular deterioration. The rich neurotropic milieu of the developing CNS wanes, as does the stem cell niches together with their vascular supply. Genetic factors and environmental stressors may further accelerate the loss of plasticity. After that, the functional recovery on disruption or degeneration is either very slow or incomplete (Tuffaha et al. 2016).

Thus, declining neural function is a combination of neurodegeneration and a loss of regeneration capacity (plasticity).

With an increasing global incidence of neurological disorders, global scientific efforts are focused on finding therapies for disorders of neural function. Cell replacement therapies using pluripotent embryonic stem cells to replace damaged or lost neural cells were attempted and encountered multiple difficulties, e.g., differentiation issues including generation of inappropriate phenotypes, immune rejection, and non-integration (Schulz et al. 2004; Robinson et al. 2005; Lau et al. 2006). Subsequently, autologous stem cell therapies were attempted, using adult neural stem cells, which also encountered cell isolation and homogeneity issues, difficulty in differentiation, and non-integration problems (Murrell et al. 2008; Khoo et al. 2011). Evoked cell and axon regeneration in lesioned and degenerating CNS were attempted using glial cells and neurotrophic factors (Lutton et al. 2012; Norazit et al. 2011b), with limited efficacy (Lutton et al. 2012; Cloutier et al. 2016). These studies and many others in the field suggest that sustained functional recovery of the nervous system would require a step back into recapitulating the initial developmental processes, where endogenous mechanisms would be revitalized to initiate self-repair.

Traditional medicinal recipes often use multiple herbs and herbal extracts to modulate one or more physiological systems in therapy. In rejuvenating the nervous system, poly-herbal preparations may be rationalized. Poly-herbal recipes presenting the correct combination of bioactive phytochemicals to modulate vascular, immune, and antioxidant mechanisms, together with neuro-stimulatory properties, is a potential therapeutic strategy.

Combinations of immune modulators, enhancers of blood perfusion, antioxidants, and other bioactive substances are available as nootropic health supplements, often based on traditional medicine recipes, e.g., Black seed oil (*Nigella sativa*), Brahmi (*Bacopa monniera*), and Gotukola (*C. asiatica*) as a combined immune suppressor/cognition enhancer/antioxidant. Similarly, health supplements combining turmeric (*C. longa*) and ginger (*Z. officinale*) or Black pepper (*Piper nigrum*) have gained popularity as herbal combinations that improve nervous system function. Curcumin bioavailability is increased when combined with piperine (the active ingredient of *Piper* spp. e.g., *nigrum*, *retrofractum*, and *longum*).

Ancient Ayurvedic recipes use similar combinations of plant materials, e.g., Thrikatu Choorna (Dry ginger (*Z. officinale*); black pepper (*P. nigrum*) and pipali (*P. longum*)) powder in a ratio of 1:1:1. The neuroprotective and memory-enhancing properties of *Z. officinale* and *C. rotundus* tubers combined (Sutalangka and Wattanathorn 2017) and the cognition enhancement demonstrated by a recipe with *A. gigas*, *Saururus chinensis*, and *Schizandra chinensis* extracts have been demonstrated (Kang et al. 2005). While robust scientific data demonstrating the safety and efficacy of some nutritional supplements, together with their mechanisms of action are available, others require further elucidation.

With growing interest in the potential health benefits of phytochemicals and their ability to enhance nervous system function, many herbs and their constituent bioactive phytochemicals have undergone scientific investigation.

*B. monniera* contains neuroactive saponins, including bacosides (Jain et al. 1993) and alkaloids, including brahmine and herpestine (Bhandari et al. 2007). An alcoholic extract containing saponins, bacosides A and B improved learning performance in rats (Singh and Dhawan 1997) and enhanced memory-related tests in mice (Zhou et al. 2009). Similarly, clinical trials showed treatment with *B. monnieri* improved learning and memory in human subjects (Stough et al. 2001; Pase et al. 2012). Improved learning and memory in healthy humans (Stough et al. 2008) and in aged subjects (Calabrese et al. 2008) and subjects with neurological disorders have been demonstrated (Stough et al. 2013). *B. monniera* is a frequently used vegetable given to children in Sri Lanka as food that improves learning and memory, similar to its ethnopharmacological use in India.

*C. asiatica* contains many phytochemicals, including brahmoside. It is reported to improve learning and memory in multiple laboratories and clinical studies (Gohil et al. 2010). *C. asiatica* is used extensively as a vegetable and in several traditional medicinal recipes in Sri Lanka.

*W. somnifera* is a popular herbal constituent in Ayurveda, used to treat general weakness, nervous exhaustion, insomnia, and amnesia, etc. *W. somnifera* extracts ameliorated the scopolamine-induced memory deficits in rats (Dhuley 2001; Kuboyama et al. 2005). Several compounds have been isolated from *W. somnifera,* including Withanolide A, a potent inducer of axonal and dendritic outgrowth, resulting in the regeneration of the cortex and hippocampus. Similarly, *C. sativa* accelerates the recovery of motor function after nerve compression injury in the peripheral nervous system, through axonal regrowth (Aziz et al. 2019). Peripheral axonal regrowth, with increased functional recovery, has been demonstrated with *G. biloba* extracts (Hsu et al. 2004). Extracts of *Eleutherococcus senticosus* have neuroprotective and neuro-regenerative activity (an outgrowth of axon and dendrites) mediated primarily through its secondary metabolites eleutherosides B, E, and isoflaxidin (Tohda et al. 2008).

## 2.7 Neuroprotection, Antioxidants, and Immune Modulation

More than 60 therapeutically useful compounds have been isolated from *C. sativa*. Among them, both delta 9-tetrahydrocannabinol and CBD are responsible for providing neuroprotection against cerebral ischemic damage (Hayakawa et al. 2007). CBD does not induce psychosis and is a compound which has neuroprotective, antioxidative, and anti-inflammatory effects (Esposito et al. 2005). It rescues pc12 cells from beta-amyloid deposits (and its derivatives) by reducing the concentration of reactive oxygen species, lipid peroxidases, and caspase 3 (Iuvone et al. 2004).

*C. longa* has a long history in traditional medicine, together with a spectrum of uses in cultural practices, including its use as an ingredient (spice) in foods, ranging from the middle-east, South Asia to far-east Asia (Hatcher et al. 2008; Niranjan and Prakash 2008; Prasad and Aggarwal 2011). The primary bioactive molecule in turmeric, curcumin (diferuloylmethane), is well investigated, using modern scientific protocols, with demonstrated ability to influence multiple signaling pathways, including survival pathways and growth factors; cytoprotective; metastatic and angiogenic pathways (Hatcher et al. 2008; Niranjan and Prakash 2008; Hewlings and Kalman 2017). Extensive experimental data on the potential therapeutic action of turmeric/curcumin are comprehensively reviewed by Aggarwal et al. (2013). Turmeric is used in traditional medicine as an immune modulator (Hatcher et al. 2008).

The active ingredient of turmeric, curcumin, is neuroprotective in ethanol-induced brain damage (Rajakrishnan et al. 1999) and ischemia–reperfusion injury (Ghoneim et al. 2002), attributed to its antioxidant activity. Neuroprotection via an antioxidant mechanism, in an ischemic injury model of the forebrain and hippocampus, has been demonstrated in rats (Al-Omar et al. 2006). *W. somnifera* root extract, similarly, reduces oxidative stress and protects neuronal cells from oxidative damages (Sankar et al. 2007). The root extract provides neuroprotection in paraquat and maneb induced (well-established oxidative stress-inducing neurotoxins) nigrostriatal dopaminergic neurodegeneration and also shown to normalize behavioral and biochemical parameters of the nigrostriatal circuitry (Prakash et al. 2013). Similarly, *A. calamus,* traditionally used to treat a range of diseases, contains a variety of secondary metabolites, including those showing anti-inflammatory and antioxidant properties (Muthuraman and Singh 2011). Methanolic extracts of *E. senticosus* have provided neuroprotection against beta-amyloid-induced oxidative stress-mediated neuron atrophy (Bai et al. 2011).

Many traditional foods, including nutraceuticals, which are commonplace in the ethnopharmacology of Sri Lanka, are consumed, providing benefits to the nervous system through their antioxidant properties and immune modulation.

## 3 Traditional Foods of Benefit to the Nervous System

As the adult nervous system begins to decline as a natural process with increasing age, the rate of this decline may be reduced through lifestyle choices. Of lifestyle factors, healthy eating is a primary element. In many eastern countries, food traditions have been maintained over many centuries, where nutraceuticals are completely integrated into diets and food choices. Most foods are prepared with varying combinations of spices, such as turmeric, nutmeg, clove, ginger, black pepper, and cinnamon, which, as discussed previously, have well established beneficial effects on the nervous system. The consumption of antioxidant containing foods, such as green tea and many other herbal infusions, fruit and vegetables, with bioactive phytochemicals such as flavonoids, are associated with protection against cerebrovascular diseases, e.g., strokes, and other neurologic disorders, such as dementia and Alzheimer's disease (Kennedy 2014). Furthermore, many vegetables, e.g., Gotukola (*C. asiatica*), Lunuwila (*B. monniera*), Vanduru Ma (*M. pruriens*), contain neuroactive substances and are available as the traditional foods in Sri Lanka. Several important plants possessing valuable active compounds, their activities on the nervous system have been listed (Table 1).

**Table 1** Plant species, neuroprotective compounds, and their actions

| Plant species | Active compounds | Activities | Reference |
|---|---|---|---|
| *Abelmoschus esculentus* | Catechin and quercetin | CNS stimulatory and antidepressant activity | Agarwal et al. (2015); Xia et al. (2018) |
| *Acanthopanax senticosus* | Isofraxidin<br>Syringin (Eleutheroside B) | Neuroprotective activity in Alzheimer's disease | Fujikawa et al. (2005) |
| *Achyranthes bidentate* | Triterpene saponins | Exhibit cognitive enhancive effects | Lin et al. (2003); Kang et al. (2005) |
| *Acorus calamus* | Phenolic content | Anti-inflammatory and antioxidant properties | Muthuraman and Singh (2011) |
| *Aloysia polystachya* | Flavonoids, Carvone | Active compound terpenoids have GABA receptors mediate anxiolytic-like activity and antidepressant-like effect | Carmona et al. (2019); Mora et al. (2005); Zeni et al. (2011); Consolini et al. (2011); Hellión-Ibarrola et al. (2006) |
| *Alpinia galangal* | Diarylheptanoids, Gingerol, quercetin, sterol glycosides | Anti-cataleptic activity via significant inhibition of mast cell degranulation | Dongre et al. (2015); Dongre et al. (2017) |
| *Angelica sinensis* | z-lingustilide,11-angeloylsenkyulide F, Confierl, Cycloastragenol, Feruliv acid | Anti-Alzheimer's activity by preventing neurotoxicity induced by β-amyloid peptide associated oxidative stress.<br>Inhibit the aggregation of β-amyloid peptide | Huang et al. (2008); Ho et al. (2009); Duan et al. (2015); Hong et al. (1994) |
| *Bacopa monnieri* | Bacoside, Jujubogenin, Pseudojujubogenin | An analgesic, antipyretic, anti-inflammatory, antioxidant with anti-lipid peroxidase activity.<br>Cognitive and memory enhancing, and neuroprotective abilities, immune suppressor | Singh and Dhawan 1982; Sundaramurthi and Packiam (2017); Yellamma and Vanaja (2016); Uabundit et al. (2010); Jain et al. 1993; Bhandari et al. (2007); Stough et al. (2001); Pase et al. (2012) |

(continued)

**Table 1** (continued)

| Plant species | Active compounds | Activities | Reference |
|---|---|---|---|
| *Bauhinia racemosa* | Kaempferol and methyl gallate | The extract has anti-cataleptic activity mast cell stabilizing and stop the histamine-releasing activity | Nirmal et al. (2011);<br>Lakdawala et al. (1980) |
| *Boerhavia diffusa* | Hydroxybenzoic hydroxybenzoic compounds | CNS stimulation, facilitating increased motor coordination and locomotor activity | Deepika et al. (2015) |
| *Cannabis longa* | Cannabigerol, Cannabidiol, cannabinol | Influence multiple signaling pathways, growth factors; cytoprotective; metastatic and angiogenic pathways<br>Antioxidant activity | Hatcher et al. (2008);<br>Niranjan and Prakash (2008); Hewlings and Kalman (2017);<br>Al-Omar et al. (2006) |
| *Cannabis sativa* | Cannabinoid | Cannabidiol and Delta 9-tetrahydrocannabinol provide neuroprotection against cerebral ischemic damage<br>Antioxidative and anti-inflammatory effects | Hayakawa et al. (2007);<br>Esposito et al. (2005);<br>Iuvone et al. (2004) |
| *Centella asiatica* | Hydrocotylin, hersaponin, histidine, threonine | Anti-Alzheimer's activity by enhancing memory and cognitive activities antioxidant properties reduce amyloid formation<br>Immune suppressor, enhancer | Veerendra Kumar and Gupta (2003);<br>Dhanasekaran et al. (2009);<br>Gohil et al. (2010) |
| *Cnidium monnieri* | Osthol | Exhibit cognitive enhancive effects | Lin et al. (2003);<br>Kang et al. (2005) |
| *Convolvulus pluricauli* | Coumarins, flavanoids | Hippocampal antioxidant activity<br>*Cognitive enhancement* | Niyati et al. (2005);<br>Balkrishna et al. (2020);<br>Bihaqi et al. (2011) |
| *Cucurbita maxima* | Flavanoids, polyphenolic compounds | A seed has CNS stimulatory properties | Doke et al. (2011) |
| *Curcuma longa* | Curcumins, Calebin-A | The active compound is curcumin; it influences multiple signaling pathways and growth factors; cytoprotective, metastatic, and angiogenic pathway. | Hatcher et al. (2008);<br>Niranjan and Prakash (2008); Hewlings and Kalman (2017);<br>Aggarwal et al. (2013); |

| | | | |
|---|---|---|---|
| | | Anti-cataleptic activity via significant inhibition of mast cell degranulation | Rajakrishnan et al. (1999); Dongre et al. (2015); Dongre et al. (2017); Dhanalakshmi et al. (2004) |
| *Cyperus rotundus* | Kaemferol, beta-sitodterol, N-hecacosana | Antioxidant ache inhibitory effects.<br>Neuroprotective and memory-enhancing properties | Sutalangka and Wattanathorn (2017) |
| *Diospyros malabarica* | Glycoside of beta-sitosterol | Possesses anti-stressor ability | Purane and Vidyadhara (2016) |
| *Eleutherococcus senticosus* | Phlorizin | Neuroprotective and neuro-regenerative activities.<br>Provide neuroprotection against beta-amyloid-induced oxidative stress-mediated neuron atrophy. | Tohda et al. (2008);<br>Bai et al. (2011) |
| *Eugenia caryophyllus* | Eugenol | Antioxidant activity<br>Antistressor activity | Asad et al. (2009);<br>Ghadermazi et al. (2017) |
| *Evodia rutaecarpa* | Dehydroevodiamine | Evodiamine present in the plant improves cognitive abilities in Alzheimer's disease. | Yuan et al. (2011) |
| *Ginkgo biloba* | Kaempferol | Extractrestores mice locomotion defects.<br>Increases dopaminergic transmission via inhibiting monoamine oxidase.<br>This pant can be used as anti-Alzheimer and anti-Parkinson's medications.<br>Peripheral axonal regrowth | Sloley et al. (2000);<br>Ahmad et al. (2005);<br>Hsu et al. (2004) |
| *Gynostemma pentaphyllum* | Quercetin | Neuroprotective effect on tyrosine hydroxylase immunopositive neurons and antioxidant activity | Choi et al. (2010) |
| *Hibiscus asper* | Flevanoid, pheniv acid | Dopaminergic neuroprotective effects through antioxidant and antiapoptotic properties.<br>Manage the neurological abnormalities in | Hritcu et al. (2011);<br>Foyet et al. (2011) |

(continued)

**Table 1** (continued)

| Plant species | Active compounds | Activities | Reference |
|---|---|---|---|
| | | Parkinson's disease.<br>Antioxidant and antiapoptotic activity. | |
| *Hyoscyamus niger* | Cleomiscosin A methyl ether, Coumarinolignans, Cleomiscosin A, Cleomiscosin B, Cleomiscosin A-9-′-acetate,<br>Cleomiscosin B-9′-acetate. | Protect against oxidative stress.<br>Monoamine oxidase inhibiting activity<br>Seeds contain significant levels of L-DOPA | Sengupta et al. (2010a, b);<br>Riederer and Jellinger (1983); Sloley et al. (2000);<br>Khatri and Juvekar (2015) |
| *Kigelia Africana* | Iridoids, isocoumarins, naphthoquinones, phenylpropanoids and phenylethanoids, furanone derivatives, flavonoids | CNS stimulatory activity than caffeine via GABA receptor-mediated mechanism | Owolabi et al. (2008) |
| *Lantana camara* | Sabiene | Ursolic acid and stearoyl glucoside show gabagenic anxiolytic activity | Awad et al. (2009) |
| *Markhamia tomentosa* | Phenylpropanoids, triterpenic acids, and anthraquinones | Hippocampal antioxidant activity | Ionita et al. (2017);<br>Ali et al. (2015) |
| *Mucuna pruriens* | L-dopa | Anticataleptic activity via dopamine and 5-HT contain in the extraction | Manyam et al. (2004b);<br>Sahu et al. (2011);<br>Singh et al. (2013a, b) |
| *Narcissus assoanus* | Galantamine | Galantamine mediated anti-Alzheimer's activity (competitive inhibitor of brain cholinesterase) | Wilkinson (2001) |
| *Nigella sativa* | Thymoquinone, dithymoquinone, thymohydroquinone, and thymol | Neuroprotective, cognition enhancer, antioxidant | Kanter et al. (2006);<br>Ghosheh et al. (1999) |
| *Ocimum sanctum* | Eugenol | Dopamine facilitated anticataleptic effect<br>Antistressor activity mediated by ocimumosides A and B and 4-allyl-1-O—D-glucopyronosyl-2-hydroxybenzene. | Aswar and Joshi (2015);<br>Bhargava and Singh (1981);<br>Prakash and Gupta (2000);<br>Misra et al. (2009);<br>Prakash and Gupta (2000) |

| | | | |
|---|---|---|---|
| *Ophiogon japanicus* | Nicotine | Exhibit cognitive enhancive effects | Lin et al. (2003); Kang et al. (2005) |
| *Oxalis corniculate* | | Show anti-Parkinson's properties by anti-oxidant activity in the brain | Aruna et al. (2016); Senthil and Kapoor (2010) |
| *Papaver somniferous* | Opium | Sedative action via GABAergic neurotransmission | Sarris et al. (2011) |
| *Passiflora quadrangularis* | Vitexin-2″-*0*-xyloside, vitexin-2″-*0*-glucoside, orientina-2″-*0*-glucoside | GABA mediated anxiolytic-like activity (showed diazepam-like sedative effects) | Gazola et al. (2015); da Silva et al. (2017); Holanda et al. (2020); Gazola et al. (2018) |
| *Piper longum* | Pierine (5-(3, 4-methylenedioxyphenyl)-1-piperidinopent-2, 4-dien-1-one) | Anti-cataleptic activity, mediated via the stabilization of mast cells. Neuroprotective effects | Dhanalakshmi et al. (2004); Hua et al. (2019) |
| *Piper nigrum* | (5-(3, 4-methylenedioxyphenyl)-1-piperidinopent-2, 4-dien-1-one) | Antioxidant activity in hippocampus and neuroprotective effect | Hua et al. (2019); Hritcu et al. (2014) |
| *Schizandrae chinensis* | Schizandrin | Protecting activity against Aβ-induced toxicity | Yang et al. (2016) |
| *Solanum xanthocarpum* | Solanocarpine, Carpesterol | Anti-cataleptic activity, mediated via the stabilization of mast cells | Dhanalakshmi et al. (2004) |
| *Syzygium aromaticum* | Eugenol, Beta-caryophyllene | Remarkable sedative effects, namely anxiolytic and anti-depressant like activity via binding of GABA and/or benzodiazepine receptors | Hossain et al. (2019) |
| *Terminalia citrina* | Chebulinic acid, gallic acid | Gabagenic anxiolytic activity and antinociceptive and anti-inflammatory effects | Das et al. (2015); Molina-Hernández et al. (2004) |
| *Tinospora cordifolia* | Cordifolioside A, Magnoflorine, tincordiside, syringin | Anti-cataleptic activity, mediated via the stabilization of mast cells | Dhanalakshmi et al. (2004); Sharma et al. (2012) |

(continued)

**Table 1** (continued)

| Plant species | Active compounds | Activities | Reference |
|---|---|---|---|
| *Uncaria rhynchophylla* | Uncarine E | Monoamine oxidase inhibiting activity | Kartika et al. (2014);<br>Fujiwara et al. (2006) |
| *Valeriana officinalis* | Valepotriates and valerenic acid | Antioxidative, neuroprotective properties,<br>Dopaminergic neuroprotective effects | Bos et al. (1998);<br>de Oliveria et al. (2008);<br>Piccinelli et al. (2004);<br>de Oliveria et al. (2008) |
| *Withania somnifera* | Glycowithanolides | Reduces oxidative stress, normalize behavioral and biochemical parameters of the nigrostriatal pathway.<br>Anti-stress activity | Sankar et al. (2007);<br>Prakash et al. (2013);<br>Kaur et al. (2001);<br>Misra et al. (2009);<br>Bhattacharya et al. (2001) |
| *Zingiber officinale* | Zingiberene, gingerols, alpha-curcumene, shogaols | Anti-cataleptic activity via significant inhibition of mast cell degranulation;<br>Antioxidant ache inhibitory effects;<br>Neuroprotective and memory-enhancing properties | Dongre et al. (2015);<br>Dongre et al. (2017);<br>Sutalangka and Wattanathorn (2017);<br>da Silveira et al. (2019) |

## 4 Conclusions

There are a plethora of phytochemicals that markedly influence neurological processes. Many of these act as immune modulators and antioxidants (often playing a similar role in the plants per se) and are of protracted benefit to the nervous system. These are commonly consumed as foods, comprising a more diffuse component of the ethnopharmacopia. Other more potent phytochemicals target the nervous system acting as deterrents or attractants for herbivores. These neuroactive phytochemicals influence the nervous system in multiple ways, such as modulating neurotransmitter biosynthesis, regulating release kinetics, act as agonists and antagonists, and in postsynaptic modulation. These neuroactive chemicals are often used in traditional medicines as polyherbal preparations. While many of these preparations await scientific elucidation, the paucity of medicines for nervous system ailments, together with the growing interest in preventative health, has vitalized awareness in foods and traditional treatments of benefit to the nervous system.

## References

Adler CH, Ahlskog JE (eds) (2000) Parkinson's disease and movement disorders: diagnosis and treatment guidelines for the practicing physician. Humana Press, Totowa, NJ, p 474

Agarwal A, Bora D, Agarwal C et al (2015) CNS stimulant and antidepressant activity of seeds of *Abelmoschus esculentus* in rats. Bull Pharmacol Res 5:47–50

Aggarwal BB, Gupta SC, Sung B (2013) Curcumin: an orally bioavailable blocker of TNF and other pro-inflammatory biomarkers. Br J Pharmacol 169(8):1672–1692. https://doi.org/10.1111/bph.2013.169

Ahmad M, Saleem S, Ahmad AS et al (2005) *Ginkgo biloba* affords dose-dependent protection against 6-hydroxydopamine-induced parkinsonism in rats: neurobehavioral, neurochemical and immunohistochemical evidences. J Neurochem 9:94–104. https://doi.org/10.1111/j.1471-4159.2005.03000.x

Ali S, El-Ahmady S, Ayoub N (2015) Phytochemicals of *markhamia* species (Bignoniaceae) and their therapeutic value: a review. Eur J Med Plants 6(3):124–142. https://doi.org/10.9734/ejmp/2015/15015

Al-Omar FA, Nagi MN, Abdulgadir MM et al (2006) Immediate and delayed treatments with curcumin prevents forebrain ischemia-induced neuronal damage and oxidative insult in the rat hippocampus. Neurochem Res 31:611–618

Aruna K, Rajeswari PDR, Sankar SR (2016) The effects of *Oxalis corniculata* l. extract against MPTP induced oxidative stress in mouse model of Parkinson's disease. J Pharm Sci Res 8(10):1136–1140

Asad M, Singh A, Dhamanigi S (2009) Anti-stress activity of hydro-alcoholic extract of *Eugenia caryophyllus* buds (clove). Ind J Pharmacol 41:28. https://doi.org/10.4103/0253-7613.48889

Aswar M, Joshi R (2015) Anti-cataleptic activity of various extracts of *Ocimum sanctum*. Int J Pharmacol Res 2:15

Awad R, Muhammad A, Durst T et al (2009) Bioassay-guided fractionation of lemon balm (*Melissa officinalis* L.) using an in vitro measure of GABA transaminase activity. Phytother Res 23:1075–1081. https://doi.org/10.1002/ptr.2712

Aziz N, Rasul A, Malik S et al (2019) Supplementation of *Cannabis sativa* L. leaf powder accelerates functional recovery and ameliorates haemoglobin level following an induced injury to sciatic nerve in mouse model. Pak J Pharm Sci 32:785–792

Bai Y, Tohda C, Zhu S et al (2011) Active components from Siberian ginseng (*Eleutherococcus senticosus*) for protection of amyloid β (25–35)-induced neurotic atrophy in cultured rat cortical neurons. J Nat Med 65:417–423. https://doi.org/10.1007/s11418-011-0509-y

Balkrishna A, Thakur P, Varshney A (2020) Phytochemical profile, pharmacological attributes and medicinal properties of *Convolvulus prostratus*—a cognitive enhancer herb for the management of neurodegenerative etiologies. Front Pharmacol 11:171. https://doi.org/10.3389/fphar.2020.00171

Bennett RN, Wallsgrove RM (1994) Secondary metabolites in plant defense mechanisms. New Phytol 127:617–633

Bhandari P, Kumar N, Singh B, Kaul VK (2007) Cucurbitacins from *Bacopa monnieri*. Phytochemistry 68:1248–1254

Bhargava KP, Singh N (1981) Anti-stress activity of *Ocimum sanctum* Linn. Ind J Med Res 73:443–451

Bhatt PM, Kothiyal P (2015) A review article on phytochemistry and pharmacological profiles of *Nardostachys jatamansi*. J Pharmacog Phytochem 3:102–106

Bhattacharya A, Ghosal S, Bhattacharya S (2001) Anti-oxidant effect of Withania somnifera glycowithanolides in chronic footshock stress-induced perturbations of oxidative free radical scavenging enzymes and lipid peroxidation in rat frontal cortex and striatum. J Ethnopharmacol 74(1):1–6. https://doi.org/10.1016/s0378-8741(00)00309-3

Bihaqi S, Tiwari M, Singh A (2011) In vivo investigation of the neuroprotective property of *Convolvulus pluricaulis* in scopolamine-induced cognitive impairments in Wistar rats. Ind J Pharmacol 43:520. https://doi.org/10.4103/0253-7613.84958

Bihaqi S, Rashaan S, Tiwari M (2016) *Convolvulus pluricaulis* as a cognition booster: relevance to Alzheimer's disease. Int J Pharmaceut Sci Drug Res 8:68. https://doi.org/10.25004/ijpsdr.2016.080201

Bos R, Woerdenbag H, van Putten F (1998) Seasonal variation of the essential oil, valerenic acid and derivatives, and valepotriates in *Valeriana officinalis* roots and rhizomes, and the selection of plants suitable for phytomedicines. Planta Med 64:143–147. https://doi.org/10.1055/s-2006-957392

Calabrese C, Gregory WL, Leo M et al (2008) Effects of a standardized bacopa monnieri extract on cognitive performance, anxiety, and depression in the elderly: a randomized, double-blind, placebo-controlled trial. J Alter Complement Med 14:707–713. https://doi.org/10.1089/acm.2008.0018

Carbonel A, Cecyn M (2019) Flavonoids as modulators of synaptic plasticity: implications for the development of novel therapeutic strategies for healthy lifestyle. doi: https://doi.org/10.5772/intechopen.84164

Carmona F, Coneglian FS, Batista PA et al (2019) *Aloysia polystachya* (Griseb.) Moldenke (Verbenaceae) powdered leaves are effective in treating anxiety symptoms: a phase-2, randomized, placebo-controlled clinical trial. J Ethnopharmacol 242:112060. https://doi.org/10.1016/j.jep.2019.112060

Choi HS, Park MS, Kim SH et al (2010) Neuroprotective effects of herbal ethanol extracts from *Gynostemma pentaphyllum* in the 6-hydroxydopamine-lesioned rat model of Parkinson's disease. Molecules 15:2814–2824. https://doi.org/10.3390/molecules15042814

Chun JN, Cho M, So I et al (2014) The protective effects of *Schisandra chinensis* fruit extract and its lignans against cardiovascular disease: a review of the molecular mechanisms. Fitoterapia 97:224–233. https://doi.org/10.1016/j.fitote.2014.06.014

Cloutier F, Kalincik T, Lauschke J et al (2016) Olfactory ensheathing cells but not fibroblasts reduce the duration of autonomic dysreflexia in spinal cord injured rats. Auton Neurosci 201:17–23. https://doi.org/10.1016/j.autneu.2016.08.015

Consolini AE, Berardi A, Rosella MA (2011) Antispasmodic effects of *Aloysia polystachya* and *A. gratissima* tinctures and extracts are due to non-competitive inhibition of intestinal contractility induced by acetylcholine and calcium. Rev Bras Farmacog 21(5):889–900. https://doi.org/10.1590/s0102-695x2011005000137

Das N, Goshwami D, Hasan MS et al (2015) Evaluation of antinociceptive, anti-inflammatory and anxiolytic activities of methanolic extract of *Terminalia citrina* leaves. Asian Pacific J Trop Dis 5:S137–S141. https://doi.org/10.1016/s2222-1808(15)60875-1

Dauer W, Przedborski S (2003) Parkinson's disease: mechanism and models. Neuron 39:889–909. https://doi.org/10.1016/s0896-6273(03)00568-3

Deepika B, Ray B, Jyothi Y et al (2015) CNS stimulant activity of aqueous extract of roots of *Boerhavia diffusa* in mice: role of brain renin angiotensin system (RAAS) in Parkinson's disease. J Chem Pharmaceut Res 7:409–412

Dhanalakshmi S, Khaserao SS, Kasture SB (2004) Effect of ethanolic extract of some anti-asthmatic herbs on clonidine and haloperidol-induced catalepsy in mice. Oriental Pharmacy Expl Med 4:95–99. https://doi.org/10.3742/opem.2004.4.2.095

Dhanasekaran M, Holcomb LA, Hitt AR et al (2009) *Centella asiatica* extract selectively decreases amyloidβlevels in hippocampus of Alzheimer's disease animal model. Phytother Res 23:14–19. https://doi.org/10.1002/ptr.2405

Dhuley JN (2001) Nootropic-like effect of ashwagandha (*Withania somnifera* L.) in mice. Phytother Res 15:524–528. https://doi.org/10.1002/ptr.874

Dillehay TD, Rossen J, Ugent D et al (2010) Early Holocene coca chewing in northern Peru. Antiquity 84:939–953

Doke P, Tare H, Sherikar A et al (2011) Central nervous system stimulant effect of the oils obtained from seeds of *Cucurbita maxima*. Int J Pharmaceutical Biol 1:30–36

Dongre PR, Bhujbal SS, Kumar D (2015) Bronchodilatory activity of *Curcuma longa, Zingiber officinale* and *Alpinia galanga* based herbal formulation (AHF). Oriental Pharmacy Exp Med 15:341–346. https://doi.org/10.1007/s13596-015-0205-7

Dongre PR, Doifode CA, Chilate VV et al (2017) Anti-inflammatory and anti cataleptic effect of herbal formulation consisting of various indigenous plants in experimental animals. Int J Pharmacognosy Phytochem Res 9:323–326. https://doi.org/10.25258/phyto.v9i2.8082

dos Santos KC, Kurtz SM, Müller SD et al (2006) Sedative and anxiolytic effects of methanolic extract from the leaves of *Passiflora actinia*. Brazilian Arch Biol Tech 49(4):565–573

Duan MH, Wang LN, Jiang YH et al (2015) *Angelica sinensis* reduced Aβ-induced memory impairment in rats. J Drug Target 24:340–347. https://doi.org/10.3109/1061186x.2015.1077848

Esposito G, De Filippis D, Carnuccio R et al (2005) The marijuana component cannabidiol inhibits β-amyloid-induced tau protein hyperphosphorylation through Wnt/β-catenin pathway rescue in PC12 cells. J Mol Med 84:253–258. https://doi.org/10.1007/s00109-005-0025-1

Fiehn O (2002) Metabolomics-the link between genotypes and phenotypes. Plant Mol Biol 48:155–171

Filipov A (1994) Medicinal plants of the Pilagá of Central Chaco. J Ethnopharmacol 44:181–193. https://doi.org/10.1016/0378-8741(94)01185-0

Foyet HS, Hritcu L, Ciobica A et al (2011) Methanolic extract of *Hibiscus asper* leaves improves spatial memory deficits in the 6-hydroxydopamine-lesion rodent model of Parkinson's disease. J Ethnopharmacol 133:773–779. https://doi.org/10.1016/j.jep.2010.11.011

Fujikawa T, Miguchi S, Kanada N et al (2005) Acanthopanax *senticosus* harms as a prophylactic for MPTP-induced Parkinson's disease in rats. J Ethnopharmacol 97:375–381. https://doi.org/10.1016/j.jep.2004.11.031

Fujiwara H, Iwasaki K, Furukawa K et al (2006) Uncaria rhynchophylla, a Chinese medicinal herb, has potent anti-aggregation effects on Alzheimer's beta-amyloid proteins. J Neurosci Res 84 (2):427–433. https://doi.org/10.1002/jnr.20891

Gazola AC, Costa GM, Castellanos L et al (2015) Involvement of GABAergic pathway in the sedative activity of apigenin, the main flavonoid from *Passiflora quadrangularis* pericarp. Rev Bras Farmacog 25:158–163. https://doi.org/10.1016/j.bjp.2015.03.009

Gazola AC, Costa GM, Zucolotto SM (2018) The sedative activity of flavonoids from *Passiflora quadrangularis* is mediated through the GABAergic pathway. Biomed Pharmacother 100:388–393. https://doi.org/10.1016/j.biopha.2018.02.002

Ghadermazi R, Amir S, Goli H (2017) Antioxidant activity of clove (*Eugenia caryophyllata* Thunb), oregano (*Oringanum vulgare* L) and sage (*Salvia officinalis* L) essential oils in various model systems Int. Food Res J24:1628–1635

Ghoneim AI, Abdel-Naim AB, Khalifa AE et al (2002) Protective effects of curcumin against ischemia/reperfusion insult in rat forebrain. Pharmacol Res 46(3):273–279. https://doi.org/10.1016/S1043-6618(02)00123-8

Ghosheh OA, Houdi AA, Crooks PA (1999) High performance liquid chromatographic analysis of the pharmacologically active quinones and related compounds in the oil of the black seed (*Nigella sativa* L.). J Pharmaceutical Biomed Analysis 19(5):757–762. https://doi.org/10.1016/s0731-7085(98)00300-8

Gohil KJ, Patel JA, Gajjar AK (2010) Pharmacological review on *Centella asiatica*: a potential herbal cure-all. Ind J Pharm Sci 72:546–556

Gresamma L (2019) To evaluate the CNS stimulant activity aqueous flower extract of *hibiscus sabdariff* Linn. In the male Wister albino rats. J Neurol Psychiatr Dis 1:104

Griffin KL, Seemann JR (1996) Plants, CO2 and photosynthesis in the 21st century. Chem Biol 3 (4):245–254. https://doi.org/10.1016/s1074-5521(96)90104-0

Hartmann T (2008) The lost origin of chemical ecology in the late 19th century. PNAS 105:4541–4546

Hatcher H, Planalp R, Cho J (2008) Curcumin: from ancient medicine to current clinical trials. Cell Mol Life Sci 65:1631–1652

Hayakawa K, Mishima K, Nozako M et al (2007) Repeated treatment with cannabidiol but not Δ9-tetrahydrocannabinol has a neuroprotective effect without the development of tolerance. Neuropharmacology 52:1079–1087. https://doi.org/10.1016/j.neuropharm.2006.11.005

Hellión-Ibarrola MC, Ibarrola DA, Montalbetti Y et al (2006) The anxiolytic-like effects of *Aloysia polystachya* (Griseb.) Moldenke (Verbenaceae) in mice. J Ethnopharmacol 105:400–408. https://doi.org/10.1016/j.jep.2005.11.013

Hellión-Ibarrola MC, Ibarrola DA, Montalbetti Y et al (2008) The antidepressant-like effects of Aloysia polystachya (Griseb.) Moldenke (Verbenaceae) in mice. Phytopathol Mediterr 15:478–483. https://doi.org/10.1016/j.phymed.2007.11.018

Hewlings S, Kalman D (2017) Curcumin: a review of its' effects on human health. Foods 6:92

Ho CC, Kumaran A, Hwang LS (2009) Bioassay guided isolation and identification of anti-Alzheimer active compounds from the root of *Angelica sinensis*. Food Chem 114(1):246–252. https://doi.org/10.1016/j.foodchem.2008.09.046

Holanda DKR, Wurlitzer NJ, Dionisio AP et al (2020) Garlic passion fruit (Passiflora tenuifila Killip): assessment of eventual acute toxicity, anxiolytic, sedative, and anticonvulsant effects using in vivo assays. Food Res Int 128:108813. https://doi.org/10.1016/j.foodres.2019.108813

Homberg U (2002) Neurotransmitters and neuropeptides in the brain of the locust. Microscopy Res Technique 56:189–209. https://doi.org/10.1002/jemt.10024

Hong GX, Qin WC, Huang LS (1994) Memory-improving effect of aqueous extract of *Astragalus membranaceus* (Fisch.) Bge. Zhongguo Zhong Yao Za Zhi 19(11):687–704

Hossain MM, Aka TD, Rahman MS et al (2019) Neuropharmacological activity of the crude ethanolic extract of *Syzygium aromaticum* flowering bud. J Nat Prod Res Ethnopharmacol l6:191–198. https://doi.org/10.15562/phytomedicine.2019.109

Hritcu L, Foyet HS, Stefan M et al (2011) Neuroprotective effect of the methanolic extract of *Hibiscus asper* leaves in 6-hydroxydopamine-lesioned rat model of Parkinson's disease. J Ethnopharmacol 137:585–591. https://doi.org/10.1016/j.jep.2011.06.008

Hritcu L, Noumedem JA, Cioanca O et al (2014) Methanolic extract of *piper nigrum* fruits improves memory impairment by decreasing brain oxidative stress in amyloid beta rat model of Alzheimer's disease. Cell Mol Neurobiol 34(3):437–449. https://doi.org/10.1007/s10571-014-0028-y

Hsu SH, Chang CJ, Tang CM et al (2004) *In vitro* and *in vivo* effects of *Ginkgo biloba* extract EGb 761 on seeded Schwann cells within poly (DL-lactic acid-co-glycolic acid) conduits for peripheral nerve regeneration. J Biomater Appl 19:163–182. https://doi.org/10.1177/0885328204045580

Hua S, Wang B, Chen R et al (2019) Neuroprotective effect of dichloromethane extraction from *Piper nigrum* L. and *Piper longum* L. on permanent focal cerebral ischemia injury in rats. J Stroke Cerebrovasc Dis 28(3):751–760. https://doi.org/10.1016/j.jstrokecerebrovasdis.2018.11.018

Huang SH, Lin CM, Chiang BH (2008) Protective effects of *Angelica sinensis* extract on amyloid β-peptide-induced neurotoxicity. Phytomedicine 15(9):710–721. https://doi.org/10.1016/j.phymed.2008.02.022

Hussein RA, El-Anssary AA (2018) Plants secondary metabolites: the key drivers of the pharmacological actions of medicinal plants. Intech Open. https://doi.org/10.5772/intechopen.76139

Ionita R, Postu PA, Beppe GJ et al (2017) Cognitive-enhancing and antioxidant activities of the aqueous extract from *Markhamia tomentosa* (Benth) K Schum stem bark in a rat model of scopolamine. Behav Brain Funct 13:5. https://doi.org/10.1186/s12993-017-0123-6

Iuvone T, Esposito G, Esposito R et al (2004) Neuroprotective effect of cannabidiol, a non-psychoactive component from *Cannabis sativa,* on beta-amyloid-induced toxicity in PC12 cells. J Neurochem 89:134–141. https://doi.org/10.1111/j.1471-4159.2003.02327.x

Jain P, Kulshreshtha DK, Bacoside A (1993) A minor saponin from *Bacopa monniera*. Phytochemistry 33:449–451

Jayaratne RK, Hettiarachchi PL, Abeysekara A (2017) A comparative pharmacognostic evaluation of anatomical and chemical characters of *Nardostachys jatamansi* (D.Don) DC. (Jatamansa) and *Valeriana mooni* Arn.Ex C.B.Clarke (Lanka thuwarala). Sri Lankan J Biol 2:9–22. https://doi.org/10.4038/sljb.v2i1.6

Jiang J, Hu C (2009) Evodiamine: a novel anti-cancer alkaloid from *Evodia rutaecarpa*. Molecules 14:1852–1859. https://doi.org/10.3390/molecules14051852

Jiang Y, Gao H, Turdu G (2017) Traditional Chinese medicinal herbs as potential AChE inhibitors for anti-Alzheimer's disease: a review. Bioorg Chem 75:50–61. https://doi.org/10.1016/j.bioorg.2017.09.004

Jirschitzka J, Schmidt G, Reichelt M et al (2012) Plant tropane alkaloid biosynthesis evolved independently in the Solanaceae and Erythroxylaceae. PNAS 109:10,304–10,309

Jirschitzka J, Mattern D, Gershenzon J (2013) Learning from nature: new approaches to the metabolic engineering of plant defence pathways. Curr Opin Biotechol 24:320–328

Kang SY, Kim YC (2007) Neuroprotective coumarins from the root of *Angelica gigas*: structure-activity relationships. Arch Pharmacol Res 30:1368–1373. https://doi.org/10.1007/bf02977358

Kang SY, Lee KY, Koo KA et al (2005) ESP-102, a standardized combined extract of *Angelica gigas, Saururus chinensis* and *Schizandra chinensis*, significantly improved scopolamine-induced memory impairment in mice. Life Sci 76:1691–1705. https://doi.org/10.1016/j.lfs.2004.07.029

Kanter M, Coskun O, Kalayc M et al (2006) Neuroprotective effects of *Nigella sativa* on experimental spinal cord injury in rats. Hum Exp Toxicol 25(3):127–133. https://doi.org/10.1191/0960327106ht608o

Kartika B, Muralidharan P, Rahman H (2014) Herbal treatment of parkinsonism: a review. Int J Pharm Sci Rev Res 5(34):185–191

Kaur P, Mathur S, Sharma M et al (2001) A biologically active constituent of *Withania somnifera* (ashwagandha) with antistress activity. Indian J Clin Biochem 16:195–198. https://doi.org/10.1007/bf02864860

Kavatagimath SA, Jalalpure SS (2016) Screening of ethanolic extract of *Diospyros malabarica* Desr.Bark for anti-diabetic and antioxidant potential. Ind J Pharm Educ Res 50:179–189. https://doi.org/10.5530/ijper.50.1.22

Kazmi I, Afzal M, Damanhouri ZA et al (2013) Anxiolytic potential of ursolic acid derivative-a stearoyl glucoside isolated from *Lantana camara* L. (Verbenaceae). Asian Pacific J Trop Med 6:433–437. https://doi.org/10.1016/s1995-7645(13)60069-3

Kennedy DO (2014) Polyphenols and the human brain: plant "secondary metabolite" ecologic roles and endogenous signaling functions drive benefits. Adv Nutr 5:515–533

Khalilzadeh E, Hazrati R, Saiah G (2016) Effects of topical and systemic administration of *Eugenia caryophyllata* buds' essential oil on corneal anesthesia and analgesia. Res Pharm Sci 11:293. https://doi.org/10.4103/1735-5362.189297

Khatri DK, Juvekar AR (2015) Propensity of *Hyoscyamus niger* seeds methanolic extract to allay stereotaxically rotenone-induced Parkinson's disease symptoms in rats. Orient Pharm Exp Med 15:387–388. https://doi.org/10.1007/s13596-015-0216-

Khoo MLM, Tao H, Meedeniya ACB et al (2011) Transplantation of neuronal-primed human bone marrow mesenchymal stem cells in hemiparkinsonian rodents. PLoS One 6:e19025

Kouli A, Torsney KM, Kuan WL (2018) Parkinson's disease: etiology, neuropathology, and pathogenesis. Pathogen Clin Aspects:3–26. https://doi.org/10.15586/codonpublications.parkinsonsdisease.2018.ch1

Kuboyama T, Tohda C, Komatsu K (2005) Neuritic regeneration and synaptic reconstruction induced by withanolide A. Br J Pharmacol 144:961–971. https://doi.org/10.1038/sj.bjp.0706122

Lakdawala AD, Dadkar NK, Dohadwalla AN (1980) Action of clonidine on the mast cells of rats. J Pharm Pharmacol 32:790–791. https://doi.org/10.1111/j.2042-7158.1980.tb13070.x

Lau AA, Hemsley KM, Meedeniya ACB (2006) Directed differentiation and characterization of genetically modified embryonic stem cells for therapy. Methods Mol Biol 329:471–484

Lin YC, Wu CR, Lin CJ (2003) The ameliorating effects of cognition-enhancing Chinese herbs on scopolamine-and MK-801-induced amnesia in rats. Am J Chin Med 31:543–549. https://doi.org/10.1142/s0192415x03001302

Lin HQ, Ho MT, Lau LS (2008) Anti-acetylcholinesterase activities of traditional Chinese medicine for treating Alzheimer's disease. Chem Biol Interact 175:352–354. https://doi.org/10.1016/j.cbi.2008.05.030

Lutton C, Young YW, Williams R et al (2012) Combined VEGF and PDGF treatment reduces secondary degeneration after spinal cord injury. J Neurotrauma 29:957–970

Manyam BV, Dhanasekaran M, Hare TA (2004a) Effect of antiparkinson drug HP-200 (*Mucuna pruriens*) on the central monoaminergic neurotransmitters. Phytother Res 18:97–101. https://doi.org/10.1002/ptr.1407

Manyam BV, Dhanasekaran M, Hare TA (2004b) Neuroprotective effects of the antiparkinson drug *Mucuna pruriens*. Phytother Res 18:706–712. https://doi.org/10.1002/ptr.1514

Mestry M, Bajaj A, Lalan K (2016) Herbal CNS stimulants. Int J Herbal Med 4:109–116

Misra DS, Maitia R, Ghosha D (2009) Protection of swimming-induced oxidative stress in some vital organs by the treatment of composite extract of *Withania somnifera*, *Ocimum sanctum* and *Zingiber officinalis* in male rat. Afr J Trad CAM 6:534–543

Molina-Hernández M, Tellez-Alcántara N, Pérez García J et al (2004) Anxiolytic-like actions of leaves of *Casimiroa edulis* (Rutaceae) in male Wistar rats. J Ethnopharmacol 93:93–98. https://doi.org/10.1016/j.jep.2004.03.039

Mondal S, Gupta M, Mazumder U et al (2005) Hepatoprotective activity of *Diospyros malabarica* bark in carbon tetrachloride intoxicated rats. Eur Bull Drug Res 13:25–30

Mora S, Díaz-Véliz G, Millán R et al (2005) Anxiolytic and antidepressant-like effects of the hydroalcoholic extract from *Aloysia polystachya* in rats. Pharmacol Biochem Behav 82:373–378. https://doi.org/10.1016/j.pbb.2005.09.007

Murrell W, Wetzig A, Donnellan M et al (2008) Olfactory mucosa is a potential source for autologous stem cell therapy for Parkinson's disease. Stem Cell 26:2183–2192

Murugaiyah V, Mattson MP (2015) Neurohormetic phytochemicals: an evolutionary–bioenergetic perspective. Neurochem Int 89:271–280

Muthuraman A, Singh N (2011) Attenuating effect of *Acorus calamus* extract in chronic constriction injury induced neuropathic pain in rats: an evidence of anti-oxidative, anti-inflammatory, neuroprotective and calcium inhibitory effects. BMC Complement Altern Med 11:24. https://doi.org/10.1186/1472-6882-11-24

Niranjan A, Prakash D (2008) Chemical constituents and biological activities of turmeric (*Curcuma longa* L.) a review. J Food Sci Technol 45:109–116

Nirmal SA, Dhasade VV, Laware RB et al (2011) Antihistaminic effect of *Bauhinia racemose* leaves. J Young Pharmacists 3:129–131. https://doi.org/10.4103/0975-1483.80301

Niyati A, Shah MB, Suhagiab BN et al (2005) A pharmacognostical study on *Convolvulus prostratus* Forssk. J Nat Remedies 5:108–114

Norazit A, Meedeniya ACB, Nguyen MN et al (2010) Progressive loss of dopaminergic neurons induced by unilateral rotenone infusion into the medial forebrain bundle. Brain Res 1360:119–121

Norazit A, Mellick G, Meedeniya ACB (2011a) Oxidative stress in Parkinson's disease; parallels between current animal models, human studies, and cells. In: Oxidative stress II. InTech Open Access Publisher. https://doi.org/10.5772/33459

Norazit A, Nguyen MN, Dickson C et al (2011b) Vascular endothelial growth factor and platelet derived growth factor modulates the glial response to a cortical stab injury. Neurosci 192:652–660

Nuñez L, D'Aquino M (2012) Microbicide activity of clove essential oil (*Eugenia caryophyllata*). Braz J Microbiol 43:1255–1260. https://doi.org/10.1590/s1517-83822012000400003

de Oliveria DM, Barreto G, De Andrade DVG et al (2008) Cytoprotective effect of *Valeriana officinalis* extract on an in vitro experimental model of Parkinson's disease. Neurochem Res 34:215–220. https://doi.org/10.1007/s11064-008-9749-y

Owolabi OJ, Amaechina FC, Eledan AB (2008) Central nervous system stimulant effect of the ethanolic extract of *Kigelia africana*. J Med Plants Res 2:20–23

Pase PM, Kean J, Sarris J et al (2012) J Alter Complement Med 18:647–652. https://doi.org/10.1089/acm.2011.0367

Piccinelli AL, Arana S, Caceres A et al (2004) New lignans from the roots of *Valerianaprionophylla* with antioxidative and vasorelaxant activities. J Nat Products 67 (7):1135–1140. https://doi.org/10.1021/np049879c

Prakash J, Gupta S (2000) Chemopreventive activity of *Ocimum sanctum* seed oil. J Ethnopharmacol 72:29–34. https://doi.org/10.1016/s0378-8741(00)00194-x

Prakash J, Yadav SK, Chouhan S (2013) Neuroprotective role of *Withania somnifera*root extract in maneb–paraquat induced mouse model of parkinsonism. Neurochem Res 38:972–980. https://doi.org/10.1007/s11064-013-1005-4

Prasad S, Aggarwal B (2011) Turmeric the golden spice, from traditional medicine to modern medicine. In: Benzie, Wachtel-Galor (eds) Herbal medicine: biomolecular and clinical aspects, 2nd edn. CRC Press/Taylor & Francis, Boca Raton, FL

Purane LM, Vidyadhara S (2016) Antistressor activity of *Diospyros malabarica* (desr.) kostel in mice and rats. Int J Pharm Sci Res 7:2320–5148

Rabiei Z, Solati K, Amini-Khoei H (2019) Phytotherapy in treatment of Parkinson's disease: a review. Pharm Biol 57:355–362. https://doi.org/10.1080/13880209.2019.1618344

Rai K, Gupta N, Dharamdasani L et al (2017) *Bacopa monnieri*: a wonder drug changing fortune of people. Int J Appl SciBiotech 5:127–132. https://doi.org/10.3126/ijasbt.v5i2.16952

Rajakrishnan V, Viswanathan PK, Rajasekharan N (1999) Neuroprotective role of curcumin from Curcuma longa on ethanol-induced brain damage. Phytother Res 13(7):571–574. https://doi.org/10.1002/(SICI)1099-1573(199911)13:7<571

Riederer P, Jellinger K (1983) Neurochemical insights into monoamine oxidase inhibitors, with special reference to deprenyl (selegiline). Acta Neurol Scand 68:43–55. https://doi.org/10.1111/j.1600-0404.1983.tb01516.x

Robinson AJ, Meedeniya AC, Hemsley KM et al (2005) Survival and engraftment of mouse embryonic stem cell-derived implants in the Guinea pig brain. Neurosci Res 53:161–168

Sahu P, Pal A, Nanda G, Champatisingh D (2011) Anticataleptic and antiepileptic activity of ethanolic extract of leaves of *Mucuna pruriens*: a study on role of dopaminergic system in epilepsy in albino rats. Ind J Pharmacol 43:197. https://doi.org/10.4103/0253-7613.77368

Sanberg PR (1980) Haloperidol-induced catalepsy is mediated by postsynaptic dopamine receptors. Nature 284:472–473. https://doi.org/10.1038/284472a

Sankar S, Manivasagam T, Krishnamurti A et al (2007) The neuroprotective effect of *Withania somnifera* root extract in MPTP-intoxicated mice: An analysis of behavioral and biochemical variables. Cell Mol Bio Lett 12:473–481. https://doi.org/10.2478/s11658-007-0015-0

Sarin CA (2017) Development and evaluation of anti-inflammatory herbal formulation. Int J Ayurvedic Herbal Med 9:323–326. https://doi.org/10.18535/ijahm/v7i4.18

Sarris J, Panossianc A, Schweitzer I et al (2011) Herbal medicine for depression, anxiety and insomnia: a review of psychopharmacology and clinical evidence. J Euroneuro 21(12):841–886

Schulz TS, Noggle SC, Palmarini GM et al (2004) Differentiation of human embryonic stem cells to dopaminergic neurons in serum free suspension culture. Stem Cells 22:1218–1238

Sengupta T, Vinayagam J, Nagashayana N et al (2010a) Antiparkinsonian effects of aqueous methanolic extract of *hyoscyamus niger* seeds result from its monoamine oxidase inhibitory and hydroxyl radical scavenging potency. Neurochem Res 36(1):177–186. https://doi.org/10.1007/s11064-010-0289-x

Sengupta T, Vinayagam J, Nagashayana N et al (2010b) Antiparkinsonian effects of aqueous methanolic extract of *hyoscyamus niger* seeds result from its monoamine oxidase inhibitory and hydroxyl radical scavenging potency. Neurochemical Res 36(1):177–186. https://doi.org/10.1007/s11064-010-0289-x

Senthil KKK, Kapoor BR (2010) Effect of *Oxalis corniculate* L. extracts on antioxidant enzymes levels in rat brain after induction of seizures by MES and PTZ. Int J Biopharm 1:58–61

Sharma U, Bala M, Kumar N, Singh B et al (2012) Immunomodulatory active compounds from *Tinospora cordifolia*. J Ethnopharmacol 141(3):918–926. https://doi.org/10.1016/j.jep.2012.03.027

da Silva JA, da Costa MJDC, da Alves MDCR, da Braga JEF, da Lima CMBL, da Pordeus LCDM (2017) Effects of the single supplementation and multiple doses of *Passiflora incarnata* L. on human anxiety: a clinical trial, double-blind, placebo-controlled, randomized. Int Arch Med 10. https://doi.org/10.3823/2276

da Silveira VM, Gomes-Rochette NF, Nunes-Pinheiro DCS et al (2019) Ginger (Zingiber officinale Roscoe). In: Nabavi SM, Silva AS (eds) Nonvitamin and nonmineral nutritional supplements, 1st edn. Academic Press, London, pp 235–239. https://doi.org/10.1016/B978-0-12-812491-8.00034-5

Singh HK, Dhawan BN (1982) Effect of *Bacopa monniera* Linn. (Brāhmi) extract on avoidance responses in rats. J Ethnopharmacol 5:205–214. https://doi.org/10.1016/0378-8741(82)90044-7

Singh HK, Dhawan BN (1997) Neuropsychopharmacology effects of the Ayurvedic nootropic *Bacopa monniera* Linn. (Brahmi). Ind J Pharmacol 29:359–365

Singh AP, Sarkar S, Tripathi M et al (2013a) Mucuna pruriens and its major constituent L-DOPA recover spermatogenic loss by combating ROS, loss of mitochondrial membrane potential and apoptosis. PLoS One 8(1):e54655. https://doi.org/10.1371/journal.pone.0054655

Singh M, Kaur M, Kukreja H et al (2013b) Acetylcholinesterase inhibitors as Alzheimer therapy: from nerve toxins to neuroprotection. Eur J Med Chem 70:165–188. https://doi.org/10.1016/j.ejmech.2013.09.050

Sloley BD, Urichuk LJ, Morley P et al (2000) Identification of kaempferol as a monoamine oxidase inhibitor and potential neuroprotectant in extracts of *Ginkgo biloba* leaves. J Pharm Pharmacol 52:451–459. https://doi.org/10.1211/0022357001774075

Son TG, Camandola S, Mattson MP (2008) Hormetic dietary phytochemicals. NeuroMolecular Med 10:236. https://doi.org/10.1007/s12017-008-8037-y

Sowndhararajan K, Kim S (2017) Neuroprotective and cognitive enhancement potentials of *Angelica gigas* Nakai root: a review. Sci Pharm 85:21. https://doi.org/10.3390/scipharm85020021

Stough C, Lloyd J, Clarke J et al (2001) The chronic effects of an extract of *Bacopa monniera* (Brahmi) on cognitive function in healthy human subjects. Psychopharmacology 156:481

Stough C, Downey LA, Lloyd J et al (2008) Examining the nootropic effects of a special extract of *Bacopa monniera* on human cognitive functioning: 90 day double-blind placebo-controlled randomized trial. Phytother Res 22:1629–1634

Stough C, Scholey A, Cropley V et al (2013) Examining the cognitive effects of a special extract of Bacopa monniera (CDRI08: Keenmnd): a review of ten years of research at Swinburne university. J Pharm Sci 16:254–258

Sundaramurthi P, Packiam KK (2017) A review on pharmacognosy and pharmacology of *Evolvulus alsinoides* (L.). Int Res J Pharmacy 8:1–4. https://doi.org/10.7897/2230-8407.087110

Sutalangka C, Wattanathorn J (2017) Neuroprotective and cognitive-enhancing effects of the combined extract of *Cyperus rotundus* and *Zingiber officinale*. BMC Complement Alternat Med 17:135. https://doi.org/10.1186/s12906-017-1632-4

Taher YA, Samud AM, El-Taher FE et al (2015) Experimental evaluation of anti-inflammatory, antinociceptive and antipyretic activities of clove oil in mice. Libyan J Med 10:28,685. https://doi.org/10.3402/ljm.v10.28685

Tohda C, Ichimura M, Bai Y (2008) Inhibitory effects of *Eleutherococcus senticosus* extracts on amyloid β (25-35)–induced neuritic atrophy and synaptic loss. J Pharmacol Sci 107:329–339. https://doi.org/10.1254/jphs.08046fp

Tuffaha SH, Budihardjo JD, Sarhane KA et al (2016) Growth hormone therapy accelerates axonal regeneration, promotes motor reinnervation, and reduces muscle atrophy following peripheral nerve injury. Plast Reconstr Surg 137:1771–1780. https://doi.org/10.1097/prs.0000000000002188

Uabundit N, Wattanathorn J, Mucimapura S et al (2010) Cognitive enhancement and neuroprotective effects of *Bacopa monnieri* in Alzheimer's disease model. J Ethnopharmacol 127:26–31. https://doi.org/10.1016/j.jep.2009.09.056

Uragoda CG (1983) History of opium in Sri Lanka. Med Hist 27:69–76

Valdez LM, Taboada J, Valdez JE (2015) Ancient use of coca leaves in the Peruvian central highlands. J Anthropological Res 71:231–258

Veerendra Kumar MH, Gupta YK (2003) Effect of *Centella asiatica* on cognition and oxidative stress in an intracerebroventricular streptozotocin model of Alzheimer's disease in rats. Clin Exp Pharmacol Physiol 30:336–342. https://doi.org/10.1046/j.1440-1681.2003.03842.x

Venkatesan R, Ji E, Kim S (2015) Phytochemicals that regulate neurodegenerative disease by targeting neurotrophins: a comprehensive review. Biomed Res Int 2015:814068. https://doi.org/10.1155/2015/814068

Weetman JA, Sharry S, Wong MB et al (2013) Increased SUMO-1 expression in the unilateral rotenone-lesioned mouse model of Parkinson's disease. J Neurosci Lett 544:119–124

Wiart C (2006) Plants affecting the central nervous system. In: Wiart C (ed) Ethnopharmacology of medicinal plants: Asia and the Pacific. Humana Press, Totowa, NJ

Wilkinson D (2001) Galantamine: a new treatment for Alzheimer's disease. Expert Rev Neurother 1:153–159. https://doi.org/10.1586/14737175.1.2.153

Wink M (2018) Plant secondary metabolites modulate insect behavior-steps toward addiction? Front Physiol 9:364. https://doi.org/10.3389/fphys.2018.00364

Xia F, Li C, Li M et al (2018) Antidepressant activity of an aqueous extract from okra seeds. RSC Adv 57:32814–32822. https://doi.org/10.1039/C8RA03201G

Yang BY, Guo JT, Li ZY et al (2016) New thymoquinol glycosides and neuroprotective dibenzo cyclooctane lignans from the rattan stems of *Schisandra chinensis*. Chem Biodivers 13 (9):1118–1125. https://doi.org/10.1002/cbdv.201500311

Yellamma K, Vanaja D (2016) *Evolvulus alsinoides* plant extract as a cognitive enhancer of special memory in AD-induced rat model. Res J Pharm Biol Chem Sci 7:2050–2054

Yuan SM, Gao K, Wang DM (2011) Evodiamine improves cognitive abilities in SAMP8 and $APP^{swe}/PS1^{\Delta E9}$ transgenic mouse models of Alzheimer's disease. Acta Pharmacol Sin 32:295–302

Zeni ALB, Zomkowski ADE, Dal-Cim T et al (2011) Antidepressant-like and neuroprotective effects of *Aloysia gratissima*: investigation of involvement of l-arginine-nitric oxide-cyclic guanosine monophosphate pathway. J Ethnopharmacol 137:864–874. https://doi.org/10.1016/j.jep.2011.07.009

Zhou Y, Peng L, Zhang W-D et al (2009) Effect of triterpenoid saponins from Bacopa monniera on scopolamine-induced memory impairment in mice. Planta Med 75(6):568–574. https://doi.org/10.1055/s-0029-1185339

# Ameliorative Effects of Shodhana (Purification) Procedures on Neurotoxicity Caused by Ayurvedic Drugs of Mineral and Herbal Origin

**Rabinarayan Acharya, Anagha Ranade, Mayur Surana, and Sharad D. Pawar**

## Contents



R. Acharya
Department of Dravyaguna Vigyan, Institute for Teaching & Research in Ayurveda, Jamnagar, Gujarat, India

A. Ranade · S. D. Pawar (✉)
Regional Ayurveda Institute for Fundamental Research, Pune, Maharashtra, India
e-mail: rophraifr.pu-ayush@gov.in

M. Surana
Vardhayu Ayurved, Nashik, Maharashtra, India

 
D. C. Agrawal, M. Dhanasekaran (eds.), *Medicinal Herbs and Fungi*,
https://doi.org/10.1007/978-981-33-4141-8_14

**Abstract** Recent reports suggest that the global burden of neurological diseases is on a constant rise. These appear to result from various metal-containing anthropogenic environmental causes, herbal medications, and dietary factors. Many metals like mercury, arsenic, and lead have shown neurotoxicity. Similarly, many herbs also have been reported to be neurotoxic.

Ayurveda drugs consist of either single drug or medicinal formulations of herbal and herbo-mineral and animal origin. The traditionally used *bhasma* (incinerated ash of organometallic compounds) preparations or formulations consisting of herbs from *visha* (poisonous) and *upavisha* (semi-poisonous) groups often pose a threat of toxicity. Medicinal plants are often a potential source of metals viz. mercury, arsenic, lead, etc. that cause neurotoxicity in modern toxicological parlance.

In order to combat the harmful effects produced as a result of single drug or formulations containing the metal/herbal neurotoxicants, Ayurveda emphasizes a process, namely *Shodhana* (purification). It is stated in the various treatises that processing of the mineral drugs mentioned in *Rasashastra* (Indian Iatrochemistry) with classically prescribed traditional purification procedures consisting of medicinal herb juices/decoctions can help reduce poisonous effects and also potentiate their therapeutic value. After critically exploring the available literature on the effect of *Shodhana* in ameliorating neurotoxicity with drugs of mineral and herbal origin mentioned in Ayurveda, it has been observed that adopting scientifically validated traditional purification techniques for poisonous medicinal drugs of plant and mineral origin can ameliorate their neurotoxic effects and enhance their therapeutic efficacy.

**Keywords** *Ayurveda* · Herbs · Metals · Minerals · Neurotoxicity · *Shodhana* (purification)

## Abbreviations

| | |
|---|---|
| AFI | Ayurvedic Formulary of India |
| $As_4S_4$ | Tetraarsenic tetrasulfide |
| CNS | Central Nervous System |
| $HgCl_2$ | Mercuric chloride |
| HgS | Mercury sulfide |
| HPTLC | High-performance thin layer chromatography |
| IQ | Intelligence quotient |
| MeHg | Methyl mercury |
| $NaAsO_2$ | Sodium arsenite |
| $NaH_2AsO_4$ | Sodium dihydrogen arsenate |
| $Pb(CH_3COO)_2$ | Lead acetate trihydrate |
| PbO | Lead oxide |
| PbS | Lead sulfide |
| PTZ | Pentylenetetrazole |
| SMA | Spontaneous motor activity |
| TEM | Transmission electron microscopy |

## 1 Introduction

Ayurveda has treatment modalities with drugs of plant and mineral origin. Among the medicinal plants, Ayurveda entails few in the group of poisonous categories referred to as *Upavisha varga. Upavisha* is a group of plant-based drugs, which are not lethal but produce moderate toxic symptoms on consumption or administration. The Schedule E1 of Drug and Cosmetic Act, 1945, of Govt. of India has included them under the poisonous herb list of indigenous medicine (Anonymous 2003). The plants under the *Upavisha* group include *Gunja* (*Abrus precatorius* Linn. (Seed)), *Shringivisha* (*Aconitum chasmanthum* Stapf. ex Holmes), *Danti* (*Baliospermum montanum* Mull. Arg.), *Arka* (*Calotropis procera* (Ait.) R. Br), *Bhanga* (*Cannabis sativa* Linn. (Except seeds)), *Jayapala* (*Croton tiglium* Linn.), *Dhatura* (*Datura metal* Linn.), *Langali* (*Gloriosa superba* Linn.), *Parasika Yavani* (*Hyoscyamus niger* Linn.), Mill, *Ahiphena* (*Papaver somniferum* Linn. (except seeds)), *Bhallataka* (*Semecarpus anacardium* Linn.), and *Kuchala* (*Stychnos nux vomica* Linn). Among these, *Vatsanabha* (*Aconitum ferox* Wall, ex Ser.) comes under the *Visha* category, i.e., poisonous category.

The drugs of mineral origin in Schedule E1 consist of *Gauripashana* (arsenic), *Hartala* (arseno sulfide), *Manahashila* (arseno sulfide), *Parada* (mercury), *Rasa Karpura* (hydrargyri subchloridum), *Tuttha* (copper sulfate), *Hingula* (cinnabar), *Sindura* (red oxide of lead), and *Girisindura* (red oxide of mercury). Thus, it can be inferred that the majority of them are organometallic compounds of arsenic, mercury, copper, and lead. Purification processes for poisonous medicinal plants and

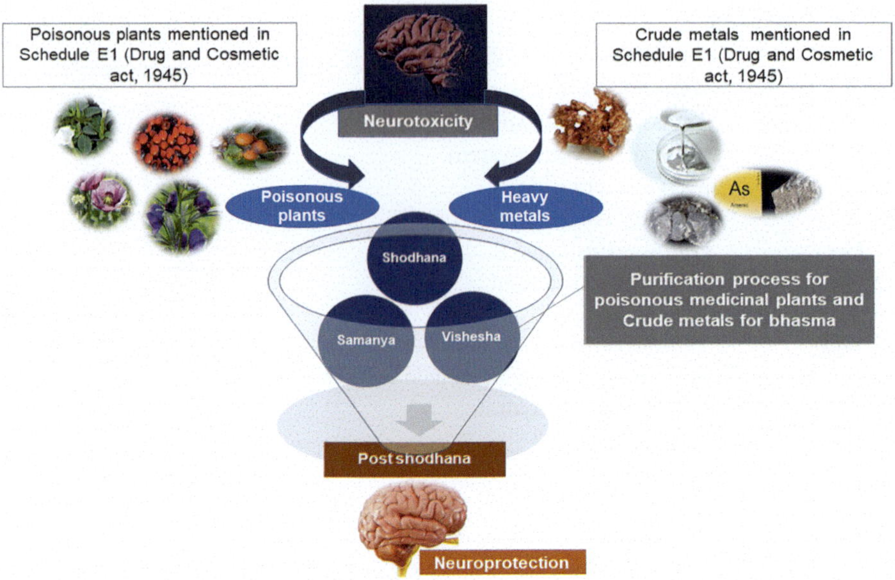


**Fig. 1** *Shodhana* (Purification) processes for poisonous medicinal plants and crude metals for *bhasma* (incinerated ash of organometallic compounds)

crude metals for *bhasma* (incinerated ash of organometallic compounds) are illustrated in Fig. 1.

With an exponential rise in the use of medicinal plants in the management of diseases, herb–drug interactions, and toxicities arising due to ingestion are a major concern in scientific society. Few toxicities arising are obvious due to the use of herbs having some myths regarding "no side effects." Moreover, lack of regulatory vigilance over procurement and sale of crude medicinal plants, suspicious practitioners who have limited knowledge about the harmful effects, self-treatment by individuals, and improper techniques of administration further add to the toxicities.

In the case of drugs of mineral origin, the *bhasma* (incinerated ash of organometallic compounds) of Ayurveda are prepared by subjecting crude form of mineral to "alchemy" procedures to alter their original forms suitable for the medication (i.e., HgS, $As_4S_4$, PbS, PbO). They appear distinct from other environmental metal contaminants ($HgCl_2$, MeHg, $NaAsO_2$, $NaH_2AsO_4$, $Pb(CH_3COO)_2$, etc.) (Liu et al. 2019). In the *Rasashastra* (Indian Iatrochemistry) literature for drugs of mineral/metallic origin, a series of procedures have been mentioned, which start from *Shodhana* (detoxification), *Marana* (incineration of metals/minerals by treating them with specifically prescribed medicinal herbs, to convert hard metal/mineral into fine and soft powder) and ends with *Amritikaran* (further treatment with metal-specific herbs) leading to the formation of nontoxic *Bhasmas.* The ancient texts have emphasized on quality testing procedures of "*Bhasma*." These tests are ultimately related to the particle size, density, and chemical and physical stability of "*Bhasma*" under high-temperature conditions (Table 1) (Savrikar and Ravishankar 2011).

**Table 1** Quality parameters for *Bhasma*

| Name of the test | Description with the meaning of test terms |
|---|---|
| *Rekhapurnatva* | When all the particles of *bhasma* enter the ridges of the fingers and do not easily come out, then it is said to be *Rekhapurnabhasma*. |
| *Varitaratva* | When all the particles of the *bhasma* float on the water surface, then it is called *Varitara*. |
| *Unam* | When a heavy grain (rice/wheat, etc.) is placed on the film of the *bhasma*, it floats over it and does not sink; then, the *bhasma* is said to be passing the *Unampariksha*. |
| *Apunarbhavatva* | After treating/incinerating with *mitrapanchaka*, if *bhasma* does not contain any free metal, then the *bhasma* is said to be *Apunarbhava*. |
| *Niruttha* | If the *Bhasma* is heated with silver up to the melting of silver and cooled, *Bhasma passes the niruttha* test if the weight of silver used remains unchanged. |
| *Dantagre Kachkachaabhava* | When a small amount of *bhasma* is chewed, if it does not feel gritty under teeth, then the *bhasma* can be said to be having fine particles. |
| *Varna* | The *bhasma* should possess a specific color after preparation as per classics. |
| *Nisvadu* | Ideal *bhasma* must be tasteless. |
| *Avami* | Ingestion of *bhasma* must not produce any nausea. |
| *Nischchandratva* | When prepared *bhasma* taken in a petri dish and observed under sunlight using a magnifying glass does not contain any luster, then it is said to be *Nishchandra* because luster indicated the presence of free metal. |

These successive procedures and *Marana* like treatments to crude minerals in *Ayurveda* render them nontoxic and favorable for the body. If these procedures are bypassed or not appropriately followed, untoward effects ought to be produced clinically (Sarkar et al. 2010). Normally, the poisonous plants and mineral drugs in their crude forms have been reported to exhibit nephrotoxicity, hepatotoxicity, neurotoxicity, dermal toxicity, etc. The present chapter focuses on the neurotoxicity aspects of medicinal plants and minerals covered in E1 Schedule of Drug and Cosmetic Act (Anonymous 2003) and ameliorative effects of *shodhana* procedure mentioned in Ayurveda on their neurotoxic properties.

## 2 Concept of *Shodhana* (Purification) in Ayurveda

Despite their adverse effects, *Vishadravya* (poisonous substances) of plant and mineral origin are used in Ayurveda owing to their efficacy in small doses as compared to nonpoisonous drugs. To use them as a therapeutic, individually or in formulations, they are subjected to some traditional processing techniques that render them safe for internal administration (Rasavagbhata 1995).

*Shodhana* (purification) is the process, which involves the conversion of any poisonous drug into beneficial, nonpoisonous/nontoxic ones. This comprises purification as well as a decrease in the levels of toxic constituents (active toxic

**Table 2** *Shodhana* of poisonous plants from E1 Schedule as mentioned in the Ayurvedic Formulary of India (AFI)

| Name in Ayurveda and plant species | *Shodhana* as mentioned in AFI |
|---|---|
| *Gunja (Abrus precatorius* Linn.) | *Swedana* is done in *Kanji* for 3 hours in *Dola Yantra.* |
| *Vatsanabha* (*Aconitum ferox* wall) | Small pieces of *Vatsanbha* are soaked in *gomutra* for 3 days, replacing the *gomutra* every day. Then they are washed and dried. |
| *Bhanga* (*Cannabis sativa* Linn.) | *Bhanga is* put in a muslin bag and washed in water till free from turbidity and then dried. |
| *Jaypal* (*Croton tiglium* Linn.) | The testa of *Jaypala* is removed, and the seeds are subjected to *Swedana in Dola Yantra* with *godugdha* for 3 h. The embryo of the seed is removed. The cotyledons dried and powdered. Trituration with *nimbu swarasa* is done for 3 days, and then it is dried in the sun and used. |
| *Dhatura* (*Datura metel* Linn) | *Dhatura* seeds are soaked in *Gomutra* for 12 h. Then they are washed with water and subjected to *Swedana* in a *Dola Yantra* containing Godugdha for 3 h. The seeds are used after removing the testa. |
| *Langali* (*Gloriosa superba* Linn.) | Small pieces of the root of *Langali* are soaked in *Gomutra* for 24 h, then washed and dried. |
| *Ahiphena* (*Papaver somniferum* Linn) | Trituration of rhizomes with ginger juice for 21 days. |
| *Kuchala* (*Stychnos nux vomica* Linn) | *Kuchala* seeds are kept in *Gomutra* for 7 days. *Fresh gomutra is* to be replaced every day. After that, it is removed and washed with water. *Swedana* in *Godugdha* with *Dola Yantra* for 3 h is done. The testa and embryo are removed; the cotyledon is roasted in ghee and powdered well. |
| *Bhallataka* (*Semecarpus anacardium* Linn) | *Bhallataka* fruits after removing the attachment of thalamus are soaked in *Gomutra* for 7 days and after that in *Godugdha* for 7 days. To reduce the oil content, the seeds are then put into a bag containing coarse brick powder with which they are rubbed carefully. Then the fruits are washed with water and dried in air. |

principles), which also results in enhanced therapeutic efficacy. *Shodhana* has been defined as a technique that is employed to separate physical and chemical impurities from the *dravya* (plant/mineral) by treatment with various herbs. It is carried out by using pharmaceutical procedures of *Swedana* (boiling), *Mardana* (trituration), etc. with particular drugs. The *shodhana* methods of poisonous plants mentioned under the E1 Schedule as per Ayurvedic formulary of India have been specified in Table 2.

*Shodhana* procedure, in general, has been classified into two types:

- *Samanya Shodhana* (common detoxification procedure).
- *Vishesha Shodhana* (specific detoxification procedure).

## *2.1* Samanya Shodhana *(Common Detoxification Procedure)*

This consists of general purification measures viz. one in which the *dravya* (plant/mineral) is washed, cut into pieces and soaked in *gomutra* (cow's urine) for some days and then removed and used further in formulations. Secondly, *swedana* (boiling in milk or decoction) is employed wherein a special arrangement is made referred to as *Dola Yantra*. The drug to be fomented, is tied in a muslin cloth, and the resultant bag-like structure (*pottali*) tied to a stick is suspended in the vessel consisting of milk/decoction in such a way that the surface of "*pottali*" just touches/dips the hot surface of the milk (Acharya 2014).

## *2.2* Vishesha Shodhana *(Specific Detoxification Procedure)*

Certain specific purification practices are recommended for a peculiar drug, e.g., *Shodhana* of *Bhallataka*, i.e., fruit of *S. anacardium* in brick powder; seeds of (*A. precatorius*) *Gunja* seeds in *Kanji* (sour gruel) and immersion of nux vomica seeds in cow's urine followed by fomentation (*swedana*) with cow's milk and roasting in cow's milk clarified butter (ghee) (Chunekar and Pandey 1999). The rhizome of *Vacha* with *panchapallava kwatha* is also an example of *vishesha shodhana* (Das 2008). The following are the various methods of specific detoxification (Sarkar et al. 2008; Murthy 2012; Upadhyaya 1999).

### 2.2.1 *Achushana* (Absorption)

Under this method, oily/greasy contents of certain toxic materials are reduced through different absorption means, e.g., *Bhallataka Shodhana* with brick powder.

### 2.2.2 *Bharjana* (Frying or Roasting)

The drug is fried with specific liquid media indigenous cow's ghee in particular on (*mandagni*) mild flame viz. e.g., *Shodhana* of *Hingu* (*Ferula narthex* exudate). *Shodhana* of *Gairika* (ochre) by frying in ghee.

### 2.2.3 *Bhavana* (Levigation)

*Bhavana* refers to the trituration of drugs with prescribed liquid media, which consist of expressed juices, decoctions for a specific time. For example, *Ahiphena* (Opium exudates) *Shodhana* with ginger juice; *Jambir* (citrus variety fruit) juice levigation to *Tamra patra* (crude copper).

### 2.2.4 *Nimajjana* (Dipping)

The *dravya* is immersed in the specified liquid media as per classics for a specific time, e.g., detoxification of Aconitum rhizome in cow's urine. Detoxification of Strychnos seeds in cow's urine.

### 2.2.5 *Parishravana* (Straining)

The solid drug is dissolved in specific liquid media as mentioned in classics and separated from insoluble impurities through straining, e.g., *Guggulu* with *triphala* decoction.

### 2.2.6 *Prakshalana* (Washing)

The drug is washed with specific liquid to remove its physical impurities, e.g., the tuber of *Puereria tuberosa* after collection is subjected to washing with water; washing of *Asparagus racemosus* (*shatavari*) roots with potable water to remove dust, mud, and insects.

### 2.2.7 *Prithakikarana* (Separation)

Physical impurities like adulteration of brick powder are removed when kept immersed in water, e.g., *Mallotus philippensis* (*Kampillaka);* removal of physical impurities in *Abhraka* (mica).

### 2.2.8 *Swedana* (Boiling Under a Liquid Bath)

The drug is boiled in prescribed liquid media through the *Dola Yantra* method, e.g., seeds of *C. tiglium* are boiled in indigenous cow's milk in *Dola Yantra; Dola Yantra* is also employed in *shodhana* of *sudha varga* (group of calcium-rich drugs).

# 3 Neurotoxic Effect of Medicinal Plants of E1 Schedule

## *3.1* Gunja (A. precatorius *Linn. (Seed))*

*A. precatorius* is a poisonous plant that possesses one of the potentially toxic principles, Abrin, a toxalbumin in its seeds that inhibits protein synthesis, thereby causing cell death (Gadadhar and Karande 2013). The seeds are crushed and taken orally for suicidal purpose. A retrospective study of poisoning cases due to *Gunja* has reported the presence of neurological symptoms like altered sensorium, tremors, and seizures in some of those cases (Karthikeyan and Amalnath 2017; Pillay et al. 2005). Since ancient times, *Gunja* seeds have been employed as poultry poison, cattle poison, and fish poisoning (Olsnese 2004).

Some rare neurological symptoms also include convulsions (Patil et al. 2016), demyelinating encephalitis (Sahoo et al. 2008; Sahni et al. 2007), raised intracranial pressure (ICP) and papilledema (Subrahmanyan et al. 2008), dural venous sinus thrombosis (Vinod et al. 2013), and pontine myelinolysis (Sarkar et al. 2017).

## *3.2* Shringivisha *(*A. chasmanthum *Stapf. Ex Holmes) and* Vatsanabha *(*A. ferox *Wall, ex Ser)*

The members of the genus *Aconitum* have considerable amounts of highly toxic aconitine and related alkaloids in their roots and tubers. Neurological symptoms arising due to the ingestion of these tubers include paraesthesia, numbness of the face, and muscle weakness (Moritz et al. 2005; Gao et al. 2018). The neurotoxic effects of aconitine are due to its action on voltage-sensitive sodium channels in axons, thereby blocking neuromuscular transmission by decreasing the release of acetylcholine (Muroi et al. 1990). Vertigo, difficulty in speech, weakness of muscles associated with twitching and spasms are also neurological symptoms reported for *Aconitum* (Modi 2005).

## *3.3* Bhanga *(*C. sativa *Linn. (Except Seeds))*

Acute intoxication with *Cannabis* in man results in the impairment of short-term memory (Solowij 1998; Miller and Branconnier 1983). This drug is often used commonly for drug abuse. Some recreational symptoms appreciated by drug addicts are euphoria, perception alterations such as time and spatial distortion, intensification of ordinary sensory experiences, and motor impairment. Some unpleasant reactions include fear, depression, etc. In the pediatric population, jerking, delirium, and anxiety have been reported (Turner et al. 2020).

## *3.4* Dhatura *(*Datura metel *Linn.)*

The seeds of this plant produce hallucinations on ingestion, causing poisoning. Consumption of any part of the plant may result in a severe anticholinergic reaction that may lead to toxicity (Boumba et al. 2005). Other toxic symptoms may comprise ataxia, impaired short-term memory, disorientation, confusion, hallucinations, psychosis, agitated delirium, seizures, coma (Alberto et al. 2001; Diker et al. 2007). *Datura* species contain copious amounts of belladonna alkaloids that can generate both local and systemic anticholinergic toxicity (Krenzelok 2010).

## *3.5* Langali (G. superba *Linn.)*

The root of this climber is poisonous, consisting of colchicines, gloriosine, and superbine. Accidental ingestion or suicidal attempts of this tuber result in specific neurological symptoms like confusion and conclusions (Rathnayaka et al. 2015). It is also reported to exhibit acute ascending polyneuropathy (Yaron et al. 2010).

## *3.6* Ahiphena (P. somniferum *Linn. (Except Seeds))*

It is the botanical source of morphine, codeine, and other alkaloids. The opioids have been used since ages as a recreational drug where overdosage results in narcotic effects. The neurotoxic symptoms due to poisoning include central nervous system depression, pinpoint pupils, and depressed respiration (Anonymous 2019).

## *3.7* Kuchala (Stychnos nux vomica *Linn)*

The seeds of this tree are a potential source of toxin named strychnine. Strychnine is known to affect the central nervous system as a competitive antagonist on the postsynaptic glycine receptors. The inhibitory effect of the spinal interneurons on the muscles is lost. This leads to twitching, muscle spasms, and these muscle spasms in a conscious patient resemble “awake” seizure known as spinal seizure somewhat like tetanus.

## 4 Reported Literature on the Effect of *Shodhana* on Poisonous Plants of E1 Schedule

### *4.1* Gunja *(*A. precatorius *Linn. (Seed Coat))*

(i) Red variety seeds of *Gunja* when processed with cow's milk for 6 h whereas, sour gruel, lemon juice, and water (as control) for 3 h following the standard of Ayurveda classics resulted in a decrease of protein contents in most of the samples after *shodhana* (Roy et al. 2016).
(ii) A study has documented the depletion of toxic alkaloid hypaphorine and abrin on the purification of *Gunja* with cow's milk or *Kanji* (Singh et al. 1998). Similarly, a decrease in hemagglutination property in post-*shodhana* samples of *Gunja* seeds has been reported (Nabar et al. 2011; Wadnerwar et al. 2017).
(iii) In acute toxicity, pre-*shodhana* red varieties of the pericarp of *Gunja* were found toxic, whereas neither mortality nor any significant alteration in behavior was observed in the pericarp of milk and *Kanji Shodhita Gunja* (Gautam 2018; Barve and Ojha 2013).
(iv) In another study of *Shodhana* in black and white varieties of *Gunja,* a decrease in weight, change in color, and decrease in the number of phytoconstituents (as per HPTLC findings) was observed (Gautam 2017).

### *4.2* Vatsanabha *(*A. ferox *Wall, ex Ser)*

It has been reported that the percentage of aconitine in *Vatsanabha* was reduced after being processed with *Gomutra* (cow's urine) as a medium (Sarkar et al. 2008). In a separate study, the total alkaloid content decreased in *Vatsanabha* after *Shodhana* (Bodhankar 2017). Also, a decrease in the number of spots representing toxic alkaloids in HPTLC in *Shodhita Vatsanabha* was observed (Arya and Kumar 2017). Sarkar and team found that *Gomutra* and *Godugdha* render *Aconite* devoid of cardiac and neuromuscular toxic effects without affecting its antipyretic activity (Sarkar et al. 2012). Also, it has been observed that the contents of less toxic substances such as aconine, hypoaconine, and benzylhypoaconine increase possibly due to conversion of the toxic aconitine into aconine or hydrolysis of the alkaloids to their respective amino alcohols after *Shodhana* (Deore et al. 2013; Handa et al. 1951; Parikh et al. 1996; Jaiswal et al. 2013).

### *4.3 Bhanga* (C. sativa *Linn. (Except Seeds))*

Ram and team reported a reduction in the tetrahydrocannabinol content after *shodhana* of *Bhanga* (cannabis leaves) with water (Ram et al. 2018). If used in

*Ashodhita* (unprocessed) form, may cause *Madakari* (intoxicant) effects; hence, Ayurveda classics have advised *Shodhana* (purifying process) of *Bhanga* before its therapeutic use. *Bhanga* leaves processed with the water-wash method have been advised for the management of pain in cancer patients (Tavhare et al. 2019).

## *4.4* Dhatura (Datura metal *Linn.)*

(i) A decrease in total alkaloid content and an increase in total protein content of seed after *Shodhana* have been reported. Also, complete removal of scopolamine and partial removal of hyoscyamine on *Shodhana* of *Dhatura* was observed, which renders its nontoxic nature (Patel et al. 2010).

(ii) It has been reported that the total alkaloid content decreases post-*Shodhana*. Also, the concentration of hyoscyamine increases upon *shodhana* treatment, whereas it was observed that with ameteline B and calystegine were absent post *shodhana.* Anti-inflammatory and anti-ulcer activity have also been observed in *Shodhita* samples, which are comparable to standard drugs (Gopalkrishna 2011).

## *4.5* Langali (G. superba *Linn.)*

(i) *Shodhana* of *Gloriosa supera* roots in *Gomutra* (cow's urine) resulted in a decrease in colchicine content post *shodhana* (Nabar et al. 2013).

(ii) In a chronic toxicity study of *Shodhita* samples of *Gloriosa*, no significant behavioral changes were observed. The results indicated that the *shodhana* process removes or reduces certain principles which tend to cause changes in some hematological and biochemical parameters (Bhide et al. 2019).

## *4.6* Kuchala (Stychnos nux vomica *Linn.)*

(i) Mitra and coworkers have reported that alkaloids like strychnine and brucine reduced in quantity when *Kuchala* seeds were processed in *Ardraka swarasa* (fresh ginger juice) (Mitra et al. 2011a).

(ii) It was observed that after processing with *Eranda taila* (castor oil), the toxic Strychnine and Brucine contents were reduced in comparison to the raw seeds of *Kuchala* (Mitra et al. 2011b). The detoxification of *Kuchala* may probably be due to the chemical conversion of strychnine and brucine into less toxic derivatives such as isostrychnine, isobrucine, strychnine N–oxide, brucine N–oxide due to enhanced N–oxidation. The level of loganic acid content of the seeds reduced during the *Shodhana* (Choi et al. 2004). There are several reports on a reduction in alkaloid content of seeds post *shodhana* (Survase et al. 2016;

Joshi and Agrawal 1977; Kumar and Sinha 2009; Cai et al. 1993; Akbar et al. 2010).

(iii) After *shodhana*, the ethanolic extracts of *kuchala* seeds were tested in Swiss Albino mice and Wistar rats for response in spontaneous motor activity (SMA), pentobarbitone-induced hypnosis, Pentylenetetrazole (PTZ) induced convulsions, diazepam-assisted protection, and morphine-induced catalepsy. The results showed inhibition of PTZ induced convulsions, catalepsy, and reduced SMA. There was a moderate degree of CNS depressant activity that, in turn, justifies the use of the traditionally processed poisonous drug therapeutically in neurological disorders like Parkinson's disease and Trigeminal neuralgia for which satisfactory treatment is still anticipated (Katiyar et al. 2010).

# 5 Neurotoxic Effect of Drugs of Mineral Origin Described in E1 Schedule

## *5.1 Arsenic*

Arsenic poisoning has been reported to cause impairment to neurological functions such as learning, short-term memory, and concentration (Bolla-Wilson and Bleecker 1987). The most frequent neurological manifestation Arsenic is peripheral neuropathy that may last for lifelong. This may lead to rapid, severe ascending weakness, similar to the Guillan–Barré syndrome, requiring mechanical ventilation (Mazumdar et al. 1992).

## *5.2 Mercury*

Mercury exposure has been reported to damage the nervous system, thereby altering brain functions leading to tremors, irritability, memory deficit, and changes in hearing or vision (Martin and Griswold 2009). Mercury, at a lower concentration, can cause neurodegenerative disorders and nerve damage in humans, namely sclerosis and Alzheimer's disease (Carocci et al. 2014).

## *5.3 Lead*

Prospective studies have reported slower development in the sensorimotor and visuomotor domains due to prenatal exposure to lead (Bellinger et al. 1987). Occupational exposure to lead results in reduced intellectual functioning in humans wherein every 10 μg/dL increments predicting an IQ decrease of 1–5 points (Khalil et al. 2009). Also, chronic lead exposure hampers visuospatial skills, executive functioning and attention, memory, motor skills, and language skills (Mason et al. 2014).

### *5.4 Copper*

Copper excess is detrimental to neuronal cells, which are sensitive to oxidative stress. Copper is well known to interact with proteins that are crucial in the development of Alzheimer's disease, amyotrophic lateral sclerosis, and Parkinson's disease (Giampietro et al. 2018). Apart from this, the development of dystonia, abnormal gait, dysarthria, and dementia characteristic of aceruloplasminemia; also, Menke's disease and Wilson's disease have been reported (Desai and Kaler 2008).

## 6 Reported Literature on the Effect of Traditional Processing on Drugs of Mineral Origin Resulting in the Preparation of *Bhasmas* of E1 Schedule

### *6.1 Herbo-Mineral Formulations Containing Arsenic*

(i) In a study, *Hartala bhasma* was found to be nontoxic, safer in the classical therapeutic dose, and did not exhibit any significant change in the expected behavior of animals (Mondal et al. 2016). In a study comprising acute and chronic toxicity evaluation of *Rasamanikya* (an arsenic preparation), no toxicity was observed at therapeutic dose levels in terms of behavior, and pathological changes in Wistar rats (Chaudhari et al. 2018).

(ii) In acute toxicity, *Rasamanikya* at the dose of 2000 mg/kg did not produce any observable toxic effects or mortality. The safety of *Rasamanikya* at therapeutic and fivefold therapeutic dose level has been revealed in the chronic toxicity study (Chaudhari et al. 2018).

### *6.2* Tamra *Bhasma (Incinerated Copper)*

(i) In a study, *Tamra bhasma* prepared without *shodhana* was found to be very toxic at a therapeutic dose level. In contrast to this, *Tamra bhasma* prepared with *shodhana* of crude Copper was found to be nontoxic. It again highlights the importance of *Shodhana* in the preparation of mineral formulations (Jagtap et al. 2013).

(ii) It has been documented that *Tamra bhasma* prepared by the traditional method from crude copper wire did not cause any behavioral change in the subchronic toxicity study in Wistar strain albino rats (Chaudhari et al. 2016; Vahalia et al. 2011).

### *6.3 Herbo-Mineral Formulations Containing Mercury*

(i) When traditionally prepared *Rasasindura* was subjected to physicochemical characterization, where it was found that *rasasindura* was chemically pure α-HgS and Hg:S in 1:1 ratio. Analysis of *Rasasindura* by transmission electron microscopy (TEM) depicted that the particles were in the nanoscale (Mukhi et al. 2017). *Rasasindura* was found to be nontoxic along with *kajjali* in a study to assess neurotoxic activity using the zebrafish behavioral model (Biswas et al. 2018).

(ii) Some mercurial preparations viz. *Aarogyavardhini vati* have inferred nontoxic nature with a battery of experimental studies at doses greater than the therapeutic human dose (Kumar et al. 2012). Many safety studies for others like *Vasanta kusumakara, Lakshmivilas rasa, Sutshekhara rasa, Smritisagara rasa, Agnitundi vati, Shwasakuthara rasa, Swarna bhasma* with mercury, etc. have been reported (Thakur et al. 2017).

### *6.4* Naga Bhasma *(Incinerated Lead)*

(i) In a preclinical study, it was observed that animals treated with Stage 1 (*Gomutra Shodhana*); Stage 2 (*Samanya Shodhana*); Stage 3 (*Vishesha Shodhana*); and Stage 4 (fully processed *Naga bhasma*) showed a gradual increase in the memory and learning. This observation was substantiated by the findings of the histopathology of the Cornu Ammonis (CA) region of the hippocampus. There was a gradual reduction in the time spent in the dark compartment (passive avoidance test) by the Wistar rats treated with Stages 1–3 of the purification of *bhasma*. Thus, we can find that, if traditional processing techniques are adopted in the case of toxic metals, the newly formed organometallic compounds are nontoxic (Quadros et al. 2016).

(ii) It has been reported that *naga bhasma* prepared by traditional procedures consisted of lead in nanocrystalline (~60 nm) lead sulfide form ($Pb^{2+}$) along with the organic contents and different nutrient elements coming from the herbs used during the processing. *Naga bhasma* prepared exhibited safety in histopathology study on rats at a dose of 6 mg/100 g/day (Singh et al. 2010).

## 7 Plants as Antidotes of Heavy Metals

Apart from a thorough knowledge of detoxification of metal origin drugs, seers of Ayurveda were also vigilant about the untoward effects caused by improper processing techniques adopted during *shodhana* and *Bhasma* preparation. They have already enlisted some common antidotes in such circumstances, which consist of *Dhanyaka* (Coriander), ginger juice, *Arjuna* (*Terminalia arjuna*), turmeric, lemon juice, aloe vera juice, etc. The text of *Basavarajeeyam* has mentioned different types of poisonous substances and its antidotes in *Visha Chikitisa* (treatment of poison)

(Dharini and Chaitra 2018). Table 3 describes the medicinal plants used as metal-specific antidotes (Jha 2000).

Toxicology in Ayurveda is described in the discipline of *Agada tantra,* which highlights poisonous drugs right from their classification, posology, therapeutical use up to antidotes due to poisoning. Acharya Chakrapani has clearly defined that the substance used to counteract the poison is called *Prativisha,* i.e., equivalent to antidote (Acharya 2014).

Further, some classical texts, *Basavarajeeyam* (Krishnamurthy 2014) and *Anupana Manjiri* (Acharya 1972), have also mentioned antidotes for poisonous medicinal plants, which have been illustrated in Table 4. Here, some other nonpoisonous plants have been attributed with properties peculiarly for this *dravya* from *Upavisha varga.*

**Table 3** Plants as antidotes of metals/mineral origin drugs

| Metals/minerals | Chelating agent/antidote |
|---|---|
| *Rasa/Parad* (mercury) | *Gandhaka* (sulfur) with *Nagvalli* (*Piper betel*) |
| *Tamra* (copper) | *Dhanyaka* (coriander) with sugar |
| *Naga* (lead) | *Swarna bhasma* (gold ash) mixed with *Haritaki* (*Terminalia chebula*) and sugar |
| *Hartala* (As2S3) | *Jeeraka* (*Carum carvi*) with sugar |
| *Manashila* (As2S3) | Cow's milk with honey |
| *Abhrak* (Biotite) | *Atasi Beeja Rasa* (juice of seeds of *Linum usitatissimum*) |
| *Sasyak* (blue vitriol) | *Jambiri nimbu* |
| *Gauripashana* (white arsenic) | *Tankan* (borax) with cow's milk with cow's ghee |
| *Makshik* (copper pyrite) | *Kullatha kwatha (Dolichus biflorus) and Dadima Kwatha (Punica granatum)* |

**Table 4** Plants as antidotes of poisonous plants of *visha* and *upavisha* category

| Poisonous Plants | Antidote |
|---|---|
| *Langali* (*Gloriosa superba* Linn.) | *Shunti* (dry ginger) (*Zingiber officinale*) |
| *Bhallataka* (*Semecarpus anacardium* Linn.) | *Palasha puspha* (*Butea monosperma*), *Kapikacchu* (Mucuna pruriens), *Eranda taila* (ricinus oil), *Haridra* (curcuma), *Bilva (Aegle marmelos) patra, Narikela, Bibhibitaki* (*Terminalia bellerica*) *mula.* |
| *Kuchala* (*Stychnos nux vomica* Linn.) | *Jambu (Syzygium cumini), Ketaki (Pandanus tectorius)* |
| *Dhatura* (*Datura metel* Linn.) | *Kamala (Nelumbo nucifera) patra choorna,* |
| *Vatsanabha* (*Aconitum ferox* Wall) | *Meghanada Patra swarasa* (leaf juice of *Amaranthus spinosus*) |
| *Ahiphena* (*Papaver somniferum* Linn.) | Ginger juice |

## 8 Conclusions

By abiding the scientifically validated traditional purification techniques for poisonous medicinal plants and metals, it is possible to ameliorate their neurotoxic effects and enhance their therapeutic efficacy. Further, in-depth studies with relevant models for assessment of neurotoxicity pre and post *shodhana* can aid in ascertaining these traditional claims.

## References


Acharya V (1972) Anupana Manjiri. 1st edn. Jamnagar. Sahitya Sanshodhan Vibhagiya Prakashan. Gujarat Ayurved University, Gujarat, India

Acharya R (2014) *Shodhana*: an Ayurvedic detoxification technique and its impact on certain medicinal plants. In: Conservation, cultivation and exploration of therapeutic potential of medicinal plants, 1st edn. Central Council for Research in Ayurvedic Sciences, p 427–450

Akbar S, Khan SA, Masood A, Iqbal M (2010) Use of *Strychnos nux-vomica* (Azraqi) seeds in Unani system of medicine: role of detoxification. Afr J Trad Complement Altern Med 7:286–290. https://doi.org/10.4314/ajtcam.v7i4.56689

Alberto K, Claudia S, Ludmilla K, Arnon B (2001) Toxic Delirium due to *Datura stramonium*. Isr Med Assoc J 3:538–539

Anonymous (2003) Drug and Cosmetic Rules 1945. Ministry of Health and Family Welfare. Govt. of India. Pdf document. p 317. http://vbch.dnh.nic.in/pdf/Rules%20and%20regulations%20of%20Drug%20and%20Cosmetics%20act.pdf.

Anonymous (2019) Toxins. Encyclopedia Britannica inc. https://www.britannica.com/science/poison-biochemistry/Plant-poisons-phytotoxins

Arya N, Kumar H (2017) Comparative physico-chemical profile of '*Vatsanabha*' (*Aconitum Ferox*, Ranunculaceae) *Mula* processed through cow's urine and cow's milk. Int J Res Ayurveda Pharm 8(5):217–222. https://doi.org/10.7897/2277-4343.085278

Barve KH, Ojha N (2013) Effective detoxification of *Abrus precatorius* Linn. Seeds by *Shodhana*. J Ayurveda Integrative Med 4(2):82

Bellinger D, Leviton A, Waternaux C, Needleman H, Rabinowitz M (1987) Longitudinal analyses of prenatal and postnatal lead exposure and early cognitive development. New Engl J Med 316 (17):1037–1043. https://doi.org/10.1056/NEJM198704233161701

Bhide B, Acharya R, Ravishankar B, Kolhe R, Nariya M (2019) Study of the safety profile of raw (*aśodhita*) and cow's urine processed (*gomūtra śodhita*) tubers of *Gloriosa superba* L. (*Lāngali*) in albino rats. Ind J Nat Prod Resour 10(3):210–217

Biswas S, Balodia N, Bellare J (2018) Comparative neurotoxicity study of mercury-based inorganic compounds including Ayurvedic medicines *Rasasindura* and *Kajjali* in zebrafish model. Neurotoxicol Teratol 66:25–34. https://doi.org/10.1016/j.ntt.2018.01.007

Bodhankar K (2017) Shodhana of Visha dravya w.s.r Vatsanabha Shodhana and Bhallataka shodhana. Int J Curr Res 9(8):56,267–56,271

Bolla-Wilson K, Bleecker ML (1987) Neuropsychological impairment following inorganic arsenic exposure. J Occup Med 29(6):500–503

Boumba VA, Mitselou A, Vougiouklakis T (2005) Fatal poisoning from ingestion of *Datura stramonium* seeds. Vet Hum Toxicol 46(2):81–82. https://pubmed.ncbi.nlm.nih.gov/15080209/

Cai BC, Yang WX, Zhu WY, Lu JC, Ye DJ (1993) Effect of processing on the extraction of alkaloids from *Strychnos*. Zhongguo Zhong Yao Za Zhi 18:23–62

Carocci A, Rovito N, Sinicropi MS, Genchi G (2014) Mercury toxicity and neurodegenerative effects. Rev Environ Contam Toxicol 229:1–18. https://doi.org/10.1007/978-3-319-03777-6_1

Chaudhari SY, Nariya MB, Galib R, Prajapati PK (2016) Acute and subchronic toxicity study of *Tamra Bhasma* (incinerated copper) prepared with and without *Amritikarana*. J Ayurveda Integr Med 7(1):23–29. https://doi.org/10.1016/j.jaim.2015.11.001

Chaudhari SY, Biradar S, Nariya M, Galib R, Prajapati PK (2018) Acute and chronic toxicity of *Rasamanikya*, an Ayurvedic arsenical formulation in rats. Ind J Pharm Sci 80(2):325–333. https://doi.org/10.4172/pharmaceutical-sciences.1000361

Choi YH, Sohn YM, Kim CY, Oh KY, Kim J (2004) Analysis of strychnine from detoxified *Strychnos nux-vomica* [corrected] seeds using liquid chromatography-electrospray mass spectrometry. J Ethnopharmacol 93:109–112. https://doi.org/10.1016/j.jep.2004.03.038

Chunekar KC, Pandey GS (1999) Bhavaprakasha Nighantu, 10th edn. Chaukhamba Bharati Academy, Varanasi, India, pp 139–141, 313–314, 568

Das G (2008) Bhaishajya Ratnavali. In: Mishra B (ed) , 19th edn. Chaukhambha Surabharati Prakashan, Varanasi, India, p 570

Deore SL, Moon KV, Khadabadi SS, Deokate UA, Baviskar BA (2013) Evaluation of toxicity of *'Vatsanabha'* (*Aconitum ferox, Ranunculaceae*) before and after *Shodhana*. J Young Pharm 5 (1):3–6. https://doi.org/10.1016/j.jyp.2013.01.001

Desai V, Kaler SG (2008) Role of copper in human neurological disorders. Am J Clin Nutr 88 (3):855S–858S. https://doi.org/10.1093/ajcn/88.3.855S

Dharini VA, Chaitra H (2018) A review on antidotes with special reference to basavarajeeyam. J Pharm Sci Innov 7(4):115–120. https://doi.org/10.7897/2277-4572.07492

Diker D, Markovitz D, Rothman M, Sendovski U (2007) Coma as a presenting sign of *Datura stramonium* seed tea poisoning. Eur J Int Med 18(4):336–338. https://doi.org/10.1016/j.ejim.2006.09.035

Gadadhar S, Karande AA (2013) Abrin immunotoxin: targeted cytotoxicity and intracellular trafficking pathway. PLoS One 8:e58304. https://doi.org/10.1371/journal.pone.005830

Gao X, Hu J, Zhang X, Zuo Y, Wang Y, Zhu S (2018) Research progress of aconitine toxicity and forensic analysis of aconitine poisoning. Forensic Sci Res 5(1):25–31. https://doi.org/10.1080/20961790.2018.1452346

Gautam DNS (2017) Comparative pharmaceutical study of white & black *Gunja* (*Abrus precatorius* L.) kernels. Int J Ayurved Herbal Med 7:2706–2712. https://doi.org/10.18535/ijahm/v7i4.10

Gautam DNS (2018) Physicochemical and acute toxicity study of *Gunja* (*Abrus precatorius*) pericarp. Res J Pharm Tech 11(5):1935–1939. https://doi.org/10.5958/0974-360X.2018.00359.1

Giampietro R, Spinelli F, Contino M, Colabufo NA (2018) The pivotal role of copper in neurodegeneration: a new strategy for the therapy of neurodegenerative disorders. Mol Pharma 15(3):808–820. https://doi.org/10.1021/acs.molpharmaceut.7b00841

Gopalkrishna SV (2011) Effect of purification (*Sodhana*) on phytochemicals, toxicity and pharmacological properties of some medicinal plants. Research and Development Cell, Thesis, Jawaharlal Nehru Technological University, Hyderabad, India

Handa KL, Chopra IC, Kohli JD, Singh K (1951) Mitigation of aconite; a preliminary note. Indian J Med Res 39:89–98

Jagtap CY, Ashok BK, Patgiri BJ, Prajapati PK, Ravishankar B (2013) Acute and subchronic toxicity study of Tamrabhasma (incinerated copper) prepared from *Ashodhita* (unpurified) and *Shodhita* (purified) *tamra* in rats. Indian J Pharm Sci 75(3):346. https://doi.org/10.4103/0250-474X.117433

Jaiswal Y, Liang Z, Yong P, Chen H, Zhao Z (2013) A comparative study on the traditional Indian *Shodhana* and Chinese processing methods for aconite roots by characterization and determination of the major components. Chem Cent J 7(1):169. https://doi.org/10.1186/1752-153X-7-169

Jha C (2000) Ayurvediya Rasashastra Chowkambha Surabharati Prakashana, Varanasi, India

Joshi D, Agrawal K (1977) Effect of purification (*Shodhana*) on the alkaloidal concentration of *kuchala* seeds. J Res Ind Med Yoga Homeopath 12:43–46

Karthikeyan A, Amalnath SD (2017) *Abrus precatorius* poisoning: a retrospective study of 112 patients. Ind J Crit Care Med 21(4):224–225. https://doi.org/10.4103/ijccm.IJCCM_320_16

Katiyar C, Kumar A, Bhattacharya SK, Singh RS (2010) Ayurvedic processed seeds of *nux-vomica*: neuropharmacological and chemical evaluation. Fitoterapia 81(3):190–195. https://doi.org/10.1016/j.fitote.2009.08.023

Khalil N, Morrow LA, Needleman H, Talbott EO, Wilson JW, Cauley JA (2009) Association of cumulative lead and neurocognitive function in an occupational cohort. Neuropsychology 23 (1):10–19. https://doi.org/10.1037/a0013757

Krenzelok EP (2010) Aspects of *Datura* poisoning and treatment. Clin Toxicol 48(2):104–110. https://doi.org/10.3109/15563651003630672

Krishnamurthy MS (2014) A reputed text of Ayurvedic therapeutic and pharmaceutical codified by "Vaidya Shree Basavaraja", Basavarajeeyam text with English translation, Visharoganidalakshana adhyaya, 1st edn. Chaukhambha Orentalia, Varanasi, India, p 621

Kumar A, Sinha BN (2009) Ayurvedic processings of *nux–vomica*: qualitative and quantitative determination of total alkaloidal contents and relative toxicity. Malays J Pharm Sci 7:83–98

Kumar G, Srivastava A, Sharma SK, Gupta YK (2012) Safety evaluation of an Ayurvedic medicine, *Arogyavardhini vati* on brain, liver and kidney in rats. J Ethnopharmacol 140(1):151–160. https://doi.org/10.1016/j.jep.2012.01.004

Liu J, Zhang F, Ravikanth V, Olajide OA, Li C, Wei LX (2019) Chemical compositions of metals in bhasmas and tibetan zuotai are a major determinant of their therapeutic effects and toxicity. Evid Based Complement Alternat Med:1697804. https://doi.org/10.1155/2019/1697804

Martin S, Griswold W (2009) Human health effects of heavy metals. Environ Sci Technol Briefs Citizens 5:1–6

Mason LH, Harp JP, Han DY (2014) Pb neurotoxicity: neuropsychological effects of lead toxicity. Biomed Res Int 2014:840547. https://doi.org/10.1155/2014/840547

Mazumdar DN, Das Gupta J, Chakraborty AK, Chatterjee A, Das D, Chakraborti D (1992) Environmental pollution and chronic arsenicosis in South Calcutta. Bull World Health Organ 70:481–485

Miller LL, Branconnier RJ (1983) Cannabis: effects on memory and the cholinergic limbic system. Psychol Bull 93:441–456

Mitra S, Shukla VJ, Acharya R (2011a) Effect of *Shodhana* (processing) on *Kupeelu* (Strychnos nux-vomica Linn.) with special reference to strychnine and brucine content. AYU 32 (3):402–407. https://doi.org/10.4103/0974-8520.93923

Mitra S, Shukla VJ, Acharya R (2011b) Role of castor oil in processing (*Shodhana*) of *Kupeelu* (*Strychnos nux vomica* Linn.) seeds: an approach of traditional Ayurveda. Int J Ayurvedic Med 2(2):62–71

Modi JP (2005) Modi's medical jurisprudence and toxicology. In: Kannan K (ed), 23rd edn., Lexis Nexis Butterworths, p 468

Mondal S, Palbag S, Gautam DN (2016) Toxicological study of arsenic containing Ayurvedic drug *Haratal Bhasma*. J Ayurved Herbal Med 2(3):82–85

Moritz F, Compagnon P, Kaliszczak IG, Kaliszczak Y, Caliskan V, Girau C (2005) Severe acute poisoning with homemade *Aconitum napellus* capsules: toxico-kinetic and clinical data. Clin Toxicol (Phila) 43:873–876. https://doi.org/10.1080/15563650500357594

Mukhi P, Mohapatra SS, Bhattacharjee M et al (2017) Mercury based drug in ancient India: the red sulfide of mercury in nanoscale. J Ayurveda Integr Med 8(2):93–98. https://doi.org/10.1016/j.jaim.2017.01.009

Muroi M, Kimura I, Kimura M (1990) Blocking effects of hypaconitine and aconitine on nerve action potentials in phrenic nerve-diaphragm muscles of mice. Neuropharmacology 29:567–572. https://doi.org/10.1016/0028-3908(90)90069-4

Murthy KRS (2012) Sharangadhara Samhita. Chaukhambha Orientalia, Varanasi, India

Nabar M, Pimpalgaokar P, Laddha K (2011) Studies on *Sodhanaprakriya* of *Gunja (Abrus precatorius* Linn) seeds. Indian J Trad Knowledge 10:693–696

Nabar MP, Mhaske PN, Pimpalgaonkar PB, Laddha KS (2013) Gloriosa superba roots: content change of colchicine during Sodhana (detoxification) process. Ind J Trad Knowledge 12 (2):277–280

Olsnese S (2004) The history of ricin, abrin and related toxins. Toxicon 44:361–370. https://doi.org/10.1016/j.toxicon.2004.05.003

Parikh KM, Doshi VJ, Salunkhe UB, Dhanvate AA (1996) Authentication of detoxification process used in traditional Indian medicine. Int Hortic Soc Acta Hortic 426:57–54. https://doi.org/10.17660/ActaHortic.1996.426.5

Patel Y, Bhat SD, Rabinarayan A, Ashok BK, Shukla VJ (2010) Role of *Shodhana* on analytical parameters of *Datura innoxia* mill and *Datura metel* Linn. Seeds. Int J Res Ayurveda Pharm 1:249–254

Patil MM, Patil SV, Akki AS, Lakhkar B, Badiger S (2016) An arrow poison (*Abrus precatorius*) causing fatal poisoning in a child. J Clin Diagn Res 10:SD03–SD04

Pillay VV, Bhagyanathan PV, Krishnaprasad R, Rajesh RR, Vishnupriya N (2005) Poisoning due to white seed variety of *Abrus precatorius*. J Assoc Physicians India 53:317–319

Quadros LS, Bangera H, Kotian SR, Bhat KM (2016) Effects of lead in various preparatory stages of *Nagabhasma* on function and histopathology of *cornu ammonis* of hippocampus. J Clin Diagn Res 10:AF01–AF04. https://doi.org/10.7860/JCDR/2016/20424.8748

Ram SG, Krishna MC, Kumar PD, Mohan MK, Babu G (2018) Conceptual review on *vijaya* (*Cannabis sativa* Linn.): a forgotten ambrosia. Int J Res Ayurveda Pharm 9(2):18–27. https://doi.org/10.7897/2277-4343.09228

Rasavagbhata (1995) Rasaratnasamuchchaya. In: Shastri A (ed). 9th edn, Chaukhambha Amarbharti Publication, Varanasi, India, p 570

Rathnayaka RN, Ranathunga PN, Fernando R (2015) Epidemiology and clinical profile of *Gloriosa superba* poisoning in Sri Lanka. Am J Toxicol 2(1):9–15

Roy S, Acharya R, Pandya P (2016) Effect of *Shodhana* (purification/processing) on powder microscopical and analytical parameters of Gunja (*Abrus Precatorius* Linn.) seeds. Int J Pharm Pharm Sci 8(11):81–85. https://doi.org/10.22159/ijpps.2016v8i11.12949

Sahni V, Agarwal SK, Singh NP, Sikdar S (2007) Acute demyelinating encephalitis after jequirity pea ingestion (*Abrus precatorius*). Clin Toxicol (Phila) 45(1):77–79. https://doi.org/10.1080/15563650601006116

Sahoo R, Hamide A, Amalnath SD, Narayana BS (2008) Acute demyelinating encephalitis due to *Abrus precatorius* poisoning—complete recovery after steroid therapy. Clin Toxicol (Phila) 46 (10):1071–1073. https://doi.org/10.1080/15563650802334671

Sarkar PK, Subha, Prajapati PK (2008) Importance of media in *Shodhana* of *Vatsanabha* (aconite). AYU 29(1):52–55

Sarkar PK, Das S, Prajapati PK (2010) Ancient concept of metal pharmacology based on Ayurvedic literature. Anc Sci Life 29(4):1–6

Sarkar PK, Prajapati PK, Shukla VJ, Ravishankar B (2012) Evaluation of effect of *shodhana* on pharmacological activities of aconite. Indian J Pharm Edu Res 46(3):243–247

Sarkar S, Basu K, Das J, Datta S (2017) *Abrus precatorius* poisoning and central pontine myelinolysis. J Pediatr Neurosci 12(4):353–355. https://doi.org/10.4103/jpn.JPN_84_17

Savrikar S, Ravishankar B (2011) Introduction to '*Rasashaastra*'-the iatrochemistry of Ayurveda. Afric J Trad Complement Alternat Med 8(5):66–82. https://doi.org/10.4314/ajtcam.v8i5S.1

Singh DG, Banerji R, Mahrotra S (1998) Effect of *Shodhana* on the toxicity of *Abrus precatorius*. Anc Sci Life 18(2):127–129

Singh SK, Gautam DN, Kumar M, Rai SB (2010) Synthesis, characterization and histopathological study of a lead-based Indian traditional drug: *Naga bhasma*. Indian J Pharm Sci 72(1):24–30. https://doi.org/10.4103/0250-474X.62232

Solowij N (1998) Cannabis and cognitive functioning. Cambridge: Cambridge University Press. https://doi.org/10.1017/CBO9780511526824

Subrahmanyan D, Mathew J, Raj M (2008) An unusual manifestation of *Abrus precatorius* poisoning: a report of two cases. Clin Toxicol (Phila) 46(2):173–175. https://doi.org/10.1080/15563650601185134

Survase MR, Taware AS, Survase MR (2016) *Kupilu* – various *Shodhana* procedures with HPLC analysis. World J Pharma Med Res 2(6):57–61

Tavhare SD, Acharya R, Reddy RG, Dhiman KS (2019) Management of chronic pain with *Jalaprakshalana* (water-wash) *Shodhita* (processed) *Bhanga* (Cannabis sativa L.) in cancer patients with deprived quality of life: an open-label single arm clinical trial. Ayu 40(1):34–43. https://doi.org/10.4103/ayu.AYU_43_19

Thakur K, Gudi R, Vahalia M, Shitut S, Nadkarni S (2017) Preparation and characterization of *Suvarna Bhasma paradamarit*. J Pharmacopuncture 20(1):36–44. https://doi.org/10.3831/KPI.2017.20.007

Turner AR, Spurling BC, Agrawal S (2020) Marijuana toxicity. Stat Pearls Publishing. https://www.ncbi.nlm.nih.gov/books/NBK430823

Upadhyaya M (1999) Ayurvedaprakasha. In: Mishra GS (ed), Chaukhambha Bharati Academy, Varanasi, India

Vahalia MK, Nadkarni KT, Sangle VD (2011) Chronic toxicity study for *Tamra Bhasma* (a generic Ayurvedic mineral formulation) in laboratory animals. Recent Res Sci Technol 3(11):76–79

Vinod KV, Thabah MM, Venkatesh T et al (2013) A rare cause of dural venous sinus thrombosis. Neurol India 61:669–670. https://doi.org/10.4103/0028-3886.125292

Wadnerwar NN, Jyotishi S, Rajput DS, Duragkar UD (2017) Effect of *Shodhana* on toxic principle of *Gunja Beeja* with reference to agglutination-an in vitro study. Ann Ayurvedic Med 6 (1–2):31–39

# St. John's Wort: A Therapeutic Herb to Be Cautioned for Its Potential Neurotoxic Effects and Major Drug Interactions

**Ayaka Fujihashi, Sindhu Ramesh, Manoj Govindarajulu, Mohammed Almaghrabi, Rishi M. Nadar, Jack Deruiter, Timothy Moore, Satyanarayana Pondugula, Dinesh Chandra Agrawal, and Muralikrishnan Dhanasekaran**

## Contents



A. Fujihashi · S. Ramesh · M. Govindarajulu · M. Almaghrabi · R. M. Nadar · J. Deruiter · T. Moore · M. Dhanasekaran (✉)
Department of Drug Development and Discovery, Harrison School of Pharmacy, Auburn University, Auburn, AL, USA
e-mail: dhanamu@auburn.edu

S. Pondugula
Department of Anatomy, Physiology and Pharmacology, Auburn University, Auburn, AL, USA

D. C. Agrawal (✉)
Department of Applied Chemistry, Chaoyang University of Technology, Taichung, Taiwan
e-mail: dcagrawal@gm.cyut.edu.tw



D. C. Agrawal, M. Dhanasekaran (eds.), *Medicinal Herbs and Fungi*,
https://doi.org/10.1007/978-981-33-4141-8_15

**Abstract** *Hypericum perforatum,* generally recognized as St. John's Wort (SJW), has been used as an alternative form of medicine for centuries in various traditions and cultures all over the world. Traditional uses for this herb include broad wound healing, gastrointestinal issues and pathologies, pain, and bronchitis, among other ailments. However, it has garnered the most attention today as a treatment for mental disorders, namely depression and bipolar disorder. Supplements of SJW are easily and widely accessible due to their commercial popularity. Though a multitude of therapeutic benefits has been reported, the efficacy of SJW remains to be documented clinically. Many studies have found it to have substantial adverse and toxic effects, and at times, potentially fatal drug interactions. The current chapter provides a brief history of *H. perforatum*, a summary of its active constituents, current and potential therapeutic uses, adverse drug effects, and drug interactions that should be taken into consideration when prophylactically and/or therapeutically using SJW with prescribed medications.

**Keywords** Drug interaction · *Hypericum perforatum* · Neuroprotection · Neurotoxicity · St. John's Wort

## Abbreviations

| | |
|---|---|
| AAPH | 2,2′-Azobis(2-amidinopropane) dihydrochloride |
| AD | Alzheimer's disease |
| AIDS | Acquired immunodeficiency syndrome |
| AUC | Area under the curve |
| Aβ | Amyloid-β |
| CDC | Centers for Disease Control and Prevention |
| CYP2C9 | Cytochrome P450 Family 2 Subfamily C Member 9 |
| CYP2E1 | Cytochrome P450 Family 2 Subfamily E Member 1 |
| CYP3A4 | Cytochrome P450 3A4 |
| DPPH | 1,1-Diphenyl-2-picrylhydrazyl |
| GABA | Gamma-aminobutyric acid |
| HIV | Human immunodeficiency virus |
| LC–MS | Liquid chromatography–mass spectrometry |
| LDL | Low-density lipoprotein |
| L-DOPA | L-3,4-Dihydroxyphenylalanine |
| MAO | Monoamine oxidase |
| MDR1 | Multidrug-resistant gene |
| MPA | Medical products agency |
| NSAIDs | Nonsteroidal anti-inflammatory drugs |
| PD | Parkinson's disease |
| PMA | Phorbol 12-myristate 13-acetate |
| SJW | St. John's Wort |
| SN | Substantia nigra |
| SSRI | Selective serotonin reuptake inhibitor |

# 1 Introduction

*Hypericum perforatum,* also known as St. John's Wort (SJW), Klamath weed, goatweed, and Tipton weed, is a flowering perennial plant in the Hypericaceae family. The plant is native to drier sites of humid and subhumid temperate parts of Europe and Asia. *H. perforatum* usually grows around 1 m in height, with branched, erect stems in the upper section and a clustered, depressed base. Stems are slender and grow superficially within the soil. The green–yellow leaves of the plant have translucent, glandular tissue dots that give the leaves a perforated appearance. Between late spring and early to midsummer, bright yellow flowers with five petals and three bundles of stamens appear (Fig. 1).

*H. perforatum* reproduces both sexually and asexually. The plant produces 33,000 seeds in a season, which are then dispersed by wind, water, or animals. It is believed that *H. perforatum,* a species with 32 chromosomes, is an allopolyploid species that resulted from the hybridization between two related species, each with 16 chromosomes, *H. maculatum* is found through western Europe extending to Siberia, and *H. attenuatum* is found in regions between Siberia and Korea and eastern China. The common name for the species finds its origin in the feast of St. John the Baptist on June 24th, which coincides with the blossoming and harvesting of the plant. The plant would be hung on doors to ward off evil spirits and protect men from harm. The name *Hypericum* is derived from Greek root words of "hyper," meaning "over" and "eikon," meaning "picture," referring to traditional uses of the plant that hung the herb over religious pictures (Benzie and Wachtel-Galor 2011). Though the plant is native to temperate regions, its ability to invade plant communities and replace native vegetation in North and South America, India, New Zealand, Australia, and South Africa has led to its classification as a noxious

**Fig. 1** "Hairy St John's Wort" by Johndal is licensed under CC BY-SA 2.0

**Fig. 2** "Demon Bloke" by Mister Bork is licensed under CC BY-NC-SA 4.0

weed. Now it is also the most common species of *Hypericum* over its native regions. The invasiveness of the plant is also problematic for livestock, as the plant causes damaging photosensitivity reactions in animals that consume the plant. Utilization of St. John's Wort for medicinal purposes extends back to the ancient physicians of Galen, Dioscorides, and Hippocrates. The herb was commonly used as a diuretic, wound treatment, and for intestinal worms. In the middle Ages, association with priests and the ability to repel demons garnered the plant the name of Fuga demonum or the Scourge of the Devil (Fig. 2).

Oral administration of *Hypericum* was commonly used to treat mental distress, bronchitis, headaches, gastritis, diarrhea, and gout. In contrast, topical administration of the oil extract of the plant was used for burns and general wound treatment, as well as ulcers and hemorrhoids (Lee 1999). Extracts were prepared by macerating the flowering tops of the plant with olive oil, palm oil, sunflower oil, or wheat–germ oil and exposed to sunlight for 40 days. Today, commercially available pharmaceutical grade preparations of *H. perforatum* are made from dried aerial parts of the plant, typically using ethanol or methanol for extraction of active components (Benzie and Wachtel-Galor 2011).

## 2 Chemical Constituents of St. John's Wort (*H. perforatum*)

The chemical composition of St. John's Wort can vary in kind and quantity by the geographical origin of the plant material, growing conditions, harvest time, use of raw or dried plant materials, method of preparation, and storage conditions (Arsić 2016). However, the most common medicinally active classes of compounds consist of naphthodianthrones, phloroglucinols, flavonoids, and phenolic compounds.

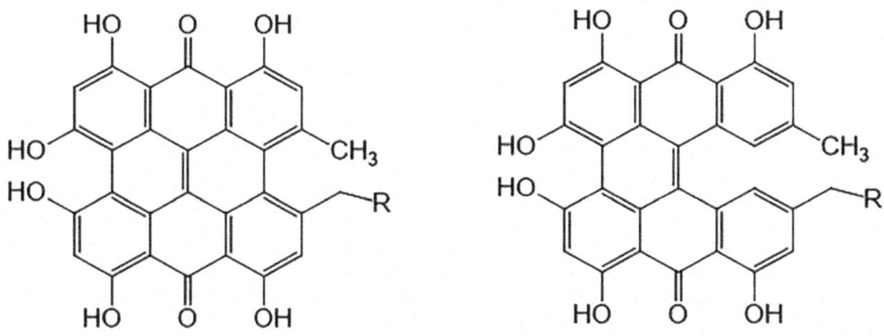


Hypericin R = H
Psuedohypericin R = OH

Protohypericin R = H
Protopseudohypericin R = OH

**Fig. 3** Chemical structures of hypericin, psuedohypericin, protohypericin, and proto-pseudohypericin

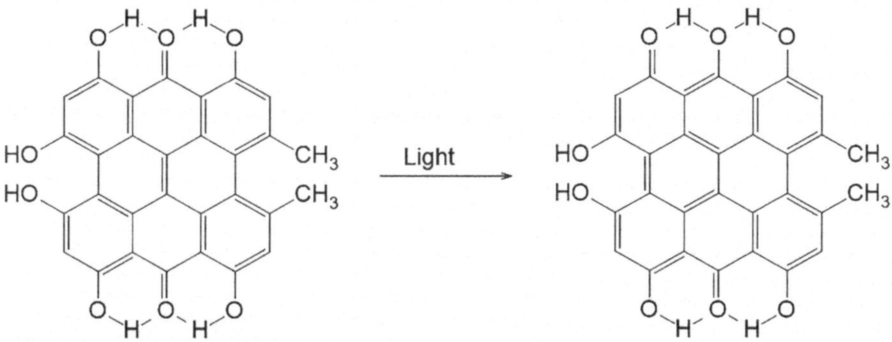


**Fig. 4** Resonant structures of hypericin

Notable compounds in the naphthodianthrone class include hypericin, psuedohypericin, protohypericin, and proto-pseudo hypericin (Fig. 3).

Hypericin is a pigment derived from anthraquinone. Found in the black dots in the flowers of *H. perforatum,* hypericin is also responsible for the red color of St. John Wort oils (Meruelo et al. 1988). Hypericin is highly photoreactive due to its structure. Four hydroxyl groups are adjacent to two carbonyl groups. Due to resonance, the hydrogen transfers between the hydroxyl groups and the carbonyl oxygen occur in the presence of fluorescent light (Fig. 4). This transfer of hydrogen atoms causes acidification of the environment (Fehr et al. 1995).

Flavonoids can be found in ranges of 7–12%, depending on the portion of the flower that is used for preparation (stem, flowers, leaves, etc.). Flavonoids are characterized by a planar structure and are notable for the free radical scavenging capabilities. Common flavonoids found in St John's Wort include flavonols, flavones, glycosides, biflavones, amentoflavone, myricetin, hyperin, oligomeric proanthocyanidins, and miquelianin (Fig. 5). Flavonol glycosides are the largest

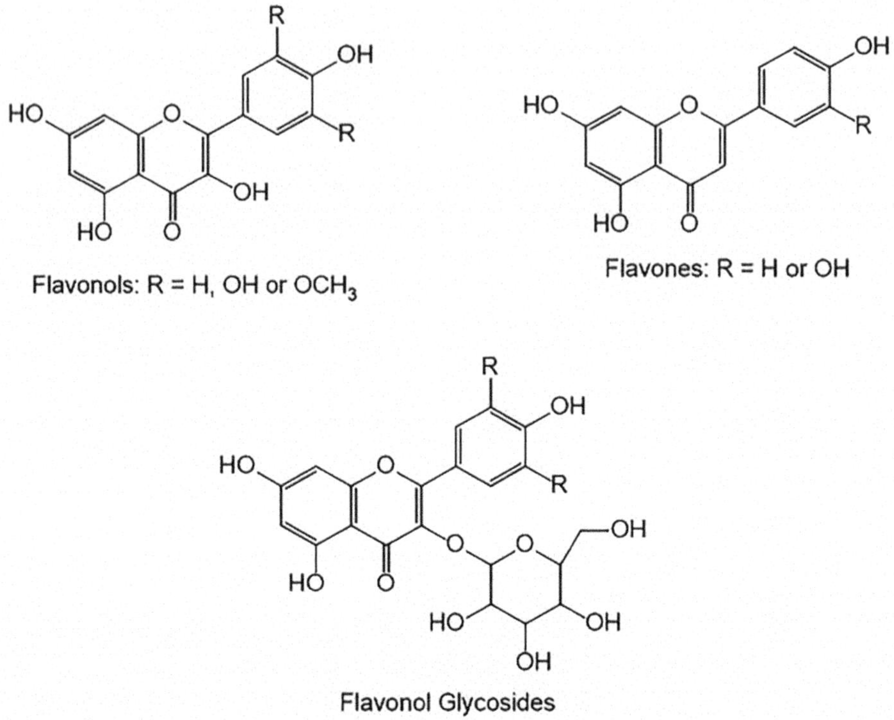


**Fig. 5** Chemical structures of flavonols, flavones, and flavonol glycosides

group of secondary metabolites obtained from *H. perforatum* (Benzie and Wachtel-Galor 2011).

Phloroglucinols include hyperforin, adhyperforin, furohyperforin, and other hyperforin analogs (Fig. 6). Hyperforin has been isolated in concentrations ranging from 2 to 4.5%; however, the instability of the compound in the presence of oxygen and light has made the total synthesis of hyperforin difficult. Essential oils found in *H. perforatum* include mono- and sesquiterpenes and range in concentrations of 0.05–0.09% (Benzie and Wachtel-Galor 2011).

Phenolic acids are biologically valuable for their antioxidant, anticancer, and anti-inflammatory activities. Using HPLC on dried crude plant material of *H. perforatum*, 22 different phenolic compounds have been identified. Notable phenolic acids obtained from St. John's Wort include chlorogenic, caffeic, p-coumaric, ferulic, p-hydroxybenzoic, and vanillic acids (Kwiecień et al. 2015) (Fig. 7).

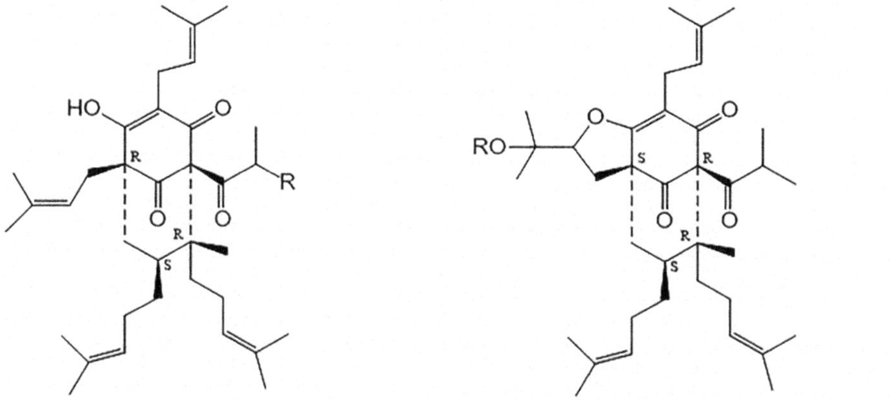

Hyperforin: R = H
Adhyperforin: R = $CH_3$

Furohyperforin: R = H
Furohyperforin hydroperoxide: R = OH

**Fig. 6** Chemical structures of hyperforin, adhyperforin, furohyperforin, and furohyperforin hydroperoxide

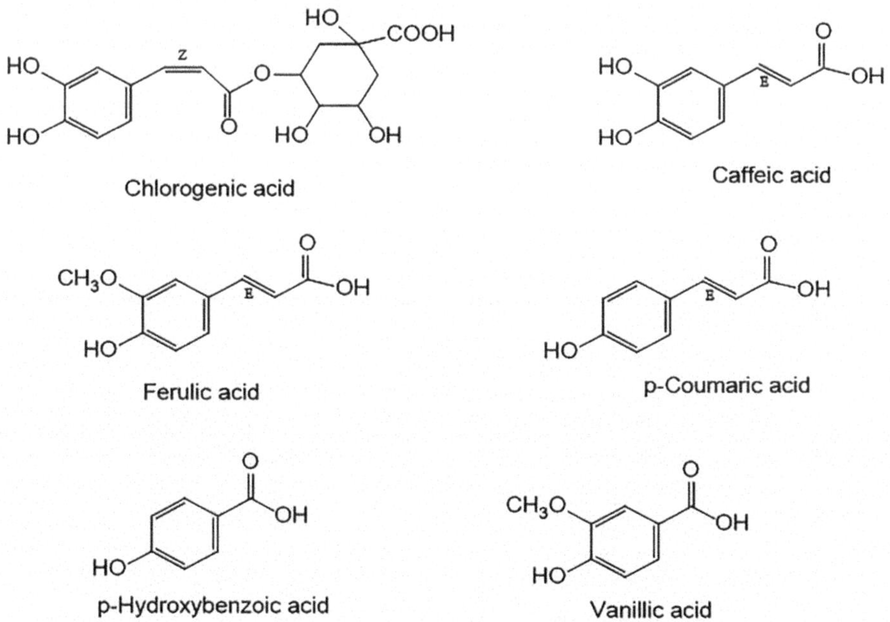


**Fig. 7** Chemical structures of chlorogenic acid, caffeic acid, ferulic acid, p-coumaric acid, p-hydroxybenzoic acid, and vanillic acid

## 3 Pharmacodynamic Effects in the Central Nervous System

St. John's Wort is primarily known for its use as a supplement in treating depression and other depressive disorders. Though depression is typically diagnosed through a patient's clinical symptoms, there have been hypotheses as to the root cause of the major depressive disorder. The monoamine hypothesis states that deficiencies of monoaminergic transmitters norepinephrine, serotonin, and/or dopamine are responsible for the main symptoms of depression, such as depressed mood, anhedonia, and fatigue. Thus, current pharmacological therapies focus on increasing the concentration of serotonin and/or norepinephrine (tricyclic antidepressants, monoamine oxidase (MAO) inhibitors, and selective serotonin reuptake inhibitors). However, such treatments are not without side effects, and there are severe cases of depression that are resistant to current treatments (Brigitta 2002). Thus, alternative therapies may be used to reduce the number or severity of side effects or to act as supplements to be used in tandem with current therapies. The efficacy of SJW in treating depression has been confirmed in a variety of clinical studies. Öztürk et al. (1996) found that extracts of *H. perforatum* were as effective as desipramine and trimipramine, two antidepressant drugs, in behavioral models including swimming time, locomotor activity, tail-flick, and hold-board experiments. Various in vitro studies support the antidepressive effects of *H. perforatum*; however, no single constituent or specific pharmacodynamic effect has been identified, primarily due to the lack of well-understood mechanisms underlying depression itself. It is believed to be the multiple compounds found in the herb that contribute to its effects. Hyperforin is one such compound, which was found to inhibit the reuptake of serotonin, dopamine, noradrenaline, GABA, and L-glutamate (Chatterjee et al. 1998). Hyperforin non-competitively inhibits the reuptake of various neurotransmitters by increasing intracellular $Na^+$ ion concentration. The elevation in $Na^+$ concentration results in a decrease in $H^+$ ion concentrations, which may be due to secondary activation of $Na^+$/$H^+$ exchanger. The sodium ionophore monensin acts as an inhibitor of the synaptosomal reuptake of monoamines. The chronic treatment induces changes on the receptor level, as is often seen with long-term usage of most antidepressants. Hypericum has been shown to downregulate $\beta_1$-adrenoceptor in the frontal cortex without changing receptor affinity. Chronic administration of hypericum has also been shown to enhance 5-HT neurotransmission by upregulating 5-$HT_{1A}$ receptors in the dorsal hippocampus and forebrain and upregulating 5-$HT_2$ receptors (Nathan 2001).

Flavonoids obtained from standard extracts of St. John's Wort are effective in protecting PC12 cells from hydrogen-peroxide-induced cell apoptosis by preventing DNA fragmentation and cell shrinkage (Lu et al. 2004). Specifically, the flavonoids quercetin and kaempferol decrease oxidation of the mitochondrial lipid membrane, providing neuroprotective action (Silva et al. 2008). A similar in vitro study has found *H. perforatum* to be effective against glutamate-induced cellular apoptosis in HT22 hippocampal cell lines. Glutamate is known to increase intracellular $Ca^{2+}$ concentration, which results in depolarization of the mitochondria membrane and subsequent accumulation of reactive oxygen species. Incubation of cells with glutamate and *H. perforatum* extracts has been shown to prevent cell death (Breyer et al. 2007). Butterweck has found that SJW is capable of inhibiting uptake of serotonin, dopamine, and norepinephrine with comparable affinity and a significant affinity for adenosine, gamma-aminobutyric acid (GABA), and glutamate receptor. In vivo extracts of *H. perforatum* were found to cause changes in neurotransmitter concentration in areas of the brain typically afflicted by depression. SJW has also been suggested in gene regulation of the hypothalamic–pituitary–adrenal axis function. Again, the pharmacological effects of SJW have been attributed to hypericin, hyperforin, and flavonoids (Butterweck 2003).

Several studies have investigated potential uses for St. John's Wort in neurological disorders beyond major depression. Specifically, the herb's ability to mitigate debilitating symptoms of neurodegenerative diseases like Alzheimer's disease (AD) and Parkinson's disease (PD) are of particular interest due to the growing global prevalence of such conditions. Alzheimer's, the most common neurodegenerative disease, is characterized by the presence of β-amyloid (Aβ) plaque depositions in the cerebral cortex. β- and γ-secretases proteolytically cleave amyloid precursor protein to produce Aβ. Aβ aggregates to form the pathologic hallmark plaques of AD. Both oligomeric and monomeric forms of Aβ are toxic to neurons, resulting in neuronal death and synaptic dysfunction. In a study conducted by Hofrichter et al. extracts of SJW were found to reduce the size and number of Aβ plaques, restore neurons, restore cognition, and activate microglia both in vitro and in vivo of APP-transgenic mice. The actions are believed to be consequences of increased "ATP Binding Cassette Subfamily C Member 1" (ABCC1) transporter activity to export Aβ across the blood–brain barrier and excrete it into the peripheral circulation (Hofrichter et al. 2013). Similar results were obtained in a study investigating the activity of "ATP Binding Cassette Subfamily B Member 1" (ABCB1) transporter in double-transgenic mice treated with SJW extracts over 60 and 120 days, which showed increased activity of ABCB1 expression and reduction in Aβ accumulation (Brenn et al. 2014).

Parkinson's disease (a movement disorder), the second highest prevalent neurodegenerative disease, is characterized by a loss of dopaminergic neurons in the substantia nigra (SN) and the presence of Lewy bodies in all affected brain regions. Currently, PD is treated primarily with L-DOPA (L-3,4-dihydroxyphenylalanine), a precursor to dopamine that can cross the blood–brain barrier. The dopamine precursor L-DOPA undergoes decarboxylated and thus converted to dopamine by dopa decarboxylase in the presence of pyridoxal phosphate to generate dopamine and alleviate the effects of PD. However, alternative therapies such as MAO inhibitors have been investigated for their antioxidant and neuroprotective effects due to the extensive side effects exhibited by current therapies. Extracts of SJW have been found to inhibit apoptotic cascades and reduce neuronal cell death in the substantia nigra against rotenone in rats by decreasing Bax levels, an apoptotic protein gene (del Rio et al. 2013).

Some studies have investigated the efficacy of *H. perforatum* in treating epilepsy; however, results reported in the literature are conflicting. One study investigating the effect of different fractions of *H. perforatum* on kindling epilepsy in chinchilla rabbits found that polar fractions of the extract were able to repress epileptic discharges. In contrast, nonpolar lipid fractions potentiated epileptic activity (Ivetic et al. 2002). Another study by Hosseinzadeh et al. (2005) confirmed *H. perforatum*'s anticonvulsant activity in pentylenetetrazole and maximum electric shock tests, suggesting a potential benefit of SJW for treating petit mal seizures. However, the use of SJW for treatment of epilepsy is generally cautioned against due to interactions with prescribed anticonvulsants used concomitantly with SJW.

## 4 Other Indications

Since St John Wort is an herbal remedy, it is classified as a dietary supplement by the Food and Drug Administration, United States of America. Thus, it is not strictly regulated for safety and efficacy compared to pharmaceutical drugs. However, St. John's Wort can be found in the form of tablets, capsules, teas, and tinctures and is widely available as a dietary supplement for treating depressive disorders and their associated symptoms such as anxiety, fatigue, poor appetite, or insomnia. Various studies, including analyses on tissue samples, animal models, and human clinical trials, appear to verify the efficacy of St John's Wort in treating depression. Because St John's Wort supplements are most heavily commercially marketed as a treatment for depression, this has been the most extensively studied claim. However, multiple studies have found several other uses and indications for the herb.

### *4.1 Antibacterial and Antiviral Effects*

According to the Centers for Disease Control and Prevention (CDC), there are 15.5 million physician office visits in the United States each year by patients who have a diagnosis of an infectious or parasitic disease. Every year, the influenza virus kills roughly 0.1% of those infected. Most current therapies rely on antibiotics, antivirals to eradicate the infectious pathogens, or focus heavily on alleviating symptoms of the disease. Extracts of *H. perforatum* have been found to exhibit antibacterial properties, with hyperforin being the main antibacterial component of the plant (Schempp et al. 1999). Hyperforin has been found to inhibit the growth of gram-positive bacteria, as well as methicillin-resistant and penicillin-resistant *S. aureus* specifically (Lyles et al. 2017). The hypericin compounds found in *H. perforatum* also have been shown to express viricidal against enveloped viruses by altering viral proteins and inhibiting the virus's ability to fuse with the cell membrane (Diwu 1995). In vitro studies support the antiviral activities of hypericin, specifically in action against enveloped viruses in human blood of acquired immunodeficiency syndrome (AIDS) (Holden 1991). Recent studies have found

3-hydroxy lauric acid present in *H. perforatum* exhibits antihuman immunodeficiency virus (HIV) activity (Maury et al. 2009).

## 4.2 Anticancer Effects

Cancer is currently one of the prominent causes of global mortality. Current therapies include surgery, radiation therapy, chemotherapy, targeted therapy, hormonal therapy, and immunotherapy. Advances in detection and treatment have increased survival rates generally; however, the treatments are not without detrimental side effects, the most common of which include pain, fatigue, nausea, and vomiting (Miller et al. 2019). St. John's Wort has been suggested as a potential alternative treatment due to the ability of its bioactive constituents to suppress cell proliferation and induce apoptosis.

Hyperforin and hypericin appear to inhibit tumor cell growth in vitro by inducing cell apoptosis through the activations of caspases and triggering the release of cytochrome c (Schempp et al. 2002). Hyperforin used in tandem with procyanidin B2 was effective in inhibiting in vitro growth of leukemia K562 and U937 cells and human brain glioblastoma cells LN229 (Hostanska et al. 2003). Hypericin has been shown to inhibit the growth of glioma, neuroblastoma, adenoma, mesothelioma, melanoma, carcinoma, sarcoma, and leukemia cells, which has been attributed to photodynamic properties of hypericin, specifically in its ability to generate superoxide radicals that kill tumor cells. Photoactivation of hypericin via white light or ultraviolet light was found to induce apoptosis in malignant cutaneous T cells and lymphoma T cells (Fox et al. 1998).

## 4.3 Anti-Inflammatory and Pain Effects

Inflammation is part of the complex biological response of body tissues to harmful stimuli, such as pathogens, damaged cells, or irritants. Blood vessels dilate, blood flow increases, and neutrophils migrate to targeted areas. Macrophages and dendritic cells release cytokines such as IL-1 and TNF-$\alpha$, and the pathogen or irritant is broken down. However, chronic inflammation can occur because of a failure to eliminate infectious organisms, recurring episodes of acute inflammation, oxidative stress, or autoimmune disorders. It can lead to a variety of diseases such as rheumatoid arthritis, lupus, asthma, and Crohn's disease. Current treatments include the use of nonsteroidal anti-inflammatory drugs (NSAIDs), steroids, or disease-modifying immunosuppressant drugs as well as instructing patients to follow an anti-inflammatory diet (Pahwa et al. 2020).

Extracts of *H. perforatum* have been shown to possess analgesic properties in various animal models of nociception. Preparations of SJW were found to be effective in increasing both the thermal and chemical pain threshold in rats (reference). Specifically, hypericin, obtained through a chloroform fraction, was effective in mediating both thermal and chemical pain, and hyperforin, obtained through a methanol fraction, was effective in thermal antinociception. The same research group also found that thermal antinociception was reversed by naloxone and phorbol 12-myristrate 13-acetate (PMA), a protein kinase C activator. In contrast, chemical antinociception was prevented by PMA (Galeotti et al. 2010). In a different study investigating the efficacy of SJW on diabetic neuropathy in Sprague–Dawley rats with streptozotocin-induced diabetes, freeze-dried seed extracts of SJW were found to verse mechanical hyperalgesia in a dose-dependent manner. The antihyperalgesic effects of SJW in diabetes may be in part due to its observed antihyperglycemic effects (Galeotti et al. 2014). In vivo studies have shown that dosing rats with St John's wort decreased blood and bowel enzymes associated with colonic inflammation and has less gastric ulcer incidences. This anti-inflammatory activity of *H. perforatum* can be attributed to quercetin and I3, II8-biapigenin, two oil extracts from SJW (Dost et al. 2009). It is believed the anti-inflammatory responses are a result of inhibition of pro-inflammatory genes (Tedeschi et al. 2003).

### 4.4 Antioxidant Activity

LC–MS separation and identification of *H. perforatum* fractions have shown significant antioxidant activity that surpasses that of synthetic antioxidants with specific ability to scavenge DPPH (1,1-Diphenyl-2-picrylhydrazyl) radical, 2,2-Diphenyl-1-(2,4,6-trinitrophenyl) hydrazyl) radical, NO radical, 2,2′-Azobis(2-amidinopropane) dihydrochloride (AAPH), and inhibition of lipid peroxidation (Orčić et al. 2011). Antioxidant properties primarily stem from the phenolic compounds found in *H. perforatum.* Specifically, the flavonoids quercetin and hyperoxide were most effective in scavenging the DPPH radical. The two compounds were not as effective in scavenging AAPH.

## 5 Adverse Drug Effects

A review of data obtained from randomized controlled studies, surveillance studies, and national and international drug safety monitoring bodies indicates *H. perforatum* to be well-tolerated (Ernst et al. 1998). Adverse reactions are reported in 1–3% of patients taking St. John's Wort, and typically these effects are mild, including reports of gastrointestinal symptoms (dyspepsia, anorexia, diarrhea, nausea, and constipation), dizziness, confusion, sedation, or tiredness. In certain clinical trials, subjects exhibited dermatologic effects, including rash, itching, and pruritis (Hammerness et al. 2003; Schrader 2000). Hypericin and pseudo-hypericin, chemical constituents of St. John's Wort, are known photosensitizing agents, and *H. perforatum* usage can lead to hypericism, or sensitivity to sunlight, by lowering light sensitivity thresholds.

Photosensitivity due to the ingestion or topical application of St. John's Wort has been reported for both humans and animals; however, presentation of symptoms of photosensitivity in humans has been rare. Lane-Brown (2000) has described three such isolated incidents of photosensitivity in humans in which all three patients exhibited bulla, erythema, and/or pain after sun exposure following oral or topical use of St. John's Wort.

Livestock is particularly vulnerable to the phototoxic effects of *H. perforatum*, especially grazing animals, light-colored skin that consumes large quantities of the plant. Symptoms following consumption include skin blisters, psychomotor agitation, and in extreme cases, death (Ernst et al. 1998).

### *5.1 Adverse Drug Effects in the Central Nervous System*

Reports of the adverse drug effects of St. John's Wort in the central nervous system are limited to isolated case studies of patients with complicated past medical histories or concomitant prescription drug use. There have been no studies exhibiting adverse effects in the central nervous system on a wide sample of patients. In a double-blind, randomized, placebo-controlled clinical trial of 200 adults diagnosed with major depression, administration of St. John's Wort extract for 4 weeks was associated with headaches for 10% or more of patients in the treatment groups with a significant difference in the incidence of headaches in treatment and placebo groups (Shelton et al. 2001). However, multiple other studies have not found the headache to be an adverse effect.

Mania induced by St. John's Wort has been described in several case studies. Most subjects exhibiting mania, seemingly induced by *H. perforatum*, had histories of major depression or bipolar disorder. A 28-year-old female with a history of depression exhibited mania induced by high doses of St. John's Wort (Fahmi et al. 2002). Another female patient with depression resulting from a traumatic brain injury exhibited hypomania after adding St. John's Wort and *Gingko biloba*, another herbal supplement, to fluoxetine and buspirone. These symptoms ceased after discontinuation of herbal supplements (Spinella and Eaton 2002). Symptoms indicative of serotonin syndrome have also been reported in association with St. John's Wort usage. In one case, a 40-year-old man with a history of anxiety disorder, depression, and selective serotonin reuptake inhibitors (SSRI-induced) mania, but not currently on any other medications, reported diaphoresis, hypertension, confusion, dyspnea, weakness, and tremors after ingesting 450 mg of St. John's Wort daily for 10 days (Parker et al. 2001).

Isolated incidences of psychosis and psychotic symptoms have been observed in patients self-administering SJW. In one case study, a 25-year-old white male with no past medical history of psychiatric disorders and a positive family history of psychiatric disorders presented to an emergency psychiatry service with symptoms of disorganized speech, paranoia, and delusions. The patient had been self-medicating for 4 g of aqueous infusions of SJW four times daily for 3 months before admission into psychiatric services to treat a *Helicobacter pylori* infection discovered by his general practitioner. The patient self-reported exacerbation of psychotic symptoms during this time (Ferrara et al. 2017).

# 6 Drug Interactions

St. John's Wort has been found to reduce the therapeutic efficacy of several prescribed medications when used concurrently. Various studies suggest the primary mechanism for drug interactions between St. John's Wort and regular prescription drugs lies in hyperforin, one of the plant's main active ingredients. The most clinically important effect of SJW is its effect on the expression of genes mediating metabolism, calcium regulation, cell proliferation, and apoptosis. Hyperforin is known to activate the pregnane X receptor, leading to the activation of genes expressing the metabolizing enzyme cytochrome P450 3A4 (CYP3A4) and the efflux transporter p-glycoprotein (Krusekopf and Roots 2005). Specifically, St. John's Wort can promote significant induction of CYP3A4 (Gurley et al. 2005), Cytochrome P450 Family 2 Subfamily E Member 1CYP2E1 (Gurley et al. 2002), and CYP2C9 activities (Yue et al. 2000), and therefore enhance the metabolism of drugs processed by these enzymes. St John's Wort is also believed to induce P-glycoprotein expression in isolated intestinal cells, lowering plasma concentrations for P-glycoprotein substrates such as digoxin, fexofenadine, and talinolol. Thus, many of the drug interactions observed with SJW are with those prescription drugs that are substrates of CYP enzymes and p-glycoprotein.

## *6.1 Anticoagulants*

In 1998, the Swedish bulletin of the Medical Products Agency (MPA) published findings of four cases of decreased efficacy of warfarin during concomitant St. John Wort usage. Patients exhibited a decreased INR value that was stabilized and returned to baseline levels before SJW usage after discontinuation of SJW (Yue et al. 2000). Though no in vitro studies have been conducted, it is believed SJW induces CYP2C9 activity, which results in an enhanced rate of warfarin inactivation.

## *6.2 Oral Contraceptives*

In a single-blind sequential trial of 16 healthy women treated with a daily low-dose oral contraceptive (norethindrone and ethinyl estradiol) and 300 mg of St. John's Wort (three times daily), there was a 13–15% reduction in the contraceptive exposure, and women exhibited an increased instance of breakthrough bleeding compared to the control. It is believed SJW increases the metabolic inactivation of norethindrone and ethinyl estradiol (Murphy et al. 2005). Vlachojannis, Cameron, and Chrubasik have obtained similar results. Concomitant use of SJW and "Ortho-Novum 1/35," an oral contraceptive, led to an increase in the clearance of norethindrone and a reduction in the half-life of ethinyl estradiol. This increased metabolism is believed to be due to increased CYP3A activity (Vlachojannis et al. 2011).

## 6.3 *Antiviral (Anti-HIV) Drugs*

SJW has also been found to decrease plasma concentrations of the HIV protease inhibitor, indinavir, by 57% in 16 healthy volunteers (Piscitelli et al. 2000). Indinavir is a substrate for CYP3A4 and P-glycoprotein. Similar results of decreased plasma concentrations of prescription drugs were found in patients taking SJW with nevirapine, a substrate of CYP3A4 and CYP2B6 (de Maat et al. 2001). In one isolated case, a patient experienced an increase in HIV RNA viral load after concomitant use of hypericum with indinavir and lamivudine (Chiba et al. 1996).

## 6.4 *Immunosuppressants*

The interactions between SJW and cyclosporine are well documented in several case studies and clinical trials (Breidenbach et al. 2000; Dresser et al. 2003; Mai et al. 2004). SJW taken at therapeutic doses with cyclosporine following heart, renal, or liver transplants showed decreased blood levels, and in some cases, patients experienced acute rejection episodes. Patients clinically improved following discontinuation of the herbal supplement. Cyclosporine is a substrate of P-glycoprotein and is metabolized by CYP3A4. Dresser et al. also believe SJW may also induce multidrug-resistant gene (MDR1) activities. In addition to cyclosporine, blood levels of tacrolimus, another immunosuppressive drug, have also been found to decrease when taken with SJW. Interactions of tacrolimus with other drugs are of special interest, for low levels can result in rejection of transplants, while high levels can lead to nephrotoxicity. In one case, a 65-year-old renal transplant recipient showed a sharp decrease in plasma levels of tacrolimus after 1 month of taking 600 mg SJW to self-treat a depressive mood disorder (Bolley et al. 2002).

## 6.5 *Statins*

Simvastatin and atorvastatin are two statins metabolized by CYP34 and are substrates for P-glycoprotein. SJW has been shown to decrease plasma levels of simvastatin and reduce the efficacy of atorvastatin (Sugimoto et al. 2001). Patients with hyperlipidemia taking atorvastatin with SJW experienced a significant increase in serum levels of low-density lipoprotein (LDL) cholesterol and total cholesterol (Andrén et al. 2007).

## 6.6 Beta-Adrenergic and Calcium Blockers

Talinolol is a beta1-adrenergic blocker and P-glycoprotein substrate. A 12-day treatment of SJW resulted in a 93% increase in clearance of talinolol, which is believed to be due to inductions or MDR1 and P-glycoprotein levels modulated by SJW (Schwarz et al. 2007). SJW has also been shown to decrease the area under the plasma drug concentration–time curve (AUC) of nifedipine and verapamil, and these two calcium channel blockers are used to treat hypertension in healthy volunteers. Both these compounds are metabolized by CYP3A4 (Tannergren et al. 2004).

## 6.7 Antidepressants

Case reports of mania induced by SJW usage as well as symptoms of central serotonin syndrome show that SJW interacts with 5-HT reuptake inhibitors, such as paroxetine, sertraline, venlafaxine, and nefazodone. Though CYP enzymes of P-glycoprotein do not metabolize these selective serotonin reuptake inhibitors (SSRI), a pharmacodynamic mechanism is believed to be responsible (Lantz et al. 1999; Prost et al. 2000). Other antidepressants that are metabolized by CYP enzymes have shown decreased efficacy when used with SJW. In a study investigating the concomitant use of 75 mg of amitriptyline, a tricyclic antidepressant, twice a day, and 900 mg hypericum extract a day for 12–14 days by 12 participants, the patients were found to have a decreased plasma AUC of amitriptyline by 22%. The same study also found that the AUC for nortriptyline decreased by 41% (Johne et al. 2002).

## 6.8 Benzodiazepines

CYP3A4 enzymes metabolize alprazolam, midazolam, and quazepam. Accordingly, SJW has shown significant interactions with these medications. Concomitant oral administration of 2 mg alprazolam and one SJW tablet was found to increase the alprazolam clearance by twofold, decrease the alprazolam plasma concentration AUC by twofold, and decrease the half-life for alprazolam from 3.9 to 2.4 h in 12 healthy volunteers (Markowitz et al. 2003). Similar results of reduced plasma levels after use with SJW were observed with midazolam in healthy volunteers

(Gurley et al. 2002). Thirteen participants taking 15 mg of quazepam after treatment with 900 mg a day of SJW for 14 days also exhibited a decreased plasma concentration and a significantly reduced AUC (Kawaguchi et al. 2004).

## 7 Future Recommendations for Prophylactic and Therapeutic Use

Though readily available in US commercial markets, St. John's Wort does not appear to be a homeopathic panacea that many manufacturers tout it to be. Despite many clinical, in vivo, and in vitro studies, key questions regarding SJW's molecular mechanisms, efficacy, and adverse effects remain unanswered. Even its therapeutic value in the treatment of depression remains in question, as pointed out by Linde et al. (2008). They identified contradictions in studies investigating the efficacy of SJW in treating major depression conducted by German-speaking and non-German-speaking countries (Linde et al. 2008). Further research is essential in elucidating major adverse impacts, especially when used concomitantly with other drugs. Isolated studies investigating adverse drug interactions must be replicated in rigorous pharmacokinetic trials to deem as sufficient evidence of its adverse effects. Studies investigating the potential neurotoxicity of SJW should have larger sample sizes to determine if observed incidences were isolated or can be generalized. Finally, research on the different formulations, different preparations, and the effects this has on its therapeutic value must also be conducted for safe usage.

## References

Andrén L, Andreasson A, Eggertsen R (2007) Interaction between a commercially available St. John's wort product (Movina) and atorvastatin in patients with hypercholesterolemia. Eur J Clin Pharmacol 63:913–916. https://doi.org/10.1007/s00228-007-0345-x

Arsić I (2016) Preparation and characterization of St. John's Wort herb extracts using olive, sunflower and palm oils. Acta Facultatis Medicae Naissensis 33:119–126. https://doi.org/10.1515/afmnai-2016-0013

Benzie IFF, Wachtel-Galor S (eds) (2011) Herbal medicine: biomolecular and clinical aspects, 2nd edn. CRC Press, Boca Raton

Bolley R, Zülke C, Kammerl M et al (2002) Tacrolimus-induced nephrotoxicity unmasked by induction of the CYP3A4 system with St John's wort. Transplantation 73:1009. https://doi.org/10.1097/00007890-200203270-00035

Breidenbach T, Hoffmann MW, Becker T et al (2000) Drug interaction of St John's wort with cyclosporin. Lancet 355:1912. https://doi.org/10.1016/s0140-6736(05)73359-6

Brenn A, Grube M, Jedlitschky G et al (2014) St. John's wort reduces Beta-amyloid accumulation in a double transgenic Alzheimer's disease mouse model-role of P-glycoprotein: St. John's wort and Alzheimer's disease. Brain Pathol 24:18–24. https://doi.org/10.1111/bpa.12069

Breyer A, Elstner M, Gillessen T et al (2007) Glutamate-induced cell death in neuronal HT22 cells is attenuated by extracts from St. John's wort (Hypericum perforatum L.). Phytomedicine 14:250–255. https://doi.org/10.1016/j.phymed.2007.02.001

Brigitta B (2002) Pathophysiology of depression and mechanisms of treatment. Dialogues Clin Neurosci 4:7–20

Butterweck V (2003) Mechanism of action of St John's wort in depression: what is known? CNS Drugs 17:539–562. https://doi.org/10.2165/00023210-200317080-00001

Chatterjee SS, Bhattacharya SK, Wonnemann M et al (1998) Hyperforin as a possible antidepressant component of hypericum extracts. Life Sci 63:499–510. https://doi.org/10.1016/s0024-3205(98)00299-9

Chiba M, Hensleigh M, Nishime JA et al (1996) Role of cytochrome P450 3A4 in human metabolism of MK-639, a potent human immunodeficiency virus protease inhibitor. Drug Metab Dispos 24:307–314

Diwu Z (1995) Novel therapeutic and diagnostic applications of hypocrellins and hypericins. Photochem Photobiol 61:529–539. https://doi.org/10.1111/j.1751-1097.1995.tb09903.x

Dost T, Ozkayran H, Gokalp F et al (2009) The effect of Hypericum perforatum (St. John's wort) on experimental colitis in rat. Dig Dis Sci 54:1214–1221. https://doi.org/10.1007/s10620-008-0477-6

Dresser GK, Schwarz UI, Wilkinson GR, Kim RB (2003) Coordinate induction of both cytochrome P4503A and MDR1 by St John's wort in healthy subjects. Clin Pharmacol Ther 73:41–50. https://doi.org/10.1067/mcp.2003.10

Ernst E, Rand JI, Barnes J, Stevinson C (1998) Adverse effects profile of the herbal antidepressant St. John's wort (Hypericum perforatum L.). Eur J Clin Pharmacol 54:589–594. https://doi.org/10.1007/s002280050519

Fahmi M, Huang C, Schweitzer I (2002) A case of mania induced by *Hypericum*. World J Biol Psychiatry 3:58–59. https://doi.org/10.3109/15622970209150602

Fehr MJ, McCloskey MA, Petrich JW (1995) Light-induced acidification by the antiviral agent Hypericin. J Am Chem Soc 117:1833–1836. https://doi.org/10.1021/ja00111a024

Ferrara M, Mungai F, Starace F (2017) St John's wort (*Hypericum perforatum*)-induced psychosis: a case report. J Med Case Rep 11(1):137. https://doi.org/10.1186/s13256-017-1302-7

Fox FE, Niu Z, Tobia A, Rook AH (1998) Photoactivated Hypericin is an anti-proliferative agent that induces a high rate of apoptotic death of normal, transformed, and malignant t lymphocytes: implications for the treatment of cutaneous lymphoproliferative and inflammatory disorders. J Invest Dermatol 111:327–332. https://doi.org/10.1046/j.1523-1747.1998.00278.x

Galeotti N, Vivoli E, Bilia AR et al (2010) A prolonged protein kinase c-mediated, opioid-related antinociceptive effect of St John's wort in mice. J Pain 11:149–159. https://doi.org/10.1016/j.jpain.2009.06.013

Galeotti N, Maidecchi A, Mattoli L et al (2014) St. John's wort seed and feverfew flower extracts relieve painful diabetic neuropathy in a rat model of diabetes. Fitoterapia 92:23–33. https://doi.org/10.1016/j.fitote.2013.10.003

Gurley BJ, Gardner SF, Hubbard MA et al (2002) Cytochrome P450 phenotypic ratios for predicting herb-drug interactions in humans. Clin Pharmacol Ther 72:276–287. https://doi.org/10.1067/mcp.2002.126913

Gurley BJ, Gardner SF, Hubbard MA et al (2005) Clinical assessment of effects of botanical supplementation on cytochrome p450 phenotypes in the elderly: St John's wort, garlic oil, *Panax ginseng* and *Ginkgo biloba*. Drugs Aging 22:525–539. https://doi.org/10.2165/00002512-200522060-00006

Hammerness P, Basch E, Ulbricht C et al (2003) St. John's wort: a systematic review of adverse effects and drug interactions for the consultation psychiatrist. Psychosomatics 44:271–282. https://doi.org/10.1176/appi.psy.44.4.271

Hofrichter J, Krohn M, Schumacher T et al (2013) Reduced Alzheimer's disease pathology by St. John's wort treatment is independent of hyperforin and facilitated by ABCC1 and microglia activation in mice. Curr Alzheimer Res 10:1057–1069. https://doi.org/10.2174/15672050113106660171

Holden C (1991) Treating AIDS with Worts. Science 254:522

Hosseinzadeh H, Karimi GR, Rakhshanizadeh M (2005) Anticonvulsant effect of *Hypericum perforatum*: role of nitric oxide. J Ethnopharmacol 98:207–208. https://doi.org/10.1016/j.jep.2004.12.007

Hostanska K, Reichling J, Bommer S et al (2003) Hyperforin a constituent of St John's wort (*Hypericum perforatum* L.) extract induces apoptosis by triggering activation of caspases and with hypericin synergistically exerts cytotoxicity towards human malignant cell lines. Eur J Pharm Biopharm 56:121–132. https://doi.org/10.1016/S0939-6411(03)00046-8

Ivetic V, Popovic M, Mimica-Dukic N et al (2002) St. John's wort (*Hypericum perforatum* L.) and kindling epilepsy in rabbit. Phytomedicine 9:496–499. https://doi.org/10.1078/09447110260572985

Johne A, Schmider J, Brockmöller J et al (2002) Decreased plasma levels of amitriptyline and its metabolites on comedication with an extract from St. John's wort (*Hypericum perforatum*). J Clin Psychopharmacol 22:46–54. https://doi.org/10.1097/00004714-200202000-00008

Kawaguchi A, Ohmori M, Tsuruoka S-I et al (2004) Drug interaction between St John's wort and quazepam. Br J Clin Pharmacol 58:403–410. https://doi.org/10.1111/j.1365-2125.2004.02171.x

Krusekopf S, Roots I (2005) St. John's wort and its constituent hyperforin concordantly regulate expression of genes encoding enzymes involved in basic cellular pathways. Pharmacogenet Genomics 15:817–829. https://doi.org/10.1097/01.fpc.0000175597.60066.3d

Kwiecień I, Szydłowska A, Kawka B et al (2015) Accumulation of biologically active phenolic acids in agitated shoot cultures of three *Hypericum perforatum* cultivars: 'Elixir', 'Helos' and 'Topas. Plant Cell Tiss Organ Cult (PCTOC) 123:273–281. https://doi.org/10.1007/s11240-015-0830-3

Lane-Brown MM (2000) Photosensitivity associated with herbal preparations off St John's wort (*Hypericum perforatum*). Med J Aust 172:302. https://doi.org/10.5694/j.1326-5377.2000.tb123965.x

Lantz MS, Buchalter E, Giambanco V (1999) St. John's wort and antidepressant drug interactions in the elderly. J Geriatr Psychiatry Neurol 12:7–10. https://doi.org/10.1177/089198879901200103

Lee MR (1999) Saint John's wort (*Hupericum perforatum*). A balm for hurt minds? Proc R Coll Physicians Edinb 29:253–257

Linde K, Berner MM, Kriston L (2008) St John's wort for major depression. Cochrane Database Syst Rev 2008:CD000448. https://doi.org/10.1002/14651858.CD000448.pub3

Lu Y-H, Du C-B, Liu J-W et al (2004) Neuroprotective effects of Hypericum perforatum on trauma induced by hydrogen peroxide in PC12 cells. Am J Chin Med 32:397–405. https://doi.org/10.1142/S0192415X04002053

Lyles JT, Kim A, Nelson K et al (2017) The chemical and antibacterial evaluation of St. John's wort oil macerates used in Kosovar traditional medicine. Front Microbiol 8:1639. https://doi.org/10.3389/fmicb.2017.01639

de Maat MMR, Hoetelmans RMW, Mathôt RAA et al (2001) Drug interaction between St John's wort and nevirapine. AIDS 15:420–421. https://doi.org/10.1097/00002030-200102160-00019

Mai I, Bauer S, Perloff ES et al (2004) Hyperforin content determines the magnitude of the St John's wort-cyclosporine drug interaction. Clin Pharmacol Ther 76:330–340. https://doi.org/10.1016/j.clpt.2004.07.004

Markowitz JS, Donovan JL, DeVane CL et al (2003) Effect of St John's wort on drug metabolism by induction of cytochrome P450 3A4 enzyme. JAMA 290:1500–1504. https://doi.org/10.1001/jama.290.11.1500

Maury W, Price JP, Brindley MA et al (2009) Identification of light-independent inhibition of human immunodeficiency virus-1 infection through bioguided fractionation of *Hypericum perforatum*. Virol J 6:101. https://doi.org/10.1186/1743-422X-6-101

Meruelo D, Lavie G, Lavie D (1988) Therapeutic agents with dramatic antiretroviral activity and little toxicity at effective doses: aromatic polycyclic diones hypericin and pseudohypericin. Proc Natl Acad Sci U S A 85:5230–5234. https://doi.org/10.1073/pnas.85.14.5230

Miller KD, Nogueira L, Mariotto AB et al (2019) Cancer treatment and survivorship statistics, 2019. CA Cancer J Clin 69:363–385. https://doi.org/10.3322/caac.21565
Murphy PA, Kern SE, Stanczyk FZ, Westhoff CL (2005) Interaction of St. John's wort with oral contraceptives: effects on the pharmacokinetics of norethindrone and ethinyl estradiol, ovarian activity and breakthrough bleeding. Contraception 71:402–408. https://doi.org/10.1016/j.contraception.2004.11.004
Nathan PJ (2001) *Hypericum perforatum* (St John's wort): a non-selective reuptake inhibitor? A review of the recent advances in its pharmacology. J Psychopharmacol 15:47–54. https://doi.org/10.1177/026988110101500109
Orčić DZ, Mimica-Dukić NM, Francišković MM et al (2011) Antioxidant activity relationship of phenolic compounds in *Hypericum perforatum* L. Chem Cent J 5:34. https://doi.org/10.1186/1752-153X-5-34
Öztürk Y, Aydin S, Beis R et al (1996) Effects of *Hypericum perforatum* L. and *Hypericum calycinum* L. extracts on the central nervous system in mice. Phytomedicine 3:139–146. https://doi.org/10.1016/S0944-7113(96)80027-4
Pahwa R, Goyal A, Bansal P, Jialal I (2020) Chronic inflammation. In: StatPearls. StatPearls Publishing, Treasure Island, FL
Parker V, Wong AH, Boon HS, Seeman MV (2001) Adverse reactions to St John's wort. Can J Psychiatr 46:77–79. https://doi.org/10.1177/070674370104600112
Piscitelli SC, Burstein AH, Chaitt D et al (2000) Indinavir concentrations and St John's wort. Lancet 355:547–548. https://doi.org/10.1016/S0140-6736(99)05712-8
Prost N, Tichadou L, Rodor F et al (2000) Interaction millepertuis-venlafaxine [St. John's wort-venlafaxine interaction]. Presse Med 29:1285–1286
del Rio MA, Sanchez-Reus MI, Iglesias I et al (2013) Neuroprotective properties of standardized extracts of Hypericum perforatum on rotenone model of Parkinson's disease. CNS Neurol Disord Drug Targets 12:665–679. https://doi.org/10.2174/1871527311312050013
Schempp CM, Pelz K, Wittmer A et al (1999) Antibacterial activity of hyperforin from St John's wort, against multiresistant Staphylococcus aureus and gram-positive bacteria. Lancet 353:2129. https://doi.org/10.1016/S0140-6736(99)00214-7
Schempp CM, Kirkin V, Simon-Haarhaus B et al (2002) Inhibition of tumour cell growth by hyperforin, a novel anticancer drug from St. John's wort that acts by induction of apoptosis. Oncogene 21:1242–1250. https://doi.org/10.1038/sj.onc.1205190
Schrader E (2000) Equivalence of St John's wort extract (Ze 117) and fluoxetine: a randomized, controlled study in mildmoderate depression. Int Clin Psychopharmacol 15:61–68. https://doi.org/10.1097/00004850-200015020-00001
Schwarz UI, Hanso H, Oertel R et al (2007) Induction of intestinal P-glycoprotein by St John's wort reduces the oral bioavailability of talinolol. Clin Pharmacol Ther 81:669–678. https://doi.org/10.1038/sj.clpt.6100191
Shelton RC, Keller MB, Gelenberg A et al (2001) Effectiveness of St John's wort in major depression: a randomized controlled trial. JAMA 285:1978. https://doi.org/10.1001/jama.285.15.1978
Silva B, Oliveira PJ, Dias A, Malva JO (2008) Quercetin, kaempferol and biapigenin from *Hypericum perforatum* are neuroprotective against excitotoxic insults. Neurotox Res 13:265–279. https://doi.org/10.1007/bf03033510
Spinella M, Eaton LA (2002) Hypomania induced by herbal and pharmaceutical psychotropic medicines following mild traumatic brain injury. Brain Inj 16:359–367. https://doi.org/10.1080/02699050110103319
Sugimoto K, Ohmori M, Tsuruoka S et al (2001) Different effects of St John's wort on the pharmacokinetics of simvastatin and pravastatin. Clin Pharmacol Ther 70:518–524. https://doi.org/10.1067/mcp.2001.120025
Tannergren C, Engman H, Knutson L et al (2004) St John's wort decreases the bioavailability of R- and S-verapamil through induction of the first-pass metabolism. Clin Pharmacol Ther 75:298–309. https://doi.org/10.1016/j.clpt.2003.12.012

Tedeschi E, Menegazzi M, Margotto D et al (2003) Anti-inflammatory actions of St. John's wort: inhibition of human inducible nitric-oxide synthase expression by down-regulating signal transducer and activator of transcription-1alpha (STAT-1alpha) activation. J Pharmacol Exp Ther 307:254–261. https://doi.org/10.1124/jpet.103.054460

Vlachojannis J, Cameron M, Chrubasik S (2011) Drug interactions with St. John's wort products. Pharmacol Res 63:254–256. https://doi.org/10.1016/j.phrs.2010.11.011

Yue QY, Bergquist C, Gerdén B (2000) Safety of St John's wort (Hypericum perforatum). Lancet 355:576–577. https://doi.org/10.1016/S0140-6736(05)73227-X

# Neurotoxic Potential of Alkaloids from Thorn Apple (*Datura stramonium* L.): A Commonly Used Indian Folk Medicinal Herb

**Malvi Choudhary, Itika Sharma, Dinesh Chandra Agrawal, Manoj K. Dhar, and Sanjana Kaul**

## Contents



**Abstract** *Datura stramonium* L., plant has been an exemplary source of folklore medicinal herb known for its mental stimulation and curative properties. The noteworthy pharmacological potential of this plant is utilized by Ayurvedic practitioners in the traditional system of Indian medicine. The major active principles contributing to remedying effects of this plant are ascribed to alkaloids, terpenoids, withanolides, steroids, flavonoids, triterpenes, phenolic compounds, carbohydrates, and tannins. These phytoconstituents bequeathed plant with anticancer, anti-inflammatory, antidiabetic, anticholinergic, antispasmodic, antioxidant,

M. Choudhary · I. Sharma · M. K. Dhar · S. Kaul (✉)
School of Biotechnology, University of Jammu, Jammu, Jammu & Kashmir, India
e-mail: sanjanakaul@jammuuniversity.ac.in

D. C. Agrawal
Department of Applied Chemistry, Chaoyang University of Technology, Taichung, Taiwan

 
D. C. Agrawal, M. Dhanasekaran (eds.), *Medicinal Herbs and Fungi*,
https://doi.org/10.1007/978-981-33-4141-8_16

antidepressant, antimicrobial, cytotoxic, insecticidal, larvicidal, and herbicidal activity. So far, 64 tropane alkaloids characterized from the *Datura* plant are of prime importance to humankind. However, a variety of tropane alkaloids are neurotoxic, and their frequent recreational abuse may lead to delirium and even death. Recent studies have been directed toward the therapeutic efficacy of these metabolites and/or their derivatives as lead molecules offering new perspectives in biological, medicinal, and pharmaceutical research. This also encourages the cultivation and conservation of this highly valuable medicinal herb for its putative health benefits and sustainable use at commercial scale.

**Keywords** *Datura stramonium* · Medicinal herb · Neurotoxic · Pharmacology · Phytoconstituents · Tropane alkaloids

## Abbreviations

| | |
|---|---|
| ACh | Acetylcholine |
| CNS | Central nervous system |
| CT | Computerized tomography |
| DPPH | 1, 1-Diphenyl-2-picrylhydrazyl |
| g | Gram |
| GPCR | G-protein coupled receptor |
| IC50 | Half maximal inhibitory concentration |
| kg | Kilogram |
| L | Liter |
| m AchR | Muscarinic acetylcholine receptor |
| M(1–5) | Mammalian subtypes |
| mg | Milligram |
| min | Minute |
| mL | Milliliter |
| MS | Mass spectrometry |
| nAChR | Nicotinic acetylcholine receptor |
| RBCs | Red blood cells |
| SA node | Sinoatrial node |
| Sp | Species |
| WBCs | White blood cells |
| μg | Microgram |

## 1 Introduction

Longer product development cycles, lack of high-throughput screening against molecular targets, uncertainty in the collection of biomaterials, and difficulties associated with the repeated isolation of known bioactive compound are the main retrospective concerns that have led to decline in the interest of pharmaceutical industries in natural-product drug discovery in the past few decades (Koehn and Carter 2005). Aside from these addressed issues, there has been a paradigm shift from synthetic molecules to natural products in the current milieu (Atanasov et al. 2015). Natural products are evolutionarily optimized compounds fashioned by natural selection. Understanding the capacity of the natural world to produce secondary metabolites resulted in the revitalization of interest for the design and synthesis of new chemotypes to improve pharmacological profiles (Lee 2004; Pye et al. 2017). These products are bestowed with molecular selectivity and are highly efficient in their cellular targets (Grisoni et al. 2018). Recently, increasing appreciation has relocated to natural product research for tailoring synthetically tractable small molecules simulating nature's chemistry (Wright 2017). During the period from 1981 to 2014, 51% of all approved therapeutic agents are either natural products or based on their skeletons (Newman and Cragg 2016). Nature in one guise or another has blessed humanity with an abundant botanical wealth for treating various human ailments since antiquity (Pandith et al. 2018). According to the WHO, the usage and demand for natural products, principally plant-based therapies have gained momentum among the world's population. The widespread acceptance and commercialization of herbal medicines have been attributed to their long-lasting curative effects, lower complications, fewer side effects, etc. (Forouzanfar and Hosseinzadeh 2018). Among the developing countries, India is one of the major global suppliers of medicinal herbs, an integral part of traditional medical practices (Tripathy et al. 2015). *Datura stramonium* L., a member of the Solanaceae family, is a well-known folklore medicinal herb. The genus name *Datura* has its origin from the Sanskrit word "dhattura" or "dhatura" (meaning divine inebriation) quoted in several classical texts (Gallego 2012). It has religious visionary importance in the Hindu godlore, where people provide thorn apple as offerings to God Shiva on religious occasions like Shivaratri and Swasthani Puja (Pandya et al. 2019). It is a widespread annual flowering plant, highly rich in tropane alkaloids (organic nitrogenated bicyclic compounds) having therapeutic potentialities (Benitez et al. 2018). The alkaloid content of *Datura* comprises a conglomeration of anticholinergic and antimuscarinic drugs like atropine, scopolamine, and hyoscyamine (Rahmoune et al. 2019). The presence of these active constituents makes this plant stand out in Ayurveda to alleviate various human ailments such as inflammation, ulcers, rheumatism, respiratory decongestion, alopecia, toothache, dental and skin infections (Bock and Norris 2016). Nevertheless, the possible toxic side effects of *Datura* poisoning include hallucinations, bizarre behavior, confusion, amnesia, delirium leading to coma, cardiac and respiratory arrest, and even death (Maheshwari et al. 2013; Gaire and Subedi 2013; Gupta et al. 2016). However, a thorough understanding of scientifically proven chemical constituents of *Datura* will stipulate its possible pharmacological and toxicological effects.

## 2 Vernacular Names of *D. stramonium* L

*Datura*, a folklore medicinal herb used in Indian systems of medicine, has been known by many common names such as Jimson weed, stinkweed, mad apple, thorn apple, dhatura tatula, Devil's snare, hell's bell, devil's trumpet, devil's weed, tolguacha, locoweed, pricklyburr, and devil's cucumber (Mishra 2018) (Fig. 1). Also, known by other names:

English Name: Thorn apple
Hindi Name: Dattura
Sanskrit: Dhastura
Tamil name: Ummattangani/Vella-Ummathai
Telugu name: Ummetta
Bengal name: Dhattura
Gujarati name: Unmatta
Kannada name: Unmatta, Dhattura
Malayalam name: Maraummam
Arabic name: Datur
Persian name: Tatur

**Fig. 1** A *Datura* plant growing in wild

## 3 Geographical Distribution

The non-anthropogenic origin of genus *Datura* was found endemic to Mexico and the southwestern USA. The Mexican territory, adjacent areas in the southwestern USA and northern Central America indicated the center of taxonomic diversity and distribution of this genus (Setshogo 2015). It is well adapted to xeric conditions of America, and higher abundance is found in tropical, subtropical, and temperate countries. In India, wastelands, old fields, eroded sites, rangelands, scrub jungles, railway lines, and overgrazed pastures are regions of its phytogeographic distribution (Anju and Ratan 2011). It is a nine species complex genus endowed with medicinal as well as poisonous properties (Jalander and Gachande 2012).

## 4 Ethnopharmacological Importance

*Datura* sp. has contributed significantly to the Indian folklore system of medicine with high utility and usage in Ayurveda. Herbal formulations made from different plant parts have been used extensively to treat human disorders since antiquity. Traditionally, its pharmacological potential lies in curing a wide range of human sufferings like burns, ulcers, alopecia, otalgia, toothache, epilepsy, rheumatism, as a central nervous system stimulant, etc. (Maheshwari et al. 2013). The whole plant expressed juice/extract is employed as a poultice in treating abscess wounds, fistulas, severe neuralgia, hemorrhoids, inflammation, pain, and bruises. The leaves of the plant have therapeutic efficacies and are used as a remedy for skin infections, headache, rheumatism, and gout. Since it is a hallucinogenic plant, its leaves are smoked for its bronchodilator action by relaxing the smooth muscles of the bronchial tube and results in the relief of asthmatic bronchial spasm and bronchitis (Soni et al. 2012; Uddin et al. 2012). In European countries, *Datura* seeds are known for its narcotic properties. Its seeds are capable of causing sedation and are useful in treating the behavior of hysterical and psychotic patients. Additionally, seeds act as a purging agent to get relief from stomach and intestinal pain due to worm infestation. The paste made from flowers of *Datura* finds wide application in the prevention of scalp infections like seborrheic dermatitis, psoriasis, alopecia, and as a prophylactic treatment to animal bites (Monira and Munan 2012). All these ethnopharmacological relevance of *Datura* plants further requires interdisciplinary and multidisciplinary approaches to derive plant-active pharmaceutical ingredients.

## 5 Phytochemistry

The genus *Datura* synthesizes a suite of variable chemotypes belonging to tropane alkaloids. These are ornithine-derived natural compounds amply present in different plant parts of the Solanaceae family and at various stages of the life cycle. Tropane alkaloids, along with its semisynthetic derivatives, are potent muscarinic receptor antagonists and are pharmaceutically active (Kukula-Koch and Widelski 2017). Up to 64 alkaloids have been detected from the *Datura* plant with atropine, hyoscyamine, and scopolamine as the major alkaloids and tigloidin, aposcopolamine, apoatropine, hyoscyamine N-oxide and scopolamine N-oxide, 3,6 ditigloyloxytropane, and 7-hydroxyhyoscyamine are few of minor alkaloids (Das et al. 2012). Pandya et al. (2019) reported the variations in the concentration of atropine and scopolamine among different plant parts. Atropine content was found maximum in roots and stems, whereas scopolamine was present in higher levels in leaves and stems. Hyoscyamine is the third major alkaloid of interest with promising pharmacological properties. These active chemical constituents are highly valued for their anticholinergic effects, thereby serving as lead structures in many pharmacopeias. Saponins, tannins, and glycosides are the main phytoconstituents present in the *Datura* plant, as studied by Shagal et al. (2012) (Fig. 2).

## 6 Neurotoxic Properties of *D. stramonium* L

Alkaloids are naturally occurring secondary plant metabolites functioning as substitutes for minerals and are also involved in protection against predators and pathogens. They are renowned cyclic nitrogen compounds that assuage human sufferings with their potential pharmacological actions and structural diversity (Dey et al. 2020). Based on structural perception, they are categorized into two divisions: (i) heterocyclic and (ii) non-heterocyclic compounds. The amino acid precursor molecule involved in the biosynthetic pathway of these alkaloid further forms the basis of classification of alkaloids into true and prototypes (Adibah and Azzreena 2019).

Tropane alkaloids are ornithine, and acetoacetate derived true alkaloids with "8-azabicyclo [3.2.1] octane skeleton" (Fig. 3), whose synthesis is ascribed mainly to botanicals of Solanaceae family like *Atropa, Datura, Hyoscyamus,* etc. The tropane alkaloids containing genera are well documented in ethnopharmacological tradition for both preventive and curative measures. The tropic acid ester, a basic structural motif of tropane alkaloids, can be enzymatically substituted by varying functional groups for the synthesis of various derivative molecules (Afewerki et al. 2019). Atropine, scopolamine, and hyoscyamine are naturally occurring bicyclic tropane alkaloids abundantly found in the Solanaceae family. They have been evaluated as potential bronco-dilators, anesthetic, antiemetics, antispasmodic, and anti-colic

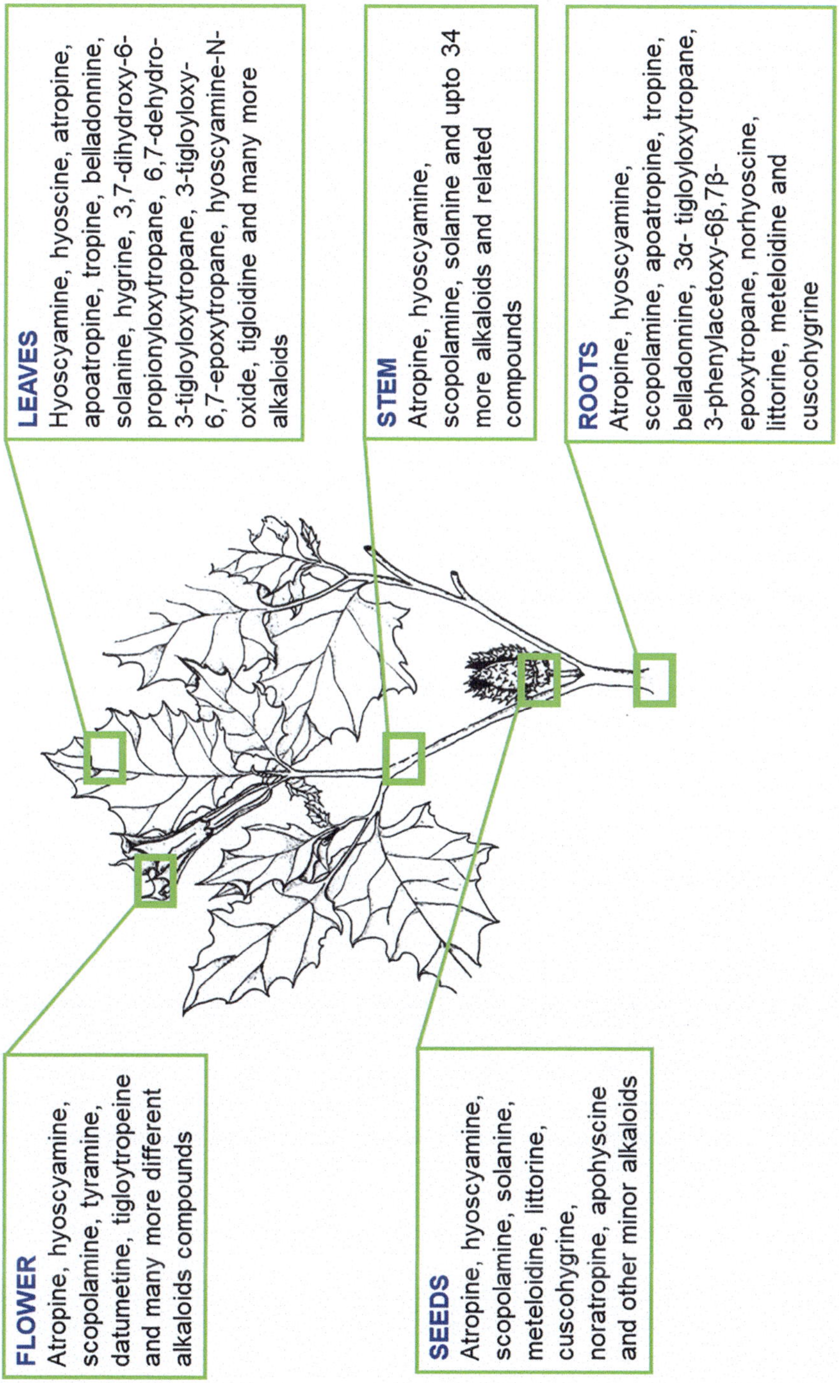


**Fig. 2** The major alkaloids present in different parts of the *Datura* plant

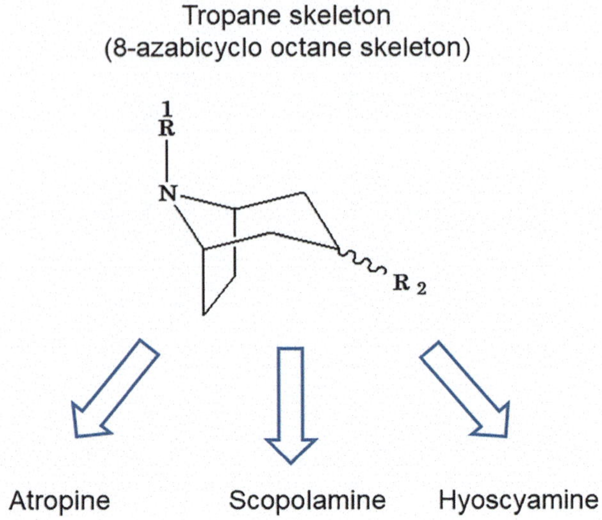


**Fig. 3** Three main alkaloids found in *Datura* plant derived from a common tropane skeleton

drugs (Grynkiewicz and Gadzikowska 2008). Also, they are widely employed as model anticholinergic drugs (Pandareesh et al. 2016; Chen and Yoon 2020).

In parallel, its structural diversity has contributed significantly to draw the attention of toxicologists to understand its toxicity level in practice. They are endowed with the ability to adversely affect the structural and functional integrity of the nervous system and have been characterized as potent "neurotoxins." Toxicant ingestion, dosage, exposure time, sensitivity, and developmental stage of an organism all interfere with the normal signaling mechanisms of neurotransmission, resulting in hallucinations and psychoactive effects (Saroya and Singh 2020). These are toxic compounds that damage the permeability of the central nervous system by crossing the blood–brain barrier (Williamson 2017).

*Datura* is a toxic herb when ingested improperly. It was introduced to the old world as an asthma herb by Romany immigrants for its bronchodilatory effect (Kohnen-Johannsen and Kayser 2019). It was found during the very first time in the Virginia state of the United States in 1676 (Wiebe et al. 2008). It is a member of the belladonna family that contains various alkaloids. When taken at high doses, these may cause tachycardia, mydriasis, restlessness, confusion, bizarre behavior, perception disorders, agitation, and subsequent amnesia. As the cost of *Datura* is lesser as compared to other drugs such as marijuana, hence it is more often used by teenagers and adults for the stupefaction purpose (Adekomi et al. 2020). Experience of varying levels of psychoactive effects is dependent on the bioavailability and predominance of belladonna alkaloids present in different plant parts. Approximately, 6 mg of atropine is present in 100 seeds of *D. stramonium*. Flowers of the plant contain the highest amount of alkaloid content, and roots comprise the lowest ratios (Pillay and Sasidharan 2019). Nevertheless, when highly toxic doses override the doses of tolerance, datura poisoning results (Karadas et al. 2011; Krenzelok 2010). Thus, in this case, alkaloid poisoning is rightly justified by the statement "the dose makes the poison." Therefore, in the medieval times, the plant parts or extracts were used as poisons, in rituals of shamanism and as magic potions in witchcraft practices (Carod-Artal 2015). The mnemonic can easily remember the clinical symptoms of poisoning: "hot as a hare, blind as a bat, dry as a bone, red as a beet, and mad as a hatter," representing hyperthermia, mydriasis, dry mouth, flushed skin, and mucous membranes, delirium (Dalefied 2017). Thus, they become a subject of illicit "recreational drugs" abuse whose illegal trafficking calls out for a social emergency.

## *6.1 Mode of Action and Clinical Presentation*

Neurotransmitters are an important class of nerve-secreted bioactive compounds that perform key regulatory roles in intercellular communication (Changeux 2020). The peripheral and autonomic nervous system (brain, peripheral plexuses, ganglia, and adrenal medulla) are involved in the release of neurotransmitter molecules. Chemically, it is a ligand that triggers stimulatory or inhibitory neuronal functions by binding and activating specific receptors present on the surface of target cells. Understanding the precise dynamics and functions of neurotransmitters is the main prerequisite for understanding the chemistry, physiology, cognitive functions, and pathologies of the brain (Fedotova et al. 2020). Acetylcholine (2-acetoxy-N, N, N-trimethylethanaminium; ACh) as a neurotransmitter was the first molecule to be identified. It is the most abundant intercellular lipophobic neurotransmitter found in both vertebrates and invertebrates. It participates in neuronal signaling, playing a fundamental role in cellular and synaptic physiology, tissue homeostasis, and is well characterized in switch network dynamics (Cox et al. 2020).

Nicotinic (nAChRs) and muscarinic ACh receptors (mAChRs) are two structurally distinct families of cell–surface receptors that are known to reproduce the physiological effects elicited by ACh. They are named so after their agonists, nicotine, and muscarine. The former function as ionotropic receptors and the latter are members of metabotropic G-protein-coupled receptors (GPCRs) (Wang et al. 2018). Recent realization of experimental data from molecular cloning studies has revealed that five distinct mammalian subtypes (M1–M5) of mAChRs are present. The cholinergic transmission of these M1–M5 receptors is based on G-protein coupling preference. M1, M3, and M5 are coupled with stimulating Gq/G11 receptors family that acts via the phospholipase C signaling pathway. M2 and M4, on the other hand, activate downstream signaling via Gi/Go-type G-proteins (Jones et al. 2012). Such a G-protein predilection results in the activity transition of key effector molecules, as shown in Fig. 4. Also, the distribution and function of these molecularly distinct mAChs subtypes have been tabulated (Table 1).

Therefore, these subtypes control a plethora of cholinergic processes such as an increase in intracellular calcium levels, contraction of smooth muscles, brain microcirculation, in pain processes by mediating vasoconstriction and vasodilation,

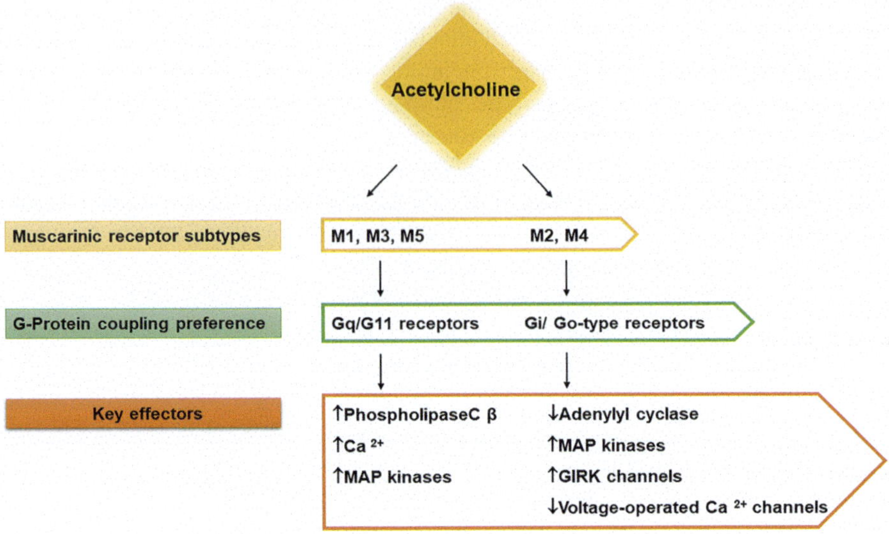


**Fig. 4** The structure and signaling pathways of mAChRs subtypes (adapted from Wess et al. 2007)

**Table 1** Distribution and function of mAChRs subtypes

| Muscarinic Receptor subtypes | Distribution | Function |
|---|---|---|
| M1 | Striatum, cortex, present postsynaptically on cell bodies and dendrites of hippocampal pyramidal neurons and granule cells | Anxiety-like behaviors and hyperlocomotion |
| M2 | Presynaptically in the cortex, basal forebrain, hippocampus, and striatum | Decreased heart rate due to hyperpolarization of the plasma membrane of excitable cells |
| M3 | Present throughout in the central nervous system | Smooth-muscle contractility and glandular secretion |
| M4 | Predominant in the striatum, low levels in cortex and hippocampus | Reducing the intracellular concentration of cAMP and prolonging opening of potassium channels |
| M5 | Postsynaptically on the dopaminergic neurons | Activators of dopamine signaling |

activation of nitric acid synthase, modulation of memory, learning processes, attention, and heart rate (Kohnen-Johannsen and Kayser 2019).

The alkaloids of *Datura* plant act as competitive and nonselective muscarinic acetylcholine receptor (mAChR) antagonists that affect the central and peripheral nervous system. They are potent anticholinergic compounds that prevent the binding of the physiological neurotransmitter, acetylcholine (Fig. 5).

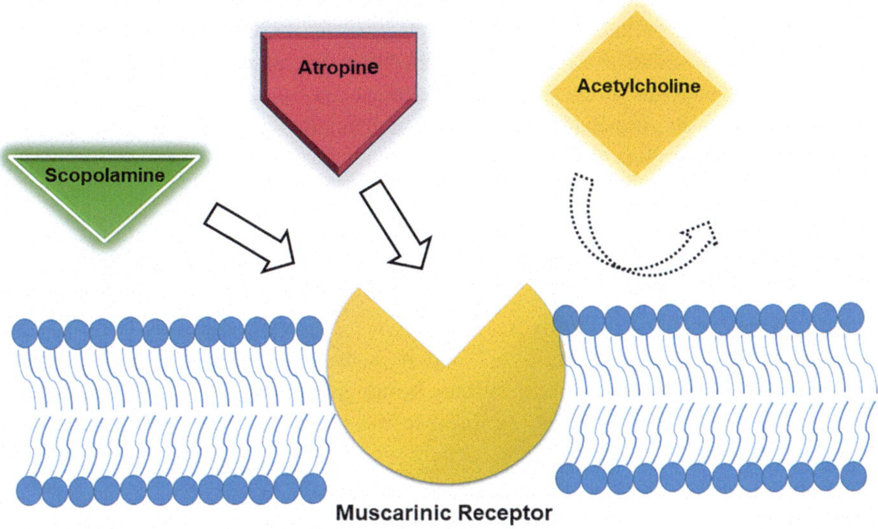


**Fig. 5** Datura alkaloids as competitive inhibitors of muscarinic receptors

These alkaloids are absorbed readily through mucous membranes when plant parts are ingested or smoked, the common route of administration. When absorbed, alkaloids cross the blood–brain barrier and cause disturbance of the central nervous system. These belladonna alkaloids block nerve impulses linked with the parasympathetic nervous system that regulate various involuntary movements including pupil dilation, heart rate, gland secretion, constriction of the bronchioles in the lungs, and movement in the alimentary canal thereby causing hallucinogens, sedation, antiemetic and amnestic effects (Kwakye et al. 2018; Pretorius and Marx 2006).

Atropine and scopolamine consumption causes blockade of parasympathetic postganglionic fibers innervating the exocrine glands thereby affecting salivary and bronchial secretions resulting in dry mouth, and other main primary toxic manifestations include mydriasis, direct ocular exposure, and tachycardia which are discussed as below:

### 6.1.1 Mydriasis

The blockade of the pupillary sphincter muscle and iris muscle leads to mydriasis, a common clinical finding of the peripheral anticholinergic syndrome. The problem in the pupillary sphincter and ciliary muscle is because of the entry of sap or dried plant material directly into the eyes, hence causing mydriasis and cycloplegia (Ijland et al. 2001). Alkaloidal content and quantity of the sap determine the persistence of mydriasis. As these alkaloids also affect the central nervous system, it can make the person agitated, confused, and disoriented (Krenzelok 2010). *Datura* foliage also contaminates the grains that affect livestock and poultry. When the corn plants are harvested and chopped, the dried *Datura* plants also get chopped and pulverized, creating dust having belladonna alkaloids that result in mydriasis and other systemic symptoms. In this case, the mydriasis is likely to be bilateral (Raman and Jacob 2005). Paralysis of the ciliary muscle, decreased vision, decreased lacrimation, visual hallucinations, increased allergies, impairment of saccadic eye movements, eye irritation, and conjunctivitis are the main symptoms involved.

### 6.1.2 Direct Ocular Exposure

In addition to mydriasis, *Datura* is also known for causing diseases such as "gardeners eye" and "corn pickers eye," which is due to the high concentration of belladonna alkaloids in the plant. When home gardeners nurture the plant in a domestic flower garden, they get exposed to the plant, and hence the disease was named as "gardener's eye." Exposure occurs when the gardeners rub their eyes with their hands. Another reason for exposure is the usage of hands or gloved hands to remove sweat from the face as a result of which sweat runs into the eyes and contamination takes place. Most people are unaware of the toxic properties of *Datura* as a result of which they do not take the necessary precautions, and exposure occurs (Krenzelok 2010).

### 6.1.3 Tachycardia

Atropine, hyoscyamine, and scopolamine are known antagonists of muscarinic acetylcholine receptors that block the receptors in the SA node, thereby causing an increase in heart rate. Tachycardia is inextricably linked to other symptoms like fever, sweating, drowsiness, fatigue, and dementia (Rajput 2014). The onset of clinical anticholinergic symptoms usually appears within 30–60 min after the ingestion of plant parts. Depending on the quantity and the exposure time, these neural symptoms may last for 24–48 h (Furbee 2009).

## 6.2 *Reported Cases of* Datura *Poisoning*

A lot of accidents have been stated to date because of the ingestion of *Datura*. *Datura* has been known as a popular poison for suicide and murder in some parts of Europe and India. The toxicological surveys give a better assessment and information about the predominant toxins found in all *Datura* species. The main central nervous system toxicity features include restlessness, hallucinations, delirium, and altered sensorium hallucination, whereas hyperthermia, sweating, pupil dilation, dry mouth and skin, urinary retention, and reduced gastric movement are common manifestations of peripheral nervous system toxicity.

The first recorded case of *Datura* poisoning dates back to 1676 when a soldier of Bacon rebellion used *Datura* leaves in salad preparation and began to hallucinate. In the 1990s and 2000s, the number of adolescents and young adults was reported to fall ill from ingesting *Datura* intentionally (Yilmaz et al. 2014). Many case studies shedding light on the accidental food contamination and ingestion of the *Datura* plant have been well documented in the literature. In 2001, a 35-year older man was found to be suffered from mydriasis after being in contact with devil's trumpet (*Datura*) while gardening (Ijland et al. 2001). This herb gained the attention of the researchers when in the same year, another case of consumption of Jimson weed by a 16-year-old male has come forward. The adolescent had taken the drug for sedative purposes. The doctors decided to consult the Poison Control Centre when a large number of cases appeared consecutively (Wenker and Suchard 2001). Another case of intoxication of an adolescent came in 2002 when his mother brought him to the hospital because of his hallucinating behavior. After analysis, doctors found the *Datura* plant to be the root cause behind the intoxication and the bizarre behavior of the adolescent (Chan 2002).

Further, a report of seven patients suffering from unilateral mydriasis has been presented by Havelius and Asman (2002). The findings suggested the accidental exposure of *Datura suaveolens* into the eyes of patients. Out of the seven patients, three of them were also suffering from cycloplegia, and one of them had also developed tachycardia. Treatment was given according to the age and the level of infection. In 2005, a healthy 76-year-old individual in the UK was also witnessed with mydriasis and identical symptoms. The patient was introduced with a unilateral fixed dilated pupil leading to blurred vision. The history of the patient also revealed the linkage of a patient with *Datura innoxia* unintentionally. However, after a few weeks, the condition of the patient has turned out to be normal by following the prescribed treatment (Raman and Jacob 2005).

Further, the case of the insobriety of a 8-year-old girl by the ingestion of the *D. stramonium* plant was reported in 2005 when she was admitted to the hospital in an unconscious condition. After a lot of examinations, the reason for the patient's condition was detected, and treatment was given accordingly (Ertekin et al. 2005). Also, a case was reported by Soneral and Connor (2005), where the stupefaction of five adolescents happened by the consumption of *D. stramonium.* Major symptoms like hallucinations, mydriasis, and disorientation were also common among the five patients. Such frequent reports of poisoning cases by *Datura* intake are a strong indication of the easy availability of this herb in the markets (Soneral and Connor 2005). Accidental toxication of horses with *D. stramonium* leaves has also been reported. The reason behind the toxication was the consumption of maize contaminated with leaves of *D. stramonium.* Due to the intoxication, the horses were witnessed with breathing issues, hyperthermia, gastric problems, loss in thirst and urination, tachycardia, and mydriasis. Horses were monitored at different stages, and accordingly, treatment followed (Binev et al. 2006).

Furthermore, Diker et al. 2007 presented cases of toxication of *D. stramonium* in two patients, one was 21-year old, and the other of 19-year old age. Both the patients showed the symptoms of being in Coma but later responded to the medications and treatment. Yet many other instances also illustrate the intentional intake of the herb. This observation was highlighted when a case of ingestion of hundred toxic seeds of *D. stramonium* by two teenagers was reported in 2007 in British Columbia. Numerous symptoms like illusions, perplexity, inexplicable and absurd speech, and dilated pupils were displayed in both the patients. An estimate of 188 cases was stated from 2001 to 2007 in Texas, whereas a report of 23 cases was submitted to the British Columbia Drug and Poison Centre from 2003 to 2006 (Spina and Taddei 2007). Also, Wiebe and his collaborators (Wiebe et al. 2008) have revealed an incident about the deliberate ingestion of the toxic *Datura* plant and hospitalization of some adolescents, which happened in Winnipeg, Manitoba in 2006. All the patients admitted were suffering from agitation and other symptoms of poisoning. In addition to this, the case of two adolescents admitted to the hospital at Geneve, Suisse, because of the ingestion of *Datura* intentionally was reviewed by Burquier et al. (2008). One of the patients had some visual problems, and the other one was reported to be suffering from amnesia. Both the patients were treated accordingly and got better after the treatment. Moreover, Krenzelok (2010) has revealed an analysis report of 1458 patients suffering from the intoxication by *Datura sp,* and the percentage of adolescents and adults was 46.2% and 41.2%, respectively. Even though deaths from the poisoning were not observed in major proportion, but it has also been found during the investigation that 72% of the total ingestion was on purpose to use the drug for insobriety. However, another report of the intoxication of *D. stramonium* seeds by three patients came in 2010.

Due to the vast availability and less legislative control on the selling of the plant, it has become a favorable psychoactive drug for adolescents since times. Proper understanding of the symptoms and appropriate treatment procedures are important for overcoming the consequences (Stella et al. 2010). Karadas et al. (2011) have reported a case of accidental ingestion of the *Datura* plant by a 65-year-old individual, and the condition of the patient improved after the treatment. In addition to this, acute, subacute, and chronic toxicity of alkaloids present in the seeds of *D. stramonium* were analyzed by Bouzidi et al. (2011). They revealed the number of doses sufficient to cause the effect in the weight of the liver, spleen, brain, and to increase the levels of RBCs, hemoglobin, and WBCs. They further described the range of doses of toxins and seeds of *Datura* sp. fatal to adults and children. Also, Bouziri et al. (2011) presented a case of accidental intake of *Datura* plant by a 3.5-year-old girl. The patient was given the treatment of diazepam, and she took a few days to come into normal condition.

Further, Amini et al. (2012) described a case series of 19 patients admitted to Imam Reza Hospital, Mashhad, Iran, due to the poisoning of *D. stramonium* from 2008 to 2011. They found children to be majorly affected by the toxication. Parenteral benzodiazepines were used to treat most of the patients without the usage of physostigmine. Also, Shea et al. (2012) presented the first case report of delirium in a 79-year-old patient who was presented with anticholinergic toxicity symptoms like hallucinations of ghosts, disorientation, slurred speech, and generalized weakness. These clinical features were compatible with alkaloid tropane poisoning. Later, the correct investigation and history of the patient revealed that the consumption of herbal tea (*Strobilanthes forrestii* Diels.) contaminated with tropane alkaloids (especially *Datura* sp) was responsible for this neurotoxicity. In addition to this, a report was presented by Devi et al. (2012) that suggested the depressant and sedative properties of *D. stramonium*. They performed numerous tests like checking the locomotory activity, free-swimming test to access the potential of aqueous extract of *D. stramonium*. They concluded that the extract was capable of producing oxidative stress, thereby resulting in damage to brain cells. Several studies based on animal models were also carried out in previous times to investigate the parameters of toxicity (Mahdeb et al. 2012).

Another study was carried out by Glatstein et al. (2012) to review the application of physostigmine in treating the patients suffering from the toxicity of the *Datura* plant. They reported a case of a 15-year-old girl admitted to the Division of Pediatric Emergency Medicine, University of Toronto, Canada, after the deliberate ingestion of the seeds of the *Datura* plant. They used physostigmine, and as a result of which, the condition of the patient got better after a few days. Hence, they suggested its usage for the treatment. Moreover, Perharic et al. (2013) presented a case of mass poisoning of 73 people living in Slovenia. Patients were found to consume the food contaminated with *D. stramonium*.

Researchers investigated the quantitative risk assessment of atropine and scopolamine present in the contaminated food and suggested the constant surveillance of food. Yilmaz et al. (2014) have further outlined a case study of a 19-year-old patient with speech issues and an unconscious situation. They further threw light on the treatment procedure. Physostigmine was used as an antidote during the therapy. In another case, Khanra et al. (2015) highlighted the psychotic effects of the hallucinogen, *Datura,* on the brain. They presented a case of a 32-year-old individual with aggressive conduct who has been continuously consuming *Datura* seeds for 3 years deliberately. As a result of the abusive effect of the plant, he was diagnosed to be suffering from paranoid schizophrenia and mental and behavioral disorder. He was treated with the antipsychotic tablet to cure his condition, and improvement was observed after a few days of treatment. Moreover, in 2015, a report of intoxication of a complete family due to *D. stramonium* turned up in a hospital in Turkey. It happened as a result of an accident when during the making of a Turkish dish-Dolma, someone mixed the pumpkin flowers with the flowers of *D. stramonium* coincidently due to their similar appearance. Symptoms were common, for example, red skin, tachycardia, hallucinations, disorientation, agitation, and hyperthermia. After proper care and prescribed treatment, all the members of the family got recovered from the poisoning (Disel et al. 2015).

In 2016, Gunaydin and his coworkers presented the case of consumption of *D. stramonium* plant by two individuals of 64- and 67-year-old age in Turkey. Both the patients returned to normal state after a few days (Gunaydın et al. 2017). Arefi et al. (2016) also presented a report depicting the intoxication of five patients admitted to the Clinical Toxicology Emergency of Baharloo Hospital, Tehran, Iran, in 2015 for a few days. Four out of the five patients were diagnosed with the symptoms of rhabdomyolysis that increased the concern of the physicians, but all the patients got cured with the help of treatment given at the hospital. Besides, a case of homicide by *Datura* poisoning during the robbery was studied by Gaff and his team (Garff et al. 2016). By the second autopsy report, it was found that the death of the patient happened by disordered cardiac rhythm because of the toxication by *Datura* sp. This study threw light on the usage of the plant in criminal activities.

Another study has been carried out to study the effect of extract of *Datura* sp., and it was found that the methanolic crude extract of *Datura metel* exhibited surgical anesthesia in dogs with increased heart and respiratory rates. As compared to the leaf extract, the seed extract was observed to be safer. According to another study, it has been shown that the ethanolic seed extract of the *Datura metel* was harmful to the health of Wistar rats in a dose-dependent manner. Chromatolysis, reduced protein synthesis due to loss of Nissl substances and nuclei, neuronal, as well as axonal injuries, were observed in the histological studies done on treated Wistar rats (Al-Snafi 2017). In addition to this, Uddin et al. (2017) have also studied the toxic effects of *D. stramonium* on rats and reached to a conclusion that the potency of the plant to induce intoxication in rats is dependent on the dose given to the treated ones.

Further, the toxicity of *D. stramonium* has been analyzed by Fu et al. (2017) by using the 1H NMR-based metabolomics approach and also found out the dose-dependent efficiency of the plant to cause a change in the behavior of tested mice. Another case study reporting voluntary ingestion of *Datura* sp. by a 22-year-old male patient was studied by Tranca et al. (2017). At the time of admission to critical care, the patient was presented with typical symptoms of *Datura* poisoning viz. mydriasis, reactive pupils, agitation, dry skin and mucosa, hyperpyrexia, and tachycardia. Similarly, Beynon and Chaturvedi (2018) also reported a case study of *datura* intoxication by an adolescent. Furthermore, misidentification in the prediction of the plant and food contamination is a common problem that is positively associated with the severity of alkaloid poisoning.

Doan et al. (2019) also performed retrospective case studies on 114 *Datura* related poisoning cases reported in the Taiwan poison control center between 1986 and 2015. Besides, Vaza and Bhalerao (2018) have reported many cases of poisoning of humans and animals by direct and indirect ingestion of plant parts/extracts. Another case came from Bursa Yuksek Ihtisas Training and Research Hospital, Department of Pediatrics, Yildirim, Bursa, Turkey, where a child of 6-year-old age was admitted due to the ingestion of the *D. stramonium* plant involuntarily. After diagnosis, he was treated firstly by performing gastric decontamination, and then physostigmine was administered to the patient (Korkmaz et al. 2019).

In addition to this, Joshi (2019) also reported a case study on *Datura* poisoning in a 70-year-old man who presented the symptoms of diminished vision, pain, redness, watering, and photophobia after instillation of *Datura* juice in his eyes. For instance, Ogunmoyole et al. (2019) studied the toxicity of *Datura* seed extract on various organs (liver, kidney, heart, and brain) of Wistar rats and concluded its significant toxicity. Further, the toxicity levels vary with various organs and depend on the solvents used for the extraction of metabolites from seed.

The studies discussing the use of *Datura sp* to cause criminal activities like murder and robbery are very prevalent and diversely reported (Pillay and Sasidharan 2019). Also, Kerchner and Farkas (2020) carried out a study in which they evaluated nearly 2500 cases of human exposure to *Datura*, conveyed to the Hungarian Toxicological Information Service from 2005 to 2017, and then they tried to link the data with the other centers present in America, Asia, and Europe.

It was found that the cases of toxication due to the application of the *Datura* plant were different in America, Asia, and Europe, being highest in Europe and lowest in America. Even it was considered among the top four toxic plants in Europe. Another study was conducted by Oboh et al. (2020) to examine the effect of alkaloid extracts made from different parts of the *Datura* plant on cognitive function in rats. Reduced memory index and increased level of anxiety-like behavior were exhibited by rats that were continuously administered with the extract. They also reported that the constituent alkaloids are responsible for causing alterations in the major biochemical parameters underlying the impairment of cognitive functions. Also, many cases depicting the suicide attempts by ingesting *Datura* sp. have been reported to date (Batool et al. 2020). Thus, the literature indicates the neurotoxicity potential of the constituent tropane alkaloids of Thorn apple: *Datura*. Hence, it was suggested to use plant products carefully so that their toxicity could be avoided.

### 6.3 *Diagnosis of* Datura *Poisoning*

To diagnose the *Datura* poisoning, it is important to learn the history of the patient, and then the head CT scan can be performed for the same. In rural areas where neuroimaging equipment is not available, the pilocarpine diagnostic test may be used for diagnosis. The use of ophthalmic drops causes pupil constriction in patients who are suffering from paralytic mydriasis because of pre- or postganglionic nerve damage. In the case of unilateral mydriasis, the unaffected eye responds immediately to pilocarpine that further supports the diagnosis procedure of mydriasis caused by ocular exposure to *Datura* (Devi et al. 2012). Gas chromatography–MS can also be used to trace the level of atropine (Pillay and Sasidharan 2019).

### 6.4 *Management of* Datura *Poisoning*

Delay in diagnosis is the most common challenge associated with *Datura* intoxication. The onset of clinical symptoms like ataxia, altered mental state, disorientation, confusion, hallucinations, psychosis, agitation, etc. should be considered as an early presentation of this poisoning. Therefore, history and consistent symptomology of the patient will support the suspicion and initial management of this toxin. Gastric lavage with activated charcoal by mouth or tube, intravenous benzodiazepine injections for control of agitation, and hyperpyrexia control are the practices followed in the early clinical presentation of poisoning in patients. Administration of benzodiazepines helps to relieve hyperpyrexia, avoidance of self-injury, and myoglobinuric renal failure, and the treatment is usually supportive.

In addition to this, for optimal patient care, there always exist specific pharmacological antagonists. In the case of minor anticholinergic activity, pharmacological treatment is not required. In the case of major anticholinergic toxicity, physostigmine inhibitors are used. They act as an antidote. Physostigmine binds acetylcholinesterase reversibly, hence, increases the amount of free acetylcholine, and function of the autonomic nervous system is restored when appropriately titrated. Consequently, physostigmine also has the proficiency to reverse the peripheral toxicity associated with *Datura*. Being a tertiary amine, it is capable of crossing the blood–brain barrier and can reverse the CNS toxicity (Bouziri et al. 2011; Disel et al. 2015). But the judicious use of physostigmine is important as the excessive amount can cause cholinergic toxicity, which is marked by adverse effects like seizures and pulmonary edema. The concentration of physostigmine used should be in a dose of 0.5–2.0 mg to adults, and it must not exceed 1 mg/min. If the patient does not respond to the dose, the second dose should be given after 20 min. Physostigmine is used in children only to treat life-threatening toxicity. When used intravenous or intramuscular, the dose should be 0.02 mg/kg, and the rate should not exceed 0.5 mg/min when given intravenously. The dose may be repeated after 5- to 10-min intervals until a dose of 2 mg has been given. In the case of injudicious use of physostigmine, atropine can be used to reverse physostigmine-induced cholinergic toxicity (Alberto et al. 2001).

Gastrointestinal decontamination was once considered as the keystone of treating patients who ingested *Datura*. Even there is no evidence of using the conventional methods of decontamination. Activated charcoal may have that capability, but it remains unvalidated. In case there is no gastrointestinal motility, activated charcoal should not be used because of the risk of the development of ileus. Once a possible diagnosis of *Datura* toxicity is completed, the pharmacological approach is to reestablish the balance in the autonomic nervous system (Salen et al. 2003). Further, public health and prevention programs initiated by poison control centers can serve as a panacea in the management of toxicity associated with the alkaloids of these plants.

## 7 Pharmacological Properties of *Datura* Species

*Datura* is an extensively growing plant and well-acknowledged to have a tremendous pharmacological perspective with high efficacy and convention in folklore medicinal herbs. The plethora of bioactivities shown by *Datura* sp. is tabulated (Tables 2, 3, 4).

**Table 2** Pharmacological properties of *Datura* sp.

| Activity | Species | Test organism | Dose | Outcomes | References |
|---|---|---|---|---|---|
| Antidiabetic activity | *Datura fastuosa* | Rat | Seed residue at 25, 50, and 75 mg/kg conc. | Glucose lowering effect was shown | Murthy et al. (2004) |
| Herbicidal activity | *Datura metel* | *Parthenium* | Aqueous and methanolic extract at different concentrations | Antagonistic effects on the growth of *Parthenium* were shown | Singh et al. (2005); Belz et al. (2007); Javaid et al. (2010) |
| Analgesic effect | *Datura fastuosa* | Mice | Aqueous extracts of seeds and leaves were tested with the administration of morphine and naloxone | Both the extracts have shown significant activity along with the administration of naloxone, but no activity was shown by extracts alone | Abena et al. (2003); Khalili and Atyabi (2004); Abbas (2013) |
| Anti-inflammatory activity | *Datura stramonium* | Rat | Plant extract of different concentrations (50, 100, and 200 mg/kg) were used | Prominent activity has been shown | Sonika et al. (2010) |
| Anti-asthmatic activity | *Datura stramonium* | Human | Elevated doses of scopolamine and atropine were implicated | Sensitization of nicotine receptors was observed, but the harmful effect on the fetus has been shown | Gaire and Subedi (2013); Al-Snafi (2017) |
| Antifertility activity | *Datura metel* | Male rat and female albino mouse | Alcoholic and acetone extracts of seeds of a plant of different concentrations were tested | Anti-implantation activity in females and a decrease in the concentration of sperms were observed | Al-Snafi (2017) |

# 8 Conclusions

*Datura*, a well-known phytomedicine, has been extensively used in ethnomedical treatment for curing a broad range of ailments. A careful analysis of major bioactive chemical constituents, particularly tropane alkaloids, reveals a spectrum of health benefits. Insight of documented scientific literature on this medicinal plant shows

**Table 3** Antimicrobial potential of different *Datura* sp.

| Activity | Species | Test organisms | Extract/concentration | Inhibited test organism | References |
|---|---|---|---|---|---|
| Antimicrobial activity | *Datura fastuosa* | *Escherichia coli, Staphylococcus aureus, Bacillus subtilis, Candida albicans, Aspergillus Niger* | Methanolic and hydrochloric acid extract (conc. 50–0.78 μg/mL) | *Bacillus subtilis* and *Staphylococcus aureus* | Dhiman et al. (2012) |
| | *Datura stramonium* | *E.scherichia coli, Staphylococcus aureus, Bacillus subtilis, Pseudomonas, Sarcina, Klebsiella* | Aqueous extract | *Sarcina* and *Staphylococcus aureus* | Shobha et al. (2014) |
| | | *Klebsiella pneumonia, Salmonella typhi, Staphylococcus aureus, Proteus vulgaris, Pseudomonas aeruginosa, Escherichia coli, Aspergillus Niger, Aspergillus fumigatus*, and *Fusarium spp* | Alcoholic extract (conc. 4 mg/0.2 mL to 8 mg/mL) | All test organisms were effectively inhibited | Reddy (2009) |
| | | *Escherichia coli, Staphylococcus aureus* | Aqueous and alcoholic extract (at 100 and 200 mg/mL concentrations) | All tested microbial species were inhibited | Gupta et al. (2013) |
| | | *Escherichia coli, Staphylococcus aureus, Pseudomonas aeruginosa, and the fungal strains Aspergillus flavus, Aspergillus Niger, Fusarium culmorum, Rhizopus stolonifera* | (a) Petroleum ether, water, ethanolic, and methanolic extract | *Pseudomonas aeruginosa* | Sharma et al. (2013) |
| | | | (b) Methanolic extract (fruit, root, stem, leaf, and callus samples) | Prominent antimicrobial activity was shown | Sharma and Sharma (2013) |

| | | | | | |
|---|---|---|---|---|---|
| | *Datura ferox* | *Bacillus subtilis, Escherichia coli, Staphylococcus aureus, Salmonella typhi, Proteus vulgaris, Aspergillus flavus, Candida albicans* | Ethanolic and aqueous extract | Ethanolic extracts had significant antimicrobial potential | Gachande and Khillare (2013); Tandon et al. (2014) |
| | *Datura innoxia* | *Staphylococcus aureus, Escherichia coli, Pseudomonas aeruginosa, Candida albicans* | Ethanolic extract (0.1 g/mL concentration) | Mild to strong inhibition was observed | Mathur et al. (2013) |
| | | *Bacillus subtilis, Escherichia coli, Staphylococcus aureus, Enterococcus faecalis, Pseudomonas aeruginosa* | Methanolic extract (conc. 2.5, 1.25, and 0.725 mg/mL) | Antimicrobial activity was displayed against all except *Escherichia coli* and *Pseudomonas aeruginosa* | Eftekhar et al. (2005) |
| | *Datura metel* | *Alternaria alternata, Aspergillus Niger, Fusarium oxysporum, Xanthomonas compestris, lactobacillus acidophilus, Pseudomonas marginalis, Pseudomonas syringae, Pseudomonas aeruginosa, Streptococcus mutans, Streptococcus salivarius, Staphylococcus aureus* | Chloroform and methanolic extract (100 and 200 mg/mL concentrations) | Prominent antimicrobial activity was displayed | Vadlapudi and Kaladhar (2012) |

(continued)

**Table 3** (continued)

| Activity | Species | Test organisms | Extract/concentration | Inhibited test organism | References |
|---|---|---|---|---|---|
| Anti-bacterial activity | *Datura stramonium* | *Pseudomonas aeruginosa*, *Klebsiella pneumonia*, *Escherichia coli* | Ethanolic extract (at conc. 5, 10, 15, 20, 25, and 30% (w/v) | Considerable antibacterial activity was observed | Banso and Adeyemo (2006) |
| | *Datura ferox* | *Bacillus megaterium*, *Bacillus cereus*, *Escherichia coli*, *Salmonella typhi*, *Staphylococcus aureus* | Aqueous extract | *Staphylococcus aureus* and *Salmonella typhi* | Jamdhade et al. (2011) |
| | *Datura innoxia* | *Escherichia coli*, *Staphylococcus aureus*, *Salmonella typhi*, *Bacillus subtilis* and *Bacillus cereus* | Extract was prepared in organic solvents (ethanol, methanol, and hexane at 100 mg/mL concentration | The leaf extracts showed better efficiency as compared to the stem and root extracts | Kaushik and Goyal (2008) |
| Antifungal activity | *Datura stramonium* | *Fusarium oxysporum*, *Rhizopus stolonifera* | Water, alcohol, and chloroform extract | Considerable antifungal activity was shown | Shinde and Dhale (2011) |
| | | *Rhizopus* sp. | Aqueous and alcoholic extract (conc. 100 and 200 mg/mL respectively) | The extracts exhibited a significant antifungal effect | Gupta et al. (2013) |
| | *Datura metel* | *Aspergillus fumigatus*, *Aspergillus flavus* | Trichloromethane extract | Significant antifungal activity was observed | Sharma (2002) |
| | | *Ascochyta rabiei* | Aqueous and methanolic extract (at varying concentrations) | Shoot extracts showed a prominent reduction in the growth of *Ascochyta rabiei* | Bajwa et al. (2008) |
| Vibriocidal activity | *Datura* sp. | *Vibrio cholera*, *Vibrio parahaemolyticus* | Ethanol, aqueous, and acetone extract | Significant vibriocidal activity against the bacterial species was observed | Sharma et al. (2009) |

**Table 4** Biological activities shown by different *Datura* sp.

| Activity | Species | Model organism | Dose | Outcome | References |
|---|---|---|---|---|---|
| Antiquorum sensing activity | *Datura metel* | *Chromobacterium violaceum, agrobacterium tumefaciens* | Ethanolic extract (at concentration 100 g/L) | The extract showed no anti-quorum sensing activity but it showed an antimicrobial effect | Adonizio et al. (2006) |
| Effect on barbituric sleeping | *Datura fastuosa* | Rat and mice | Aqueous extracts | Threshold barbituric sleeping considered | Abena et al. (2003) |
| Effect on rectal temperature | *Datura fastuosa* | Rat and mice | Temperature was determined after the administration of the extract | Extract was shown to reduce the temperature at 800 mg/kg concentration of extract | Abena et al. (2003) |
| Effect on apomorphine stereotypes | *Datura fastuosa* | Rat and mice | Apomorphine (1 mg/kg) was injected after introducing the seed or leaves extract at 400 and 800 mg/kg concentration | No modification has been shown on apomorphine stereotype by the extract | Abena et al. (2003) |
| DPPH scavenging assay | *Datura fastuosa* | 1,1-diphenyl1-picrylhydrazyl radical scavenging assay; iron-chelating model | Methanolic and hydroalcoholic extracts of seeds were prepared from 2 to 1000 μg/mL concentration | Scavenging of the free radicals has occurred. The $IC_{50}$ values for hydroalcoholic and methanolic extracts were 25.78 and 28.34 μg/mL, respectively | Dhiman et al. (2012) |
| Radical scavenging activity | *Datura stramonium, Datura metel, Datura innoxia* | DPPH radical scavenging activity | Fine extraction of the plant material was done with N-hexane by using soxhlet apparatus. | *Datura innoxia* had the quenching potential of 60%, and *Datura stramonium* had the quenching capacity of 30% | Ramadan et al. (2007) |
| DPPH radical scavenging | *Datura metel* | DPPH radical scavenging activity | N-hexane, ethyl acetate, chloroform, and methanolic extracts of plant materials were prepared at four concentrations ranging from 25 to 100 mg/mL concentrations | The chloroform extract showed maximum inhibition and minimum inhibition was displayed by n-hexane extract | Sangeetha et al. (2014) |
| Egg-laying effect | *Datura stramonium* | Two-spotted spider mite *T. urticae* Koch female | At 2500 and 25,000 mg/L concentration, leaf and seed extracts were sprayed | The extracts showed considerable deterrent effects on egg-laying capacity | Kumral et al. (2010) |

anticholinergic, antiasthmatic, anticancer, antiepileptic, antidiabetic, anti-inflammatory, antidepressant, and analgesic effect. However, the deliberate/accidental ingestion of the *Datura* plant or its parts results in neuropathological manifestations by damaging the functional dynamics of muscarinic acetylcholine receptors. This results in the outgrowth of ideas among the medicinal chemists for the conversion of these neurotoxic alkaloids into model anticholinergic drugs entities based on molecular probing for evaluating novel mechanisms of biological action. During the past two decades, awareness of pharmacodynamics and toxicokinetic properties of *Datura* alkaloids has generated a wealth of data in the search for more selective muscarinic acetylcholine receptors agonists and antagonists. Intensive research on mAChR-mutant animal models has not only generated new information on physiological and potential pathophysiological roles of individual mAChR subtypes. Besides, recapitulating the mACh signaling pathways for better understanding the role of these alkaloids in neurological and psychiatric diseases. Thus, much more comprehensive studies based on preclinical and clinical trials are required to unravel the potential anticholinergic metabolites of this medicinal herb. Also, in-depth investigations on its toxicity levels, bioavailability, and mechanical aspects are still needed to be explored for the development of key bioactive constituents for rational designing of drug molecules. This will further bring a major paradigmatic shift resulting in the development of novel muscarinic drugs with a high degree of selectivity, efficiency, and reduced side effects.

## References

Abbas DA (2013) Analgesic, anti-inflammatory and antidiarrhoeal effects of *Datura stramonium* hydroalcoholic leaves extract in mice. Int J Recent Res Appl Stud 14(1):193–199

Abena AA, Miguel LM, Mouanga A, Assah TH, Diatewa M (2003) Evaluation of analgesic effect of *Datura fastuosa* leaves and seed extracts. Fitoterapia 74(5):486–488. https://doi.org/10.1016/S0367-326X(03)00124-2

Adekomi AD, Tijani AA, Ghazal KO (2020) Some effects of the aqueous leaf extract of *Datura metel* on the frontal cortex of adult wistar rats (*Rattus norvegicus*). Eur J Anat 14(2):83–89

Adibah KZ, Azzreena MA (2019) Plant toxins: alkaloids and their toxicities. GSC Biol Pharm Sci 6 (2):21–29. https://doi.org/10.30574/gscbps.2019.6.2.0003

Adonizio AL, Downum K, Bennett BC, Mathee K (2006) Antiquorum sensing activity of medicinal plants in southern Florida. J Ethnopharmacol 105(3):427–435. https://doi.org/10.1016/j.jep.2005.11.025

Afewerki S, Wang JX, Liao WW, Cordova A (2019) The chemical synthesis and application of tropane alkaloids. Alkaloids Chem Biol 81:151–233. https://doi.org/10.1016/bs.alkal.2018.06.001

Alberto K, Claudia S, Ludmilla K, Arnon B (2001) Toxic delirium due to *Datura stramonium*. Isr Med Assoc J 3:538–539

Al-Snafi AE (2017) Medical importance of *Datura fastuosa* (syn: *Datura metel*) and *Datura stramonium*—a review. IOSR J Pharm 7(2):43–58

Amini M, Khosrojerdi H, Afshari R (2012) Acute *Datura stramonium* poisoning in east of Iran-a case series. Avicenna J Phytomed 1(4):274–276

Anju D, Ratan L (2011) Phytochemical and pharmacological status of *Datura fastuosa* Linn. Int J Res Ayurveda Pharm 2(1):145–150

Arefi M, Barzegari N, Asgari M, Soltani S, Farhidnia N, Fallah F (2016) *Datura* poisoning, clinical and laboratory findings—report of five cases. Rom J Leg Med 24(4):308–311. https://doi.org/10.4323/rjlm.2016.308

Atanasov AG, Waltenberger B, Pferschy-Wenzig EM, Linder T, Wawrosch C, Uhrin P et al (2015) Discovery and resupply of pharmacologically active plant-derived natural products: a review. Biotechnol Adv 33(8):1582–1614. https://doi.org/10.1016/j.biotechadv.2015.08.001

Bajwa R, Shafique S, Shafique S (2008) Fungitoxicity of aqueous and solvent extracts of *Datura metel* against *Ascochyta rabiei*. Mycopathologia 6(1–2):17–22

Banso A, Adeyemo S (2006) Phytochemical screening and anti-microbial assessment of *Abutilon mauritianum, Bacopa monnifera* and *Datura stramonium*. Biokemistri 18(1):39–44. https://doi.org/10.4314/biokem.v18i1.56390

Batool A, Batool Z, Qureshi R, Raja NI (2020) Phytochemicals, pharmacological properties and biotechnological aspects of highly medicinal plant: *Datura stramonium*. J. Plant Sci 8(2):29–40. https://doi.org/10.11648/j.jps.20200802.12

Belz RG, Reinhardt CF, Foxcroft LC, Hurle K (2007) Residue allelopathy in *Parthenium hysterophorus* L.—does parthenin play a leading role? J Crop Prot 26(3):237–245. https://doi.org/10.1016/j.cropro.2005.06.009

Benitez G, March-Salas M, Villa-Kamel A et al (2018) The genus *Datura* L.(Solanaceae) in Mexico and Spain–Ethnobotanical perspective at the interface of medical and illicit uses. J Ethnopharmacol 219:133–151. https://doi.org/10.1016/j.jep.2018.03.007

Beynon SJ, Chaturvedi S (2018) *Datura* intoxication in an adolescent male: a challenge in the internet era. J Pediatr Child Health 54(1):84–87. https://doi.org/10.1111/jpc.13726

Binev R, Valchev I, Nikolov J (2006) Clinical and pathological studies of jimson weed *(Datura stramonium)* poisoning in horses. Trakia J Sci 4(3):56–63

Bock JH, Norris DO (2016) Introduction to forensic plant science. In: Bock JH, Norris DO (eds) Forensic plant science. Academic Press, London, pp 1–22. https://doi.org/10.1016/b978-0-12-801475-2.00001-4

Bouzidi A, Mahdeb N, Kara N (2011) Toxicity studies of alkaloids of seeds of *Datura stramonium* and synthesis alkaloids in male rats. J Med Plants Res 5(15):3421–3431

Bouziri A, Hamdi A, Borgi A et al (2011) *Datura stramonium* L. poisoning in a geophagous child: a case report. Int J Emerg Med 4:31. https://doi.org/10.1186/1865-1380-4-31

Burquier VDG, Narring F, Broers B (2008) Intoxication volontaire par *Datura* et circumstances de consommation: à propos de deux cas chez l'adolescent. Presse Med 37(6):982–985. https://doi.org/10.1016/j.lpm.2008.01.019

Carod-Artal FJ (2015) Hallucinogenic drugs in pre-Columbian Mesoamerican cultures. Neurologia (English Edition) 30(1):42–49. https://doi.org/10.1016/j.nrleng.2011.07.010

Chan K (2002) Jimson weed poisoning—a case report. Perm J 6(4):28–30

Changeux JP (2020) Discovery of the first neurotransmitter receptor: the acetylcholine nicotinic receptor. Biomol Ther 10(4):547–562. https://doi.org/10.3390/biom10040547

Chen NW, Yoon YK (2020) Scopolamine, a toxin-induced experimental model used for research in Alzheimer's disease. CNS Neurol Disord Drug Targets 19:85–93. https://doi.org/10.2174/1871527319666200214104331

Cox MA, Bassi C, Saunders ME, Nechanitzky R et al (2020) Beyond neurotransmission: acetylcholine in immunity and inflammation. J Intern Med 287(2):120–133. https://doi.org/10.1111/joim.13006

Dalefied R (2017) Poisonous plants. In: Dalefied R (ed) Veterinary toxicology for Australia and New Zealand. Elsevier, Amsterdam, pp 485–572. https://doi.org/10.1016/b978-0-12-420227-6.00023-2

Das S, Kumar P, Basu SP (2012) Review article on phytoconstituents and therapeutic potential of *Datura stramonium* Linn. J Drug Del Ther 2(3):4–7

Devi MR, Bawari ME, Paul SB, Sharma GD (2012) Characterization of the toxic effects induced by *Datura stramonium* L. leaves on mice: a behavioral, biochemical and ultrastructural approach. Asian J Pharm Clin Res 5(3):143–146

Dey P, Kundu A, Kumar A, Gupta M et al (2020) Analysis of alkaloids (indole alkaloids, isoquinoline alkaloids, tropane alkaloids). In: Nabavi et al (eds) Recent advances in natural products analysis. Elsevier, Amsterdam

Dhiman A, La R, Bhan M, Dhiman B, Hooda A (2012) Plebeian assessment of anti-microbial and in vitro anti-oxidant zest of *Datura fastuosa* L. seeds. J Pharm Sci Innov 1(4):49–53

Diker D, Markovitz D, Rothman M, Sendovski U (2007) Coma as a presenting sign of *Datura stramonium* seed tea poisoning. Eur J Int 18(4):336–338. https://doi.org/10.1016/j.ejim.2006.09.035

Disel NR, Yilmaz M, Kekec Z, Karanlik M (2015) Poisoned after dinner: dolma with *Datura stramonium*. Turkish J Emerg 15(1):51–55. https://doi.org/10.5505/1304.7361.2015.70894

Doan UV, Wu ML, Phua DH, Mendez Rojas B, Yang CC (2019) *Datura* and *Brugmansia* plants related antimuscarinic toxicity: an analysis of poisoning cases reported to the Taiwan poison control center. Clin Toxicol 57(4):246–253. https://doi.org/10.1080/15563650.2018.1513527

Eftekhar F, Yousefzadi M, Tafakori V (2005) Antimicrobial activity of *Datura innoxia* and *Datura stramonium*. Fitoterapia 76(1):118–120. https://doi.org/10.1016/j.fitote.2004.10.004

Ertekin V, Selimoglu MA, Altinkaynak S (2005) A combination of unusual presentations of *Datura stramonium* intoxication in a child: Rhabdomyolysis and fulminant hepatitis. J Emerg 28(2):227–228. https://doi.org/10.1016/j.jemermed.2004.11.006

Fedotova MV, Kruchinin SE, Chuev GN (2020) Hydration features of the neurotransmitter acetylcholine. J Mol Liq 304:112757. https://doi.org/10.1016/j.molliq.2020.112757

Forouzanfar F, Hosseinzadeh H (2018) Medicinal herbs in the treatment of neuropathic pain: a review. Iran J Basic Med Sci 21(4):347–358. https://doi.org/10.22038/IJBMS.2018.24026.6021

Fu Y, Si Z, Li P, Li M, Zhao H et al (2017) Acute psychoactive and toxic effects of *D. metel* on mice explained by 1 H NMR based metabolomics approach. Metab Brain Dis 32(4):1295–1309. https://doi.org/10.1007/s11011-017-0038-9

Furbee B (2009) Neurotoxic plants. In: Dobbs MR (ed) Clinical neurotoxicology: syndromes, substances, environments. Elsevier, Philadelphia, PA, pp 523–542. https://doi.org/10.1016/B978-032305260-3.50053-8

Gachande BD, Khillare EM (2013) In-vitro evaluation of *Datura* species for potential anti-microbial activity. Biosci Discov 4(1):78–81

Gaire BP, Subedi L (2013) A review on the pharmacological and toxicological aspects of *Datura stramonium* L. J integr Med 11(2):73–79. https://doi.org/10.3736/jintegrmed2013016

Gallego MJ (2012) *Datura* L. In: Castroviejo S, Aedo C, Liinz M, Munoz Garmendia F, Nieto Feliner G, Paiva J, Benedí C (eds) Flora Iberica 11. CSIC, Madrid, Spain, pp 216–224

Garff LE, Delannoy Y, Mesli V, Hedouin V, Tournel G (2016) Forensic features of a fatal *Datura* poisoning case during a robbery. Forensic Sci Int 261:17–21. https://doi.org/10.1016/j.forsciint.2016.02.028

Glatstein MM, Alabdulrazzaq F, Garcia-Bournissen F, Scolnik D (2012) Use of physostigmine for hallucinogenic plant poisoning in a teenager: case report and review of the literature. Am J Ther 19(5):384–388. https://doi.org/10.1097/MJT.0b013e3181f0cbb4

Grisoni F, Merk D, Consonni V, Hiss JA et al (2018) Scaffold hopping from natural products to synthetic mimetics by holistic molecular similarity. Comms Chem 1(1):44. https://doi.org/10.1038/s42004-018-0043-x

Grynkiewicz G, Gadzikowska M (2008) Tropane alkaloids as medicinally useful natural products and their synthetic derivatives as new drugs. Pharmacol Rep 60(4):439–463

Gunaydın YK, Yıldız CG, Işıkoğlu V et al (2017) *Datura stramonium* poisoning: two case reports. J Emerg Med Case Rep 8:4–6. https://doi.org/10.5152/jemcr.2016.1471

Gupta AK, Ahirwar NK, Shinde N et al (2013) Phytochemical screening and anti-microbial assessment of leaves of *Adhatoda vasica*, *Azadirachta indica* and *Datura stramonium*. UK J Pharm Biosci 1(1):42–47. https://doi.org/10.20510/ukjpb/1/i1/91118

Gupta A, Kumar S, Mahindroo N, Saini RV (2016) Bioactive fraction from *Datura stramonium* Linn. Promotes human immune cells mediated cytotoxicity towards lung and breast cancer cells. Pharm J 8(5):435–439. https://doi.org/10.5530/pj.2016.5.4

Havelius U, Asman P (2002) Accidental mydriasis from exposure to Angel's trumpet (*Datura suaveolens*). Acta Ophthalmol Scand 80(3):332–335. https://doi.org/10.1034/j.1600-0420.2002.800319.x

Ijland M, Poels EF, Verhagen WI (2001) Bilateral mydriasis due to Angel's trumpet. J Neuro Ophthalmol 26(2):133–135. https://doi.org/10.1076/noph.26.2.133.10303

Jalander V, Gachande BD (2012) Effect of aqueous leaf extracts of *Datura* sp. against two plant pathogenic fungi. Int J Food Agric Vet Sci 2(3):131–134. http://www.cibtech.org/jfav.htm

Jamdhade MS, Survase SA, Kare MA, Bhuktar AS (2011) Antibacterial activity of genus *Datura* L. in Marathwada, Maharashtra. J Phytology 2:42–45

Javaid A, Shafique S, Shafique S (2010) Herbicidal effects of extracts and residue incorporation of *Datura metel* against *Parthenium* weed. Nat Prod Res 24(15):1426–1437. https://doi.org/10.1080/14786410903075440

Jones CK, Byun N, Bubser M (2012) Muscarinic and nicotinic acetylcholine receptor agonists and allosteric modulators for the treatment of schizophrenia. Neuropsychopharmacology 37 (1):16–42. https://doi.org/10.1038/npp.2011.199

Joshi RS (2019) Corneal toxicity due to *Datura inoxia*. J Ophthalmic Vis Res 14(3):366

Karadas S, Selvi Y, Sahin M et al (2011) *Datura stramonium* intoxication: report of a case with psychiatric symptoms. Dusunen Adam 24(2):152–154. https://doi.org/10.5350/DAJPN2011240210

Kaushik P, Goyal P (2008) *In vitro* evaluation of *Datura innoxia* (thorn-apple) for potential antibacterial activity. Ind J Microbiol 48(3):353–357. https://doi.org/10.1007/s12088-008-0020-7

Kerchner A, Farkas A (2020) Worldwide poisoning potential of *Brugmansia* and *Datura*. Forensic Toxicol 38(1):30–41. https://doi.org/10.1007/s11419-019-00500-2

Khalili NM, Atyabi SM (2004) Evaluation of analgesic effect of *Datura stramonium* seed extract in hot plate and formalin tested on male rats. J Appl Res Med Aromat Plants 20(3):309–322

Khanra S, Khess CR, Srivastava N (2015) Chronic non-fatal *Datura* abuse in a patient of paranoid schizophrenia: a case report. Addict Behav 43:39–41. https://doi.org/10.1016/j.addbeh.2014.12.002

Koehn FE, Carter GT (2005) The evolving role of natural products in drug discovery. Nat Rev Drug Discov 4(3):206–220. https://doi.org/10.1038/nrd1657

Kohnen-Johannsen KL, Kayser O (2019) Tropane alkaloids: chemistry, pharmacology, biosynthesis and production. Molecules 24(4):796. https://doi.org/10.3390/molecules24040796

Korkmaz MF, Bostanc M, Onur H, Cagan E (2019) *Datura stramonium* poisoning: a case report and review of the literature. Eur Res J 5(1):186–188. https://doi.org/10.18621/eurj.392041

Krenzelok EP (2010) Aspects of *Datura* poisoning and treatment. Clin Toxicol 48(2):104–110. https://doi.org/10.3109/15563651003630672

Kukula-Koch WA, Widelski J (2017) Alkaloids. Pharmacognosy, Academic Press, In, pp 163–198. https://doi.org/10.1016/B978-0-12-802104-0.00009-3

Kumral NA, Cobanoglu S, Yalcin C (2010) Acaricidal, repellent and oviposition deterrent activities of *Datura stramonium* L. against adult *Tetranychus urticae* (Koch). J Pest Sci 83(2):73–180. https://doi.org/10.1007/s10340-009-0284-7

Kwakye GF, Jimenez J, Jimenez JA, Aschner M (2018) Atropa belladonna neurotoxicity: Implications to neurological disorders. Food Chem Toxicol 116:346–353. https://doi.org/10.1016/j.fct.2018.04.022

Lee KH (2004) Current developments in the discovery and design of new drug candidates from plant natural product leads. J Nat Prod 67(2):273–283. https://doi.org/10.1021/np030373o

Mahdeb N, Bouzidi A, Kara N, Benouadah Z, Soufane S (2012) Acute and subacute toxicity studies of alkaloids of seeds and synthetic alkaloids of *Datura stramonium* in female rats. Pharmacol 3 (11):598–604

Maheshwari NO, Khan A, Chopade BA (2013) Rediscovering the medicinal properties of *Datura* sp.: a review. J Med Plants Res 7(39):2885–2897. https://doi.org/10.5897/JMPR11.1657

Mathur P, Singh A, Srivastava VR, Singh D, Mishra Y (2013) Anti-microbial activity of indigenous wildly growing plants: potential source of green antibiotics. Afr J Microbiol Res 7 (29):3807–3815

Mishra S (2018) *Datura Stramonium* (common name: jimson weed) medicinal uses, side effects and benefits. World J Pharm Res 12(7):1011–1019. https://doi.org/10.20959/wjpr201812-12710

Monira KM, Munan SM (2012) Review on *Datura metel*: a potential medicinal plant. Glob J Res Med Plants Indig Med 1(4):123–132

Murthy BK, Nammi S, Kota MK et al (2004) Evaluation of hypoglycemic and antihyperglycemic effects of *Datura metel* (Linn.) seeds in normal and alloxan-induced diabetic rats. J Ethnopharmacol 91(1):95–98. https://doi.org/10.1016/j.jep.2003.12.010

Newman DJ, Cragg GM (2016) Natural products as sources of new drugs from 1981 to 2014. J Nat Prod 79(3):629–661. https://doi.org/10.1021/acs.jnatprod.5b01055

Oboh G, Ademiluyi A, Agbebi O (2020) Behavioral and biochemical indices of neurotoxicity in jimson weed administered rats. Afr J Biomed Res 23(1):87–96

Ogunmoyole T, Adeyeye RI, Olatilu BO et al (2019) Multiple organ toxicity of *Datura stramonium* seed extracts. Toxicol Rep 6:983–989. https://doi.org/10.1016/j.toxrep.2019.09.011

Pandareesh MD, Anand T, Khanum F (2016) Cognition enhancing and neuromodulatory propensity of *Bacopa monniera* extract against scopolamine induced cognitive impairments in rat hippocampus. Neurochem Res 41:985–999. https://doi.org/10.1007/s11064-015-1780-1

Pandith SA, Dar RA, Lattoo SK et al (2018) *Rheum australe*, an endangered high-value medicinal herb of North Western Himalayas: a review of its botany, ethnomedical uses, phytochemistry and pharmacology. Phytochem Rev 17(3):573–609. https://doi.org/10.1007/s11101-018-9551-7

Pandya J, Rajhans S, Mankad A (2019) A review on pharmacological study of *Datura stramonium* L. Int J Res Advent Technol 7(3):1545–1549. https://doi.org/10.32622/ijrat.732019171

Perharic L, Kozelj G, Druzina B, Stanovnik L (2013) Risk assessment of buckwheat flour contaminated by thorn-apple (*Datura stramonium* L.) alkaloids: a case study from Slovenia. Food Addit Contam 30(2):321–330. https://doi.org/10.1080/19440049.2012.743189

Pillay VV, Sasidharan A (2019) Oleander and *Datura* poisoning: an update. Indian J Crit Care Med 23(4):250–255. https://doi.org/10.5005/jp-journals-10071-23302

Pretorius E, Marx J (2006) Datura stramonium in asthma treatment and possible effects on prenatal development. Environ Toxicol Pharmacol 21(3):331–337. https://doi.org/10.1016/j.etap.2005.10.006

Pye CR, Bertin MJ, Lokey RS et al (2017) Retrospective analysis of natural products provides insights for future discovery trends. Proc Natl Acad Sci 114(22):5601–5606

Rahmoune B, Zerrouk IZ, Bouzaa S et al (2019) Amino acids profiling in *Datura stramonium* and study of their variations after inoculation with plant growth promoting Rhizobacteria. Biologia 74:1373–1383. https://doi.org/10.2478/s11756-019-00287-y

Rajput H (2014) Effects of *Atropa belladonna* as an anti-cholinergic. Nat Prod Chem Res 1:670. https://doi.org/10.4172/2329-6836.1000104

Ramadan MF, Zayed R, El-Shamy H (2007) Screening of bioactive lipids and radical scavenging potential of some Solanaceae plants. Food Chem 103(3):885–890. https://doi.org/10.1016/j.foodchem.2006.09.040

Raman SV, Jacob J (2005) Mydriasis due to *Datura inoxia*. Am J Emerg 22(4):310–311. https://doi.org/10.1136/emj.2003.013060

Reddy BU (2009) Anti-microbial activity of *Datura stramonium* L. and *Tylophora indica* (Burm. F.) Merr. Pharmacologyonline 1:1293–1300

Salen P, Shih R, Sierzenski P, Reed J (2003) Effect of physostigmine and gastric lavage in a *Datura stramonium*- induced anticholinergic poisoning epidemic. Am J Emerg Med 21(4):316–317. https://doi.org/10.1016/S0735-6757(03)00036-6

Sangeetha S, Deepa M, Sugitha N, Mythili S, Sathiavelu A (2014) Anti-oxidant activity and phytochemical analysis *of Datura metel.* Int J Drug Dev Res 6(4):280–285

Saroya SA, Singh J (2020) Introduction to psychoactive medicinal plants. In: Saroya SA, Singh J (eds) Psychoactive medicinal plants and fungal neurotoxins. Springer, Singapore. https://doi.org/10.1007/978-981-15-2313-7_1

Setshogo MP (2015) A review of some medicinal and or hallucinogenic Solanaceous plants of Botswana: the genus *Datura* L. Int J Appl Res Nat Prod 1(2):15–23

Shagal MH, Modibbo UU, Liman AB (2012) Pharmacological justification for the ethnomedical use of Datura stramonium stem-bark extract in treatment of diseases caused by some pathogenic bacteria. Int Res Pharm Pharmacol 2(1):016–019

Sharma GL (2002) Studies on antimycotic properties of *Datura metel.* J Ethnopharmacol 80 (2003):193–197. https://doi.org/10.1016/S0378-8741(02)00036-3

Sharma P, Sharma RA (2013) Comparative anti-microbial activity and phytochemical analysis of *Datura stramonium* l. plant extracts and callus in vitro. Eur J Med Plants 3(2):281–287. https://doi.org/10.9734/EJMP/2013/2284

Sharma A, Patel VK, Chaturvedi AN (2009) Vibriocidal activity of certain medicinal plants used in Indian folklore medicine by tribals of Mahakoshal region of Central India. Indian J Pharmacol 41(3):129–133. https://doi.org/10.4103/0253-7613.55212

Sharma RA, Sharma PA, Yadav AN (2013) Anti-microbial screening of sequential extracts of *Datura stramonium* L. Int J Pharm 5(2):401–404

Shea YF, Chow TY, Chiu PK et al (2012) Delirium due to herbal tea contaminated with tropane alkaloid. J Clin Gerontol Geriatr 3(3):110–112. https://doi.org/10.1016/j.jcgg.2012.06.003

Shinde V, Dhale DA (2011) Antifungal properties of extracts of *Ocimum tenuiflorum* and *Datura stramonium* against some vegetable pathogenic fungi. J Phytology 3(12):41–44

Shobha G, Soumya C, Shashidhara KS, Moses V (2014) Phytochemical profile, antibacterial and antidiabetic effects of crude aqueous leaf extract of *Datura stramonium.* Pharmacophore 5 (2):273–278

Singh HP, Batish DR, Pandher JK, Kohli RK (2005) Phytotoxic effects of *Parthenium hysterophorus* residues on three Brassica species. Weed Biol Manag 5(3):105–199. https://doi.org/10.1111/j.1445-6664.2005.00172

Soneral SN, Connor NP (2005) Jimson weed intoxication in five adolescents. WMJ-MADISON 104(7):70–72

Soni P, Siddiqui AA, Dwivedi J, Soni V (2012) Pharmacological properties of *Datura stramonium* L. as a potential medicinal tree: an overview. Asian Pac J Trop Biomed 2(12):1002–1008. https://doi.org/10.1016/S2221-1691(13)60014-3

Sonika G, Manubala R, Deepak J (2010) Comparative studies on anti-inflammatory activity of *Coriandrum sativum*, *Datura stramonium* and *Azadirachta indica.* Asian J Exp Biol Sci 1 (1):151–154

Spina SP, Taddei A (2007) Teenagers with jimson weed (*Datura stramonium*) poisoning. CJEM 9 (6):467–469. https://doi.org/10.1017/S1481803500015530

Stella L, Vitelli MR, Palazzo E et al (2010) *Datura stramonium* intake: a report on three cases. J Psychoactive Drugs 42(4):507–512. https://doi.org/10.1080/02791072.2010.10400713

Tandon C, Priti M, Manodeep S (2014) Anti-bacterial compound in different species of *Datura*—a review. J Acad Res 8(21):2090–2098. https://doi.org/10.5897/AJMR2014.6805

Tranca DA, Szabo R, Cocis M (2017) Acute poisoning due to ingestion of *Datura stramonium*—a case report. Rom J Anaesth Int Care 24(1):65–68

Tripathy V, Basak BB, Varghese TS, Saha A (2015) Residues and contaminants in medicinal herbs—a review. Phytochem Lett 14:67–78. https://doi.org/10.1016/j.phytol.2015.09.003

Uddin G, Rauf A, Akhtar S (2012) Studies on chemical constituents, phytochemical profile and pharmacological action of *Datura alba.* Middle-East J Med Plants Res 1(1):14–18. https://doi.org/10.5829/idosi.mejmpr.2011.1.1.1104

Uddin F, Hossain A, Das R, Rahman Ahmad M, Akanda R, Islam S (2017) Evaluation of toxic effects of *Datura* leaves (*Datura stramonium*) in rat. Int J Agric Environ Res 3(4):3486–3497

Vadlapudi V, Kaladhar DSVGK (2012) Anti-microbial study of plant extracts of *Datura metel* L. against some important disease causing pathogens. Asian Pac J Trop Dis 2(1):94–97. https://doi.org/10.1016/S2222-1808(12)60130-3

Vaza JS, Bhalerao SA (2018) Phytochemistry, medicinal values and pharmacological potential of *Datura stramonium* L. review. Int J Pharm Drug Anal 6(7):540–544

Wang H, Jing M, Li Y (2018) Lighting up the brain: genetically encoded fluorescent sensors for imaging neurotransmitters and neuromodulators. Curr Opin Neurobiol 50:171–178. https://doi.org/10.1016/j.conb.2018.03.010

Wenker K, Suchard J (2001) Outbreak of *Datura* ingestion at a juvenile correctional facility. Cal J Emerg Med 2(3):37–38

Wess J, Eglen RM, Gautam D (2007) Muscarinic acetylcholine receptors: mutant mice provide new insights for drug development. Nat Rev Drug Discov 6(9):721–733

Wiebe TH, Sigurdson ES, Katz LY (2008) Angel's trumpet (*Datura stramonium*) poisoning and delirium in adolescents in Winnipeg, Manitoba: summer 2006. J Paediatr Child Health 13 (3):193–196. https://doi.org/10.1093/pch/13.3.193

Williamson ME (2017) Herbal neurotoxicity: an introduction to its occurrence and causes. In: Pelkonen et al (eds) Toxicology of herbal products. Springer, Cham, Switzerland, pp 345–362. https://doi.org/10.1007/978-3-319-43806-1_14

Wright GD (2017) Opportunities for natural products in 21st century antibiotic discovery. Nat Prod Rep 34(7):694–701. https://doi.org/10.1039/c7np00019g

Yilmaz MS, Isik B, Ongar M, Solakoglu AG, Baba M, Kavalci G, Salt O (2014) Delirium due to *Datura stramonium* ingestion: a case report. Sci Adv 2(10):523–527. https://doi.org/10.9734/AIR/2014/10167

# Medicinal Plants in Uganda as Potential Therapeutics against Neurological Disorders

**Savina Asiimwe, Godwin U. Anywar, Esezah Kyomugisha Kakudidi, and Patience Tugume**

**Contents**



**Abstract** According to the World Health Organization, the global burden of neurological disorders (NDs) is projected to increase from 92 at present to 103 million Disability Life Adjusted Years (DALYs) by 2030. However, NDs are not prioritized in modern healthcare systems considering the funds allocated and the limited availability of specialized modern health practitioners. The traditional health practitioners are more accessible and culturally acceptable yet not fully recognized in the formal health sector. It stems from fears related to integration arising from mutual suspicion between the two sectors. In this chapter, we present complementary and alternative therapies that could potentially narrow the treatment gap in the management of ND in Uganda. Specifically, plant species from the Ugandan context are presented from ethnobotanical studies. Twenty-three plant species from 14 families were collated from studies conducted between 1995 and 2020 to treat ND majorly epilepsy. Medicinal plants have bioactive compounds such as polyphenols and alkaloids, which potentially delay neurodegeneration, improve memory and cognitive function, and thus potential sources of new drugs in the management of ND.

S. Asiimwe · G. U. Anywar · E. K. Kakudidi · P. Tugume (✉)
Department of Plant Sciences, Microbiology & Biotechnology, Makerere University, Kampala, Uganda
e-mail: ptugume@cns.mak.ac.ug



D. C. Agrawal, M. Dhanasekaran (eds.), *Medicinal Herbs and Fungi*,
https://doi.org/10.1007/978-981-33-4141-8_17

**Keywords** Neurological disorders · medicinal plants · ethnobotanical · herbalists · Uganda

## Abbreviations

| | |
|---|---|
| AD | Alzheimer's disease |
| AIDS | Acquired immunodeficiency syndrome |
| ALS | Amyotrophic lateral sclerosis |
| ASD | Autistic spectrum disorder |
| AUD | Alcoholism/alcohol use disorder |
| CNS | Central nervous system |
| DALYs | Disability life adjusted years |
| FASD | Fetal alcohol spectrum disorder |
| FDA | Food & Drug Authority |
| GBD | Global Burden of Disease |
| HIV | Human immunodeficiency virus |
| MDD | Major depression disorder |
| MNS | Mental, neurological, and substance use |
| ND | Neurological disorders |
| OM | Orthodox medicine |
| PD | Parkinson's disease |
| PSD | Poststroke depression |
| SSA | Sub Saharan Africa |
| TCAM | Traditional, complementary, and alternative medicine |
| THP | Traditional health practitioners |
| WHO | World Health Organization |
| YLDs | Years lived with disability |

## 1 Introduction

Medicinal plant species have been essential sources of both preventive and curative therapies for humans and livestock since ancient times (Lulekal et al. 2008). Historical accounts of traditional medicinal plants depict their use from 5000 to 4000 BC in China, and 1600 BC by Syrians, Babylonians, Hebrews, and Egyptians. Ethnobotanical knowledge regarding the use of medicinal plant species has evolved through adaptive processes. This knowledge is handed down through generations by cultural transmission (Berkes et al. 2008). Communities worldwide possess different indigenous experiences in the use of plants to treat various ailments (Omoruyi et al. 2012). Herbal medicines are experiencing a greater resurgence, as many more people are turning their attention from modern drugs towards herbal or alternative medicines (Williams and Ahmad 1999). Traditional medicines (TMs) are used globally as these depend on locally available plant species, plant-based products, and the capitalization of traditional knowledge repositories (Awas and Demissew 2009). Also, cultural acceptability, affordability, and efficacy against certain types of diseases and are considered to have fewer side effects compared to modern medicines (Kakudidi et al. 2016; Tugume et al. 2016; Anywar et al. 2020). It is not surprising, therefore, that 80% of the population in the developing world depends on plant remedies for primary health care, since pharmaceuticals are highly-priced (Cragg and Newman 2001; 2002).

## 2 Methods

This chapter constitutes a literature review on medicinal plants used in the management of neurological disorders. Information was gathered from published sources, including scholarly journal articles, books, reports, and conference proceedings. The search engines and databases used included: Web of Science, Research4Life, Pub Med, Google Scholar, and ScienceDirect. The following keywords were used in different combinations: medicinal plants, neurological disorders, healthcare, traditional medicine, central nervous system, neurotoxicity, and neurodegeneration. The searches were then refined to look for specific information on the uses of plants for neurological disorders (NDs). Also, phytochemical and pharmacological activities, burden, and incidence of ND were considered. The scientific names of the plant species were updated and harmonized using theplantlist.org version 1.1 accessed on March 1, 2020. The plant families were also updated using the angiosperm phylogeny group IV.

## 3 Categorization, Incidence, and the Burden of Neurological Disorders

A neurological disorder is a term used to describe a disease that affects the central nervous system (CNS) because of physical injury to the brain or nerves (Rizk et al. 2018). The ND often leads to debilitating conditions that result in the progressive degeneration or death of neurons, which cause problems with movement or with mental functions (Helzner et al. 2008).

The World Health Organization (WHO) has often focused on infectious and noncommunicable diseases. Neurological disorders have not been prioritized in the modern healthcare systems considering the funds allocated and fewer health workers treating them. Yet, ND equally presents a substantial disease burden in the developing world (Institute of Medicine 2010).

Mathers and Loncar (2005) noted that ND caused a much higher burden than malignant neoplasms, digestive, and respiratory diseases constituting 6.3% of the global disease burden in 2005. It was equivalent to 92 million DALYs and projected to increase to 103 million by 2030. NDs are the fifth leading cause of death globally and are responsible for 5.53% of total mortality. Over 80% of ND occur in low middle-income countries, with about 6.8 million dying each year (Mathers and Loncar 2005; Kakudidi et al. 2016) irrespective of their sociodemographic and economic status. However, other studies showed the prevalence of ND in developing countries was slightly lower than in industrialized countries (Prince 2000; Hendrie et al. 2001). The variation is due to failure to detect mild dementia due to lack of access to doctors and inadequate diagnostic capabilities in developing countries.

The cost of ND treatment is more than $100 billion annually (Ratheesh et al. 2017). Some NDs often affect the elderly and are a major cause of morbidity, mortality, and severely constrain the social welfare systems (Sonibare and Ayoola 2015). The prevalence of ND presents an immense economic and social burden to the health care systems worldwide, especially in low-income countries taking into account increased life expectancy and limited neurological services (Ahmad et al. 2015; Rizk et al. 2018). For instance, much as the government of Uganda has taken steps toward improving and strengthening mental health services at the grassroots (Nsereko et al. 2011), the treatment gap continues to widen, considering the patient to health worker ratio of 1.13:100,000 (WHO 2016). In Uganda, about 4% of the total health budget is allocated to mental, neurological, and substance use (MNS) disorders, indicating insufficient resources to tackle the enormous burden (Institute of Medicine 2010).

There are more than 600 diseases of the nervous system. The common ones are epilepsy, Alzheimer's disease (AD), Parkinson's disease (PD), Huntington's disease, cerebrovascular diseases, neuro-infections, traumatic disorders, and psychiatric diseases (Hiruma-Lima et al. 2006). Neurological disorders may be caused by trauma, e.g., skull fracture and spinal cord injury; cerebral vascular accidents, e.g., stroke; neoplasia; infections, e.g., meningitis; metabolic disorders, e.g., diabetic neuropathy; genetic disorders, e.g., Down's syndrome; environmental factors, e.g., heavy metal encephalopathies; poor nutrition; immunological factors, e.g., multiple sclerosis (Rizk et al. 2018), and HIV/AIDS-associated neurological disorders (Rizk et al. 2018; Anywar et al. 2020). A brief description of the most prevalent neurological disorders is given below.

Stroke is the abrupt failure in functioning of the brain in a localized area when there is interrupted blood flow leading to cell damage. The effects of stroke include paralysis, speech disorder, amnesia, comatose, or death (Park et al. 2009). Stroke is the leading cause of deaths accounting for 67.3% among all neurological disorders (Feigin et al. 2017). In 2009, a review of 1227 studies from ten African countries showed that 0.3% of the population had suffered from stroke. It was estimated that by 2013 10.8% would suffer from stroke with 9.6% survival rate (Adeloye 2014). In Uganda, the prevalence of stroke in adults was estimated at 14.3% (Nakibuuka et al. 2015) while poststroke depression (PSD) affected 31.5% of patients within 6 months after the onset of stroke (Gyagenda et al. 2015). Additionally, stroke was reported at 6.8% in children admitted to Mulago hospital with sickle cell anemia (Munube et al. 2016).

Alzheimer's disease (AD) is a progressive, irreversible neurological disorder characterized by profound memory loss enough to interfere with social and occupational functioning (Akhondzadeh and Meleki 2007), unusual behavior, personality changes, and loss of cognitive ability (Parihar and Hemnani 2004). AD is the most common cause of severe mental deterioration (dementia) in the elderly (Akhondzadeh and Meleki 2007; Poorgholam et al. 2018). The people affected by AD alone increased from 26.6 million in 2006 to 36 million in 2014 (Ratheesh et al. 2017) and contributed 60–70% of ND cases in 2016 (WHO 2016). Severe cases of AD lead to incontinence, complete loss of memory for the sense of time and place, leading to total dependence of patients on other people (Alzheimer's Association 2010). WHO projected the incidence of AD and other dementias to increase by 66% by 2030 (Mathers and Loncar 2005). There is paucity of data on AD in Sub-Saharan Africa (SSA). Our search revealed that in Nigeria, the incidence of dementia was 2.19% while dementia-related AD was at 1.15% (Sosa-Ortiz et al. 2012). In rural Uganda, the prevalence of AD and other related dementias was estimated to be 20% (Mubangizi et al. 2020).

Depression is a medical illness that affects the mood and manifests as sadness, fatigue, loss of interest, and appetite. It is a significant cause of disability leading to severe social and economic deficits (Saki et al. 2014) with lifetime prevalence at 21% of the population in some developed countries (Kamalipour et al. 2008). The global prevalence of major depression disorder (MDD) was averaged at 4.7% (Ferrari et al. 2013). The Global Burden of Disease (GBD) 1990 study reported that depressive disorders were the fourth leading cause of burden, accounting for 3.7% of DALYs (Murray et al. 2007). By 2000 MDD was the third leading cause of burden (4.3% of DALYs), and the principal cause of disability accounting for 13.4% DALYs in women and 8.3% in men. The GBD 2000 update estimated that the 12-month prevalence of major depressive episode (MDE) was 1.6% in males and 2.5% in females (Üstün et al. 2004). Tomlinson et al. (2009) reported a prevalence of 9.7% among older persons (≥65 years) in South Africa whereas Rochat et al. (2011) reported antenatal depression to affect 47% of women. Aggregated data from 14 districts in Uganda among the population aged ≥15 years reported the prevalence of major depressive disorders at 29.3% (Kinyanda et al. 2011).

Dementia is a malfunction of the CNS that leads to gradual death of brain cells. It causes a decline in the cognitive domain (attention, memory, executive function, visual–spatial ability) and language, especially in the elderly (Al-Qazzaz et al. 2014). Globally 24.3 million people were estimated to have dementia, with 4.6 million cases occurring annually (Mathers and Loncar 2005). It was projected to double every 20 years to 81.1 million by 2040, with 71% in developing countries. Similarly, WHO (2007) estimated 35.6million people to have dementia annually. In 2010, the prevalence of dementia among people aged ≥50 years in Africa was estimated at 2.4% translating to 2.76 million (Rhiannon et al. 2012). Dementia and depression affected 13.2% of Ugandans aged ≥60 years (Namuli 2015). Given the increasing number of dementia patients, the cost of treatment is enormous and expected to rise.

Parkinson's disease (PD) is a chronic neurodegenerative disease after Alzheimer's (Dorsey et al. 2007; Elbaz et al. 2016). It is characterized by the loss of dopaminergic neurons leading to decreased levels of dopamine in the striatum and disruption of the motor control (Obeso et al. 2010). WHO projected the incidence of

PD to be 11.5% by 2030 (Mathers and Loncar 2005). Generally, the population of PD patients in SSA is relatively small. Existing country variations are difficult to make due to methodological differences in studies (Uduak et al. 2018). Most people with PD in SSA, remain undiagnosed and untreated (Dotchin et al. 2007). Kaddumukasa et al. (2015) noted paucity of evidence and data on PD awareness in Uganda.

Epilepsy is a chronic disorder associated with alterations in the mental processes, the state of consciousness, or involuntary movements leading to convulsions. A convulsion is an abnormal discharge of a group of neurons in the CNS, more specifically the brain. The convulsion may occur in different clinical forms depending on the discharge rate and its spread (Saki et al. 2014). Epilepsy affected approximately 70 million people worldwide, of whom 90% were from developing countries (Singh and Trevick 2016), and about 10 million people in Africa (Sahoo et al. 2011). The epilepsy prevalence rate in Uganda was estimated at 13.3% (Nakibuuka et al. 2015).

Mental, neurological, and substance use (MNS) disorders are increasingly affecting the youth due to the abuse of substances such as cannabis, alcohol, cocaine, heroin, pethidine, tobacco, khat, and kuber (Awuzu et al. 2014). Caring for MNS patients poses a formidable challenge for individuals, families, governments, and societies around the world (Institute of Medicine 2010). In SSA, MNS disorders account for 19% of years lived with disability (YLDs) (Whiteford et al. 2013). According to WHO, MNS disorders accounted for 5.2% of all DALYs in Uganda (WHO 2017).

Autistic spectrum disorder (ASD) is a neurodevelopmental condition characterized by impairment in communication skills, stereotyped behavior, and social relations (Matson and Boisjoli 2008). ASD affects approximately 1 in 100 individuals (Baird et al. 2006). In 2010, the estimated prevalence was 0.76% equating to 52 million cases of ASD. Globally, ASD accounted for more than 58 DALYs per 100,000 population (Baxter et al. 2015). The prevalence of ASD among children with developmental disorders in Egypt and Tunisia was 33.6% and 11.5%, respectively (Seif Eldin et al. 2008), and 11.4% in Southeastern Nigeria (Bakare et al. 2012). There is currently scarcity of data in Uganda on the prevalence of ASD, although the prevalence of nodding disease akin to ASD was reported at 6.8–16.7% in Northern Uganda (Iyengar et al. 2014; Colebunders et al. 2016).

Multiple sclerosis (MS) is a chronic inflammatory disease of the CNS which causes differential dysfunction of the motor, sensory, visual, and autonomic systems. MS affects an estimated 2.3 million people worldwide with women being affected more than twice as often as men (Doshi and Chataway 2017). WHO projected the incidence of MS to increase to 9.14% by 2030 (Mathers and Loncar 2005). In SSA the age-standardized prevalence of MS ranged between 2.4 and 3.8 per 100,000 population (GBD 2016 Multiple Sclerosis Collaborators (2019)).

Alcohol use disorder (AUD) is a chronic condition characterized by uncontrolled alcohol consumption resulting in cognitive deficiencies, uncoordinated movements, and adverse social; occupational; or health consequences (Harper 2007; Brust 2010). A related condition is fetal alcohol spectrum disorder (FASD), which results from alcohol consumption during pregnancy. It is characterized by attention deficit, hyperactivity, impaired cognitive function, and mood disorders (Barr et al. 2006). AUD kills approximately 3.3 million people worldwide (Lim et al. 2012; WHO 2014) and is a leading risk factor for premature death and disability (Endal 2013). In South Africa, alcohol abuse accounted for 7.1% of all deaths and 7% of total DALYs (Schneider et al. 2007). In Uganda, the prevalence of AUD was reported at 17% (32% men, 7% women) (Roberts et al. 2011).

## 4 Modes of Diagnosis and Management of Neurological Disorders

Traditional, complementary, and alternative medicine (TCAM) uses diagnostic systems incompatible with conventional medicine due to differences in the philosophy of illness concerning causes/concepts of diseases. Diagnosis of neurological disorders and the treatment prescribed by traditional healers reflect the indigenous beliefs and cultural interpretations of the problem peculiar to each local culture or community (Gureje et al. 2015). Diagnosis often takes place in a shrine where the admitted patients are also hospitalized. The diagnostic tools, methods, and materials for neurological disorders in TCAM vary according to culture in contrast to the orthodox medicine that observes signs of illness (Owuor et al. 2005).

The African traditional therapies for neurological disorders involve phytotherapy, psychotherapy, and spiritopsycotherapy (Hirt and M'pia 2008) although there is an overlap of the therapies but the latter two are beyond the scope of this chapter. Phytotherapy involves the use of herbs (Mbwayo et al. 2013) although Ndetei (2007) noted that traditional psychotherapy is more effective than psychotherapy as practiced in modern medicine. Another reported method is a craniotomy, practiced by the Kisii in Kenya (Ovuga et al. 1999). Treatment techniques include administration of medication through body cuts; nasal drops or sniffing herbal medicine; oral administration, bathing/washing with the drug; and burning or inhaling the herbs.

## 5 Traditional Use of Medicinal Plants for Neurological Disorders

Traditional healing practices have been used in many countries long before orthodox medicine (Gureje et al. 2015). Traditional health practitioners (THPs) mainly involve the use of medicinal plants, which form a significant component of the natural therapeutic remedies and are also food supplements (Dike et al. 2012; Duke 2019). Traditional medicine continues to occupy an important position in African societies because it is a pillar of culture and human socialization (Kokwaro 1993; Owuor et al. 2005). Traditional health practitioners exist alongside modern health practitioners (Ovuga et al. 1999) and are often the first ones to be contacted for any ailment, not only for healing but also to obtain an explanation for the causes (Mbwayo et al. 2013). Communities often refer patients with illnesses to specific THP because they are considered competent and adopt culturally acceptable methods. Therefore, THP cannot be disregarded in the management of ND (Haimanot 1985; Abbo et al. 2019).

TCAM and orthodox medicine (OM) have similar objectives of healing or preventing disease. The TCAM view of health differs from OM regarding diseases and causes/concepts of disease (Sofowora 1996; WHO 1999). According to the TCAM concept, there are five causes of disease: (i) physical ailments, (ii) psychological, (iii) astral, (iv) spiritual, and (v) esoteric (Lombo 1979). However, OM recognizes physical ailments. Thus, the concept of disease in TCAM is broader and more holistic than in OM (Sofowora 1996; Mbwayo et al. 2013; Abbo et al. 2019; Ningsangrenla and Rao 2019).

Scientists have demonstrated the contribution of medicinal plants to the treatment of many infectious, noninfectious, and noncommunicable diseases in many parts of the world (Okigbo and Mmeka 2006; Anywar et al. 2020). Therefore, neurological disorders would not be an exception. Various publications have recorded numerous herbal extracts that have been used to treat ND (Sonibare and Ayoola 2015; Ningsangrenla and Rao 2019).

## 6 Role of Traditional Healers in Neurological Disorders Healthcare

Traditional health practitioners (THPs) usually live in the same community where they practice, speak the same language, and have extensive knowledge of the social environment within which they operate, as well as resources used (Abbo et al. 2019). THPs are more accessible than modern medicine practitioners and moreover the latter are fewer. However, the contribution of THP is not fully recognized in the formal health sector due to fears related to integration arising from mutual suspicion between the two sectors. Relatedly there is a concern expressed by biomedical and religious establishments regarding the "unscientific" and "unorthodox" practices of traditional healers (Patel 2011).

An estimated 30–60% of people with ND-related problems visit THP in Uganda (Nsereko et al. 2011; Akol et al. 2018). Nsereko et al. (2011) attribute the high percentage of clients to beliefs about the causal factors; nature of service delivery: testimonies from those who have benefited from the services and awareness of the availability; social-economic status; severity of the condition, and avoidance of stigma experienced in modern health centers.

The high ND burden, the large treatment gap, and the inadequate modern medical health facilities (Silberberg and Katabira 2006) require copartnership between traditional healers and OM health practitioners in ND care (Patel 2011). The complementarity of therapies used would lessen the gap that requires training the THP to diagnose and handle patients properly to eliminate the "unscientific" and "unorthodox" practices, as well as removing suspicions between the two sectors (Patel 2011). Additionally, with a shortage of trained physicians to handle ND, the nurses and community health workers would require more training on how to diagnose and treat the common neurological conditions that present to their clinics (Silberberg and Katabira 2006).

## 7 Ethnopharmacological Knowledge on Neurological Disorders

Herbal medicines are largely effective alternatives to conventional drugs because of the presence of various phytochemicals (Sasidharan et al. 2011; Ahmad et al. 2016). Clinical trials performed in mild to moderate AD dementia, have led to more effective and safe alternatives from natural sources, for instance, galantamine from *Leucojum* spp., *Narcissus* spp. and *Galanthus* spp. (Heinrich and Teoh 2004; Cummings et al. 2014).

Ethnopharmacology has been recognized as a source of knowledge-driven drug discovery from plants (Farnsworth 1990; Heinrich 2010). This recognition is based on local or traditional knowledge of plants' pharmacological or toxicological properties (Cordell and Colvard 2005; Heinrich 2010). Some of the drugs discovered through ethnomedical knowledge have been proven safe and effective due to the presence of naturally optimized phytochemicals (Ayaz et al. 2019). With over 250,000 plant species worldwide, the potential for new drug discovery is enormous.

The number of synthetic drugs approved for neurodegenerative diseases has not increased for some time. Only one new ND drug for amyotrophic lateral sclerosis (ALS) called Radicava, was approved by the Food and Drug Authority (FDA) (Cruz 2018). The previously clinically approved medical drugs for ND have had limited efficacy and are associated with side effects (Watkins et al. 1994; Husain et al. 2007; Ayaz et al. 2019) and therefore calls for urgent development of new plant-based ND drugs. Figure 1 provides a summary of the management of ND in Uganda.

## 8 Plants Species Used for the Treatment of Neurological Disorders in Uganda

A total of 23 plant species distributed in 14 families were collated from studies done in Uganda between 1995 and 2020 that are used to treat ND. The dominant family Leguminosae had nine species while the rest of the families had one species each. Epilepsy is the major neurological disorder treated using herbal medicine in Uganda (Table 1).

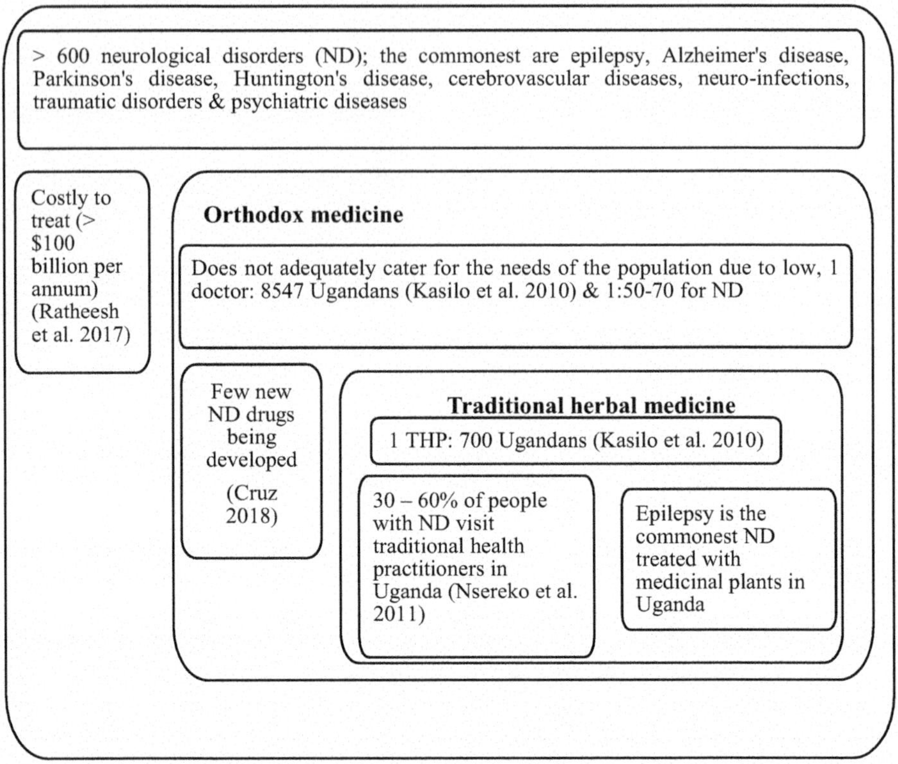


**Fig. 1** Orthodox and traditional complementary and alternative medicine in the management of neurological disorders

## 9 Phytochemical and Pharmacological Activity of Plant Species

There is paucity of studies on neuroprotective activity of plants traditionally used to manage ND in Uganda. However, some medicinal plant species traditionally used to treat epilepsy (Table 1) have been pharmacologically tested in various studies and showed anticonvulsant and antiepileptic activities in vivo (Table 2). The anticonvulsant and antiepileptic activities are attributed to the presence of alkaloids, saponins, flavonoids, terpenoids, and coumarins. For instance, alkaloids showed potent anti-epileptiform activity in vitro on rat hippocampal slices through interaction with the voltage-dependent $Na^+$ channel (Ameri 1998). Flavonoids prevented seizures induced by Pentylenetetrazole (PTZ) through central benzodiazepine receptors (Medina et al. 1990). Terpenoids such as α-Terpineol exhibited protection against PTZ, picrotoxine, and maximal electroshock-induced seizures in mice (de Sousa et al. 2006, 2007). Jalsrai et al. (2010) reported the saponin protective effects against PTZ-induced seizures.

**Table 1** Plants species used for the care of epilepsy in Uganda

| Family | Medicinal plant | Parts used | Methods of preparation | References |
|---|---|---|---|---|
| Amaranthaceae | *Chenopodium opulifolium* Schrad. Ex W. D. J. Koch. and Ziz | Root | Powder mixed with water and drunk or mixed with petroleum jelly for body smearing | Tabuti et al. (2003) |
| | *Dysphania ambrosioides* (L.) Mosyakin and Clemants (Syn. *Chenopodium ambrosioides*) | Leaves | Maceration, infusion drunk | Kamatenesi et al. (2011) |
| Asteraceae | *Microglossa pyrifolia* (Lam.) Kuntze | Leaves | Decoction bathed/ powder sniffed or applied in the nose | Kamatenesi et al. (2011) |
| Apocynaceae | *Carissa spinarum* L. Syn: *Carissa edulis* (Forssk)Vahl | Root | Smoke powder/infusion for bathing | Oryema et al. (2010) |
| Combretaceae | *Terminalia glaucescens* Planch. ex Benth. | Root | Decoction used for bathe; powder added to petroleum for body smearing | Tabuti et al. (2003) |
| Crassulaceae | *Kalanchoe crenata* (Andrews) Haw. | Stem, leaves | Decoction drunk | Tabuti et al. (2003) |
| Lamiaceae | *Hoslundia opposita* Vahl | Root | Decoction drunk | Okello and Ssegawa (2007), Oryema et al. (2010) |
| Leguminosae | *Acacia hockii* De Willd | Root | Powder mixed in water and drunk or bathed | Tabuti et al. (2003) |
| | *Albizia coriaria* Oliv. | Root | Decoction drunk | Oryema et al. (2010), Anywar et al. (2020) |
| | *Albizia malacophylla* (A. Rich) Walp. | Root | Inhale smoke of burnt root; apply powder on incisions | Tabuti et al. (2003), Oryema et al. (2010) |
| | *Antiaris toxicaria* Lesch. | Leaves | Infusion bathed | Tugume et al. (2016) |
| | *Dichrostachys cinerea* (L.) Wight & Arn.) | Root, leaves | Powder infusion drunk | Katende et al. (1995) |
| | *Erythrina abyssinica* DC. | Root, bark | Decoction drunk | Anywar et al. (2020) |
| | *Indigofera emarginella* A. Rich. | Leaves | Decoction drunk | Ssegawa and Kasenene (2007) |

(continued)

**Table 1** (continued)

| Family | Medicinal plant | Parts used | Methods of preparation | References |
|---|---|---|---|---|
| | *Sesbania sesban* (L.) Merr | Leaves | Infusion drunk | Ssegawa and Kasenene (2007) |
| | *Tamarindus indica* L. | Roots | Herbal bath | Tugume et al. (2016) |
| Myricaceae | *Morella salicifolia* (Hochst. Ex A. Rich.) Verdc. and Polhill | Leaves, bark, root | Powder added to food or drinks | Kamatenesi et al. (2011) |
| Passifloraceae | *Adenia cissampeloides* (Planch. ex Hook.) Harms | Leaves | Infusion bathed | Tabuti et al. (2003) |
| Peraceae | *Clutia abyssinica* Jaub. and Spach | Leaves | Decoction drunk and bathed | Hamill et al. (2000), Asiimwe et al. (2014) |
| Poaceae | *Arundinaria alpina* K. Schum | Root | Infusion bathed | Tugume et al. (2016) |
| Polygalaceae | *Securidaca longipedunculata* Fresen | Root | Decoction drunk | Okello and Ssegawa (2007) |
| Rutaceae | *Zanthoxylum chalybeum* Engl. | Bark, root, leaves | Decoction drunk | Anywar et al. (2020) |
| Verbenaceae | *Lantana camara* L. | Fruits | Fruit extract drunk | Kamatenesi et al. (2011) |

*Albizia coriaria* and *Albizia malacophylla* were documented as used for treating epilepsy in Uganda (Table 1), but a closely related species *Albizia glaberrima* used for nervous system disorders and epilepsy in West Africa was pharmacologically tested and found to be active. The aqueous leaf extract of *A. glaberrima* showed a dose-dependent anxiolytic/muscle relaxant at low doses and sedative-hypnotic/anticonvulsant activities at high doses in mice (Adebesin et al. 2015). *Carissa spinarum* showed bioactive principles with anticonvulsant activity in mice (Ya'u et al. 2008, 2015). The aqueous root extract of *Securidaca longipedunculata* exhibited anticonvulsant, anxiolytic, and sedative activities in mice in a dose-dependent manner (Adeyemi et al. 2010; Okomolo et al. 2011). Phytocompounds with anticonvulsant activity have been isolated from *Hoslundia opposita* (Adesina and Ette 1982). Although *Erythrina abyssinaica* is used for treating epilepsy in Uganda, anticonvulsant properties of the alkaloids (+)-erythravine and (+)-11-α-hydroxy-erythravine were isolated from the flowers of a related species *Erythrina mulungu* (Faggion et al. 2011). Similarly, *Erythrina mysorensis* had anticonvulsant activity in epileptic mice models (Nagaraja et al. 2012). The flavonoid, abyssinone V-4′ from *Erythrina droogmansiana* also showed anticonvulsant

**Table 2** Bioactive compounds in neuroprotection against epilepsy

| Plant species | Known constituents | References |
|---|---|---|
| *Acacia hockii* | Alkaloids, flavonoids, steroids, terpenoids, cardiac glycosides, phenolics, saponins, anthracene derivatives, pigments | Kamau et al. (2016), Lagnika et al. (2016) |
| *Adenia cissampeloides* | Tannins saponins, phlobatannins, flavonoids, anthraquinones, terpenoids, steroids, alkaloids, carbohydrates, glycosides | Njoku and Obi (2009) |
| *Albizia coriaria* | Benzyl alcohol, acacic acid lactone, lupenone, lupeol, betulimic acid | Ganza (2014), Byamukama et al. (2015) |
| *Albizia malacophylla* | Saponins, flavonoids, phenolic glycosides, alkaloids, triterpenoids | He et al. (2020) |
| *Antiaris toxicaria* | Anthracene glycosides, tannins, flavonoids, alkaloids, saponins, reducing sugars, and triterpenoids | Mante et al. (2017) |
| *Chenopodium opulifolium* | Polyphenolics (flavonoids, caffeic acid, tannins) | Ajayi et al. (2017) |
| *Clutia abyssinica* | Alkaloids, anthraquinones, flavonoids, glycosides, phenolic compounds, saponins, steroids, terpenes, and tannins | Mergia et al. (2015) |
| *Dichrostachys cinerea* | Tri-terpenoids, saponins, alkaloids, β-amyrin, β-sitosterol | Stark et al. (2013) |
| *Dysphania ambrosioides* Syn. *Chenopodium ambrosioides* | Stigmasterol, β-sitosterol, octadecanoic acid, scopoletin and 1-piperoylpiperidine | Shah and Khan (2017) |
| *Erythrina abyssinica* | Alkaloids | Fahmy et al. (2018) |
| *Hoslundia opposita* | Hoslunddiol, euscaphic acid, 5,7-dimethoxy-6-methyl flavone | Mujovo et al. (2008) |
| *Indigofera emarginella* | Terpenoids, flavonoids, nitro-group containing compounds, steroids | Rahman et al. (2018) |
| *Kalanchoe crenata* | Sterols, saponins, phenolics, bufadienolide, flavonoids | Nguelefack et al. (2006) |
| Kalanchoe *densiflora* | Tannins, saponins, terpenoids, flavonoids, cardiac glycosides, cotyledoside | Bruneton (1999), Kirui et al. (2014) |
| *Lantana camara* | Bicyclogermacrene, δ-3-Carene, Caryophyllene oxide, α-Cubebene, β-Caryophyllene, β-Cubebene, δ-Elemene, β-Elemene, Germacrene B, Germacrene D, Humulene, linalool, Longicyclene, δ-Muurolene, Sabinen, β Sesquiphellandrene; Lantadene A, B | Bruneton (1999), dos Santos et al. (2015) |

(continued)

**Table 2** (continued)

| Plant species | Known constituents | References |
|---|---|---|
| *Microglossa pyrifolia* | α-Humulene, α-pinene, germacrene D, (E)-β-ocimene, (E)-β-caryophyllene, thymol | Boti et al. (2007), Bum et al. (2009) |
| *Morella salicifolia* | Myricanol; diarylheptanoids | Jahng and Park (2018), Dickey et al. (2019) |
| *Securidaca longipedunculata* | Indole alkaloid, securinine, ergotalkaloids, flavonoids, alkaloids | Taiwe and Kuete (2014) |
| *Sesbania sesban* | Steroids, amino acids, triterpenoids, saponins, vitamins, pyruvic acid, glycosides, carbohydrates | Gomase (2012) |
| *Tamarindus indica* | Tannins, polyphenols, flavonoids, cardiac glycosides, amino acids, terpenoids, leteolin, carbohydrates, lipids, ubiquinol (coenzyme $Q_{10}$), carotenes, uric acid, lipoic acid, glutathione, ascorbic acid | Tawfik (2011), Boccardi et al. (2016), Rahman (2016) |
| *Terminalia glaucescens* | Terminalin A, glauconoic acid | Bum et al. (2009) |
| *Yushania alpina* Syn *Arundinaria alpina* | Phytosterol | Nongdam and Tikendra (2014) |

activity in mice at doses of 25 and 100 mg/kg against generalized convulsions induced by picrotoxine (Talla et al. 2016).

Limited pharmacological studies on the CNS properties of *Acacia* spp. have been done despite its wide use in ethno-medicine. *Acacia hockii* was documented as used for treating epilepsy in Uganda (Table 1), but pharmacological studies of closely related species have shown moderate or minimal activity on ND. For instance, the bark extracts of *A. sieberiana* and *A. xanthophloea* showed dose-dependent sedative activity (Stafford et al. 2005). *A. juliflora* and *A. nubica* leaf extracts showed tonic seizure prevention capacity (Hebbar et al. 2004). *A. nilotica*, *A. raddiana*, *A. seyal*, *A. sieberiana*, *A. modesta*, *A. farnesiana*, *A. catechu*, and *A. cyanophylla* inhibited acetylcholinesterase (AChE) a potential target in the treatment of Alzheimer (Tengku and Raja 2013; Ghribia et al. 2014; Thangavelu and Ramasamy 2015). Catechin isolated from *A. catechu*, and administered at doses of 10, 20, and 40 mg/kg significantly enhanced the learning capacity and retention of memory in mice (Biradar et al. 2012). Pharmacological studies need to be done on *A. hockii* to validate its use in the management of epilepsy in Uganda.

A decoction of *Microglossa pyrifolia* significantly protected mice against induced seizures, maximal electric shock, and picrotoxin in mice at a dose of 260 mg/kg. The decoction also delayed onset of seizures in mice injected intraperitoneally with isonicotinic hydrazide acid (Bum et al. 2009). The inhibition of seizures suggested possession of anticonvulsant activity (Salih and Mustafa 2008) and the involvement of glycine receptors (Findly et al. 2002). The anticonvulsant effects of *M. pyrifolia* may have resulted from antagonistic action on N-methyl-D-aspartate (NMDA) receptors or reversion antioxidant–oxidant imbalance (Pahuja et al. 2012).

Essential oils from different plant species reduced symptoms in patients of Alzheimer's and Parkinson's diseases (Jimbo et al. 2009; Dobetsberger and Buchbauer 2011). In vivo tests using Winstar rats showed that *L. camara* was not toxic at 2000 mg/kg dose (OECD 2002). However, lantadene A and B showed toxicity by causing diarrhea, vomiting, and cyanosis in humans and other animals (Bruneton 1999). The crude leaf extract of *Sesbania sesban* stimulated the CNS activity compared to caffeine a reference drug (Naik et al. 2011).

In vivo and in vitro models reported neuroprotective activities of *Antiaris toxicaria* in stroke and potassium cyanide-induced neurotoxicity respectively (Ilesanmi et al. 2017). Ilesanmi et al. (2017) further reported its neuroprotective activity by raising the action of Na+/K+-ATPase, acetylcholinesterase, and tyrosine hydroxylase. The aqueous extract of *A. toxicaria* also delayed onset and altered duration of seizures in mice and rats at a dose of 200–800 mg/kg comparable to the standard drugs Carbamazepine and Nifedipine at (30 mg/kg) (Mante et al. 2017). *Arundinaria alpina* (Bamboo) leaf extract showed no toxic effects in rats after 90 days of administration at doses of 1.43, 2.87, and 4.30 g/kg per day hence considered safe for human consumption (Lu et al. 2005). The leaf and bark extract of *Tamarindus indica* exhibited cholinesterase inhibitory activity with $IC_{50}$ of 188.67 μg/mL and 176.30 μg/mL for acetylcholinesterase and butyrylcholinesterase enzyme respectively (Rahman 2016).

Traditionally herbal medicines are considered safe for human use. However, they have both useful and harmful effects on humans and other animals based on the dose (Sekhar et al. 2012). For instance, cotyledoside extracted from *Kalanchoe densiflora* showed neurotoxic effects in humans and sheep at a dose of 7 g/kg characterized by diarrhea, bradycardia, and arrhythmia (Bruneton 1999). Despite the toxic effects of some plant secondary metabolites, they are important in drug development. It is therefore imperative to conduct extensive toxicological studies on traditionally used medicinal plants to avoid any adverse effects on the body (Nicholson et al. 2002).

## 10 Conclusions

Traditional medicinal knowledge used by indigenous communities is not only useful for the conservation of cultural traditions and community health care but also contributes to future drug development. Globally, there is diverse knowledge of medicinal plants used to manage neurological disorders. This chapter has collated information on medicinal plant use especially in the management of epilepsy in Uganda. The generated information revealed that some knowledge exists in society although few scientific studies on phytochemistry and pharmacology have been done to validate their use in traditional medicine. Thus, there is a need for extensive ethnobotanical surveys to capture more information on management of other ND using medicinal plants, as well as phytochemical and pharmacological studies. Also, there is a need for the Ministry of Health and other stakeholders in the health sector to partner with THP to reduce the treatment gap in the management of ND.

## References

Abbo C, Odokonyero R, Ovuga E (2019) A narrative analysis of the link between modern medicine and traditional medicine in Africa: a case of mental health in Uganda. Brain Res Bull 145:109–116

Adebesin IF, Akindele AJ, Adeyemi OO (2015) Evaluation of neuropharmacological effects of aqueous leaf extract of *Albizia glaberrima* (Leguminosae) in mice. J Ethnopharmacol 160:101–108. https://doi.org/10.1016/j.jep.2014.11.040

Adeloye D (2014) An estimate of the incidence and prevalence of stroke in Africa: a systematic review and meta-analysis. PLoS One 9:e100724. https://doi.org/10.1371/journal.pone.0100724

Adesina SK, Ette EI (1982) The isolation and identification of anticonvulsant agents from *Clausena anisata* and *Afraegle paniculata*. Fitoterapia 53(3):63–66

Adeyemi OO, Akindele AJ, Yemitan OK et al (2010) Anticonvulsant, anxiolytic and sedative activities of the aqueous root extract of *Securidaca longepedunculata* Fresen. J Ethnopharmacol 130:191–195. https://doi.org/10.1016/j.jep.2010.04.028

Ahmad S, Ullah F, Ayaz M et al (2015) Antioxidant and anticholinesterase investigations of *Rumex hastatus* D. Don: potential effectiveness in oxidative stress and neurological disorders. Biol Res 48:20. https://doi.org/10.1186/s40659-015-0010-2

Ahmad S, Ullah F, Sadiq A et al (2016) Chemical composition, antioxidant and anticholinesterase potentials of essential oil of *Rumex hastatus* D. Don collected from the north west of Pakistan. BMC Complement Alternat Med 16:29. https://doi.org/10.1186/s12906-016-0998-z

Ajayi AM, Tanayen JK, Magomere A, Ezeonwumelu JOC (2017) Antinociceptive and anti-inflammatory effects of aqueous extract of *Chenopodium opulifolium* schrad leaves. J Intercult Ethnopharmacol 6:14–21. https://doi.org/10.5455/jice.20161208013809

Akhondzadeh S, Meleki J (2007) Herbal medicines in the treatment of psychiatric and neurological disorders. Iran J Psychiatry 1:119–138

Akol A, Moland KM, Babirye JN, Engebretsen IMS (2018) "We are like co-wives": traditional healers' views on collaborating with the formal child and adolescent mental health system in Uganda. BMC Health Serv Res 18:258. https://doi.org/10.1186/s12913-018-3063-4

Al-Qazzaz NK, Bin ASH, Ahmad SA et al (2014) Role of EEG as biomarker in the early detection and classification of dementia. Scientific World J 2014:1. https://doi.org/10.1155/2014/906038

Alzheimer's Association (2010) 2010 Alzheimer's disease facts and figures. Alzheimers Dement 6:158–194. https://doi.org/10.1016/j.jalz.2010.01.009

Ameri A (1998) Inhibition of stimulus-triggered and spontaneous epileptiform activity in rat hippocampal slices by the Aconitum alkaloid mesaconitine. Eur J Pharmacol 342:183–191

Anywar G, Kakudidi E, Byamukama R et al (2020) Indigenous traditional knowledge of medicinal plants used by herbalists in treating opportunistic infections among people living with HIV/AIDS in Uganda. J Ethnopharmacol 246:112205. https://doi.org/10.1016/j.jep.2019.112205

Asiimwe S, Namutebi A, Borg-Karlsson A et al (2014) Documentation and consensus of indigenous knowledge on medicinal plants used by the local communities in Western Uganda. J Nat Prod Plant Resour 4(1):34–42

Awas T, Demissew S (2009) Ethnobotanical study of medicinal plants in Kafficho people, southwestern Ethiopia. In: Proceedings of the 16th international conference of Ethiopian studies. NTNU-Trykk Press, Trondheim, Norway, pp 711–726

Awuzu EA, Kaye E, Vudriko P (2014) Prevalence of cannabis residues in psychiatric patients: a case study of two mental health referral hospitals in Uganda. Subst Abuse 8:1–5. https://doi.org/10.4137/SART.S13254

Ayaz M, Ullah F, Sadiq A et al (2019) Natural products-based drugs: potential therapeutics against Alzheimer's disease and other neurological disorders. Front Pharmacol 10:1417. https://doi.org/10.3389/fphar.2019.01417

Baird G, Simonoff E, Pickles A et al (2006) Prevalence of disorders of the autism spectrum in a population cohort of children in South Thames: the special needs and autism project (SNAP). Lancet 368:210–215. https://doi.org/10.1016/S0140-6736(06)69041-7

Bakare M, Ebigbo P, Ubochi V (2012) Prevalence of autism spectrum disorder among Nigerian children with intellectual disability: a stopgap assessment. J Health Care Poor Underserved 23:513–518. https://doi.org/10.1353/hpu.2012.0056

Barr HM, Bookstein FL, O'Malley KD et al (2006) Binge drinking during pregnancy as a predictor of psychiatric disorders on the structured clinical interview for DSM-IV in young adult offspring. Am J Psychiatry 163(6):1061–1065. https://doi.org/10.1176/ajp.2006.163.6.1061

Baxter AJ, Brugha TS, Erskine HE et al (2015) The epidemiology and global burden of autism spectrum disorders. Psychol Med 45:601–613. https://doi.org/10.1017/S003329171400172X

Berkes F, Colding J, Folke C (2008) Navigating social-ecological systems: building resilience for complexity and change. Cambridge University Press, Cambridge, UK

Biradar SM, Tarak KC, Kulkarni VH et al (2012) Antiamnesic and antioxidant effect of Acacia catechu-catechin in normal, aged and scopolamine challenged cognitive deficit mice. J Pharmacol and Toxicol 7(5):231–241

Boccardi V, Baroni M, Mangialasche F, Mecocci P (2016) Vitamin E family: role in the pathogenesis and treatment of Alzheimer's disease. Alzheimers Dement (NY) 2(3):182–191. https://doi.org/10.1016/j.trci.2016.08.002

Boti JB, Koukoua G, N'Guessan TY, Casanova J (2007) Chemical variability of Conyza sumatrensis and Microglossa pyrifolia from Côte d'Ivoire. Flavour Fragr J 22:27–31. https://doi.org/10.1002/ffj.1743

Bruneton J (1999) Toxic plants dangerous to humans and animals. Intercept Limited

Brust J (2010) Ethanol and cognition: indirect effects, neurotoxicity and neuroprotection: a review. Int J Environ Res Public Health 7:1540–1557

Bum EN, Taiwe GS, Nkainsa LA et al (2009) Validation of anticonvulsant and sedative activity of six medicinal plants. Epilepsy Behav 14:454–458

Byamukama R, Ganza B, Namukobe J et al (2015) Bioactive compounds in the stem bark of *Albizia coriaria* (Welw. Ex Oliver). Int J Biol Chem Sci 9:1013. https://doi.org/10.4314/ijbcs.v9i2.37

Colebunders R, Hendy A, Mokili JL et al (2016) Nodding syndrome and epilepsy in onchocerciasis endemic regions: comparing preliminary observations from South Sudan and the Democratic Republic of the Congo with data from Uganda. BMC Res Notes 9:182

Cordell GA, Colvard MD (2005) Some thoughts on the future of ethnopharmacology. J Ethnopharmacol 100:5–14

Cragg GM, Newman DJ (2001) Natural product drug discovery in the next millennium. Pharm Biol 39:8–17

Cragg GM, Newman DJ (2002) Drugs from nature: past achievements, future prospects. Adv Phytomed 1:23–37

Cruz MP (2018) Edaravone (Radicava): a novel neuroprotective agent for the treatment of amyotrophic lateral sclerosis. Pharm Ther 43:25

Cummings JL, Morstorf T, Zhong K (2014) Alzheimer's disease drug-development pipeline: few candidates, frequent failures. Alzheimers Res Ther 6:37

Dickey C, Jinwal U, Calcul L et al (2019) Myricanol derivatives and uses thereof for treatment of neurodegenerative diseases. US Patent No. 20180265442

Dike IP, Obembe OO, Adebiyi FE (2012) Ethnobotanical survey for potential anti-malarial plants in South-Western Nigeria. J Ethnopharmacol 144:618–626. https://doi.org/10.1016/j.jep.2012.10.002

Dobetsberger C, Buchbauer G (2011) Actions of essential oils on the central nervous system: an updated review. Flavour Fragr J 26:300–316. https://doi.org/10.1002/ffj.2045
Dorsey ER, Constantinescu R, Thompson JP et al (2007) Projected number of people with Parkinson disease in the most populous nations, 2005 through 2030. Neurology 68:384–386
dos Santos RC, de Melo Filho AA, Chagas EA et al (2015) Chemical composition, antimicrobial and anti-acetylcholinesterase activities of essential oil from *Lantana camara* (Verbenaceae) flowers. J Med Plants Res 9:922–928
Doshi A, Chataway J (2017) Multiple sclerosis, a treatable disease. Clin Med (Northfield Il) 17:530
Dotchin CL, Msuya O, Walker RW (2007) The challenge of Parkinson's disease management in Africa. Age Ageing 36:122–127. https://doi.org/10.1093/ageing/afl172
Duke JA (2019) Handbook of edible weeds. CRC Press, Boca Raton, FL
Elbaz A, Carcaillon L, Kab S, Moisan F (2016) Epidemiology of Parkinson's disease. Rev Neurol (Paris) 172:14–26. https://doi.org/10.1016/j.neurol.2015.09.012
Endal D (2013) Global burden of disease figure show: alcohol grows as risk factor for death and disability. http://www.add-resources.org/
Faggion SA, Cunha AOS, Fachim HA et al (2011) Anticonvulsant profile of the alkaloids (+)-erythravine and (+)-11-α-hydroxy-erythravine isolated from the flowers of *Erythrina mulungu* Mart ex Benth (Leguminosae–Papilionaceae). Epilepsy Behav 20:441–446. https://doi.org/10.1016/j.yebeh.2010.12.037
Fahmy NM, Al-Sayed E, El-Shazly M, Singab AN (2018) Comprehensive review on flavonoids biological activities of Erythrina plant species. Ind Crop Prod 123:500–538. https://doi.org/10.1016/j.indcrop.2018.06.028
Farnsworth NR (1990) The role of ethnopharmacology in drug development. In: Bioactive compounds from plants, Ciba Foundation Symposium 154. Wiley Interscience, New York
Feigin VL, Abajobir AA, Abate KH et al (2017) Global, regional, and national burden of neurological disorders during 1990–2015: a systematic analysis for the Global Burden of Disease Study 2015. Lancet Neurol 16:877–897
Ferrari AJ, Somerville AJ, Baxter AJ et al (2013) Global variation in the prevalence and incidence of major depressive disorder: a systematic review of the epidemiological literature. Psychol Med 43:471–481
Findlay GS, Wick MJ, Mascia MP et al (2002) Transgenic expression of a mutant glycine receptor decreases alcohol sensitivity of mice. J Pharmacol Exp Ther 300(2):526–534
Ganza B (2014) Isolation and characterization of the bioactive compounds in the stem bark of Albizia Coriaria. http://hdl.handle.net/10570/3351
GBD 2016 Multiple Sclerosis Collaborators (2019) Global, regional, and national burden of multiple sclerosis 1990–2016: a systematic analysis for the Global Burden of Disease Study 2016. Lancet Neurol 18:269–285. https://doi.org/10.1016/S1474-4422(18)30443-5
Ghribia L, Ghouilaa H, Omrib A et al (2014) Antioxidant and anti–acetylcholinesterase activities of extracts and secondary metabolites from *Acacia cyanophylla*. Asian Pac J Trop Biomed 4: S417–S423
Gomase PV (2012) *Sesbania sesban* Linn: a review on its ethnobotany, phytochemical and pharmacological profile. Asian J Biomed Pharm Sci 2:11
Gureje O, Nortje G, Makanjuola V et al (2015) The role of global traditional and complementary systems of medicine in treating mental health problems. Lancet Psychiatry 2:168–177. https://doi.org/10.1016/S2215-0366(15)00013-9
Gyagenda JO, Ddumba E, Odokonyero R et al (2015) Post-stroke depression among stroke survivors attending two hospitals in Kampala Uganda. Afr Health Sci 15:1220–1231
Haimanot RT (1985) Parkinson's disease in Ethiopia: a prospective study of 70 patients. East Afr Med J 62:571–579
Hamill FA, Apio S, Mubiru NK et al (2000) Traditional herbal drugs of southern Uganda, I. J Ethnopharmacol 70:281–300. https://doi.org/10.1016/S0378-8741(00)00180-X
Harper C (2007) The neurotoxicity of alcohol. Hum Exp Toxicol 26:251–257

He Y, Wang Q, Ye Y et al (2020) The ethnopharmacology, phytochemistry, pharmacology and toxicology of genus Albizia: a review. J Ethnopharmacol 257:112677

Hebbar SS, Harsha VH, Shripathi V, Hegde GR (2004) Ethnomedicine of Dharwad district in Karnataka, India—plants used in oral health care. J Ethnopharmacol 94:261–266

Heinrich M (2010) Ethnopharmacology in the 21st century-grand challenges. Front Pharmacol 1:8

Heinrich M, Teoh HL (2004) Galanthamine from snowdrop—the development of a modern drug against Alzheimer's disease from local Caucasian knowledge. J Ethnopharmacol 92:147–162

Helzner EP, Scarmeas N, Cosentino S et al (2008) Survival in Alzheimer disease. Neurology 71:1489–1495. https://doi.org/10.1212/01.wnl.0000334278.11022.42

Hendrie HC, Ogunniyi A, Hall KS et al (2001) Incidence of dementia and Alzheimer disease in 2 communities: Yoruba residing in Ibadan, Nigeria, and African Americans residing in Indianapolis, Indiana. JAMA 285:739–747. https://doi.org/10.1001/jama.285.6.739

Hirt HM, M'pia B (2008) Natural medicine in the tropics. Marianum Press, Kisubi, Uganda

Hiruma-Lima CA, Calvo TR, Rodrigues CM et al (2006) Antiulcerogenic activity of Alchornea castaneaefolia: effects on somatostatin, gastrin and prostaglandin. J Ethnopharmacol 104:215–224. https://doi.org/10.1016/j.jep.2005.09.007

Husain GM, Mishra D, Singh PN, Rao ChV KV (2007) Ethnopharmacological review of native traditional medicinal plants for brain disorders. Pharmacog Rev 1:20–28

Ilesanmi OB, Akinmoladun AC, Olayeriju OS et al (2017) Modulation of key biochemical markers relevant to stroke by Antiaris africana leaf extract following cerebral ischemia/reperfusion injury. Afr J Tradit Complement Altern Med 14:253–264

Institute of Medicine (2010) Mental, neurological, and substance use disorders in sub-Saharan Africa: reducing the treatment gap, improving quality of care: summary of a joint workshop. The National Academies Press, Washington, DC

Iyengar PJ, Wamala J, Ratto J et al (2014) Prevalence of nodding syndrome—Uganda, 2012–2013. MMWR Morb Mortal Wkly Rep 63:603

Jahng Y, Park JG (2018) Recent studies on cyclic 1, 7-diarylheptanoids: their isolation, structures, biological activities, and chemical synthesis. Molecules 23:3107. https://doi.org/10.3390/molecules23123107

Jalsrai A, Grecksch G, Becker A (2010) Evaluation of the effects of Astragalus mongholicus Bunge saponin extract on central nervous system functions. J Ethnopharmacol 131:544–549. https://doi.org/10.1016/j.jep.2010.07.031

Jimbo D, Kimura Y, Taniguchi M et al (2009) Effect of aromatherapy on patients with Alzheimer's disease. Psychogeriatrics 9:173–179

Kaddumukasa M, Kakooza A, Kaddumukasa MN et al (2015) Knowledge and attitudes of Parkinson's disease in rural and urban Mukono District, Uganda: a cross-sectional, community-based study. Park Dis 2015:196150

Kakudidi E, Kirimuhuzya C, Anywar G et al (2016) Medicinal plants used in the management of noncommunicable diseases in Uganda. In: Tsay et al (eds) Medicinal plants—recent advances in research and development, 1st edn. Springer Nature, Singapore

Kamalipour M, Akhoundzadeh SH, Rezazadeh SHA (2008) Herbal medicines in the treatment of depression and anxiety. J Med Plants 7(4):1–7

Kamatenesi MM, Acipa A, Oryem-Origa H (2011) Medicinal plants of Otwal and Ngai sub counties in Oyam District, northern Uganda. J Ethnobiol Ethnomed 7:7. https://doi.org/10.1186/1746-4269-7-7

Kamau JK, Nthiga PM, Mwonjoria JK et al (2016) Anti-inflammatory activity of methanolic leaf extract of *Kigelia africana* (lam.) Benth and stem bark extract of Acacia hockii De wild in mice. J Dev Drugs 5:1–8

Katende A, Ssegawa P, Tengnäs B (1995) Useful trees and shrubs of Uganda. Technical Handbook series 10. Regional Soil conservation Unit/SIDA, Nairobi

Kinyanda E, Woodburn P, Tugumisirize J et al (2011) Poverty, life events and the risk for depression in Uganda. Soc Psychiatry Psychiatr Epidemiol 46:35–44

Kirui SC, Kiprop AK, Kiplagat FK et al (2014) Chemical assessment and antimicrobial activity of solvent extracts from Kalanchoe densiflora. Int J Educ Res 2:1–22

Kokwaro JO (1993) Medicinal plants of East Africa. In: Medicinal plants of East Africa. Kenya Literature Bureau, Nairobi, Kenya

Lagnika L, Tchatchedre M, Amoussa AMO et al (2016) Phytochemical assesment, in vitro antibacterial and antioxidant activities Acacia hockii De wild. Biomedicine 3:1–8

Lim SS, Vos T, Flaxman AD et al (2012) A comparative risk assessment of burden of disease and injury attributable to 67 risk factors and risk factor clusters in 21 regions, 1990–2010: a systematic analysis for the Global Burden of Disease Study 2010. Lancet 380:2224–2260. https://doi.org/10.1016/S0140-6736(12)61766-8

Lombo J (1979) The healing powers of herbs, with special reference to obstetrics and gynaecology. In: Sofowora EA (ed) African medicinal plants. University of Ife Press, Ife, Nigeria, pp 23–31

Lu B, Wu X, Tie X et al (2005) Toxicology and safety of anti-oxidant of bamboo leaves. Part 1: acute and subchronic toxicity studies on anti-oxidant of bamboo leaves. Food Chem Toxicol 43:783–792

Lulekal E, Kelbessa E, Bekele T, Yinegar H (2008) An ethnobotanical study of medicinal plants in Mana Angetu District, southern Ethiopia. J Ethnobiol Ethnomed 4. https://doi.org/10.1186/1746-4269-4-10

Mante PK, Adongo DW, Woode E (2017) Anticonvulsant effects of antiaris toxicaria aqueous extract: investigation using animal models of temporal lobe epilepsy. BMC Res Notes 10:1–9

Mathers CD, Loncar D (2005) Updated projections of global mortality and burden of disease, 2002–2030: data sources, methods and results. Geneva World Heal Organ

Matson JL, Boisjoli JA (2008) Strategies for assessing Asperger's syndrome: a critical review of data based methods. Res Autism Spectr Disord 2:237–248

Mbwayo AW, Ndetei DM, Mutiso V, Khasakhala LI (2013) Traditional healers and provision of mental health services in cosmopolitan informal settlements in Nairobi, Kenya. Afr J Psychiatry 16:134–140

Medina JH, Paladini AC, Wolfman C et al (1990) Chrysin (5, 7-di-OH-flavone), a naturally-occurring ligand for benzodiazepine receptors, with anticonvulsant properties. Biochem Pharmacol 40:2227–2231

Mergia E, Terefe G, Teklehaymanot T, Shibeshi W (2015) Phytochemical screening and *in vitro* antitrypanosomal activity of aqueous and methanol leaf extract of *Clutia* abyssinica (Euphorbiaceae) against *Trypanosoma congolense*. Pharmacologia 6(3):79–87. https://doi.org/10.5567/pharmacologia.2015.79.87

Mubangizi V, Maling S, Obua C, Tsai AC (2020) Prevalence and correlates of Alzheimer's disease and related dementias in rural Uganda: cross-sectional, population-based study. BMC Geriatr 20:48

Mujovo SF, Hussein AA, Meyer JJM et al (2008) Bioactive compounds from *Lippia javanica* and *Hoslundia opposita*. Nat Prod Res 22:1047–1054

Munube D, Katabira E, Ndeezi G et al (2016) Prevalence of stroke in children admitted with sickle cell anaemia to Mulago hospital. BMC Neurol 16:175

Murray CJL, Lopez AD, Black R et al (2007) Global burden of disease 2005: call for collaborators. Lancet (London, England) 370:109

Nagaraja TS, Mohamood R, Krishna V et al (2012) Anticonvulsant activity of *Erythrina mysorensis* bark extract in an animal model of epilepsy. J Pharmacol Pharmacother 3:62

Naik N, Tare HL, Sherikar AK et al (2011) Central nervous system stimulant effect of extracts obtained from the barks of *Sesbania sesban*. Int J Inst Pharm Life Sci 1:77–92

Nakibuuka J, Sajatovic M, Nankabirwa J et al (2015) Stroke-risk factors differ between rural and urban communities: population survey in Central Uganda. Neuroepidemiology 44:156–165

Namuli JD (2015) A case of Alzheimer's dementia in Uganda. In: Musisi S, Jacobson S (eds) Brain degeneration and dementia in sub-Saharan Africa. Springer, New York, pp 247–254

Ndetei DM (2007) Traditional healers in East Africa. Int Psychiatry 4:85–86

Nguelefack TB, Nana P, Atsamo AD et al (2006) Analgesic and anticonvulsant effects of extracts from the leaves of *Kalanchoe crenata* (Andrews) Haworth (Crassulaceae). J Ethnopharmacol 106:70–75
Nicholson JK, Connelly J, Lindon JC, Holmes E (2002) Metabonomics: a platform for studying drug toxicity and gene function. Nat Rev Drug Discov 1:153–161
Ningsangrenla L, Rao PSS (2019) Traditional healing practices and perspectives of mental health in Nagaland. J North East India Stud 9:33–56
Njoku VO, Obi C (2009) Phytochemical constituents of some selected medicinal plants. African J Pure Appl Chem 3:228–233
Nongdam P, Tikendra L (2014) The nutritional facts of bamboo shoots and their usage as important traditional foods of Northeast India. Int Sch Res Notices 2014:679073
Nsereko JR, Kizza D, Kigozi F et al (2011) Stakeholder's perceptions of help-seeking behaviour among people with mental health problems in Uganda. Int J Ment Health Syst 5:5
Obeso JA, Rodriguez-Oroz MC, Goetz CG et al (2010) Missing pieces in the Parkinson's disease puzzle. Nat Med 16:653–661
OECD (2002) Test No. 423: Acute oral toxicity-acute toxic class method. OECD Publishing
Okello J, Ssegawa P (2007) Plants used by communities of Ngai sub county, Apac District, Northern Uganda. Afr J Ecol 45:76–83
Okigbo RN, Mmeka EC (2006) An appraisal of phytomedicine in Africa. Curr Appl Sci Technol 6:83–94
Okomolo FCL, Mbafor JT, Bum EN et al (2011) Evaluation of the sedative and anticonvulsant properties of three Cameroonian plants. Afr J Tradit Complement Altern Med 8:181–190
Omoruyi BE, Bradley G, Afolayan AJ (2012) Ethnomedicinal survey of medicinal plants used for the management of HIV/AIDS infection among local communities of Nkonkobe municipality, eastern Cape, South Africa. J Med Plants Res 6:3603–3608
Oryema C, Bukenya-Ziraba R, Omagor N, Opio A (2010) Medicinal plants of Erute county, lira district, Uganda with particular reference to their conservation. Afr J Ecol 48:285–298
Ovuga E, Boardman J, Oluka EGAO (1999) Traditional healers and mental illness in Uganda. Psychiatr Bull 23:276–279
Owuor BO, Mulemi BA, Kokwaro JO (2005) Indigenous snake bite remedies of the Luo of western Kenya. J Ethnobiol 25:129–141
Pahuja M, Mehla J, Gupta YK (2012) Anticonvulsant and antioxidative activity of hydroalcoholic extract of tuber of Orchis mascula in pentylenetetrazole and maximal electroshock induced seizures in rats. J Ethnopharmacol 142(1):23–27
Parihar MS, Hemnani T (2004) Alzheimer's disease pathogenesis and therapeutic interventions. J Clin Neurosci 11:456–467
Park D-H, Borlongan CV, Willing AE et al (2009) Human umbilical cord blood cell grafts for brain ischemia. Cell Transplant 18:985–998
Patel V (2011) Traditional healers for mental health care in Africa. Glob Health Action 4:7956
Poorgholam P, Yaghmaei P, Hajebrahimi Z (2018) Thymoquinone recovers learning function in a rat model of Alzheimer's disease. Avicenna J Phytomed 8:188
Prince M (2000) Methodological issues for population-based research into dementia in developing countries. A position paper from the 10/66 dementia research group. Int J Geriatr Psychiatry 15:21–30
Rahman M (2016) In-vitro evaluation of cholinesterase inhibitory and antioxidant activities of tamarindus indica for the treatment of Alzheimer's disease. A dissertation submitted to the Department of Pharmacy, East West University in the partial fulfillment for the degree of Bachelor of Pharmacy
Rahman TU, Zeb MA, Liaqat W et al (2018) Phytochemistry and pharmacology of genus indigofera: a review. Rec Nat Prod 12:1–13
Ratheesh G, Tian L, Venugopal JR et al (2017) Role of medicinal plants in neurodegenerative diseases. Biomanufacturing Rev 2:2

Rhiannon G-C, Adeloye D, Chan KY et al (2012) An estimate of the prevalence of dementia in Africa: a systematic analysis. J Glob Health 2
Rizk MZ, Fouad GI, Aly HF (2018) Neurological disorders: causes and treatments strategies. Int J Public Mental Health Neurosci 5(1):1–10
Roberts B, Ocaka KF, Browne J et al (2011) Alcohol disorder amongst forcibly displaced persons in northern Uganda. Addict Behav 36:870–873
Rochat TJ, Tomlinson M, Bärnighausen T et al (2011) The prevalence and clinical presentation of antenatal depression in rural South Africa. J Affect Disord 135:362–373
Sahoo N, Manchikanti P, Dey SH (2011) Herbal drug patenting in India: IP potential. J Ethnopharmacol 137:289–297. https://doi.org/10.1016/j.jep.2011.05.022
Saki K, Bahmani M, Rafieian-Kopaei M et al (2014) The most common native medicinal plants used for psychiatric and neurological disorders in Urmia city, northwest of Iran. Asian Pacific J Trop Dis 4(Suppl):S895–S901. https://doi.org/10.1016/S2222-1808(14)60754-4
Sasidharan S, Chen Y, Saravanan D et al (2011) Extraction, isolation and characterization of bioactive compounds from plants' extracts. Afr J Tradit Complement Altern Med 8(1):1–10
Schneider M, Norman R, Parry C et al (2007) Estimating the burden of disease attributable to alcohol use in South Africa in 2000. South African Med J 97:664–672
Seif Eldin A, Habib D, Noufal A et al (2008) Use of M-CHAT for a multinational screening of young children with autism in the Arab countries. Int Rev Psychiatry 20:281–289. https://doi.org/10.1080/09540260801990324
Sekhar JC, Sandhya S, Vinod KR et al (2012) Plant toxins-useful and harmful effects. Hygeia-J Drugs Med 4:79–90
Shah H, Khan AA (2017) Phytochemical characterisation of an important medicinal plant, Chenopodium ambrosioides Linn. Nat Prod Res 31:2321–2324
Silberberg D, Katabira E (2006) Neurological disorders. In: Jamison DT (ed) Disease and mortality in sub-Saharan Africa, 2nd edn. The International Bank for Reconstruction and Development/ The World Bank, Washington, DC
Singh A, Trevick S (2016) The epidemiology of global epilepsy. Neurol Clin 34:837–847
Sofowora A (1996) Medicinal plants and traditional medicine in Africa. Karthala 2:365
Sonibare MA, Ayoola IO (2015) Medicinal plants used in the treatment of neurodegenerative disorders in some parts of Southwest Nigeria. African J Pharm Pharmacol 9:956–965
Sosa-Ortiz AL, Acosta-Castillo I, Prince MJ (2012) Epidemiology of dementias and Alzheimer's disease. Arch Med Res 43:600–608
de Sousa DP, Gonçalves JCR, Quintans-Júnior L et al (2006) Study of anticonvulsant effect of citronellol, a monoterpene alcohol, in rodents. Neurosci Lett 401:231–235
de Sousa DP, Quintans L Jr, de Almeida RN (2007) Evolution of the anticonvulsant activity of α-terpineol. Pharm Biol 45:69–70
Salih MAM, Mustafa AA (2008) A substance in broad beans (Vicia faba) is protective against experimentally induced convulsions in mice. Epilepsy Behav 12(1):25–29
Ssegawa P, Kasenene JM (2007) Medicinal plant diversity and uses in the Sango bay area, southern Uganda. J Ethnopharmacol 113:521–540. https://doi.org/10.1016/j.jep.2007.07.014
Stafford GI, Jäger AK, van Staden J (2005) Activity of traditional south African sedative and potentially CNS-acting plants in the GABA-benzodiazepine receptor assay. J Ethnopharmacol 100:210–215. https://doi.org/10.1016/j.jep.2005.04.004
Stark TD, Mtui DJ, Balemba OB (2013) Ethnopharmacological survey of plants used in the traditional treatment of gastrointestinal pain, inflammation and diarrhea in Africa: future perspectives for integration into modern medicine. Animals 3:158–227
Tabuti JRSR, Dhillion SS, Lye KA et al (2003) Traditional herbal drugs of Bulamogi, Uganda: plants, use and administration. J Ethnopharmacol 85:19–44. https://doi.org/10.1016/s0378-8741(03)00161-2
Taiwe GS, Kuete V (2014) Neurotoxicity and neuroprotective effects of African medicinal plants. Toxicol Surv African Med Plants 2014:413–444. https://doi.org/10.1016/B978-0-12-800018-2.00014-5

Talla E, Yaya Gbaweng J, Mokale L et al (2016) Chemical constituents from *Erythrina droogmansiana* (Fabaceae), radical scavenging and antibacterial potential of some extracts and compounds. Nat Prod An Indian J 12:12–20

Tawfik MK (2011) Coenzyme Q10 enhances the anticonvulsant effect of phenytoin in pilocarpine-induced seizures in rats and ameliorates phenytoin-induced cognitive impairment and oxidative stress. Epilepsy Behav 22:671–677

Tengku N, Raja S (2013) Inhibition of acetylcholinesterase and NADH oxidase by *Acacia farnesiana*. Panacea J Pharm Pharm Sci 1:15–17

Thangavelu L, Ramasamy R (2015) *In vitro* acetyl cholinesterase inhibitory assay of *Acacia catechu* willd ethanolic seed extract. Pharm J 7(5):280–282. https://doi.org/10.5530/pj.2015.5.5

Tomlinson M, Grimsrud AT, Stein DJ et al (2009) The epidemiology of major depression in South Africa: results from the south African stress and health study. S Afr Med J 99:367–373

Tugume P, Kakudidi EK, Buyinza M et al (2016) Ethnobotanical survey of medicinal plant species used by communities around Mabira central Forest reserve, Uganda. J Ethnobiol Ethnomed 12:1–28. https://doi.org/10.1186/s13002-015-0077-4

Uduak W, Bandmann O, Walker R (2018) Parkinson's disease in sub-Saharan Africa: a review of epidemiology, genetics and access to care. J Mov Disord 11:53–64. https://doi.org/10.14802/jmd.17028

Üstün TB, Ayuso-Mateos JL, Chatterji S et al (2004) Global burden of depressive disorders in the year 2000. Br J Psychiatry 184:386–392. https://doi.org/10.1192/bjp.184.5.386

Watkins PB, Zimmerman HJ, Knapp MJ et al (1994) Hepatotoxic effects of tacrine administration in patients with Alzheimer's disease. JAMA 271:992–998

Whiteford HA, Degenhardt L, Rehm J et al (2013) Global burden of disease attributable to mental and substance use disorders: findings from the global burden of disease study 2010. Lancet 382:1575–1586. https://doi.org/10.1016/S0140-6736(13)61611-6

WHO (1999) WHO monographs on selected medicinal plants (Vol. 1). Geneva

WHO (2007) Neurological disorders affect millions globally: WHO report. https://www.who.int/mediacentre/news/releases/2007/pr04/en/. Accessed 24 Mar 2020

WHO (2014) Management of substance abuse unit. Global status report on alcohol and health, 2014. World Heal Organ

WHO (2016) What are neurological disorders? In: Online Q&A. https://www.who.int/features/qa/55/en/. Accessed 24 Mar 2010

WHO (2017) Health statistics and information systems. http://www.who.int/healthinfo/global_burden_disease/estimates/en/index2.html%0A

Williams JT, Ahmad Z (1999) Priorities for medicinal plants research and development in Pakistan. Medicinal and Aromatic Plants Program in Asia (MAPPA). IDRC, SARO, New Delhi, India

Ya'u J, Yaro AH, Abubakar MS et al (2008) Anticonvulsant activity of *Carissa edulis* (Vahl) (Apocynaceae) root bark extract. J Ethnopharmacol 120:255–258. https://doi.org/10.1016/j.jep.2008.08.029

Ya'u J, Yaro AH, Malami S et al (2015) Anticonvulsant activity of aqueous fraction of *Carissa edulis* root bark. Pharm Biol 53:1329–1338. https://doi.org/10.3109/13880209.2014.981280

# Ayurvedic Ideology on *Rasapanchak*-Based Cognitive Drug Intervention

**Anagha Ranade, Mayur Surana, Shivani V. Dhokne, Sudesh Gaidhani, and Sharad D. Pawar**

## Contents



A. Ranade · S. D. Pawar (✉)
Regional Ayurveda Institute for Fundamental Research, Pune, Maharashtra, India
e-mail: rophraifr.pu-ayush@gov.in

M. Surana
Vardhayu Ayurved, Nashik, Maharashtra, India

S. V. Dhokne
Dr. D.Y. Patil Institute of Pharmaceutical Sciences and Research, Pimpri, Pune, Maharashtra, India

S. Gaidhani
Central Council for Research in Ayurvedic Sciences (Ministry of AYUSH, Govt. of India), New Delhi, India

 
D. C. Agrawal, M. Dhanasekaran (eds.), *Medicinal Herbs and Fungi*,
https://doi.org/10.1007/978-981-33-4141-8_18

**Abstract** Ayurveda uniquely describes health as a state of mental and spiritual well-being apart from physical well-being. It has imparted an etiology of mental origin to various diseases. The basic memory functions are regarded to be a combination of the power of acquisition (*Grahana*), retention (*Dharana*), and recollection (*Smarana*).

The fundamental doctrine of "*Tridosha*" states that *Vata*, *Pitta*, and *Kapha* are the three (energy) or biological rhythms regulating the physical and mental functioning of the human body. *Pitta* is responsible for understanding and grasping knowledge due to its *Satva*, *Aashukari*, and *Tikshna* property. *Vata* is responsible for the association of ideas. *Kapha* provides stability, which is necessary for the retention of memory.

The seers have stated that a drug exhibits its action either by its *Rasa*, *Guna*, *Veerya*, *Vipaka*, and *Prabhava*. This pentad ultimately influences the alteration in the status of *Dosha*. Similarly, *Satva* corresponding to *Pitta*, *Raja* corresponding to *Vata*, and *Tama* corresponding to *Kapha* are stated to regulate the functions of the brain and intellect in humans. It is the proper combination of the pentad above in medicinal plant drugs that aids in altering the deranged mechanisms of cognitive functions as per Ayurveda. *Sheeta veerya*, along with *madhura rasa* and *madhura vipaka*, improves retention power and *katu* rasa, *katu vipaka*, and *ushna veerya* drugs enhance the ability to acquire new information and data interpretation. The concept of *Veerya* is close to the "Yin and Yang" theory of Chinese medicine. Many medicinal plants viz. *Guduchi*, *Yashtimadhu*, *Brahmi*, and so on have been entitled as *medhya rasayana*, i.e., these are bestowed with enhanced properties to correct the cognitive dysfunction. Apart from these, various other drugs based on the pentad principle mentioned above have been said to participate in augmenting functions of different levels of brain and intellect.

Thus, it is the need of the hour to understand the rationale of Ayurveda's ideology behind the management of cognitive dysfunctions. It will undoubtedly help researchers to understand and find leads in various disorders of the nervous system. The chapter will explore the rationale of Ayurveda in the management of cognitive disorders.

**Keywords** Ayurveda · Cognition · *Medhya rasayana* · Memory · *Rasapanchak*

## Abbreviations

| | |
|---|---|
| Ach | Acetylcholine |
| AChE) | Acetylcholinesterase |
| AD | Alzheimer's disease |
| AS | *Argyreia speciosa* |
| BDNF | Brain-derived neurotrophic factor |
| BM | *Bacopa monnieri* |
| CAT | Catalase |
| DAD | Diallyl disulfide |
| DAT | Diallyl trisulfide |
| EPM | Elevated plus-maze |
| GABA | Gamma aminobutyric acid |
| GSH | Glutathione |
| LIF | Leukaemia inhibitory factor |
| MAO-A | Monoamine oxidase A |
| NO | Nitric oxide |
| PDE | Phosphodiesterase |
| PTU | Propylthiouracil |
| RAM | Radial arm maze |
| ROS | Reactive oxygen species |
| SOD | Superoxide dismutase |
| STZ | Streptozotocin |
| TGF-β3 | Transforming growth factor-beta 3 |
| TH-IR | Tyrosine hydroxylase immunoreactive |
| TNF-α | Tumour necrosis factor |

## 1 Introduction

Cognition is a collective term used for gaining knowledge and comprehension. The cognitive processes are of varied types ranging from simple tasks such as sensation, perception, and attention to complex tasks like language, learning, memory, thoughts, and decision making (Smith and Kelly 2015).

Ayurveda uniquely describes health as a state of mental and spiritual wellbeing apart from physical wellbeing. It has imparted an etiology of mental origin to various diseases. Memory, i.e., *Smriti*, is an important factor in the case of the perception of knowledge. The memory functions are described under "*Medha*" in Ayurveda. *Medha* is defined as the capacity to retain and comprehend information. *Medha* is further classified into *Dharana shakti*, i.e., power of retention, *Grahana shakti*, i.e., power of grasping, *Smarana shakti*, i.e., the power of recollection and *Vivek Shakti*, i.e., power of discrimination (Ray and Ray 2015). Thus, the term "*Medhya*" is the property that helps to enhance these functions. Ayurveda entails various herbs that possess this "*medhya*" property.

Ayurvedic ideology on "*Raspanchaka*"-based cognitive drug intervention has been illustrated in Fig. 1. In Ayurveda, the pharmacological effect of *Ausadha dravya* (drug) has been explained in *Dravyaguna* based on the theory of *Rasapanchaka*. It comprises of *Rasa* (taste), *Vipaka* (post digestion effect), *Guna* (quality), *Veerya* (potency), and *Prabhava* (special effect). All of them form an integral part of Ayurvedic pharmacodynamics and pharmacokinetics. The seers have stated that any drug exhibits its action either by its *Rasa, Guna, Vipaka, Veerya, and Prabhava* (Sastry and Chaturvedi 1983). *Veerya* has been described as an entity that possesses higher potency than Rasa, Vipaka, in establishing the drug action. *Charak* abides by the term "*Shakti*," which is the potency/energy with which the drug initiates its action and is consistently present in the body. Though herbs have been used extensively as memory boosters, the proper usage of them depends on their *rasapanchak* and *veerya* in particular.

A thorough review of medicinal plants indicated to possess *medhya* property was done from an important lexicon named *Bhavprakash Nighantu* (Table 1). Also, a comprehensive literature search was carried out using different search engines to compile the researchers reported on the cognitive effects of functions of these plants mentioned in the above said *Nighantu*. Out of them, in vitro and in vivo pharmacological activities of 24 medicinal plant-based drugs and gold have been reported in Table 2 given below.

## 2 Reported Cognition-Related Activities of Medicinal Plants Entitled as *Medhya* in *Bhavprakash Nighantu*

### *2.1* Bacopa monnierri (Brahmi)

*Brahmi* suppresses neuronal oxidative stress and inhibits the acetylcholinesterase activity by ameliorating neurodegeneration. Several studies were carried out on *brahmi* extracts and its active components. Bacoside A, Baccoside B, Bacosaponins, Betullinic acid are known to have a significant role in neuroprotection. Several studies have also evidenced the antioxidant property of *Brahmi* (*Bacopa monnieri*) (BM), widely known for its nootropic activity to treat epilepsy and for memory enhancement and improvement of brain function. The hypothesis that BM extract could offset neurotoxicant-induced oxidative dysfunctions in the developing brain in a rotenone mouse model was tested (Shinomol et al. 2012). The ethanolic extract of BM leaves (BME) was used to treat, and its effect was measured by assessment of lipid peroxidation, ROS generation, Hydroperoxide levels, protein carbonyls, GSH, SOD, CAT, and activities of cholinergic enzymes. Several shreds of evidence support this significant offset given by BME, viz. the normalized levels of oxidative markers (malondialdehyde, ROS levels, and hydroperoxides) and restoration of

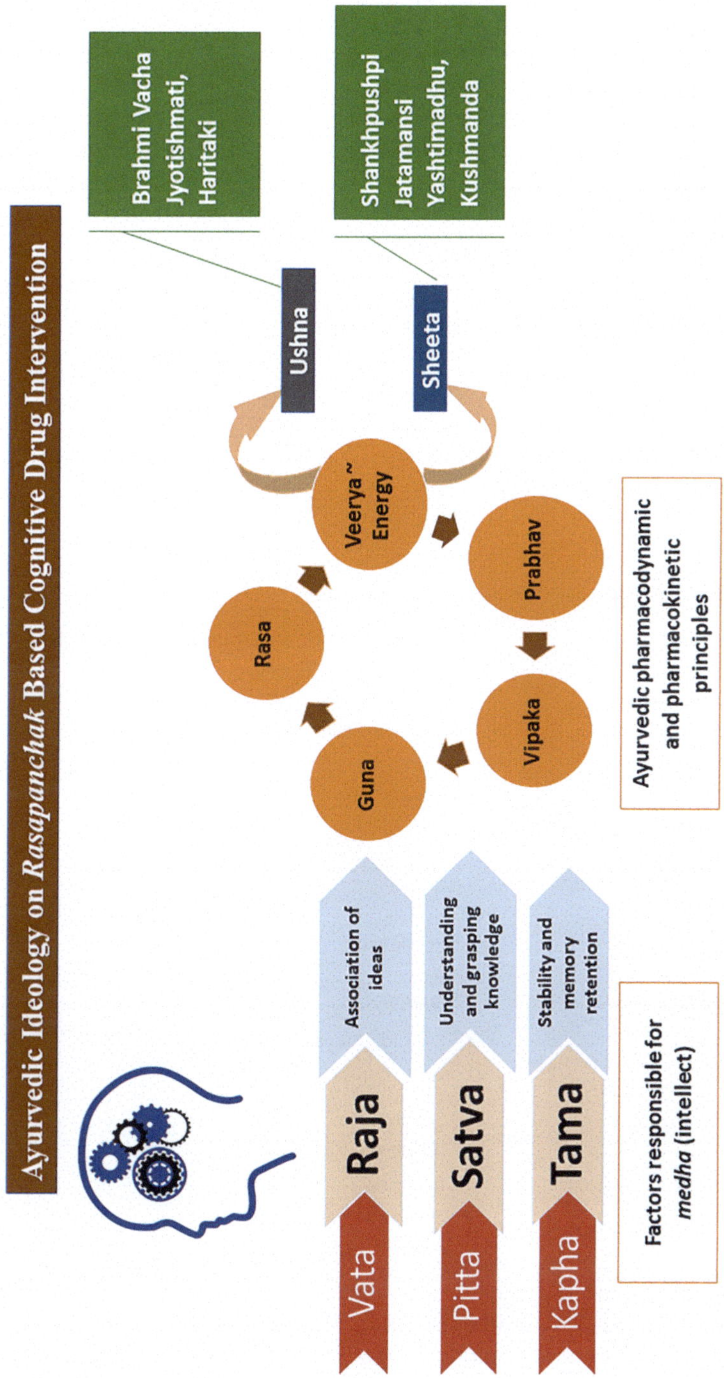


**Fig. 1** Ayurvedic ideology on "Raspanchak"-based cognitive drug intervention

**Table 1** List of plants in Bhavprakash Nighantu as "*medhya*"

| Sl. No | Botanical name of the plant and Family | Vernacular name | Plant parts | Ayurvedic indications |
|---|---|---|---|---|
| 1. | *Bacopa monnierri* | *Brahmi* | Whole | *Medhya* |
| 2. | *Centella asiatica* | *Mandukaparni* | Whole | *Medhya* |
| 3. | *Glycyrrhiza glabra* | *Yashtimadhu* | Rhizome/ stolon | *Medhya* |
| 4. | *Convolvulus pluricaulis* | *Shankhapushpi* | Whole | *Medhya* |
| 5. | *Terminalia chebula* | *Haritaki* | Fruit/ pericarp | *Medhya* |
| 6. | *Tinospora cordifolia* | *Guduchi* | Stem | *Medhya* |
| 7. | *Celastrus paniculatus* | *Jyotishmati* | Seed oil | *Medhya* |
| 8. | *Acorus calamus* | *Vacha* | Rhizome | *Medhya* |
| 9. | *Nardostachys jatamansi* | *Jatamansi* | Rhizome | *Medhya* |
| 10. | *Benincasa hispida* | *Kushmanda* | Fruit | *Medhya* |
| 11. | *Piper longum* | *Pippali* | Fruit | *Medhya* |
| 12. | *Cuminum cyminum* | *Jeera* | Fruit | *Medhya* |
| 13. | *Carum bulbocastrum* | *Kala Jeera* | Fruit | *Medhya* |
| 14. | *Nigella sativa* | *Upakunchika* | Fruit | *Medhya* |
| 15. | *Allium sativum* | *Lahsun* | Bulb | *Medha hito* |
| 16. | *Semicarpus anacardium* | *Bibba* | Ripe fruit | *Medhya* |
| 17. | *Nardocstachys jatamansi* | *Jatamansi* | Rhizome | *Medhya* |
| 18. | *Taxus baccata* | *Sthauneyaka* | Leaves | *Medhakar* |
| 19. | *Clitorea ternatea* | *Aparajita, vishnukranta* | Whole plant | *Medhya, smruti, buddhi* |
| 20. | *Asparagus racemosus* | *Shatavari* | Root | *Medhagni pushtida* |
| 21. | *Argyria nervosa* | *Vriddhadaru* | Root | *Medhakara* |
| 22. | *Sphaeranthus indicus* | *Mundi* | Whole plant | *Medhya* |
| 23. | *Convolvulus pluricaulis* | *Shankhpushpi* | Whole plant | *Medhya, manas rogahar, smrutida* |
| 24. | *Pupalia lappacea* | *Nagadamani* | Whole plant | *Matiprada* |
| 25. | *Punicum granatum* | *Svadu dadima* | Beeja | *Medha* |
| 26. | *Gold* | – | – | *Medha, smruti, mati* |
| 27. | *Rice variety kedara*) (rice grown in wetlands) | – | Beeja | *Medha* |
| 28. | *Hordeum vulgare* | *Yava/Barley* | Seed | *Medha vardhaka* |
| 29. | *Corchorus capsularis* | *Nadishak* | Leaves | *Medhya* |
| 30. | *Corchorus fascicularis* | *Chanchuki* | Leaves | *Medhya* |

depleted GSH levels, standardized protein carbonyl content, restored activity levels of cytosolic antioxidant enzymes, neurotransmitter function, and dopamine levels. Thus, it suggested that BM can be used as a prophylactic agent for neurodegenerative disorders. Ahirwar and Tembhre (2016) studied assessment of acetylcholinesterase inhibition by *Bacopa monneiri* and acephate in the hippocampus of chick

**Table 2** Drugs from *Bhavprakash* whose role in cognition deficits have been reported

| Sl. No | Botanical name of the plant and family | Sanskrit name | *Rasa* | *Veerya* |
|---|---|---|---|---|
| 1. | *Bacopa monnierri* | *Brahmi/Aindri* | *Tikta* | *Ushna* |
| 2. | *Acorus calamus* | *Vacha* | *Katu* | *Ushna* |
| 3. | *Argyria nervosa* | *Vriddhadaru* | *Katu Kashaya* | *Ushna* |
| 4. | *Benincasa hispida* | *Kushmanda* | *Madhura* | *Sheeta* |
| 5. | *Boerhavia diffusa* | *Punarnava* | *Madhura, Katu* | *Ushna* |
| 6. | *Celastrus paniculatus* | *Jyotishmati* | *Katu* | *Ushna* |
| 7. | *Centella asiatica* | *Mandukaparni* | *Tikta* | *Sheeta* |
| 8. | *Convolvulus pluricaulis* | *Shankhapushpi* | *Tikta* | *Sheeta* |
| 9. | *Desmodium gangeticum* | *Shalaparni* | *Madhura* | *Ushna* |
| 10. | *Glycyrrhiza glabra* | *Yashtimadhu* | *Madhura* | *Sheeta* |
| 11. | *Nardocstachys jatamansi* | *Jatamansi* | *Tikta* | *Sheeta* |
| 12. | *Hedychium spicatum* | *Shati* | *Katu* | *Ushna* |
| 13. | *Terminalia chebula* | *Haritaki* | *Kashaya* | *Ushna* |
| 14. | *Allium sativum* | *Rasona* | *Katu* | *Ushna* |
| 15. | *Clitorea ternatea* | *Aparajita,* | *Katu* | *Sheeta* |
| 16. | *Asparagus racemosus* | *Shatavari* | *Tikta Madhura* | *Sheeta* |
| 17. | *Nigella sativa Linn.* | *Upakunchika* | *Katu* | *Ushna* |
| 18. | *Cuminum cyminum* | *Jeera* | *Katu* | *Ushna* |
| 19. | *Hordeum vulgare* | *Yava/Barley* | *Kashaya* | *Ushna* |
| 20. | *Punicum granatum* | *Svadu dadima* | *Madhura* | *Ushna* |
| 21. | *Piper longum,* | *Pippali* | *Katu* | *Ushna* |
| 22. | *Tinospora cordifolia* | *Guduchi* | *Tikta* | *Ushna* |
| 23. | *Sphaeranthus indicus* | *Mundi* | *Madhur* | *Ushna* |
| 24. | *Semecarpus anacardium* | *Bhallataka* | *Katu* | *Ushna* |

brain for the impediment of Alzheimer's disease. They found out that the ethanolic extract of BM leaves and twigs inhibits acetylcholinesterase (AChE), which catalyzes and breaks down the ACh. This study concludes that BM was a competitive and non-competitive inhibitor of AChE. Thus it helps to improve the ACh level. *Brahmi* has been reported to have a neuroprotective effect against β-amyloid protein and glutamate-induced neurotoxicity in primary cortical cultured neurons. In Alzheimer's disease, the neuronal dysfunction due to amyloid-β and tau protein accumulation is responsible for the disease progression (Dubey and Chinnathambi 2019). Overproduction of reactive oxygen species increased neuroinflammation, and proteinopathy leads to disease-associated behavioral deficits and dementia. This effect was determined by measuring the neuronal cell viability. The study exhibited the GSH and lipid peroxidation inhibition activity (Limpeanchob et al. 2008).

Flies (*Drosophila melanogaster*) are nowadays widely used as a model for neurodegenerative diseases. In one study, BM inhibited the rotenone-induced toxicity in *Drosophila*. BM significantly reduces the level of endogenous oxidative markers like protein carbonyl content, malondialdehyde, and hydrogen peroxide. This approach confirms the utility of the *Drosophila* model for neuroprotection

(Hosamani and Muralidhar 2009). The whole plant extract of BM has been shown to inhibit the aluminium-induced oxidative stress in the hippocampus of the rat brain (Jyoti and Sharma 2006). BM has a positive effect on ischemia-induced brain injury. Neurobehavioral parameters like rotarod, hanging wire, pole fall, elevated plus maze, passive avoidance test, and biochemical estimations like nitric oxide, SOD, CAT provide evidence to this. *Brahmi* is mainly used in the Alzheimer's model of rats and mice, as well as of amnesia, where the bacosides play a vital role in synaptic function.

## *2.2* A. calamus (Vacha)

*A. calamus* was traditionally used for the improvement of memory and cognition in old age (Pandey 2016). *A. calamus* has shown a potential role in Alzheimer's disease. Cyclic nucleotides play a significant role in cellular function in neurodegenerative diseases like Alzheimer's/dementia. The bioactive compound of *A. Calamus* is found to be phosphodiesterases (PDE) inhibitor (Kumar et al. 2015). It helps in the regulation of cyclic nucleotide levels. The cyclic nucleotide is responsible for synaptic plasticity, neurotransmitter regulation, and transcriptional regulation of various neurotrophic factors. Extract of *A. calamus* attenuates cognitive function by enhancing anticholinesterase activity (Oh et al. 2004).

## *2.3* Argyreia speciosa (Vriddhadaru)

*Argyreia speciosa*, commonly known as "Elephant creeper" and *Vriddhadaru* is well known for brain tonic, nootropic, antianxiety, and anticonvulsant activity. The ethanolic extract of the root of *A. speciosa* (AS) was studied against global cerebral ischemic reperfusion injury in Wistar rat. Where due to comprehensive cerebral ischemic reperfusion injury, cerebral infarction, and brain edema increases, and it results in a neurological deficit, sensorimotor dysfunction, and decreased neurobehavioral function. *A. speciosa* markedly attenuates all the changes caused by global cerebral ischemic reperfusion injury to the normal state (Jaiswal and Tailang 2018). The mechanism behind this neuroprotective behavior may be due to its antioxidant and anti-inflammatory properties. Aqueous extract of AS has been evaluated in mice for aging, scopolamine, and diazepam-induced memory deficits (Hanumanthachar et al. 2007). To assess the short-term and long-term memory, elevated plus maze and passive avoidance tests were used. Due to possible mechanisms, it elicits the anti-amnesic effect, where amnesia was induced by diazepam, scopolamine, and aging. It also inhibits the acetylcholinesterase enzyme to elevate the memory and behavioral function (Joshi and Parle 2006).

## *2.4* Benincasa hispida (Kushmanda)

In the Ayurvedic system of medicine, *Benincasa hispida* (BH) is the main ingredient of *kushmanda lehyam*, which is widely used in nervous disorders. The species belongs to the Cucurbitaceae family and is commonly known as ash guard. Triterpenes, sterols, and glycosides have been isolated from *B. hispida*. In one study, the fruit pulp extract of *B. hispida* was used to evaluate the cognition facilitating effect in colchicine-induced experimental rat model of Alzheimer's disease (AD) (Roy and Mazumdar 2016). A behavioral study performed using a shuttle box with two identical compartments, separated with hurdle by which retention of the acquired active avoidance response was assessed. Acetylcholine (ACh), choline acetyltransferase (ChAT) levels, and antioxidants (SOD, CAT, GSH) levels were estimated. *Kushmanda* has been found to reverse the colchicine-induced cognitive deficits significantly. The aqueous pulp extract of *Kushmanda* provides antioxidant protection against the colchicine-induced oxidative stress. It is an anti-depressant, anticonvulsant, anti-inflammatory, and antioxidant properties prove it to be a potent agent in neurodegenerative and neuroinflammatory diseases. There is one study showing the antidepressant-like effect of the methanolic extract of *Kushmanda* fruit in mouse forced swim test. The methanolic extract of fruit shows antidepressant activity by increasing the levels of norepinephrine, dopamine, and serotonin, and decreasing the levels of GABA and by inhibiting MAO-A (Dhingra and Joshi 2012).

## *2.5* Boerhaavia diffusa (Punarnava)

The *Boerhaavia diffusa* plant contains a large number of compounds, including punarnavine, boeravinone, hypoxanthine 9-L-arabinofuranoside, ursolic acid, punarnavoside. Hydroalcoholic extract of *B. diffusa* has shown neuroprotective effect against MPTP-induced neurodegeneration in rats (Chakkaravarthi 2015). The chemotherapeutic potential of *B. diffusa* was investigated by Khan et al. (2013). The methanolic extract of *B. diffusa* roots and its different fractions, including liriodendrin rich fraction, was investigated for a possible role in anticonvulsant activity (Kaur and Goel 2009). A comparative study of hydroalcoholic extract and polyherbal formulation of *B. diffusa* for their anti-stress activity using cold-restraint stress model was conducted by Desai et al. (2011). Hydroalcoholic extract and polyherbal formulation showed significant comparative anti-stress activity.

## *2.6* Celastrus paniculatus (Jyotishmati)

*Celastrus paniculatus* (Celastraceae) is a hardy shrub well known for its medicinal properties like nootropic or brain tonic. The oil expressed from the seeds of

*C. paniculatus* is valued for improving intellect and memory. The water extract of the seed of *C. paniculatus* enhances the memory process in rat and mice evidenced in the elevated plus-maze and sodium nitrite-induced amnesia models (Bhanumathy et al. 2010). The water-soluble seed extract of *C. paniculatus* and organic extracts showed neuroprotective activity on neuronal cells of the rat forebrain against hydrogen peroxide-induced oxidative injury (Godkar et al. 2003), and glutamate-induced injury (Godkar et al. 2006). Revanna et al. (2019) studied the behavioral and neuroprotective efficacy of *C. paniculatus* on the hippocampus of the rat brain in kainic acid-induced neurodegenerative diseases. The seed oil of *C. paniculatus* was used to analyze behavioral tests like hole board test, shuttle box test, Morris water maze test, and blood parameters, e.g., RBC, WBC, hemoglobin, platelet count; and serum parameters like serum uric acid, serum cortisol, and serum cholinesterase. The change in normal behaviors due to the induction of kainic acid was significantly reversed by the treatment of *C. paniculatus* as a mark of neuroprotection. Lekha et al. (2010) estimated the cognitive enhancement and neuroprotective effect of *Celastrus paniculatus* seed oil on male Wistar rats. The radial-arm maze paradigm was used to study the effect on learning and memory. Where, on treatment with seed oil, decreased AChE, and enhanced radial-arm maze activities were noted. Bhagya et al. (2016) determined the neuroprotective effect of *C. paniculatus* on chronic stress-induced cognitive impairment. In this study, open field and elevated plus maze (EPM) tests were used to analyze anxiety-like behavior, whereas partially baited radial arm maze (RAM) and T-maze tests were used to evaluate spatial learning and memory capabilities.

## *2.7* Centella asiatica (Mandukaparni)

*Centella asiatica* (Apiaceae) is a medicinal plant widely used as antioxidant, anti-inflammatory, and neuroprotective. Commonly known as Indian Pennywort or *gotukola* or *thankuni*, the plant is used as traditional Chinese and Ayurvedic medicine to improve cognitive function and reverse cognitive impairments. In an in vitro assay, crude extracts of *C. asiatica* containing madecassoside and asiaticoside were found to be neuroprotective and neuritogenic. *C. asiatica* is well known for its neuroprotective effect in in vivo and in vitro. Gray et al. (2018) determined that *C. asiatica* attenuates hippocampal mitochondrial dysfunction and improves memory and executive function in β-amyloid overexpressing mice. Water extract of *C. asiatica* attenuates β-amyloid (Aβ)-induced spatial memory deficits in mice and improves neuronal health. Using object location memory task, conditioned fear response, odor discrimination reversal learning test, and memory function were executed. Treatment with aqueous extract of *C. asiatica* improved hippocampal mitochondrial function and increased hippocampal and cortical expression of mitochondrial genes. Gene expression of the transcription factor NRF2, as well as its antioxidant target enzymes increased with improved behavioral performances. Teerapattarakan et al. (2018) estimated the neuroprotective effect of a standardized extract of *C. asiatica* ECa233 in rotenone-induced parkinsonism rats. The locomotory activity was executed using an open field test, and biochemical parameters like SOD, CAT, MDA were measured. Standardized extract of *C. asiatica* ECa233 can protect against rotenone-induced motor deficits and dopaminergic neuronal death. The response was mediated through the protection of mitochondrial complex I activity, the effects of antioxidants, and the enhancement of antioxidant enzyme expression and enhanced locomotion. Chiroma et al. (2019) assessed the protective effect of *C. asiatica* against D-galactose and aluminium chloride-induced rats. T-maze spontaneous alternation, modified elevated plus maze, and novel object recognition tests are different cognitive paradigms that were used to evaluate spatial and non-spatial learning and memory. Viswanathan et al. (2019) analyzed the protection of the mouse brain from paracetamol-induced stress by *C. asiatica* methanol extract. Treatment of methanolic extract of *C. asiatica* increased the scavenging activity, and this was mediated by inhibiting expression of pro-inflammatory cytokines and increasing the expression of the anti-inflammatory cytokine, which also protected mouse primary astrocytes.

## *2.8* Convolvulus pluricaulis (Shankhapushpi)

*Convolvulus pluricaulis*, also known as *Shankhapushpi* (Family: Convolvulaceae), is used as the best nerve tonic (*medhya rasayana*) to treat nervous disorders and aging (Chunekar and Pandey 2018). *C. pluricaulis* has ameliorated microtubule-associated protein tau-induced neurotoxicity in the *Drosophila* model of Alzheimer's disease (AD) (Kizhakke et al. 2017). In the rat brain cerebral cortex, it prevented the impairment of cholinergic and antioxidant activity, which was induced by aluminium (Bihaqi et al. 2009). It was assessed to attenuate scopolamine-induced amnesia and exhibited neuroprotective activity against tumor necrosis factor-alpha (TNF-$\alpha$) and prostaglandin E2 in rats. *C. pluricaulis* showed protective action against neuroinflammation-associated depressive behavior induced by chronic unpredictable mild stress in rats (Gupta and Fernandes 2019). For behavioral assessment, open field test, sucrose preference test, and forced swim test were used.

## *2.9* Desmodium gangeticum (Shalaparni)

*Desmodium gangeticum*, commonly known as *Salvan*, belonging to the Leguminosae family, is used as a bitter tonic and in nervous debility. It possesses antioxidant, anti-inflammatory, neuroprotective activity in neurological disorders. Joshi and Parle (2006) assessed the anti-amnesic activity of *D. gangeticum* in mice. Aqueous extract of *D. gangeticum* showed nootropic potential in mice. For evaluation of learning and memory exteroceptive behavioral models such as pole climbing test, Morris water maze, elevated plus maze, passive avoidance paradigms were used.

## *2.10* Glycyrrhiza glabra (Yashtimadhu)

The roots and rhizomes of *Glycyrrhiza glabra* (Leguminosae) are used as a sedative and anxiolytic agents. Glycyrrhizin, the main constituent of *Glycyrrhiza glabra*, possesses antioxidant, anti-inflammatory, and antiviral activity. Dhingra et al. (2004) studied the memory-enhancing activity of *Glycyrrhiza glabra* in mice. Glycyrrhizin also attenuates isoflurane-induced cognitive deficits in neonatal rats via its anti-inflammatory activity (Wang et al. 2015). Morris water maze was used to examine the effects of isoflurane and glycyrrhizin treatment on memory performance. Glycyrrhizin treatment prevented the deficits in spatial memory induced by neonatal exposure to isoflurane. Glabridin, a major active isoflavone from *Glycyrrhiza glabra* (licorice), has shown a neuroprotective effect in vitro and in vivo models. Glabridin significantly inhibited the staurosporine-induced cytotoxicity and apoptosis of cultured rat cortical neurons and reduced the DNA laddering caused by staurosporine. To facilitate cell survival, it suppressed the elevated Bax protein and caspase–3 pro-enzyme and decreased bcl-2 induced by staurosporine in cultured rat cortical neurons. It also inhibits superoxide production in cultured cortical neurons by modulation of multiple pathways associated with apoptosis (Yu et al. 2008). Gonzalez-Reyes et al. (2016) analyzed glycyrrhizin ameliorates oxidative stress and inflammation in the hippocampus and olfactory bulb in lithium/pilocarpine-induced epilepticus in rats. *G. glabra* suppressed the induction of the proinflammatory cytokines interleukin-1 beta (IL-1) and tumor necrosis factor-alpha.

## *2.11* Hedychium spicatum (Shati)

*Shati (Hedychium spicatum)* methanol extract is known for its nootropic and immunomodulatory activity due to the presence of saponin (Shete and Bodhankar 2010).

## *2.12* Nardostachys jatamansi (Jatamansi)

*Nardostachys jatamansi* (family Valerianacae) possesses anticonvulsant, anti-Parkinson's, tranquilizing, hepatoprotective, and neuroprotective activities. The common part used is a rhizome. Its oil acts as a bitter tonic, stimulant, antispasmodic, antiepileptic with major constituents like sesquiterpenes and coumarins (Bhatt and Kothiyal 2015). Salim et al. (2003) observed that pretreatment with an alcoholic extract of *N. jatamansi* protected rats against focal ischemia caused by middle cerebral artery occlusion. German et al. (1978) noted the tranquilizing activity of sesquiterpene valeranone isolated from *Nardostachys jatamansi* rhizomes. Salim S et al. (2003) recorded the protective effect of *Nardostachys jatamansi* in rat cerebral ischemia. Ahmad et al. (2006) estimated attenuation by *Nardostachys jatamansi* of 6-hydroxydopamine-induced Parkinsonism in rats. The study found that the ethanolic extract of *Nardostachys jatamansi* roots can slow the neuronal injury in a 6-OHDA-rat model of Parkinson's, which was exhibited by an increased density of tyrosine hydroxylase immunoreactive (TH-IR) fibers in the ipsilateral striatum of the lesioned rats. *Nardostachys jatamansi* improved behavioral outcomes in streptozotocin (STZ)-induced cognitive impairment in rats. Passive avoidance and Morris water maze were the standard behavioral assessment parameters used to test *Nardostachys jatamansi* efficacy.

## *2.13* Terminalia chebula (Haritaki)

*Terminalia chebula* Retz (Combretaceae) is commonly known as black or chebulic myrobalan. Fruits of *Terminalia chebula* have been used for neuropathy, memory loss, epilepsy, and depression. Chang and Lin (2012) showed the neuroprotective effect *T. chebula* and its potential to bind with GABA receptors. Manyam (1999) reported *T. chebula* delayed aging progression and improved cognitive deficits. Kumar et al. (2017) investigated the protective effect of *T. chebula* against seizures, seizure-induced cognitive impairment, and oxidative stress in the rats.

## *2.14* Tinospora cordifolia (Guduchi)

Rawal et al. (2004) evaluated the effect of *Tinospora cordifolia* on the free radical generation and lipid peroxidation during oxygen–glucose deprivation in rat hippocampal slices. The electron paramagnetic resonance spectroscopy revealed strong free radical scavenging properties against reactive oxygen and nitrogen species. In another study, the levels of proinflammatory cytokines like NOS, COX, IL-6, IL-1$\beta$, and TNF-$\alpha$ were found to be downregulated by *T. cordifolia* flower ethanolic extract in lipopolysaccharide-induced inflammation of neurons in rats (Prakash et al. 2017). The neuroprotective activity of *T. cordifolia* is attributed to the presence of several phytochemicals (berberine, flavonoids, alkaloids, glycosides, diterpenoid lactones). The ethanolic extract of aerial parts of *T. cordifolia* showed a significant reduction in levels of dopamine and complex I activities. It helped in reducing oxidative stress and restoration of locomotion in 6-hydroxy dopamine-induced Parkinson's diseased rats (Kosaraju et al. 2014). Sharma and Kaur (2018) observed the neurodegenerative potentials of butanol extract fraction of *T. cordifolia* with the modulation of neuronal differentiation, apoptosis, and homeostasis pathways in in vitro excitotoxicity induced by glutamate in neuronal cells. Also, it showed the neuroprotection activity in rats (Mishra and Kaur 2013; Mishra et al. 2016).

## *2.15* Allium sativum (Rasona)

Garlic (*Allium sativum*) (Alliaceae) is used as a food additive. It contains biological active sulfur compounds, such as diallyl disulfide (DAD) and diallyl trisulfide (DAT), for its nutraceutical values. *A. sativum* has been shown to improve the cognitive function of the brain and has neuroprotective activity. Wichai et al. (2019) recorded the neuroprotective effect of *A. sativum* against amyloid β protein-induced memory impairment in mice due to its antioxidant activities and scavenging reactive oxygen species. Hazzaa et al. (2020) investigated the neuroprotective potential of *A. sativum* against monosodium glutamate-induced excitotoxicity, i.e., impact on short-term memory, gliosis, and oxidative stress. Implications for short term memory was analyzed by using open field maze and T-maze. The prevention of neurotoxicity occurred due to the enhanced antioxidant activity and reduced lipid peroxidation.

## *2.16* Asparagus racemosus (Shatavari)

Neuroprotective effect of *Asparagus racemosus* root extract via the enhancement of brain-derived neurotrophic factor and estrogen receptor in ovariectomized rats was studied by Lalert et al. (2018). They observed that the ethanol extract of *A. racemosus* root could improve learning and memory impairment due to the enhancement of BDNF and ERs up-regulation, which was induced by ovariectomy. It also arrested aging, increase longevity, and improve mental function. Asparacoside, shatavarin IV, V and XI, racemosol, and asparagamine are the major constituents of *A. racemosus* responsible for its neuroprotective action. Ojha et al. (2010) reported that oral administration of methanolic extract of *A. racemosus* root could enhance memory and protect against the scopolamine-induced amnesia in rodents. Parihar and Hemnani (2004) reported that mice pre-treated with *A. racemosus* root acetone extract for 2 weeks could prevent the neuronal damage induced by kainic acid.

## *2.17* Nigella sativa (Upakunchika)

*Nigella sativa* belongs to the family Ranunculaceae. The black seeds are also known as black cumin, black caraway seed, and Habbatul barakat. Thymoquinone is an active constituent of this plant. Akhtar et al. (2014) recorded neuroprotective property of *N. sativa* loaded oral provesicular lipid formulation in in vitro and ex vivo models. Asiaei et al. (2017) evaluated the neuroprotective effects of *N. sativa* extract upon the hippocampus in PTU-induced hypothyroidism juvenile rats. It was found that *N. sativa* oil enhanced the spatial working memory performance of rats in a radial arm maze model (Khairul et al. 2013). The hydro-alcoholic extract of *N. sativa* prevented scopolamine-induced spatial memory deficits and decreased the AChE activity as well as oxidative stress in the brain tissues of rats (Hosseini et al. 2015).

## *2.18* Clitorea ternatea (Aparajita)

The roots, seeds, and leaves of *Clitorea ternatea* Linn (Fabaceae) are widely used as a brain tonic to promote memory and intelligence. It is also reported to be having anti-depressant, anticonvulsant, anti-inflammatory properties. Talpate et al. (2014) explored the nootropic and neuroprotective potential of *C. ternatea* leaves ethanol extract in the experimental model for diabetes mellitus induced cognitive decline. Y-maze, Morris water maze, and the radial arm maze tests were used to evaluate its effect on spatial working memory, spatial reference memory, and spatial working-reference memory. Acetylcholinesterase, lipid peroxide, superoxide dismutase (SOD), total nitric oxide (NO), catalase (CAT), and glutathione (GSH) assays were performed to study the neuroprotective effects. Rai et al. (2001) investigated that *C. ternatea* Linn root extract treatment during the growth spurt period enhanced learning and memory in rats.

## *2.19* Cuminum cyminum (Jeera)

*Cuminum cyminum* Linn (Apiaceae), commonly known as cumin, is mainly used in cuisines to increase the taste and flavor of food. Cuminol, cymine, phellandrene, carvone, and cuminique alcohol are the main constituents of cumin. *C. cyminum* possesses antioxidant, anti-inflammatory, and antiepileptic properties. *C. cyminum* is a potent immunomodulator which, in a dose-dependent manner, increased T-cells (CD4 and CD8) count and Th1-predominant immune response (Chauhan et al. 2010). Koppula and Choi (2011) reported that *C. cyminum* extract could attenuate scopolamine-induced memory loss and stress-induced urinary biochemical changes in rats. A study on the mouse model of Parkinson's disease *C. cyminum* extract attenuated MPTP-induced oxidative stress and behavioral impairments (Kim et al. 2016).

## *2.20* Aurum (Gold)

The neuroprotective effect of gold nanoparticles composites in Parkinson's disease model was studied by Kaikai Hu et al. (2018). Ostergaard et al. (2009) reported that bio-released gold ions modulate the expression of neuroprotective and hematopoietic factors after brain injury. They affirmed that gold ions modulate neurotrophic factor by increasing the expression of neurotrophin (NT)-4, transforming growth factor-beta 3 (TGF-β3), leukemia inhibitory factor (LIF), and metallothionein I + II (MT-I + II).

## *2.21* Hordeum vulgare (Yava)

*Hordeum vulgare* is a natural source of vitamins and minerals and possess antioxidant and anti-depressive action. The antidepressant effect is due to the presence of α-tocopherol and ascorbic acid. Yamaura et al. (2012) recorded antidepressant-like effects of young green barley leaf in the mouse forced swimming test.

## *2.22* Piper longum (Pippali)

*Piper longum* has antidepressive, antiepileptic, anti-inflammatory, anticonvulsive, and antioxidant properties. Hua et al. (2019) observed that piperine is neuroprotective in rats with cerebral ischemic injury. According to Go et al. (2018), *P. longum* decreased cognitive impairment and improved hippocampus function in the aged mice. It was found that, with *P. longum* administration, the phosphorylated levels of N-methyl-D-aspartate receptor subtype 2B (NR2B), calmodulin-dependent protein kinase II alpha (caMKIIα), and extracellular signal-regulated kinase 1/2 (ERK1/2) increased in the hippocampus of the aged mice. Due to an increase in CA-3 specificity, *P. longum* is known as a potent marker of neuronal plasticity, learning, and memory.

## *2.23* Punica granatum (Dadima)

*Punica granatum* (Pomegranate) possesses the antioxidant capacity, increases behavioral performance, and reduces amyloid plaque deposition in the hippocampus by the consumption pomegranate pulp or fruit juice (Akhtar et al. 2015). Morzelle et al. (2016) studied the neuroprotective effects of pomegranate peel extract after chronic infusion with amyloid-β peptide in mice. They observed that pomegranate peel extract reduced amyloid plaque density, increased the expression of neurotrophin BDNF, and reduced the activity of acetylcholinesterase enzyme. Also, it reduced lipid peroxidation and the concentration of the proinflammatory cytokine TNF-α. In another study, pomegranate extract inhibited the nuclear factor of activated T-cell activity and microglial activation in in vitro, and a transgenic mouse model of Alzheimer's disease (Rojanathammanee et al. 2013).

### *2.24* Semecarpus anacardium (Bhallataka)

Shukla et al. (2000) reported neuroprotective effects of *Semecarpus anacardium* in stress-induced neurodegeneration in albino rats.

### *2.25* Sphaeranthus indicus (Mundi)

Neuroprotective effect of petroleum ether, methanolic, and aqueous extracts of *Sphaeranthus indicus* flower heads on lipofuscinogenesis and fluorescence product in the brain of D-galactose-induced aging accelerated mice were reported by Ambikar and Mohanta (2013).

## 3 Discussion

The advancement in neuroscience in the last decade has helped us to understand the behavioral pattern, biochemical, and molecular mechanisms involved in learning and memory processes. It has provided us the base for the various preclinical behavioral and pharmacological models for the assessment of drugs used for the treatment of psychiatric and neurodegenerative disorders associated with cognitive dysfunction (Alzheimer's disease, Parkinson's disease, dementia, and Down's syndrome, etc.). The response of cognitive drugs is evaluated by using series of tests, including exteroceptive (active and passive avoidance behavioral task) and interoceptive (chemical-induced amnesia or cognitive dysfunction) aversive stimuli models for understanding behavioral and pharmacological changes in learning and memory processes. The exteroceptive aversive stimuli models help in understanding the effect of drugs on a short-term or long-term memory and learning processes. For example, plus maze, shuttle box, and Morris water maze tests are used for the assessment of drugs acting on long term memory, short term memory, and learning processes, respectively. The interoceptive aversive stimuli models, which help in task-specific regions in memory and learning processes with nucleus basalis magnocellularis and hippocampal lesions are neurotoxins like kainic acids, colchicine, *N*-methyl-D-aspartic acid, and so on (Kulkarni 2005).

## 4 Pharmacological Action Through Ayurvedic Perspective

The fundamental doctrine of "*Tridosha*" states that *Vata*, *Pitta*, and *Kapha* are the three (energy) or biological rhythms regulating the physical and mental functioning of the human body. Similarly, *Satva* corresponding to *Pitta*, *Raja* corresponding to *Vata* and *Tama* corresponding to *Kapha* are stated to regulate the functions of the brain and intellect in humans. Thus, in the case of intelligence, i.e., *Medha*, *Pitta* regulates the function of understanding and grasping knowledge. *Vata* is responsible for the co-ordination of ideas, language, and response. *Kapha* is known for its property of stability, which is necessary for the retention of memory (Dwivedi and Mehra 2012).

The models used in the assessment of the cognitive effects of above Ayurvedic plants are based on the modern medicine perspectives. Almost all medicinal plants in Ayurveda and other traditional systems have been screened for AChE inhibitory activity. Where results observed are not about a particular *Veerya*. Also, not all these plants have shown positive results in Alzheimer's clinically. Thus, there is a need to rethink the use of appropriate models for screening.

In order to know their grasping power-enhancing property, some different models can also be recommended, as these drugs will help in restoring the *Sadhak pitta* function, which is learning and comprehension. Also, the use of single herbs must be as per given in Classics like in the case of '*Medhya Rasayanas*' of Charak Samhita (Srivastav and Saxena 2019).

The neurodegeneration phenomenon can be correlated to the deranged *Vata* function, which is supposed to occur as per classics in the geriatric population. Thus, the conduction of impulses and synaptic function loss can be corrected by administering drugs that alleviate the *Vata* function and promote nutrition by *Madhura rasa*, *Ushna veerya* dravyas. These will help in the correction of co-ordination and comprehension capacities of the individual. But if irritability is more along with memory loss, it will be more advisable to administer *madhura*, *tikta*, and *Sheeta veerya* dravyas. General cognition related models aim to directly target these functions for all the available plants that are reported to help promote intellect. But, *Ayurveda* has laid specific criteria dependent on age, *Dosha* (*Vata*, *Pitta or Kapha* bioenergies) involved, Dhatu (Tissue level), *Prakriti* (constitution) and etiological factors viz. sleeping in the daytime, faulty dietary habits, and behavioral conduct. All these criteria influence the *Agni* (metabolic status) of the individual. These thus manifest the pathology in *Tamas and Rajas guna* derangement, which leads to functional deficits of intellect and memory. Therefore, pharmacological models should be developed to assess these etiological factors, which will then give impeccable leads to drug mechanisms in Ayurveda.

The *Medhya* drugs used in children will differ as they already have *Kapha* predominance. Thus, they will need more of a *dravya*, enhancing the function of *Pitta* related to learning capabilities, comprehension, grasping, language. The dosage forms and posology will also be different. Ayurveda recommends almost all *Medhya* drugs in the form of *Ghrita* (ghee) or oil preparations, which is logical as they can cross the blood-brain barrier. Also, routes of administration influence drug action. These *ghrita* are recommended right from birth along with gold to enhance intellectual functions (Tiwari 2002).

## 5 Conclusions

To conclude, the models for assessment of cognition viz. Morris maze, Elevated plus maze, Passive avoidance test in modern parlance can partially answer the cognitive effects of medicinal plants because the basis of Ayurveda pharmacology has a holistic base consisting of *rasapanchaka* concept. Some models target anxiety-like behavioral aspects, some target spatial learning, but the meaning of "*medhya*" in Ayurveda has a further depth and is very complex. As discussed above, it is more related to the quality of tissues and psyche (*satva sarata)*, ultimately, which resides upon the homeostasis of the particular individual. Thus, cognition therapy of Ayurveda using herbs is also decided upon the *Prakriti* of the individual, his/her metabolism capacity, and so on. Therefore, probing into the concept of *Rasapanchak* and the development of tools for assessment of drug mechanisms targeting "*Medhya*" karma is essential for obtaining new insights in neuropharmacology.

## References


Ahirwar S, Tembhre M (2016) Assessment of acetylcholinesterase inhibition by *Bacopa monneiri* and acephate in hippocampus of chick brain for impediment of Alzheimer's disease. Pharm Pharmacol Int J 4:413–417. https://doi.org/10.15406/ppij.2016.04.00088

Ahmad M, Yousuf S, Khan MB et al (2006) Attenuation by *Nardostachys jatamansi* of 6-hydroxydopamine-induced parkinsonism in rats: behavioral, neurochemical and immunohistochemical studies. Pharmacol Biochem Behav 83(1):150–160. https://doi.org/10.1016/j.pbb.2006.01.005

Akhtar M, Imam SS, Ahmad MA et al (2014) Neuroprotective study of *Nigella sativa*—loaded oral provesicular lipid formulation: *in vitro* and *ex vivo* study. Drug Deliv 21:487–494. https://doi.org/10.3109/10717544.2014.886640

Akhtar S, Ismail T, Fraternale T, Sestili P (2015) Pomegranate peel and peel extracts: chemistry and food features. Food Chem 174:417–425. https://doi.org/10.1016/j.foodchem.2014.11.035

Ambikar DB, Mohanta GP (2013) Neuroprotective effect of petroleum ether, methanolic and aqueous extracts of flower heads of *Sphaeranthus indicus* on lipofuscinogenesis and fluorescence product in brain of D-galactose induced aging accelerated mice. Orient Pharm Exp Med 13:301–306. https://doi.org/10.1007/s13596-013-0110-x

Asiaei F, Fazel A, Rajabzadeh AA, Hosseini M (2017) Neuroprotective effects of *Nigella sativa* extract upon the hippocampus in PTU-induced hypothyroidism juvenile rats: a stereological study. Metabolic Brain Dis 32:1755–1765. https://doi.org/10.1007/s11011-017-0025-1

Bhagya V, Christofer T, Rao BSS (2016) Neuroprotective effect of *Celastrus paniculatus* on chronic stress-induced cognitive impairment. Ind J Pharmacol 48:687–693. https://doi.org/10.4103/0253-7613.194853

Bhanumathy M, Harish MS, Shivaprasad HN, Sushma G (2010) Nootropic activity of *Celastrus paniculatus* seed. Pharm Biol 48:324–327. https://doi.org/10.3109/13880200903127391

Bhatt M, Kothiyal P (2015) A review article on phytochemistry and pharmacological profiles of *Nardostachys jatamansi* DC- medicinal herb. J Pharmacog Phytochem 3(5):102–106

Bihaqi SW, Sharma M, Singh AP, Tiwari M (2009) Neuroprotective role of *Convolvulus pluricaulis* on aluminum induced neurotoxicity in rat brain. J Ethnopharmacol 124:409–415. https://doi.org/10.1016/j.jep.2009.05.038

Chakkaravarthi N (2015) Neuroprotective effect of hydro-alcoholic extract of *Boerhaavia diffusa* Linn against MPTP-induced neurodegeneration in rats. Dissertation submitted in partial fulfillment of the Master of Pharmacy in Pharmacology. http://repository-tnmgrmu.ac.in/id/eprint/4938

Chang CL, Lin CS (2012) Phytochemical composition, antioxidant activity, and neuroprotective effect of *Terminalia chebula* Retzius extracts. Evid Based Complement Alternat Med:125247. https://doi.org/10.1155/2012/125247

Chauhan PS, Satti NK, Suri KA, Amina M, Bani S (2010) Stimulatory effects of *Cuminum cyminum* and flavonoid glycoside on cyclosporine-a and restraint stress induced immune-suppression in Swiss albino mice. Chem Biol Interact 185(1):66–72. https://doi.org/10.1016/j.cbi.2010.02.016

Chiroma SM, Hidayat Baharuldin MT, Mat Taib CN et al (2019) Protective effect of *Centella asiatica* against D-galactose and aluminium chloride induced rats: behavioral and ultrastructural approaches. Biomed Pharmacother 109:853–864. https://doi.org/10.1016/j.biopha.2018.10.111

Chunekar KC, Pandey GS (2018) Bhavaprakashnighantu. Reprint, Chaukhambha Krishnadas Academy, Varanasi, p 439

Desai SK, Desai SM, Navdeep S et al (2011) Antistress activity of *Boerhaavia diffusa* root extract and a polyherbal formulation containing *Boerhaavia diffusa* using cold restraint stress model. Int J Pharm Pharmaceut Sci 3(1):130–132

Dhingra D, Joshi P (2012) Antidepressant-like activity of *Benincasa hispida* fruits in mice: possible involvement of monoaminergic and gabaergic systems. J Pharmacol Pharmacother 3(1):60–63. https://doi.org/10.4103/0976-500X.92521

Dhingra D, Parle M, Kulkarni SK (2004) Memory enhancing activity of *Glycyrrhiza glabra* in mice. J Ethnopharmacol 91(2–3):361–365. https://doi.org/10.1016/j.jep.2004.01.016

Dubey T, Chinnathambi S (2019) Brahmi (*Bacopa monnieri*): an Ayurvedic herb against the Alzheimer's disease. Arch Biochem Biophys 676:108153. https://doi.org/10.1016/j.abb.2019.108153

Dwivedi KN, Mehra B (2012) Mode of action of medhya drugs: a review. Anc Sci Life 32:83

German R, Rucker G, Tautges J, Wenzl H, Graf E (1978) Isolation and pharmacodynamic activity of the sesquiterpene valeranone from *Nardostachys jatamansi* DC. Arzneimittelforschung 28 (1):7–13

Go J, Park T, Han G, Park H et al (2018) Piperlongumine decreases cognitive impairment and improves hippocampal function in aged mice. Int J Mol Med 42:1875–1884. https://doi.org/10.3892/ijmm.2018.3782

Godkar P, Gordon RK, Ravindran A, Doctor BP (2003) *Celastrus paniculatus* seed water soluble extracts protect cultured rat forebrain neuronal cells from hydrogen peroxide-induced oxidative injury. Fitoterapia 74:658–669. https://doi.org/10.1016/S0367-326X(03)00190-4

Godkar P, Gordon RK, Ravindran A, Doctor BP (2006) *Celastrus paniculatus* seed oil and organic extracts attenuate hydrogen peroxide and glutamate-induced injury in embryonic rat forebrain neuronal cells. Phytomedicine 13(1–2):29–36. https://doi.org/10.1016/j.phymed.2003.11.011

Gonzalez-Reyes S, Santillan-Cigales JJ, Jimenez-Osorio AS et al (2016) Glycyrrhizin ameliorates oxidative stress and inflammation in hippocampus and olfactory bulb in lithium/pilocarpine-induced status epilepticus in rats. Epilepsy Res 126:126–133. https://doi.org/10.1016/j.eplepsyres.2016.07.007

Gray NE, Zweig JA, Caruso M et al (2018) *Centella asiatica* attenuates hippocampal mitochondrial dysfunction and improves memory and executive function in β-amyloid overexpressing mice. Mol Cell Neurosci 93:1–9. https://doi.org/10.1016/j.mcn.2018.09.002

Gupta GL, Fernandes J (2019) Protective effect of *Convolvulus pluricaulis* against neuroinflammation associated depressive behavior induced by chronic unpredictable mild stress in rat. Biomed Pharmacother 109:1698–1708. https://doi.org/10.1016/j.biopha.2018.11.046

Hanumanthachar J, Navneet K, Jyotibala C (2007) Evaluation of nootropic effect of *Argyreia speciosa* in mice. J Health Sci 53:382–388. https://doi.org/10.1248/jhs.53.382

Hazzaa SM, Abdelaziz SAM, Eldaim MAA, Daim MAA (2020) Neuroprotective potential of *Allium sativum* against monosodium glutamate-induced excitotoxicity: impact on short-term memory, gliosis, and oxidative stress. Nutrients 12:1028. https://doi.org/10.3390/nu12041028

Hosamani R, Muralidhar (2009) Neuroprotective efficacy of *Bacopa monnieri* against rotenone-induced oxidative stress and neurotoxicity in *Drosophila melanogaster*. NeuroToxicol 30:977–985. https://doi.org/10.1016/j.neuro.2009.08.012

Hosseini M, Mohammadpour T, Karami R et al (2015) Effects of the hydro-alcoholic extract of *Nigella Sativa* on scopolamine-induced spatial memory impairment in rats and its possible mechanism. Chin J Integr Med 21:438–444. https://doi.org/10.1007/s11655-014-1742-5

Hu KBS, Chen XMS, Chen WMS et al (2018) Neuroprotective effect of gold nanoparticles composites in Parkinson's disease model. Nanomedicine 14(4):1123–1136. https://doi.org/10.1016/j.nano.2018.01.020

Hua S, Liu J, Zhang Y, Li J, Zhang X, Dong L, Zhao Y, Fu X (2019) Piperine as a neuroprotective functional component in rats with cerebral ischemic injury. Food Sci Nutr 7(11):3443–3451. https://doi.org/10.1002/fsn3.1185

Jaiswal BS, Tailang M (2018) Protective effect of ethanolic extract from the root of *Argyreia speciosa* against global cerebral ischemic reperfusion injury in rats. J Drug Deliv Ther 8 (6):8–15. https://doi.org/10.22270/jddt.v8i6.1990

Joshi H, Parle M (2006) Antiamnesic effect of *Desmodium gangeticum* in mice. Yakugaku Zasshi 126(9):795–804. https://doi.org/10.1248/yakushi.126.795

Jyoti A, Sharma D (2006) Neuroprotective role of *Bacopa monniera* extract against aluminium-induced oxidative stress in the hippocampus of rat brain. NeuroToxicol 27(4):451–457. https://doi.org/10.1016/j.neuro.2005.12.007

Kaur M, Goel RK (2009) Anti-convulsant activity of *Boerhaavia diffusa* plausible role of calcium channel antagonism. Evid Based Complement Alternat Med:310420. https://doi.org/10.1093/ecam/nep192

Khairul M, Sahak A, Mohamed AM et al (2013) *Nigella sativa* oil enhances the spatial working memory performance of rats on a radial arm maze. Evid Based Complement and Alternat Med:1–5. https://doi.org/10.1155/2013/180598

Khan SM, Ansari IA, Ahamd A et al (2013) Chemotherapeutic potential of *Boerhavia diffusa* Linn; a review. J Appl Pharma Sci 3(1):133–139. https://doi.org/10.7324/JAPS.2013.30126

Kim J, Kopalli SR, Koppula S (2016) *Cuminum cyminum* Linn (Apiaceae) extract attenuates MPTP-induced oxidative stress and behavioral impairments in mouse model of Parkinson's disease. Trop J Pharm Res 15(4):765–772. https://doi.org/10.4314/tjpr.v15i4.14

Kizhakke AP, Olakkaran S, Antony A, Tilagul SK, Hunasanahally GP (2017) *Convolvulus pluricaulis* (Shankhapushpi) ameliorates human microtubule-associated protein tau (hMAPτ) induced neurotoxicity in Alzheimer's disease Drosophila model. J Chem Neuroanatomy 95:115–122. https://doi.org/10.1016/j.jchemneu.2017.10.002

Koppula S, Choi DK (2011) *Cuminum cyminum* extract attenuates scopolamine-induced memory loss and stress-induced urinary biochemical changes in rats: a noninvasive biochemical approach. Pharm Biol 49(7):702–708. https://doi.org/10.3109/13880209.2010.541923

Kosaraju J, Chinni S, Roy PD, Kannan E, Antony AS, Kumar MN (2014) Neuroprotective effect of *Tinospora cordifolia* ethanol extract on 6-hydroxy dopamine induced parkinsonism. Indian J Pharmacol 46(2):176–180. https://doi.org/10.4103/0253-7613.129312

Kulkarni SK (2005) Handbook of experimental pharmacology, 3rd edn. Vallabh Prakashan, Delhi, India

Kumar A, Sharma V, Singh VP et al (2015) Herbs to curb cyclic nucleotide phosphodiesterase and their potential role in Alzheimer's disease. Mech Ageing Dev 149:75–87. https://doi.org/10.1016/j.mad.2015.05.009

Kumar R, Arora R, Agarwal A, Gupta YK (2017) Protective effect of *Terminalia chebula* against seizures, seizure-induced cognitive impairment and oxidative stress in experimental models of seizures in rats. J Ethnopharmacol 215:124–131. https://doi.org/10.1016/j.jep.2017.12.008

Lalert L, Kruevaisayawan H, Amatyakul P, Ingkaninan K, Khongsombat O (2018) Neuroprotective effect of *Asparagus racemosus* root extract via the enhancement of brain-derived neurotrophic factor and estrogen receptor in ovariectomized rats. J Ethnopharmacol 225:336–341. https://doi.org/10.1016/j.jep.2018.07.014

Lekha G, Kuma BP, Rao SN, Arockiasamy I, Mohan K (2010) Cognitive enhancement and neuroprotective effect of *Celastrus paniculatus* Willd seed oil (Jyothismati oil) on male wistar rats. J Pharm Sci Tech 2(2):130–138

Limpeanchob N, Jaipan S, Rattanakaruna S, Phrompittayarat W, Ingkaninan K (2008) Neuroprotective effect of *Bacopa monnieri* on beta-amyloid-induced cell death in primary cortical culture. J Ethnopharmacol 120(1):112–117. https://doi.org/10.1016/j.jep.2008.07.039

Manyam BV (1999) Dementia in Ayurveda. J Altern Complement Med 5(1):81–88. https://doi.org/10.1089/acm.1999.5.81

Mishra R, Kaur G (2013) Aqueous ethanolic extract of *Tinospora cordifolia* as a potential candidate for differentiation based therapy of glioblastomas. PLoS One 8(10):e78764. https://doi.org/10.1371/journal.pone.0078764

Mishra R, Manchanda S, Gupta M, Kaur T, Saini V, Sharma A, Kaur G (2016) *Tinospora cordifolia* ameliorates anxiety-like behavior and improves cognitive functions in acute sleep deprived rats. Sci Rep 6:25,564. https://doi.org/10.1038/srep25564

Morzelle MC, Salgado JM, Telles M, Mourelle D, Bachiega P, Buck HS, Viel TA (2016) Neuroprotective effects of pomegranate peel extract after chronic infusion with amyloid-β peptide in mice. PLoS One 11(11):e0166123. https://doi.org/10.1371/journal.pone.0166123

Oh MH, Houghton PJ, Whang WK, Cho JH (2004) Screening of Korean herbal medicines used to improve cognitive function for anti-cholinesterase activity. Phytomedicine 11(6):544–548. https://doi.org/10.1016/j.phymed.2004.03.00gol1

Ojha R, Sahu AN, Muruganandam AV, Singh GK, Krishnamurthy S (2010) *Asparagus recemosus* enhances memory and protects against amnesia in rodent models. Brain Cogn 74(1):1–9. https://doi.org/10.1016/j.bandc.2010.05.009

Ostergaard M, Larsen A, Stoltenberg M, Penkowa M (2009) Bioreleased gold ions modulate expression of neuroprotective and hematopoietic factors after brain injury. Brain Res 1307:1–13. https://doi.org/10.1016/j.brainres.2009.10.028

Pandey A (2016) Therapeutic potential of Vachadi ghrita (a medicated ghee) on CNS and other ailments—a review. Atreya Ayurveda Foundation Publication 36–46. info:eu-repo/semantics/article

Parihar MS, Hemnani T (2004) Experimental excitotoxicity provokes oxidative damage in mice brain and attenuation by extract of *Asparagus racemosus*. J Neural Transm (Vienna) 111:1–12. https://doi.org/10.1007/s00702-003-0069-8

Prakash R, Sandhya E, Ramya N, Dhivya R, Priyadarshini M, Sakthi Priya B (2017) Neuroprotective activity of ethanolic extract of *Tinospora cordifolia* on LPS induced neuroinflammation. Transl Biomed 8(4):135. https://doi.org/10.21767/2172-0479.100135

Rai KS, Murthy KD, Karantha KS, Rao MS (2001) *Clitoria ternatea* (Linn) root extract treatment during growth spurt period enhances learning and memory in rats. Ind J Physiol Pharmacol 45 (3):305–313

Rawal A, Muddeshwar M, Biswas S (2004) Effect of *Rubia cordifolia*, *Fagonia cretica* Linn and *Tinospora cordifolia* on the free radical generation and lipid peroxidation during oxygen-glucose deprivation in rat hippocampal slices. Biochem Biophys Res Commun 324 (2):588–596. https://doi.org/10.1016/j.bbrc.2004.09.094

Ray S, Ray A (2015) Medhya Rasayanas in brain function and disease. Med Chem 5(12):505–511. https://doi.org/10.4172/2161-0444.1000309

Revanna G, Pujari S, Subramanian V, Rao SR (2019) Effects of *Celastrus paniculatus* Willd and *Sida cordifolia* Linn in kainic acid-induced hippocampus damage in rats. Ind J Pharma Edu Res 53(3):537–544. https://doi.org/10.5530/ijper.53.3.86

Rojanathammanee L, Puig KL, Combs CK (2013) Pomegranate polyphenols and extract inhibit nuclear factor of activated T-cell activity and microglial activation *in vitro* and in a transgenic mouse model of Alzheimer disease. J Nutr 143(5):597–605. https://doi.org/10.3945/jn.112.169516

Roy C, Mazumdar D (2016) A nootropic effect of *Benincasa hispida* on ACH and CHAT activity in colchicine induced experimental rat model of Alzheimer's disease: possible involvement. World J Pharm Res 3(9):1282–1296

Salim S, Ahmad M, Zafar KS, Ahmad AS, Islam F (2003) Protective effect of *Nardostachys jatamansi* in rat cerebral ischemia. Pharmacol Biochem Behav 74(2):481–486. https://doi.org/10.1016/S0091-3057(02)01030-4

Sastry K, Chaturvedi G (1983) Agnivesha, Caraka Samhita, elaborated Vidyothini hindi commentary, vol 1, 11th edn. Chowkamba Bharathi Academy, Varanasi

Sharma A, Kaur G (2018) *Tinospora cordifolia* as a potential neuroregenerative candidate against glutamate induced excitotoxicity: an *in vitro* perspective. BMC Complement Altern Med 18:268. https://doi.org/10.1186/s12906-018-2330-6

Shete RV, Bodhankar SL (2010) *Hedychium spicatum*: evaluation of its nootropic effect in mice. Res J Pharmacog Phytochem 2(5):403–406

Shinomol GK, Mythri RB, Bharath MMS, Murlidhara (2012) *Bacopa monnieri* extract offsets rotenone-induced cytotoxicity in dopaminergic cells and oxidative impairments in mice brain. Cell Mol Neurobiol 32(3):455–465. https://doi.org/10.1007/s10571-011-9776-0

Shukla SD, Jain S, Sharma K, Bhatnagar M (2000) Stress induced neuron degeneration and protective effects of *Semecarpus anacardium* Linn and *Withania somnifera* Dunn in hippocampus of albino rats: an ultrastructural study. Ind J Exp Biol 38(10):1007–1013

Smith AD, Kelly A (2015) Cognitive processes. The encyclopedia of adulthood and aging:1–4. https://doi.org/10.1002/9781118521373.wbeaa213

Srivastav N, Saxena V (2019) Role of medhyarasayanas (nootropic drugs) in developmental disabilities of children. Int J Health Sci Res 9(6):315–322

Talpate KA, Bhosale UA, Zambare MR, Somani RS (2014) Neuroprotective and nootropic activity of *Clitorea ternatea* Linn (Fabaceae) leaves on diabetes induced cognitive decline in experimental animals. J Pharm Bioallied Sci 6(1):48–55. https://doi.org/10.4103/0975-7406.124317

Teerapattarakan N, Benya-aphikul H, Tansawat R, Wanakhachornkrai O, Tantisira MH, Rodsiri R (2018) Neuroprotective effect of a standardized extract of *Centella asiatica* ECa233 in rotenone-induced parkinsonism rats. Phytomedicine 44:65–67. https://doi.org/10.1016/j.phymed.2018.04.028

Tiwari PV (2002) Vriddhajivaka. Suthrasthana, chapter 18. In: Kashyap Samhita. 1st edn. ChaukhambaVisvabharati

Viswanathan G, Mohan V, Radhakrishnan N, Sasikumar A, Prasanna A, Nair R, Baby S (2019) Protection of mouse brain from paracetamol-induced stress by *Centella asiatica* methanol extract. J Ethnopharmacol 236:474–483. https://doi.org/10.1016/j.jep.2019.03.017

Wang W, Chen X, Zhang J, Zhao Y, Li S, Tan L, Gao J, Fang X, Luo A (2015) Glycyrrhizin attenuates isoflurane-induced cognitive deficits in neonatal rats via its anti-inflammatory activity. Neuroscience 316:328–336. https://doi.org/10.1016/j.neuroscience.2015.11.001

Wichai T, Pannangrong W, Welbat J, Chaichun A, Sripanidkulchai K, Sripanidkulchai B (2019) Effects of aged garlic extract on spatial memory and oxidative damage in the brain of amyloid-ß induced rats. Songklanakarin J Sci Technol 41:311–318

Yamaura K, Nakayama N, Shimada M et al (2012) Antidepressant-like effects of young green barley leaf (*Hordeum vulgare* L.) in the mouse forced swimming test. Pharm Res 4(1):22–26. https://doi.org/10.4103/0974-8490.91030

Yu XQ, Xue CC, Zhou ZW, Li CG, Du YM, Liang DJ, Zhou SF (2008) *In vitro* and *in vivo* neuroprotective effect and mechanisms of glabridin, a major active isoflavan from *Glycyrrhiza glabra* (licorice). Life Sci 82(1–2):68–78. https://doi.org/10.1016/j.lfs.2007.10.019

# Neurotoxic Medicinal Plants of Indian Himalayan Regions: An Overview

**Supriya Sharma, Ashish Raina, Dinesh Chandra Agrawal, Manoj K. Dhar, and Sanjana Kaul**

## Contents




**Abstract** The Indian Himalayan region plays an important role in providing products of great therapeutic importance to humankind. The plants thriving in the



S. Sharma · A. Raina · M. K. Dhar · S. Kaul (✉)
School of Biotechnology, University of Jammu, Jammu, Jammu & Kashmir, India
e-mail: sanjanakaul@jammuuniversity.ac.in

D. C. Agrawal
Department of Applied Chemistry, Chaoyang University of Technology, Taichung, Taiwan





D. C. Agrawal, M. Dhanasekaran (eds.), *Medicinal Herbs and Fungi*,
https://doi.org/10.1007/978-981-33-4141-8_19

Himalayas are being used as curative agents by tribal people since ages. The unique topographic and climatic conditions of this region make it a prolific source of medicinal plants. The presence of diverse metabolites in these plants attributes toward their great therapeutic potential. In the last few years, a tremendous increase in the demand for natural therapies has been observed all around the globe. Despite the great therapeutic potential, some of these medicinal plants have subacute and chronic toxicities associated with them. The toxins produced by them can adversely affect human health. It varies from a mild allergic reaction to a severe medical complication that may lead to death. Neurotoxicity adversely affects the nervous system. The signs and symptoms of neurotoxicity are depression, anxiety, paralysis, headache, hallucination, impaired vision, numbness or weakness in limbs, behavioral changes, etc. It may be caused due to an accidental intake, drug abuse, overdose, chemotherapy, organ transplant, etc. Neurotoxins kill or disrupt the activity of neurons, thereby adversely affecting the nervous system. The cerebellum is a part of the brain that has been reported to be most affected by neurotoxins. Some of the neurotoxic medicinal plants of the Indian Himalayan region include *Aconitum chasmanthum*, *Datura stramonium*, *Conium maculatum*, *Catharanthus roseus*, *Cannabis sativa*, *Ricinus communis*, *Gloriosa superba*, *Mandragora officinalis*, *Melia azedarach*, *Podophyllum hexandrum*, etc. This book chapter aims to provide an overview of neurotoxic medicinal plants of the Indian Himalayan region and their neurotoxins. A special emphasis has been laid on the mechanism of neurotoxicity. A few case studies have also been added to explain the adverse effects of neurotoxins.

**Keywords** Indian Himalayan region · Medicinal plants · Neurotoxins · Neurotoxicity

## Abbreviations

| | |
|---|---|
| ANS | Autonomic nervous system |
| CBD | Cannabidiol |
| CBC | Cannabichromene |
| CBG | Cannabigerol |
| CNS | Central nervous system |
| COP-1 | Coatomer protein 1 |
| ER | Endoplasmic reticulum |
| FDA | Food and Drug Administration |
| GABA | Gamma-aminobutyric acid |
| IHR | Indian Himalayan Region |
| β-ODAP | β-N-Oxalyl-l-α,β-diaminopropionic acid |
| OAP | L-3-Oxalylamino-2-aminopropionic acid |
| PNS | Peripheral nervous system |
| RIP | Ribosome-inactivating protein |
| rRNA | Ribosomal ribonucleic acid |
| THC | Tetrahydrocannabinol |

## 1 Introduction

Himalayan medicinal plants play a prolific role in the livelihood of the rural population in the region. Tribal people have used some of these plants as curative agents against various ailments for ages. Nowadays, their importance has been recognized at the international level. The arc-shaped Himalayan region extends along the entire Northern boundary of India. It covers a vast area along the Northern frontiers of India, including (from west to east): Jammu and Kashmir, Himachal Pradesh, Uttar Pradesh, Sikkim, and Arunachal Pradesh. The Himalayan region mainly comprises of the mountain chain covering a surface area of 8 million $km^2$ (approximately) and has a length of 3000 km (Fig. 1).

Despite its location near the tropics, it represents a complete transaction from tropical to temperate conditions. It covers over 18% of the dynamic area of the Indian subcontinent and harbors a plethora of medicinal plants continually being used for their therapeutic properties (Malik et al. 2015).

Highly diverse topographic and climatic conditions of the Indian Himalayan Region (IHR) make it a repository of highly valuable medicinal plants (Gupta et al. 2018). The majestic Indian Himalayan ranges can be divided into three categories based on the latitudinal (geographical) division: (1) The Himalayan Ranges which further includes: The Great Himalayas (Inner Himalayas), The Middle or the lesser Himalaya (Himachal or lower Himalayas), The Shiwalik (Outer Himalayas); (2) The Trans Himalayas and (3) The Eastern Hills or The Purvanchal. The Western Himalayan region of India covers Panjab Himalayas, Kashmir and

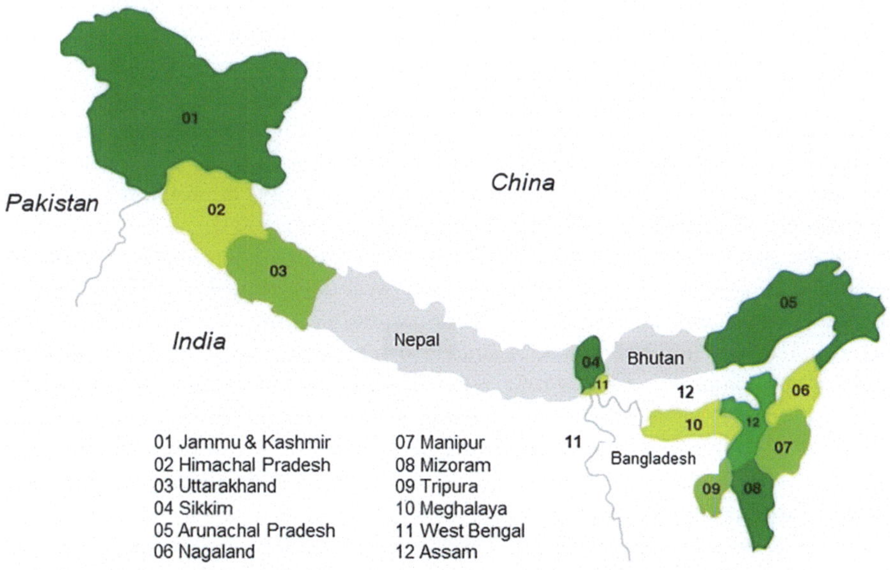


**Fig. 1** The Himalayan region along the Northern boundary of India (map not to scale)

Himachal Himalayas, and Kumaun Himalayan regions, and the Eastern Himalaya covers Sikkim Himalayas based on the rivers flowing through them.

The varying altitude of IHR makes it a rich source of diverse medicinal plants. As per an estimate, there are over nearly 1743 species of plants with therapeutic value in the IHR (Malik et al. 2015). The medicinal plants of IHR are a rich source of bioactive compounds with profound pharmacological activities. About 65% of the Indian population is dependent on the traditional system of medicine. Indian tribal and ethnic people are dependent on the plants of IHR to gain economic values and primary health care benefits. Due to the low cost and good results, the usage of these medicinal plants is still practiced in different communities in India. Owing to their nutraceutical and ethnopharmacological importance, they can be utilized for fulfilling nutritional and medicinal needs.

Although medicinal plants of IHR have tremendous therapeutic applications, however, some of them have been reported to produce several toxins. The toxins produced by these plants can adversely affect human health in different ways. These toxins might be hepatotoxic, neurotoxic, nephrotoxic, etc., and result in relatively mild allergic reactions to serious medical complications that include death. Since they are lethal to animals or livestock, therefore they can be used as an effective tool against them. These plant resources of the IHR can be explored for the development of self-defense by using various biotechnological tools. Furthermore, their usage for the development of multi-system targeted warfare defensive agents may have several advantages over other warfare agents such as nuclear, microbial, and chemical warfare agents.

Neurotoxicity is a state where there is physical damage to the brain due to the exposure to neurotoxins. Neurotoxins kill neurons or disrupt their activity, thereby adversely affecting the normal functioning of the brain (Fatima and Nayeem 2016). The signs and symptoms of neurotoxicity include depression, anxiety, paralysis, headache, hallucination, impaired vision, numbness or weakness in limbs, behavioral changes, etc., it may be caused due to an accidental intake, drug abuse, chemotherapy, organ transplant, etc. It is the cerebellum, a part of the brain that has been reported to be most affected by the neurotoxins. Since it is responsible for the processing of information to conduct muscle activity, neurotoxins result in altered reflexes, confusion, and loss of body control.

Among toxic plants, neurotoxic plants have gained the interest of the scientific community nowadays. Literature studies have well documented neurotoxic effects of various medicinal plants of IHR. Some of the toxic medicinal plants of IHR include, *Aconitum chasmanthum*, *Arisaema triphyllum*, *Datura stramonium*, *Delphinium brunonianum*, *Convallaria majalis*, *Conium maculatum*, *Catharanthus roseus*, *Cannabis sativa*, *Coriaria myrtifolia*, *Ricinus communis*, *Gloriosa superba*, *Heracleum canescens*, *Hyoscyamus niger*, *Karwinskia humboldtiana*, *Mandragora officinalis*, *Melia azedarach*, *Podophyllum pelatum*, *Silybum marianum*, *Solanum xanthocarpum*, etc. (Carod-Artal 2015; Gupta et al. 2018). Among them, *A. chasmanthum*, *D. stramonium*, *C. roseus*, *C. sativa*, *R. communis*, *G. superba*, *K. humboldtiana*, *M. officinalis*, *M. azedarach*, *P. somniferum*, *P. hexandrum*, etc.

have profound neurotoxic effects. These plants include anticholinergic, nicotinic, convulsant, or epileptogenic and hallucinogenic plants (Diaz 2016).

Literature studies have well documented neurotoxic effects of these plants. Direct contact of a highly toxic plant, *Aconitum chasmanthum,* found in the meadows of the Himalaya with skin causes numbness (Ameri 1998). It is due to the presence of a toxic diterpenoid alkaloid, aconitine present in it. Both cardiotoxic and neurotoxic effects of this neurotoxin have been reported. Accidental consumption of this plant has resulted in adverse effects like aconite poisoning (Chan 2009). Likewise, eating a few berries of *Melia azedarach* can result in the death of humans (Oelrichs et al. 1983). It also results in the paralytic attacks in birds that eat too many seeds.

This book chapter sheds light on some of the highly neurotoxic plants of the Indian Himalayan Region. The exceptional geographic and climatic conditions of this region make it a treasure home to a wide range of plants. The chapter covers the detailed mechanism of the neurotoxicity caused by Himalayan medicinal plants. Furthermore, case studies on the neurotoxic effects of these plants have also been included.

## 2 Mechanism of Neurotoxicity

The nervous system comprises the central nervous system (CNS) and peripheral nervous system (PNS). The CNS consists of the brain, and the spinal cord and PNS consist of the network of neurons (afferent sensory and efferent motor nerves) arising from CNS to all parts of the body, thus regulating bodily functions. A part of PNS is the autonomic nervous system (ANS) that is not in the conscious control of organisms and includes the functioning of vital organs. The ANS is further divided into a sympathetic and parasympathetic nervous system that regulates the same organ in opposite ways, like upregulating and downregulating circulation and digestion, etc. The damage to any part of the nervous system leads to various disorders and damages, which may lead to specific symptoms like anxiety, numbness, tingling, depression, impaired vision, impotence, paralysis, and in severe condition can lead to organ failure that can be fatal. Neurotoxicity is the state of physical damage caused by potent neurotoxins, which include substances that alter the normal functioning or kill the neurons, causing an adverse effect on the nervous system. This neurotoxicity may be caused by pesticides, metal ions, food additives, coloring substances, and even natural products (Fatima and Nayeem 2016).

Medicinal plants have long been used for their pharmacological properties imparted by the presence of secondary metabolites but often cause clinical toxicity, thus causing damage to the body. The neurological pathology of medicinal plants is accredited to the potent neurotoxins that often induce acute and rarely chronic neurotoxicity among individuals. The mechanism by which neurotoxin causes toxicity is diverse. Neurotoxins generally disrupt the activity of the nervous system (central and peripheral) and neuromuscular junctions by either activating or inactivating the important ion channels and neuronal pumps leading to inhibition of neuronal signal transduction (Kristanc and Kreft 2016). It causes adverse effects, often injuries that either appears immediately or delayed like limb weakness, loss of memory, headache, sexual dysfunction, vision loss, mental and behavioral problems, etc. (Choudhury et al. 2011). These harmful effects may even lead to the death of human beings.

The transmission of information across the synaptic cleft in neurons occurs with the help of molecules called neurotransmitters, which act as messengers of the nervous system. Few examples of neurotransmitters are acetylcholine, adrenaline/ noradrenaline, dopamine, serotonin, histamine, glutamate, gamma-aminobutyric acid (GABA), etc. Many neurotoxins exert their effect by binding to various receptors acting either agonist (full or partial) or antagonist to neurotransmitters, thus modulating their normal functioning. Some neurotoxins exert their toxicity by inhibiting the enzyme that downregulates the concentration of neurotransmitters such as monoamine oxidases and cholinesterases (Wink 2000). Some alkaloids such as α-chaconine, α-solanine, tomatine, 3 β-hydroxylupanine, and cepharanthine have been found to exert toxic effects by interfering with the membrane integrity of the cell (Wink 2000).

The mechanism of neurotoxicity induced by different neurotoxins has been briefly explained. Ricin is a type 2 ribosome-inactivating protein (RIP) derived from *Ricinus communis* L. (Euphorbiaceae), also known as castor bean (Polito et al. 2016). It belongs to the A-B family of toxins bearing heterodimeric A-chain and B-chain forming its structure. Ricin A-chain is a globular 267 amino acid long structured by 8 α-helices and 8 β-strands and is responsible for the catalytic activity of the molecule. Ricin B-chain is 262 amino acid long organized by two homologous domains responsible for the binding activity of the molecule. Both A-chain and B-chain are bound together by a disulfide bridge (Doan 2004; Polito et al. 2019). Although, the castor bean is used in folk medicine for a variety of pharmacological properties such as to promote lactation, anti-rheumatic activity, for gastropathy, constipation, inflammation, fever, ascites, bronchitis, cough, skin diseases, to induce uterine contraction as an abortive, etc., but, it is a highly potent neurotoxin often resulting to death in case of intoxication (Scarpa and Guerci 1982; Ladda and Kamthane 2014).

The cellular uptake of ricin occurs by binding to cell surface receptors where it recognizes N-acetylgalactosamine, β-1,4-linked galactose, or mannose residues on the cell membrane. Eventually, the holotoxin is internalized by receptor-mediated endocytosis. After ricin uptake, the molecule is first delivered to early endosomes from where it is sent back either to the cell surface or is proteolytically degraded by lysosomes delivered via late endosomes. Few non-degraded ricin is routed to Golgi complex and cycled to the endoplasmic reticulum (ER) via coatomer protein 1 (COP-1)-coated vesicles dependent or independent pathway. After reaching ER, two chains get separated, and A-chain is cycled to cytosol with help from cytosolic chaperones, ribosome-mediated refolding, and interactions with ribosomal and proteasomal factors which also helps A-chain to regain its active form. Active A-chain having catalytic activity is an N-glycosidase that removes a universally-conserved adenine at position 4324 of 28s ribosomal RNA of the large subunit of the eukaryotic ribosome. This interaction leads to damage of 28s rRNA and inhibition of protein synthesis. Moreover, ricin also leads to damage via inflammation, DNA damage, and apoptosis (Sowa-Rogozinska et al. 2019; Polito et al. 2019).

Abrin is a toxic plant protein obtained from seeds of *Abrus precatorius* L. (Leguminosae), commonly known as jequirity bean or rosary pea. It belongs to type 2 RIPs and has structure and properties similar to ricin (Zhu et al. 2019). Structurally, abrin consists of two chains A and B linked to each other by a disulfide linkage. The A- chain comprises 251 amino acid residues, has a molecular weight of 30 kDa, and consists of three folding domains bearing catalytic activity. The B-chain is the heptamer comprising of 268 amino acids long chain (molecular weight 35 kDa) with two highly conserved saccharides binding site (Dickers et al. 2003; Olsnes et al. 1974). Abrin's mechanism of toxicity is similar to ricin. Abrin binding to the cell is facilitated by B-chain of the molecule, which has a high affinity for mannose cell receptors on a variety of cells, particularly reticuloendothelial cells. After the binding, internalization of the molecule takes place via endocytosis. The A-chain inside the cell is cycled to ribosomes, where it catalytically halts ribosomal function by removing the adenine residues from positions 4 and 324 of 28S rRNA of the larger subunit. It inhibits protein synthesis and ultimately leads to cell death. In vitro experiments depict abrin as more potent than ricin. Abrin is also known as arrow poison because it was used on the tip of the arrow or spear in olden days to induce poisoning to make sure the fatality of the purpose (Dickers et al. 2003; Patil et al. 2016).

Swansonine is an indolizidine alkaloid, is a natural toxin found in flowering plants such as *Astragalus*, *Oxytropis*, *Ipomea* sp., *Swansona canescens* frequently in the rangelands of the western United States, South America, Asia, and also in Europe. Locoweed is more often ingested by domesticated animals while grazing, which causes a severe neurological condition called as locoism (Panter et al. 2014; Kristanc and Kreft 2016).

Swansonine is structurally similar to mannose. It functions by inhibiting lysosomal mannosidase and Golgi mannosidase II which are crucial to glycosylation of many proteins, including membrane receptors, hormones, and a variety of enzymes. This inhibition causes composite oligosaccharides and glycoprotein to accumulate in the lysosomes, which results in vacuolation of the cell types, including neurons (Kristanc and Kreft 2016). The likely mechanism by which swansonine inhibits the enzyme is the cis-glycol functional group, which identifies the enzyme glycosylation cation, and the eight hydroxyl groups increase similarity with the mannose cation. Finally, the intramolecular hydrogen bond increases the specificity of swansonine and affinity with the enzyme (Wu et al. 2016). Swansonine is soluble in water and excreted via urine if ingested. The toxicity is reversible if the plant is not ingested for 5–6 days, and cell repair begins soon as the poison clearance takes place from the body. However, the damage to the nervous system results in permanent neurological and behavioral disorders such as gait abnormalities, ataxia, abnormal posture, etc. Swansonine also causes a severe effect on the reproductive success of the animals (Wu et al. 2016).

Colchicine is a neutral lipophilic alkaloid found in plants like *Colchicum luteum*, *Colchicum autumnale*, and *Gloriosa superba*. It is a highly toxic natural compound having molecular formula $C_{22}H_{25}NO_6$. The minimum lethal dose of colchicine is reported to range between 7–28 mg (Finkelstein et al. 2010; Ade and Rai 2010). Colchicine acts by binding to intercellular protein tubulin inhibiting its alpha and beta forms to polymerize and form a microtubule. In addition to altered protein assembly in the Golgi apparatus, this microtubular disruption results in altered cell shape, decreased endocytosis and exocytosis, weakened cellular motility, and arrest of mitosis (Ben-Chetrit and Levy 1998). The arrest of mitosis is caused due to failure of chromosomal separation, a microtubule-dependent process, thus inhibiting cell division. The auto-toxic response of this microtubule disruption occurs in neuronal cells, due to accumulation of toxic cellular product which fails to get removed because of inhibition of important microtubular functions such as cytoplasmic motility and extracellular secretion of hormones and neurotransmitters leading to multi-organ failure in certain cases. This disruption also leads to toxic action on myocardial cells, decreased expression of adhesion molecules on neutrophils, histamine release from mast cells, inhibiting the secretion of insulin, among many others (Finkelstein et al. 2010; Gooneratne et al. 2014).

# 3 Neurotoxic Plants

Various medicinal plants obtained from IHR have a potential role in curing the neurological disorder. But a slight increase in the level of their dosage results in their adverse neurotoxic effects that are mostly irreversible. Many researchers have well documented the neurotoxicity of medicinal plants and neurotoxic syndromes triggered by them. Carod-Artal (2015) reported that *Mandragora officinalis*, *Datura stramonium*, *Conium maculatum*, *Coriaria myrtifolia*, *Ricinus communis*, *Gloriosa superba*, *Catharanthus roseus*, *Karwinskia humboldtiana*, and *Podophyllum hexandrum* are some of the medicinal plants that cause acute or chronic neurotoxicity in human beings. Acute poisoning caused by their extracts, fruits, seeds, roots, etc. is relatively common and their most common toxicity is gastrointestinal, followed by neurotoxicity. It often affects children and young people. An overview of some of the neurotoxic plants of IHR has been given below (Table 1).

## *3.1* Aconitum chasmanthum

*Aconitum chasmanthum* is commonly known as Gaping Monkshood, and in Hindi, it is referred to as Mohri. It is distributed throughout IHR at altitudes ranging from 2300 to 4300 m above sea level. It is a highly toxic plant resulting in neurotoxicity and cardiotoxicity. In order to reduce its toxicity, it is processed before use. However, inadequate processing, as well as consumption of dose larger than its

**Table 1** Neurotoxic components of medicinal plants of Indian Himalayan Region

| Medicinal plant | Family | Common name | Neurotoxic component |
|---|---|---|---|
| *Aconitum chasmanthum* | Ranunculaceae | Gaping monkshood | Aconitine |
| *Abrus precatorius* | Fabaceae | Rosary pea | Abrin |
| *Acorus calamus* | Acoraceae | Sweet flag | β-asarone |
| *Melia azedarach* | Meliaceae | Chinaberry tree | Meliatoxins |
| *Cannabis sativa* | Cannabaceae | Marijuana | Cannabinoids |
| *Datura stramonium* | Solanaceae | Thorn apple | Tropane alkaloids |
| *Gloriosa superba* | Colchicaceae | Flame lily | Colchicine |
| *Catharanthus roseus* | Apocynaceae | Pink periwinkle | Vinca alkaloids |
| *Podophyllum hexandrum* | Berberidaceae | Indian podophyllum | Podophyllotoxin |
| *Ricinus communis* | Euphorbiaceae | Castor bean | Ricin |
| *Lathyrus sativus* | Fabaceae | Grass pea | β-*N*-oxalyl-L-α,-β-diaminopropionic acid |
| *Mandragora officinalis* | Solanaceae | Mandrake | Tropine alkaloids |
| *Conium maculatum* | Apiaceae | Hemlock | Conium and other alkaloids |
| *Ipomea carnea* | Convulvulaceae | Bush morning glory | Swainsonine |

recommended dose, increases the risk of poisoning. Tribal people have been reported to use its dried pulverized roots by mixing with butter as an ointment on the abscess and boils (Ameri 1998). "Naswar" is prepared by mixing it with tobacco.

The toxicity of the plant can be attributed to aconitine, which is a highly toxic diterpenoid alkaloid known to suppress the inactivation of voltage-dependent $Na^+$ channels by binding to neurotoxin binding site 2 of the α-subunit of the channel protein (Gupta et al. 2018). Its toxicity can be depicted from the fact that even skin contact with it causes numbness in people (Chan 2009). Hydrolyzation of aconite alkaloids is done by soaking and boiling to convert it into less toxic or non-toxic derivatives.

## *3.2* Acorus calamus

*Acorus calamus* is one of the highly valued medicinal and economically important plant species used since the time of Ayurveda, despite its neurotoxicity. It belongs to the family Acoraceae and is commonly known as Sweet flag and Bach in Hindi. It is widely distributed near the marshy places or along the streams and water springs. In IHR, it has been reported from foothills located near the marshy places and the riverbanks of various regions of Himachal Pradesh like Kinnaur, Kullu, Mandi, Shimla, Kangra, Sirmour, Khajjair, Simoli and Zundi, Uttarakhand like Chamoli,

Batarni-Dhara, Doon valley Dehradun, Almora, Haldwani and Nainital, Sikkim, Jammu and Kashmir (Mittal et al. 2009).

This plant is used as an important component in over 77 ayurvedic preparations like Aswagandharistam, Dhanwantharaishtam, Nimbamrithasavam, Hinguevachadi chroornam, Cheriyanarayanathailam, Gugguluthiktha kamquatham, etc. Moreover, its usage in the modern medicinal system has been well documented. The plant produces a brownish, yellow-colored, essential oil commonly called the 'Calamus oil,' which is widely used in modern medicines as tonics or tablets, food industry for making frozen desserts, yogurts, cakes, confectionery and desserts (Venskutonis and Dagilyte 2003).

The importance of the calamus oil derived from *Acorus calamus* is due to the presence of phenolic ether, asarone. It is the most active constituent of the essential oil and constitutes up to 96% of the total constituents. Asarone is commonly found in three isomeric forms viz. alpha (α), beta (β), and gamma (γ). Among these three forms, β-Asarone is considered the most characteristic component of the oil. Despite great medicinal and industrial applications of its essential oils Food and Drug Administration (FDA) banned its usage (Thirach et al. 2003). It is due to the carcinogenic and neurotoxic properties of β-asarone, the active constituent.

## *3.3* Abrus precatorius

It is commonly called Rosary pea or Chanoti in Hindi, and it grows wild in most parts of India. It is native to India, with a wide distribution from the Himalayas down to Southern India. Nowadays, it grows in all tropical regions of the world (Solanki and Zaveri 2012). In India, it was used in Hindu culture from very early times as medicine. It has been reported to exhibit a broad range of therapeutic effects, like antibacterial, antifungal, antiserotonergic, antitumor, analgesic, anti-inflammatory, antispasmodic, antidiabetic, anti-migraine, anti-inflammatory, healing wounds, throat scratches, and sores, etc.

Almost all parts of the plant are toxic, but the highest concentrations of toxins are found in the seeds. The color of the seeds varies from black to red. The most common color of the seeds is glossy red. A black band is found at the end of the seed which makes it attractive (Jang et al. 2010). The neurotoxicity of the seeds is because of the presence of a toxic plant substance, abrin. Due to high toxicity, it was used as an arrow poison in olden days. Generally, its toxicity initially involves gastrointestinal manifestations. However, there are a few reports where altered sensorium and seizures have been reported as the initial evidence (Patil et al. 2016). Its toxic effects on the CNS include headache, dilated pupils, hallucinations, drowsiness, tetany, and seizures. Furthermore, a rare complication of *A. precatorius* poisoning causes demyelination (Sahoo et al. 2008). Patil et al. (2016) reported rare evidence of its neurological manifestation in the pediatric age group.

Abrin affects the normal functioning of ribosomes that involves halting protein synthesis and leading to cellular death. Among human beings, a dosage of 0.1–1 μg/

kg proves to be fatal. Ingestion of intact whole seeds delays the onset of toxicity, but an increase in toxicity decreases the time to onset the symptoms have been observed after crushing of seeds (Reedman et al. 2008).

## *3.4* Melia azedarach

It is commonly known as Chinaberry tree or Bakain in Hindi. It is distributed throughout IHR, ranging from North-Western to North-Eastern Himalayas (Biplov et al. 2015; Chauhan et al. 2017). Various biological activities like insecticidal, antibacterial, antifungal, antimalarial, anticancer, antiviral, and many more have been reported from it (Gupta et al. 2018). Despite its promising biological activities, *M. azedarach* has been reported to possess certain toxic effects. All parts of the tree are poisonous. Due to its toxicity, its bark and fruit extract are used to kill parasitic roundworms. The tree is so poisonous that even eating as few as 5–6 berries can result in the death of human beings. Moreover, birds that eat its seeds become paralyzed. The toxicity is due to the presence of some terpenoids, such as meliatoxins (Carpinella et al. 2003). Tetranortriterpenes of the limonoid class have been reported to exhibit acute nervous symptoms resulting in the death of animals/mammals (Oelrichs et al. 1983; Isman 2006).

## *3.5* Cannabis sativa

It is commonly called Marijuana or bhang in Hindi. It is distributed in a variety of habitats and altitudes that ranges from sea level to the alpine foothills of the Himalayas. It has been cultivated and used for 5000–6000 years. Therefore, it is difficult to pinpoint the origin of this species (El Sohly et al. 2017). Furthermore, it is an economically important crop and was used in folk medicine as well as a source of textile fiber since the dawn of time (Andre et al. 2016). This plant has been reported to contain more than 560 (approximately) constituents, out of which 400 (approximately) are toxic. The major psychotropic effects are attributed to toxic constituent 9-tetrahydrocannabinol (THC), a cannabinoid. People are using it for centuries for the effects of its psychoactive resins. Besides, tobacco-like preparation (marijuana) of the leaves and flowers is also used for a long time. The resin extracted from the top of flowering plants contains much higher THC concentration and is widely used for psychoactive effects (Fitzerald et al. 2013).

*Cannabis sativa* is divided into three phenotypes (Galal et al. 2009): phenotype I (drug-type), with THC $>0.5\%$ and cannabidiol (CBD) $<0.5\%$ (THC/CBD $\gg 1$); an intermediate phenotype II (intermediate-type), with CBD as the major cannabinoid (THC/CBD $\sim 1$); phenotype III (fiber-type or hemp), with low THC content (THC/CBD $\ll 1$). Various environmental factors play an important role in determining the amount of cannabinoid present in various parts of the plant at different growth stages. But CBD: THC ratio is under genetic control (De Meijer et al. 2003).

Besides THC, there are cannabinoids with several medicinal functions, such as cannabidiol (CBD), cannabichromene (CBC), and cannabigerol (CBG). Various, non-cannabinoid constituents belonging to diverse classes of natural products are responsible for the therapeutic potential of the plant. Their potential applications in the treatment of several serious illnesses, such as glaucoma, depression, neuralgia, multiple sclerosis, Alzheimer's, and alleviation of symptoms of HIV/AIDS and cancer have been documented (El Sohly et al. 2017).

## *3.6* Datura stramonium

The troublesome weed *D. stramonium* is known as Jimson weed. It is also known as thorn apple, angel's trumpet, locoweed, etc. It is native to deserts of America, Europe, Asia, and Africa and is also mainly distributed in the IHR from Kashmir to Sikkim (up to 2–700 m altitudes). Since it is a plant that has both poisonous as well as medicinal properties, but it is tremendously used in ethnomedicine and modern medicine (Chander and Sharma 2020). It has been reported that all parts of the plant are poisonous if ingested by humans or livestock. Although its small quantity is used for medicinal purposes, overdosage can result in severe toxicity. Traditionally, it was used for curing various human ailments, such as ulcers, wounds, inflammation, rheumatism, sciatica, bruises and swellings, fever, asthma, and bronchitis, etc. About 64 tropane alkaloids have been detected from *D. stramonium*. Among them, two major tropane alkaloids are atropine and scopolamine (Berkov et al. 2006).

The incidence of *D. stramonium* poisoning is very common in India, and usually, it involves seeds. In India, the seed of *D. stramonium* is believed to be associated with the God Shiva among Hindus, which can promote misuse of the plant on religious occasions, such as Shivaratri and Swasthani Puja (Gaire 2008). Mostly, its poisoning is sporadic, with a cluster of cases being reported mostly among adolescents. However, some of the cases of unintentional poisoning by *D. stramonium* have also been reported where it was accidentally taken, or an overdose of decoction prepared from herbal prescription was consumed (Devi et al. 2011). Some of the common symptoms of its poisoning include delirium, agitation and seizures, mydriasis, blurred vision, photophobia, dry mouth, and mucous membranes, extreme thirst, tachycardia, nausea, and vomiting, decreased bowel sounds, difficulty in swallowing and speaking, hyperthermia, hypertension, loss of consciousness, and coma (Gaire 2008). Therefore, it is mostly studied for its toxicological properties.

The neurotoxicity is due to the presence of tropane belladonna alkaloids including, hyoscyamine, hyoscine, atropine, and scopolamine, etc., produced in various parts of the plant (Deng 2005). Atropine has more exciting properties, whereas scopolamine has more relaxing and hallucinogenic properties. These tropane belladonna alkaloids may result in anticholinergic poisoning if taken in large concentrations. Tropane alkaloids contain a methylated nitrogen atom (N-$CH_3$) and include anticholinergic drugs atropine, scopolamine, and hyoscyamine (Devi et al. 2011).

## *3.7* Gloriosa superba

It is a flowering plant commonly known as Climbing lily and Kalihari in Hindi. It is a native of tropical Africa and is now found naturally throughout much of tropical Asia, including India, Sri Lanka, Malaysia, and Burma. In IHR, it has been reported from Northern to Eastern Himalayas. Various parts of the plant are widely used in ethnomedicine for a long time (Khan et al. 2007). Its seed and tuberous rhizome are very useful medicinally for the treatment of gout, wounds, snakebite. *G. superba* contains alkaloids, colchicines, gloriosine, 3-desmethyl colchicine, β-lumicolchicine, N-formyldesacetyl-colchicine, 2-desmethyl colchicine, and compounds such as chelidonic acid and salicylic acid that plays an important role in its therapeutic potential (Duke 1985).

Indians use it commonly on windowsills to deter snakes. Also, this species is believed to have various magical properties (Malayandi et al. 2019). Its leaves are used as an ingredient in arrow poisons due to the toxicity of its phytoconstituents. Mostly, *G. superba* is implicated in cases of self-poisoning. Colchicine is present in the seeds and tubers in highly concentrated amounts that mediate toxicity. Besides antimitotic effects, it also results in the death of neuronal cells. Binding nature of colchicine with tubulin results in disruption of tubulin polymerization, which further delays the onset of toxic encephalopathy, thereby disrupting neuronal cells (Malayandi et al. 2019). Gooneratne et al. (2014) reported toxic encephalopathy due to the poisoning of *G. superba* alkaloid, colchicine. The encephalopathy might be attributed to a direct-effect of colchicines that was probably mediated through its effect on microtubular transport.

## *3.8* Catharanthus roseus

*C. roseus* was earlier called as *Vinca rosea*. It is commonly known as Madagascar periwinkle and Sadabahar in Hindi. This plant species is endemic to Madagascar (Malagasy Republic) and is cultivated as an ornamental plant in gardens throughout the world (Garcia 1954). Also, it is widely distributed in Indian Himalayas. The plant has been reported to possess great therapeutic potential, especially as a cancer fighter. Folkloric usage in lowering blood pressure and blood sugar has been documented. The plant is a rich source of alkaloids that are important for its therapeutic potential. Neurosedative properties of its alkaloids have been reported. Among alkaloids, vinca alkaloids are of great importance. Four major vinca alkaloids, vinblastine, vinorelbine, vincristine, and vindesine obtained from this plant, are in clinical use. They are the second-most-used class of cancer drugs (Moudi et al. 2013). Apart from antitumor activity, these alkaloids play a vital role in treatment for acute lymphoblastic leukemia, Hodgkin's disease, non-Hodgkin's lymphomas, multiple myeloma, Wilm's tumor, Ewing's sarcoma, and small cell carcinoma of the lung.

Despite useful activities, the usage of vinca alkaloid, vincristine, is limited due to its neurotoxic effects. The adverse effects are manifested with reduced motility of the intestines resulting in constipation and peripheral neuropathy (Moudi et al. 2013). The early symptoms such as numbness and tingling of hands and feet are accompanied by loss of deep tendon reflexes. Under extreme conditions, muscle weakness develops, which is more marked in the distal muscles of the hands and feet. Besides, some other manifestations of its neurotoxicity involve ocular palsies, hoarseness of voice, and autonomic neuropathy in the form of postural hypotension and atony of the urinary bladder (Legha 1986). The neurotoxicity is dose-related and is usually reversible on the interruption of the therapy. However, the recovery rate is slow and takes several months. Furthermore, no specific antidote has been established against its neurotoxicity.

The vinca-alkaloids are amongst the most important drugs that induce peripheral neurotoxicity. Since they are widely used for various malignancies such as ovarian and breast cancer, and hematological cancers, thus, chemotherapy-induced neuropathy is common. Furthermore, among patients who already have neuropathic symptoms (due to diabetes mellitus, hereditary neuropathies, or earlier treatment with neurotoxic chemotherapy) are more vulnerable to the development of chemotherapy-induced peripheral neuropathy.

Accidental massive overdose with vincristine may cause CNS toxicity and may lead to death. Cortical blindness, athetosis, ataxia, and parkinsonian-like symptoms caused by vincristine are infrequent and usually reversible. Vincristine is used especially in non-Hodgkin's lymphoma, Hodgkin's disease, and leukemia. The most important dose-limiting adverse effect of vinblastine is bone-marrow suppression, which usually precedes neurotoxicity. Vindesine and vinorelbine cause mild neurotoxicity resulting in loss of deep tendon reflexes (Verstappen et al. 2003). Likewise, many other neurotoxic complications are caused by chemotherapy in cancer patients due to vinca alkaloids.

## *3.9* Podophyllum hexandrum

It is an herbaceous and rhizomatous species of the genus *Podophyllum* that has endangered status in India. It is distributed in restricted pockets of Indian Himalayan regions, especially Northern Himalayas such as Garhwal, Kashmir, etc. Like other plants as discussed above, it is being used in several systems of traditional medicine such as Ayurveda for curing various ailments like cold, constipation, biliary fever, plague, bacterial infections, mental disorders, allergies, etc. The medicinal value of plants is due to various phytoconstituents, but lignans obtained from rhizome are of great importance (Bhadula et al. 1996). This plant is valued all over the world for its medicinal importance.

Podophyllotoxin is important lignin isolated from its rhizome and acts as a precursor for chemotherapeutic semi-synthetic drugs. It has an important biological activity in blocking mitosis. Since *P. hexandrum* has been reported to contain three times more amount of podophyllotoxin than its counterpart *P. pelatum*, therefore, it is more exploited (Alam et al. 2009). Moreover, its chemical synthesis is also considered as complicated (Giri and Narasu 2000). Due to this reason, its population size is decreasing and is now among threatened species of IHR.

Although widely used as a therapeutic agent, podophyllotoxin has also been reported to cause peripheral and autonomic neuropathy (Carod-Artal 2015). Chang et al. (1992) conducted a study to report the neurotoxicity of podophyllotoxin. Etoposide, a podophyllotoxin-based chemotherapeutic drug, is used in the treatment of lung cancer and leukemia (Verstappen et al. 2003). A high-dose therapy of etoposide results in headache, seizures, and somnolence in patients who have undergone bone marrow transplants (Leff et al. 1988).

## *3.10* Ricinus communis

This plant is native to India and is widely distributed in IHR (Sharma et al. 2013; Aziz et al. 2016). The plant has gained importance due to the presence of various phytoconstituents like triterpenoids, flavonoids, lignin, tannins, alkaloids, glycosides, etc. in its leaves, root, and seeds. Among these compounds, alkaloids have significant importance. Castor beans are a rich source of castor oil and contain about 40–60% oil that is affluent in fatty acid triglycerides, predominantly the component called ricinolein (Prince et al. 2011). Despite, castor beans also hold ricin, a poison due to which more level of focus is given on this plant.

Ricin is referred to as a toxic glycoprotein found in seeds of *R. communis*. It belongs to a heterogeneous group of proteins, lectins, and shares the property of binding to glycoproteins and glycolipids of the cell surface. It is one of the most potent and lethal substances known and is considered as an important bioweapon, a highly attractive agent for bioterrorism activities. Besides, it presents great potential in cell-based research as well as an anticancer agent. Moreover, it is being explored for nanoparticle formulations in tumor therapy (Franke et al. 2019). It acts as a state-of-the-art tool in neurobiology for selectively destroying neuronal populations. With the specificity offered by it, it can be used as a potential conjugating molecule for monoclonal antibodies in monoclonal antibody technology to get improved selectivity in the treatment of human malignancy (Ilercil et al. 1994). However, some studies have reported that their neurotoxicity might be dose-limiting in human therapy, while others have not predicted it in preclinical animal trials. They can also agglutinate red and malignant cells or stimulate cell proliferation. Besides, it is also used for the study of the physiological consequences on CNS premotor neurons of the loss of their neuronal target during the execution of well-defined motor tasks (De la Cruz et al. 1995). Due to these properties, ricin has long been known in the pharmacopeia and used since ages for both medical as well as criminal purposes (Aziz et al. 2016). Ricin interferes irreversibly with the protein synthesis by catalytically inactivating the 60S rRNA subunit. It works in such an efficient manner that even a single molecule is enough to kill a cell. *R. communis* apart from ricin also contains several allergenic compounds such as castor oil, agglutinin, ricinin, nudiflorin, etc. (Franke et al. 2019).

## *3.11* Lathyrus sativus

It is commonly known as Grasspea is a high-yielding legume consumed as a food in Northern India. It is drought-resistant and is valued for its high protein content. This crop can adapt to extreme conditions, apart from being largely grown in various states if India, it is also found in IHR ranging from Northwestern to Eastern Himalayan regions (Kumari 2001; Talukdar 2013). However, the presence of the neurotoxin, β-*N*-oxalyl-L-α,β-diaminopropionic acid (β-ODAP) in seeds has hindered its development as an important legume crop. Studies have shown that consumption of its large quantities for prolonged periods causes irreversible paralysis. But some low-toxin lines have been developed nowadays that make it safer for both animal and human foods (Yan et al. 2006). Consumption of its seeds for more than 3 months causes neurolathyrism. The agent (β-ODAP) responsible for this cortical motor neuron disease is present in seed as the free amino acid (Murti et al. 1964; Rao et al. 1964). The higher intake of this legume crop, especially under drought conditions by poorly nourished people, is the characteristic setting for neurolathyrism epidemics (Yan et al. 2006).

Lathyrism is considered as one of the oldest neurotoxic diseases known to humankind. It results from excessive consumption of the *L. sativus* and its related species (Spencer and Schaumburg 1983). It is an irreversible, non-progressive spastic paraparesis that is associated with poorly understood degenerative changes in the spinal cord. This disease is prevalent in Central India, where the poorer sections of the people eat its seeds. Moreover, it has also been reported from other countries, viz. Spain, France, other Mediterranean countries, and the United States, where other species of Lathyrus are grown (Murti et al. 1964). There are two distinct types of lathyrism, neurolathyrism, and osteolathyrism. Neurolathyrism is characterized by nervous disorders such as hyperirritability, weakness, and paralysis of leg muscles and convulsions. Rao et al. 1964 contributed to the isolation and characterization of neurotoxins from this plant. The first evidence that *L. sativus* neurotoxicity might be caused by a compound, L-3-oxalylamino-2-aminopropionic acid (OAP) was given by Parker et al. (1979). Although the experiment was performed on a primate (young male squirrel monkeys, *Saimiri sciureus*) with a mature blood-brain barrier, it supported a possible role for OAP in the etiology of human neurolathyrism. But the etiology of this crippling disease is still unclear.

## *3.12* Mandragora officinalis

It is found in the high Himalayas and is the sole representative of the genus Mandragora in India. It is commonly known as Himalayan mandrake. In India, it is found only in two states, Arunachal Pradesh and Sikkim. It has been reported to be used by various tribes of these states to cure different ailments in their daily lives (Maity et al. 2019). It is used to cure gastric and respiratory problems. It is also used to get rid of joint pains. There are reports where it was being given to patients before surgery to reduce the sensitivity of the brain and for its soporific properties (Blakemore and Jennett 2001). This plant contains some important alkaloids, e.g., tropine alkaloids such as hyoscine and anisodamine that are highly toxic despite their important medicinal uses (Zhang et al. 1994). Their side effects due to overdose include drowsiness, hallucinations, etc. The unique chemical properties of its root have made it famous for its magical powers (Fowler 2015).

Its tropine alkaloids cause anticholinergic plant poisoning. Ingestion of its seed in salads or stews or brewed into teas adversely affects CNS and PNS. Their tropine alkaloids inhibit acetylcholine receptors in the brain resulting in fever, tachycardia, blurred vision, vertigo, etc. It may later result in confusion, dysarthria, hallucinations, and fatal coma (Diaz 2016).

## *3.13* Conium maculatum

It is commonly known as Hemlock or Dog's poison. It is traditionally being used in the treatment of spasmodic disorders and to relieve nervous excitation, rheumatic pains, stomach-aches, nervousness, and restlessness. It is also used externally to treat herpes, erysipelas, breast tumors, etc. (Madaan and Kumar 2012). The plant contains phytoconstituents like piperidine alkaloids, coniine, 2-phenyl-3,4,5,6-tetrahydropyridine, and 5-hydroxy-2-pentylpiperidine; fatty constituents, palmitic, stearic, oleic, petroselinic, linolenic and arachidic acids; unsaponifiable matter containing β-sitosterol and lupeol; flavonoids; coumarins, polyacetylenes; volatile and nonvolatile oils.

Piperidine alkaloids possess nicotinic effects like coniine, N-methyl coniine, conhydrine, and pseudoconhydrine and γ-coniceine (Karakasi et al. 2019). Coniine is neurotoxic for humans and causes death via respiratory paralysis. It's poisoning usually occurs by the ingestion of the plant, but it may also be caused by the ingestion of meat of the animal feeding it (Konca et al. 2014). Although every part of the plant is toxic, the highest alkaloid concentration exists in the seeds (Leete 1994). Ataxia and headache are observed in the early stage of poisoning, while increased salivation, tachycardia, and pupillary dilation develop due to the effects of the plant on the autonomic ganglia.

### *3.14* Ipomea carnea

It is commonly known as Bush Morning Glory and is widely used in different traditional medical systems such as Ayurveda, Siddha, and Unani. It has been reported to possess medicinal properties such as antioxidant, antidiabetic, antimicrobial, wound healing, anti-inflammatory, immunomodulatory, cardiovascular, hepatoprotective activity (Bhalerao and Teli 2016). Besides, its stem is also used for making paper. Therapeutical importance of the plant indicates that it has potential bioactive phytochemicals, convolvulinolate (11-hydroxy-pentadecane acid), ipurolic acid (3,11- dihydroxy-tetradecane acid), 7-hydroxy-decane acid, acacetin-7-galactoside, ipomotocin, 2-epi-lentiginosine, swainsonine, N-methyltrans-4-hydroxy-L-prolineat, calystegines, etc. (Singh et al. 2020).

Swainsonine is an etiological agent responsible for neurotoxicity of this plant, and it causes a chronic neurotoxicity syndrome, locoism. Since it has close structural similarity to mannose, it can inhibit the lysosomal mannosidase and Golgi mannosidase II which induces an enzymatic dysfunction resulting in accumulation of complex oligosaccharides in lysosomes (Kristanc and Kreft 2016). Other abnormalities, such as abnormal posture, ataxia, and posterior limb paresis, may also occur. The cerebellum is mainly affected by Ipomoea intoxication (Bhalerao and Teli 2016).

## 4 Case Studies: Neurotoxins of Medicinal Plants

A few case studies on the neurological manifestations of poisoning caused by various medicinal plants and their active constituents have been described below.

### *4.1 Colchicine Poisoning*

Gooneratne et al. (2014) reported the case of a 28-year-old woman who had eaten tubers of *G. superba* to end her life after a domestic dispute. The case highlighted a delayed toxic encephalopathy that could be attributable to the direct effect of its major toxic constituent, colchicines. Colchicine might have affected its microtubular transport. MR scanning of the patient's brain showed bilateral symmetrical T2 basal ganglia hyperintensities in the caudate and lentiform nuclei. Generally, patients with colchicine poisoning develop confusion and seizures and get concomitant renal or liver dysfunctioning. However, in their report, the patient had no significant metabolic derangement. Moreover, another unusual feature of encephalopathy observed was the delayed onset, reminiscent of that following anoxic encephalopathy and carbon monoxide poisoning.

Since, the active toxic compound of the plant is alkaloid colchicine, which inhibits its polymerization into microtubules by binding to tubulin. Studies on various experimental and animal models have shown that colchicine causes the neuronal cell death of certain neuronal populations. Besides, it may induce an autotoxic response, leading to neuronal death of certain populations due to the accumulation of a toxic cellular product, which usually is transported by a microtubule-dependent process. This might be the explanation for the delayed onset of encephalopathy in the patient.

### *4.2 Hemlock Poisoning*

Konca et al. (2014) reported a case of a 6-year-old girl who was admitted to hospital with a burning sensation in the mouth, hypersalivation, tremor in hands, and walking imbalance after ingestion of poisonous plant *C. maculatum* (Hemlock). Since the plant resembles common vegetables, she accidental ingested it. Initially, the patient was fully conscious but agitated, and the amount of saliva was too much for her to swallow. Her neurological examination revealed that she had a tremor in both hands, loss of motor function in both lower limbs, dilation in both pupils, limitations in the movement of tongue, ataxia and decreased response to light.

Various piperidine alkaloids present in the plant are a kind of peripheral neurotoxins that affect neuromuscular junctions and create nicotinic effects in the autonomic ganglia. It has been reported that ataxia and headache are the symptoms observed in the early stage of poisoning (Ferah et al. 2006). The plant affects autonomic ganglia resulting in hypersalivation, tachycardia, and pupillary dilation. Further, increased cholinergic stimuli may increase muscle weakness or result in paralysis, bradycardia, and central nervous system depression.

### *4.3 Abrin Poisoning*

Patil et al. (2016) reported a case where a child accidentally ingested *A. precatorius* seeds. The child was presented with altered sensorium and convulsions. It is a rare case and has not been reported earlier in the pediatric age group. Before the convulsions, there was no history of fever, vomiting, or pain abdomen, or rashes, but the history suggested of pica for the last 2–3 months.

In this case, the presence of pica and the beautiful nature of seeds must have attracted the child for the ingestion of its seeds. These seeds are rich in abrin, which is a highly toxic plant substance. Out of two polypeptides chains present in abrin, one removes adenine residue from 28s rRNA and stops protein synthesis. In contrast, another chain facilitates the entry of a molecule in the cells by acting on the cell membrane. However, delayed and unusual initial neurologic manifestation reported in this case rather than classical initial hemorrhagic gastroenteritis might have occurred due to the ingestion of intact seeds of *A. precatorius*.

### 4.4 Aconitine Poisoning

Tak et al. (2016) reported a case of a 55-year-old male who was presented with a history of nausea, vomiting, palpitation paresthesis, and profuse sweating in 2 h after ingestion of Aconitum root. He procured it from Ayurvedic pharmacy, consumed 5 g of it, and became symptomatic within 1 h. Examination revealed shock with irregular pulses.

Aconitum is a fast-acting toxin with aconitine, and related alkaloids as its main constituents and its pure 2 mg may cause death. Its toxicity is characterized immediately after ingestion with symptoms like the burning sensation of the lips, tongue, mouth, and throat. Furthermore, it may result in numbness of the oral cavity and throat. Besides, salivation, nausea, vomiting, dizziness, syncope, palpitation, and diarrhea may occur (Chan 2009). These neurotoxins adversely affect the voltage-sensitive sodium channels of the cell membranes of myocardium, nerves, and muscles of excitable tissues. Once aconitine binds with the voltage-sensitive sodium channels, it persistently activates the sodium channels, which become refractory to excitation. Arrhythmia-like situation is triggered due to delayed after-depolarization and early after-depolarization. Moreover, such properties are also in part due to its anticholinergic effects mediated by the vagus nerve.

### 4.5 Scopolamine Poisoning

Dufayet et al. (2020) reported a rare case of scopolamine-facilitated robbery, via inhalation. The victim was a 58-year-old man without any medical problems. After an unknown man tried to apply a piece of cloth on his face, he gradually lost consciousness. Clinical evaluation revealed the disturbance of consciousness with bilateral mydriasis, hypertension, and tachycardia. Scopolamine was the only substance found in both whole blood and urine samples using liquid chromatography tandem mass spectrometry. Since the half-life of scopolamine is short, therefore its poisoning is rare. But this is the first case of an inhaled scopolamine robbery attempt that had been reported with quantitative toxicological findings.

Scopolamine is a tropane alkaloid found in *D. stramonium* and some other plants and is a non-selective competitive inhibitor of acetylcholine receptors. Some common symptoms of its poisoning, anticholinergic encephalopathy, include agitation, confusion, hallucination, convulsions, and even coma. Generally, the symptoms regress within 2 days, but some fatal cases have also been noted (Le Garff et al. 2016; Lusthof et al. 2017).

## 5 Conclusions

Medicinal plants have long been used for their pharmacological properties without stressing much on the toxic nature of these plants. There always have been high public interest and acceptance for herbal products due to the misconception that "natural always means safe," but this is not entirely true. Nature does possess its defense mechanism to safeguard itself by producing phytochemicals that can act as potent toxins to the consumer. Improper intake, misidentification, improper dosage, and lack of expertise are the major causes of poisoning by them. The primary intention of this chapter is to stress the neurotoxic potential of some medicinal plants of the Indian Himalayan region, emphasizing their toxicity, mechanism of action, and clinical studies to recapitulate the available data on these medicinal plants. The fact much research has been carried out in unraveling the toxicological potential of medicinal plants, but there remains a lot of a grey area left, which needs to be filled soon. Furthermore, the generation of antidotes against neurotoxins, used as biorational drugs in medicine and agriculture, has always been the interest. There are opportunities for carrying out intensive research in a multidisciplinary approach to investigate the real insights in the toxico-kinetics and mechanism of action of these toxins.

## References

Ade R, Rai MK (2010) Colchicine, current advances and future prospects. Nusantara Biosci 2 (2):90–97. https://doi.org/10.13057/nusbiosci/n020207

Alam MA, Gulati P, Gulati AK et al (2009) Assessment of genetic diversity among *Podophyllum hexandrum* genotypes of the North-Western Himalayan region for podophyllotoxin production. Indian J Biotechnol 8(4):391–399

Ameri A (1998) The effects of *Aconitum* alkaloids on the central nervous system. Prog Neurobiol 56(2):211–235. https://doi.org/10.1016/s0301-0082(98)00037-9

Andre CM, Hausman JF, Guerriero G (2016) *Cannabis sativa*: the plant of the thousand and one molecules. Front Plant Sci 7:11–17. https://doi.org/10.3389/fpls.2016.00019

Aziz S, Rabniwaz A, Ghani KS (2016) Phytochemical and biological screening of *Ricinus communis* seed oil grown wild in Jammu & Kashmir. J Pharmacog Phytochem 5(3):89–92

Ben-Chetrit E, Levy M (1998) Colchicine: update. Semin Arthritis Rheum 28(1):48–59. https://doi.org/10.1016/s0049-0172(98)80028-0

Berkov S, Zayed R, Doncheva T (2006) Alkaloid patterns in some varieties of *Datura stramonium*. Fitoterapia 77(3):179–182. https://doi.org/10.1016/j.fitote.2006.01.002

Bhadula SK, Singh A, Lata H et al (1996) Genetic resources of *Podophyllum hexandrum* Royle, an endangered medicinal species from Garhwal Himalaya, India. Plant Genet Res News Lett 106 (27):26–29

Bhalerao SA, Teli NC (2016) *Ipomoea carnea* Jacq.: ethnobotany, phytochemistry and pharmacological potential. Int J Curr Res Biosci Plant Biol 3(8):138–144. https://doi.org/10.20546/ijcrbp.2016.308.021

Biplov CS, Biswakarma S, Shukla G et al (2015) Documentation and utilization pattern of ethnomedicinal plants in Darjeeling Himalayas, India. Int J Usuf Mngt 16(2):3–11

Blakemore C, Jennett S (2001) The Oxford companion to the body. Oxford University Press, New York. https://doi.org/10.1093/acref/9780198524038.001.0001

Carod-Artal FJ (2015) Adverse neurological effects caused by the ingestion of plants, seeds, and fruits. In: Watson R et al (eds) Bioactive nutraceuticals and dietary supplements in neurological and brain disease, 1st edn. Academic Press, San Diego, pp 215–219. https://doi.org/10.1016/B978-0-12-411462-3.00023-0

Carpinella MC, Defago MC, Valladares G et al (2003) Anti-feedant and insecticide properties of a limonoid from *Melia azedarach* (Meliaceae) with potential use for pest management. J Agric Food Chem 51(2):369–374. https://doi.org/10.1021/jf025811w

Chan TY (2009) Aconite poisoning. Clin Toxicol (Phila) 47:279–285. https://doi.org/10.1080/15563650902904407

Chander H, Sharma G (2020) Some rare ethnomedicinal plants of lower foothills of North-Western Himalaya in Himachal Pradesh. In: Pant et al (eds) Ethnobotany and biodiversity conservation. Indu Book Service Private Limited, New Delhi, pp 123–143

Chang LW, Yang CM, Chen CF et al (1992) Experimental podophyllotoxin (bajiaolian) poisoning: I. Effects on the nervous system. Biomed Environ Sci 5(4):283–292

Chauhan PS, Bisht S, Ahmed S (2017) Traditional and ethnobotanical uses of medicinal trees in district Tehri Garhwal (Western Himalayas). Int J Ayurvedic Herbal Med 7(1):2442–2448

Choudhury C, Bawari M, Sharma GD (2011) Assessment of neurotoxic potential of a medicinal plant extract on mice. Ecoscan 1:299–304

De la Cruz RR, Pastor AM, Delgado-Garcia JM (1995) The neurotoxic effects of *Ricinus communis* agglutinin-II. J Toxicol Toxin Rev 14(1):1–46. https://doi.org/10.3109/15569549509089967

De Meijer EPM, Bagatta M, Carboni A et al (2003) The inheritance of chemical phenotype in *Cannabis sativa* L. Genetics 163:335–340

Deng F (2005) Effects of glyphosate, chlorsulfuron, and methyl jasmonate on growth and alkaloid biosynthesis of jimson weed (*Datura stramonium* L.). Pestic Biochem Physiol 84(2):16–26. https://doi.org/10.1016/j.pestbp.2004.09.007

Devi MR, Bawari M, Paul SB et al (2011) Neurotoxic and medicinal properties of *Datura stramonium* L.—review. Assam Univ J Sci Technol 7(1):139–144

Diaz JH (2016) Poisoning by herbs and plants: rapid toxidromic classification and diagnosis. Wilderness Environ Med 27(1):136–152. https://doi.org/10.1016/j.wem.2015.11.006

Dickers KJ, Bradberry SM, Rice P et al (2003) Abrin poisoning. Toxicol Rev 22(3):137–142. https://doi.org/10.2165/00139709-200322030-00002

Doan LG (2004) Ricin: mechanism of toxicity, clinical manifestations, and vaccine development, a review. J Toxicol Clin Toxicol 42(2):201–208. https://doi.org/10.1081/CLT-120030945

Dufayet L, Alcaraz E, Dorol J et al (2020) Attempt of scopolamine-facilitated robbery: an original case of poisoning by inhalation confirmed by LC-MS/MS and review of the literature. Forensic Toxicol 38(1):264–268. https://doi.org/10.1007/s11419-019-00492-z

Duke JA (1985) Handbook of medicinal herbs. CRC Press, Boca Raton, FL, p 677

El Sohly MA, Radwan MM, Gul W et al (2017) Phytochemistry of *Cannabis sativa* L. In: Kinghorn A et al (eds) Phytocannabinoids, vol 103. Springer, Cham, pp 1–36. https://doi.org/10.1007/978-3-319-45541-9_1

Fatima N, Nayeem N (2016) Toxic effects as a result of herbal medicine intake. In: Larramendy M et al (eds) Toxicology-new aspects to this scientific conundrum, vol 26. Tech Open, London, pp 193–207. https://doi.org/10.5772/64468

Ferah AU, Dilek MO, Ercument Y et al (2006) Hemlock poisoning (case report): die like Socrates. Resuscitation 70(2):337–340. https://doi.org/10.1016/j.resuscitation.2006.06.125

Finkelstein Y, Aks SE, Hutson JR et al (2010) Colchicine poisoning: the dark side of an ancient drug. Clin Toxicol 48(5):407–414. https://doi.org/10.3109/15563650.2010.495348

Fitzerald KT, Bronstein AC, Newquist KL (2013) Marijuana poisoning. Top Companion Anim Med 28(1):8–12. https://doi.org/10.1053/j.tcam.2013.03.004

Fowler CJ (2015) Plant sources of anti-muscarinics. BJU Int 115:4–7

Franke H, Scholl R, Aigner A (2019) Ricin and *Ricinus communis* in pharmacology and toxicology-from ancient use and "papyrus Ebers" to modern perspectives and "poisonous plant of the year 2018". Naunyn Schmiedeberg's Arch Pharmacol 392:1181–1208. https://doi.org/10.1007/s00210-019-01691-6

Gaire BP (2008) Monograph on *Datura stramonium*. Pokhara University Press, Kaski, Nepal, pp 1–114

Galal AM, Slade D, Gul W et al (2009) Naturally occurring and related synthetic cannabinoids and their potential therapeutic applications. Recent Pat CNS Drug Discov 4(2):112–136. https://doi.org/10.2174/157488909788453031

Garcia F (1954) A botany symposium on medicinal plants. In: Proceed 8th Pacific sci congr natl res council of the Philippines, IV(A), p 182

Giri A, Narasu ML (2000) Production of podophyllotoxin from *Podophyllum hexandrum*: a potential natural product for clinically useful anticancer drugs. Cytotechnology 34:17–26. https://doi.org/10.1023/A:1008138230896

Gooneratne IK, Weeratunga P, Caldera M et al (2014) Toxic encephalopathy due to colchicines—*Gloriosa superba* poisoning. Pract Neurol 14(5):357–359. https://doi.org/10.1136/practneurol-2013-000753

Gupta SM, Manikyaprabhu K, Dwibedi SK et al (2018) Himalayan toxic plants of defense importance. Acta Sci Med Sci 2(3):44–48

Ilercil O, Laske DW, Walbridge S et al (1994) Neurotoxicity of immunotoxins. Mol Chem Neuropathol 21(2–3):379–386. https://doi.org/10.1007/BF02815363

Isman MB (2006) Botanical insecticides, deterrents, and repellents in modern agriculture and an increasingly regulated world. Annu Rev Entomol 51(1):45–66. https://doi.org/10.1146/annurev.ento.51.110104.151146

Jang DH, Hofmann RS, Nelson LS (2010) Attempted suicide, by mail order: *Abrus precatorius*. J Med Toxicol 6(4):427–430. https://doi.org/10.1007/s13181-010-0099-1

Karakasi MV, Tologkos S, Papadatou V et al (2019) *Conium maculatum* intoxication: literature review and case report on hemlock poisoning. Forensic Sci Rev 31(1):23–36

Khan H, Ali Khan M, Hussan I (2007) Enzyme inhibition activities of the extracts from rhizomes of *Gloriosa superba* Linn (Colchicaceae). J Enz Inhib Med Chem 22(6):722–725. https://doi.org/10.1080/14756360601164853

Konca C, Kahramaner Z, Bosnak M et al (2014) Hemlock (*Conium maculatum*) poisoning in a child. Turkish J Emerg Med 14(1):34–36. https://doi.org/10.5505/1304.7361.2013.23500

Kristanc L, Kreft S (2016) European medicinal and edible plants associated with subacute and chronic toxicity part I: plants with carcinogenic, teratogenic and endocrine-disrupting effects. Food Chem Toxicol 92:150–164. https://doi.org/10.1016/j.fct.2016.04.007

Kumari V (2001) Field evaluation of grass pea (*Lathyrus sativus* L.) germplasm for its toxicity in the northwestern hills of India. Lathyrus Lathyrism News Lett 2:82–84

Ladda PL, Kamthane RB (2014) *Ricinus communis* (castor): an overview. Int J Res Pharmacol Pharmacother 3(2):136–144

Le Garff E, Delannoy Y, Mesli V et al (2016) Forensic features of a fatal *Datura* poisoning case during a robbery. Forensic Sci Int 261:17–21. https://doi.org/10.1016/j.forsciint.2016.02.028

Leete EB (1994) Becoming a hospice volunteer. Am J Hosp Palliat Care 11(2):27–32. https://doi.org/10.1177/104990919401100209

Leff RS, Thompson JM, Daly MB et al (1988) Acute neurologic dysfunction after high-dose etoposide therapy for malignant glioma. Cancer 62(1):32–35. https://doi.org/10.1002/10970142(19880701)62:1<32::AIDCNCR2820620108>3.0.CO;2-X

Legha SS (1986) Vincristine neurotoxicity. Med Toxicol 1:421–427. https://doi.org/10.1007/BF03259853

Lusthof KJ, Bosman IJ, Kubat B et al (2017) Toxicological results in a fatal and two non-fatal cases of scopolamine-facilitated robberies. Forensic Sci Int 274:79–82. https://doi.org/10.1016/j.forsciint.2017.01.024

Madaan R, Kumar S (2012) Screening of alkaloidal fraction of *Conium maculatum* L. aerial parts for analgesic and anti-inflammatory activity. Ind J Pharm Sci 74(5):457–460. https://doi.org/10.4103/0250-474X.108423

Maity D, Ghosh J, Midday M (2019) Traditional use of *Mandragora caulescens*, a poorly known medicinal plant of Sikkim Himalaya. Explor Anim Med Res 9(2):204–206

Malayandi S, Rajendran K, Palaniyandi SA et al (2019) *Gloriosa superba*, a source of the bioactive alkaloid colchicine: chemistry, biosynthesis and commercial production. In: Swamy MK et al (eds) Medicinal plants: chemistry, pharmacology and therapeutic applications, 1st edn. CRC Press, Boca Raton, FL, pp 3–9

Malik ZA, Bhat JA, Ballabha R et al (2015) Ethnomedicinal plants traditionally used in health care practices by inhabitants of Western Himalaya. J Ethnopharmacol 172:133–144. https://doi.org/10.1016/j.jep.2015.06.002

Mittal N, Ginwal H, Varshney V (2009) Pharmaceutical and biotechnological potential of *Acorus calamus* Linn.: an indigenous highly valued medicinal plant species. Pharmacog Rev 3 (5):93–103

Moudi M, Go R, Yien CY et al (2013) Vinca alkaloids. Int J Prev Med 4(11):1231–1236

Murti VV, Seshadri TR, Venkitasubramanian TA (1964) Neurotoxic compounds of the seeds of *Lathyrus sativus*. Phytochemistry 3(1):73–78. https://doi.org/10.1016/S0031-9422(00)83997-7

Oelrichs PB, Hill MW, Vallely PJ et al (1983) Toxic tetranortriterpenes of the fruit of *Melia azedarach*. Phytochemistry 22(2):531–534. https://doi.org/10.1016/0031-9422(83)83039-8

Olsnes S, Refsnes K, Pihl A (1974) Mechanism of action of the toxic lectins abrin and ricin. Nature 249:627–631. https://doi.org/10.1038/249627a0

Panter KE, Welch KD, Gardner DR (2014) Poisonous plants: biomarkers for diagnosis. In: Gupta RC (ed) Biomarkers in toxicology. Elsevier, Oxford, pp 563–589. https://doi.org/10.1016/C2012-0-01373-7

Parker AJ, Mehta T, Zarghami NS et al (1979) Acute neurotoxicity of the *Lathyrus sativus* neurotoxin, L-3-oxalylamino-2-aminopropionic acid, in the squirrel monkey. Toxicol Appl Pharmacol 47(1):135–143. https://doi.org/10.1016/0041-008X(79)90080-2

Patil MM, Patil SV, Akki AS et al (2016) An arrow poison (*Abrus precatorius*) causing fatal poisoning in a child. J Clin Diagn Res 10(3):SD03–SD04. https://doi.org/10.7860/JCDR/2016/18234.7439

Polito L, Bortolotti M, Maiello S et al (2016) Plants producing ribosome-inactivating proteins in traditional medicine. Molecules 21(11):1560. https://doi.org/10.3390/molecules21111560

Polito L, Bortolotti M, Battelli MG et al (2019) Ricin: an ancient story for a timeless plant toxin. Toxins 11(6):324. https://doi.org/10.3390/toxins11060324

Prince ES, Parameswari P, Mahaboob KR (2011) Protective effect of *Ricinus communis* leaves extract on carbon tetrachloride induced hepatotoxicity in albino rats. Iran J Pharm Sci 7 (4):269–278

Rao SL, Adiga PR, Sarma PS (1964) The isolation and characterization of β-N-oxalyl-L-α, β-diaminopropionic acid: a neurotoxin from the seeds of *Lathyrus sativus*. Biochemistry 3 (3):432–436. https://doi.org/10.1021/bi00891a022

Reedman L, Shih RD, Hung O (2008) Survival after an intentional ingestion of crushed *Abrus* seeds. Western J Emerg Med 9(3):157–159

Sahoo R, Hamide A, Amalnath SD et al (2008) Acute demyelinating encephalitis due to *Abrus precatorius* poisoning - complete recovery after steroid therapy. Clin Toxicol 46 (10):1071–1073. https://doi.org/10.1080/15563650802334671

Scarpa A, Guerci A (1982) Various uses of the castor oil plant (*Ricinus communis* L.) a review. J Ethnopharmacol 5(2):117–137. https://doi.org/10.1016/0378-8741(82)90038-1

Sharma J, Gairola S, Gaur RD et al (2013) Ethnomedicinal plants used for treating epilepsy by indigenous communities of sub-Himalayan region of Uttarakhand, India. J Ethnopharmacol 150 (1):353–370. https://doi.org/10.1016/j.jep.2013.08.052

Singh R, Navneet, Mansotra DK et al (2020) Ethnobotanical and pharmacological aspects of *Ipomoea carnea* Jacq. and *Celosia cristata* Linn. In: Joshi PC et al (eds) Current status of researches in biosciences. Today & Tomorrow's Printers and Publishers, New Delhi, India, pp 493–506

Solanki A, Zaveri M (2012) Pharmacognosy, phytochemistry and pharmacology of *Abrus precatorius* leaf: a review. Int J Pharm Sci Rev Res 13(2):71–76

Sowa-Rogozinska N, Sominka H, Nowakowska-Gołacka J et al (2019) Intracellular transport and cytotoxicity of the protein toxin ricin. Toxins 11(6):350. https://doi.org/10.3390/toxins11060350

Spencer PS, Schaumburg HH (1983) Lathyrism: a neurotoxic disease. Neurobehav Toxicol Teratol 5(6):625–629

Tak S, Lakhotia M, Gupta A et al (2016) Aconite poisoning with arrhythmia and shock. Indian Heart J 68:S207–S209. https://doi.org/10.1016/j.ihj.2015.08.010

Talukdar D (2013) Antioxidant potential and type II diabetes related enzyme inhibition properties of raw and processed legumes in Indian Himalayas. J Appl Pharm Sci 3(3):13–19. https://doi.org/10.7324/JAPS.2013.30303

Thirach S, Tragoolpua S, Punjaisee S et al (2003) Antifungal activity of some medicinal plant extracts against *Candida albicans* and *Cryptococcus neoformans*. Acta Hort (597):217–221. https://doi.org/10.17660/ActaHortic.2003.597.31

Venskutonis PR, Dagilyte A (2003) Composition of essential oil of sweet flag (*Acorus calamus* L.) leaves at different growing phase. J Essent Oil Res 15(5):313–318. https://doi.org/10.1080/10412905.2003.9698598

Verstappen CC, Heimans JJ, Hoekman K et al (2003) Neurotoxic complications of chemotherapy in patients with cancer. Drugs 63(15):1549–1563. https://doi.org/10.2165/00003495-200363150-00003

Wink M (2000) Interference of alkaloids with neuroreceptors and ion channels. In: Atta-ur-Rehman (ed) Studies in natural products chemistry, vol 21. Elsevier, Amsterdam, pp 3–122. https://doi.org/10.1016/S1572-5995(00)80004-6

Wu C, Han T, Lu H et al (2016) The toxicology mechanism of endophytic fungus and swainsonine in locoweed. Environ Toxicol Pharmacol 47:38–46. https://doi.org/10.1016/j.etap.2016.08.018

Yan ZY, Spencer PS, Li ZX et al (2006) *Lathyrus sativus* (grass pea) and its neurotoxin ODAP. Phytochemistry 67(2):107–121. https://doi.org/10.1016/j.phytochem.2005.10.022

Zhang Z, Lu A, William GD (1994) Solanaceae. In: Wu ZY et al (eds) Flora of China (Verbenaceae through Solanaceae). Science Press, Beijing, and Missouri Botanical Garden Press, St. Louis, pp 300–332

Zhu W, Yu H, Wang Y et al (2019) Integration of transcriptomics, proteomics and metabolomics data to reveal the biological mechanisms of abrin injury in human lung epithelial cells. Toxicol Lett 312:1–10. https://doi.org/10.1016/j.toxlet.2019.04.033

# Neuroprotective Effects of *Portulaca oleracea* and *Portulaca quadrifida* Linn

**Manal Buabeid, Sabrina Ait Gacem, Ayaka Fujihashi, Ayeh Trish, Rishi M. Nadar, Manoj Govindarajulu, and Muralikrishnan Dhanasekaran**

## Contents



**Abstract** *Portulaca oleracea* and *Portulaca quadrifida* Linn. belong to the family Portulacaceae. Both plant species possess several similar medicinal effects, allowing one to act as a substitute for the other in traditional herbal practice. These species possess an important neuroprotective activity and can modulate stress due to their significant anticonvulsant, antianxiety, and antidepressant properties. Besides, neuroprotection ability, *P. oleracea* is known for its wound healing, anti-ulcerogenic, antitumor, antidiabetic, antiaging, antinociceptive, and muscle relaxant activities. While *P. quadrifida* has similar uses of *P. oleracea*, in addition, it is useful in treating cough, asthma, urinary discharges, hemorrhoids, ulcers, gastrointestinal

M. Buabeid (✉) · S. A. Gacem
College of Pharmacy and Health Sciences, Ajman University, Ajman, UAE
e-mail: m.buabeid@ajman.ac.ae

A. Fujihashi · R. M. Nadar · M. Govindarajulu · M. Dhanasekaran
Department of Drug Discovery and Development, Auburn University, Auburn, USA

A. Trish
College of Science and Mathematics, Auburn University, Auburn, USA



D. C. Agrawal, M. Dhanasekaran (eds.), *Medicinal Herbs and Fungi*,
https://doi.org/10.1007/978-981-33-4141-8_20

ailments, dysentery, and has anti-inflammatory property. The present chapter discusses the important neuroprotective activity and other therapeutic benefits of two *Portulaca* species.

**Keywords** *Portulaca oleracea* · *Portulaca quadrifida* · Portulacaceae · Neuroprotection

## Abbreviations

| | |
|---|---|
| AChE | Acetylcholinesterase |
| AD | Alzheimer's disease |
| BLA | Blood lactic acid |
| CNS | Central nervous system |
| DOPA | 3, 4-Dihydroxyphenylalanine |
| EPO | Erythropoietin |
| *H. pylori* | *Helicobacter pylori* |
| HIF-1 | Hypoxia-inducible factor-1 |
| LDH | Lactate dehydrogenase |
| PD | Parkinson's disease |
| PFK | Phosphofructokinase |
| PK | Pyruvate kinase |
| SUN | Serum urea nitrogen |
| WHO | World Health Organization |

## 1 Introduction

For centuries, people have been relying on both conventional and ancient traditional remedies for different medical practices. Many people prefer the ancient traditional systems of medicine due to the belief that modern drugs are associated with several side effects compared to herbal remedies (Pandey et al. 2013). The traditional herbal preparations represent a substantial amount of the global drug market. Still, the lack of quality control, standardization, and documentation made the acceptance of these remedies as alternatives to conventional medications difficult.

Plants have been used for years as traditional remedies in different parts of the world. Two important plant species that have been studied for their diverse medical effects are *Portulaca oleracea,* also known as purslane (Fig. 1a, b), and *Portulaca quadrifida* Linn., commonly known as "chickweed." Both plant species belong to the family Portulacaceae (Burkill 1995; Petropoulos et al. 2016). Both species are annual wild plants common in warm tropical regions and exert similar medicinal effects, allowing one to act as a substitute for the other in traditional herbal practice (Dymock et al. 1980).

*P. oleracea* is an annual green plant known to have different names such as purslane, baqlahamqa, pigweed, parsley, and global panacea, the term given by the

**Fig. 1** (**a**) *Portulaca oleracea* (Attribution: Mauricio Mercadante). (**b**) *Portulaca quadrifida* Linn. (Attribution: https://indiabiodiversity.org)

World Health Organization (WHO) for this widely used medicinal plant (Masoodi et al. 2011; Xu et al. 2006). *P. oleracea* is claimed to be of Asian origin distributed in tropical and high temperate regions. It is also spread in the Middle East, as well as the Mediterranean region (Masoodi et al. 2011; Zhu et al. 2010). *P. oleracea* is usually consumed as a leafy vegetable, especially in the Mediterranean diet (Simopoulos 2004; Nebel and Heinrich 2009).

*P. quadrifida* Linn. is an annual and a small diffused plant common in tropical areas (Kirtikar and Basu 2001). This plant is usually found among rocks or sandy alkaline soil. The leaves of this plant are used in salad, as well, for many other purposes (Masoodi et al. 2011; Kar and Borthakur 2008). The present chapter deals with the neuroprotective activity and several other therapeutic properties of *P. oleracea,* and *P. quadrifida* Linn. (Fig. 2).

## 2 Active Constituents of *P. oleracea* and *P. quadrifida* Plants

The active constituents of these plants have been explored in several research studies, and several components have been isolated from these plants (Tables 1 and 2). The isolated components of these plants were found to be responsible for many beneficial properties and effects, which encouraged their traditional use for medicinal practices in several parts of the world (Chowdhary et al. 2013).

*P. oleracea* L. contains a variety of components, and each contributes to one of the properties exerted by the plant. Flavonoids, for example, are polyphenolic compounds that are known for their antioxidant properties. The highest levels of flavonoids are present in the root, followed by the stem and leaves of *P. oleracea* (Lim and Quah 2007; Xu et al. 2006). Besides, mature *P. oleracea* plants had higher phenolic compound levels and antioxidant activities than immature plants (Kamal-Uddin et al. 2012). It was observed in previous studies that total phenolic and

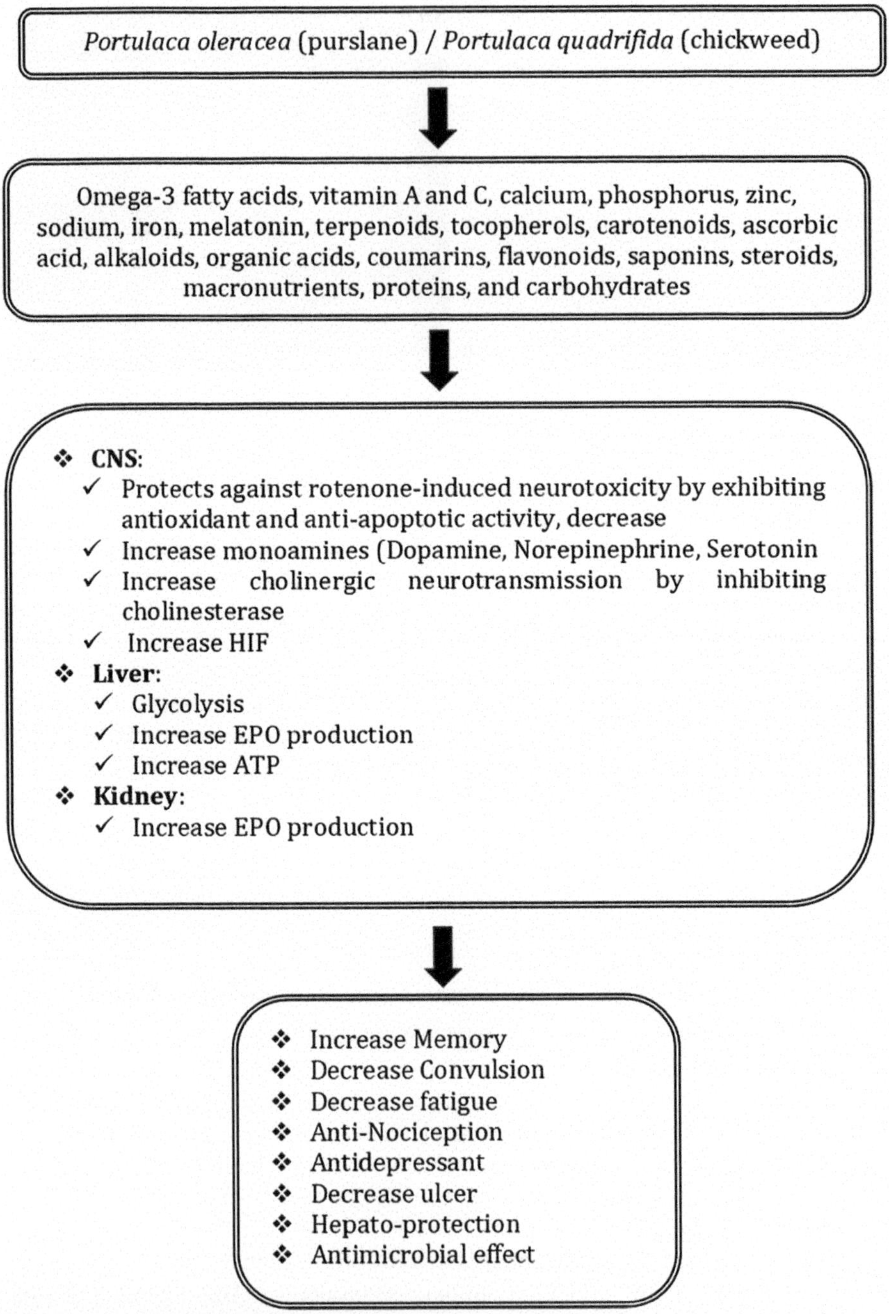


**Fig. 2** Neuroprotective and other therapeutic effects of *P. oleracea* and *Portulaca quadrifida* Linn.

**Table 1** Active constituents of *Portulaca oleracea* plant

| Plant species | Active constituents | References |
|---|---|---|
| *Portulaca oleracea* L. | Omega-3 fatty acids | Iranshahy et al. (2017)<br>Omara-Alwala et al. (1991)<br>Oliveira et al. (2009) |
| | Vitamin A and C, calcium, phosphorus, zinc, sodium, and iron | Iranshahy et al. (2017)<br>Simopoulos et al. (1992)<br>Mohamed and Hussein (1994) |
| | Antioxidants such as melatonin | Simopoulos et al. (2005) |
| | Terpenoids, tocopherols, carotenoids, and ascorbic acid | Xin et al. (2008)<br>Liu et al. (2000)<br>Szalai et al. (2010) |
| | Alkaloids | Iranshahy et al. (2017)<br>Xu et al. (2016)<br>Li et al. (2017)<br>Xiang et al. (2005)<br>Tian et al. (2014)<br>Jiao et al. (2015) |
| | Organic acids and coumarins | Wang et al. (2017)<br>Sicari et al. (2018)<br>Awad (1994) |
| | Flavonoids | Xu et al. (2006)<br>Erkan (2012)<br>Zhu et al. (2010) |
| | Saponins | Iranshahy et al. (2017) |
| | Steroids | Iranshahy et al. (2017) |
| | Macronutrients such as proteins and carbohydrates | Mohamed and Hussein (1994) |

flavonoid content were measured in several extracts such as ethyl acetate, methanol, dichloromethane, and n-hexane. The methanol extract of *P. oleracea* was reported to have a strong correlation between its flavonoids content and the antioxidant activity (Salehi et al. 2013).

Alkaloids such as dopamine, norepinephrine, dopa, and N-*trans*-feruloyltyramine, including DOPA compounds (3, 4-dihydroxyphenylalanine) belong to the family of alkaloids. It was found that the content of these two compounds is higher in the leaves than the stem and seeds (Chen et al. 2003). It was observed that different extraction methods provide different content of alkaloids. For example, using methanol in the extraction of *P. oleracea* plant provides a high level of dopamine. In contrast, when distilled water was used to perform the extraction, the higher content of norepinephrine was obtained (Yue et al. 2005).

*P. oleracea* L. contains other essential ingredients such as omega-3 fatty acids, vitamin A and C, calcium, phosphorus, zinc, sodium, iron, terpenoids, tocopherols,

**Table 2** Active constituents of *Portulaca quadrifida* plant

| Plant species | Active constituents | Reference |
|---|---|---|
| *Portulaca quadrifida* L. | Omega-3 fatty acids | Trupti et al. (2018) |
| | Triterpenoids, terpenoids | Verma et al. (2016)<br>Narasimha et al. (2018)<br>Trupti et al. (2018) |
| | Alkaloids | Verma et al. (2016)<br>Trupti et al. (2018)<br>Narasimha et al. (2018)<br>Jagtap et al. (2019)<br>Mulla and Swamy (2010) |
| | Flavonoids | Narasimha et al. (2018)<br>Jagtap et al. (2019)<br>Mulla and Swamy (2010) |
| | Saponins | Verma et al. (2016)<br>Narasimha et al. (2018)<br>Jagtap et al. (2019) |
| | Steroids | Verma et al. (2016)<br>Jagtap et al. (2019) |
| | Amino acids and proteins | Narasimha et al. (2018)<br>Jagtap et al. (2019) |
| | Tannins | Verma et al. (2016)<br>Narasimha et al. (2018)<br>Jagtap et al. (2019) |
| | Glycosides | Jagtap et al. (2019) |

carotenoids, alkaloids, organic acids, flavonoids, saponins, steroids, and macronutrients (Table 1).

*P. quadrifida* Linn., have several constituents that play a role in the pharmacological effects. As shown in Table 2, the plant species contain constituents similar to those of *P. oleracea* L. such as alkaloids, steroids, terpenoids, and tannins (Verma et al. 2016). Several studies have identified other components that have a high medicinal value, such as flavonoids, saponins, triterpenoids, glycosides, and amino acids in *P. quadrifida* extracts (Table 2) (Narasimha et al. 2018; Jagtap et al. 2019).

## 3 Neuroprotective Activity of Two *Portulaca* Species

*P. oleracea* is said to remove free radicals, which are unstable atoms that bind and react with other atoms to complete their outer shell in order to become stable (Abdel Moneim 2013). Oxygen molecules divide into single atoms with unpaired electrons, making them unstable free radicals that try to bond with other molecules to reach stability. This process can lead to oxidative stress (Medicalnewstoday.com 2020).

Oxidative stress can have many adverse effects on the body. It can lead to several diseases and contribute to aging as the free radicals break cells down over time. As the body ages, the ability to fight the effects of these radicals is decreased, resulting in the accumulation of more free radicals and more oxidative stress that lead to cell damage and several diseases such as Alzheimer's, Parkinson's, and other dementias, as well as, cardiovascular diseases due to clogged arteries (Medicalnewstoday.com 2020; Sawada and Carlson 1987). *P. oleracea* has shown a positive effect in removing such radicals (Abdel Moneim 2013).

Parkinson's disease (PD) is a long-term degenerative disease that affects the central nervous system (CNS) and is associated with cell death in the brain. Seventy percent of dopamine-secreting neurons are affected, and there is a decreased activity of the mitochondrial respiratory chain (Complex I) (Davie 2008; Hoglinger et al. 2005). Neuroprotective and several other therapeutic effects of *P. oleracea* and *P. quadrifida* Linn. have been illustrated in Fig. 2.

*P. oleracea* was found to have an impact in removing free radicals and inhibiting neuron apoptosis. It was also found to affect antagonizing dopamine depletion and complex-I inhibition in rat striatum, which indicates that it can be a promising agent for neuroprotective effects and can be used for Parkinson's disease (Abdel Moneim 2013).

In a study by Abdel-Moneim, the neuroprotective activity of the aqueous extract of *P. oleracea* against brain damage triggered by the administration of rotenone in albino rats was evaluated (Abdel Moneim 2013). Rotenone is known to inhibit the mitochondrial complex I, which results in the production of oxygen free radicals that stimulate oxidative stress, a known cause of apoptosis in neurons. It was observed that this extract might inhibit the metabolism of dopamine and apoptosis induction in rats, making this plant a good option for future therapy of Parkinson's disease and other neurodegenerative disorders.

In another study, it was observed that *P. oleracea* extract enhanced dopamine levels in the cerebellum, thalamus, cerebral cortex, and hypothalamus. Also, it increased the levels of serotonin and norepinephrine in the cerebellum, medulla oblongata, thalamus, hypothalamus, and cerebral cortex (Abdel Moneim et al. 2012). In contrast to this, levels of dopamine, serotonin, and norepinephrine in the spinal cord decreased, while the levels of acetylcholinesterase increased in all parts of the brain except in the cerebellum. One explanation for this effect on neurotransmitters could be linked to the fatty acids and melatonin contents in the plant (Abdel Moneim et al. 2012).

Xu and Shan (2014), in their study, showed that polysaccharides from *P. oleracea* plants could enhance the swimming time to fatigue of mice and enhance glycogen levels. Also, it decreased serum urea nitrogen (SUN) and blood lactic acid (BLA), suggesting that this plant can reduce exhaustion induced by forced swimming in mice.

The alkaloidal extract of *P. oleracea* significantly inhibited the activity of acetylcholinesterase (AChE), an enzyme that catalyzes the breakdown of acetylcholine and other choline esters that act as neurotransmitters. The use of acetylcholinesterase inhibitors is considered a promising strategy for Alzheimer's disease (AD) treatment. Therefore, *P. oleracea* could be used for both the treatment and prophylaxis of AD (Yang et al. 2012; Tiwari et al. 2012).

Concerning the protection of the nerve tissues and cells from damage, *P. oleracea* increased the expression level of hypoxia-inducible factor-1 (HIF-1) (Wang et al. 2007). HIF-1 is mainly regulated by cellular oxygen concentration; thus, under hypoxic conditions, this factor activates transcription of genes encoding erythropoietin, glucose transporters, and other genes whose protein products increase oxygen delivery or facilitate metabolic adaptation to hypoxia (Semenza 2000).

Wang and coworkers reported that *P. oleracea* stimulated glycolysis and erythropoietin (EPO) production, a glycoprotein cytokine secreted by the kidney in response to cellular hypoxia, which is also associated with nerve tissue protection (Wang et al. 2007). Erythropoietin (EPO) reduces the possibility of hypoxic–ischemic damage, and it also exhibits neuroprotective activity in the brain. The ethanolic extract of *P. oleracea* plant stimulated the endogenous erythropoietin expression by stabilizing HIF-1α and HIFs in mice brain. It declined the serum enolase levels in hypoxic mice as well as the activity of caspase-3 in the neurons, which reduced the damage resulted from the hypoxic condition (Wanyin et al. 2012).

It was observed in in vivo and in vitro study that extracts of *P. oleracea* affected hypoxia in the nerve tissues of male mice as it enhanced the levels of ATP, pyruvate kinase (PK), phosphofructokinase (PFK), and lactate dehydrogenase (LDH) of hypoxic brain cortices, as well as increment of ATP available for the neurons and EPO expression when compared with the control group (Wang et al. 2007).

The administration of *P. oleracea* plant extract resulted in several other effects, such as a significant decrease in calcium concentration in the brain cortex (Abdel Moneim 2013); the increased onset time of convulsions, and a decreased locomotor activity (Radhakrishnan et al. 2001); polysaccharides from the extract decreased serum urea nitrogen and blood lactic acid and increased hepatic glycogen content (Xu and Shan 2014).

In a two-month clinical trial carried out to investigate the impact of *P. oleracea* plant extract on psychological symptoms, risperidone and biperiden were administered to the schizophrenic patients. The results showed that risperidone improved the mental conditions of chronic schizophrenic patients (Parvin et al. 2013).

The other species, *P. quadrifida,* exhibited several effects on both the central and peripheral nervous systems evaluated by in vivo muscle relaxant activity, spontaneous motor activity, anticonvulsant activity, and antinociceptive activity. Also, spontaneous motor activity and antinociceptive activity were observed to be decreased (Patil et al. 2012; Magaji et al. 2008). *P. quadrifida* plant extract contains antioxidants that play an important role in reducing stress-induced conditions, making it an excellent candidate as a nutritional supplement to treat several conditions associated with free radicals and oxidative stress.

Shahwan and Gacem (2018) investigated the impact of *P. quadrifida* extract on prooxidant changes, and to identify the effect of active constituents on stress. The study revealed that *P. quadrifida* extract showed a significant resistance towards the oxidative metabolism triggered by restraint-induced stress. The post-extract treatment (curative) was found to be more effective in restoring the altered oxidative metabolism compared to the pre-extract treatment (prophylactic). The antioxidant property was attributed to the presence of β-carotene, Vitamin C, and polyphenolic compounds. Thus, *P. quadrifida* could be used as a dietary supplement to combat various neurodegenerative diseases.

In another study, it was found that the aqueous extract of *P. quadrifida* has many significant anticonvulsant, antianxiety, and antidepressant activities (Vanga et al. 2018). The antianxiety effect of this plant extract is comparable to that of the standard drug diazepam. The aqueous extract of this plant showed a significant reduction in convulsions compared with the control mice group. Also, all phases have shown a decreased duration indicating the anticonvulsant effect, which is comparable to the standard anticonvulsant drug phenytoin. The antidepressant activity, anticonvulsant, and the antianxiety activities were attributed to the presence of alkaloids and glycosides (Vanga et al. 2018).

## 4 Other Therapeutic Effects of Two *Portulaca* Species

*P. oleracea* L. has been used for many years in different parts of the world for many medicinal purposes in traditional practice due to its unique healing effects. In the past, this plant was used in traditional Chinese healing practices to counteract toxicity and reduce heat (The Pharmacopoeia Commission of PRC 2015).

*P. oleracea* L. is also used for many medicinal purposes like relieving insect bites, fever, eczema, acne, bronchitis, and coughs (Zhang et al. 2002). It is also known for its wound healing (Rashed et al. 2003), anti-ulcerogenic (Karimi et al. 2004), antitumor (Ahangarpour et al. 2016; Zhou et al. 2015; Lee et al. 2014; Shen et al. 2013), antidiabetic (Li et al. 2009), antiaging (Kim et al. 2015), and antinociceptive and muscle-relaxant activities (Trupti et al. 2018; Jain et al. 2014). Also, *P. oleracea* plants possess several other therapeutic effects.

### *4.1 Anti-Inflammatory and Analgesic Properties*

Several studies reported an anti-inflammatory activity for *P. oleracea* and in some recent studies, several novel alkaloids have been isolated that exert anti-inflammatory effect (Xu et al. 2016; Li et al. 2016; Meng et al. 2016; Zhou et al. 2015; Lee et al. 2012; Chan et al. 2000).

### *4.2 Antibacterial*

The methanolic extract of *P. oleracea* plants possesses antimicrobial properties. The plant contains several flavonoids, tannins, and glycosides responsible for the antibacterial activity (Jagtap et al. 2019; Xia et al. 2015). Also, *P. quadrifida* has antibacterial activity against bacteria *Helicobacter pylori* (H pylori), causing gastric ulcers (Omara-Alwala et al. 1991).

## 4.3 Antioxidant

The antioxidant effect is due to the presence of certain constituents such as glutathione, Vitamin C, Vitamin A, β-carotene, and B-complex vitamins (Gu et al. 2015; Kamal-Uddin et al. 2014; ErKan 2012; Siriamornpun and Suttajit 2010; Yang et al. 2009; Lim and Quah 2007).

## 4.4 Protection Against Cardiovascular Diseases

*P. oleracea* is known for its antihypertensive properties (Dkhil et al. 2011), as well as its cardiovascular protective effects, which can be attributed to the presence of omega-3 fatty acids that result in the protective effect by decreasing the thickness of blood (Kamal-Uddin et al. 2014; Liu et al. 2000).

## 4.5 Hepatoprotective Activity

Hepatoprotective activity was observed in some investigational studies in which several doses were used to observe the pharmacological activity, and it was concluded that a dosage higher than 400 mg/kg body weight could provide significant hepatoprotective activity in mice (Trupti et al. 2018; Shafi and Tabassum 2015).

*P. quadrifida* is also known for its various medicinal effects similar to those found in *P. oleracea.* It is also said to be useful in cough, asthma, urinary discharges, hemorrhoids, ulcers, abdominal complaints, dysentery, and for its use as an anti-inflammatory (Siriamornpun and Suttajit 2010; Hoffman et al. 2004; Kirtikar and Basu 2001). *P. quadrifida* is used in some countries such as Nigeria as a local application for swelling (Patil et al. 2012; Kirtikar and Basu 1984). This plant was also found to have an anti-colorectal cancer activity (Khasim and Kumar 2016). Other several medicinal properties of *P. quadrifida* have been reported such as:

## 4.6 Anthelmintic Activity

*P. quadrifida* was found to possess anthelmintic properties, especially against *Pheretima posthuma* (Narasimha et al. 2018).

### *4.7 Antifungal and Antimicrobial Activity*

*P. quadrifida* Linn. was used for many years in Indian traditional medicine (Ayurveda) in the treatment of athlete's foot, as well as for its activity against *Candida albicans* and *Aspergillus fumigates* (Kar and Borthakur 2008; Hoffman et al. 2004).

## 5 Safety Aspects

The safety of both *P. quadrifida* and *P. oleracea* L. was studied in previous studies, and it was found that these plants are nontoxic. The application of leaf extracts of *P. quadrifida* against the brine shrimp indicated that the leaf extract had LC50 values >100 μg/mL, indicating that leaf extract is nontoxic compared to Potassium dichromate. Thus, the leaves of these plants safe for human consumption and can be used as a nutraceutical (Kumaran and Naidu 2018). In separate in vitro studies, *P. oleracea* did not show any cytotoxic effects and was considered as pharmacologically inactive (LC50 values were more than 100 μg/mL) (Tan et al. 2013; Shabsoug et al. 2008). In contrast to these, some other studies revealed that the methanolic extract showed moderate toxicity as well as hepatic and renal toxicities (Milad et al. 2017; Musa et al. 2007).

Most of the clinical trials have not reported any serious adverse effects, but the most common side effects observed in some studies included thyroiditis, facial nerve palsy, and skin rash, and constipation.

## 6 Conclusions

Both *P. oleracea* and *P. quadrifida* species exert similar medicinal effects, allowing for one to act as a substitute for the other in the traditional herbal medicine. Extracts of both plant species possess an important neuroprotective activity and can modulate stress as they possess a significant anticonvulsant, antianxiety, and antidepressant activity. The phytochemical constituents responsible for the anticonvulsant effect of *P. quadrifida* are not yet fully known, needing detailed studies. The pharmacological activities of these plants have been observed through in vitro and animal studies. Therefore, further supporting clinical trials are necessary for future research. Also, most of the studies have been carried out on crude extracts of these plants; therefore, many constituents remain unknown and need further research.

## References


Abdel Moneim A (2013) The neuroprotective effects of Purslane (*Portulaca oleracea*) on rotenone-induced biochemical changes and apoptosis in brain of rat. CNS Neurol Disord Drug Targets 12:830–841. https://doi.org/10.2174/18715273113129990081

Abdel Moneim AE, AlNasr I, Dkhil MA et al (2012) Neuronal activities of *Portulaca oleracea* in adult rats. J Med Plant Res 6:3162–3168. https://doi.org/10.5897/JMPR11.1616

Ahangarpour A, Lamoochi Z, Moghaddam H et al (2016) Effects of *Portulaca oleracea* ethanolic extract on reproductive system of aging female mice. Int J Reprod Biomed 14:205–212. https://doi.org/10.29252/ijrm.14.3.205

Awad N (1994) Lipid content and anti-microbial activity of phenolic constituents of cultivated *Portulaca oleracea* L. Bull Fac Pharm 32(1):137–142

Burkill HM (1995) The useful plants of west tropical Africa, 2nd edn. Royal Botanic Gardens, Richmond, UK

Chan K, Islam MW, Kamil M et al (2000) The analgesic and anti-inflammatory effects of *Portulaca oleracea* L. subsp. sativa (haw.) Celak. J Ethnopharmacol 73:445–451. https://doi.org/10.1016/S0378-8741(00)00318-4

Chen J, Shi YP, Liu JY (2003) Determination of noradrenaline and dopamine in Chinese herbal extracts from *Portulaca oleracea* L. by high-performance liquid chromatography. J Chromatogr 1003(1):127–132. https://doi.org/10.1016/S0021-9673(03)00786-6

Chowdhary CV, Meruva A, Elumalai RA (2013) A review on phytochemical and pharmacological profile of *Portulaca oleracea* Linn. (purslane). Int J Res Ayurveda Pharm 4:34–37. https://doi.org/10.7897/2277-4343.04119

Davie CA (2008) A review of Parkinson's disease. Br Med Bull 86(1):109–127. https://doi.org/10.1093/bmb/ldn013

Dkhil MA, Moniem AA, Al-Quraishy S et al (2011) Antioxidant effect of purslane (*Portulaca oleracea*) and its mechanism of action. J Med Plants Res 5:1589–1593

Dymock W, Warden CJ, Hooper D (1980) Portulacaceae. Pharmacographia Indica 1:158–159

Erkan N (2012) Antioxidant activity and phenolic compounds of fractions from *Portulaca oleracea* L. Food Chem 133:775–781. https://doi.org/10.1016/j.foodchem.2012.01.091

Gu JF, Zheng ZY, Yuan JR et al (2015) Comparison on hypoglycemic and antioxidant activities of the fresh and dried *Portulaca oleracea* L. in insulin resistant HepG2 cells and streptozotocin-induced C57BL/6J diabetic mice. J Ethnopharmacol 161:214–223. https://doi.org/10.1016/j.jep.2014.12.002

Hoffman BR, Delas A, Blanco K et al (2004) Screening of anti-bacterial and antifungal activities of ten medicinal plants from Ghana. Pharm Biol 42(1):13–17. https://doi.org/10.1080/13880200490504925

Hoglinger G, Lannuzel A, Khondiker M et al (2005) The mitochondrial complex I inhibitor rotenone triggers a cerebral tauopathy. J Neurochem 95(4):930–939. https://doi.org/10.1111/j.1471-4159.2005.03493.x

Iranshahy M, Javadi B, Iranshahi M et al (2017) A review of traditional uses, phytochemistry and pharmacology of *Portulaca oleracea* L. J Ethnopharmacol 205:158–172. https://doi.org/10.1016/j.jep.2017.05.004

Jagtap S, Gujar K, Ghare A (2019) Phytochemical screening and anti-microbial activity of *Portulaca quadrifida* Linn. Asian J Pharm Clin Res 12(3):78–81. https://doi.org/10.22159/ajpcr.2019.v12i3.27587

Jain S, Parihar S, Pandey N (2014) Medicinal plants with neuropharmacological properties from Indian origin. Int J Pharm Pharm Sci 6(10):36–40

Jiao ZZ, Yue S, Sun HX et al (2015) Indoline amide glucosides from *Portulaca oleracea*: isolation, structure, and DPPH radical scavenging activity. J Nat Prod 78:2588–2597. https://doi.org/10.1021/acs.jnatprod.5b00524

Kamal-Uddin MK, Juraimi AS, Ali ME et al (2012) Evaluation of antioxidant properties and mineral composition of purslane (*Portulaca oleracea* L.) at different growth stages. Int J Mol Sci 13:10,257–10,267. https://doi.org/10.3390/ijms130810257

Kamal-Uddin MD, Juraimi AS, Hossain MR et al (2014) Purslane weed (*Portulaca oleracea*): a prospective plant source of nutrition, omega-3 fatty acid, and antioxidant attributes. Sci World J 95:1019. https://doi.org/10.1155/2014/951019

Kar A, Borthakur SK (2008) Wild vegetables of Karbi-Anglong district, Assam. Nat Prod Radiance 7(5):448–460

Karimi G, Hosseinzadeh H, Ettehad N (2004) Evaluation if the gastric anti-ulcerogenic effects of *Portulaca oleracea* L. extracts in mice. Phytother Res 18:484–487. https://doi.org/10.1002/ptr.1463

Khasim SM, Kumar EP (2016) In vivo anticancer screening of *Portulaca quadrifida* Linn on colon cancer. World J Pharm Res 5(1):1175–1197

Kim JY, Oh HM, Kwak SC et al (2015) Purslane suppresses osteoclast differentiation and bone resorbing activity via inhibition of Akt/GSK3β-c-Fos-NFATc1 signaling in vitro and prevents lipopolysaccharide-induced bone loss in vivo. Biol Pharm Bull 38:66–74. https://doi.org/10.1248/bpb.b14-00567

Kirtikar KR, Basu BD (1984) Indian medicinal plants, vol 11. Oriental Enterprises, Uttaranchal Dehra Dun, pp 172–177

Kirtikar JR, Basu BD (2001) Indian medicinal plants, vol 2. Oriental Enterprises, Uttaranchal Dehra Dun, pp 333–335

Kumaran RT, Naidu N (2018) In-vitro cytotoxicity of assay of leaves of *Portulaca quadrifida* using brine shrimp assay. Int J Pharmacognosy 5(7):437–439

Lee AS, Kim JS, Lee YJ et al (2012) Anti-TNF-α activity of *Portulaca oleracea* in vascular endothelial cells. Int J Mol Sci 13:5628–5644. https://doi.org/10.3390/ijms13055628

Lee AS, Kim KH, Park C et al (2014) *Portulaca oleracea* extracts protect human keratinocytes and fibroblasts from UV induced apoptosis. Experimental dermatology Exp Dermatol 23:13–17. https://doi.org/10.1111/exd.12396

Li FL, Li QW, Gao DW et al (2009) Preparation and antidiabetic activity of polysaccharide from *Portulaca oleracea* L. Afr J Biotechnol 8:569–573

Li CY, Meng YH, Ying ZM et al (2016) Three novel alkaloids from *Portulaca oleracea* L. and their anti-inflammatory effects. J Agric Food Chem 64:5837–5844. https://doi.org/10.1021/acs.jafc.6b02673

Li CY, Ying ZM, Gao MZ et al (2017) Two new similar alkaloids from *Portulaca oleracea* L. Nat Prod Res 31:1792–1798. https://doi.org/10.1080/14786419.2017.1292507

Lim YY, Quah EL (2007) Antioxidant properties of different cultivars of *Portulaca oleracea*. Food Chem 103:734–740. https://doi.org/10.1016/j.foodchem.2006.09.025

Liu L, Howe P, Zhou YF et al (2000) Fatty acids and β-carotene in Australian Purslane (*Portulaca oleracea*) varieties. J Chromatogr 893:207–213. https://doi.org/10.1016/S0021-9673(00)00747-0

Magaji MG, Anuka JA, Abdu-Aguye I et al (2008) Preliminary studies on anti-inflammatory and analgesic activities of *Securinega virosa* (Euphorbiaceae) in experimental animal models. J Med Plan Res 2(2):039–044

Masoodi MH, Ahmad B, Mir SR et al (2011) *Portulaca oleracea* L. a review. J Pharm Res 4:3044–3048

Medicalnewstoday (2020) Free radicals: how do they affect the body. Available via: https://www.medicalnewstoday.com/articles/318652. Accessed 25 Feb. 2020

Meng YH, Ying ZM, Xiang Z et al (2016) The anti-inflammation and pharmacokinetics of a novel alkaloid from *Portulaca oleracea* L. J Pharm Pharmacol 68:397–405. https://doi.org/10.1111/jphp.12526

Milad I, Behjat J, Mehrdad I et al (2017) A review of traditional uses, phytochemistry and pharmacology of *Portulaca oleracea* L. J Ethnopharmacol 205:158–172. https://doi.org/10.1016/j.jep.2017.05.004

Mohamed A, Hussein A (1994) Chemical composition of Purslane (*Portulaca oleracea*). Plant Foods Hum Nutr 45:1–9. https://doi.org/10.1007/BF01091224

Mulla SK, Swamy P (2010) Preliminary pharmacognostical and phytochemical evaluation of *Portulaca quadrifida* Linn. Int J Pharmtech Res 2(3):1699–1702

Musa K, Ahmed A, Ibrahim G et al (2007) Toxicity studies on the methanolic extract of *Portulaca oleracea* L (Fam. Portulacaceae). J Biol Sci 7:1293–1295. https://doi.org/10.3923/jbs.2007.1293.1295

Narasimha RY, Prasada RM, Supriya A et al (2018) Preliminary phytochemical analysis and anthelmintic activity of ethanolic extract of *Portulaca quadrifida* whole plant. International journal of current research in. Life Sci 7(6):2173–2175

Nebel S, Heinrich M (2009) Ta chórta: a comparative ethnobotanical-linguistic study of wild food plants in a Graecanic area in Calabria, southern Italy. Econ Bot 63:78–92. https://doi.org/10.1007/s12231-008-9069-9

Oliveira I, Valentão P, Lopes R et al (2009) Phytochemical characterization and radical scavenging activity of *Portulaca oleracea* L. leaves and stems. Microchem J 92:129–134. https://doi.org/10.1016/j.microc.2009.02.006

Omara-Alwala TR, Mebrahtu T, Prior DE et al (1991) Omega-three fatty acids in purslane (*Portulaca oleracea*) tissues. J Am Oil Chem Soc 68(3):198–199. https://doi.org/10.1007/BF02657769

Pandey MM, Rastogi S, Rawat AK (2013) Indian traditional ayurvedic system of medicine and nutritional supplementation. Evid Based Complement Alternat Med 2013:376327. https://doi.org/10.1155/2013/376327

Parvin N, Farzaneh S, Vardanjani LR et al (2013) The effects of *Portulaca oleracea* L (purslane) on psychologic symptoms of schizophrenic patients. Eur Psychiatry 28(1):423. https://doi.org/10.1016/S0924-9338(13)75770-5

Patil AG, Joshi VS, Koli SP et al (2012) Pharmacognistical and phytochemical analysis of *Portulaca quadrifida* Linn. Res J Pharm Biol Chem Sci 3(1):90–100

Petropoulos SA, Karkanis A, Martins N et al (2016) Phytochemical composition and bioactive compounds of common Purslane (*Portulaca oleracea* L.) as affected by crop management practices. Trends Food Sci Technol 55:1–10. https://doi.org/10.1016/j.tifs.2016.06.010

Radhakrishnan R, Zakaria MM, Islam MW et al (2001) Neuropharmacological actions of *Portulaca oleracea* L v. sativa (hawk). J Ethnopharmacol 76:171–176. https://doi.org/10.1016/S0378-8741(01)00230-6

Rashed AN, Afifi FU, Disi AM (2003) Simple evaluation of the wound healing activity of a crude extract of *Portulaca oleracea* L. (growing in Jordan) in Mus musculus JVI-1. J Ethnopharmacol 88:131–136. https://doi.org/10.1016/S0378-8741(03)00194-6

Salehi P, Asghari B, Esmaeili MA et al (2013) Glucosidase and amylase inhibitory effect and antioxidant activity of ten plant extracts traditionally used in Iran for diabetes. J Med Plant Res 7:257–266

Sawada M, Carlson J (1987) Changes in superoxide radical and lipid peroxide formation in the brain, heart, and liver during the lifetime of the rat. Mech Ageing Dev 41(2):125–137. https://doi.org/10.1016/0047-6374(87)90057-1

Semenza G (2000) Expression of hypoxia-inducible factor 1: mechanisms and consequences. Biochem Pharmacol 59(1):47–53. https://doi.org/10.1016/S0006-2952(99)00292-0

Shabsoug B, Khalil R, Abuharfeil N (2008) Enhancement of natural killer cell activity in vitro against human tumor cells by some plants from Jordan. J Immunotoxicol 5:279–285. https://doi.org/10.1080/15376510802312027

Shafi S, Tabassum N (2015) Toxicity evaluation of hydro-alcoholic extract of *Portulaca oleracea* (whole plant) in Swiss albino mice. Int J Pharm Pharm Sci 7(2):506–510

Shahwan M, Gacem AS (2018) Neuroprotective antioxidant effect of *Portulaca quadrifida* Linn. Leaves extract on immobilization stress-induced changes in rat's brain. Asian J Pharm Clin Res 11(10):378. https://doi.org/10.22159/ajpcr.2018.v11i10.27209

Shen H, Tang G, Zeng G et al (2013) Purification and characterization of an antitumor polysaccharide from *Portulaca oleracea* L. Carbohydr Polym 93:395–400. https://doi.org/10.1016/j.carbpol.2012.11.107

Sicari V, Loizzo MR, Tundis R et al (2018) *Portulaca oleracea* L. (Purslane) extracts display antioxidant and Hypoglycaemic effects. J Appl Bot Food Qual 91:39–46

Simopoulos AP (2004) The traditional diet of Greece and cancer. Eur J Cancer Prev 13:219–230. https://doi.org/10.1097/01.cej.0000130011.99148.07

Simopoulos AP, Norman HA, Gillaspy JE et al (1992) Common purslane: a source of omega-3 fatty acids and antioxidants. J Am Coll Nutr 11(4):374–382. https://doi.org/10.1080/07315724.1992.10718240

Simopoulos AP, Tan DX, Manchester LC et al (2005) Purslane: a plant source of omega-3 fatty acids and melatonin. J Pineal Res 39(3):331–332. https://doi.org/10.1111/j.1600-079X.2005.00269.x

Siriamornpun S, Suttajit M (2010) Microchemical components and antioxidant activity of different morphological parts of Thai wild purslane (*Portulaca oleracea*). Weed Sci 58(3):182–188. https://doi.org/10.1614/WS-D-09-00073.1

Szalai G, Dai N, Danin A et al (2010) Effect of nitrogen source in the fertilizing solution on nutritional quality of three members of the *Portulaca oleracea* aggregate. J Sci Food Agric 90:2039–2045. https://doi.org/10.1002/jsfa.4049

Tan GCS, Wong KM, Pearle-Won GQ et al (2013) In vitro cytotoxic and antiproliferative effects of *Portulaca oleracea* methanol extract on breast, cervical, colon and nasopharyngeal cancerous cell lines. Sains Malays 42(7):927–935

The Pharmacopoeia Commission of PRC (2015) The pharmacopeia of the People's Republic of China (part 1). China Medical Science and Technology Press, Beijing, pp 49–50

Tian JL, Liang X, Gao PY et al (2014) Two new alkaloids from *Portulaca oleracea* and their cytotoxic activities. J Asian Nat Prod Res 16:259–264. https://doi.org/10.1080/10286020.2013.866948

Tiwari P, Dwivedi S, Singh MP et al (2012) Basic and modern concepts on cholinergic receptor: a review. Asian Pac J Trop Dis 3(5):413–420. https://doi.org/10.1016/S2222-1808(13)60094-8

Trupti PD, Chitra CK, Pratik PD et al (2018) Phytochemical analysis of *Portulaca oleracea* and *Portulaca quadrifida* extracts using gas chromatography–mass spectrometry. Asian J Pharm Clin Res 11(9):204–207. https://doi.org/10.22159/ajpcr.2018.v11i9.26662

Vanga R, Srujana CY, Jyothi K (2018) Pharmacological evaluation of *Portulaca quadrifida*. MOJ Anat Physiol 5(6):335–341. https://doi.org/10.15406/mojap.2018.05.00221

Verma SC, Hazra K, Devgan M et al (2016) Phytochemical and high-performance liquid chromatography analysis of extract of *Portulaca quadrifida* Linn. Asian J Pharm Clin Res 9 (3):163–164

Wang W, Gu L, Dong L et al (2007) Protective effect of *Portulaca oleracea* extracts on hypoxic nerve tissue and its mechanism. Asia Pac J Clin Nutr 16(1):227–233

Wang HQ, Zhang LY, Wang YY (2017) Isolating and identifying organic acids from *Portulaca oleracea* and determining their anti-cyanobacterial activity. Pol J Environ Stud 26:441–445. https://doi.org/10.15244/pjoes/64465

Wanyin W, Liwei D, Lin J et al (2012) Ethanol extract of *Portulaca oleracea* L. protects against hypoxia-induced neuro damage through modulating endogenous erythropoietin expression. J Nutr Biochem 23:385–391. https://doi.org/10.1016/j.jnutbio.2010.12.015

Xia L, Li JM, Liu B et al (2015) Separation and identification of four new compounds with antibacterial activity from *Portulaca oleracea* L. Molecules 20:16375–16387. https://doi.org/10.3390/molecules200916375

Xiang L, Xing DM, Wang W et al (2005) Alkaloids from *Portulaca oleracea* L. Phytochemistry 66:2595–2601. https://doi.org/10.1016/j.phytochem.2005.08.011

Xin HL, Xu YF, Hou YH et al (2008) Two novel triterpenoids from *Portulaca oleracea* L. Helv Chim Acta 91:2075–2080. https://doi.org/10.1002/hlca.200890221

Xu Z, Shan Y (2014) Anti-fatigue effects of polysaccharides extracted from *Portulaca oleracea* L. in mice. Indian J Biochem Biophys 51:321–325

Xu X, Yu L, Chen G (2006) Determination of flavonoids in *Portulaca oleracea* L. by capillary electrophoresis with electrochemical detection. J Pharm Biomed Anal 41(2):493–499. https://doi.org/10.1016/j.jpba.2006.01.013

Xu L, Ying ZM, Wei WJ et al (2016) A novel alkaloid from *Portulaca oleracea* L. Nat Prod Res 31:902–908. https://doi.org/10.1080/14786419.2016.1253081

Yang Z, Liu C, Xiang L et al (2009) Phenolic alkaloids as a new class of antioxidants in *Portulaca oleracea*. Phytother Res 23:1032–1035. https://doi.org/10.1002/ptr.2742

Yang Z, Zhang D, Ren J et al (2012) Acetylcholinesterase inhibitory activity of the total alkaloid from traditional Chinese herbal medicine for treating Alzheimer's disease. Med Chem Res 21 (6):734–738. https://doi.org/10.1007/s00044-011-9582-8

Yue ME, Jiang TF, Shi YP (2005) Simultaneous determination of noradrenaline and dopamine in *Portulaca oleracea* L. by capillary zone electrophoresis. J Sep Sci 28:360–364. https://doi.org/10.1002/jssc.200400045

Zhang XJ, Ji YB, Qu ZY et al (2002) Experimental studies on antibiotic functions of *Portulaca oleracea* L. in vitro. Chin J Microecol 14:277–280

Zhou Y, Xin H, Rahman K et al (2015) *Portulaca oleracea* L.: a review of Phytochemistry and pharmacological effects. Bio Med Res Int 2015:1–11. https://doi.org/10.1155/2015/925631

Zhu HB, Wang YZ, Liu YX et al (2010) Analysis of flavonoids in *Portulaca oleracea* L. by UV–Vis spectrophotometry with comparative study on different extraction technologies. Food Anal Method 3:90–97. https://doi.org/10.1007/s12161-009-9091-2

The manufacturer's authorised representative in the EU is Springer Nature Customer Service Centre GmbH, Europaplatz 3, 69115 Heidelberg, Germany. If you have any concerns regarding our products, please contact ProductSafety@springernature.com

Printed and bound by CPI Group (UK) Ltd, Croydon, CR0 4YY
30/09/2026
02225241-0015